AF517784

# SOIL CONSERVATION

***For Sustainable Agriculture***

# SOIL CONSERVATION

## *For Sustainable Agriculture*

**Hough Hammond Bennett**

**AGROBIOS (INDIA)**

*Published by:*
**AGROBIOS (INDIA)**
Behind Nasrani Cinema
Chopasani Road, Jodhpur
Ph.: 0291-2643993
E.Mail: agrobiosindia@gmail.com

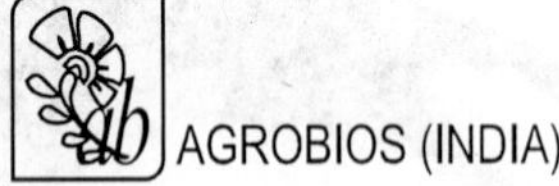


AGROBIOS INDIA (2023)



ISBN (13): 978-81-7754-104-5

***Published by:*** Dr. Updesh Purohit for Agrobios (India), Jodhpur
***Laser Typeset at:*** Yashee Computers, Jodhpur
***Cover Design by:*** Reena
***Printed at:*** Agrobios Digitals, Jodhpur

# PREFACE

In fifteen decades, Americans have transformed a wilderness into a mighty nation. In all the history of the world, no people ever built so fast and yet so well. This will be a land of liberty, they said in the beginning, and as they hacked the forest, drove their ploughshares deep into the earth, and spread their herds across the ranges, they sang of the land of the free that they were making. All that they finally built upon this continent is founded in that faith—that here there would be opportunity and independence and security for any man.

Those things are the power and the hope of this democracy. And they have sprung, very largely, from the goodness of our land, its capacity to produce rewardingly. Yet with astonishing improvidence, Americans have plundered the resource that made it possible to realize their dream.

Moving across this country in the greatest march of occupation ever known, they have exploited and abused this soil. As a result, our vital land supply has been steadily sapped by the heavy drain of soil erosion.

Since the first crude plow uprooted the first square foot of sod, and since man's axe first bit into virgin forest, erosion of the soil has been a problem. It is as old as history. Down through the ages it has influenced the lives of men and the destinies of nations and civilizations. In the United States today, no problem is more urgent.

Millions of acres of our land are ruined, other millions of acres already have been harmed. And not mere soil is going down the slopes, down the rivers, down to the wastes of the oceans. Opportunity, security, the chance for a man to make a living from the land—these are going too. It is to preserve them—to sustain a rewarding rural life as a bulwark of this nation, that we must defend the soil.

This nation is still producing bountiful crops. But many thousands of farmers already feel the pinch of erosion. Tens of thousands of them are finding it increasingly difficult to eke out a living on eroded land almost regardless of agricultural prices.

In other words, even in this young nation, pressure on the land already has become acute in many localities. Many areas have been damaged to such an extent by erosion that not enough productive soil is left for the present population. In Puerto Rico, portions of the Southern Piedmont, and the Rio Grande Valley, for example, erosion already has crowded many people off the land and brought others to the level of precarious subsistence farming. Some of this land can be stabilized, and some severely impoverished areas can be improved, but many land users must seek better soil elsewhere if they are to remain in the business of farming or ranching. Today the nation has an abundance of land, but not enough *good* land. Probaby, if there had not been so much good land in the beginning, there would not have developed the early idea that the productive soil of America was limitless and inexhaustible. This erroneous appraisal of the land resource, passed along as a tradition, accounts for much of our costly steep-land tillage, overgrazing, and failure to defend vulnerable soil from the ravage of erosion.

The present plight of the land brings to mind the extremes to which other countries with small areas of arable soil must go in order to make maximum use of every available acre. In southern France, for example, certain poor soils are utilized under a rotation of fish culture with grain production. In Italy, under the program of the *Bonifica Integrale*, many areas of severely gullied steep slopes are being smoothed down with explosives in order to reclaim them for agricultural use. Always, where populations have increased and agricultural lands have been exploited and wasted, people have looked beyond their borders for additional land. This urge has brought about conquests, wars, and migrations to new lands. Permanent agriculture has been achieved in only a few regions, for the most part of relatively small size, throughout the world. Some parts of the world, blessed with gentle rains, favorable soil, moderate slopes, native skills, and inherent love for the soil, have been held securely. Elsewhere—in Peru, Guatemala, Mexico, the Philippine Islands, parts of Europe, and China—people of primitive culture in ancient civilizations, bench-terraced and, in some instances, irrigated steeply sloping land. The investment of human labor in such enterprises reached fabulous proportions: about $18,000 or more an acre, on the basis of present costs for human labor, went into the walled terraces and irrigation works of the Incas in the Andes Mountains.

A permanent agriculture, then, is possible, even where the land is highly vulnerable to erosion, when people are willing to pay the price of protecting it. Where the price has not been paid, civilizations have disintegrated and disappeared. If necessary for survival, the American people undoubtedly would bench-terrace all their tilled land, as did the Incas, but it would be done at an undreamed of cost. Fortunately, American agriculture is now in a stage where heavy costs may be avoided by consistently working with, rather than against, natural forces, and by provident action based on a thorough diagnosis of the present problems of land use.

All our experience has demonstrated that erosion can be controlled in a practical way. The need is for forthright, determined, nation-wide action. Today's necessity for public action is the outgrowth of yesterday's failure to look more carefully to our land. Foresight in the last century, during our march of agricultural occupation, would have produced a different result. Today we are simply retracing our steps across this land in a march of agricultural conservation. "Soil Conservation" is primarily concerned with this second march.

Looking across the background of events which led up to the present national program of soil and water conservation, a number of points stand out in the perspective of the author as prominent mileposts along the way. Among these are the findings of the soil survey of Fairfield County, South Carolina, which, in 1911, disclosed that 90,000 acres of formerly cultivated land had been so cut to pieces by gullies that it had to be classed as *rough gullied land*, and that an additional 46,000 acres of formerly rich bottomland had been converted into swampy *meadow* land because the streams, gorged with the products of erosion, had lost their original channel capacities. This was probably the first survey of a large area in America which pointed specifically and quantitatively to the wholesale ravages of unrestrained soil erosion.

Another outstanding event was the publication of two bulletins giving the results of measurements of the rate of soil and water losses from definite types of land. One of these bulletins was published by the Agricultural Experiment Station of Missouri, in 1923 (Erosion and Surface Runoff under Different Soil Conditions, *Research Bull.* 63); and the other by the Spur branch of the Agricultural Experiment Station of Texas (Factors Influencing Runoff and Soil Erosion, *Bull.* 411).

Another step toward national recognition of the seriousness of erosion was the educational campaign carried on by the U. S. Department of Agriculture in the late 1920's. In this effort, Soil Erosion, A National Menace, *Circ.* 33, published in 1928, played an important part. This program of education aroused nation-wide interest, and on Dec. 18, 1928, the House of Representatives adopted, without a dissenting vote, the Buchanan Amendment to the Agricultural Appropriation Bill for the fiscal year 1930, appropriating $160,000:

> To enable the Secretary of Agriculture to make investigations, not otherwise provided for, of the causes of soil erosion and the possibility of increasing the absorption of rainfall by the soil in the United States, and to devise means to be employed in the preservation of soil, the prevention or control of destructive erosion and the conservation of rainfall by terracing or other means, independ-

ently or in cooperation with other branches of the Government, State agencies, counties, farm organizations, associations of business men, or individuals.

The Agricultural Bill, including this amendment, was subsequently adopted by the Congress.

This amendment followed a detailed inquiry by its author, Representative James P. Buchanan of Texas, into the problem of soil erosion and its prevention before the Subcommittee of the House Committee on Appropriations (Hearings on the Agricultural Appropriation Bill for 1930, 70th Cong., 2d Sess., pp. 310–330, Nov. 21, 1928).

As the result of this and subsequent appropriations, ten soil-erosion experiment stations were established on important types of farm land in various parts of the country. Information quickly acquired through these investigations, together with that derived from the work of the state experiment stations, proved a vigorous stimulant to the educational program. This information spread rapidly around the world and undoubtedly had much to do with the present widespread interest in the problem of soil conservation in various parts of Africa, Australia, and other countries.

On Sept. 19, 1933, the Soil Erosion Service was established in the Department of the Interior with $5,000,000 provided under the authority of the National Industrial Recovery Act. Almost immediately soil and water conservation demonstration projects were established in a large number of erosion problem areas throughout the country. Behind these original projects, and every one established later, was the basic, underlying principle that permanent results in conservation of the soil could not be attained except through the coordinated treatment of the different kinds of land that make the farms, ranches, and watersheds of the nation, in accordance with their specific needs and adaptabilities. By making use of the various proved methods for meeting these needs, and applying them so that one measure would support another and so that what was done in one field or on one farm would contribute protection to another field or farm, remarkably good results were attained.

The great dust storm of May 12, 1934, stimulated national interest in the problem of erosion. This spectacular dust cloud was the first one in history big enough to retain its identity as it swept across the country from the Great Plains to beyond the Atlantic Coast. It blotted out the sun over a large part of the nation and sifted through the windows of New York skyscrapers. When that happened it began to dawn on the public that something had gone wrong with the land resource of the nation.

Another milepost of historic importance was the passage of the Soil Conservation Act of Apr. 27, 1935, in which Congress definitely committed the national government to the policy of soil conservation. The preamble of that act reads:

> It is hereby recognized that the wastage of soil and moisture resources on farm, grazing, and forest lands of the Nation, resulting from soil erosion, is a menace to the national welfare and that it is hereby declared to be the policy of Congress to provide permanently for the control and prevention of soil erosion and thereby to preserve natural resources, control floods, prevent impairment of reservoirs, and maintain the navigability of rivers and harbors, protect public health, public lands and relieve unemployment.

This was an historic event, as was the inclusion in the Omnibus Flood Control Act of 1936, of a specific provision for erosion-control and water-retardation work over upstream agricultural lands to supplement the control effect obtained with large engineering installations downstream, along the channels of major streams. A number of other acts including erosion-control provisions have also been passed recently. Altogether, Congress has made possible the launching of a great and beneficent program for the conservation of the nation's most indispensable asset—its agricultural lands.

Another epochal step in the history of soil conservation in the United States was the passage by a large number of states of Soil Conservation District Laws, beginning with the enactment of the first act of this kind in Arkansas in 1937. Under these state authorizations, districts have been established in many parts of the country, comprising many millions of acres; work in them is progressing rapidly and effectively.

Thus, within a few short years, the problem of erosion control and soil conservation has been moved out from a position of comparative obscurity to become the objective of a national policy and the basis for an expanding program of work on the land.

Much success and much progress have been achieved under the national program since its inception. Farmers generally, as well as the public, have approved this type of work, although its full value cannot be accurately determined this early. Appraisal of the importance of the land to national welfare demands consideration of the future and recognition of the land as a resource that needs to be defended forever in order that it may remain productive and continue to support the population.

"Soil Conservation" explores the wide ramification of the land problem into many fields—physics, chemistry, and biology, economics and sociology, climate, soils, ecology, geology, engineering, and others. The point is stressed that lasting accomplishment—a permanent agriculture—can be achieved only by coordinating the knowledge of many sciences toward a common objective. Major emphasis is placed on (1) the erosion process; (2) physical effects of erosion on land, vegetation, and agriculture; (3) economic, social, and human-welfare aspects of the erosion problem; (4) relation of erosion to floods and siltation; (5) conservation action (work on the land); (6) techniques, plans, and programs for soil and water conservation; (7) search for new and improved methods for defense of the soil resource and for conservation of rainfall, wildlife, and other resources dependent on the land; and (8) results obtained through (*a*) direct and coordinate application of conservation techniques and (*b*) education. More concisely, the primary objective of this volume is to present a comprehensive statement of the *science and practice of soil and water conservation*. Simplification of presentation is sought through division of the material under two major groupings: Part 1: The Problem; Part 2: Soil Conservation.

If this volume seems to center about the program of the Soil Conservation Service, it is because the author is most familiar with that program, having had intimate association with its conception, inception, and development. It is fully understood that other organizations, Federal, state, and private, as well as individuals, have contributed a vast amount of fundamental information, education, and work on the land to the development and furtherance of various phases of the conservation movement, including defense of the soil through better land use. The term *national program* used frequently throughout the text may be open to possible misinterpretation. It should be explained that the word *national* as used here is not synonymous with *Federal*, nor does the term *national program* imply activities of the national government alone. On the contrary, it is used to designate the whole soil-conservation movement, in which not only the national government, but also the states and many private organizations and individuals, are taking a vital part.

The author's experience of thirty-six years' work studying land problems and conducting land surveys in the United States and other countries, as well as the experience of other specialists and travelers, has been drawn upon in the preparation of the text.

The toll of soil wastage by erosion would have been much greater but for the beneficent effects of soil-building crops, crop rotations, cover crops, improved varieties of crops, etc., developed through the activities of many patient, persistent, and able specialists and land users in this and other countries. To this large group of workers the nation owes a debt of gratitude that can scarcely be paid. The author makes full acknowledgment to these men and women for the wealth of material that has been drawn upon in the preparation of this volume. Limitation of space has imposed severe restrictions on specific citations and credits; it also has necessitated laying aside much material pertinent to a complete treatise on the subject, as well as the details of the programs of other agencies engaged in related fields of conservation. Only indirect acknowledgment can be made of the work of a long list of those who have made valuable contributions in the field of soil and water conservation and better land use by referring to two recently published bibliographies: Bibliography on Soil Erosion and Soil and Water Conservation, U. S. Dept. Agr. *Misc. Pub.* 312, 1938; and A Selected Bibliography on Management of Western Ranges, Livestock, and Wildlife, U. S. Dept. Agr. *Misc. Pub.* 281, 1938.

A few of those to whom the author owes much for their painstaking assistance in compiling and preparing technical data and editing manuscript are: Louise M. Phillips, Gordon K. Zimmerman, George A. Barnes, R. H. Davis, Lois Olson, C. W. Thornthwaite, C. R. Enlow, John F. Preston, G. C. Dobson, Ernest G. Holt, W. R. Van Dersal, T. B. Chambers, C. L. Hamilton, F. J. Crider, E. A. Norton, H. E. Middleton, W. C. Lowdermilk, A. J. Pieters, G. W. Musgrave, C. F. Sharpe, J. A. Bonsteel, R. A. Winston, C. E. Ramser, Angus McDonald, S. W. Cosby, W. A. Rockie, A. H. Paschal, H. G. Lewis, E. J. Carpenter, R. L. Boke, E. A. Kinnear, Arnold Davis, Morrill Tozier, S. B. Detwiler, and Jess Davidson.

HUGH HAMMOND BENNETT.

Washington, D. C.,
*November*, 1939.

# CONTENTS

## PART 2

## SOIL CONSERVATION

# PART ONE
# The Problem

# Chapter I. The Problem in the United States

## The Virgin Land

The earliest settlers arriving on the North American continent found a land richly endowed by nature and virtually unexploited by man. Except in an inconsequential way, the aborigines had done little to cultivate this land or change its virgin character. Before the gaze of the transplanted Europeans lay a vast wilderness teeming with apparently inexhaustible stores of game, fish, fur, timber, grass, and soil. Across the broad expanses of the country, from semitropical to boreal climates, from humid to arid regions, spread an infinite variety of soil types, topographic and climatic conditions, and vegetative types and patterns. Almost everywhere the fertile land supported some kind of vegetation—trees or grasses, shrubs or chaparral. In all but a few scattered barren areas, dense stands of perennial plants sheltered the ground from the elements and enriched it with their decaying material. Soil nourished vegetation, and vegetation protected soil, in a compact of mutual advantage and growth.

Rains fell on the land, and snows melted with the changing seasons; but water tended to move slowly over the ground surface, checked and kept clear by the tangled canopy of vegetative growth. The deep, humus-charged, granular topsoil, perforated even into the subsoil by decaying plant roots and burrowing earthworms, insects, and animals, soaked up the raindrops, which filtered down to nourish the growth of vegetation or to replenish underground reservoirs and springs. Rivers ran clear, except in flood, when abrasive rushing waters tore soil from the banks, sometimes muddying even the Missouri and the Mississippi. Generally speaking, however, the natural circulation of waters was a uniform and orderly process. Flood heights and silt-laden streams were the spasmodic exceptions in a land of prevailing harmony and balance. Topsoil was removed from the land surface no faster than it was built up from beneath by the slow, complex processes of nature.

## The March of Occupation

Into this virgin land the eager colonists entered with energy and enthusiasm. They began a transformation of the earth's surface that is probably without parallel in the history of the world. The occupation of the continent was accomplished not through steady infiltration of population into undeveloped regions but rather through a remarkably rapid advance over a wide front by farmers, stockmen, prospectors, miners, trappers, loggers, explorers, and adventurers. Along the line of advance, there was little thought of conservation or of depleting resources. With a

Fig. 1.—The trees, underbrush, and mat of leafmold in this Maryland area of virgin land have anchored this soil safely to the landscape. Nearly all the rainfall is absorbed. *(Photograph by Soil Conservation Service.)*

country of immense potential wealth beckoning for development, it is small wonder that the emphasis lay, unconsciously, on speedy exploitation.

For a time, the advance was checked by an extensive belt of forest land (Fig. 1) bordering the Atlantic and eastern Gulf. Cutting, girdling, and burning acres of timber required hard work and patience. For more than a hundred years, settlement was largely confined to a relatively narrow strip lying along the Eastern seaboard. Eventually, however, with the pressure of new population from Europe, the westward trek began. Moving slowly at first, the pioneers set to work on the virgin forests with axe and fire. Little by little, the land was cleared for farmsteads; vast quantities of timber were burned at the "logrollings" of the settlers; trails and roadways were opened into the heart of the continent.

By about 1830, most of the better land east of the Mississippi was occupied. Then, across the grasslands lying west of the river, the migration moved at a much swifter pace. Over vast areas the prairie sod of the tall-grass country was broken with amazing speed. Before long the short-grass country of the Great Plains was also invaded, first by buffalo hunters, then by ranchmen, and finally by farmers. By 1890, most of the better lands lying within the borders of the United States had been settled. After little more than 200 years from the beginning of the westward movement, the last frontier was dissolved in the Pacific Ocean.

History records the daring of the early pioneers, who blazed the path for a new nation, and the heroic suffering of the settlers who followed. In many respects it is a thrilling story of adventure, hardihood, and courage. The march across the continent was accomplished in the face of great odds and sometimes at tragic costs in human misery. It was followed by a rapid and energetic development in agriculture, industry, commerce, transportation, and communications, which finally produced one of the greatest nations on the face of the earth.

## Depletion of Natural Resources

Both the march of land occupation and the ensuing national development were accompanied, however, by a prodigious wastage of the resources with which nature originally stocked the land. The white inhabitants of this country, in their "conquest of the wilderness" and their "subjugation of the West," piled up a record of heedless destruction that nearly staggers the imagination. Slopes once clothed with mighty forests now lie bare and stark. Formerly rich lands are riddled with gullies (Fig. 2). Level plains country that once supported lush stands of native "short" grasses is overgrown with weeds or covered with shifting sands left in the wake of dust storms.

What caused this tragic transformation? What happened to the bountiful land that inspired early explorers to enthusiastic comment and rhapsodic description? The answer lies largely in a false philosophy of plenty, a myth of inexhaustibility, which prevailed generally for many years and persists, in some quarters, even at the present time.

Yet in the time of our forefathers this was a normal, perhaps an inevitable, reaction to environmental conditions. Nearly everywhere the early settlers faced rich farm and grazing lands (Fig. 3) which stretched away as far as the eye could see. There was every reason to conclude that the agricultural domain was limitless and inexhaustible. Free land extended to the far horizons. Occupation was encouraged by the Federal Government, by the territories and states, by railroads and land companies. Go West, and take your choice of free land or cheap land, was the advice to young and old. It was a dramatic era—an era of empire building.

Fig. 2.—This maze of gullies, which crosses an entire county in the Southern Coastal Plain and cuts through portions of two adjoining counties, has permanently destroyed more than a hundred thousand acres of some of the best land in the locality. The original gully began about 60 years ago in a cultivated field. Stewart County, Georgia. (*Photograph by Soil Conservation Service.*)

Fig. 3.—Sweetwater County, Wyoming, presented this scene to the Hayden Expedition in 1870. Thick grasses grew nearly everywhere at that time, with trees in the stream bottoms. Streams flowed clear and soil erosion was of no importance.

For a hundred years the settlers came by ox cart, covered wagon, and railway. For a hundred years this hurried settlement of the West absorbed the rapidly increasing population of the nation. It was only natural that those who took over the West, like those who occupied the East, should look upon the grasses of the virgin prairies, plains, and tablelands and upon the trees of the valleys as nothing more than obstacles in the way of human progress.

The fact remains, however, that much of the land was abused, mined, and ravaged. In the turmoil of national growth, abundant resources were reduced to a state of impoverishment or near-extinction. Buffaloes were slaughtered by the thousands, merely for their hides. Trappers took their harvest of pelts without restriction or restraint. Protective forests were cut from sloping hillsides and entire watersheds; immense areas of grassland were broken or bared by the onrushing settlers with their plows and their livestock. Minerals were extracted, and their wealth dissipated in a surge of exploitation.

The story of the passenger pigeon is characteristic. The last of these birds died in a Cincinnati zoo in September, 1914. Ornithologists say that once this species was one of the most abundant game birds ever known in any country. Their flights frequently darkened the skies; the branches of trees are said to have been broken off by the very weight of their numbers settling to roost. Yet within a few generations, the legions of this species have been effaced from the earth. Should man dwell upon this planet for millions of years, he would never behold another passenger pigeon.

In like manner, other valuable resources have been exhausted and continue to be exhausted. What the final result will be in terms of the national economy no one can predict with accuracy. One fact, however, is eminently clear. The potential wealth and living standard of this nation, or of any nation, depend ultimately on its store of natural resources. If, through carelessness and neglect, these resources are wasted beyond a certain point, the whole structure of national achievement must be impaired.

## Soil Erosion

Out of the long list of nature's gifts to man, none is perhaps so utterly essential to human life as soil. And topsoil is the most vital part of soil (made up of topsoil plus the layers beneath). Lying at an average depth of about 7 or 8 inches over the face of the land, this upper layer of the soil is the principal feeding zone of the plants, which provide food for human or livestock consumption, fiber for clothing, and timber for shelter. Soil constitutes the physical basis of our agricultural enterprise; it is a *sine qua non* in the production of practically all food (except fish), of all fiber (without exception), and of all wood (without exception). Under many conditions, however, it is the most unstable of all major natural resources.

Water or wind, in moving across the ground surface, exerts an abrasive force which picks up soil particles and carries them away in suspension.

In a natural, undisturbed environment, the dense cover of vegetation (Fig. 4) retards this surface transposition of soil to a pace so slow, generally, that new soil is formed from the parent materials beneath as rapidly as the finished product (topsoil) is carried away from above. Under such conditions, the removal of topsoil is known as *normal erosion*, sometimes referred to as *geological erosion* or the *geologic norm of erosion*. It is a normal process, proceeding with the tediousness of centuries. It abrades at one place and builds (aggrades) at another. In slowly sculpturing the highlands of the world, it contributes material for the development of alluvial plains, valley fills, and aeolian deposits.

Fig. 4.—Native grasses provide an excellent protection for the soil of this Oklahoma area. (*Photograph by Soil Conservation Service.*)

Where the land surface is bared of protective vegetation—as it must be under cultivation—the soil is exposed directly to the abrasive action of the elements. Transposition processes of an extremely rapid order are set in motion. Stripped of the protective cover that normally anchors soil to the landscape, this indispensable material frequently is moved a thousand times faster than under natural conditions. This accelerated phenomenon of soil removal is known as *soil erosion*. Unless steps are taken to check its progress, it becomes the most potent single factor in the deterioration of productive land.

This volume is primarily concerned with processes, effects, and controls involved with soil erosion—the accelerated cutting away of land resulting from man's occupation and use. Fundamentals having to do with this abnormal process and its rate of progress, as well as with methods for its control or prevention, cannot be understood, evaluated, or adequately planned for, however, without an understanding of basic processes and rates pertaining to the geologic norm of soil removal. The final answers to all the questions involved cannot be given now; nevertheless, much valuable information has been acquired through the instrumentalities of research, surveys, and practical work on the land.

## Wastage of Soil in the United States

National habits of waste in this country have nowhere been exhibited more flagrantly than in the use of agricultural land. From erosion surveys, soil surveys, and other measurements of soil losses, it is estimated that

Fig. 5.—Severe erosion in this Oklahoma field where the crop rows ran downhill. (*Photograph by Soil Conservation Service.*)

erosion in the United States already has ruined, or seriously impoverished, approximately 282 million acres. From an additional 775 million acres, erosion has stripped away varying proportions of the fertile topsoil. Considering only cropland, it is estimated that erosion has ruined about 50 million acres for further practical cultivation. Another area of cropland approximating 50 million acres is bordering on the same condition. Nearly 100 million acres more, still largely in cultivation, has been severely damaged by the loss of from one-half to all the topsoil. On at least another 100 million acres of cropland, erosion is getting actively under way.

Examining the details, it must be admitted that there was in this country a large aggregate area of inferior land to begin with, much of which was so obviously poor that it has never, to this day, been touched by the plow. Vast areas in the West were too rough, too steep, too stony, or too arid for any use other than limited grazing or forestry. In the East there were large tracts too steep, too stony, too wet, or too sandy for plowing.

Nevertheless, many fundamental principles of sound land use were grossly violated—tens of thousands of acres of excessively steep land were plowed. Within 10 years, 20 years, 50 years, more or less, the soil has been washed away, and a condition of permanent, or near-permanent, ruin is the result. Even where the land itself was not too steep for a limited and conservative agriculture, farmers waded in with plow and axe, stripped nature's protective cover from the ground (Fig. 5), and indiscriminately, without regard for the future, laid bare the rich soil to the erosive forces of wind and water. This was not done maliciously but simply without thought of the consequences.

## Soil an Indispensable Resource

Lack of foresight and restraint, starting in the early Colonial period and continuing through the present day, has created in this country a land problem of tremendous implications. What makes the situation so grave is the irreplaceable nature of soil. Once this valuable asset leaves a field, it is as irretrievably lost as if consumed by fire, as far as that particular field is concerned. It cannot be hauled back economically even though temporarily stranded only a short distance down the slope. A thousand tons would be required to cover one acre to a depth of 7 inches.

Soil is reproduced from its parent material so slowly that we may as well accept as a fact that, once the surface layer is washed off, land so affected is, from the practical standpoint, generally in a condition of permanent impoverishment. As nearly as can be ascertained, it takes nature, under the most favorable conditions, including a good cover of trees, grass, or other protective vegetation, anywhere from 300 to 1,000 years or more to build a single inch of topsoil. When 7 inches of topsoil is allowed to wash away, therefore, at least 2,000 to 7,000 years of nature's work goes to waste. The time involved may be much longer; the building of the second inch may require many more years than the building of the first inch at the surface, and so on downward. Studies of old eroded areas abandoned to trees or other types of vegetation indicate that the building of soil by the natural process generally proceeds from the surface downward.

## Widespread Damage of Erosion

Probably the most reliable picture of the extent of erosion in this country can be obtained from the results of a nation-wide reconnaissance survey, made by the Soil Erosion Service of the United States Department of the Interior during the summer of 1934. The figures gathered at that time, together with more detailed studies in numerous problem areas, make it possible to estimate with reasonable accuracy the amount of land affected in varying degrees. The survey revealed that erosion is much more widespread in the United States than had previously been imagined. Some evidence of its damaging effects was found over more than a billion acres of crop and grazing land.

In addition to the 50 million acres of cropland now virtually useless for further production, because it has been stripped of topsoil or riddled with gullies, another 150 million acres of arable land has declined far enough to make farming difficult or unprofitable. Over an additional area of nearly 680 million acres of all kinds of land, traces of water erosion are now discernible; and on much of this land the damage is constantly increasing in severity. Finally, a large area, located in the Great Plains from Texas to North Dakota and in other parts of the West, is characteristically subject to wind erosion, wherever exposed through the activities of man. The survey indicates that this form of erosion is active in some degree over more than 200 million acres of farm and grazing land.

## Annual Losses of Soil

Available measurements indicate that at least 3,000,000,000 tons of solid material is washed out of the fields and pastures of America every year. It is estimated that about 730,000,000 tons of solid matter is discharged annually into the Gulf of Mexico by the Mississippi River alone.[1] These materials come largely from the farms of the Mississippi Basin; as alluvial deposits, they form land richer than the flood plains of the Nile. But the sediment entering the oceans represents merely a fraction of the soil washed out of fields and pastures. The greater part is piled up or temporarily lodged along lower slopes, often damaging the soil beneath; or it is deposited over rich, alluvial stream bottoms or in channelways, harbors, reservoirs, irrigation ditches, and drainage canals.

## Losses of Plant Food

The vast quantity of our soil wasted every year contains 92,172,300 tons of the five principal elements of plant food (phosphorus, potassium, nitrogen, calcium, and magnesium), as computed from the average of analyses of 389 samples of surface soils collected throughout the country (1.55 per cent potash, 0.15 per cent phosphoric acid, 0.10 per cent nitrogen, 1.56 per cent lime, 0.84 per cent magnesia).[2] Of this total, 43,361,000 tons consists of phosphorus, potassium, and nitrogen, the principal

[1] Lower Mississippi River Delta, Louisiana Dept. Cons., *Geol. Bull.* 8, 1936.

[2] Bennett, H. H., and Chapline, W. R. Soil Erosion a National Menace, *U. S. Dept. Agr., Circ.* 33, 1928.

ingredients of commercial fertilizer. According to estimates,[1] approximately 668,000 tons of phosphorus, potassium, and nitrogen was used in the United States during the fiscal year ending June 30, 1934. The same authority estimates the value of commercial fertilizers sold in the United States during the calendar year 1934 at $158,500,000.

In other words, erosion removes from the country's fields and pastures every year available and potential plant food amounting to about sixty times the available plant food returned to the soil in various forms of commercial fertilizers, assuming 1934 to be a fairly representative year

Fig. 6.—Subsoil. Hard, dry, and baked by the sun, this land has little value for productive purposes. The topsoil was blown off this Texas field to the depth of plowing. (*Photograph by Soil Conservation Service.*)

with respect to fertilizer usage. Furthermore, erosion removes not only the plant food itself but actually the entire body of the soil—plant nutrients, humus, beneficial microscopic organisms, and all other constituents. Plant food can be restored to soil worn lean by cropping or leaching; but when the soil itself is washed into the streams and oceans, nature can rebuild its counterpart only after centuries of activity.

## Subsoil Farming

Where farmers operate on land stripped of its vitally important surface layer of topsoil (Fig. 6), they frequently have but the slimmest

[1] Mehring, A. L., and Smalley, H. R. "A Survey of Plant Food Consumption in the United States for the Year Ending June 30, 1934." National Fertilizer Association, Washington, D. C., 1935.

opportunity to make a satisfactory living, whether prices are up or down. They have, in short, been reduced to the status of subsoil farmers; and subsoil farming too often is the equivalent of bankrupt farming on bankrupt land. According to measurements on a variety of types of important agricultural lands, the productivity of these eroded slopes, as compared with the original soil, has been reduced by 35 to 97 per cent.

In the last analysis, soil is the raw material of agriculture—the primary source of its output. Without it a farmer has no more chance of producing a satisfactory crop than an automobile manufacturer has of turning out motor cars without steel or rubber. Nevertheless, many of these subsistence or submarginal farmers, operating on erosion-impoverished land, continue to produce something and, in the aggregate, enough to compete, in a price-depressing way, with those farmers who are tilling good land.

## Indirect Damage of Erosion

Aside from the destruction and impoverishment of farm land, erosion has indirect consequences which threaten the permanence of investments amounting to billions of dollars in navigation, power, municipal water supply, and irrigation developments. Products of surface wash and gully excavation are carried away by storm waters and often deposited in stream channels or reservoirs. Yawning gullies concentrate rainfall and discharge it at maximum speed to gorge the channelways of tributaries and trunk streams with destructive floods. As erosion strips away the absorptive top layers of soil, rainwater finds more difficulty in penetrating the frequently less permeable layers lying beneath. In many sections of the country, the effects of drought on plant life have been aggravated by progressive soil removal.

## Cost of Erosion

Conservative estimates indicate that the annual monetary cost of erosion in the United States amounts to at least $400,000,000 in terms of lost productivity alone. This loss already totals probably not less than $10,000,000,000; and unless erosion is effectively curbed, the probable future costs will be equally gigantic. The annual $400,000,000 direct loss would, within fifty years, accumulate to not less than $20,000,000,000; and since unrestrained erosion progresses at an increasing rate (the subsoil usually being more erodible than the topsoil), the cost may extend to $30,000,000,000 or beyond. To this would have to be added huge losses due to (1) clogging of great reservoirs and shoaling of stream channels with the sedimentary products of erosion; (2) the abandonment of irrigated areas dependent on reservoirs; (3) the virtual abandonment of large agricultural sections; (4) the economic devastation of large western areas dependent on grazing; and (5) the disintegration of rural communities and transfer of large farm populations to relief rolls or to new means of livelihood. Furthermore, soil wasted by erosion is irreplaceable; accordingly, from the long-time view, from the standpoint of a nation that would be permanent, the real value of land is indeterminate.

## History of Erosion

Accelerated erosion of the soil is not merely a recent threat to human security; it is as old as agriculture. In all probability, it began when the first rain struck the first furrow turned by a crude implement in the hands of prehistoric man. It has been going on ever since, wherever man's efforts to gain a livelihood from the soil have led him to remove the natural cover of protective vegetation.

History is largely a record of man's efforts to wrest the land from nature, because man relies for sustenance on the products of the soil. Yet too frequently man's conquest of the land has been disastrous; over extensive areas, his culture of the earth has resulted in extreme impoverishment or complete destruction of the very soil resources upon which he is dependent. When this has occurred on a wide scale, the consequence has been the disappearance of civilization from the affected region. Recent archeological evidence indicates that erosion doubtless played a large part in undermining and obliterating many ancient civilizations in Africa, in the Near East, and in Central Asia.

## Susceptibility of the United States

In 1934, the United States had, according to available statistics, approximately 414 million acres of cropland, including idle fields and plowable pastures. Of this total area, about 100 million acres was in corn, 28 million in cotton, and 20 million in potatoes, tobacco, sorghums, and other clean-tilled crops that leave the soil exposed to wind and rain during the growing season. In Europe, excluding Russia, there is approximately the same number of acres of cropland as in the United States, but only 65 million acres is devoted to row crops, as compared with 150 million acres in this country. In other words, the United States, with about the same area of cultivated land as western Europe, exposes two and a half times as many acres to the more serious forms of erosion. Moreover, the rains over much of Europe do not come in such sudden, dashing downpours as over most of that part of the United States where clean tillage is practiced. This combination of dashing rains and vast acreages in row crops in the United States is largely responsible for the present terrific erosion damage.

## Disappearance of the Frontier

In short, little has been left undone to accentuate the gravity of the erosion problem over a large portion of continental United States. The plain truth is that Americans, as a people, have never learned to love the land and to regard it as an enduring resource. They have seen it only as a field for exploitation and a source of immediate financial return. In the days of expanding frontier it was customary, when land was washed, cropped, or grazed to a condition of impoverishment, to pull up stakes and move on to fresher fields and greener pastures. Today such easy migration

Fig. 7.—Sand dune in the Great Plains formed by wind erosion. (*Photograph by Soil Conservation Service.*)

is no longer possible. The country has expanded to the full limits of its boundaries, and erosion is causing a progressive shrinkage of the tillable area (Fig. 7). The early frontier psychology of land treatment must be abandoned once and for all. In its place a new frontier has appeared. A restricted area of land—an indispensable area, subject to still further restriction by the inroads of uncontrolled erosion—has taken the place of a former abundance of land. Now, man must move rapidly over this diminishing area in order to clear away not trees or prairie grasses but old methods of wasteful land use and substitute therefor new methods of conservation that will provide security for the soil and for those living by the soil.

## Necessity for Erosion Control

As matters now stand, control of erosion is the first and most essential step in the direction of correct land utilization on about 75 per cent of the present and potential cultivated area of the nation. Looking to the future, it appears, on the basis of past experience, that within the next hundred years at least 100,000,000 acres of the remaining valuable agricultural land will become severely impoverished through erosion unless adequate protection is provided. This progressive impoverishment, if permitted to continue unchecked, eventually will reduce the fertile area of farm land from the present total of about 450,000,000 to not more than about 150,000,000 acres. Such an area might easily prove insufficient for the maintenance of a satisfactory national standard of living. And beyond a hundred years, which is not a long time in the life of a nation (or should not be), rain and wind will still be moving soil from vulnerable areas.

In other words, accelerated soil erosion presents the nation, not merely the individual farmer, with a physical land problem of enormous importance to the continuing welfare of agriculture in particular and the entire social structure in general. Moreover, beyond this whole land problem exists the intimate physical relationship of eroding land to mounting flood heights and to damaging silt deposits. No permanent solution of these latter two problems appears possible without better control of runoff all the way from the crests of ridges down across the watersheds, where floods originate and silt loads are picked up, on to the very channelways of streams and rivers. Control of runoff means control of erosion; the one necessarily involves the other.

## Techniques of Erosion Control and Water Conservation

Over a period of many generations, farmers in America and in foreign lands have developed, largely by a process of trial and error, a number of devices and practices for the control of erosion and the retardation of runoff. In recent years, agricultural technicians have studied these measures, tested their effectiveness under varying conditions, and made some valuable improvements. Today there is a practical solution, or a partial solution at least, for virtually every erosion problem with which the farmer or stockman may be confronted.

## Responsibility for Soil Defense

Conservation of the soil, in a national sense, requires the adoption of sound land-use principles and practices by agriculture as a whole. The attainment of this objective involves the widespread use of physical measures of land defense and the adjustment of certain economic and social forces tending to encourage exploitation of the soil.

The responsibility for such a national program falls upon both the nation and the individual. National responsibility involves the protection of society's interest in a natural resource of vital importance to the whole people. Government functions properly in discharging this responsibility. Equally strong, however, is the interest of the individual in the land that he owns. National action may be led and aided by government, but the soil must be conserved ultimately by those who till the land and live by its products. Without a widespread recognition of this latter responsibility, any governmental program of soil conservation must be doomed to eventual futility and failure.

# Chapter II. Erosion and Civilization

## Erosion, A Phase of Agricultural History

The history of soil erosion is an integral part of the history of agriculture. Primitive farming, the world over, is closely adapted to the environment in which it develops and has little permanent effect on the land itself. Accelerated erosion begins when the demand for land exceeds the primitive supply; agricultural expansion begins with the competition for land thus developed. Such competition is almost as old as history itself. The earliest account of it comes from the Bible.[1] Cain was a tiller of the soil, and Abel the keeper of the sheep. Cain brought as an offering the fruit of the ground, but the Lord was wroth, whereas Abel's gift of the firstlings of the flock met with favor and respect. Then followed the slaying of the herdsman by the jealous farmer. Up through the ages this rivalry of farmer and herdsman has continued. Eventually, in a modified form, it appeared in the Great Plains of North America.

Farming of the north European type was introduced by the colonists along the Atlantic seaboard and spread westward with the agricultural frontier. In the West, cattle were introduced from Spain via Mexico, and the range pushed northward and eastward. In the Great Plains the two systems eventually came into conflict. The result has been some of the most serious erosion that has occurred on this continent. Farming, overgrazing, and low rainfall conspired to create the "Dust Bowl."

Like agriculture in general, erosion has its roots in the past, and its processes are regionally interdependent because many of them are set in motion by the introduction of new crops and farm practices. Agriculture advanced by borrowing and adapting to the local environment farm practices that had been developed in other regions. When the colonists came to the Eastern seaboard of the United States they cultivated native crops, but their methods of cultivation were those designed for use under conditions of the gentle rainfall and the comparatively smooth topography of English fields. No adjustments were made in these methods to conform to soil or slope conditions.

[1] Gen. 4:1–12.

The agricultural traditions of both Spain and England were hybrid in nature. The Spanish practices were derived from North Africa, through the Carthaginians and the Moors, who had adopted earlier Phoenician crops and farm practices. Many of these crops and practices had been developed still farther east and were introduced into Phoenicia by way of Syria and Mesopotamia. In the later middle ages, England was faced with the danger of food shortage and for the improvement of agriculture turned to Roman sources. Roman practices were based in part on the experiences of older agricultural areas at the eastern end of the Mediterranean and in Mesopotamia.

The earliest known writings on the subject of agriculture date back to about 3000 B.C. They come from the lower delta of the Tigris and Euphrates Rivers, an area that in prehistoric and early historic times served as a center of dispersal of agricultural knowledge.

## The Beginning of Erosion History

### Mesopotamia: The Cradle of European Agriculture

In Mesopotamia, normal erosion from the Assyrian highlands and deposition in the lower reaches of the rivers followed the retreat of the great ice sheets. The alluvial deposits at the mouths of the rivers were exceptionally fertile. Here, exploitive agriculture, accompanied by the cutting of the forests at the headwaters of the rivers, began before the dawn of history. Erosion was accelerated by man's activities, but the exact extent cannot be determined definitely. In flowing over their delta, the Tigris and the Euphrates had a natural tendency to shift their courses. During the period of Babylonian supremacy, they were close enough to permit the construction of irrigation canals which headed in the Euphrates and drained into the Tigris. The irrigation waters spread the sediment carried by the rivers; and where it was deposited, the land gradually became higher. In consequence, when the rivers were forced to shift their courses they moved away from each other. The increased difficulty of irrigation and the decline in agriculture that resulted were definitely determined by the use of the land.

In early postglacial times, the Tigris and the Euphrates flowed into the Persian Gulf near Hit and Samarra, some 600 miles north of the present shoreline. The period of continuous tradition began about 3000 B.C. By this time, Ur and Eridu, which tradition reports as originally seaports, had become inland cities. Since then the shoreline has moved some 200 miles farther south (Fig. 8). The Karun River, flowing westward from the Persian highlands (Fig. 9), also contributed its silt to the Persian Gulf and built up a bar which extended eastward from Basra and protected lower Mesopotamia from the inroads of the sea. The Tigris and Euphrates

Rivers had deposited their silt farther north, and the water behind the bar was free from sediment but dark in color and rich in chemical matter. It was here that the Chaldeans settled in prehistoric times. The soil was extraordinarily productive but subject to inundation during the rainy

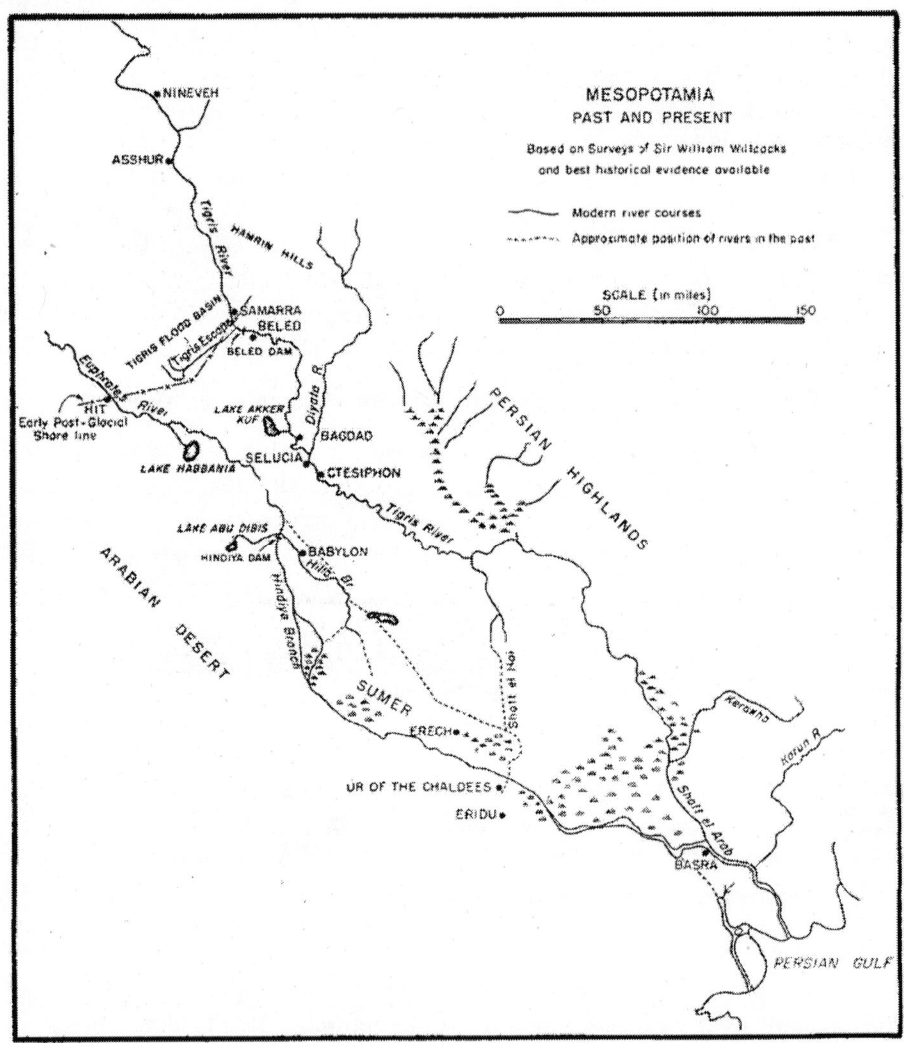


Fig. 8.—Mesopotamia—past and present. (*Revised by Soil Conservation Service.*)

season. For protection the Chaldeans built many miles of immense dikes 100 feet or more in width. Similar environmental conditions are encountered today but farther to the south along the Shatt el Arab.

As irrigation agriculture spread to the land farther up the rivers, the problems of flooding and sedimentation became more serious. The powerful communities farther north in the delta resorted to the only means that they knew for protecting themselves against floods and at the same time preventing the accumulation of sediment in their canals. This consisted of completely shutting off the water from certain branches of the rivers by earthen dams. While protecting the area farther upstream, those in the lower delta were subjected to increased sedimentation and higher floods.

From Bagdad south for a distance of nearly 200 miles, the level of the Euphrates is higher than that of the Tigris, and the ancient irrigation canals drained eastward from the Euphrates to the Tigris. In the lower reaches, the Shatt el Hai carried the main volume of the Tigris. Here, too,

Fig. 9.—Remains of a stone structure (center) surrounded by skeletonized land. Southwestern Persia. (*Courtesy Reed Haythorne.*)

the rivers were close enough to permit drainage between them, and the intervening area was honeycombed by irrigation canals. The remains of their courses can be found today, but it is evident that all are not of the same date. The conquering armies of Babylon brought back with them hosts of slaves, whose chief duty was to keep the irrigation channels free of silt. In spite of an abundance of labor, the ditches filled with sediment and were obliterated by subsequent cultivation.

As the height of both cultivated fields and river beds was gradually increased, the tendency of the rivers to shift their courses became more pronounced. The Shatt el Hai built up its bed until it became higher than the land to the east of it, and eventually the main channel of the Tigris shifted eastward. This increased the distance between the rivers in some places by more than 50 miles. The development of irrigation in the lower delta today requires the restoration of the former Hai channel.

FIG. 10.—Above, abandoned parts of the city of Samarra, Mesopotamia, and surrounding fields are now covered with blowing sand, as shown in foreground. Below, the summer palace of Harun-al-Rashid, Mesopotamia, surrounded by a waste of water- and wind-eroded land. (*Courtesy H. I. Cozens.*)

Along the Euphrates, the irrigated fields lay to the east of the river. Irrigation waters were drained to the east, and the flood waters allowed to escape to the west. Babylon was protected from floods by two large natural depressions, the Habbania and the Abu Dibis. The Hindiya Canal provided additional protection by diverting the water to the west of

Babylon. During the dry season it was barricaded by a temporary dam, which was constructed and destroyed yearly, a task requiring the labor of 10,000 slaves. Sedimentation was most rapid along the Hilla, or main branch of the Euphrates. Today the main stream flows through the former Hindiya flood outlet. This shift has increased the distance between the Tigris and the Euphrates as much as 35 miles in places, and the difficulty of constructing connecting irrigation canals has been increased proportionally.

Near Beled, north of Bagdad, the flood waters of the Tigris were turned into a large basin with a hard bed of conglomerate underlying a

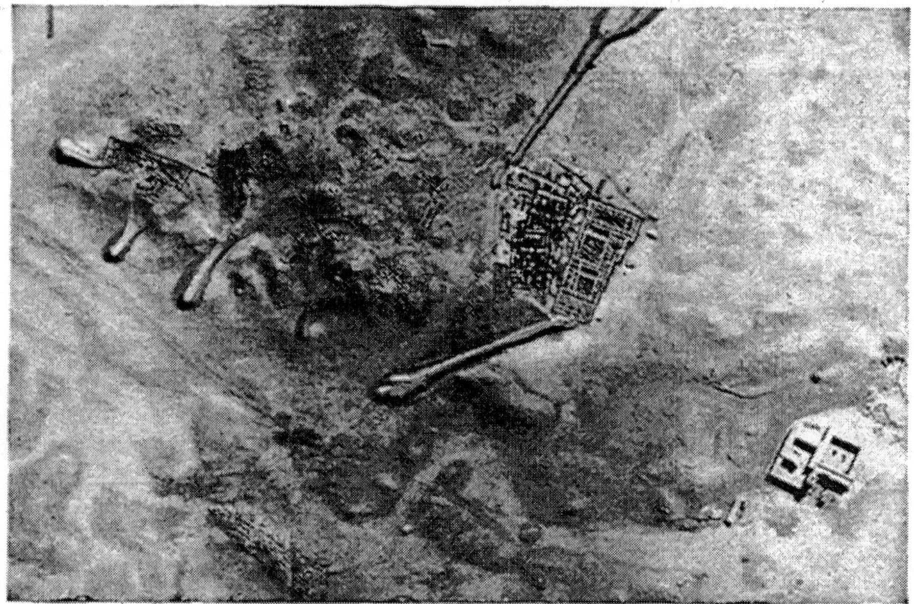

Fig. 11.—Ruins of Tel Asmar, Mesopotamia, and the irrigation works that once surrounded it (faint lines). (*Courtesy Reed Haythorne.*)

thin alluvial deposit. The three heads of the famous Nahrwan Canal, whose construction is commonly attributed to Nimrod, led from the upstream side of the dam. The dam was maintained for 3,000 years and was not destroyed until about 600 or 700 years ago, during the later caliphate days (Fig. 10). Since its destruction, the channel of the Tigris has shifted to the east, and the lands near the head canal, formerly among the most fertile of the Tigris Valley, have been so badly cut by ravines and gullies that, according to Sir William Willcocks,[1] their restoration today is not financially practicable.

[1] Willcocks, Sir William. "Irrigation of Mesopotamia." London and New York, 1917. Accompanied by atlas.

In the past, the outlet basins and canals along the Euphrates provided effective protection against floods; but along the Tigris, floods were never satisfactorily controlled. The three flood basins of the Tigris and Euphrates Rivers had a combined capacity of nearly 5 million acre-feet of water, of which about a fourth was available for irrigation during the dry seasons, but even this was insufficient for the delta when there was heavy irrigation upstream. The competition for water supply was a major cause in the constant struggles between Babylonia and Assyria for political

FIG. 12.—It was necessary to dig about 15 feet or more to reach the tops of the ruins of the ancient city of Khorsabad, Mesopotamia. The covering material consists of wind-blown soil. Khorsabad is supposed to have lain buried for 2,600 years. (*Courtesy Reed Haythorne.*)

supremacy and resulted in repeated abandonment of cultivated land. It was during the periods when Babylon was strong enough to control the whole of Mesopotamia that agriculture was most flourishing in the lower delta. After 500 B.C., when Persia controlled both rivers in their entirety, Babylonian agriculture again prospered.

Wherever irrigation channels were neglected or destroyed by invading armies, the adjacent fields became deserts. To the northeast of Bagdad is a desert of blowing sand in which stand the ruins of a group of cities and extensive irrigation canals. Their existence clearly demonstrates that the area was once productive and prosperous. Among these cities was Tel Asmar which is now being excavated by the Oriental Institute of the University of Chicago (Fig. 11). Vegetation could not survive after abandonment of irrigation, and dust storms increased in violence. The ruins of ancient cities are today covered with wind-borne debris, and the expeditions of the Oriental Institute encountered in their explorations storms that carried dust to heights of over 15,000 feet. After a 30-hour sandstorm it is not uncommon to find that inches of sand have been blown into the excavations.[1]

Assyria, to the north of Bagdad, is a country favored more by position than by local environment. Its rainfall, less than 20 inches annually, comes chiefly in the winter, and droughts are frequent. As the political power of ancient Assyria grew, population increased, and the need for cultivated land and extensive pastures became more imperative. Summer sandstorms were frequent even in early times (Fig. 12), and winter rains washed soil from the sloping lands, leaving bare rock exposed. This may explain in part the need for the westward expansion that began in the second millennium B.C. and for the control of the trade routes from Assyria to the Mediterranean.

## Routes of Communication between Mesopotamia and the Mediterranean

The first trade route to develop between Mesopotamia and the Mediterranean was that from Antioch to the great bend of the Euphrates. The country through which it passed was lacking in sedentary populations, but the mountain regions bear witness to ancient forests that were known to the Assyrians. The cutting of this timber, before the beginning of written history in Mesopotamia, is in part responsible for the height and violence of the floods of the Euphrates and for the sedimentation in the lower delta. The chief products transported by the Euphrates were building stones, asphalt, and lumber from the Taurus, Amanus, and Lebanon Mountains.

The longer, though more practical, route lay along the piedmont south of the Taurus Mountains and stretched in a northward-bending arc from Assur and Nineveh, at the head of navigation on the Tigris, to Antioch. Because food for men and fodder and pasture for animals could be secured along the way, this was the route most commonly followed by armies. The slightly heavier rainfall of the piedmont was supplemented by streams fed by melting snow from the mountains to the north. By 2000 B.C., the Babylonian Empire had extended to the Mediterranean Sea. As a result of this expansion, improved methods for conserving water by the con-

[1] Wilson, John A. Letter to W. C. Lowdermilk, Mar. 17, 1937.

struction of reservoirs and irrigation canals were introduced into Syria and were accompanied by agricultural advancement southward toward the desert of northern Arabia.

The piedmont was a much coveted trade route and was conquered and reconquered by Babylonians, Assyrians, Egyptians, Persians, Greeks, and Romans. Today much of the area is barren, unproductive, and swept by spring floods and summer dust storms. The cities have been deserted, and the soil washed from the fields. Most of the deterioration of the land must be attributed to neglect resulting from the disastrous effects of successive conquests and to the decline in trans-Syrian trade following the downfall of the Roman Empire.

Farther south lay the third important trade route, leading from Antioch, Damascus, or the cities of Transjordania through Palmyra to Dura on the Euphrates. The importance of the desert cities did not emerge until later Hellenic times; and under Roman protection, trade developed rapidly and agriculture thrived locally. To a large extent both were artificially created and lasted only until Roman protection ceased. When the trade with the East shifted to the sea routes, cities declined, fields were abandoned, and wind erosion increased. The ruins of cities and aqueducts have been covered by sand, but the character of the region as a whole has been changed but little. According to Alois Musil,[1] the desert oases of today could support a population as dense as that of the past if a strong political power was established.

## Early Agriculture in Palestine, Phoenicia, and Syria

From the beginning of history, the lands at the eastern end of the Mediterranean Sea have been the home of settled agricultural people. Forests covered the higher slopes of the Amanus and Lebanon Mountains and of Mount Hermon. The rain falling in the highlands sank into the ground, percolated through the limestone bedrock, and emerged lower down the slopes as permanent springs. Because trees were scarce on the lowlands, the mountain forests were regarded with religious reverence. To Abraham and other early nomadic chieftains, the planting of trees near a spring was an act of virtue. The destruction of groves by fire was a punishment sent from heaven, and the destruction of fruit or olive trees during war was an unpardonable offense.

The mountains at the east end of the Mediterranean originally were forested, but cultivation and animal husbandry gradually encroached on the woodland. Today all of the Mediterranean lands show a lower proportion of forests than their climate and mountain relief would suggest.

[1] Musil, Alois. "Northern Neǧd," Am. Geog. Soc. *Oriental Explorations and Studies* 5, 1928 (see App. X, pp. 304–319).

The Hebrews were at first a nomadic, pastoral people, but after settling in the Promised Land of Canaan they adopted the agricultural traditions of their predecessors. The limited lowland areas were reserved for crops; the pasturing of horses and cattle there was forbidden by law until the time of David and Solomon, when more extensive pasture lands were added to the country by conquest. Sheep and goats were relegated to the hill pastures and were a serious factor in preventing the natural regeneration of the forests.

The rainfall of the region is concentrated during the winter months and is followed by a long dry summer period. In Palestine, a single storm often changes a trickling wadi into a torrent. When Barak was battling against the troops of Sisera,[1] Jehovah saved the Israelites by sending a sudden flood which overwhelmed the troops of Sisera and mired their chariots. Recently H. V. Morton[2] encountered similar flood conditions in traveling from Antioch to Selucia. Two miles out of Antioch his car was mired and had to be towed out of the mud by the oxen of a near-by plowman. The next day he set out again on horseback, in hopes of fording the river. This time he was stopped by the flood waters of a small tributary to the Orontes and was forced to postpone the trip until the flood subsided.

The removal of the forests increased the number and violence of the floods. As the water ran off the cleared slopes, it carried with it the soil. Today treeless slopes, with bare rock exposures, form one of the most characteristic features of the landscape.

Trees like those described in the Bible are still found in places and indicate that climatically the area is capable of reforestation; but on many of the slopes practically all of the soil has been washed away, and complete revegetation is impossible. Weitz,[3] in discussing plans for the reforestation of Palestine, recommends the planting of native trees and carobs in the hills which, "owing to the erosion of the slopes which has gone on for centuries—ever since the Jews were exiled from Palestine—and to the continuous felling of fruit and forest trees," have been brought to their present denuded state.

The erosion of the highlands has been accompanied by rapid sedimentation along the lower river courses. In excavating at the base of the mountains near the site of ancient Antioch (Fig. 13), archaeologists have had to dig through as much as 28 feet of water-borne debris to uncover the remains of the former splendor.[4] At the mouth of the Orontes, the

[1] Judges 4:15, 5:21.

[2] Morton, H. V. "In the Steps of St. Paul." New York. 1936.

[3] Weitz, J. An Afforestation Plan for Palestine, *Palestine and Middle East Econ. Mag.*, Vol. 8, pp. 391–394, 1937.

[4] In a letter to H. H. Bennett, Oct. 8, 1934, William A. Campbell said: "The site of ancient Antioch is located between a range of low mountains, an offshoot of the Casian Range, and the Orontes River, so that the depth of the deposition varies from 7 m. to

sediment carried down by the river has converted the old port of Selucia into dry land, and the present city is located 5 miles north of the river mouth. The sediment carried down by the Orontes today can be identified far out in the Mediterranean Sea (Fig. 14).

Fig. 13.—Excavating ancient Antioch, Syria, from beneath a covering of erosion debris. (*Courtesy Reed Haythorne.*)

In the past, erosion and deposition were controlled, to a large extent, by the use of erosion-control practices. The construction and cultivation of irrigated terraces comprised Phoenicia's chief contribution to the agricultural development of the Mediterranean lands. Traditionally, the Greeks acquired their knowledge of irrigation from Phoenicia, where Hercules and Cadmus, the founder of Boeotia, derived their skill. In Palestine, the remains of irrigated terraces are found around Carmel,

---

8.50 m. along the mountain range to .10 m. to .25 m. in the plain along the river. Practically all of the deposit has been caused by the uncontrolled erosion of the mountain range. Following the great earthquakes of the sixth century A.D. and the Persian invasion of 540 A.D. large areas of the city were abandoned, and the debris of the demolition gradually covered by water transported soil and gravel from the mountains to the depths noted above. This deposition has been frequently broken through by the trenches of natives searching for building stone and for marble to burn into lime. Such activities are marked by irregular strata of chipped stone.

"The valley and the district over the ancient city is at present intensively farmed within the limitations set by primitive implements and agricultural practices. The surface of the soil is merely scratched by a home-made wooden plow pulled by oxen, the grain

Gilboa, and Samaria, all of which were renowned for their fertility in ancient times. On the slopes of Judea, broad terraces were constructed for barley fields as well as for the olive and vine. The Hebrews had one distinct advantage over other Mediterranean peoples. Like the Babylonians, to the east, their agricultural principles were incorporated in their

FIG. 14.—Eroded uplands near Antioch, Syria, partly terraced for agricultural purposes. Note sediment of Orontes River extending far into the Mediterranean (light color). (*Courtesy Geographical Review.*)

religious law. To some extent this may explain why agriculture was able to flourish in Palestine even under unfavorable environmental conditions.

The political unrest of the second century A.D. resulted in the destruction not only of the engineering works but also of the agricultural traditions of Palestine. Trees were cut, winter rains washed the soil from the

crops are reaped and threshed by hand, and alternate strips of land are left fallow each year. Only the vegetable crops are fertilized and these by manure of all kinds and decomposed garbage from the town. The covering of the formerly inhabited areas by the erosion of the mountain has added to the area of tillable soil available, but there is no attempt to control the erosion or the rapid run-off of water following a rain. Consequently, crops near the mountain are frequently washed out or covered by water-transported soil, and during the dry season they suffer a good deal from lack of moisture. There is no doubt that the agriculture of the district has regressed greatly since ancient times due to the uncontrolled water erosion and to the inferior agricultural methods of the present inhabitants."

slopes, reservoirs were destroyed, and the neglected irrigation channels filled with sediment. The former productivity of the cultivated slopes cannot be completely restored, even with the more advanced scientific knowledge of today.

Probably the worst destruction at the eastern end of the Mediterranean occurred near Antioch. The countryside is drained by the Orontes River and its tributaries. These tributaries rising in the adjacent wooded mountains furnished a water supply even during the dry summer months. After the forests were cut, an extensive system of terraces and irrigation and drainage canals, combined with scientific dry farming, effectively reduced erosion. Traces of five terraces can still be seen, and some of the old water channels are today being used by the more intelligent farmers. The productivity of the area was renowned throughout the ancient world. Oil and choice wines were exported by way of the Orontes River. In addition to its agricultural assets, Antioch was the terminus of the routes to Babylonia, India, and China and was an outstanding commercial center. According to Libanius of Antioch,[1] the city in 360 A.D. had a population of 400,000. All the forces of nature contributed to her benefit—soil, streams, and a temperate climate—and farmers could be seen driving their plow oxen almost to the summits of the mountains. The fall of Antioch shortly thereafter caused the decline of both commerce and agriculture. In 1934 the population of Antioch, including the army post, was but 28,000. The surrounding hills are bare and rocky, and fields are cultivated by farmers who use a primitive type of plow which merely scratches the surface of the earth.

## Erosion in Greece

Knowledge of terracing and irrigation was introduced into Greece from the lands at the eastern end of the Mediterranean, but the date of their introduction has been lost in a maze of mythology. When Hercules cleaned the Augean stables and slew the Hydra, he was confining the untamed waters into channels so that they would serve rather than destroy the works of man. This is the first account of irrigation in Greece. Before this, the Greeks had been a pastoral people. With the improvement of agriculture they were gradually converted into farmers. Their hilltops were forested and covered with soil, and the fertile valleys produced a surplus of grain. Forests similar to those of Greece during the age of fable still can be seen on some of the smaller islands of the Aegean Sea; but in Cyprus and Crete, the stepping stones between Greece and Phoenicia, the forests have been cut down, the hills have been overgrazed, the soil has been washed from the slopes, and the marshy areas at the mouths of the rivers have increased in size.

[1] Bouchier, E. S. "A Short History of Antioch," p. 13. London. 1921.

As population increased, more farming land was needed. Gradually the forests of Greece were cut down and converted into fields. By the ninth century B.C. the land problem had become acute, and Lycurgus of Sparta formulated the first Greek land tenure laws. About 600 B.C., corresponding laws, attributed to Solon, were enacted in Athens. As more farm land was needed, colonies were established. Most of these, like Libya and Sicily, were grain producers; the island colonies served as way stations along the routes of the grain trade. With the founding of colonies, the Greek states experienced a gradual shift from grain production to the growing of some of the more valuable crops, such as the olive and the vine. Although grain could be produced more cheaply in the colonies than in Greece, the grain trade was constantly endangered by war, piracy, or severe storms. Because of this, the Greek states were always near the margin of starvation. In Thessaly and Boeotia, where there were proportionally larger areas of level land, grain production continued. In the other states, the land had been deteriorating progressively under cultivation until grain could no longer be grown at a profit. Such deterioration was noted by Xenophon,[1] who, in the fifth century B.C., commented on the profit that he was able to secure by purchasing old, worn-out farms, restoring them, and later reselling.

Direct evidence of erosion in ancient times is lacking in Greek literature, but there are references to current agricultural practices. The arts of dry farming and irrigation had developed long before the Greek states became sea powers. Fields were commonly cultivated and left fallow in alternate years, the fallow land being plowed three to five times during the winter. The "thrice-fallowed field" is mentioned by Homer, and Hesiod called the fallow "the guardian against death and ruin." Deep plowing, also, probably was used at the time of Homer, and the Thessalonians employed a special digging tool, the *mischum*, which served much the same purpose as the subsoil plow in this country.

Plowing was regarded as the essence of good farming. The furrows were plowed in straight lines and then cross plowed obliquely. So frequent were the plowings and so close the furrows that it was considered unnecessary to harrow a well-plowed field. Straight furrows were a sign of good plowing. This seems to have originated among the Persians, and Xenophon holds up as an example to the Greek farmers the straight furrows of Cyrus. Such practices left the soil particularly susceptible to washing. As in the semiarid sections of North America, the chief concern in Greece was adequate moisture for crops; the condition of the soil was overlooked. Theophrastus summed up this attitude by saying, "It is the weather rather than the soil that determines the harvest."[2]

[1] Xenophon. "The Economist," XX, 22–26.

[2] Semple, E. C. "Geography of the Mediterranean Region," p. 297. New York. 1931.

Many of the rivers of ancient Greece were characteristically muddy and deposited large amounts of sediment. The Greeks understood the principles of deposition and correctly attributed to this cause the locking of islands to the mainland. Hecataeus of Miletus (500 B.C.) was probably the first to describe Egypt as the "gift of the Nile," a description current to the present day. With this as background, it is not surprising that the Greeks underestimated the attendant dangers of erosion in the highlands and of the increase in malarial swamplands near the river mouths.

When Alcmaeon, after slaying his mother, fled from Arcadia to escape the vengeance of the gods, the Pythian priestess advised him to seek the newest land that had been formed if he were to escape the avenging spirit. Accordingly, he settled on the delta of the Achelous River. The sediment being deposited at his time had already given rise to the tradition that all of the Echinadian Islands, near its mouth, would soon be tied to the mainland. The process was never completed. In the second century A.D. Pausanias[1] commented on the clear water of the river. He attributed this to the shift from cultivation to grazing along its headwaters, a shift resulting from ravaging of the region and the destruction of irrigation works by war. To prove his point, he contrasted this with conditions along the Meander River which "flowing through the lands of Phrygia and Caria, which are plowed every year . . . has in a short time turned the sea between Priene and Miletus into dry land." Similarly, the city of Tarsus in Cilicia, once visited by Cleopatra's fleet, is now 10 miles inland.

The Greeks, however, did little to check the growing evil. The philosophizing on the subject of agriculture, which became popular in the later periods, referred generally to the practices of preceding generations as the ideal and contained few exhortations to farmers to experiment further.

Occasionally, conservation of soil and water was practiced but usually for other purposes. The trade on which the states thrived was dependent on wooden ships. Government control of the lumber trade, therefore, developed at an early date for the purpose of preserving the forests and maintaining a local supply of wood. Frequently, high mountains were sacred to the gods, and forests were spared because of religious motives. To a small extent these forests helped to regulate the runoff and reduced both the flood hazard and soil erosion.

Erosion of the sloping lands of Greece continues to the present. During his three archaeological expeditions to Corinth, Professor T. L. Shear[2] found it necessary to remove over 21,000 tons of erosion debris from the

[1] Pausanias. "Description of Greece," VIII, 24.

[2] Shear, T. L. Excavations at Corinth in 1925, *Am. Jour. Archaeology* (2d ser.), Vol. 29, pp. 381–397; also, Excavations in the Theater District and Tombs of Corinth in 1928, *Am. Jour. Archaeology* (2d ser.), Vol. 32, pp. 474–495.

theater alone. In August, 1906, a heavy rain washed such great quantities of mud into the excavations at Peirene that the Greek government had to clear it away and build a wall above the diggings to prevent a repetition of the catastrophe. Again in 1925, a week's time was required to remove debris that had washed into the excavations and to construct protective dikes. At another time, work at the theater in Corinth was delayed a week in order to protect a bit of mosaic pavement that was in danger of being washed away.

At Athens, also, there is evidence of erosion. On the top of the Acropolis there is no soil at all. Only on the south and west sides, where the theater and the Byzantine wall have formed a barrier, is there any accumulation of soil. It seems unlikely that the bare, rocky hills of Greece could today support a culture comparable to that of the past.

## Erosion in Ancient Italy

Early records reveal that the Greeks, who in the eighth century B.C. settled in the country that is now Italy, found there wooded mountains and fertile lowlands. Even then, Greece was dependent on a foreign food supply, and the forests of Italy supplied timber to maintain the navies. Sicily was known as the "granary of Greece"; later, grain cultivation spread to the peninsula of Italy.

Among the early Greek settlements in Italy, Sybaris[1] was outstanding. For 200 years no Hellenic city could compare with it in wealth and splendor. Although it was located in the lowland between the mouths of the Sybaris and Crati Rivers, it was renowned for its salubrious climate and productive fields. In 510 B.C., Sybaris was attacked and conquered by the neighboring Crotonians. They destroyed the irrigation works, released the waters of the Crati River, and flooded the city. By this time, the forests of the adjacent mountains had been cut down, and floods, which were characteristic of the region, could no longer be controlled. Each flood brought with it sediment which was deposited in the lowlands. The Crati and Sybaris Rivers originally had separate mouths. Today they unite 3 miles from the sea. As channels were blocked, the rivers overflowed and created marshes which formed breeding grounds for the malaria-carrying Anopheles mosquito. The plains became uninhabitable, and the farmers were forced to cultivate fields higher up the slope, thereby increasing both erosion and flood hazards. For hundreds of years the lowlands were abandoned, and even the site of ancient Sybaris was unknown.

The destruction of the Sybarite culture was accelerated by conquest that released the forces of nature; but throughout Italy deforestation and the cultivation of the slopes resulted in more frequent floods. Fields, once

[1] Ringland, A. Mussolini's Sybarites, *Am. Forests,* Vol. 39, pp. 291–297, 334, 1933.

cultivated, were converted into uninhabitable swamps, and the farmers were forced to cultivate land higher up the slopes where there was greater danger of erosion.

Throughout the Mediterranean region, the winter storms were followed by sweeping floods. Vergil describes such a flood, probably from experience encountered on his father's farm: "The swift moving flood from the mountain stream overwhelms the fields, lays low the smiling crops, wipes out the work of the oxen, and sweeps the forest with it headlong."[1] Deforestation also contributed to the severity of and increase in number of floods. According to Semple,[2] deforestation caused an increase in the floods of the Tiber River, which eventually led to proposals, at the beginning of the Christian era, to create artificial lakes to hold back the flood waters. They were not constructed, because they would have required the flooding of some of the most fertile lands of the area. Instead, an additional channel was dug for the lower Tiber to permit the rapid drainage of flood water.

In the Italian peninsula, the rainfall was greater than that of Greece, and the climate was better suited to grain production. Rome was at first a nation of small farmers, and the size of the farms was determined by the amount of land that could be well cultivated by the landowner and his *familia*, which consisted of his immediate family and possibly a few slaves. Roman literature began with the discussion of the theory of agriculture based on the experience on small farms. When Cincinnatus was called to become dictator, he was found plowing his 4 *jugera* (about 2½ acres) of land on the Vatican Hill.

Erosion proceeded slowly; and with a nation of careful farmers, there was an opportunity of developing control measures before the land had become ruined. As in Greece, the first concern was to secure an adequate water supply. Irrigation and dry farming were the accepted practices. In the irrigation channels, sediment accumulated rapidly, and the clearing of the ditches was one of the farmer's most important duties. On feast days, when other farm labor was prohibited, the clearing of drainage canals was permitted. The Romans also recognized the relation between cultivation and soil removal. Frontinus, in the first century A.D., considered that the water of the Anio Canal, which flowed through cultivated land, was too muddy for any use other than watering the Roman gardens.[3]

Rome, at first, was able to produce most of its agricultural needs locally and, in consequence, was a nation of farmers rather than traders. Good farming was regarded as the highest of accomplishments. To attain this

[1] Vergil. Aeneid, II, 305.

[2] Semple, E. C. "The Geography of the Mediterranean Region," pp. 110–111. New York. 1931.

[3] Frontinus. "De Aquis Romae," I–II.

goal, the Roman authors searched the literature of the past, and of other countries, in order that farmers might profit by wider experience. Cato, Varro, and Pliny built upon the work of their predecessors, and punctilious care was given to documentation. Columella, in addition, drew upon the experience of farmers in his native Spain. Vergil's writings were regarded by his contemporaries as a compilation of the best practices of his own times and of preceding generations.

All these authors consistently urged that farmers experiment further, that practices be adapted to individual needs, and that the success of crops demanded that the nature of the soil and the slope be considered. This to some extent enabled the Romans to avoid the less satisfactory practices of the past, but straight furrows continued to be a mark of good farming. Pliny,[1] who lived from A.D. 23 to 79, emphasizes this by advising the plowman "never stop to take breath in the middle," or the furrow will be crooked. But other recommendations are included: "Upon a hillside furrows are to be drawn transversely, only. . . . It is a good plan, too, to leave a channel every now and then if the nature of the spot requires it, by making furrows of a larger size, to draw off the water into drains." Here is one of the very early recommendations for contouring and what amounted to a form of field terracing, such as is used today.

Lucerne, or alfalfa, was particularly favored as a forage crop. The Romans traced its origin to Media and its introduction into the Mediterranean region to the wars with the Persians. Lucerne was valued because its deep roots broke up the soil, but it was early recognized that "it fertilized the ground as well as any manure." The same properties were recognized in beans, lupine, vetch, and other leguminous crops. The bean and vetch were regarded as exceptions to the general rule of frequent plowing, since they could be sown without turning the soil at all, thus saving labor and at the same time reducing the susceptibility of the soil to washing.

Lucerne normally yielded from four to seven cuttings a year and sometimes as many as nine. According to Pliny, a single sowing would last for 30 years. Columella considered 10 years the average life of a field but commented on the fact that the one sowing might last for 20 years. In either case, it added nitrogen and humus to the soil, served as a soil binder, reduced the frequency of plowing, and by so doing reduced erosion. Vergil also mentioned a simple form of rotation consisting of fallow, grain, and legume. Although he considered the fallow more desirable, he conceded that, if it were not practical, grain could be sown on the land directly after vetches or beans, since "all of these have a tendency to make the soil more fertile." Lucerne was introduced via Rome into

[1] Pliny. "Natural History," XVIII, 49.

northern Europe. In the United States, it was introduced by the Spaniards under the name alfalfa, derived from the Moors.

The period of conquest and empire building brought with it a change of agricultural conditions in Italy. North Africa and Egypt together contributed two-thirds of the Roman grain supply. Against this, the Roman farmers could not compete. During the period of the Gracchi in the second century B.C., the distribution of cheap or free grain was introduced. Once established, this custom could not be broken, even though its evils were recognized. By the time of Julius Caesar, over

FIG. 15.—Overgrazed and eroded mountain slopes in northern Italy. (*Courtesy Italian Government.*)

200,000 people in Rome alone were receiving free grain. Cheap grain from Africa caused the ruin of the small farm in Italy, despite all efforts of the agrarian party to revive it. On the large landholdings, or *latifundia*, other crops were found to be more profitable than grain. According to Mommsen,[1] the raising of cattle became more profitable in Italy than any type of crop production (Fig. 15); second in rank were vegetable gardens and olives; and fodder crops and grain production yielded the poorest returns.

After the Second Punic War (200 B.C.), the vine became increasingly important in Roman agriculture. As olive and vine culture increased, the Romans turned to the countries that had been producing them for a longer period in order to discover the best methods of cultivation. In this

[1] Mommsen, T. "The History of Rome," Vol. I, p. 445. New York. 1887. Trans. by W. P. Dickson.

the Carthaginians excelled; and in consequence the Romans translated the entire 28 volumes of Mago, the Carthaginian, into Latin. These were the only Carthaginian books ever deemed by the Romans worthy of translation. With the fall of Rome, these books were completely lost, and our knowledge of Mago's works is derived from the Roman authors who quoted him.

A shift from grain production to animal husbandry might be expected as the size of the farms increased and labor became scarcer. The olive and the vine, however, both required more intensive care. The gradual increase in their production would indicate an actual change in the character of the land. At the beginning of the Christian era it was current opinion that the land of Italy was worn out, and the deterioration of field agriculture was acknowledged. Although Pliny dissented from this opinion, it has been widely accepted by modern as well as Roman authorities. The soils of the country were notably durable and had remarkable recuperative powers. Apparently, the period of decline in Roman agriculture served as a fallow period for the countryside at large and permitted the region to recover for the use of later generations.

## North Africa: Erosion on the Desert Margin

Carthage, from which Rome obtained her most valuable information on vine and olive culture, had an entirely different historical background. At first, the cities of North Africa thrived on trans-Saharan trade exclusively. Carthage, situated at the east end of the Atlas Mountains, was surrounded by land of such fertility that the Phoenicians abandoned their commercial traditions and developed an agricultural colony; this agriculture, however, was based on the productivity of the land around Carthage and not on that of the continent to the south. Cultivation was of a highly intensive type, and no more land was tilled than could be cared for adequately. The profits from this type of agriculture were large, and the erosion hazard slight.

Early descriptions of North Africa reflected the plantation system of agriculture. Herodotus,[1] writing in the fifth century B.C., says that

> "Libya is not to be compared for goodness of soil to Asia or Europe, except for the Cynips region, which is named after the river that waters it. The land is equal to any country in the world for cereal crops, and is nothing like the rest of Libya. For the soil here is black and springs of water abound; so that there is nothing to fear from drought; nor do the heavy rains (and it rains in that part of Libya) do harm when they soak into the ground. The returns of the harvest come up to the measure which prevails in Babylonia. The soil is likewise good in the country of the Euesperite [Berenice], for there the land brings forth in the best years a hundredfold. But the Cynips region yields three hundredfold."

[1] Herodotus, IV, 198.

Yields as great as those described by Herodotus were occasionally found for single plants, but so exceptional were they that they were brought to Italy as curiosities.

After the Third Punic War (149–146 B.C.), the cultivation of grain for export increased rapidly around Carthage and throughout North Africa, partly because of the growing requirements of Italy and partly because the establishment of peace prevented further depredations by the nomads of the deserts to the south. The area of wheat cultivation was never large. In Cyrenaica, it extended inland for a distance of about 50 miles; back

FIG. 16.—The oasis village of Salah, southern Algeria, enveloped by shifting sand dunes. (*Courtesy Geographical Review.*)

of Carthage the area was more extensive, but much of the land along the Libyan coast produced no grain. The introduction of the camel made possible the cultivation of wheat farther inland, around the desert oases. As cultivation was extended toward the desert, the hazard of wind erosion increased. As long as adequate supervision was maintained and an abundant supply of slave labor was available to care for the aqueducts and irrigation canals, the land was protected against wind erosion by a crop cover. Neglect of the aqueducts was followed by land abandonment and increased erosion. To some extent, agriculture survived the Vandal invasions, but much land was abandoned, and the labor supply declined. Then followed Moorish invasions, and thereafter famines are reported more frequently (Fig. 16).

With the establishment of Byzantine power about 400 A.D., an effort was made to restore to Africa the prosperity that it experienced under Rome. The recovery was not complete, partly because the civil administrators so oppressed the people that they were willing to aid any invaders. A more important cause of failure was the wastage of the land that followed the destruction of the irrigation works. The desert gradually began to resume the appearance that it had before the founding of Carthage. After the decline of the Byzantine Empire, the cities fell to ruin, sands from the desert covered them with shifting dunes, the walls of the wells caved in, and the irrigation works were destroyed by invaders. In less than a hundred years the work of seven centuries was completely destroyed. So complete was the ruin that another thousand years elapsed before an effort again was made to cultivate the land.

Today, some areas are in much the state that they were at the time the Phoenicians settled there in the sixth century B.C.[1] In others, cultivation destroyed the humus that had required centuries for its formation; the soil was loosened by plowing; and wind storms converted the cultivated fields into expanses of dune-covered waste, thus artificially extending the desert area and making it permanently unfit for cultivation. This theory is held by Gautier,[2] the great authority of today on Saharan conditions. The chief factors that contributed to the extension of agriculture in North Africa were improved transportation resulting from the introduction of the camel and the establishment of peace under the Romans.

It is also probable that centuries of irrigation lowered the water table to such an extent that irrigation was no longer possible. The opening of new wells may have tapped the artesian water supply and, although making cultivation possible in new areas, reduced the acreage of cropland in others. This suggestion is advanced by Ball[3] for the oases of the Libyan

[1] In interpreting conditions in the past, there is a tendency to misconstrue statements. The meanings of words often change during a period of a century or more. For example, the word "granary" to the Romans frequently implied "storehouse." When an area is spoken of as a granary of Rome, it may have produced crops other than grain. In the Byzacene region of modern Tunisia, there is evidence that little grain was ever cultivated. The products were the olive, the vine, and fruit trees. The groves and forests mentioned by the Romans were probably orchards, and grain was raised only in limited areas where irrigation was possible. That trees were not the natural vegetation except in the mountains or particularly well-watered sections of the plains is indicated by the description of the areas by Sallust. It was after his time that agriculture was introduced into the area. With the abandonment of the orchards at the end of the Roman dominion, the land gradually reverted to its natural state, and current descriptions of the area are almost exact duplicates of those of Sallust.

[2] Gautier, E. F. "The Sahara: The Great Desert." New York. 1935. Trans. by Dorothy Ford Mayhew.

[3] Ball, J. Problems of the Libyan Desert, *Geog. Jour.*, Vol. 70, pp. 21–38, 105–128, 209–224, 1927.

deserts. He points out that the remains of Roman cultivation are not all of the same date. This would imply shifting of population in relation to a changing water supply even in ancient times. Wherever the land had been cultivated the sand was particularly susceptible to blowing after it was abandoned. Drifting continued actively until sufficient humus and vegetation could accumulate to hold it fast.

## The End of Roman Agricultural Supremacy

The fall of the Roman Empire was a major factor in the decline of agriculture in North Africa and the countries at the eastern end of the Mediterranean. In Europe, the effects were equally far-reaching. The agricultural hegemony of Rome was completely destroyed. The population was seriously reduced by wars; the peace, so essential to agricultural development, was ended. Even the Roman literature on agriculture, including the translations of the works of Mago the Carthaginian, was lost. During the Middle Ages, the traditions of Roman agriculture were retained only in a few of the monasteries. Not until the sixteenth century was there a general renaissance of agriculture in Europe.

## Erosion in the East

### The Rise and Fall of Persian Agriculture

In the ancient world, agricultural progress followed the lines of communication. From Babylonia, agricultural practices spread northward into Assyria and eastward to Phoenicia, Greece, and Carthage. In the Mediterranean region, agriculture attained its highest development under the Roman Empire. To the east, Persia was Babylonia's nearest neighbor. The Persians, who introduced the horse into Mesopotamia, were originally a nomadic, pastoral people wandering over the steppes of western Asia, and through them the first contact with the Far East was established. In this way peaches were introduced from China, and chickens from India. In Elam, on the western edge of the Persian plateau, rivers and fertile valleys were more numerous, and delta lands more extensive. Here the Persians became a settled agricultural people, and their farming practices were closely associated with those of Babylonia. In the sixth century B.C., Cyrus, the king of Anshan, united the wandering tribes of Persia for the first time. In 30 years, the ruler of the small Elamite kingdom became the master of western Asia. Cyrus was not only a conqueror but also a patron of agriculture. Xenophon, his great admirer, held up as an example to the Greeks the agricultural precepts of Cyrus.

The work begun by Cyrus was continued by his successors until the Persian Empire included Media, Babylonia, and Assyria and extended westward to the Mediterranean and Greece. This was a period of notable interchange of ideas between the East and the West. Persia adopted the agricultural practices of the West and, in turn, introduced into the Mediterranean region new crops, such as lucerne, the almond, and the peach. The result was a rapid development of agriculture which continued even after the decline of the Persian Empire. The population of Persia continued to increase until the beginning of the Christian era. Since then, it has remained almost constant. About 500 B.C., during the reign of Darius, the magnificent city of Persepolis was built, in a remote section of the highlands, as a residence for the kings. But the real centers of the nation continued to be Susa, Babylon, and Ecbatana.

Much of central Persia and adjacent Turkestan is desert. Irrigation and dry farming are possible only in limited areas at oases, near permanent streams, on slopes of wooded highlands, and in the mountainous areas to the north where the rainfall is heavier. The desert agriculture had a tendency to shift from place to place with the changing water supply. According to tradition, land newly irrigated could be cultivated for five generations before abandonment became necessary. The rivers brought with them sediment which gradually raised the height of the irrigated fields and increased both the difficulty of irrigation and the tendency of the rivers to shift their courses. The old fields were abandoned, and the people moved to a new location. After the fields were abandoned, dust storms increased. In places, as at Merv, the level of the land was lowered 5 to 15 feet by wind erosion. At Paikent, in contrast, the old city was covered by blowing dust, and now only the citadel mound and the top parts of the city wall can be seen above the sand.

In the mountains scattered over the desert area there are evidences of former forests; hillsides, now abandoned, have been stripped of their soil. The Mongolian invasions, which began about the twelfth century A.D., were accompanied by rapid cutting of the forests. After each storm, torrents swept down the slopes, carrying with them sediment which was deposited over the cultivated fields below. In the region between the Persian-Baluchistan boundary and Quetta can be seen mile after mile of terraced hillsides that were not abandoned until the thirteenth century.

In some areas, recovery from the first invasions of the Mongols was possible because of the added incentive provided by overland trade. Samarkand, in Turkestan, was destroyed by Genghis Khan in 1221; but by the time of Tamerlane, a century and a half later, the city and the surrounding country were flourishing again. The complete economic ruin of the country came gradually as trade shifted from land to sea routes.

### Desert Agriculture and Erosion in the Tarim Basin

The Tarim Basin, in Central Asia, was known to the Chinese by 177 B.C., at which time the agricultural Khotanese were conquered by the

Yue-Chi, Mongolian invaders from the northeast. Because of its strategic position at the junction of trade routes between China, India, and the Mediterranean, cities grew, and important centers of religion and learning developed under the rule of the Chinese. The agricultural possibilities of the region were exploited to their utmost to supply the local population and the passing caravans until the reconquest by the Mongols in 220 A.D. In the eighth century, when the area was conquered by the Tibetans, the Tarim Basin was completely cut off from Chinese trade. Even during the early part of Chinese occupancy, agriculture shifted from place to place, but, as cultivation spread, demands on the water supply increased, and agriculture became more precarious.

Probably the highest development was attained during the second and early part of the third centuries A.D. The rivers, which rise in the surrounding rim of mountains, flow toward the interior of the Tarim Basin, and their water is eventually dissipated in the desert. If the rivers become longer or shorter, the size of the desert decreases or increases proportionally. After the third century, rivers gradually became shorter, and the desert area increased in size. In some places, the sediment deposited by irrigation built up alluvial plains 10 feet or more in depth, under which are to be found the remains of former cultivation. Old cemeteries and roads passing through farming land usually lie considerably below the level of the surrounding fields. At Yotkan, the remains of early cultivation lie, in some places, as much as 20 feet below the present surface. The alluvium is lighter in color and easily distinguished from the buried soil or culture strata below. The demands of the numerous attendants of the shrine at Yotkan undoubtedly accelerated the destruction. Uzun Tati, a place of less importance, was inhabited until about the close of the thirteenth century and, as Pien, was described by Marco Polo.

Where agricultural abandonment is attributable to the failure of the water supply, wind erosion has been conspicuous. In both Turkestan and Persia, travelers such as Pumpelly, Stein, and Hedin have observed numerous flat-topped, debris-covered mounds 5 to 25 feet above the level of the surrounding land. Further investigations show that these were the sites of old towns and that their surface represents the original level of the surrounding country. The lowering of the surface was caused by wind erosion, and only where the loess soil was covered by debris too heavy to be removed by the wind was the erosion checked. This situation has been found throughout arid and semiarid Asia from Lop Nor through Persia (Fig. 17).

Sir Aurel Stein,[1] in his more recent visits to the Tarim Basin, has found that conditions are now in process of reversal; that the streams that were formerly decreasing in length are again extending their courses forward

[1] Stein, M. A. "On Ancient Central-Asian Tracks." London. 1931.

toward the desert; and that areas abandoned for centuries again can be cultivated. This may be due in part to changes in the ground-water circulation, to the shifting of the lower courses of the streams, to the recurrence of periods of heavier than average rainfall, or to the slow processes of natural recovery.

Fig. 17.—Ruined dwelling on disastrously eroded land near Lou Lan, central Asia. The laborer on the top of the platform indicates the original level of the land and the extent of soil removal by wind. (*Courtesy Geographical Review.*)

## The Agricultural Renaissance in England and the Attendant Erosion Problems

### The Beginning of Modern Wet-land Agriculture

Roman methods of farming were introduced into all parts of the Empire and were accompanied by a general improvement of agriculture. When the Roman legions conquered Britain, they found there a well-developed type of upland agriculture. The fields were square and laid out in checkerboard fashion (Fig. 18). On the downhill side, fields were separated by "lynchets," or banks, which followed the contours. On the slopes, the fields were bordered by "balks" (ditches lined by low banks) running up and down the hillside. The lynchets developed gradually as a result of the continuous turning of the soil down slope in plowing and by sheet wash from the fields. Air photographs show that the soil along the lynchets differs in color from that of the cultivated land above, and soil investigations show that they contain topsoil washed from above. The Celtic field system first appeared about 500 B.C. and continued for a period of about 1,000 years. Although the Roman agricultural system was introduced, it modified but did not replace the Celtic system. Both traditions were obliterated completely during the Saxon invasions.

The Saxon invasions after the fourth century A.D. caused a radical change in agriculture. The open-field system, with long narrow fields, was introduced and continued until the eighteenth century and, in a few places, until the present time. Not only was a new system introduced,

FIG. 18.—Eroded Celtic fields in southern England abandoned about 400 A.D. (*Crown Copyright Reserved. Courtesy of the Controller of His Majesty's Stationery Office and the Director General, Ordnance Survey.*)

but the old eroded fields were abandoned. Many of them have not been cultivated since the time of the Saxon conquest. The Saxon fields were generally located in the lowlands. These facts have been revealed by archaeological air surveys.[1] As yet, these surveys have been limited to the south of England, but similar agricultural changes probably have

[1] Crawford, O. G. S. Air Survey and Archaeology, Ordnance Survey *Prof. Papers*, (New ser.) No. 7, 1928.

occurred in the north also. Near the Harper Adams Agricultural College farm in northern Shropshire is found evidence of Roman cultivation 5 feet below the present surface of the land, and the slopes above the fields are badly eroded. The abandonment of the steeper fields either showed wisdom on the part of the English farmers or was forced on them because the fields were too badly eroded to produce crops successfully.

The growth of urban population after the fourteenth century created the need for an increase in the supply of food and raw materials for manufacturing, but the rural population had declined as a result of wars and plagues such as the Black Death. Unlike that of the Mediterranean, the agriculture of Northern Europe depended entirely on the natural water supply, and at first only level or nearly level land was cultivated. This primitive type of wet-land agriculture was unable to supply the growing needs. Its improvement was initiated by the great ecclesiastics of the time, who had retained on their land estates some of the Roman practices and had studied the agricultural writings of Cato, Varro, and Columella. The rhymed English version of Palladius came from the religious house at Colchester; the *Rules* for the management of a manor have been attributed to Robert Grosseteste, Bishop of Lincoln; and Robert of Henley, who wrote widely on agricultural topics, was a Dominican. Their ideas were superimposed upon the Saxon type of agriculture and in some cases retarded rather than advanced it. In Northern Europe, for example, the uncultivated fallow was a hangover of an early type of agriculture in which the land was cropped until it was exhausted and then allowed to remain idle until it recovered naturally. The length of the fallow period had been gradually reduced until land was left idle for only one year at a time and one year in three. In the Mediterranean region the plowed fallow had been adopted as a means for conserving moisture. The two types of fallow were easily confused, and the value that the Romans placed on the fallow was responsible for its persistence in Northern Europe long after wet-land agriculture had advanced to the stage in which it was no longer necessary.

During the Middle Ages, the manorial, open-field system prevailed throughout Europe. Since the tenants did not possess their fields in perpetuity, there was little incentive for improvement. Flocks were pastured on the commons, and their manure was largely lost to agriculture. Crop yields were low, and it was prophesied that the population of the world eventually would be limited by the productive capacity of the land. This was the basis of the Malthusian theory, which was propounded about the beginning of the nineteenth century.

## Development of Modern Agriculture

The roots of modern English agriculture were laid in Flanders, where the first of the peasant revolts occurred. These were followed by other agricultural rebellions both in England and on the Continent. The weavers of Flanders required more wool from England, and this resulted in an early movement for enclosing land in order to provide better pastures and an improved wool supply. The later enclosures (1650–1850) were essentially to increase crop production.

Flemish influence was also responsible for the introduction of crop rotation into England. Plattes, who was experimenting with rotations in 1683, was of Flemish extraction, and Sir Richard Weston came in contact with Flemish methods when he was forced to flee from England during the civil wars. His "Legacie" was regarded by Arthur Young as the greatest single contribution to the improvement of British agriculture. Of the early rotations, the four-course Norfolk rotation of the seventeenth century was the most famous. This consisted of wheat, root crop, barley, and clover—the fallow being completely eliminated. Rotations were greatly improved thereafter and were adapted to meet the needs of the different types of environment found in England.

The eighteenth century was a period of rapid improvement in plowing and planting, particularly in England. Jethro Tull was the father of this movement and introduced and popularized deep plowing, "horse-hoeing" husbandry, and drill planting. Among the outstanding agricultural leaders of the time were Lord Townshend, Bakewell of Dishley, Coke of Norfolk, Arthur Young, and Sir John Sinclair. All were concerned primarily with lowland agriculture in which drainage rather than erosion was the major problem, but their work created a general interest in experimental farming and led to the adoption of sound agricultural practices which form part of the basis of modern soil conservation.

One erosion-control measure, however, seems to have merited considerable attention: the broad-wheeled wagon, which was designed for use on badly eroded farm lands as well as on the poor roads that were prevalent in England during the eighteenth century. It solved the problem of transportation and prevented wagon-wheel ruts from enlarging and developing into gullies. There are other evidences that sheet wash was not uncommon, even on the comparatively level English fields. In order to provide adequate drainage, the fields were ridged. The drainage channels between the ridges filled with sediment washed from the fields, and the clearing of these ditches presented a serious problem. Both Young and Sinclair realized that the richest soil was washed from the fields and recommended that it be collected periodically and spread over the land as fertilizer.

## Erosion and Its Control in Scotland

In Scotland, the problem of soil erosion received more attention than in England because of the limited amount of level land. But the general improvement of agriculture in Scotland lagged. The open-field system was retained until after 1695, when the first general enclosure act was passed. Until then there was no incentive for the improvement of the land itself. The awakening of agriculture during the last quarter of the eighteenth century came largely as a result of the efforts of Sir John Sinclair to popularize good farming. The works of Tull and his followers were received with far greater enthusiasm than in England. As late as 1780, most of the Scottish agriculture was of a crude, subsistence type and far inferior to the English; but by 1815 it had improved to such an extent that it was commonly regarded as the best in Europe.

An increase in the use of soil-conserving practices was an important phase of the general agricultural improvement, and it was from Scotland, rather than from England, that the United States derived most of its early soil-conserving practices. Although the square or oblong field was generally recommended to simplify the problems of plowing, the size of the field was determined by the slope of the land. Sinclair,[1] writing in the early nineteenth century, discussed at some length the problem of ridging on sloping lands. Ridges running up and down the slope were unequivocally condemned. Horizontal ridges had the advantage of holding the soil but did not provide the necessary drainage. In some places, this difficulty was overcome by carrying the ridge around small hills, beginning at the bottom and spiraling upward. The method most highly recommended by Sinclair was the construction of ridges from the top down and sloping diagonally to the right. This provided the necessary drainage but retarded the runoff and reduced soil erosion. For land of "excessive steepness," a plow with a shifting moldboard was recommended. This was commonly known as the *turn-wrest* plow. Deep plowing met with universal approval. It was regarded as almost comparable to the hoe-and-spade cultivation that had produced luxuriant verdure on the small garden patches of earlier times. In order to popularize good plowing, annual competitions were established in various parts of the country. The first of these was initiated in the county of Clackmannan in 1780. It is interesting to note that the practices employed in Scotland are identical with those tested by Washington, Jefferson, and Randolph in the United States.[2]

In large areas of the Highlands, a poor form of subsistence farming supplemented by equally poor animal husbandry persisted. The cattle were almost wild, and the sheep so small, it is said, that the carcass usually weighed less than 25 pounds. The higher pastures were never

[1] Sinclair, Sir John. "An Account of the Systems of Husbandry Adopted in the More Improved Districts in Scotland." Vol. I, pp. 165–170. Edinburgh. 1814.

[2] Hall, A. R. Early Erosion-control Practices in Virginia, U. S. Dept. Agr. *Misc. Pub.* 256, 1937.

properly or fully utilized. Population was sparse, and rentals were necessarily low. The introduction of commercial sheep farming brought a radical change. The sheep introduced from southern Scotland and England were larger in size, hardier, and produced more wool. Cheviots were introduced into the Highlands about 1750, and after 1760 came a steady influx of sheepherders. Around Callander, the number of sheep increased from 1,000, in 1770, to 18,000, in 1790. Throughout the Highlands, rentals increased, crofters were forced to leave their farms, and the raising of cattle practically ceased. The deforestation of large areas in Scotland dates back to the introduction of commercial sheep farming. By 1850, the pastures had begun to show signs of deterioration, and thereafter more frequent mention is made of "depleted grazings."

Continuous grazing precluded the regeneration of the forests. In southern Scotland, where commercial sheep raising was an established industry before the twelfth century, the dangers of promiscuous grazing were recognized, and the old *Leges Forestarum* of the time of Richard Coeur de Lion contained provision for limiting forest grazing. As grazing continued, the plant cover deteriorated. Succulent grasses and heather declined, and bracken and moor-mat grass increased. In other places, injudicious burning was practiced in an effort to maintain a good stand of grass. Where the ground was left bare, appreciable quantities of the dry surface soil were blown away by high winds. This has been observed on some of the agricultural experiment plots. In 1694, a sandstorm occurred that converted part of the Culbin farming land into a miniature Sahara for a period of 200 years.[1] Farm lands near the seacoast are particularly susceptible to damage of this type when the natural vegetation is removed.

Removal of the vegetation also leads to water erosion and the formation of "screes," or small landslides, during heavy rains. Furthermore, grazing, trampling, or the rubbing of sheep against the sides of the screes often causes a break to become so enlarged that an entire hillside is affected. According to Fenton,[2] "within historic time sheep farming, both directly and indirectly, has altered the appearance of a large part of Scotland."

Since 1930, the problem of erosion in Scotland has received considerable attention, especially from ecologists, because of the close relation between erosion and the type of plant cover. Wherever heather and the better pasture grasses have been superseded by bracken, there is serious danger of erosion. One phase of erosion in Scotland has no counterpart in the United States. The erosion of the upland peat has caused deposition

[1] Batten, H. M. The Tragedy of Culbin Sands, *Nat. History*, Vol. 38, pp. 143–148, 1936.

[2] Fenton, E. Wyllie. The Influence of Sheep on the Vegetation of Hill Grazings in Scotland, *Jour. Ecology*, Vol. 25, pp. 424–430, 1937.

lower down, and the soil thus covered has become too acid to support good herbage.

Situations similar to those in Scotland have been found in the hill pastures of England and Wales, but in neither country is the problem so serious. A new erosion hazard, however, threatens England today. The opening of the New World provided England with an increased food supply before it became necessary to cultivate the more erodible sloping lands. The nationalism of today and the demand for self-sufficiency are promoting an increase in the area devoted to wheat, beets, potatoes, and other foodstuffs. Because of the shallowness of the soil profile and the sloping surface of the land in the areas not previously cultivated, it seems probable that the productive capacity will be reduced materially by erosion, unless measures are adopted for its prevention.

## The Palatinate and Its Relation to Erosion Control in the United States

Early agriculture in the United States, in general, followed the English traditions, but, locally, other influences dominated. During the Colonial and early national period, the German settlers in Pennsylvania were regarded as the best farmers in the entire country. This superiority may be attributed directly to the influence of farm practices developed in the Palatinate. After the Treaty of Westphalia in 1695, the Palatinate at first served as a refuge for persecuted religious sects from other parts of Europe. Among these were the essentially agricultural Mennonites. Many were of Swiss origin and had left their own country because they objected to its policy of supplying mercenary soldiers to any country that needed their services. Later, when the Catholics came into power, religious intolerance forced the Protestant sects to leave the Palatinate. Before the Revolution, there were over 100,000 Germans and Swiss in Pennsylvania alone, and smaller groups had settled in other colonies. Wherever they settled, the type of farming practiced was notably superior to that of the surrounding communities. In Europe, persecution had forced them to move frequently, and in America they sought permanent homes. To maintain the productivity of the land was of paramount importance. The German farms were comparatively small and intensively cultivated, and, consequently, land exhaustion, followed by abandonment and a westward movement of population, was practically unknown.

## Erosion in the Americas

### The Highland Terraces of Peru

In the United States, the introduction of European methods of plowing and cultivating was directly responsible for a rapid increase in erosion.

The original colonies were established during the agricultural revolution in England and at the very beginning of the period of agricultural development in Scotland. Few techniques for controlling erosion had been developed in Europe, and these had not been generally accepted by the rank and file of the farmers. Wheat and barley, the staple grain crops of Great Britain, were commonly sown broadcast. Corn, the chief native grain crop, was domesticated in America, and the method of planting differed radically from the European system. Corn was planted in hills, and only hoe or spade cultivation was used. Straight rows and clean cultivation

FIG. 19.—Terraced slopes in Peru. (*Courtesy Aerial Explorations, Inc.*)

were impossible. With the introduction of the plow, the hill crop was converted into a row crop. Other crops, such as tobacco and potatoes, experienced a similar change in methods of cultivation. The farmers from the English lowlands failed to recognize the erosion hazard that accompanied row cultivation on sloping land.

In Peru, where many of our native crops were domesticated, the lands were sloping, and the Incas and their predecessors had developed highly efficient methods of soil conservation (Fig. 19). The Indians of our eastern seaboard had acquired crops from the highlands of Peru and Mexico, but their agriculture was less highly developed since means of direct communication were lacking.

In Peru, only the limited areas in the valley floors were by nature well adapted to agriculture; and in many of these the rocks and stones brought down by the mountain glaciers had to be removed before the land could be cultivated. The mountain slopes were once covered with forests or a dense scrub vegetation. Some of the ravines that are too deep or narrow for cultivation have reverted to forest; and in the Panticalla Valley thousands of acres that are now tree covered were once terraced and cultivated.[1] As population increased, it became necessary to clear the slopes. This was done before the beginning of continuous tradition, and the earliest terraces show signs of carefully planned construction. The care used in conserving soil indicates that the ancient Peruvians had

Fig. 20.—Retaining wall of ancient terrace at Ollantaytambo, Peru, showing nicety with which stones were fitted together by the pre-Incas centuries ago. (*Courtesy O. F. Cook.*)

experienced soil erosion, but no evidence has been found as to its exact date or extent.

The Incaic or pre-Inca terraces were supported by stone retaining walls (Fig. 20) which were commonly 8 to 14 feet in height and occasionally as much as 50 feet. Behind the walls, the soil was filled in by hand. The lower levels consisted of stones and gravel, and the upper 2 or 3 feet of good agricultural soil. The terraces were irrigated by means of aqueducts which brought the water from sources that were often miles away. Where the soil was loose, the aqueducts were paved with stone; and in other sections they had to be cut from solid rock. The water was led from terrace to terrace through stone-faced outlets or over rough rocks and caught below in rock-paved basins. Near Macchu Picchu, the water from

[1] Cook, O. F. Agriculture and Native Vegetation in Peru, Washington Acad. Sci. *Jour.*, Vol. 6, pp. 284–293, 1916.

one terrace was spread among a number of outlets and fell like a shower on the terrace below. The system employed by the Incas and their predecessors effectively held the soil on the steep slopes. Today, thousands of acres of Incaic terraces are still being cultivated and constitute the main agricultural land in many areas (Fig. 21).

Even the cultivation of the valleys presented serious obstacles. In order to control floods, the rivers were confined by retaining walls. These gradually were built higher and moved closer and closer to the stream channel. The old walls were left standing but covered with soil and then

FIG. 21.—Ancient staircase terraces, still in use, at Ollantaytambo, southern Peru. (*Courtesy O. F. Cook.*)

cultivated. The valley floor was divided into broad terraces supported by stone walls. The topsoil for the terraces, both in the valleys and on the hills, was frequently brought in from long distances. According to tradition, the soil for the Inca gardens at Cuzco was brought from near Quito, a distance of 700 miles.

The intensive agriculture of Peru began its development in pre-Inca times, and its origin is still unknown. It is evidently the result of definite planning, and apparently units of considerable size were developed at one time. The result has been the most effective method of controlling erosion developed anywhere in the world though probably the most expensive so far as labor is concerned. The disintegration of much of this highly developed agriculture may be directly attributed to the devastations accompanying the Conquest and not to any intrinsic flaws in the methods employed.[1]

[1] Cook, O. F. Staircase Farms of the Ancients, *Nat. Geog. Mag.*, Vol. 29, pp. 474–535, 1916.

## Sedimentation in the Maya Lowlands

Among the Maya of Yucatan and Central America, corn was the chief crop but was produced on nearly level lowlands. The methods of cultivation employed are not definitely known, but the dense population of the past probably required more intensive farming than is practiced today. During the sixth century A.D., the old Maya empire was one of the most densely populated areas of its size (49,220 square miles of habitable land) to be found anywhere in the world. Ricketson[1] estimates the population

Fig. 22.—Complete denudation of formerly forested hillsides at Cajabon in eastern Guatemala, a result of repeated cutting and burning, followed by rapid erosion of exposed surface. (*Courtesy O. F. Cook.*)

conservatively at 13,300,000—a population comparable in density to that of England between 1821 and 1831.

The descendants of the Maya today employ the milpa system,[2] under which the land is cleared, cultivated for two or three years, and then abandoned. The yields of the second year usually amount to half of those of the first. Subsequent cultivation, within periods of 8 to 10 years, is said to be impractical because of still lower yields. Declining yields are due to three principal causes, according to the observations of the author: encroachment by weeds, depletion of the immediately available supply of

[1] Ricketson, O. G. Uaxactun, Guatemala—Group E—1926–1931, Carnegie Inst. *Pub.* 477, Pt. I, p. 23, Washington, 1937.

[2] Cook, O. F. Milpa Agriculture, a Primitive Tropical System, Smithsonian Inst. *Ann. Rept.* for 1919, pp. 307–326, Washington, 1921.

plant food, and erosion. The *rastrojos*, or lands that have been recleared from second growth after about 8 to 15 years of abandonment, frequently produce satisfactory yields, according to the standards of the patch cultivators of the jungle, provided the soil is deep, of gentle slope, and not subject to erosion. The yields from abandoned lands that have suffered seriously from erosion have been much lower upon reoccupation after years of jungle fallow. Among the most severely eroded areas are the red lands covered with pine forests in the drainage basin of the Chamelecon River in western Honduras. Although these were once occupied by the Maya, as evidenced by ruins, they are now largely unfit for cultivation. The predominant red clay is very stony and shallow, with frequent outcrops of the parent mica schist. Such land frequently occurs immediately alongside jungle-covered virgin areas, of similar origin and topographic features, which possess a deep red clay soil, often more than 3 feet in thickness. This would indicate that other pinelands of the region, as well as similar areas in eastern Guatemala, were ruined by erosion resulting from cultivation and were then abandoned (Fig. 22).

Professor O. F. Cook has the following to say about erosion in Guatemala:[1]

> "That the ancient occupations of the humid mountain regions of eastern Guatemala by agricultural civilizations were very prolonged or were repeated in several prehistoric ages is indicated by the very severe erosion which this region has suffered.
>
> "There is practically no erosion at all on slopes covered with dense tropical forest. The soil is never loosened by frost, but is held in place by the matted roots of trees and thatched over with a compact layer of fallen leaves.
>
> "Terracing of the land shows that agriculture was extensively practiced in former times in regions now unoccupied. Two principal forms of prehistoric stone terraces, built evidently for agricultural purposes, may be recognized in the Central American region, in addition to the narrow terraces of earth. . . . There are (1) narrow, high terraces to hold drainage water and prevent erosion in the narrow valleys or on steep slopes of mountains and (2) broad, low terraces apparently leveled to keep rain water from running off rather than to apply irrigation" (Fig. 23).

The corn requirements during the height of development of Maya empires made necessary the clearing and cultivation of extensive areas of wooded slopes. In 1931, C. Wythe Cooke,[2] of the United States Geological Survey, advanced the theory that the accelerated erosion caused by the

[1] Cook, O. F. Vegetation Affected by Agriculture in Central America, U. S. Dept. Agr. *Bull.* 145, April, 1909.

[2] Cooke, C. W. Why the Maya Cities of Peten District, Guatemala, Were Abandoned, Washington Acad. Sci. *Jour.*, Vol. 21, pp. 283–288, 1931. See also: Morley, Sylvanus Griswold. The Inscriptions at Copan, Carnegie Inst. *Pub.* 219, pp. 452–457, Washington, 1920.

clearing of the slopes had resulted in increased sedimentation and the gradual conversion of the clear-water lakes of the old Maya empire into swamps. In 1932, he accompanied the Carnegie Institution Expedition to Uaxactun, where he examined the sediments deposited in Joventud

Fig. 23.—Terrace culture in the table lands of Guatemala, between Totonticapan and Quezaltenango, used largely for growing wheat, the high altitude making corn more precarious. The culture is maintained by cutting off the margin of each terrace and spreading the material over the surface of the terrace below. (*Courtesy O. F. Cook.*)

Bajo. The silt of the later deposits was found to be black carbonaceous clay, representing surface soil washed from the adjacent uplands. The indications are that sedimentation destroyed water transportation, diminished the fish and game supply, polluted the water supply, and made the continued cultivation of the uplands impossible.[1]

[1] Cooke, C. W. Possible Solution of the Maya Mystery, *Sci. Monthly*, Vol. 37, pp. 362–365, 1933.

Archaeological investigations during the last few years further support the theory of progressive erosion and sedimentation and indicate that there was no sharp break between the old and later Maya empires but a gradual migration northward as the older cultivated lands were destroyed by erosion and sedimentation and then abandoned. The once flourishing agricultural region with its dense population is now covered with impenetrable swamps and dense jungles of recent growth.

## Conclusion

The current erosion crisis in North America can be definitely attributed to the exploitation of the land that followed European settlement. Although the processes of erosion operate locally, the economic effects have become international in scope. The causes, however, lie in the distant past and in those countries in which our crops and farm practices originated. Throughout the history of the world, the advancement of agriculture has been made possible by the introduction of new methods of farming from foreign areas. Since these were not developed locally, or specifically adapted to the new environment, each introduction presented a potential erosion hazard. Erosion in the United States and Canada differs from that in the Old World chiefly in its magnitude and the rapidity with which it developed. Other new countries, such as the Union of South Africa and Australia, are experiencing many of the erosion problems that plague North America and parts of South America. Like the United States, these countries have limited agricultural inheritances. From the standpoint of erosion, the heritage may be increased by an investigation of conditions in other countries with longer agricultural histories.

# Chapter III. Results of Erosion

Only in the past few years have the far-reaching effects and implications of the soil-erosion problem in the United States received widespread recognition. As recently as 10 years ago, soil erosion was commonly regarded as little more than a spasmodic phenomenon confined to those farms where the land for some unexplained reason was subject to excessive washing. Except in the Southeastern States, the average farmer seldom consciously did more to check the wastage than follow the generally ineffective and casual practice of throwing a little brush or rubbish into some of his gullies. With this, he usually dismissed the problem until the appearance of the next gully or group of gullies. The average city dweller knew nothing about the problem and so never gave it a thought. Generally speaking, the idea that sheet erosion was steadily undermining the physical basis of the agricultural enterprise of the nation never occurred to the American farmer, and academicians everywhere failed to sense the cause of widespread land decline in some of the older parts of the world or to relate soil depletion to the migrations of populations from such impoverished areas. The notion that soil wastage was anything more than a purely agricultural problem—a matter of the mechanics of farming—was generally rejected as the most apparent kind of nonsense.

Outside a few localities in the United States where the problem of erosion literally forced its attention on land users, as in parts of the Cotton Belt, little or no recognition was given the problem, and practically nothing was done about it. Some awakening came, finally, through an educational and investigational program inaugurated by the United States Department of Agriculture and some of the states,[1] about 12 years ago. Then, on May 12, 1934, the nation witnessed a disturbance that was completely without precedent in American history. Dust clouds arising over sun-parched fields in western Kansas, Texas,

[1] Bennett, H. H. Cultural Changes in Soils from the Standpoint of Erosion, *Jour. Am. Soc. of Agronomy*, Vol. 23, June, 1931; and National Program of Soil and Water Conservation, *Jour. Am. Soc. of Agronomy*, Vol. 23, May, 1931.

Bennett, H. H., and Chapline, W. R. Soil Erosion a National Menace, U. S. Dept. Agr. *Circ.* 33, 1928.

Oklahoma, and eastern Colorado, were lifted into the pathways of high air currents and carried eastward across two-thirds of the continent. Soil from the Great Plains, for the first time since the coming of white man, darkened the sun over the nation's capital, sifted through the screens of tall office buildings in New York City, and moved on for hundreds of miles over the Atlantic Ocean. People on the eastern seaboard were shocked into a realization that something had gone wrong with the land to the west. All over the country, people began to inquire about this new phenomenon and to discuss its significance.

Since that time, thousands of newspaper stories and magazine articles have been written about soil erosion. The subject has been treated on public platforms and over the radio. Motion pictures have brought the story with amazing effectiveness to even the most strictly urban communities. Meanwhile, governmental, scientific, and educational institutions have been actively searching for further knowledge about the character and extent of erosion damage. The United States Department of Agriculture, state colleges of agriculture, and state agricultural experiment stations have brought forth a wealth of new information about this process which a little while ago was so little understood. Contributions to the science of soil erosion are now coming from many parts of the world. Today it is possible to formulate a statement of the soil erosion problem in more accurate terms—in terms quite different from those which would have been used a decade ago.

Soil erosion is now generally recognized in the United States as a powerful and destructive force which, directly or indirectly, affects the lives of every man, woman, and child. It is understood now not as a freak of nature which occasionally turns up on isolated farms and ranches but as an almost continually active process which attacks countless fields, whole watersheds, and broad farming communities. Some evidence of its damaging effects has been discovered over an area comprising approximately a billion acres of crop, grazing, forest, and abandoned lands of continental United States. It has extended into nearly every major farming region and has affected, in some degree, the production of nearly every staple crop as well as many specialty crops. Erosion and eroded land have introduced a relatively new and disturbing factor into the whole complicated cycle of water circulation, drainage, and storage.

## Area Affected

It may never be possible to gather completely accurate statistics of the total land area in the United States that has been affected by soil erosion. The task of making a detailed, nationwide survey is so vast that many damaged areas doubtless would increase in extent, or depth of soil removal, before the study could be completed. There is now available, however, much information pertaining to the extent and seriousness of the problem, gathered through erosion surveys and soil surveys as well as from investigations conducted at erosion experiment stations located in important problem areas. By combining and weighing all this information, it is possible to estimate with a reasonable degree of accuracy the areal extent of erosion damage.

An aggregate area of at least 50 million acres of former cropland in this country has been ruined by erosion (Fig. 24). This is a large area—about

FIG. 24.—Formerly cultivated good land ruined by gullying. Poor subsoil material washing from such areas damages good bottomland downstream. Local control of such wasted land provides protection for remote areas. North Carolina. (*Photograph by Soil Conservation Service.*)

the size of the whole state of Nebraska—and it is definitely out of the picture for purposes of further practical crop production. Most of this land already has been abandoned; all of it is submarginal, although a small portion still supports a few people in conditions of essential poverty. Most of the area was formerly productive land, producing good crops, occasionally bumper crops. Now, it is so completely stripped of topsoil or so badly dissected by gullies that farming is either an economically unprofitable venture or a definite physical impossibility.

Probably another 50 million acres of cropland, some of it still in cultivation, is bordering on a condition of practical uselessness. Most of it is already submarginal and should be retired from cultivation. It is not now properly appraised as cropland.

An additional cropland area of about 100 million acres, now in use, has been severely impoverished by losses of topsoil and gullying. Nearly all of it has lost a fourth or more of its original topsoil; some of it has lost as much as three-fourths, and some all of the topsoil. Unless immediate steps are taken, the larger part of this land will fall below the marginal line in a relatively short time. Finally, there is an indeterminate area, probably amounting to well over 100 million acres, where erosion just recently has begun to make active progress or may be expected to do so in the very near future.

Fig. 25.—Erosion by a heavy rain following fire, northern Idaho. (*Photograph by Soil Conservation Service, 1936.*)

Erosion is not confined, however, to cropland alone. It occurs also on an enormous aggregate area of overgrazed range in western United States. Fires and close grazing have produced severe erosion on much woodland (Figs. 25 and 26).

Including these eroded areas with the cropland areas, and excluding only mountains, desert, and badland sections, a total of some 282,000,000 acres has been ruined or severely impoverished, and an additional area of 775,000,000 acres is being affected to some extent by erosion at the present time. Thus, more than half of the 1,904,000,000 acres comprising the total land area of the United States already has been affected by erosion in some degree.

The effect of this damage on the supply of land for agricultural production must, of course, take into account land not now in cultivation but which could, with reasonable adjustments, be brought into use for crops.

On the basis of Census statistics, about 413,000,000 acres were actually in cultivation in 1930, including idle or fallow land. In addition to this actual cropland, however, an area of approximately 109,000,000 acres of plowable pasture is immediately available for productive use. Perhaps 50,000,000 acres of grazing land in Western areas might also be brought into cultivation with reasonable ease, although much of it would be subject to serious wind or water erosion. The total of actual and imme-

FIG. 26.—Rill, gully, and sheet erosion on steep overgrazed slope, San Joaquin Valley, California. (*Photograph by U. S. Forest Service, 1933.*)

diately potential cropland in the United States, therefore, may be estimated at some 572,000,000 acres. How much additional land could be brought into use, in event of need, by extensive clearing, drainage, irrigation, and other adjustments it is difficult to determine with any degree of accuracy on the basis of available information.

Excluding the 100,000,000 acres which, for the most part, already have been ruined or nearly ruined as cropland, the process of soil erosion already has damaged or is now an active threat to more than one-third of the immediately arable land in the United States.

Table 1 presents a summarization of the approximate extent of erosion in the United States, by major groups of land, based on degree of seriousness.

In the humid sections of the country, only a small portion of the cultivated land is flat enough to be entirely free from soil washing—much less than was supposed as recently as 5 years ago. The large bulk of sloping land in the corn and cotton belts, in the Palouse section of the Pacific Northwest, in the tobacco areas, and in the interior valleys and basins of California and other Western States may be classed as definitely erodible by water. The rolling grainlands of the humid regions are also subject to impoverishing erosion, although a cover of grain is much more resistant to erosion than the "row" (cultivated) crops. The vast central

TABLE 1.—ESTIMATED EXTENT OF EROSION DAMAGE IN THE UNITED STATES, BY MAJOR GROUPS[1]

| *Condition of Erosion* | *Area, Acres* | |
|---|---|---|
| Damaged, all types: | | |
| Essentially ruined or severely damaged | 282,000,000 | |
| Moderately damaged or beginning | 775,000,000 | |
| Total area damaged | | 1,057,000,000 |
| Damaged, cropland (area harvested, crop failure, idle or fallow): | | |
| Essentially ruined for cultivation | 50,000,000 | |
| Severely damaged | 50,000,000 | |
| One-half to all topsoil gone | 100,000,000 | |
| Erosion beginning | 100,000,000 | |
| Total area cropland damaged | 300,000,000 | |
| Not now damaged: | | |
| Land in forest, swamp, marsh, etc. | | 702,000,000 |
| Damage not defined: | | |
| Desert, badlands, Western mountain areas, etc. | | 145,000,000 |
| Total land area | | 1,904,000,000 |

[1] Based primarily on erosion surveys.

plains region—the Great Plains area—extending from the Panhandle of Texas, northward through Oklahoma, Kansas, Colorado, and Nebraska, into the Dakotas and eastern Montana and Wyoming, is only locally troubled by water erosion, because of the very moderate slope of much of the land and the low supply of rainfall. The aggregate area of rolling land subject to water erosion is, nevertheless, enough to constitute a land problem of no little importance in the Plains. But the greater problem of the area is wind erosion. Wherever the land is continuously cultivated, overgrazed, or left bare of cover, whether sloping or level, it is generally susceptible to blowing in times of protracted drought. The degree of severity ranges all the way from minor disturbances of the surface soil to the major dust storms that sweep across the country.

Table 2 presents estimates of actual and potential soil erosion in the United States, all land considered.

SOME EXAMPLES OF EROSION-DEPLETED LAND. Of the enormous acreage of seriously eroded land in the United States, certain areas stand

TABLE 2.—PRESENT AND POTENTIAL SOIL EROSION IN THE UNITED STATES

| | Land utilization (1930 census), acres | Land plowable, without extensive clearing, drainage, or irrigation operations (estimated), acres | Eroding and potentially erodible land in crops or available for crops without clearing, drainage, or irrigation (estimated), acres | Eroding and potentially erodible land of all types (estimated), acres |
|---|---|---|---|---|
| Land in farms: | | | | |
| Farmsteads, lanes, waste | 45,000,000 | | | 25,000,000 |
| Crop failure | 13,000,000 | 13,000,000 | 11,050,000 | 11,050,000 |
| Idle or fallow | 41,000,000 | 41,000,000 | 34,850,000 | 34,850,000 |
| Cropland harvested | 359,000,000 | 359,000,000 | 305,150,000 | 305,150,000 |
| Plowable pasture | 109,000,000 | 109,000,000 | 92,650,000 | 92,650,000 |
| Nonplowable pasture | 270,000,000 | | | 229,500,000 |
| Woodland pasture | 85,000,000 | | | |
| Woodland not pastured | 65,000,000 | | | |
| | 987,000,000 | | | |
| Land not in farms: | | | | 345,850,000[1] |
| Private forest, not grazed | 151,000,000 | | | |
| Public forest, not grazed | 57,000,000 | | | |
| Public forest, grazed | 106,000,000 | | | |
| Private forest, grazed | 143,000,000 | | | |
| Private grazing land | 94,000,000 | 50,000,000 | 42,500,000 | 79,900,000 |
| Public grazing land | 235,000,000 | | | 199,750,000 |
| Cities, towns, roads, parks, railroads, desert, waste | 131,000,000 | | | 8,500,000 |
| | 917,000,000 | 572,000,000 | 486,200,000 | 1,332,200,000 |
| Total land area | 1,904,000,000 | | | |

[1] This figure represents: (1) 85 per cent of the estimated 229,000,000 acres of plowable land in forests or woodlands (194,650,000 acres), plus (2) 80 per cent of the estimated 189,000,000 acres of grazing land, and potential grazing land, now in forests or woodland (151,200,000 acres).

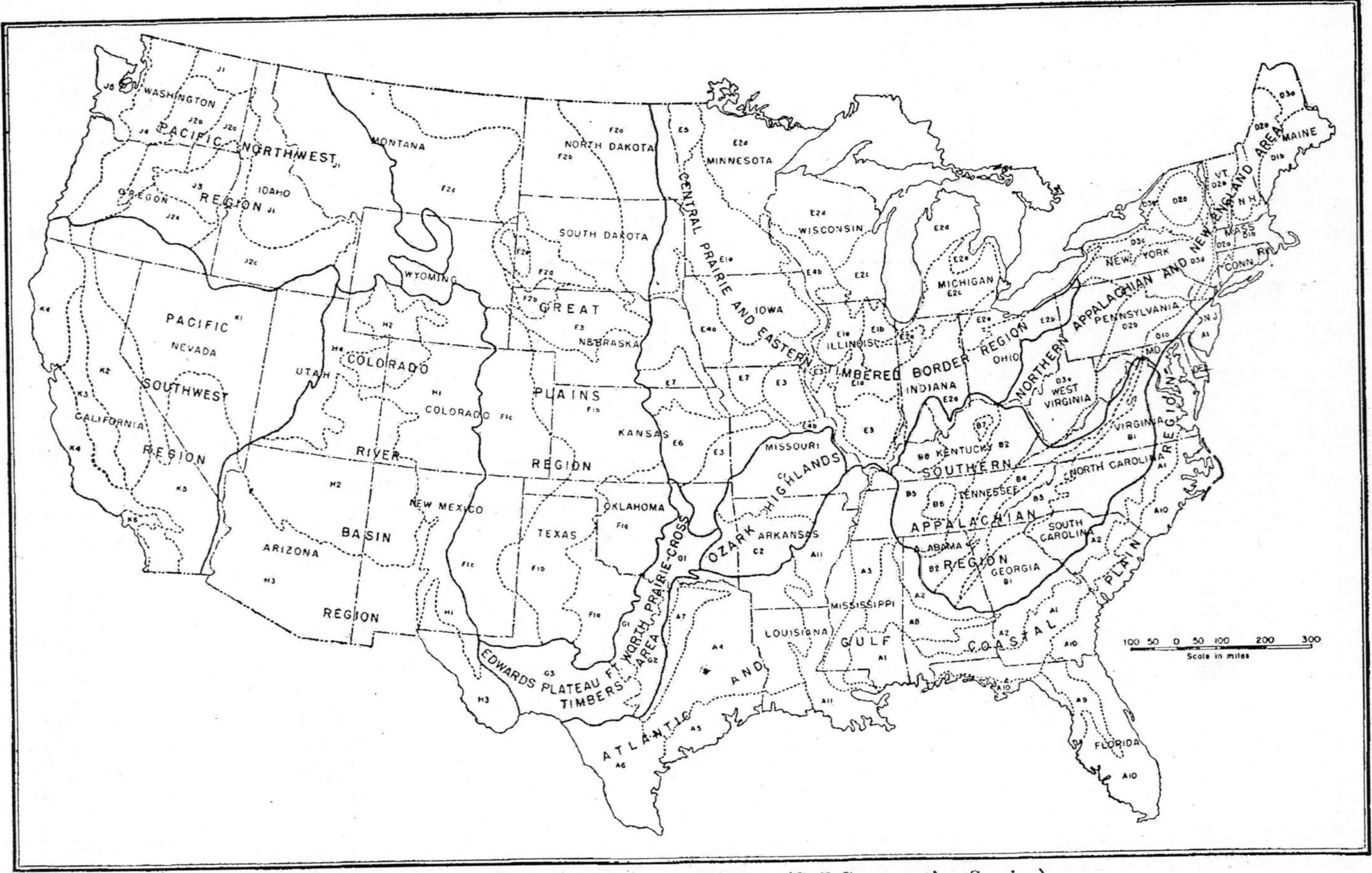


MAP. 1.—Problem areas in soil conservation. (*Soil Conservation Service.*)

The problem areas are roughly differentiated on the basis of the environmental complex, involving present and potential conditions with respect to erosion and water lossses. The principal determinative features are present degree of erosion, susceptibility to erosion, topography, climate, soil, and present and past vegetative conditions.

A, ATLANTIC AND COASTAL PLAIN REGION
- 1, Middle Coastal Plain
- 2, Upper Coastal Plain
- 3, Loess Area (Lower Mississippi Section)
- 4, Interior West Gulf Coastal Plain
- 5, Gulf Coastal Prairie
- 6, Rio Grande Plain Area
- 7, Texas Black Belt
- 8, Mississippi-Alabama Black Belt and Clay Hill Area
- 9, Florida Rolling Sandy Lands
- 10, Flatwoods
- 11, Lower Mississippi Alluvial Plain, Terrace, and Crowleys Ridge Area

B, SOUTHERN APPALACHIAN REGION
- 1, Piedmont
- 2, Appalachian Mountains and Plateau
- 3, Blue Ridge Mountains Area
- 4, Appalachian Valley and Ridges
- 5, Highland Rim Area
- 6, Nashville Basin
- 7, Lexington Bluegrass Area
- 8, Kentucky-Indiana Sandstone-Shale Hills

C, OZARK HIGHLANDS
- 1, Northern Ozarks
- 2, Southern Ozarks

D. NORTHERN APPALACHIAN AND NEW ENGLAND AREA
- 1, Piedmont
  - a, Unglaciated Piedmont
  - b, Glaciated Piedmont
- 2, Appalachian Mountains and Valleys
  - a, Adirondacks-Green and White Mountains
  - b, Ridges and Valleys
- 3, Appalachian Plateau
  - a, Aroostook Section
  - b, Eastern Great Lakes-St. Lawrence Valley
  - c, Finger Lakes-Mohawk Section
  - d, Delaware-Susquehanna Section
  - e, Allegheny-Ohio Section

E, CENTRAL PRAIRIE AND EASTERN TIMBERED BORDER REGION
- 1, Dark Prairie Area
  - a, Mixed Till Plain
  - b, Shale Till Plain
- 2, Eastern Timbered Border
  - a, Indiana-Ohio Plain
  - b, Lake Erie Plain
  - c, Southern Michigan-Wisconsin Plain
  - d, Northern Great Lakes Timbered Section
  - e, Great Lakes Plain
- 3, Claypan Prairies
- 4, Loess Area
  - a, Missouri River Section
  - b, Lower Missouri and Upper Mississippi Section
- 5, Red River Area
- 6, Residual Limestone and Shale Plain
- 7, Lower Missouri Dark Brown Till Section

F, GREAT PLAINS REGION
- 1, Southern Plains
  - a, Red Plains
  - b, Dark Brown Plains
  - c, Brown Plains
- 2, Northern Plains
  - a, Black Plains
  - b, Dark Brown Plains
  - c, Brown Plains
  - d, Pierre Shale Area
  - e, Black Hills Area
- 3, Nebraska Sand Hills

G, EDWARDS PLATEAU-FORT WORTH PRAIRIE-CROSS TIMBERS AREA
- 1, West Cross Timbers
- 2, Fort Worth Prairie and East Cross Timbers
- 3, Edwards Plateau

H, COLORADO RIVER BASIN REGION
- 1, Rocky Mountains
- 2, Colorado Plateau
- 3, Southwestern Mesas, Mountains and Basins
- 4, Wasatch Mountain Areas

J, PACIFIC NORTHWEST REGION
- 1, Rocky Mountains
- 2, Columbia Plateau
  - a, Palouse
  - b, Columbia River Plains
  - c, Snake River Plains
  - d, Oregon Plateau
- 3, Blue Mountains
- 4, Cascade Mountains
- 5, Pacific Coast Area

K, PACIFIC SOUTHWEST REGION
- 1, Great Basin Area
- 2, Sierras
- 3, California Valley
- 4, Coast Ranges and Lowlands
- 5, Sonoran Area
- 6, Southern California Coastal Plain Area

out in bold relief.[1] Throughout hundreds of important farming localities, erosion has now progressed so far that the rehabilitation of numerous areas, both small and large, is practically out of the question. These "sore spots" of American agriculture lie scattered across the length and breadth of the nation; taken together, they constitute a grim reminder of the power and destructiveness of rain and wind (see Map 1, *Problem Areas in Soil Conservation*, and Map 2, *General Distribution of Erosion*).

The Piedmont Plateau and upper coastal plain sections of the Southeast comprise one of the oldest agricultural regions in the nation. Cultivation began on some of this land about the middle of the seventeenth century, and erosion has been a problem on much of it from the beginning. As early as 1685, the first William Byrd of Virginia wrote about a heavy rain on his farm land that washed away tobacco hills "with all the top of the manured land."[2] For 200 years and more, rainwater has gouged away at Southeastern farms, cutting ever deeper into an enormous total area of the land surface.

Today, a number of counties in this region contain some of the most severely eroded land within the boundaries of continental United States. Recently a survey of erosion conditions was made in the old plantation belt of the southern Piedmont,[3] including within its scope 6,759,288 acres, of which 543,491 acres were flat bottomland. The survey shows that erosion in the past has caused some degree of damage on 95 per cent of the upland portion of this broad section. On 72.5 per cent of the upland, representing nearly all the cleared land of the area, the process is still actively under way, some 2,729,772 acres, or about 44 per cent, having reached the stage of gullying. Approximately 22 per cent of the formerly cultivated area, abandoned at one time or another principally because of erosion, has grown up to voluntary stands of "oldfield" pine, so that much of the washing on such land has been arrested.

Even on land having a slope of 3 per cent or less (representing 18.6 per cent of the upland area), some of it recently cleared, erosion is visible in many places. However, less than 3 per cent of the land of this moderate slope has been affected by gullying, and less than 3 per cent has lost all

[1] Bennett, H. H. The Wasting Heritage of the Nation, *Sci. Monthly*, Vol. 27, pp. 97–124, 1928; Relation of Erosion to Vegetative Changes, *Sci. Monthly*, Vol. 35, pp. 385–415, 1932. "The Soils and Agriculture of the Southern States." The Macmillan Company. New York. 1921.

[2] Hall, A. R. Early Erosion-control Practices in Virginia, U. S. Dept. Agr. *Misc. Pub.* 256, 1937.

[3] Georgia Land Use Problems, Georgia Exper. Sta. *Bull.* 191, pp. 99–107, 1935. This was the first comprehensive erosion survey ever made of a large area. It was conducted by G. L. Fuller of the University of Georgia, cooperating with the United States Department of Agr. and the Georgia Agricultural Experiment Station. The basic principles of the survey were developed cooperatively by G. L. Fuller and H. H. Bennett.

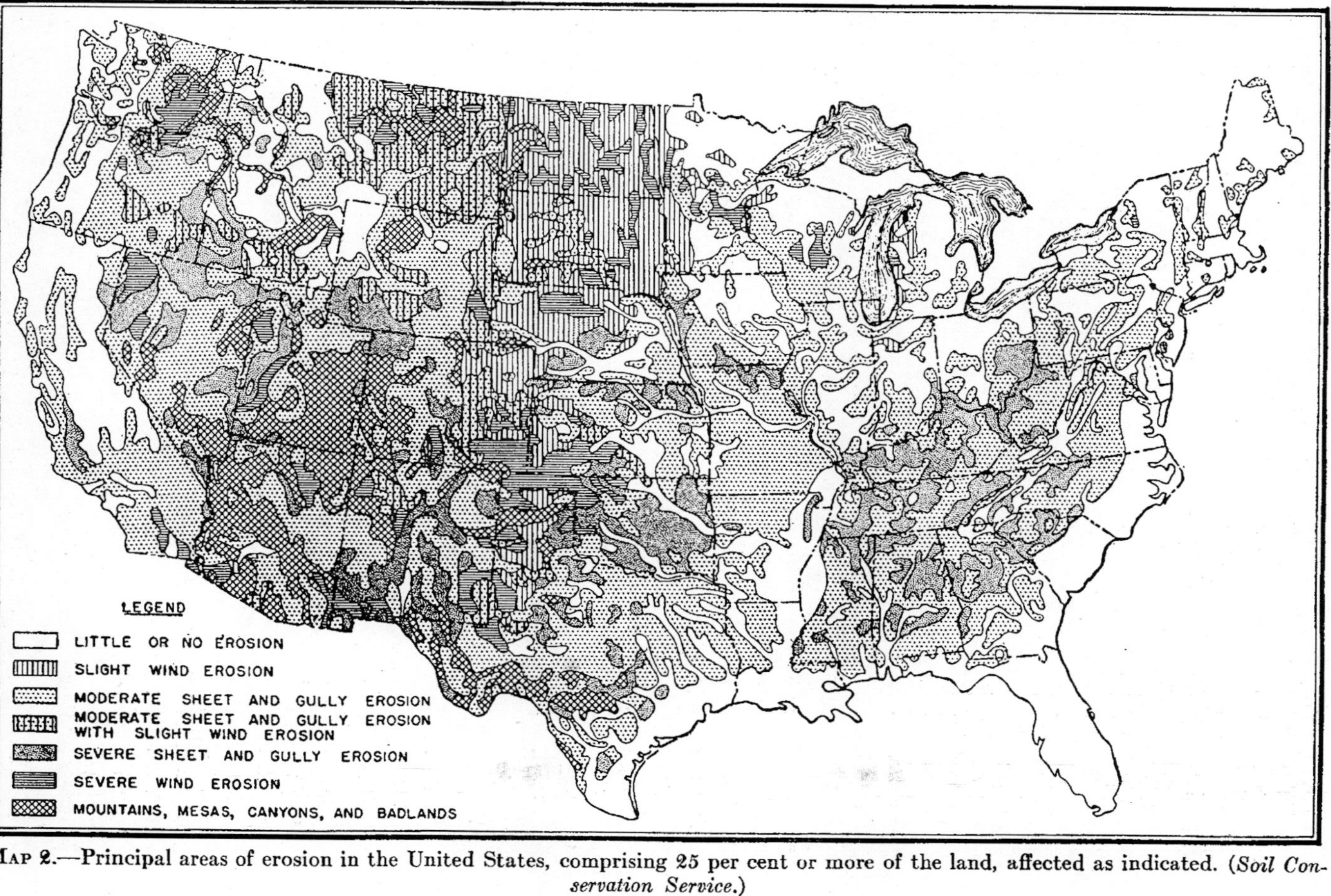


MAP 2.—Principal areas of erosion in the United States, comprising 25 per cent or more of the land, affected as indicated. (*Soil Conservation Service.*)

of the surface soil. But on slopes ranging from 3 to 7 per cent (representing 56.4 per cent of the upland area) 42.6 per cent is gullied, and 54.7 per cent has suffered from sheet washing, with only 2.7 per cent having suffered no erosion. On slopes ranging from 7 to 12 per cent (representing 17.7 per cent of the upland area), 82.2 per cent of the upland is affected by gullying, and 11.3 per cent by sheet washing.

Fig. 27.—Providence Cave; a gully in southwestern Georgia said to have started some 50 years ago. Now more than 100 feet deep. (*Photograph by Soil Conservation Service, 1936.*)

Summing up the results of this Piedmont erosion survey, covering parts or all of 35 counties, more than a million acres have little or no value for agriculture, owing to the ravages of man-induced erosion. Of the upland area, 44 per cent, or about 2,734,000 acres, is definitely submarginal for crop production, as the result of long-continuing erosion. An additional 28 per cent, or 1,750,000 acres, is well worth saving for crop use and can be saved through the execution of a complete erosion-control program. The area not seriously damaged, chiefly because of mild slope (0 to 3 per cent), amounts to about one and a half million acres.

What has happened in Stewart County, Georgia, is a striking example of the extremes to which erosion can sometimes progress when land is farmed without proper precautions. Approximately 70,000 acres—one-fourth of the land in this small county—had been essentially ruined by gullying and deep sheet erosion twenty-five years ago. Furthermore, some of the gullies are probably without counterpart on the North American

Fig. 28.—End result of accelerated erosion in a former field of rich Black Belt land, west-central Alabama. Both the black topsoil and the brownish subsoil washed off to level of parent material. Originally Houston clay, now Selma chalk. (*Photograph by Soil Conservation Service, 1936.*)

continent—huge, yawning chasms 100 feet or more in depth (Fig. 27). One of the worst is Providence Cave, which is said to have started with the drip from a barn roof some fifty years ago.[1] Already this enormous earth scar is said to have engulfed a schoolhouse, two farm buildings, and much good farm land.

Across the States of Mississippi and Alabama runs a broad strip of land known as the *Black Belt*. Once the dark-colored prairie soil of this region made it one of the richest farming sections in the Southeastern States. For generations, the Black Belt supported thriving cotton plantations and a prosperous agricultural economy. But since the turn of the century, cotton production in this area has become increasingly difficult;

[1] Soil Survey, Stewart County, Georgia. Field Operations, Bur. of Soils, U. S. Dept. of Agr., 1913.

at the present time, much of the former cropland is in grass. Over most of the area, it is now virtually impossible to collect even a sample of the original productive topsoil. The soil has vanished. Even the subsoil has washed from hundreds of acres, down to the chalky substratum of whitish calcareous beds (Selma Chalk Formation), representing the parent material (Fig. 28).

Severe erosion damage, however, is by no means confined to the Southeast. Other agricultural sections have suffered to an almost equal degree. Oklahoma, for example, one of the country's newest agricultural states, shows a record that is almost appalling. The land was thrown open to

FIG. 29.—A badly eroded farm, redlands of central Oklahoma. (*Photograph by Soil Conservation Service, 1936.*)

extensive agricultural settlement at various intervals during the past fifty years.[1] A general survey[2] has shown that some soil has been washed or blown from 13 million of the 16 million acres in cultivation and that more than 6 million acres have reached the stage of gullying. In all probability, this is the most outstanding example of rapid land decline on a wide scale to be found in the annals of human history (Fig. 29).

The Pacific Northwest, including the States of Washington, Oregon, and Idaho, is another of the nation's youngest farming sections. Only within the past 60 years has agriculture been practiced in these states on any extensive scale. But already the ravages of erosion, both by wind and by water, have been extensive. According to survey data, about 36 per

[1] McDonald, Angus. Erosion and Its Control in Oklahoma Territory, U. S. Dept. Agr. *Misc. Pub.* 301, 1938.

[2] The Oklahoma Soil Erosion Survey, 2d Southwest Soil and Water Conservation Conference *Proc. Rept.*, Stillwater, Okla., pp. 9–10, 1930.

cent of the farm land in the intermountain zone of the Pacific Northwest is subject to severe soil blowing, and an approximately equal area has lost one-half or more of its original topsoil as the result of accelerated runoff from cultivated land that a little while ago was perfectly stabilized by the natural cover of bunch grasses. Although this damage as yet is not generally comparable to that which has taken place in the older parts of the country, it nevertheless represents a spread of erosion that would not have been considered possible 60 years ago. Actually, land already is being abandoned in this new region, largely as the result of erosion that

FIG. 30.—Erosion resulting from melting of snow on summer-fallowed land. Palouse region southeastern Washington. Light-colored elevations in background stripped of dark soil by erosion since beginning of agriculture. January, 1934. (*Photograph by Soil Conservation Service.*)

has either stripped off the productive soil (Fig. 30) or slashed the slopes with gullies.

Lying in the heart of the Corn Belt, Indiana has frequently been cited as one of the country's most stable agricultural states. Yet it now appears that over 40 per cent of the farming land in Indiana is suffering more or less seriously from erosion. An area of almost 2 million acres has been stripped of more than three-fourths of its original topsoil or is so badly gullied as to be virtually worthless for continued crop production; and sheet erosion is active on an additional 7½ million acres (Fig. 31). From the standpoint of erosion damage, Indiana is still far from being the most severely eroded part of the older farm areas of the east, at least in point of extent; but not many parts of the country have been more completely ruined than a number of rather large areas in some of the limestone counties of the southern part of the state. The same can be said of numerous areas in southern Iowa; northern Missouri; southeastern Nebraska; northeastern Kansas; southern Illinois; southwestern Wisconsin; south-

FIG. 31.—Severe erosion, southeastern Indiana. (*Photograph by Soil Conservation Service, June, 1938.*)

FIG. 32.—Severe erosion of valley land induced by overgrazing. New Mexico. (*Photograph by Soil Conservation Service.*)

eastern Ohio; parts of the bean district of California; many localities in Pennsylvania, Maryland, and New Jersey; and large areas throughout

Kentucky, Tennessee, Arkansas, Texas, and northern Louisiana. Land devastation has proceeded so far in many parts of New Mexico and

Fig. 33.—Erosion of rich potato land cultivated up and down slope. Aroostook County, Maine. April, 1936. (*Photograph by Soil Conservation Service.*)

Fig. 34.—Result of heavy rain accompanying hurricane September, 1938. Tobacco field near Hartford, Conn. Barn in background blown down at same time. (*Photograph by Soil Conservation Service.*)

Arizona that even the details of topography have been altered: Where there were no gullies under original conditions toward the close of the century, nearly every important valley and tributary valley of some localities embracing thousands of square miles is now incised with one or more gullies (Fig. 32). Like conditions are found over large sections in Utah, western Colorado, and parts of Idaho and Nevada. Even Maine has local erosion problems of a serious nature (Fig. 33), as have the other New England States (Fig. 34) and New York.

In a single county in southeastern Ohio (Muskingum), an aggregate area of about 100,000 acres of formerly cultivated land has been largely or entirely stripped of topsoil, with most of it now producing only poverty

Fig. 35.—Gully on Redding farm along line *AB*, Map 3. (*Photograph by Soil Conservation Service.*)

grass, goldenrod, bushes, and other vegetation of little or no value except that it has done much to check further erosion (in this instance, after the soil has gone downstream).[1]

An illustration of a severely eroded area in the southeastern Ohio hill country is the Redding farm, 1½ miles southwest of Mt. Sterling, Muskingum County. This land, cleared about one hundred years ago, formerly had the reputation of being one of the best farms of that section. Later, careless cultivation, such as continued production of corn and cultivation without consideration of the contour, developed a serious erosion problem (Fig. 35). A detailed survey made in 1933 shows that nearly 54 acres of the upland portion of the farm have been essentially destroyed by deep sheet washing and gullying (Map 3), that nearly 7 acres have lost from 3 to 12 inches of soil, and that 8 acres of bottomland

[1] Conrey, G. W., Cutler, J. S., and Paschall, A. H. Soil Erosion in Ohio, Ohio Agr. Exper. Sta. *Bull.* 589, p. 9, 1937.

have been covered with about 7 feet of inferior material washed down from the eroding slopes. Only one acre was uneroded. With less than 8 acres of cultivable land left, this 70-acre farm was abandoned several years ago, and the house and farm buildings have fallen with decay.

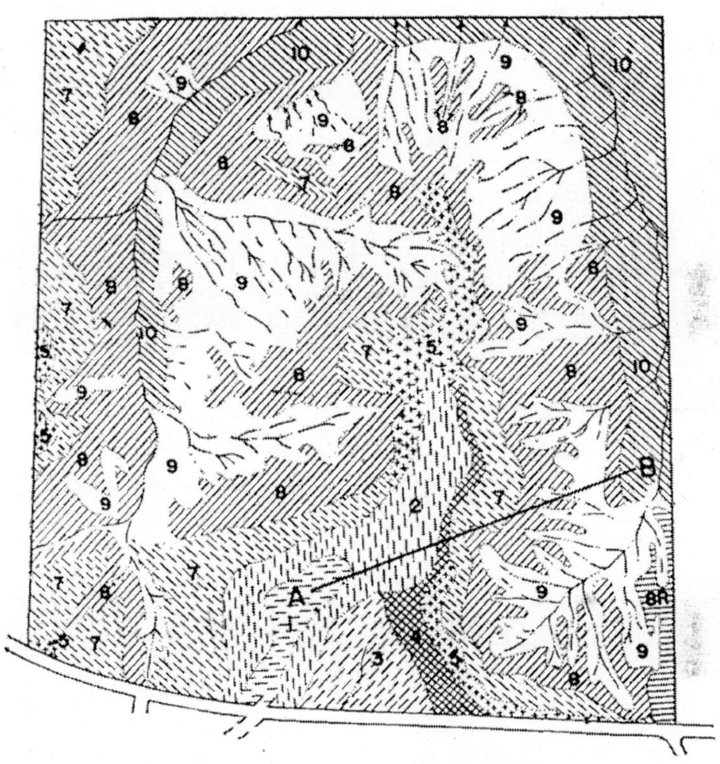

(1) No erosion, 10 in. topsoil intact, level upland 1 acre

(2) 3.5 in. topsoil removed, gentle slope 4 acres

(3) 6 in. topsoil removed, moderate slope 1.5 acres

(4) 10 in. topsoil removed, moderate slope 1.2 acres

(5) 15 in. soil and subsoil removed, moderate slope 2.5 acres

(7) 2.5 ft. soil and subsoil removed, moderately steep 10 acres

(8) 3.5 ft. soil and subsoil removed, steep and gullied 20.5 acres

(8R) Severely eroded, abandoned to 2d growth forest, steep 0.8 acre

(9) 5 to 30 ft. soil and subsoil removed, steep and gullied 20 acres

(10) Overwash from adjacent uplands, poorer than formerly 8.5 acres

Map 3.—Erosion survey of Redding farm, 1½ miles southwest of Mt. Sterling, Muskingum County, Ohio, 1933.

Graph 1 shows the effect of erosion on soil, subsoil, and underlying parent rock of a cross section of the Redding farm from level, uneroded upland to the stream bottom on the east (line *AB*, Map 3).

A soil survey of Fairfield County, South Carolina,[1] made in 1911, showed 90,000 acres of formerly cultivated land so cut to pieces that it was mapped as unplowable *Rough Gullied Land* and that an additional 46,000 acres of once rich stream bottom were classed as essentially worthless *Meadow*. The latter was so classified because of the frequency of overflow and swampy conditions resulting from the clogging of the channels with the products of erosion. Since that survey, much additional land in the county has gone out of cultivation because of severe erosion.

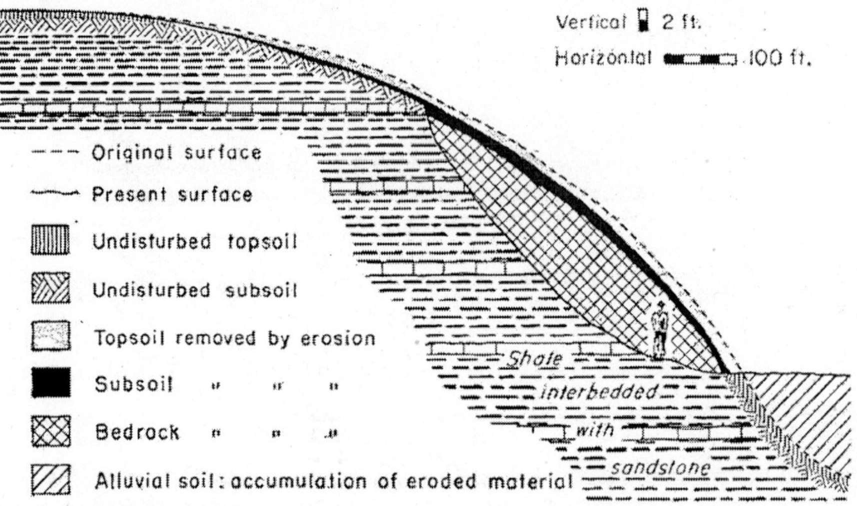


GRAPH 1.—Profile of slightly and severely eroded cross section of Redding Farm from crest to, and including, silted stream bottom. Cross section corresponds to line *AB*, Map 3.

Fortunately, these severely gullied areas have grown over partly or entirely with voluntary oldfield pine.

A survey of erosion conditions in 20 counties of the Southern Great Plains area (southeastern Colorado, southwestern Kansas, and the Panhandle areas of Oklahoma and Texas) shows that of 16,312,377 acres surveyed, nearly 98 per cent have been affected by accelerated erosion (mostly wind erosion) to a degree ranging from slight to very serious, with approximately 53 per cent affected to a serious degree.

## Erosion and Farming Operations

The ease with which farming operations can be conducted depends largely on the smoothness of the ground surface and the tractability of the soil. Rough, hummocky ground; stiff, unyielding clay soils; and shallow, stony areas always have been avoided by farmers wherever other land was available. One of the most immediate results of soil removal by erosion

[1] Soil Survey, Fairfield County, South Carolina. Field Operations, Bur. of Soils, U. S. Dept. of Agr., 1911.

is the frequent, undesirable change in the topographic details of the ground surface. When a field is cut by numerous rills, the difficulties of plowing are considerably increased even though the ground is smoothed off (and the washes forgotten) by the next cultivation (Fig. 36). When deep gullies appear, however, the work of the farmer in that particular field is multiplied. In such advanced stages of erosion, the effects are not amenable to easy repair or forgetfulness. Usually, such fields are abandoned because of the impracticability of crossing the ravines.

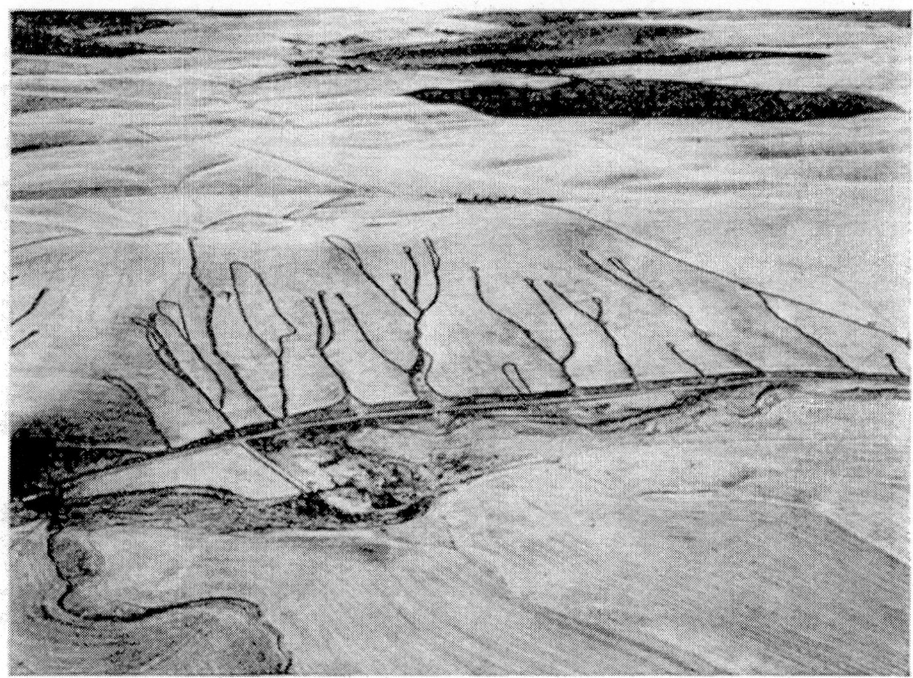

FIG. 36.—Gullies plowed in on steep wheat fields in Palouse region of Washington to facilitate operation of harvesting machinery. Photo made after the wheat was cut, August, 1937. (*Photograph by Soil Conservation Service.*)

As the productive layer of topsoil washes away with every heavy rain, the farmer gradually approaches the subsoil with his plowing, and the subsoil frequently consists of tough, stubborn clay (Fig. 37) or loose, droughty, sandy material; rotten rock; or hardpan. Subsoil farming is not only an unprofitable venture in most instances; all too often, it is also a backbreaking, expensive, or fruitless task.

EFFECTS OF SOIL LOSS ON YIELD. Over the past thirty or forty years, the average yield per acre for most of the country's staple crops has been maintained at a fairly even level. In some instances, the per-acre yield has shown a slight increase; in others, a moderate decrease. Meanwhile, erosion has been marching steadily across the nation's cultivated land, extending over an ever widening area, and cutting constantly deeper into productive soil.

When soil is removed bodily from a field, both available and potential plant food, along with inert mineral material and everything composing the body of the soil, is carried away. As erosion progresses, as stiff clay of relatively low absorptive capacity is approached and finally exposed, the ability of the land to supply the moisture necessary for plant growth is generally impaired, the beneficial activity of microscopic organisms lessened, and efficient tillage rendered more difficult. Nearly every

Fig. 37.—Erosion-exposed clay subsoil of Shelby loam, sun-baked, cracked, and dry—exceedingly poor soil.

important soil type in the United States is made up of a series of definite soil layers, or horizons—of relatively productive top layers, with less productive horizons beneath. Experiments conducted on a number of important types of farm land throughout the country have shown definitely that yields on topsoil run anywhere from about $1\frac{1}{2}$ to 33 times as large as on similarly treated areas that had been artificially desurfaced down to subsoil. Even where the subsoil was fertilized and supplied with organic matter, lower yields were produced. The effect of erosion on yields is treated in greater detail under Chap. IX (Part 1).

Erosion Deposits on Agricultural Land. The effects of erosion on crop and grazing lands are not confined to those fields and pastures which are stripped of their productive topsoil by water or wind. Material, first from the topsoil, then from erosion-exposed subsoil, carried in suspension by running water is frequently dropped on low-lying areas where the land flattens and the rate of flow is diminished. Wind-assorted sand—the heavier residue left after dust storms—blown from an unprotected field or pasture may be spread over adjoining lands for many miles around (Fig. 38). In the instance of deposition by both water and wind, it is the larger, less fertile particles that usually are dropped on agricultural lands; the finer, lighter, and more fertile material generally is carried long distances or transported out to sea. Moreover, the subsoil material discharged from raw gullies is usually of relatively unproductive character

Fig. 38.—Highway and adjacent land covered by wind-drifted sand. Western Oklahoma. (*Photograph by Soil Conservation Service*, 1935.)

and impoverishes lower slopes and alluvial plains over which it is laid down by runoff or overflow water. Coarse sand, gravel, and even boulders are often washed from eroding uplands and deposited over productive lowlands. The total amount of damage that has been inflicted on American farm lands by deposits of erosion debris is not subject to accurate measurement, but the aggregate area unfavorably affected, including those bottomlands which have been made swampy or semiswampy by increased overflow from stream channels choked with the products of erosion, amounts to probably 7 or 8 million acres of formerly good cultivated or cultivable land. During floods or violent dust storms, damage by deposition of the products of erosion sometimes reaches tremendous proportions (Fig. 39). The Ohio River flood of January, 1937, for example, laid down 17 million tons of sand in varying depths over 26,000 acres of rich alluvial lands between Pittsburgh and Cairo. Hundreds of acres of bottom soil valued at $100 an acre were ruined, at least temporarily, by heavy deposits of loose sand of low productivity. Of the 812,836 acres covered by the 1937 flood on the Ohio, 723,743 represented for the most part good alluvial farm land, and 89,093 acres urban areas. Approximately 400,000 acres, or 55 per cent, of the farm land were affected by the deposition of 47 million tons of soil material. Nearly 4 million tons were deposited in towns and cities.

Altogether, the cost of erosional deposition to the American farmer within recent years has amounted to many millions of dollars.

FIG. 39.—Four and a half inches of soil deposited in corn field, Ohio River bottom, flood of 1935. (*Photograph by Soil Conservation Service.*)

**EROSION DEPOSITS IN RESERVOIRS, WATERWAYS, AND HARBORS.** Eroded soil is not only a menace to agricultural land; it is piling up in immense quantities behind dams, in waterways, and in harbors. Huge deposits of silt are rapidly mounting in hundreds of American reservoirs, crowding out the water that these costly structures were originally built to hold and cutting short their span of useful life (Fig. 40). In many instances, however, erosional damage to reservoirs is not limited to the actual reduction in storage capacity. Silt deposits tend to spread out the stored water and expose a broader surface to losses by evaporation. In the more arid regions, such as the Southwest, exposures of storage water to the warm, dry air may constitute an exceedingly serious problem.

It is estimated that about 8,600 reservoirs in the United States are of more than immediate local importance. The total value of these structures has been conservatively placed at two billion dollars. About one-fifth of these reservoirs, representing approximately 75 per cent of the total valuation, are structures whose value depends almost entirely on their ability to store water. When this capacity is lost, the investment simply has to be written off the books. In other words, erosion conceivably could damage American reservoirs to the extent of $1,500,000,000. Furthermore, since it is impracticable to clear out a major reservoir, once it is filled with silt, not only is the storage capacity lost, but, in the practical sense, the dam site, a valuable resource in itself, is disposed of.

FIG. 40.—Reservoir filled with the products of erosion, Dan River, Schoolfield, Virginia. June, 1936. (*Photograph by Soil Conservation Service.*)

Although such wastage may not be permitted by an enlightened nation to proceed so far, erosion, nevertheless, is reducing rapidly the value of many costly structures built for purposes of irrigation, flood control, power production, municipal water supply, and recreation. Detailed surveys conducted by the Soil Conservation Service have shown that reservoir silting is a grave problem which is not confined to any one section of the country.[1]

For example, one privately owned reservoir on New River, in southwestern Virginia, was found to be 80 per cent filled with silt after twenty years of operation. A lake created for municipal water supply near Waco, Tex., had lost nearly 20 per cent of its original storage capacity in less than six years after construction. A recreation pond near Galva, Ill., was

[1] Eakin, Henry M. Silting of Reservoirs, U. S. Dept. Agr. *Tech. Bull.* 524, 1936.

more than half filled with the debris of erosion in a period of twelve years. Thirteen power reservoirs in the Southeast have filled to the top of the dam within an average of 29½ years following completion of the dam. During the 1938 California flood, some of the reservoirs lost more than one-third of their capacity by deposition of silt, gravel, and boulders. These are some of the worst examples, but they do not by any means close the list of reservoirs severely affected by sedimentation.

Silt deposition is also a troublesome problem along both natural and artificial waterways. Erosion debris is clogging numerous small streams

FIG. 41.—Drainage ditch northwestern Mississippi, filled with erosion debris before the bonds for the establishment of the drainage district were paid off. March, 1936.

and shoaling the channels of navigable waterways; it is also reducing the carrying capacity of drainage canals (Fig. 41) and impairing the effectiveness of irrigation systems. It is forcing many streams from their natural courses, thus causing damage along the banks and accentuating the problem of flood control. In many of the country's harbors and along a number of important streams, navigation is constantly threatened by shoaling, and expensive dredging operations are being necessitated by huge accumulations of soil, sometimes carried from fertile uplands hundreds of miles away. By the processes of soil erosion, soil transportation, and soil deposition, one of the nation's most valuable assets has been transformed along numerous streams into a definite liability and a menace.

EFFECTS OF EROSION ON FLOODS. The history of floods on the North American continent extends back before the days of Jamestown and Plymouth Rock. Hernando de Soto found the Mississippi River in flood when he explored the lower valley in the sixteenth century. That flood was a thoroughly natural occurrence, of course. Every stream throughout the world that is bordered with a strip of alluvium, whether a mere ribbon or a vast plain like that of the Mississippi bottoms, has been subject to overflow for undeterminable periods. Alluvium is the product of deposition from overflow water; it is formed in no other way. So, the fact that de Soto found the Mississippi in flood nearly four centuries ago contributes nothing to our knowledge of flood changes in America; it would have been helpful, however, if he had erected at the time a permanent bench mark recording the precise crest of that inundation. (With such information it would be possible now to compare the height of floods along the Mississippi under conditions of a virgin country, with flood crests that have marched down that mighty waterway since the intense alteration of surface conditions over so much of the American landscape by farmers, stockmen, and lumbermen—provided precipitation records for the entire watershed had also been recorded.)

The mechanical composition of the alluvium of thousands of streams characteristically subject to floods reveals a highly significant fact. The alluvium laid down (above the level of stream-bed deposits) by floods of the preagricultural period is generally composed of relatively fine soil particles, and the material is comparatively uniform in texture; on the other hand, the alluvium deposited by more recent floods over the preagricultural sediments is ordinarily coarser and much more diverse through the profile. The line of separation frequently is so sharp that it can be photographed readily (Fig. 42). In other words, the floods that occurred under virgin land conditions must have been more of the nature of gentle inundations which laid down, for the most part, uniformly fine material; in contrast, the floods of recent years appear to have been more on the order of rushing, tearing torrents, capable of transporting heavier, coarser particles to be deposited alternately with the fine materials of intervening moderate overflows.

Another important fact revealed by recent studies of alluvial profiles is the greatly increased rate of flood-plain aggradation that has taken place since the beginning of extensive land-use activities. Generally, recent deposition along these streams has been so rapid that there has not been time between overflows for the development of a distinct dark-colored, humus-charged topsoil layer, such as usually characterizes the upper part of the buried deposits. Furthermore, the depth of the recent alluvial deposits in many instances is as great as, and in some instances greater than, that of the preagricultural deposits, although the former were laid down within periods of little more than 40, 50, or 100 years, whereas the latter probably required thousands of years for their accumulation.

Within the past forty years or so, floods in nearly every section of the country seem to have increased in frequency, volume, and velocity. For example, in early 1937, a deluge swept the whole Ohio Valley, breaking records for height of flow (flood crests) at many points. The previous spring, record-breaking floods struck with appalling destructiveness over much of the northeastern section of the country, leaving difficult and costly problems to be solved, not only immediately along channelways but out in the fields and pastures from which so much of the flood waters come. Nearly every year unprecedented floods are recorded along various streams of the country.

FIG. 42.—Profile on the watershed of a tributary of Coon Creek, Wis. Since the beginning of farming about 60 years ago, 5 feet of soil material has washed down over the alluvial bottoms. The time required to deposit the 8 feet of material now buried below the dark line, which marks the surface of the soil at the time of settlement, may have exceeded 10,000 years. The old alluvium of the preagricultural stage has the uniform character of deposits laid down by moderate floodwaters. The new alluvium has the diverse character of deposits laid down by violent floodwaters. (*Photograph by Soil Conservation Service.*)

Meanwhile, as floods have become more destructive, erosion has reached its most serious proportions in the history of the country. The mere fact that these two phenomena appear to have developed coincidentally does not, of course, necessarily prove a causal relationship. Determination of the precise connection between soil erosion and floods must await further research. It is now possible, however, to state rather definitely at least a few underlying facts.

Soil erosion is caused by accelerated runoff following removal of the natural vegetative cover; runoff, in turn, is speeded up as absorptive topsoil is removed to expose relatively impervious subsoil over which water moves with increased speed in the direction of stream channels. Millions of gullies form new tributaries to natural drainageways; they concentrate runoff water and discharge it into streams at maximum speed. Other new tributaries are roadside ditches and drainage canals. Even the hundreds of millions of furrows between crop rows that run up and down slopes are gutters draining rainwater rapidly off the land.

The natural drainage system of the country, the branching network of rivers and their tributaries, was carved out by nature to dispose of runoff that formerly flowed or trickled down from naturally vegetated slopes. At the present time, these channels are being called on, over short periods, to carry enormously increased volumes of water. To make matters worse, some of them are in no condition to take care of even a normal burden. Erosion debris washed down from upland fields and pastures has accumulated in numerous stream channels, reducing their carrying capacity and helping to raise the crests of floods.

In the light of these facts, it becomes a little easier to understand why, from year to year, levees are being built higher and higher, and dams larger and larger. This may be only a partial explanation; possibly other factors are involved. Obviously, frozen ground of bared fields and soil wet with melting snow or continuing rains have been important factors in many floods; but widespread soil erosion and unwise farming practices, which accentuate its progress, nevertheless are among the major influences lying behind the nation's growing flood problem (see Chap. XV, Part 2.).

EROSION AND WATER SUPPLIES. Although the more humid regions, and to a considerable extent the dry land areas also, have been plagued in recent years with increasingly destructive floods, the drier sections of the country have been the frequent victims of that contrasting phenomenon known as *drought*. It may seem strange that a process that accentuates the problem of dealing with surplus quantities of water should also have an intensifying effect on the equally vital problem of water shortage. Yet erosion plays just such an apparently contradictory dual role. The relationship of erosion to droughts is not so plain as is its connection with floods, but it can be demonstrated with a reasonable degree of clarity.

Man at present is powerless to prevent periods of subnormal precipitation. On the basis of recent evidence,[1] it appears that the influence of specially planted vegetation on continental rainfall is either negligible or nonexistent. Other human endeavors to stimulate the release of atmospheric moisture have proved equally fruitless. The severity with which any drought is felt, however, depends largely on the available supplies of

[1] Holzman, B. Sources of Moisture for Precipitation in the United States. U. S. Dept. Agr. *Tech. Bull.* 589, 1937.

moisture stored both within the soil and above the ground. If these supplies are adequate, farmers can weather all but the most prolonged dry periods without excessive suffering or loss. On the other hand, when moisture stores are low, even a brief drought sometimes can wreak heavy destruction on plant growth.

The depletion of surface stores of water has been treated at some length in connection with the problem of reservoir silting. Small storage ponds, like large reservoirs, are losing capacity to hold water because of uncontrolled erosion. A survey made by the Soil Conservation Service of four representative farm reservoirs in South Dakota showed that the original storage capacity of one had been decreased by about 15 per cent and of three by more than 25 per cent in periods of approximately twenty-five to thirty years.[1] It seems highly probable that other farm ponds throughout the Great Plains might show an equal or even greater degree of silting damage. Furthermore, erosional debris in some of the country's larger irrigation reservoirs is seriously reducing the capacity of these structures to store surplus water for prolonged droughts. Abnormal demands on water in those reservoirs subject to serious silting can, especially in times of drought, easily inflict hardship on thousands of irrigation farmers operating below the dams. Mounting stores of silt are a poor preparation for months of subnormal rainfall.

Perhaps even more important than storage of water above the ground are the supplies beneath the surface. Most of the rainwater that does not flow off immediately from eroded slopes ordinarily filters down through the soil pores and other openings into the substrata, some being lost as evaporation. Where the intake of surface water exceeds the soil's capacity for retaining moisture in the upper horizons, the excess proceeds downward, as water or percolation, until it strikes an impervious layer of subsoil or rock or a permeable layer or opening leading to an outlet. Above any impermeable layer, percolating water accumulates and either feeds off slowly to springs, streams, or artesian wells or is trapped, depending on underground conditions. Water thus held is available for pumping where the depth is not too great. Thirsty plant roots, reaching down into the substrata, are nourished by this underground reservoir, where it is not too deep, as well as by the moisture held in the upper part of the soil. It has been estimated that subsurface water storage normally exceeds by several hundred times the total capacity of all the country's reservoirs. If this vast quantity of water were so distributed that it could be tapped and made available to the various farming regions on a basis of need, the nation might never be faced with another acute problem of drought.

[1] Connaughton, Mark P. Advance report on the sedimentation survey of Lake Hayes, Hayes, South Dakota. Soil Conservation Service, 1938.

Glymph, Louis M., Jr. Advance resport on the sedimentation survey of Lake Hurley, Gettysburg, South Dakota. Soil Conservation Service, 1938.

What is more important, because of immediate practicability, is the storing within the large internal voids or pore space, characteristic of most soils, of more of the rainfall. Although it may not be practicable to fill all such cavities or internal spaces (constituting the great reservoir of the soil) with water, it generally can be utilized far more effectively than the average American farmer has appreciated—and to his great advantage, with respect to both erosion control and efficient utilization of rainfall. This can be done by holding much of the rain—in some localities, all of it—on the land with practical measures, such as contouring, terracing, subsoiling, and strip cropping. Thus held, more of the water that falls on the land—sometimes all of it—can be stored (absorbed by the soil) for subsequent crop use. Not only do these methods increase the available supply of soil moisture that is stored, but they frequently make it easier for plant roots to penetrate the soil in search of moisture. The reduced absorptive capacity of land as the result of exposing dense clay subsoil by erosion, together with the increased difficulty that plants encounter in taking moisture from such clay when dry, hard, and resistant to penetration, is essentially the equivalent of a reduction of the rainfall, so far as the growth of plants is concerned.

It is important to emphasize the fact that underground water supplies are subject to depletion just as definitely as are the stores in surface reservoirs and that soil erosion frequently is a major cause of the difficulty. As brought out in the chapter on infiltration, erosion tends to prevent the accumulation of soil and underground water (water table) by interfering with the normal processes of intake and percolation. When a drop of rain strikes vegetation, it is shattered in such a way that a spray of clear water develops, part of which adheres to the plant surface as clear water. Although a fraction is lost by evaporation, the greater part almost immediately finds its way into the numberless channels that perforate the soil. But when a raindrop strikes bare soil, especially pulverized soil, the force of impact picks up fine particles, thus making muddy water. As this muddy water sinks into the ground, the fine particles tend to filter out at or near the surface and to form a thin, muddy film which chokes the pores of the soil. As a result, only part of the water enters the reservoir of the soil; the remainder flows over the surface—downhill.

It has been estimated that during the drought of 1934, the water table dropped in some parts of the Great Plains as much as 20 feet. The larger part of this reported loss probably was due directly to rainfall shortage, but some of it doubtless was the result, directly or indirectly, of accelerated soil erosion. In various parts of the country, gullies have so deeply incised the pervious materials of many aggraded valleys as effectively to drain out the accumulated water of the substrata, frequently damaging the vegetation and sometimes even killing trees. Drainage works also have contributed locally to subsidence of the water table.

Over large parts of the country, water may be considered the beginning and end of agriculture. Throughout the Great Plains and over wide areas in the West and Southwest, the key to successful farming is pretty largely an adequate supply of moisture. Already erosion has made sharp inroads into that supply; in the future its depredations may be even greater. If erosion caused no other type of damage, this alone would justify a program of prevention and control.

Erosion on the Western Range. Perhaps nowhere have the harmful effects of erosion on water supplies and vegetation been more acutely felt than on the Western range. This vast expanse of territory, lying between the eastern border of the Great Plains and the strip of country bordering on the Pacific Ocean, comprises about 40 per cent of the country's total land surface. In early days, it was primarily a grass country, where the native stands in many localities grew tall and lush. It is still a grass country, but over wide areas the original cover has been crowded out by less desirable plants or has disappeared completely. Studies and surveys show that fully half the range country is now suffering a serious depletion of forage plants after only a few decades of grazing.

On the Western range, soil erosion and declining cover are now complementary processes, the one speeding up the other in a costly process of vanishing resources. Depletion or severe diminution of the coverage of vegetation, however, probably preceded the first evidence of advancing erosion. Overgrazing has been a problem on the ranges of the Western States since the days in the nineteenth century when cattlemen and sheepmen fought for exclusive use of the grass and water holes. Today that vast range, once capable of supporting twenty-two million animal units, has an estimated carrying capacity of only about eleven million. Yet the area was being misused recently to the extent of asking it to support more than seventeen million animal units, representing an overuse of nearly 55 per cent.

In the range country, as elsewhere, exposed or thinly covered soil is a natural prey of erosion. As erosion exposes the unabsorptive lower layers of soil, as the ground surface is puddled over by muddy water, the process of moisture infiltration is disturbed. More water rushes over the ground at increasingly faster rates, carrying larger loads of soil. With this goes that part of the land which is so essential for a healthy growth of grass or other forage. In short, soil, grass, and water are inextricably related; they all decline together. Estimates and surveys reveal that fully 80 per cent of the Western range is affected by erosion to a greater or lesser degree; a total area of around 575 million acres is slowly or rapidly losing its ability to produce an adequate cover.

Erosion Damage to Transportation Systems. One of the most remarkable achievements of the United States has been the creation of an intricate network of transportation facilities—highways, secondary roads, and railway systems stretching across the country and reaching into nearly every town and hamlet. Yet millions of miles of these costly structures are faced with the hazards of erosion. Highway and railroad embankments frequently have been undermined or washed out by the cutting of uncontrolled water. At times traffic is impeded or blocked by erosion debris spread across the roadways. Along many thousands of miles, roadside ditches have become a menace to both the roadway itself and those who use the roads, as they cut wider and deeper with the erosion produced both by direct rainfall and runoff from adjoining lands. The aggregate cost of erosion damage to American transportation systems is not definitely known. A study of the problem in a typical county—Vernon County, Wisconsin—shows that erosion and flood damage are sometimes responsible for as much as 40 per cent of the annual cost of highway maintenance. Estimates from local authorities indicate an erosion damage to the highway system of Texas, not including repairs to country roads, amounting to $1,566,540 for an eight-month period beginning in September, 1935.[1]

The indirect damage attributable in part or entirely to erosion, such as bridges washed out and culverts choked with silt and rock, cannot be determined, but it probably runs into millions of dollars annually.

Effects of Wind Erosion. The widespread dust storm of May 12, 1934, was the first occurrence of its kind since white men arrived on the North American continent. No other dust storm arising from agricultural land had reached such proportions, although local storms of the same type have frequently occurred in a number of Western areas. This one is estimated to have moved 300,000,000 tons of soil from the Great Plains region.

Since that time, other storms have rolled half across the country to dissipate themselves over the Atlantic Ocean or the Gulf of Mexico. In early 1937, a dust storm, originating in the Texas-Oklahoma Panhandle region, traveled northeasterly across many states and on into Canada. Extensive disturbances of this kind bring to mind the dust-laden winds that characterize parts of the border of the Sahara. The Harmattan, for example, is such a wind. It blows southward from the western Sahara area into Nigeria for as much as 6 months in the year, often depositing nearly a foot of soil around dwellings during the course of its annual visit.[2]

In spite of their spectacular nature, however, dust storms probably have not represented the most damaging form of wind erosion. Sand and

[1] Rowalt, E. M. Soil Defense in the Piedmont. U. S. Dept. Agr. *Farmers' Bull.* 1767, 1937.

[2] Forests in Relation to Climate, Water Conservation, and Erosion. Union of South Africa Dept. Agr. and Forestry *Bull.* 159, 1935.

silt marching ahead of wind near the ground cover thousands of acres of range and cropland, smothering plant life and eliminating whatever chances there may have been for a harvest or a cover of grass. Often the abrasive force of these wind-driven soil particles is so great as to shear off growing plants just above the surface of the land. In many instances, the shifting, blowing soil has covered highways and formed troublesome fence-line drifts and sand dunes which stand as a constant menace to adjoining farm lands.

Surveys indicate that the productivity of about ten million acres has been essentially depleted by wind erosion alone and that a much larger area has been damaged in varying degrees.

To the land, wind erosion has become a deadly force; to the people living in sections of the country subjected to severe wind erosion, it has become both an economic and a social menace. Uncontrolled, it impoverishes or destroys the source of their major industries—grazing and agriculture. In the more acutely affected localities, it appears to be causing an increase of pulmonary troubles among human beings, the condition sometimes being referred to as "dust pneumonia."[1] In addition, it is blanketing homes, offices, and stores with dust, covering farm machinery and buildings with sand, endangering the lives of motorists, and increasing the wear of automobiles and farm machinery.

Damage to land and plant life by wind erosion is not restricted to arid or semiarid areas, but extends to every state. The effects are local, however, and confined principally to sandy land.

Economic and Social Consequences of Erosion. In a social and economic as well as a physical sense, the impact of erosion on the fabric of national agriculture has been far-reaching and profound.

It is almost axiomatic that any process that destroys the essential productivity of the soil can exercise only an adverse effect on an economy that depends fundamentally on the soil. Erosion is such a process, perhaps the most vicious. Other factors being equal, its adverse effect on agricultural economy will be reflected usually in progressive deterioration of productive land and in lower farm returns. It frequently leads eventually to submarginality and abandonment and in extreme cases to rural migration, disruption of the tax base, general community disintegration, and similar maladjustments of an economic and social nature. This is not to say, of course, that soil erosion is either the sole or a primary cause of all such maladjustments. But certainly it must be grouped with other factors of a physical, social, and economic nature that exercise a debilitating effect on agriculture generally.

[1] Brown, Earle G., *et al.* Dust Storms and Their Possible Effect on Health, with Special Reference to the Dust Storms in Kansas in 1935. Reprint 1707, *Public Health Reports*, Government Printing Office, 1936.

Temporarily, its ill effects may be totally or partially offset by other factors of a physical or economic nature. The application of fertilizers, for example, may serve for a time to extend the crop-productive life of land against the inroads of erosion; or, similarly, high prices may for a time compensate for high costs of production on land depleted by erosion. In the former case, however, failure to control erosion will result eventually in the complete removal not only of the naturally productive surface soil but of the subsurface soil which has been artificially fertilized. In the latter case, uncontrolled erosion eventually will make it uneconomic to attempt to produce from the land regardless of prices, since it is physically impossible to grow crops on bedrock or fields riddled by gullies.

In some instances, entire communities have been abandoned largely as the result of serious soil depletion, because throughout rural America the welfare of most communities depends largely or entirely on the successful operation of farm lands. In a nation as youthful as the United States, any number of communities, no matter how few, abandoned either directly or indirectly as a result of soil erosion, is too many. Yet the inroads of erosion on the soil wealth of the country have been so rapid within the past few years that many rural communities are even now face to face with potential land bankruptcy and abandonment, with countless areas—farms and parts of farms—already retired from cultivation or abandoned.

The winter of 1936 and spring of 1937 witnessed a considerable migration of farming population from parts of the Great Plains region because of the hazards and toll of drought and wind erosion. Some years ago, there was a similar migration, though not so concentrated or so readily identified as such, from portions of the Piedmont section of the Southeast, from various localities in the more erodible sections of the Atlantic-Gulf Coastal Plain, and from parts of the southeastern Ohio hill country. To a major degree, erosion has been a causal factor in almost every agricultural migration in the United States in recent years.

Experience has shown that soil exploitation and its progeny soil erosion are not easily halted, once started. As the body of the soil and its productivity are depleted concurrently, the land must be used harder and harder to maintain production. For example, the farmer may be forced to cultivate erosion-exposed stubborn clay more intensively or to extend cultivation to areas that hitherto he had considered unfit for crop use. Inevitably, the fallacy of this procedure is proved, economically and physically; unsound production finally strips the land of its more productive soil and then slashes it with gullies, leaving it impoverished or worthless for further crop use and frequently irreclaimable for cultivation.

Often an unsound credit policy has been the motivating agency in extending the period of use—rather, of soil exploitation. Where land has grown poorer and net income has dwindled correspondingly, the demand for loans and credit has mounted. In turn, excess credit and the extension of loans on such collateral have tended to prolong the use of land definitely submarginal. Throughout the country, however, banks and other credit organizations are beginning to base loans on normal land values determined by earning power. Fertile topsoil is beginning to receive recognition as the real backbone of the farm and the principal capital of the farmer.

Tenancy may be traced in many cases to soil impoverishment. The most recent survey shows that more than 42 per cent of the farm operators of the United States are tenants. More than one-fourth of all persons gainfully employed in agriculture are paid farm laborers, and fewer than half of the farmers of the nation own the land that they till. In 1880, according to statistical records, one out of every four farmers was a tenant.[1] Today more than two out of every five are tenants. Furthermore, the equity in the land of many so-called *owners* is so small that it scarcely merits the designation of ownership.

The average American tenant farmer moves every three or four years, and in many sections of the country one-year leases run as high as 50 per cent. Brief occupancy by tenants has encouraged mining of the soil and paved the way for serious erosion. To the average tenant, the knowledge that he will soon move to another farm discourages care for the land that he is currently farming. Without the incentive and pride of ownership, he allows the soil to wash or blow away and the buildings to fall into disrepair.

Once, tenancy in the United States was a sound system of land tenure. It was considered a step up the ladder from farm laborer to farm owner. The position of the tenant farmer was completely consistent with American tradition. This century, however, has seen tenancy become more and more the end, rather than the beginning, of farm ownership.

Soil exploitation and erosion have hurried the rate of transition from marginal to submarginal farming operations in many localities. As productive topsoil gives way to unproductive subsoil, yields diminish, and farm budgets fall out of balance. Farm mortgages and increased demands for credit have been successive, immediate results. Unable to meet obligations, many owners are becoming tenants. In some sections, 20 to 30 per cent of the land is in the hands of credit agencies, and tenants now operate approximately 45 per cent of all American farm land.

The transition from marginal to submarginal farming operations, through the process of erosion, has had other deleterious effects. In some cases, it has increased the tax burden, for, with diminished net returns from land declining in productivity and value, the farmer must neverthe-

[1] Report of the President's Committee for Farm Tenancy. National Resources Committee, 1937.

less pay taxes at the same base until his property is revalued downward for tax purposes. Then the community feels another effect of erosion, for the almost inevitable result of a lowered taxing base is a higher taxing rate on land still able to pay.

Impoverished and gullied farms are not promising markets for farm machinery, electricity, or the many other industrial products normally needed by the rural population. Beset by taxes, mortgages, credit payments, and declining soil productivity, the farmer cannot be the ally of urban industries in maintaining a healthy business cycle. Moreover, under the circumstances, he is virtually forbidden to take a normal, active part in local community life. The farm that is incapable of giving the family a living income provides no incentive for social betterment. On the contrary, it means early termination of education for the children, preclusion of church and club activities, and prohibition of entertainment and intellectual luxuries for the adults.

As the area of arable land dwindles on a farm, because of erosion, the operator may be forced to curtail his agricultural activities or seek additional land by purchase or rental. Often he can increase or maintain production output by better husbandry on his better acres—and this offers more opportunity than is generally supposed. But the difficulties are not always so simple. Land of reasonable productivity may not be available in the community, or there may be neither funds nor credit for an outlet in this direction. In this way, and others, uncontrolled erosion stalemates land users with debt, forces them to migrate or enter some other business, or leaves them stranded on wasted land as submarginal patch farmers or public dependents.

Under some conditions, such as those existing in parts of the Great Plains, certain types of farming may not be adaptable to the soil or climate. Erosion may take too heavy a toll, for example, where wheat is the main crop. The only practical remedy may be to return the land to grass and for the farmer to get into the livestock business. Frequently, this will require a larger area of land. If additional land cannot be obtained by purchase or lease, it may be better in the long run for the farmer to sell or lease his holdings and seek a livelihood elsewhere.

Summarizing the effects of soil erosion, it can be said that the process, if uncontrolled, impoverishes not only the land but those who live on and by the land, as well as communities and urban areas dependent in part or entirely on the welfare of the farmer.

# Chapter IV. Principles, Processes, and Types

The wearing away of the land surface by running water, wind, waves, and moving ice generally has been looked upon as a normal geological process. Failure to distinguish between the timeless process of normal erosion and the rapid process of accelerated erosion resulting from human disturbance of natural conditions of the land surface explains in a large degree the general failure to recognize the vast difference in the effects of these contrasting types of soil removal. In no small measure this failure to differentiate between normal erosion, under which soil is built up and maintained in balance favorable to growth of most plants, and accelerated erosion, under which soil is torn down and wasted, accounts for the indifference that has prevailed so long with respect to the large and increasing area of land impoverished or destroyed as the result of human occupancy.

## Normal Erosion

In its broadest sense, normal erosion, sometimes referred to as *geological erosion*, or *natural erosion*, represents the erosion characteristic of the land in its natural environment, undisturbed by human activity. It is a continual process of surface planation, a normal geological activity which, across the ages, has contributed to the sculpturing of mountains, plateaus, valleys, canyons, plains, and mesas and to the building of alluvial plains and deltas, coastal plains, lacustrine beds, submerged continental shelves, high piedmont plains, valley fills, aeolian deposits, alluvial cones and fans, and colluvial aprons.

Normal erosion, assisted by the complex process of rock weathering, aids both in the formation of soil and in its distribution from place to place. It occurs in a natural, undisturbed environment, where vegetation, with its canopy, stems, ground cover of vegetative litter, and underground network of binding roots, together with the absorptive, stable character of normal, humus-bound soil, retards the transposition of surface soil by rain, wind, and gravitational movement to a pace no more rapid, generally, than the pace at which new soil is formed from the parent materials beneath.

Under the impact of forces associated with the natural soil environment—the collective influence of vegetation, microorganisms, climate, and corollary physical and chemical activities—the soil is so processed, normally, as to establish within its mass characteristics that give it marked resistance to surface removal. Mellow, granular, and spongelike, the topsoil absorbs rainfall. Hidden conduits—root holes and the burrowings of insects, earthworms, and other animals—perforating both surface and subsurface layers carry water into the deep substrata; and infiltration is further assisted by such structural openings as the soil pores, cracks, cleavages, or fractures that usually puncture the profiles of normally developed soils. Various constituents, including important plant-food

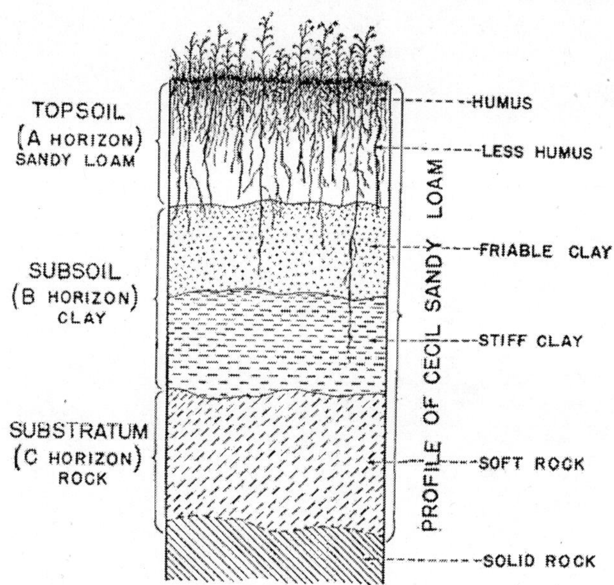

FIG. 43.—Profile of a virgin soil.

elements, are slowly removed in solution or transposed to lower depths as the result of leaching induced by this internal circulation of water. Under natural conditions, however, this removal of plant-food elements is balanced, in part at least, by the upward movement of soluble constituents, derived from the "raw" sublayers through plant roots and stems, into leaves and twigs. Subsequently, the leaves and twigs return to the earth as ground litter and, finally, through decay, are returned to the soil. In addition to these physicochemical activities, percolating water further alters the soil profile, especially of the more stable, less sloping areas, by the slow physical transposition of fine soil particles to lower depths (from the *A* into the *B* horizon). In time, sublayers thus increased in density may serve to quicken runoff by impeding infiltration. Nevertheless, some degree of balance persists regardless of these natural alterations through the soil profile; and topsoil is developed and maintained. Figure 43 depicts the simpler aspects of an undisturbed soil profile.

Erosion, normally proceeding under a protective cover of vegetation, goes on so slowly that it probably is beneficial as a rule, seldom harmful in effect. Without it there would be, undoubtedly, much more severely leached land, more waterlogged land, and more land with unfavorable hardpan or claypan.

As a matter of convenience, that phase of surface wearage having to do with the abrasion of consolidated rocks on which there is little or no soil is designated *rock erosion* and ascribed to the process of normal or geological erosion. Such activity is the principal contributor to the development of rock gorges and badlands.

## Soil Erosion

The vastly accelerated process of soil removal brought about by human interference with the normal equilibrium between soil building and soil removal is designated *soil erosion*. This frequently is called *accelerated erosion* and sometimes *abnormal erosion*.

Under the artificially created conditions resulting from the removal of the protective cover of vegetation, soil is displaced bodily much faster than it can be formed. Unless adequate measures are taken to guard against this abnormal, highly accelerated phenomenon of soil removal, it becomes the most potent single factor contributing to the deterioration of productive land.

Normally, erosion proceeds on bared areas (whether bared by axe or plow, grazing, fire, invasion of rodents, or whatever cause) at an increasing rate as the upper, more absorptive layers of soil are successively removed. As pointed out, the humus-charged, granular topsoil is generally more resistant to erosion than the less absorptive, less stable layers beneath. It is permeated with millions of structural air spaces and with openings produced by decaying plant roots, burrowing earthworms, insects, larger animals, and the fracturing caused by shrinkage on drying. The combined effects of the spongelike organic matter and the activities of microorganisms feeding on this organic material keep topsoil granular, absorptive, and cohesive. Heavy rains tend to seal over the soil pores of bared fields, and the larger openings into the body of the soil are choked with muddy injections. Eventually, underlying layers deficient in organic matter are exposed. Such exposed material, whether of clay or of coarser grain, is more erodible as a rule. With few exceptions, the exposed sublayers (*B* or *C* horizon) are distinctly more erodible—more susceptible to removal by water or wind—than the layers of the topsoil (*A* horizon). Exposed clay subsoil absorbs water so slowly that heavy rains produce rapid runoff and thus further speed the rate of erosion. Runoff water concentrates in greater volume and moves with mounting speed and erosive effect as gullies are cut deeper into the body of the earth.

In this manner, erosion, biting into the land, proceeds by a vicious straight-line process to impoverished soil → destroyed soil and land ruin; it may lead to farm abandonment → disintegration of rural communities → increased relief loads → decline of the nation.

In the practical sense, the result is as indicated; but in the absolute sense, across long periods of time, the line bends back: Volunteer vegetation comes in on the skeletonized, abandoned areas and increases in density; slowly, exceedingly slowly, soil rebuilds; and on it, vegetation progresses in the direction of climax types. Forest probably will return eventually, perhaps like the original stands; grass coverage may be reestablished in time, such as that which once clothed the prairies and plains; and mixed dry-land forage of the kind that stockmen, pioneering western United States, found over much of the semiarid and arid country may again invade depleted areas.

But such completion of the circle—washed-out land, regenerated soil, and again rich coverage of vegetation—involves long periods of time, not now determinable—generations, undoubtedly, possibly much longer. There is evidence, as previously noted, indicating that under natural conditions soil builds back from the raw materials of the substrata so slowly that not less than 300 to 1,000 years or more may be required in some regions to replace 1 inch of true topsoil. If soil builds downward from the surface, as it probably does, much longer might be required to rebuild the second inch, still longer to restore the third, and so on until 6 to 8 or 9 inches of normal topsoil depth is thus recreated through nature's slow but persistent efforts.

## Erosion by Water

Water and wind are the active forces of soil erosion, differing in the nature of their action and in outward manifestation but similar in the sense that both remove and transport surface soil. With respect to transporting capacity, the efficiency of both agencies is greatly increased with increase in velocity. Both present major problems having to do with land defense and preservation, and both call for similar methods of control, the basic essential of which is to reduce the erosive effect by slowing the rate of runoff or velocity of the wind with obstructions across their respective lines of travel. A fundamental difference is that slope is essential to erosion by water, whereas it has no direct causal effect on erosion by wind.

Water erosion is the transposition of soil by rainwater, including melted snow, running rapidly over exposed land surfaces. It is conditioned by factors of slope, soil type, land use, and amount and intensity of rainfall and is confined to sloping areas where the land is of a kind susceptible to washing and where land-use practice has stripped the surface of protective vegetation. It is a progressive process intensified by degree of slope and aggravated by cultivation, by overgrazing, and sometimes by burning and the activities of rodents. In general, it may be broadly defined under three major types, which, although closely related, are by no means mutually exclusive. Two or more of them may occur simultaneously in the same field; one may develop into another. But for the purposes of

| | A<br>Virgin Soil,<br>Per Cent | B<br>Three-fourths Topsoil Gone<br>Per Cent | C<br>All Topsoil Gone,<br>Per Cent |
|---|---|---|---|
| Content organic matter | 7.4 | 1.7 | 0.8 |
| Content nitrogen | 0.3 | 0.1 | 0.06 |
| Content clay | 9.0 | 10.2 | 26.8 |

FIG. 44.—Sheet erosion. All of original dark soil removed from light-colored areas (C). Dark-colored areas (B) still retain some of the darker topsoil. (*Photograph by Soil Conservation Service.*)

discussion, the general category of water erosion may conveniently be subdivided into sheet erosion, rill erosion, and gully erosion.

SHEET EROSION. Sheet washing is the more or less even removal of soil in thin layers over an entire segment of sloping land. It is the least conspicuous and the most insidious type of erosion. Frequently it causes the color of the land to change gradually from dark to light (Figs. 44 and 45), as the removal of dark-hued, humus-charged topsoil exposes relatively light-colored, humus-deficient subsoil. Usually this change in color is accompanied by a progressive decline in yield.

Unprotected land varies widely in its susceptibility to sheet washing, the differences depending principally on topographic features, climatic environment, and the character of the soil. Steep and fairly steep lands and those subjected to heavy or intense rainfall are most likely to be problem areas. But the vulnerability of any field or pasture to sheet erosion is conditioned to an important degree on the inherent erodibility of the soil itself. Areas where a loose, shallow layer of surface soil overlies a dense subsoil of low permeability are peculiarly susceptible to this form of water erosion. It is also likely to prevail on soils of high silt content, fragile sandy soils, stiff clays, and all soils deficient in organic matter. Actually, sheet erosion takes place to some extent wherever water flows across unprotected sloping land, as evidenced by the fact that runoff from such areas is always muddied, in some degree, with the products of

Fig. 45.—Sheet erosion has removed all dark soil in this Kansas field, down to light-colored, limy subsoil. The process has been accentuated by downhill cultivation. (*Photograph by Soil Conservation Service.*)

erosion and is never so clear as runoff from wooded and grass-covered areas.

Similarly, nearly all bare areas are subject to wind erosion when dry, particularly where the soil is loose. A fundamental difference of some importance is that when soil is removed from a field by water, the same agency cannot retransport any of it to the place of origin, as sometimes happens under the impact of shifting wind.

Sheet washing proceeds so slowly as a rule that farmers often fail to notice its effects or even to realize the cause of the color changes in their sloping fields or the appearance and expansion of spots of relatively unproductive subsoil or bedrock exposed by erosion. Eventually, some of them realize, of course, that changes have come about; but many are inclined to look upon such alterations in the soil as the result of an undefinable natural change.

Figure 46 illustrates how sheet erosion moves soil from unprotected areas. Here, the protection afforded by a single gravel or fragment of sandstone allowed columns of soil several inches high to be formed through the simple process of cutting away the surrounding uncovered, unprotected soil by running water. (Such effects can be duplicated in many places by inserting a bottle cap in the soil. It will be found that with such protection of the surface, even on moderate slopes, some of the more erodible soils, especially sandy soils, are whittled away by sheet washing at a surprising rate.)

Sheet erosion grades so imperceptibly into rill erosion that the two cannot everywhere be sharply differentiated. Some grooving of the soil goes on in connection with much or most of the erosion commonly assigned

Fig. 46.—Left, pedestals one-half natural size formed by removal of surrounding soil material (sand) through sheet erosion. Tops protected by fragments of stone. Right, pedestals one-fourth natural size formed by removal of surrounding soil material (sandy loam) through sheet erosion. Tops protected by gravel. Beltsville, Md.

to the category of sheet washing; but in the broader sense, sheet erosion involves the removal of a thin sheet of soil more or less uniformly from the entire extent of an exposed area of uniform character.

*Surface sealing* of the soil commonly is a corollary of the less violent forms of water erosion (sheet erosion), especially on dry soils suddenly wetted. Dry soil in a finely pulverized or loose condition, when wetted by rain, frequently forms a crust. This action varies somewhat with different soils; but in general, the action is as follows:

The raindrops strike the soil with considerable force. The larger ones disturb the immediate surface, especially where the soil is loose or fragile. Disturbance may be either by compaction or by a geyser-like loosening of the material probably something like the splash of a drop of milk falling in a saucer (Fig. 47), depending on the character of the soil and its condition with respect to looseness. Although the drops may be shattered and broken into smaller drops, generally they form small globules of water embedded in the surface. Dry soil resists wetting largely because of air contained in the small pores and on the surface of the particles, so that for a brief interval only a small portion of the surface material is actually wetted. The time required for wetting depends largely on the moisture content of the soil. The drier the soil the slower it wets. Under the beating of rain, however, the surface soon becomes moistened and, frequently, puddled or packed together. Under such conditions, the soil beneath still contains considerable quantities of trapped (sealed-in) air which, by

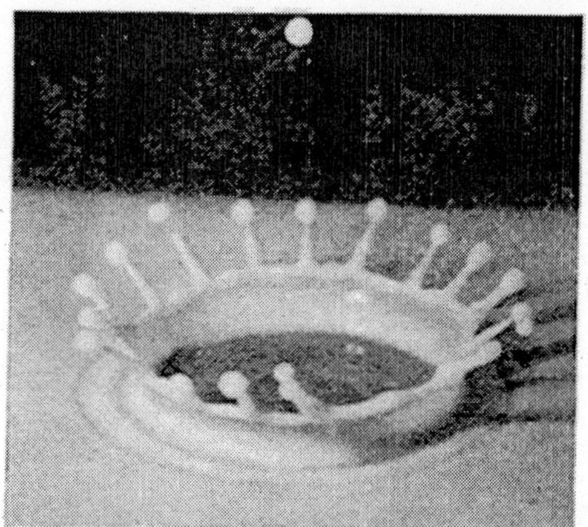

FIG. 47.—The splash of a drop of milk falling in a saucer. (*Courtesy Professor Harold E. Edgerton, Massachusetts Institute Technology.*)

actually filling the openings through which it must travel, quite effectively retards downward movement of water. The surface tension of the water in contact with the soil air further resists movement downward. If merely the surface is wetted, a crust is formed which on drying may become even more resistant to penetration by later rains, thus increasing runoff. When conditions are such that water is readily absorbed by the soil, small particles are carried with it. By a process of straining out, these particles plug the soil pores, thus sealing the surface against free penetration of water.

If the soil is dry and cloddy, capillary forces pull water into the clods, trapping and compressing considerable quantities of air. As the moisture decreases the strength of the cementation, the compressed air and unequal stresses break up the clods, causing them to slack into a pasty mass. Spreading out, portions of this slacked, more or less viscous material are drawn into the spaces between the clods so that the surface soon becomes much less permeable and, if unprotected by vegetation, quickly assumes a sealed condition resistant to rapid penetration of water.

The physical processes thus temporarily sealing the soil surface collectively produce the same effect as puddling, and the result is lowered infiltration and increased runoff. Frequently, this surface layer is so impervious that the soil at a depth of an inch or two will be dry, even with rain continuing. As rain proceeds, however, the abrasive force of the increased runoff produces increased scouring and cutting, which under certain conditions wears through the skinlike layer and starts trenching or rilling. Concentration in slight depressions, or as the result of deflection by obstacles in the pathway of flowing water, also produces channels or rills that serve to accelerate the erosive effect of the continuing rain.

Fig. 48.—Rill erosion, crisscross pattern. Ohio. (*Photograph by Soil Conservation Service.*)

Soil thus surface sealed, especially where not incised by rill erosion, hardens (crusts) on drying, frequently to such degree that where the material beneath is soft or loose, sprouting seeds are unable to push through the cemented layer. Under conditions of this kind, particularly on soils that do not crack or fracture readily, a physical situation is developed that favors excessive erosion with subsequent rainfall.

Muddied water not only serves to clog soil openings with material filtered from the percolating suspension, thus accelerating runoff and erosion, but frequently, if not generally, has a greater cutting effect than clear water, especially where the suspension carries sharp-edged mineral particles, such as sand derived from the breaking down of near-by quartz-bearing rocks.

Rill Erosion. Instead of flowing evenly over a sloping field, runoff water generally tends to concentrate in streamlets of sufficient volume and velocity to generate increased cutting power. The result of this form of runoff is rill erosion, which, in contrast to sheet washing, is characterized by small but well-defined incisions left in the land surface by the cutting action of the water. The trenching typically is straight lined, approximately; but frequently the incisions join in intricate crisscross patterns (Fig. 48).

From the practical standpoint, rill erosion is that type of accelerated erosion which produces small channels that can be obliterated by ordinary methods of tillage.

It is more apparent than sheet washing but almost as often neglected. The small incisions are easily plowed over, so that most farmers are likely

FIG. 49.—Eroded soil deposited over snow. Washington. (*Photograph by Soil Conservation Service.*)

to forget the tiny gashes or to minimize their importance once they are smoothed out with agricultural implements. With respect to extent of damage, this form of soil washing is about as serious as sheet washing. Being more conspicuous, however, more farmers realize the significance of the problem.

Where slopes are smooth, runoff water ordinarily concentrates in rill-producing streamlets only where there is an intense rainfall and a relatively small amount of percolation. Rill erosion is consequently most common in regions of rather intense precipitation and on lands of low absorptive capacity. Soils with a high silt content are especially vulnerable, although the process usually occurs during heavy rains on all areas where loose soil overlies dense subsoil. The sudden melting of snow, such as takes place in the region of the Palouse (southeastern Washington and adjacent parts of Idaho and Oregon) under the impact of warm winds from the Pacific (chinooks), produces very severe rill erosion. Frequently, the eroded soil is deposited on lower lying accumulations of snow to form peculiar patterns (Fig. 49).

Gully Erosion takes place either where the concentrated runoff from a slope increases sufficiently in volume and velocity to cut deep incisions (gullies) in the land surface or where the concentrated water continues cutting the same groove long enough to develop such incisions. Usually, gullies follow sheet erosion or result from the neglect of rills. But frequently, they have their beginning in slight depressions of the land surface where runoff water normally concentrates. Often they develop in natural field depressions or in ruts left by the wheels of farm machinery driven up- and downhill over soft ground. They frequently form also in the trails of livestock and along furrows running up and down the slope.

Ordinarily, these erosion-produced channels carry water only during or immediately after rains or following the melting of snow. Gullies usually cannot be obliterated by normal tillage; most of them cannot be crossed by farm machinery.

Once the gullying process has started, the shape of the incision is generally influenced by the relative stiffness, or resistance, of the soil strata and underlying rock material. For example, over much of the loessial soil regions and alluvial valley fills of the West, both surface soil and subsoil are commonly friable and easily cut by flowing water. Under such conditions, gullies tend to develop vertical walls which result from undermining and collapse of the banks. They extend at the heads and along the sides from water going in over the banks, and they branch readily. Many of the barrancas and "washes" that have cut so deeply into portions of the landscape of various parts of the West and Southwest are typical examples of this U-shaped type of gully.[1]

Where the subsoil is resistant to rapid cutting because of its heavy texture or toughness, and especially where the underlying geological material, or substratum, is not essentially softer than the horizon above, gullies develop sloping banks and take a V-shaped form (Fig. 50). This type is commonly found in areas where the surface soil is underlain by a stiff clay. Although V-shaped gullies usually develop less rapidly than other types, they frequently present an equally serious hazard.

Still another type, representing, at least temporarily, a combination of the U-shaped and V-shaped gullies, prevails in certain localities. This third type develops first as a V-shaped gully. But after the water has cut below the resistant subsoil, it strikes an underlying stratum of loose or soft rock material. Undercutting and caving then occur as the result of

[1] Ireland, H. A., Sharpe, C. T., and Eargle, D. H. Principles of Gully Erosion in the Piedmont of South Carolina. U. S. Dept. Agr. *Tech. Bull.* 633, 1938.

Fig. 50.—Gullies of V shape on limestone soil, northwestern Georgia. (*Photograph by Soil Conservation Service.*)

Fig. 51.—South Carolina Piedmont gully which began as V-shaped ravine. The change to U shape, with vertical walls, was caused by undercutting of soft, decomposed granitic rock. (*Photograph by Soil Conservation Service.*)

this differential erosion, and the incision changes from a V-shaped to a U-shaped channel. In the more advanced stages, these gullies closely resemble the barrancas of the West. Gullies of this type have developed within half a century to depths of 40 to 100 feet in the southern Piedmont region (Fig. 51), especially in localities where decomposed granitic rock (*C* horizon) underlies a subsoil (*B* horizon) of brittle, red clay; in rolling sections of the Atlantic-Gulf coastal plain and in western valley-fill areas where firm or moderately firm clay overlies a soft sandy stratum (Fig. 52);

Fig. 52.—Undercutting type of western gully. New Mexico. (*Photograph by Soil Conservation Service.*)

and in the areas of loessial soils where soft, unstable silt underlies a comparatively firm *B* horizon of higher clay content, as in the Memphis silt loam along the lower Mississippi.

The undercutting, or U-gully, is one of the most destructive types as well as one of the most difficult to control. Ordinary dams are not a sure means of checking their growth; special procedures are necessary. Diversion of water from the heads and sides is generally essential to success. Following this, stabilization can be effected, as a rule, by establishing protective vegetation in the ravines.

A peculiar type of gully, common to soils having tough clay or claypan subsoil, such as the shallow loessial soils overlying intractable clay (as the Grenada silt loam of western Mississippi and Tennessee (Fig. 53) and the claypan soils of the West, especially those of high sodium content, expands laterally about as rapidly as in the direction of the head (Fig. 54).

This process seemingly is governed by the fact that the resistant, tough subsoil, where exposed, tends to direct runoff toward the less stable

FIG. 53.—Gully of lateral-extension type on Grenada silt loam soil, northwestern Mississippi.

FIG. 54.—Gully erosion on soil having tough clay subsoil, southwestern California.

material of the sides, where bank erosion, undercutting, and caving cut back the enclosing walls.

Another type of gully erosion is that characterizing the exceedingly plastic, heavy clays of the humid region, such as the subsoil of the Iredell,

Susquehanna, Lowell, and White Store soils.[1] These tough clays when wet are so adhesive that clear water seems almost incapable of prizing the grains apart. If only clean water runs over the exposed clay, there may be very little erosion save that which removes the loosened fragments formed when the clay dries, shrinks, and scales or cracks into blocky fragments. Muddy water, however, especially that containing sand, gradually abrades the clay material to form broad, shallow gullies, with gently sloping sides (Fig. 55). A somewhat similar type common to the

Fig. 55.—Broad, shallow gully on tough clay. South Carolina. (*Photograph by Soil Conservation Service.*)

Black Belts of central Texas and central Alabama and Mississippi (principally on the Houston soils) is even shallower and broader. Continuing sheet erosion develops washes of this latter type, often in situations where there was little or no depression in the original surface.

Where sandy soil overlies a heavy material, as clay or silty clay, gully erosion in arid and semiarid regions usually is not so disastrous as where the textural stratigraphy is the reverse, provided the sublayer is not composed of such excessively loose material as uncemented sand or gravel. This is due to the fact that the sandy soils, even those containing much clay, of low-rainfall areas do not, as a rule, develop the peculiar columnar

[1] These and other soil series and types referred to in this volume are described in the soil survey reports of the U. S. Department of Agriculture Bureau of Chemistry and Soils; in Soils of the United States, "Atlas of American Agriculture," U. S. Department of Agriculture, 1936; and in various other publications referred to.

structure so characteristic of clay soils and soils of high silt content. The vertical splitting associated with the development of such columnar structure speeds up gully expansion by blocking off and causing to fall in, after undercutting, large masses of soil from the sides and also hastens the development of lateral gullies along the lines of splitting. In the former instance, gully control would be largely dependent on the installation of efficient dams; in the latter, both the sides and the heads of the gullies

FIG. 56.—Left, pothole gully scouring, which sometimes results in the development of a corkscrew curvature in gully bottom, such as that to the right. (*Photograph by Soil Conservation Service.*)

would have to be protected from the disastrous effects resulting from direct entrance of water. Without such protection, dams probably would not be very durable.

The effects of such *structural erosion* are most commonly seen, in the United States, in association with the trenching of valley fills and alluvial plains of the Southwest.

Other peculiarities of gully erosion are directly or indirectly related to soil character, such as the tunnel type of gully development, where rodent holes and holes left by the decay of plant roots enlarge and finally cave in.

The highly friable, very fine sandy soils found in parts of the Great Basin region are peculiarly susceptible to *tunnel erosion*. In the great areas of loess in Northern China, the development of such tunneling is commonly associated with structural fissures, like those responsible for the lateral gully formation (and sometimes tunneling) referred to above.

The grooved, or shelving, type of *badland erosion* produced by unequal removal of material from exposed strata of varying resistance (*differential erosion*) is responsible for the short, steep gullies and peculiar erosion patterns met with in the Dakotas and many parts of the West.

Some of the structural peculiarities involved with the sculpturing of gullies are not fully understood. For example, it is not clear why ordinary

FIG. 57.—Waterfall erosion at head of gully. (*Photograph by Soil Conservation Service.*)

V-shaped gullies, such as those commonly found on the red granitic lands of the southern Piedmont, tend to follow relatively straight lines, whereas another type of V-gully, that common to deep moderately plastic clay soils of shale origin, found in parts of the Southern Appalachian region (as in the vicinity of Rome, Ga.), assumes an almost corkscrew curvature near the bottom, with frequent potholes at the bottom. Figure 56 illustrates the *pothole type of gully cutting*, which probably accounts for the corkscrew curvature in gully bottoms.

*Waterfall erosion* contributes to gully extension by undercutting the sublayers, principally at the heads of the trenches (Fig. 57). It also contributes to the development of tributary gullies and to some extent to lateral expansion. Gully growth by this activity proceeds principally as the result of differential cutting back of the vari-textured layers exposed in the sides. Especially is the process active in soils having soft or unstable substrata, functioning, in effect, after the manner of normal gully undercutting in soils of this type.

This type of erosion is also distinctively effective in cutting away soil at the base of neglected field terraces, particularly where the channels have filled to such extent that runoff overtops the benched strips.

EFFECT OF SOIL FRAGMENTATION ON THE EROSION PROCESS. Some soils, especially clays and clay loams of arid and semiarid regions having a high content of water-soluble salts, crack or scale on the surface of erosion-

FIG. 58.—Gully bank erosion by surface scaling or fragmentation. Arizona. (*Photograph by Soil Conservation Service.*)

exposed subsoil in such manner as frequently to supplement the direct effect of erosion. Under such conditions, the slightest ground disturbance, such as a puff of wind, the tramping of stock, or a shower of rain, frequently causes the loosened or partly loosened fragments to fall off. These tend to accumulate in the bottom of gullies to be swept away with the next rain or to lodge on shelving positions or behind fragments of vegetation on steep slopes to be set in motion as small avalanches by any disturbance of the talus material.

The sides of gully walls in some of the common valley clays of the Western grazing region, such as the valley phase of Reeves silty clay in west Texas and New Mexico,[1] flake or crack off on drying to hasten lateral

[1] Soil Survey, Trans-Pecos Area, Texas. Bur. Chemistry and Soils, U. S. Dept. of Agr. (Series 1928) No. 35.

spread of water-eroded trenches. Rains dislodge from gully walls many of the partly loosened fragments, and with drying thereafter the surface cracks or flakes again to continue the process through alternate wet and dry conditions. It frequently happens that holes are thus cut through the inter-gully walls to assist in lateral growth of the ravines (Fig. 58).

Rock Erosion Phase of the Geologic Process. Those more violent effects of erosion characterized by such examples as the badlands of the Dakotas and the "painted desert" areas of the Southwest usually

Fig. 59.—Rock erosion. (*Photograph by Soil Conservation Service.*)

have been assigned to the geologic process. Since man has had no causative association with this type of erosional activity, so far as can be determined, the classification in that sense is correct.

From the geologic viewpoint erosion has been going on, of course, since the land was first uplifted from the sea. It would seem, therefore, that in the arid parts of the Southwestern States, where, locally, vegetative cover probably never has been very effective under existing climatic conditions, erosion may be far older than the advent of primitive man or even the appearance of the buffalo. Grazing by all forms of animal life, extant and extinct, may have played a part. The rate of erosion within these areas, of course, may have been accelerated in more recent time by the activities of man.

In so far as the immediate erosion problem is concerned, however, there is no pressing necessity to go behind the records that have been written on the land as the result of man's occupation and use of the land. For this reason, these advanced effects of long-continuing erosion, along with those produced by the abrasion of stream flow and wind through indeterminable time, such as the mighty canyons of the world and the fantastic rock carvings of western United States, will be assigned to the *rock erosion* phase of the geologic process (Fig. 59).

ASSORTING EFFECT OF THE EROSION PROCESS. To some degree, especially on sandy soils subjected to rains of moderate intensity, runoff picks up and carries away the finer, lighter particles by a process of *elutriation*, or *selective erosion*, leaving behind the larger, heavier particles. Thus, heavy sandy loam is sometimes changed to light sandy loam or, perhaps, even to sand; and loam may be changed to light loam or even to fine sandy loam. With clay soils and soils of granular or fragmental structure, however, there seems to be much less tendency toward textural changes by such selective removal of the finer material. The indications are that under intense rains, the selective effect of runoff is not so pronounced, the tendency being more toward cutting away of soil bodily.

A conspicuous effect of selective erosion is seen in the formation of *erosion pavement*,[1] where, as the fine material is gradually removed from gravelly or stony soil, more and more of the coarse constituents accumulate at the surface. In time, the surface becomes paved or cobbled, and such pavement tends to reduce or even prevent further erosion except where runoff from heavy rains is sufficient to move the gravel and so cut into the soil beneath.

The assorting effect of water on the solid suspension is far more pronounced when runoff, with its load of soil, enters a stream, although the process is much the same as the direct selective action on land. Stream flow carries the finer products of erosion much greater distances than the coarser, heavier particles. As a result, the less productive coarse material —sand, gravel, and sometimes cobbles—frequently is left as *overwash* which damages valley lands. Also, much coarse bed-load material is left in the stream channel, on bars and riffles, and some of it finds its way into reservoirs.

An example will illustrate how violently rushing flood water assorts the finer, more productive material from soil washed into streams, to leave behind coarser and much less productive alluvium:

In September, 1936, as the result of a destructive flood produced by a 9½-inch rain, a layer of coarse sand, ranging up to 24 inches in depth, was deposited over the much finer textured soil of portions of the alluvial plain of Eighteen-mile Creek, near Clemson College, South Carolina

[1] Shaw, C. F. Erosion Pavement, *Geog. Rev.*, Vol. 19, No. 4, October, 1929.

(Fig. 60). Mechanical analyses show in this instance, as in numerous others throughout the country, an increase in coarseness of texture upward through the profile (Table 3), as progressive deposition has raised the surface of the alluvial plain. The material at 90 inches, representing the surface soil under virgin conditions, contains 17 per cent of all grades of sand; the intermediate layer contains 83 per cent; and the top layer (the deposit by the 1936 flood), 93 per cent. With respect to rates of deposition, investigation of the whole profile, covering the character and depth of deposits before and after agricultural occupation, shows that rate

**Fig. 60.**—Productive alluvial soil buried under nonproductive coarse sand by a flood resulting from a 9½-inch rainfall on Sept. 29, 1936. Eighteen-mile Creek, near Clemson College, S. C. (*Photograph by Soil Conservation Service.*)

of sedimentation has increased greatly since the land was brought into cultivation. As regards fertility, chemical analyses show that the coarse, light-colored material of the 1936 flood contains only 0.10 per cent organic matter, 0.01 per cent phosphoric acid, and 0.007 per cent nitrogen as compared with 2.98 per cent organic matter, 0.16 per cent phosphoric acid, and 0.09 per cent nitrogen contained in the sediments of the preagricultural, or virgin, stage. Stated differently, the original dark-colored, fine-textured, productive, virgin soil of the preagricultural stage shows a content of approximately thirty times as much organic matter, thirteen times as much nitrogen, and sixteen times as much phosphoric acid as the recent deposit of almost sterile sand.

The content of nitrogen, phosphoric acid, and potash in the alluvium corresponding to virgin conditions within the watershed of this stream appears to be about the same as the average amount of those constituents in the predominant upland soils of the watershed (mainly soils derived from granitic rocks, such as Cecil sandy loam and Cecil clay loam). The

TABLE 3.—CHEMICAL AND PHYSICAL ANALYSES[1] OF SURFACE, INTERMEDIATE,[2] AND DEEP ALLUVIAL DEPOSITS[3] THROUGH THE PROFILE OF THE ALLUVIAL PLAIN OF EIGHTEEN-MILE CREEK, NEAR CLEMSON COLLEGE, S. C.[4]

| | *Depth of Light-colored Surface Layer Deposited by Flood of Sept. 29, 1936 0 to 24 In. Per Cent* | *Depth of Layer Immediately beneath Surface Deposit 25 to 32 In. Per Cent* | *Depth of Dark-colored Layer Probably Representing Virgin Soil at Beginning of Agricultural Operations in Watershed 90 to 93 In. Per Cent* |
|---|---|---|---|
| Organic matter | 0.10 | 1.47 | 2.98 |
| Nitrogen | 0.007 | 0.04 | 0.09 |
| Phosphoric acid | 0.01 | ..... | 0.16 |
| Potash | 2.41 | ..... | 2.44 |
| Fine gravel | 4.5 | 0.2 | 0.0 |
| Sand (coarse, medium, fine, and very fine sands) | 93.4 | 83.5 | 17.1 |
| Silt and clay (diameter silt 0.05 to 0.005 mm., clay 0.005 to 0 mm.) | 2.0 | 15.3 | 80.1 |
| Ultra fine (colloid, diameter 0.002 to 0 mm.) | 1.0 | 7.6 | 32.5 |

[1] Analyses by Bureau of Chemistry and Soils; samples collected by Soil Conservation Service following record flood of Sept. 29, 1936.

[2] The layers between the depth of 33 and 90 inches were highly diverse texturally but averaged finer grained than the layers above and coarser than those below.

[3] Original flood-plain deposits.

[4] About 200 yards below Anderson-Clemson College Highway, south side of creek.

topsoil of Cecil sandy loam and Cecil clay loam in the southern Piedmont contains an average of 0.10 per cent nitrogen, 0.12 per cent phosphoric acid, and 1.27 per cent potash (average of horizons $A_1$ and $A_2$), according to analyses of carefully selected virgin samples.[1] However, the content of phosphoric acid is a little higher and of potash considerably higher in the old alluvium than in the virgin upland (or parent soils), indicating some possible selective removal of these constituents by water erosion.

Available information indicates that water erosion generally tends to remove soil bodily rather than by a process of selecting either the finer

[1] Soils of the United States, "Atlas of American Agriculture," U. S. Department of Agriculture, pp. 55, 56, 58, 1936.

particles or the more soluble constituents. It is obvious, however, that there is some selective action with respect to removal of the finer particles, particularly from the coarser grained soils. Such assorting effect probably is considerably more pronounced in the process of wind erosion than with water erosion (see "Erosion by Wind," page 116).

Mechanical analyses of samples of washoff from the same type of soil, at the soil and water conservation experiment stations, show considerable selective action in the removal of coarse and fine material, particularly as between light and heavy washoffs, for some soils and very little difference for others.[1] For example, Table 4 shows that where the ground was covered with sweetclover, washoff from Vernon fine sandy loam carries much more clay and much less sand than from the heavier runoff and washoff from the area devoted to cotton. The results show also that the material eroded from the cotton area contains practically the same relative amounts of sand, silt, and clay as the parent soil. But the lighter washoff from sweetclover shows a marked difference from the soil in place, with respect to content of these three classes of material.

TABLE 4.—SAND, SILT, AND CLAY IN WASHOFF FROM COTTON AND SWEETCLOVER, RESPECTIVELY, GROWING ON VERNON FINE SANDY LOAM, 9.75 PER CENT SLOPE[1]

| | Runoff, per cent precipitation | Erosion, tons per acre | Washoff | | | Soil in place | | |
|---|---|---|---|---|---|---|---|---|
| | | | Sand, per cent | Silt, per cent | Clay, per cent | Sand, per cent | Silt, per cent | Clay, per cent |
| Cotton | 14.7 | 68.52 | 74.7 | 17.32 | 6.8 | 74.3 | 17.1 | 7.5 |
| Sweetclover | 9.0 | 0.56 | 15.1 | 38.9 | 36.3 | 73.6 | 16.1 | 9.1 |

[1] Soil and water conservation experiment station, Guthrie, Okla., 1932.

The various other types of land investigated (Kirvin fine sandy loam, northeast Texas; Colby silty clay loam, western Kansas; Houston black clay, central Texas; Shelby silt loam, north-central Missouri; Cecil sandy clay loam, west-central North Carolina; and Marshall silt loam, southwestern Iowa) do not show such marked selective effect of erosion in the removal of fine material as in the instance of the Vernon fine sandy loam

[1] Middleton, H. E., Slater, C. S., and Byers, H. G. The Physical and Chemical Characteristics of the Soils from the Erosion Experiment Stations, U. S. Dept. Agr. *Tech. Bull.* 430, 1934.

of Oklahoma. Results obtained on Houston clay at Temple, Tex., for example, show very little difference as between the mechanical composition of the soil in place and the washoff derived from this soil. The results obtained from Nacogdoches fine sandy loam of Texas, however, show considerable selective effect as between the comparative amounts of fine material removed in light and heavy washoff, much like that exhibited by the material eroded from Vernon fine sandy loam.

With respect to selective removal of organic matter, Middleton, Slater, and Byers say: "Practically in every case the organic content of the washoff is greater than in the soil composite."[1]

Duley and Miller, in discussing their studies of the effects of erosion on Shelby loam, say: "Chemical analyses showed that the amounts of nitrogen, phosphorus, calcium, and sulphur in the eroded material from corn or wheat land may equal or exceed the amounts taken off in the crops."[2]

The effects of dam failures and cloudbursts show that there is no limit to the size of the soil particle and almost none to the size of individual boulders that can be moved by flowing water, provided the velocity and volume are great enough.

According to a principle of hydraulics, the weight (or volume) of a particle moved by flowing water varies as the sixth power of the velocity; or, in other words, if the velocity is doubled, the weight (or volume) of a particle that can be moved is multiplied by 64. This principle, or law, is based on idealized conditions which never occur in nature. The corresponding law applicable to natural conditions has not yet been determined. Available data and observations indicate, however, that the weight of particles that flowing water will transport under natural conditions varies more nearly as the fifth power of the velocity, or possibly slightly less. In other words, if the velocity is doubled, the weight (or volume) of the particle that can be moved is multiplied by 32 or a number slightly less.

It must be remembered that this refers only to size of the particle and not to the total volume of material that can be moved under similar conditions. On this question, scientific and engineering literature throws even less light, but it may be reasonably safe to state that, by analogy, the quantity of material that can be moved by flowing water under natural conditions would vary approximately as the fourth power; or if the velocity is doubled, the quantity will be multiplied by 16.

[1] Middleton, H. E., Slater, C. S., and Byers, H. G. The Physical and Chemical Characteristics of the Soils from the Erosion Experiment Stations, U. S. Dept. Agr. *Tech. Bull.* 430, 1934.

[2] Duley, F. L., and Miller, M. F. Erosion and Surface Runoff under Different Soil Conditions, Univ. Mo. *Research Bull.* 63, 1923.

Dubuat[1] finds through his investigation of the transporting power of water that the velocity at which movement of various material just starts is as follows:

| *Material* | *Velocity Required to Start Movement, Feet per Second* |
|---|---|
| Potter's clay | 0.27 to 0.35 |
| Coarse angular sand | 0.7 to 1.1 |
| River gravel, size of aniseed | 0.35 to 0.53 |
| River gravel, size of peas | 0.62 to 0.7 |
| River gravel, size of common beans | 1.1 to 1.55 |
| Rounded pebbles 1 in. in diameter | 2.1 to 3.2 |
| Angular flints, about 1½ in. | 3.2 to 4.0 |

The increased power of water to move a particle of soil or a pebble, when the velocity is increased, will vary greatly with the physical situation. If the particle or pebble lies free on the surface, unsupported by vegetation or other physical stabilizers, movement will tend to accord with this formula. Under conditions of support, however, resistance to movement may partly or completely counterbalance the increased impact of running water produced by increased velocity.

## Erosion by Wind

In its arena of activity, erosion by wind presents a problem of equal gravity to that of water erosion. Often it occurs in areas where water erosion also is active, but in any one locality the two types rarely assume an equal degree of importance. Soil washing attains its most serious proportions on land with a considerable degree of slope and an intensive rainfall; soil blowing, on the other hand, becomes an acute problem on both level and sloping areas of low rainfall.

Unlike erosion by water, wind erosion is not easily divisible into forms or subtypes. One example of soil blowing ordinarily differs from another in degree rather than in kind, in so far as present knowledge of the wind erosion process goes. Research in the aerodynamics of soil movement, now under way, indicates that a number of fundamentals involved with the process may prove helpful, when understood, in developing techniques for control of wind erosion. In severity, wind erosion may range all the way from a slight disturbance of the surface soil over a small area to the huge dust storms that sweep across many states, remove countless tons of soil, and constitute a major catastrophe.[2]

[1] Dubuat-Nançay, L. G. "Principes d'hydraulique." Paris. 1786.

[2] Free, E. E. The Movement of Soil Material by the Wind, U. S. Dept. Agr. Bur. Soils *Bull.* 68, 1911.

Under conditions of normal ground cover and natural soil equilibrium, wind erosion, like water erosion, proceeds with the slowness of a normal geologic process. Air currents, with their momentary eddies and deflections, pick up soil material in small quantities, transport it from one area to another, and aid in the development of new soils. But in relatively flat and gently undulating, treeless regions, like the Great Plains, where the sweep of wind is almost unbroken by topographic irregularities, any violent disturbance of the natural vegetative cover brings on a tremendous acceleration of the wind erosion process.

When the grass cover is removed by plowing, the original stability of the soil is greatly reduced. Cultivated soil, depleted of the binding effect of grass roots and of the spongy organic matter that normally accumulates under a cover of grass, becomes less cohesive. After periods of subnormal rainfall, it turns into a dry powdery substance, lying loosely over the surface of the land. This loose, dry material is easily swept up by the wind and transported over long distances. The coarser, heavier particles left behind are blown along near the surface to accumulate about obstacles in their path, such as clumps of vegetation, ground hummocks, houses, farm implements, and fences.

Soils vary considerably in their resistance to wind erosion, depending generally on their structure, size of particles, and content of organic matter. Neither coarse sands nor heavy clays, however, are immune. Actually, the former are more susceptible and usually begin to blow immediately after plowing. The finer textured soils, especially those of granular structure, generally show the greatest resistance. These sometimes remain undisturbed through years of cultivation, although, when subjected continuously to diminution of organic matter under cultivation, the granules break down eventually, and the deflocculated particles are susceptible to wholesale removal by wind.

The action of wind, as well as water, on the soil is something like that of a sieve. Wind picks up the lighter, more fertile soil particles and lifts them into the pathways of high air currents, which often carry them for hundreds and, at times, thousands of miles. The coarser, less fertile particles skip and roll along the surface until they pile up in drifts behind obstacles. A comparison of the soil material blown away by a dust storm with that left behind reveals in striking fashion this sifting nature of the wind erosion process.

Early in 1937, a dust storm originating in the Texas-Oklahoma Panhandle country traveled northeasterly across many states and on into Canada. Soil material laid down by the storm on snow and ice in Iowa was collected and compared with samples taken from a small dune formed by the same wind disturbance near Dalhart, Tex. Analyses (Table 5) show that the dust carried a distance of more than 500 miles contains ten times as much organic matter, nine times as much nitrogen, nineteen times as much phosphoric acid, and 45 per cent more potash as compared with the dunesand piled up in the general locality of the dust storm's source. The transported material is seen to be of much finer texture, containing no sand whatever, as compared with nearly 92 per cent sand in the residuary, drifting dune left behind.

TABLE 5.—SUMMARY OF CHEMICAL AND PHYSICAL ANALYSES OF VIRGIN SOIL, DUNESAND, AND DUST DERIVED FROM CULTIVATED SOIL AS RESULT OF STORM ON AND PRECEDING FEB. 6, 1937.[1]

| | *Unplowed Grassland, near Dalhart, Tex.* | *Dunesand, near Dalhart, Tex. (Formed on and Immediately Preceding Feb. 6, 1937)* | *Dust, near Clarinda, Iowa (Collected from Surface of Snow, Feb. 8, 1937)* |
|---|---|---|---|
| | *Per Cent* | *Per Cent* | *Per Cent* |
| Organic matter | 1.06 | 0.33 | 3.35 |
| Nitrogen | 0.06 | 0.02 | 0.19 |
| Phosphoric acid | 0.04 | trace | 0.19 |
| Potash | 2.05 | 1.77 | 2.58 |
| Sand (coarse, medium, fine, and very fine sands) | 79.2 | 91.8 | 0.0 |
| Silt and clay (diameter silt 0.05 to 0.005 mm., clay 0.005 to 0 mm.) | 19.6 | 7.5 | 97.0 |
| Ultra fine (colloid, diameter 0.002 to 0 mm.) | 8.1 | 5.2 | 33.4 |

[1] Samples of grassland and new dune collected by H. H. Finnell, Soil Conservation Service; dust sample collected by R. A. Norton, soil and water conservation experiment station, near Clarinda, Iowa; analyses by Bureau of Chemistry and Soils.

As compared with virgin grassland in the general vicinity of the storm's origin, analyses indicate a similar assorting effect with respect to removal of both soil particles and chemical constituents. Table 5 shows that the unaffected grass-covered soil carries 79.2 per cent coarse material (total sands) as compared with no sand in the dust and 19.6 per cent fine material (silt and clay) as compared with 97 per cent in the dust. The indicated selective trend as regards chemical constituents is as follows: The dust contains more than three times as much organic matter and nitrogen, respectively, than the virgin soil, nearly five times as much phosphoric acid, and 26 per cent more potash.

A sample deposited by the same February storm at Hays, Kans., about 300 miles from the area of disturbance, was found to contain 3.7 per cent sand, 67.8 per cent silt, and 25.3 per cent clay, indicating, in comparison with the sample collected at Clarinda, Iowa, 500 miles from the source of material, that the coarser materials gradually settle out as dust storms move away from their point of origin. Most of the coarse material in this Kansas sample was of the class "very fine sand," with a diameter range from 0.1 to 0.05 mm., but part of it was of larger size 0.4 per cent fine and 0.3 per cent medium sand. The content of organic matter was practically the same as that in the sample collected at Clarinda, Iowa (3.34 per cent); whereas both the phosphoric acid and the potash were a little lower (0.14 and 2.46 per cent, respectively).

TABLE 6.—CHEMICAL ANALYSES OF UNPLOWED GRASSLAND, DUNESAND, AND DUST FROM NEAR DALHART, TEX.; HAYS, KANS.; AND CLARINDA, IOWA[1]

| | *Unplowed Grassland, near Dalhart, Tex.* | *Dunesand, near Dalhart* | *Dust from Hays, Kans.* | *Dust from Clarinda, Iowa* |
|---|---|---|---|---|
| | *Per Cent* | *Per Cent* | *Per Cent* | *Per Cent* |
| $SiO_2$ | 87.24 | 91.35 | 69.87 | 66.31 |
| $TiO_2$ | 0.24 | 0.15 | 0.55 | 0.63 |
| $Al_2O_2$ | 6.07 | 4.37 | 11.53 | 13.93 |
| $Fe_2O_3$ | 1.14 | 0.79 | 2.88 | 4.24 |
| MnO | 0.03 | 0.02 | 0.07 | 0.10 |
| CaO | 0.34 | 0.31 | 3.15 | 1.98 |
| MgO | 0.25 | 0.14 | 1.08 | 1.43 |
| $K_2O$ | 2.05 | 1.77 | 2.46 | 2.58 |
| $Na_2O$ | 0.62 | 0.25 | 1.18 | 0.92 |
| $P_2O_5$ | 0.04 | trace | 0.14 | 0.19 |
| $SO_3$ | 0.03 | 0.03 | 0.30 | 0.18 |
| Ignition loss | 1.84 | 0.84 | 6.60 | 7.26 |
| Total | 99.89 | 100.02 | 99.81 | 99.75 |
| Nitrogen | 0.06 | 0.02 | 0.20 | 0.19 |
| Carbon dioxide | None | None | 1.31 | 0.71 |
| Organic matter | 1.06 | 0.33 | 3.34 | 3.35 |

[1] Analyses by Bureau of Chemistry and Soils.

Table 6 shows the chemical composition of the grassland and dunesand in Texas and of the dust samples collected in Kansas and Iowa.

Samples of various other wind-blown deposits indicate similar selective removal of the finer, richer components of the soil.

Although this was a major dust storm, its density was not so great as that of some of the others that have struck out of the Plains since 1933. The amount of material falling at Ames, Iowa, during this storm of Feb. 7, 1937, is reported at 34.2 tons per square mile; at Marquette, Mich., 14.9 tons; and in New Hampshire, 10 tons per square mile. The sheet of ice and snow that covered much of Iowa at the time of the February storm was colored ash-gray by the material. The high content of ultra fine particles had the effect of making the ice exceedingly slippery, and injury to livestock from slipping and falling was reported by many farmers.

Various measurements of the amount of dust in the air reported by observers indicate that more than 100 tons of dust per square mile fell in parts of the country affected by the great dust storm of May 12, 1934, which covered a vast territory extending from the vicinity of the Rocky Mountains to several hundred miles out over the Atlantic.

In Europe, the amount of material that fell in Westphalia during the dust storm of 1859, supposed to have originated in the Sahara (a *sirocco*), is reported at 85.8 tons per square mile.[1]

Free defines the complete removal of the fine material of the soil ("deflation") as "the aeolian analogue of the removal of water-borne silt by the rivers." He also ascribes the sandy or stony character of deserts to this phase of wind erosion and points out that "desert pavements" are formed principally by the removal of the fine components of a gravelly or stony soil through wind action. This leaves the larger material to form a pavementlike cover, just as water erosion develops erosion pavement.

Some of the more important differences between the processes of water and wind erosion are as follows: Water erosion takes place downslope, and the transported products, concentrated along drainage lines, are carried to or part way to the sea, whereas wind erosion, not being controlled by gravity so completely as water erosion, proceeds in any direction, affecting both level and sloping land, according to the direction of air currents. The lighter wind-borne materials, traveling long distances, are laid down any- and everywhere and usually so thinly as not to constitute a hazard by accumulation in stream channels and reservoirs. True, the coarser wind-eroded residuary materials are likely to blow faster downhill than across level areas. Such ground drift represents a hazard to anything in its path. Except in the instance of coarse-textured soils, wind does not transport material from land in a condition of saturation or near saturation, which frequently favors erosion by water.

The popular conception of wind erosion is that it is confined to arid and semiarid regions. Although, to be sure, the most severe damage occurs in such regions, the process frequently affects seriously the more sandy soils of humid regions. Along the Atlantic seaboard, for example, sand transported by wind from cultivated fields sometimes chokes drainage ditches and frequently damages young plants. Vegetable growers in parts of the Atlantic States, such as the vicinity of Charleston, S. C., sometimes practice strip cropping as a means of reducing abrasion by wind-driven sand. Muck and peat lands, also, are sometimes strip cropped in the Great Lakes region to prevent the light soil particles from blowing away.

In some localities, for example in western Michigan, relatively unproductive sand, transported as ground drift, frequently covers up the lower

[1] Free, E. E. The Movement of Soil Material by the Wind, U. S. Dept. Agr. Bur. of Soils *Bull.* 68, 1911.

lying areas of agricultural land. On many farms where muck is the most productive land, this type of damage is very serious, frequently causing abandonment. In addition to areas now blowing, much land of sandy character, in this and other parts of the Great Lakes region, is potentially subject to this hazard. If present farm practices are continued, exhaustion of the binding organic matter will cause serious soil movement by wind.

TYPE OF DUST STORMS. Some attempt has been made recently to classify dust storms. For example, "black dusters," as they are sometimes called in the Panhandle country of Texas and Oklahoma, are said to be associated with northwest winds traveling at a rather high velocity and lasting 10 hours or more. These storms have covered areas 600 miles in length and 400 miles in width, with an average ceiling height of 7,500 feet. Strong southwest winds are said to produce lighter colored dust storms, which, although not so spectacular as black dusters, may cover even larger areas and attain an average ceiling height of 10,000 feet. By comparing the density of dust at various heights, it has been estimated that a storm covering 5,000 square miles may transport in the air more than seven million tons of soil, or the equivalent of an inch of soil over 46,000 acres. It has also been estimated that severe black dusters covering 175,000 square miles may transport more than 200 million tons of soil. Mechanical analyses of a large number of samples collected in the Southern Plains area show that the soil of recent wind drifts (local accumulation) contains about 25 per cent more sand and about 38 per cent less silt and clay than the surface soil of uncultivated areas of the same type of cultivated land on which the erosion took place.[1] Comparison of the analyses of the wind-drifted soil and the noneroded soil indicates that of the sandy types, such as those which cover large areas in the Plains country and certain other parts of the United States, three-fourths of the soil shifted by wind is drifted material accumulated against fences and other obstructions, with approximately one-fourth removed beyond the boundary of the fields from which the samples were collected. On the finer textured soils, a high proportion is removed beyond the boundaries of the fields. Thus, a severe storm transporting millions of tons of dust would leave from two to four times that amount piled as troublesome drifts along fence lines, over highways, around buildings, and on croplands.

The small dust storms that play over dry, plowed fields under the force of gentle winds, and the spiraling "dust clouds" (Fig. 61) of whirlwinds so frequently seen over the western landscape, generally do little individual damage. But such small blows increase in size with increased wind velocity; frequently, the local dust clouds combine to develop big dust storms, such

[1] Daniel, Harley A., and Langham, Wright H. The Effect of Wind Erosion and Cultivation on the Total Nitrogen and Organic Matter Content of Soils in the Southern High Plains, *Jour. Am. Soc. of Agronomy*, Vol. 28, August, 1936.

as settlers in the plains described decades ago.[1] Such dust clouds, although frequently traveling many miles and causing considerable annoyance, are not comparable in size to the major "dusters"—the big black dust clouds (Fig. 62)—that, from time to time since early in 1934, have driven from the heart of the continent across to the Atlantic or out over the Gulf. Nevertheless, the big blows sometimes represent a composite of dust lifted from numerous fields in the path of the wind.

**Examples of Normal and Accelerated Rates of Erosion.** The normal rate of erosion from Cecil sandy clay loam, for example, on a 10-per cent slope, in the middle southern Piedmont, is probably somewhere near 0.002 ton soil loss per acre per annum. This is the average

Fig. 61.—Local wind erosion by whirlwinds blowing, spirally, the loose, dry soil. (*Photograph by Soil Conservation Service.*)

annual rate at which soil has been lost from such land over a 5-year period under a good cover of forest, protected from fire. Measurements over a longer period may change this figure slightly, but it probably is not very far from correct, since the measurements on which it is based have covered a considerable range of rainfall conditions and relate to the same soil and slope, under the same undisturbed forest cover.

Such a rate of soil loss means that if there should be no rebuilding of soil from beneath, about 575,000 years would be required to wash off the 7 inches of topsoil normally found on this type of land, under the native cover of hardwoods. At such slow rate of planation, soil probably rebuilds from beneath fast enough to balance surface removal. At any rate, a layer

[1] Throckmorton, R. I., and Compton, L. L. Soil Erosion by Wind, Kansas State Board Agr. *Rept.*, December, 1937.

approximately 7 inches deep has been maintained for an undeterminable period in the southern Piedmont, under forest conditions essentially identical with those on the area where the measurements were made.

The time required to lose the 7-inch layer of surface soil from the same kind of land under the accelerated rate of erosion conforming with the annual production of a clean-tilled crop (cotton) would be, according to present measurements, 46 years, or about 1/12,500 times as long as under the protection of a cover of hardwoods. Where the land is completely bare of vegetation from year to year, only about 18 years would be required to take off the full depth of topsoil. In other words, under a good forest

FIG. 62.—Major type of dust storm near Lamar, Colorado, 1937. (*Photograph by Soil Conservation Service.*)

cover the topsoil of this particular type of land would last, without replacement by building from beneath, about 32,000 times as long as under the condition of a bared surface and, under grass, about 4,500 times as long.

Measurements on other types of land in other parts of the country indicate the same general trend (see Table 14, Chap. V, Part 1). In some localities, grass appears to be as effective as forest in holding soil.

The rates of erosion referred to here and, for the most part, elsewhere in this volume pertain largely to the effects of sheet washing, both normal and accelerated. Some rill erosion has taken place on the areas from which the measurements have been made, but no gully erosion. As yet, very little quantitative information is available with respect to rates of gully erosion. Observation, however, indicates that where gullying is in progress, soil is removed at a faster rate than where the planation involves sheet washing only, or sheet and rill washing. Usually, sheet erosion or rill erosion, or both, accompany the gullying process.

A rough appraisal of the rates of erosion under conditions of active gullying might be as follows: Generally speaking, land subjected to continuing severe gullying has attained the death stage of decline.

## Interception of Rainfall by Vegetation

Interception of rain by vegetation has two indirect effects on rates of infiltration, runoff, and erosion: (1) Part of the rain adheres to the trunks, stems, and leaves of plants as a film or as drops of water (Fig. 63), to evaporate without coming in contact with the ground; and (2) the larger drops frequently are shattered into small drops or spray which have a diminished impact effect on exposed soil. The first effect must be considered in fixing values of effective rainfall, and the latter in interpreting and appraising infiltration, runoff, and erosion on vegetated areas.

Fig. 63.—Part of rainfall intercepted by grass. (*Photograph taken during rain by E. L. Crandall.*)

# Chapter V. Rates of Erosion and Runoff

An adequate interpretation of the process of soil erosion involves an understanding of the numerous physical, physicochemical, and biologic factors, both natural and artificially introduced, that have to do with any modification of the forces and conditions that govern or materially influence rates and amounts of water intake by the land, runoff, and soil removal. Although available data with respect to a number of these pertinent factors are not complete by any means, enough is known of the significance of various aspects of many of them for at least an approximate understanding of the erosion process.

## Effects of Exposure

When the natural cover of vegetation is removed, soil is exposed immediately to the accelerated transporting effect of rainwater or wind. From the beginning of the erosion process, soil openings tend to clog with fine material filtered from suspension in the runoff. This reduces infiltration and thereby increases the off flowage and accentuates the erodibility of most soils. Continuing erosion eventually strips off the more resistant, loamy surface layer, to expose less stable and, usually, less permeable and more erodible sublayers. Plowing further vitiates, or destroys, the natural permeability of virgin soil by disrupting its normal granular or loamlike structure and by obstructing or closing obscure passageways, such as cracks, fractures, and the tunnelings of earthworms, insects, rodents, and plant roots. Continuing cultivation aids in the dissipation of the humus content—the life-giving, spongelike, binding material—through processes of attrition and decomposition.

Thus, the use of land involves its exposure to moderate or violent changes, according to the nature of the land, the biologic and climatic environment, and the way it is used. From the standpoint of land utility, increased rates of soil removal and accompanying losses of rainfall, induced by these artificially produced exposures, are by far the most serious consequences of man's occupation and use of the land.

## Diversity of Modifying Factors

It would be difficult to conceive a physical agency involving more modifying factors than soil erosion. The interdependent processes of runoff, infiltration, absorption, and erosion are profoundly affected not only by slope, climate, soil, and cover of vegetation and vegetative litter but also by the changed condition of the surface soil or sublayers resulting from methods of land utilization and, subsequently, by the condition and character of the different sublayers successively exposed by continuing planation. These and numerous other factors, separately and interdependently, introduce into the dynamics of this four-phased process (from the practical standpoint, runoff, infiltration, absorption, and erosion function as a composite physical agency affecting soil removal and water wastage) an almost inconceivable number of variables, many of them highly significant in relation to fundamental principles involved with rates of soil and water losses as well as practical measures for reducing such losses.

Type and condition of cover, degree of slope, kind and condition of soil, and climate all markedly affect erosion and runoff. Cover affects rates of soil removal and runoff more than any other single factor. Declivity of the land and length of slope powerfully affect hydrologic processes intimately and inseparably involved with soil planation and runoff. Intensity of rainfall, condition of the ground with respect to moisture content, and openness of the soil (permeability) are contributing factors of great importance. When the ground is frozen, erosion obviously is less rapid than at other times. When wet soil is frozen, the tendency is for more nearly complete closing of the openings, and consequent reduced infiltration capacity, than where dry soil is frozen. Soil type alone introduces an almost endless variety of conditions that appreciably or profoundly influence rates of infiltration and absorption and, therefore, of runoff and soil denudation. The texture of the surface material, its inherent structure, and consistence, markedly affect rates of water intake, as do also similar characteristics expressed in the frequent widely divergent sublayers through the soil profile, down to and including the horizon of parent material. Obviously, rainwater passes down into and through coarse sand or loose gravelly soil much more rapidly than through dense clay. Again, the infiltration and absorption capacities of sand, sandy loam, fine sandy loam, loam, silt loam, and other relatively pervious materials vary widely with the depth to sublayers of relatively impervious material, such as dense clay, claypan, hardpan, and solid rock. The hundreds of distinct soil types occurring on slopes of varying gradient and length are subjected to a multiplicity of uses, to various intensities of rainfall, and to alternate freezing and thawing. This wide range of variants cannot be ignored, either in research pertaining to the erosion process or in the development of practical methods for the control of erosion. Moreover, types of farming, grazing practices, and methods of cultivation introduce numerous other modifying conditions constituting pertinent runoff and erosion variants.

## Deficiency of Information

After considering the long list of the more evident factors, it may be somewhat appalling to the specialist, interested either in the principles of the erosion process or in the development of measures for control and prevention, that still other categories of less obvious variables must be considered. For example, much less is known about the mechanical relation of gravitational creep, sliding, solifluction, soil granulation, fragmentation, dispersion, alternate freezing and thawing, and sedimentation to processes and rates of erosion and runoff than about those more obvious effects of such physical factors as slope, soil type, and cover. Too little information is available with respect to the causative relationships between rates of runoff and the development of rill erosion and its various peculiar patterns, such as straight-line rills, crisscross rills, and rills that divide in the downslope direction in contrast to the normal tendency to branch upslope. Rates of differential erosion, such as the processes involved with the fantastic patterns of land and rock sculpturing in regions of low rainfall, have not been measured. Baffling contrasts revealed in the amounts of material eroded from cross sections of varying widths, where slope, soil type, ground condition, cover, precipitation, and cultural treatment are identical, are evidence of inadequate knowledge concerning the hydraulics of loading (the taking of soil material into suspension) and of unloading (sedimentation) on the part of thin and thick sheets of flowing water. Moreover, very little has been done in the direction of determining rates and interrelationships of evaporation, absorption, and internal circulation of soil-contained moisture under varying characteristics and conditions of soil, slope gradient, soil treatment, and exposure to sunlight. The relation of the effects of animal life and of microorganisms on the soil to water intake and retention, and to runoff and erosion, has been studied only in an incidental way. Not nearly enough study has been made of the exact processes involved with water intake and retention by different soils, subjected, under different conditions of slope, tilth, and cover, to long periods of rain of various intensities. For example, not much is really known with respect to the precise meaning of the term *saturated soil*, either as to when and how an upland soil becomes saturated or as to how long, under varying conditions, it remains soaked with water.

Even less is known about rates and dynamics of wind erosion under the varying influences of a great number of factors similar to those referred to on page 127.

Without going into further detail here, it becomes evident that much additional information, such as can be obtained only through a comprehensive program of research, is needed, not only along the line of functional processes and relationships but in the way of precise measurements of rates of soil removal, water loss, and water retention under a multiplicity of physical, biologic, chemical, and climatic conditions.

Regardless of this emphasis on deficiencies of present knowledge pertaining to the dynamics and rates of erosion, runoff, infiltration, etc., much valuable information has been acquired, and new facts are coming to light almost every day. Considering the tardiness with which programs of education, research, and work on the land have got under way, present progress is highly encouraging.

## Soil and Water Losses from Important Types of Farm Land Under (*A*) Exposure and (*B*) Protection

Data on soil and water losses obtained at soil and water conservation experiment stations, through other experimental work, and by surveys, supplemented by field observations and the experience of farmers, reveal the almost universal effectiveness of close-growing vegetation in accelerating infiltration, slowing down runoff, and protecting soil from erosion by wind and water. In the thousands of measurements of the effects of individual rains heavy enough to produce runoff, this has been true with a consistency that would be monotonous were not the information of the greatest importance in the interest of programs having to do with conservation of water and land. The general trend can well be illustrated by citation of some of the summaries from approximately 50,000 separate measurements, over periods ranging from three to eleven years, of the effects of individual rains (and thawing of snow), under a great variety of characteristics and conditions with respect to soil, slope, cover, cultural methods, and climate.

The results as typified in examples presented in the following tables and discussions are generally consistent throughout. Exceptions to the usual trend have been so rare as to be of little importance, especially since most or all of them have been the result of unusual conditions, such as the effects of drifted snow suddenly melting or excessive erosion or runoff due to excavations by rodents.

Summary of Comparative Rates of Runoff and Erosion from 13 Important Types of Land under: (1) Cultivation and (2) Protective Cover. Losses of soil and of water from 13 important varieties of

TABLE 7.—SOIL AND WATER LOSSES UNDER CULTIVATION AND PROTECTIVE COVER, FROM 13 IMPORTANT SOILS[1]

| Soil,[2] location, and years of measurement | Annual rainfall for period, inches | Slope, per cent | Clean tilled | | | Dense cover | | | Time to remove 7 inches soil | | Approx. reg'nl area, mill'n acres |
|---|---|---|---|---|---|---|---|---|---|---|---|
| | | | Crop | Soil loss annually, tons per acre | Water loss annually, per cent precipitation | Cover | Soil loss annually, tons per acre | Water loss annually, per cent precipitation | Clean tilled, years | Dense cover, years | |
| Cecil s. cl. l., Statesville, N. C. 1932–1936...... | 48.42 | 10.0 | Cotton | 25.08 | 10.47 | Mixed grasses | 0.014 | 0.29 | 46 | 82,150 | 41 |
| Nacogdoches f.sdy. l., Tyler, Tex. 1932–1936...... | 42.34 | 10.0 | Cotton | 8.18 | 16.48 | Bermuda grass | 0.006 | 0.53 | 160 | 214,200 | 48 |
| Kirvin f. sandy loam, Tyler, Tex. 1932–1936...... | 41.77 | 8.75 | Cotton | 30.07 | 22.23 | Bermuda grass | 0.05 | 0.91 | 46 | 27,400 | |
| Shelby silt loam, Bethany, Mo. 1931–1935...... | 34.79 | 8.0 | Corn | 68.78 | 28.31 | Alfalfa | 0.25 | 7.49 | 16 | 4,540 | 18 |
| Muskingum silt loam, Zanesville, Ohio. 1934–1936 | 35.46 | 12.0 | Corn | 73.23 | 41.95 | Bluegrass | 0.04 | 6.5 | 16 | 28,900 | 31 |
| Clinton silt loam, LaCrosse, Wis. 1933-1935..... | 34.12 | 16.0 | Corn | 88.66 | 20.84 | Bluegrass | 0.03 | 2.82 | 11 | 33,600 | 22 |
| Dubuque silt loam, LaCrosse, Wis. 1934–1935...... | 36.53 | 30.0 | Corn | 81.44 | 19.12 | Bluegrass | 0.22 | 6.83 | 11 | 4,110 | |
| Vernon f. sandy loam, Guthrie, Okla. 1930–1935 | 33.12 | 7.7 | Cotton | 24.29 | 14.22 | Bermuda grass | 0.032 | 1.23 | 50 | 38,200 | 37 |
| Houston blk. cl., Temple, Tex. 1931–1936...... | 32.76 | 4.0 | Corn | 23.83 | 14.22 | Bermuda grass | 0.03 | 0.04 | 34 | 26,700 | 8 |
| 1933–1936...... | 34.90 | 2.0 | Corn | 10.62 | 13.38 | | 0.10 | 0.95 | 75 | 8,010 | |
| Marshall silt loam, Clarinda, Iowa. 1933–1935...... | 26.82 | 9.0 | Corn | 18.82 | 8.64 | Bluegrass | 0.06 | 0.97 | 48 | 15,200 | 23 |
| Palouse silt loam, Pullman, Wash. 1932–1935 | 21.74 | 30.0 | Wheat, fallow | 8.52 | 9.40 | Mixed grasses | 1.45 | 3.71 | 130 | 740 | 8 |
| | 21.74 | 30.0 | Bare | 27.82 | 25.03 | Spr. wheat | 1.65 | 3.23 | 38 | 650 | |
| Colby silty cl. loam, Hays, Kans. 1930–1935 | 20.36 | 5.0 | Kafir | 11.74 | 16.20 | Native grass | 0.029 | 0.30 | 91 | 36,900 | 10 |
| Miles clay loam, Spur, Tex. 1926–1937.......... | 20.73 | 2.0 | Cotton | 7.03 | 15.53 | Buffalo grass | 1.29 | 4.34 | 150 | 880 | |
| Averages....... | ...... | ...... | ....... | 33.87 | 18.40 | ........ | 0.35 | 2.68 | | | |

[1] Representative of the principal types of erodible farm land within an area of approximately 250,000,000 acres.

[2] s. cl. l. = sandy clay loam; f. sdy. l. = fine sandy loam; blk. cl. = black clay; silty cl. loam = silty clay loam.

agricultural land representative of the principal types of erodible farm land within a number of problem areas comprising approximately two hundred and fifty million acres are shown in Table 7. These data cover periods of investigation, at the several soil and water conservation experiment stations, ranging from 3 to 11 years. The results constitute the best possible information available with respect to relative losses of soil and of water under the contrasting conditions of clean tillage and dense cover.

Without exception, it is seen that the average annual soil losses from the areas devoted to clean-tilled crops are many times greater than the corresponding losses from the areas heavily covered with protective vegetation, the excess of losses ranging from about 6 to 2,955 times as much. The water losses as immediate runoff range, for the cultivated or exposed areas, from about 2.5 to 355 times the corresponding losses from the areas protected with permanent cover.

The data show also wide differences in relative erodibility of these diverse types of land. For example, average annual soil losses from the Shelby, Muskingum, Dubuque, and Clinton types range from 68 to 88 tons per acre under clean cultivation, whereas the corresponding losses from the Nacogdoches, Miles, and Palouse types range from 7 to 8 tons an acre. In the same way, water losses as runoff from the Dubuque, Clinton, Shelby, and Muskingum are the highest, ranging from 19 to 42 per cent of the total precipitation. Lowest runoff is revealed for the Marshall, Palouse, and Cecil types, the amounts ranging from 8 to 10 per cent of the total precipitation.

The wide differences as between rates of erosion and runoff for the various types undergoing the same type of treatment involve, principally, differences of soil, slope, and amount and intencity of rainfall, other modifying factors having been controlled as effectively as possible. For example, both soil and water losses from Nacogdoches fine sandy loam (devoted to cotton) were much less than for the Kirvin fine sandy loam (devoted to cotton) located on the same farm (8 tons per acre annually, as against 30 tons per acre), although the slope of the former was 1.25 per cent steeper. Shelby silt loam, on the other hand, lost 69 tons of soil per acre annually from an 8 per cent slope, as against 89 tons lost from Clinton silt loam having a slope twice as steep, both areas having been used for corn, and both receiving almost identical amounts of precipitation in storms of similar intensities.

What is of more significance, from the standpoint of land users, is that the soil losses from corresponding slopes of Nacogdoches, Kirvin, Shelby, and Clinton types covered continuously with good stands of grass or alfalfa were only 1/1,363, $\frac{1}{601}$, $\frac{1}{273}$, and 1/2,955 as great, respectively, as where clean-tilled crops (cotton and corn) were grown. Similarly, the corresponding water losses as immediate runoff from these important types of farm land were only $\frac{1}{31}$, $\frac{1}{24}$, $\frac{1}{4}$, and $\frac{1}{7}$ as great, respectively, as those measured on the areas used continuously for cotton and corn.

In a dense cover of vegetation, then, the farmer has the most powerful practical weapon known for reducing erosion and runoff. It is observed in Table 7 that under continuous corn the time indicated as necessary for Shelby silt loam on an 8 per cent slope to lose 7 inches of topsoil (about the average depth of this type of land) is 16 years; whereas under grass, the corresponding time indicated as necessary to remove an equivalent layer of soil is 4,540 years. Kirvin fine sandy loan of 8.7 per cent slope would lose its topsoil, according to the indication of available measurements, in 46 years under cotton, as against 27,400 years under Bermuda grass.

The average results of the many thousands of measurements of the effects of individual rains on these 13 important types of farm land show that under the exposure of clean tillage and fallow, soil is lost at a rate about ninety-five times as rapidly as the corresponding loss from areas safeguarded with a dense cover of vegetation and that nearly seven times as much of the rainfall is lost as runoff from cultivated land as from land protected with vegetation.

## Losses from Representative Erodible Land, Rolling Section of Corn Belt (Lower Missouri Dark-brown Till Section) (Shelby Silt Loam)

Effect of Cover. Measurements made at Bethany, Mo., on one of the most extensive and important soils in the rolling glacial-till section of the Corn Belt (8 per cent slope of Shelby silt loam, probably representative of about the average erodibility of the region) show that the average annual loss of soil by erosion from land devoted continuously to corn has been at the rate of 68.8 tons an acre annually. At the same time, the loss of water as runoff was 28.3 per cent of the total precipitation (snow and rain). From exactly the same kind of soil, having the same declivity and receiving the same rainfall, the soil loss under alfalfa has been at the rate of only 0.25 ton per acre annually, along with a runoff of but little more than 7 per cent of the entire precipitation. In other words, alfalfa has been 275 times more effective than corn in the retention of soil (grass has been 237 times more effective). The efficiency of alfalfa in relation to rainfall-retention on this type of land has been approximately 3.8 times that of ground where corn is grown continuously. For the same period, the same type of land kept bare of all vegetation has lost an average of 112.8 tons of soil per acre annually, or 451 times as much as from comparable land devoted to alfalfa. It is interesting that the water loss from this area of bare fallow was only slightly greater (2.89 per cent) than the runoff from the corn area, although the soil loss was nearly twice as much.

This part of the Corn Belt—the Lower Missouri Dark-brown Till section (see *Problem Area* map)—comprises some 18 million acres.

Effectiveness of Rainfall. During the five-year period 1931-1935, when the mean annual precipitation was 34.8 inches, 399 rains fell, 116 of which (29 per cent) caused runoff from land used continuously for corn. Of the 42.2 inches of rainfall in 1931, 46 rains totaling 32.57 inches caused runoff from corn ground. The runoff from these 46 rains, representing 26 rain periods, was 12.3 inches, or 29.1 per cent of the rainfall for the year, which fell in 92 rains. Seventeen of these 46 rains measured more than 1 inch each, amounting in total to 21.91 inches of precipitation. Many light summer showers were completely dissipated, and others nearly so, by almost immediate evaporation from the heated soil. Under certain conditions, some of the heavier rains, even where falling at a fairly intense rate, were completely absorbed, whereas under other conditions lighter rains of about equal intensity produced runoff.

In 1931, 29.14 per cent of all the rain that fell on the continuous corn area was lost as immediate runoff. That year, 46 rains, totaling 9.65 inches, or 22.85 per cent of the total precipitation, produced no runoff from this area of corn ground. More than half of this 22.85 per cent fraction of the rainfall was lost as evaporation. Thus, the immediate loss of rainfall during this year amounted to at least 40.5 per cent of the total precipitation (29 per cent runoff plus 11.5 per cent evaporation from light showers). Since part of the remainder was lost as percolation and part as evaporation other than transpiration (in addition to the evaporation loss from the light showers), the corn crop apparently had for its use considerably less than half of the total rainfall for the year—probably something under 40 per cent.

In contrast, only 2 per cent of the rainfall of 1931 was lost from the alfalfa field. Thus it appears that this densely growing crop had for its use, on the same slope and type of soil as that used for the corn, and with the same rainfall, in the neighborhood of 68 per cent of the total precipitation for that year. This estimate is based on the 2 per cent lost immediately as runoff, plus 30 per cent estimated loss by evaporation and percolation.

Thus, it is necessary in any accurate appraisal of the effectiveness of rainfall to consider much more than annual averages of rainfall or the seasonal distribution of rains. That part which is left in the soil for plant use is vitally important. For example, 20 inches of rain in the Great Plains may produce just as much wheat where proper water-conservation measures are employed as 50 inches of similarly distributed rain on land of equal slope and productivity in Pennsylvania, if 60 per cent of the latter is permitted to go to waste as runoff and excessive percolation and evaporation through failure to provide adequate safeguards for the storing of moisture in the reservoir of the soil.

It was found at the Bethany, Mo., station that the alfalfa soil following drought regained its normal moisture content more quickly than corn ground. During the severe drought of 1934, the alfalfa area cracked to depths of 2 to 4 feet, but the soil of the corn area cracked only moderately. This tendency of the soil to fracture deeply under alfalfa probably

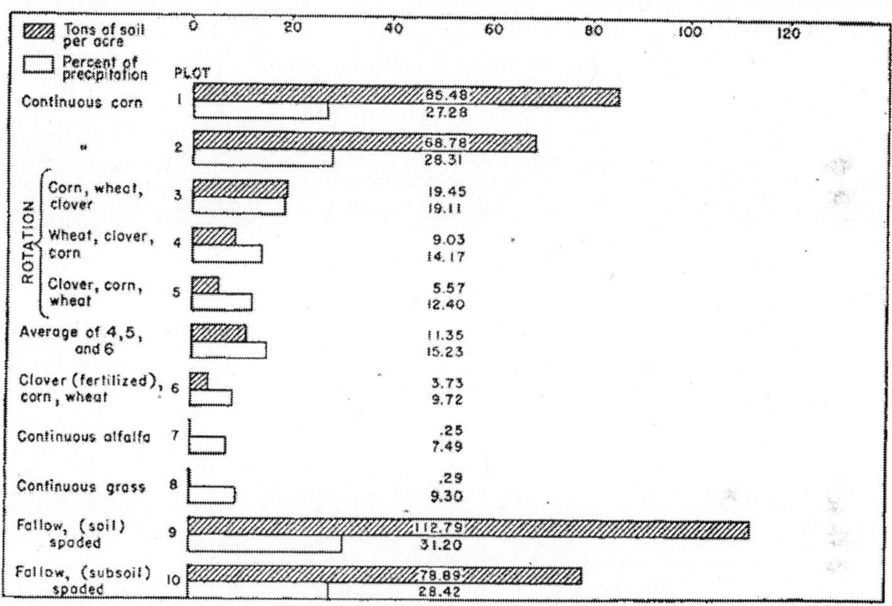


Graph 2.—Average annual soil and water losses, 1931–1935, inclusive, from 8 per cent slope of Shelby silt loam, at the soil and water conservation experiment station, Bethany, Mo. Average annual precipitation, 34.79 inches. (Plot 1 = 6 by 145.2 feet; plots 2 to 10, inclusive, = 6 by 72.6 feet.) (*Soil Conservation Service.*)

accounts partly for its much higher infiltration capacity as compared with that of the same soil under corn.

Effect of Rainfall Intensity. The results of erosion and runoff given above, in connection with effective precipitation, are based on a 5-year period (Graph 2), except as noted. This, of course, is not long enough for the establishment of absolutely accurate averages. For example, a single intensive rain of 3.75 inches that fell on these controlled areas on Apr. 3 to 4, 1934 (3.10 inches of which fell at an average rate of 2.62 inches an hour), caused a soil loss from corn ground (Graph 3) that was nearly as great as the entire loss caused by all of the 56 rains that fell on the same field in 1933, that is, approximately 46, as against nearly 50 tons an acre (Graph 4). From this exceptional rain, 68 per cent of the water was immediately lost as runoff, whereas only 29.7 per cent of

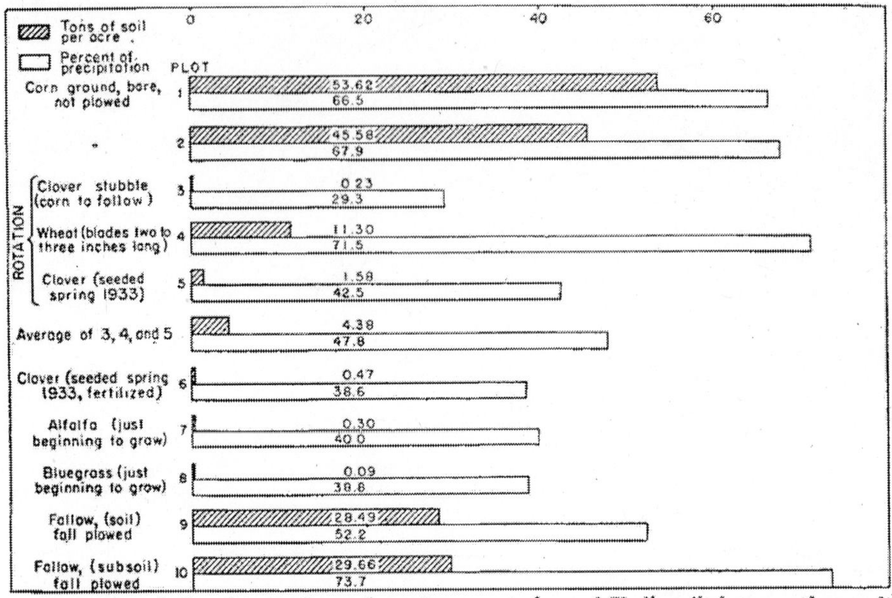


GRAPH 3.—Soil and water losses from 8 per cent slope of Shelby silt loam as the result of one rain of 3.75 inches, Apr. 3–4, 1934, at the soil and water conservation experiment station, Bethany, Mo. Rainfall reached intensity of 4.71 inches per hour for a period of 7 minutes; 3.10 inches fell at rate of 2.62 inches per hour. (Plot 1 = 6 by 145.2 feet; plots 2 to 10, inclusive, = 6 by 72.6 feet.) (*Soil Conservation Service.*)

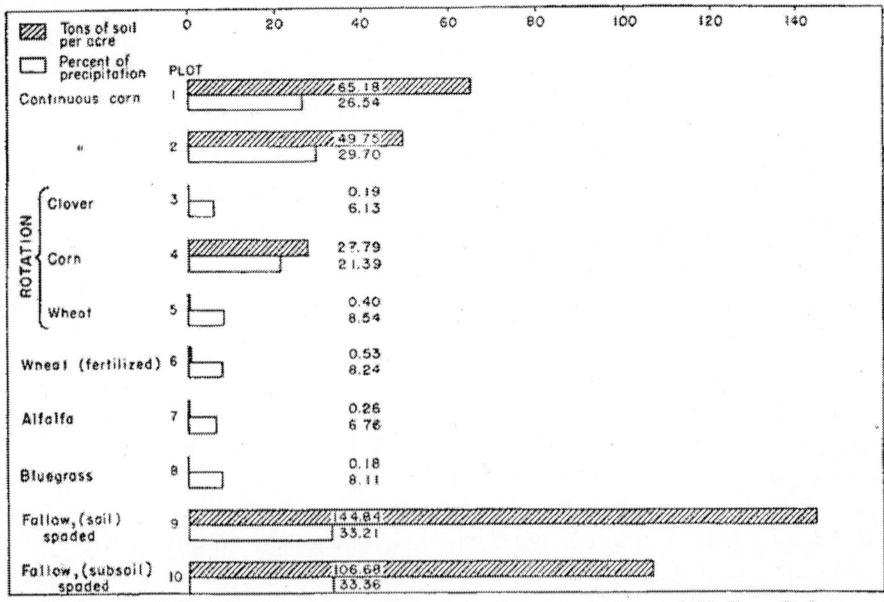


GRAPH 4.—Soil and water losses, 1933, from 8 per cent slope of Shelby silt loam, at the soil and water conservation experiment station, Bethany, Mo. Precipitation, 32.43 inches. (Plot 1 = 6 by 145.2 feet; plots 2 to 10, inclusive, = 6 by 72.6 feet.) (*Soil Conservation Service.*)

the total rainfall of 1933 was lost from the same area of corn land. The average results, nevertheless, have been sufficiently consistent from year to year, not only for the Bethany, Mo., station but on other soils of other localities, to be considered highly significant.

Graph 2 shows the details of soil and water losses from an 8 per cent slope of Shelby silt loam at Bethany, Mo., as annual averages for a five-year period. Graph 5 shows the corresponding losses from a 3.7 per cent slope of similar soil at Columbia, Mo., as annual averages for a 13-year period.

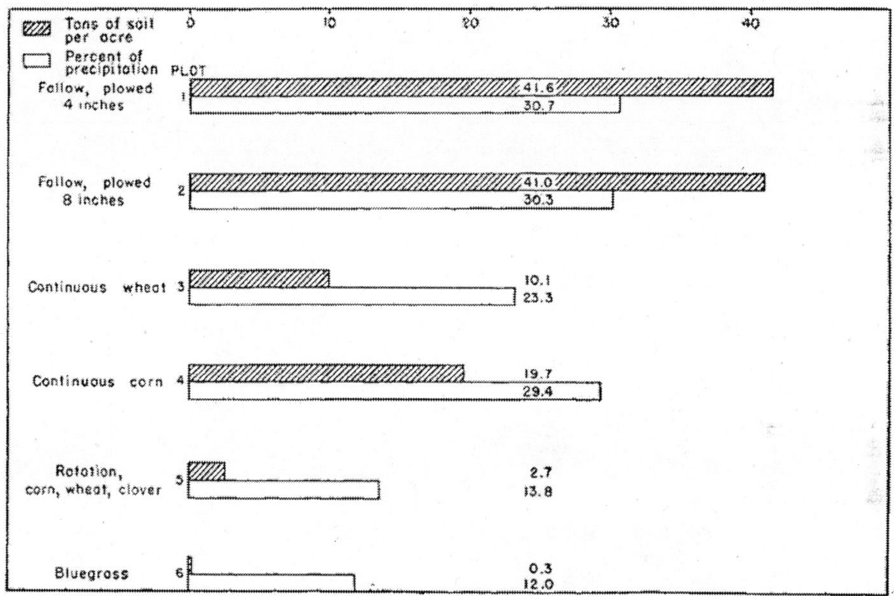


GRAPH 5.—Average annual soil and water losses, 1918–1931, inclusive, from 3.68 per cent slope of Shelby loam at the Agricultural Experiment Station, Columbia, Mo. Average annual precipitation, 40.37 inches. (Plots 1 to 6, inclusive, = 6 by 90.75 feet.)

EFFECT OF EROSION ON DURABILITY OF TOPSOIL. In the locality where these data were obtained, the soil depth on 8 per cent slopes of Shelby silt loam is approximately 7 inches. The rate of soil removal by erosion under continuous corn production, as revealed in these measurements, would require, theoretically, about 16 years to remove the surface layer down to a distinctly different and much less productive sublayer. Actually, a somewhat longer period would be required for this much soil to be washed off, because some of the subsurface material would be brought up and mixed with the topsoil by plowing, thus increasing somewhat the life depth of the soil.

Comparative theoretical periods indicated as necessary to erode a 7-inch layer of soil from Shelby silt loam of 8 per cent slope are presented below for comparable bare areas and areas in corn, alfalfa, and a rotation of corn, wheat, and clover:

INDICATED TIME NECESSARY TO ERODE 7 INCHES OF SOIL FROM SHELBY SILT LOAM, 8 PER CENT SLOPE

| | *Years* |
|---|---|
| Alfalfa | 4,540 |
| Rotation | 100 |
| Corn | 16 |
| Bare ground | 10 |

Graph 6 illustrates the comparative rates of soil and water losses from similar areas of Shelby silt loam in alfalfa, crop rotation, continuous corn, and bare fallow.

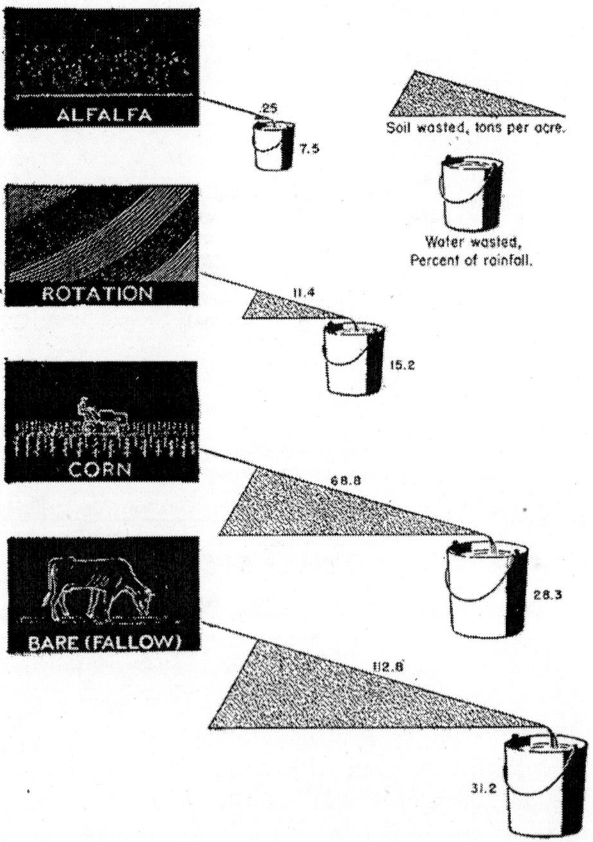


GRAPH 6.—Annual losses of soil and water from same kind of land (Shelby silt loam, 8 per cent slope), with same rainfall, from alfalfa, rotated fields, corn, bare land. (Soil and water conservation experiment station, Bethany, Mo., 1931–1935.) (*Soil Conservation Service.*)

## Losses from Representative Erodible Land of the Missouri Valley Loessial Region (Marshall Silt Loam)

From another variety of silt loam, the extensive Marshall silt loam of the Missouri River section of the loessial area covering about 23 million acres (part of the Central Prairie Region, see *Problem Area* map), the losses measured over a period of 3 years 1933–1935) at the erosion experiment station between Clarinda and Shenandoah in southwestern Iowa have been, from a 9.0 per cent slope, at the approximate rate of 18.82 tons of soil per acre per annum and 8.64 per cent of the precipitation, from continuous corn, and 0.06 ton of soil and 0.97 per cent of the precipitation from bluegrass (see Graph 7).

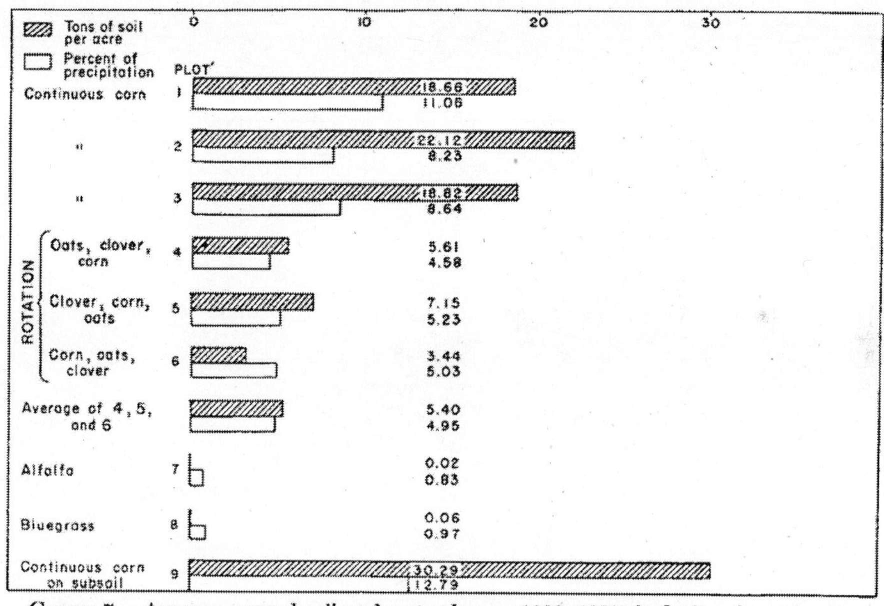


GRAPH 7.—Average annual soil and water losses, 1933–1935, inclusive, from 9 per cent slope of Marshall silt loam, at the soil and water conservation experiment station, Clarinda, Iowa. Average annual precipitation, 26.82 inches. Plot 1 = 6 by 36.3 feet; plot 2 = 6 by 145.2 feet; plots 3 to 9, inclusive, = 6 by 72.6 feet. (*Soil Conservation Service.*)

The variety of silt loam on which these measurements were made differs from the Shelby silt loam in such characteristics as a higher content of silt and lower content of sand and fine gravel and in having a higher pore space and, therefore, greater infiltration capacity and resistance to erosion.

Comparative periods indicated as theoretically necessary to remove 7 inches of topsoil from Marshall silt loam of 9 per cent slope are shown below for comparable areas in grass, a three-year crop rotation, and corn:

INDICATED TIME NECESSARY TO ERODE 7 INCHES OF SOIL FROM MARSHALL SILT LOAM, 9 PER CENT SLOPE

| | Years |
|---|---|
| Grass | 15,200 |
| Rotation | 170 |
| Corn | 48 |

Graph 8 illustrates the comparative rates of erosion and runoff from similar areas of Marshall silt loam of 9 per cent slope used for grass, corn, and a rotation of corn, oats, and clover.

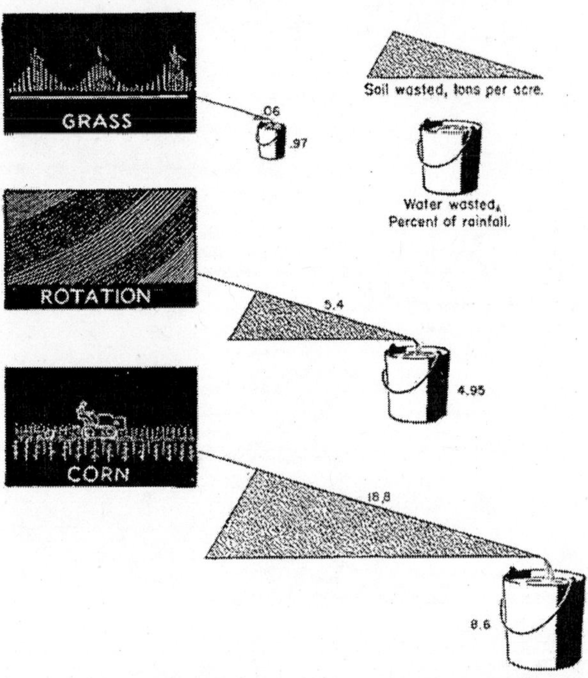


GRAPH 8.—Annual losses of soil and water from same kind of land (Marshall silt loam, 9 per cent slope), with same rainfall, from grass, rotated fields, corn. (Soil and water conservation experiment station, Clarinda, Iowa, 1933–1935.) (*Soil Conservation Service.*)

## Losses from Representative Erodible Land of the Wisconsin-Minnesota-Illinois-Iowa Nonglaciated Area (Clinton Silt Loam)

Soil and water losses from the Clinton silt loam, 16 per cent slope, for the 3-year period 1933–1935 are shown for various cultural treatments in Graph 9. The measurements, made at the soil and water conservation experiment station at La Crosse, Wis., pertain to land that is considered representative of about the average condition of the more erodible of the important farming soils of the Lower Missouri and Upper Mississippi section of this unglaciated (predominantly loessial) region of approximately 22 million acres (see *Problem Area* map). Graph 10 shows the

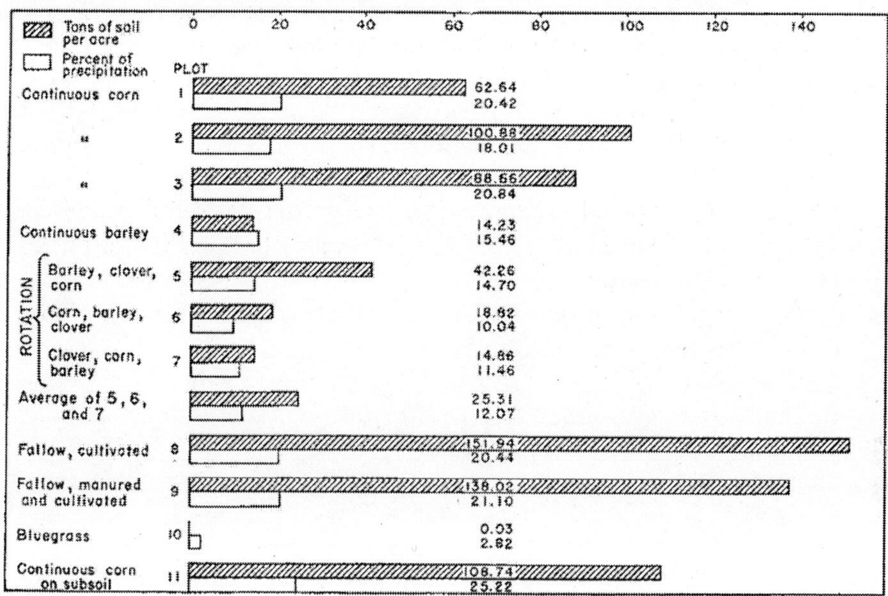


GRAPH 9.—Average annual soil and water losses, 1933–1935, inclusive, from 16 per cent slope of Clinton silt loam, at the soil and water conservation experiment station, La Crosse, Wis. Average annual precipitation, 34.12 inches. (Plot 1 = 6 by 36.3 feet; plot 2 = 6 by 145.2 feet; plots 3 to 11, inclusive, = 6 by 72.6 feet). (*Soil Conservation Service.*)

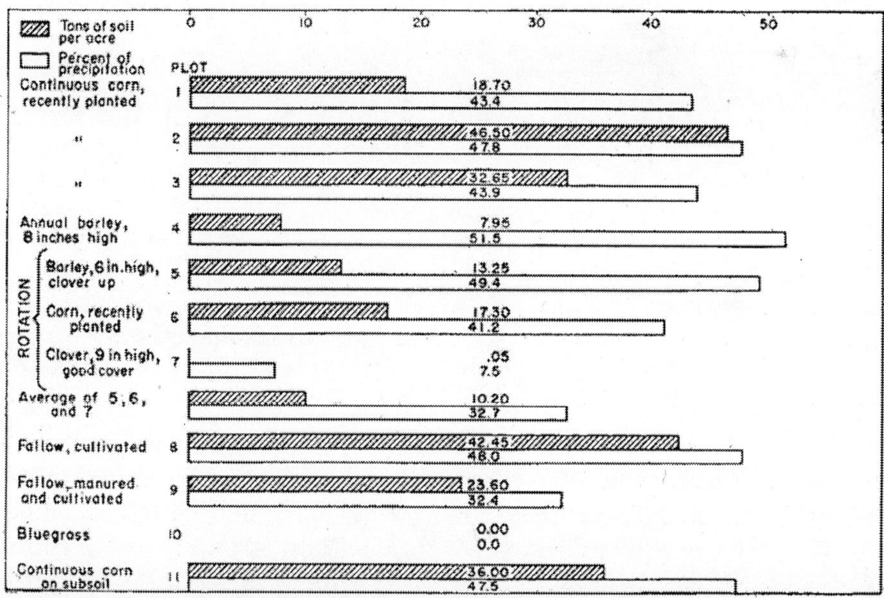


GRAPH 10.—Soil and water losses from 16 per cent slope of Clinton silt loam as the result of one rain of 1.6 inches, May 22, 1936, at the soil and water conservation experiment station, La Crosse, Wis. Rainfall reached intensity of 4.8 inches per hour for a period of 5 minutes. (Plot 1 = 6 by 36.3 feet; plot 2 = 6 by 145.2 feet; plots 3 to 11, inclusive, = 6 by 72.6 feet.) (*Soil Conservation Service.*)

losses from the same areas represented by Graph 9, resulting from a single intensive rain (1.6 inches) on May 22, 1936. At one stage, this downpour reached an intensity of 4.8 inches of rain per hour for a period of 5 minutes. The soil swept from the corn ground by this one rain amounted to nearly two-fifths of the average annual loss over a three-year period, and the rate of runoff was more than twice the average rate of water loss for the same period. Five days before this heavy rain, the areas on which the measurements were made received a rainfall of 0.20 inch.

In a number of respects, Clinton silt loam is very similar to the Marshall silt loam referred to above. One of the principal differences is its lighter color and lower average content of organic matter.

Below are shown the indicated lengths of time that would be required to wash off 7 inches of topsoil from Clinton silt loam of 16 per cent slope under four markedly different types of treatment: (1) grass; (2) corn; (3) a rotation of corn, barley, and clover; and (4) without a cover of vegetation:

INDICATED TIME NECESSARY TO ERODE 7 INCHES OF TOPSOIL FROM CLINTON SILT LOAM, 16 PER CENT SLOPE

| | *Years* |
|---|---|
| Grass | 33,600 |
| Rotation | 40 |
| Corn | 11 |
| Bare (fallow) | 7 |

## Losses from Representative Erodible Land of Moderate Slope, Winter Wheat Belt, East-central Great Plains (Colby Silty Clay Loam)

Average annual soil and water losses from one of the most important soils of the winter wheat section of the eastern Great Plains (Colby silty clay loam) are shown in Graph 11, which gives the measurements made at the Hays, Kans., soil and water conservation experiment station for the 6-year period 1930–1935. The gradient of the land is 5 per cent; the annual rainfall for the period was 20.36 inches. The loss of 11.74 tons of soil per acre annually from a cultivated crop, kafir, exceeded the loss from wheat (2.25 tons an acre) by more than five times; and, at the same time, the water loss was much greater, or 16 per cent, as against 10 per cent. The soil loss from native grass, under comparable conditions, was only $\frac{1}{391}$ of that from the clean-tilled crop, and the water loss was only $\frac{1}{54}$ as great. This land is representative of a large total area of erodible soil in the east-central part of the Great Plains—that is, in the northerly part of the Dark-brown Plains section of the Southern Plains (see *Problem Area* map).

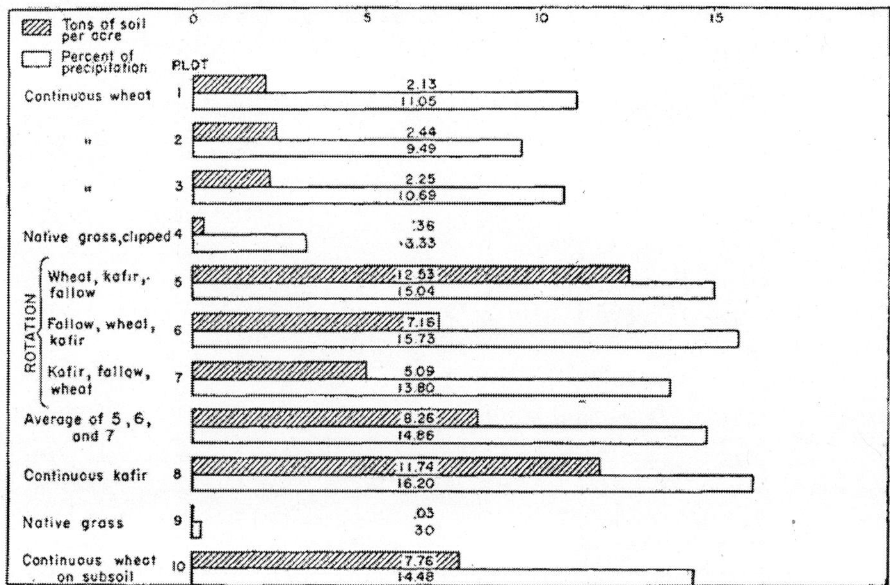


Graph 11.—Average annual soil and water losses, 1930–1935, inclusive, from 5 per cent slope of Colby silty clay loam, at the soil and water conservation experiment station, Hays, Kans. Average annual precipitation 20.36 inches. (Plot 1 = 6 by 36.3 feet; plot 2 = 6 by 145.2 feet; plots 3 to 10, inclusive, = 6 by 72.6 feet.) (*Soil Conservation Service.*)

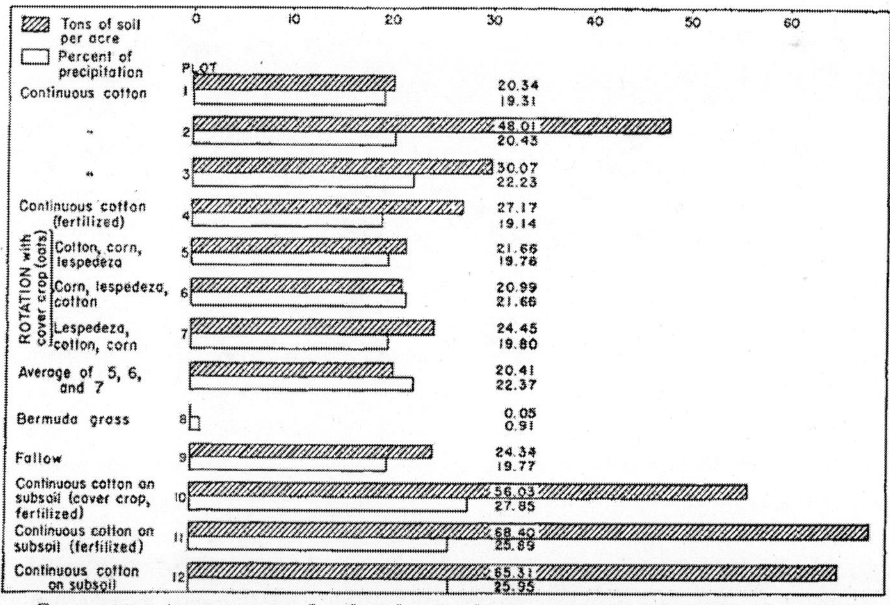


Graph 12.—Average annual soil and water losses, 1932–1936, inclusive, from 8.75 per cent slope of Kirvin fine sandy loam at the soil and water conservation experiment station, Tyler, Tex. Average annual precipitation, 41.77 inches. (Plot 1 = 6 by 36.3 feet; plot 2 = 6 by 145.2 feet; plots 3 to 12, inclusive, = 6 by 72.6 feet.) (*Soil Conservation Service.*)

## Losses from Representative Erodible Land of Arkansas-Louisiana-East Texas Sandy Lands Region (Interior West Gulf Coastal Plain) (Kirvin Fine Sandy Loam)

In Graph 12 are shown the details of soil and water losses from Kirvin fine sandy loam (8.75 per cent slope), which is representative, or approxi-

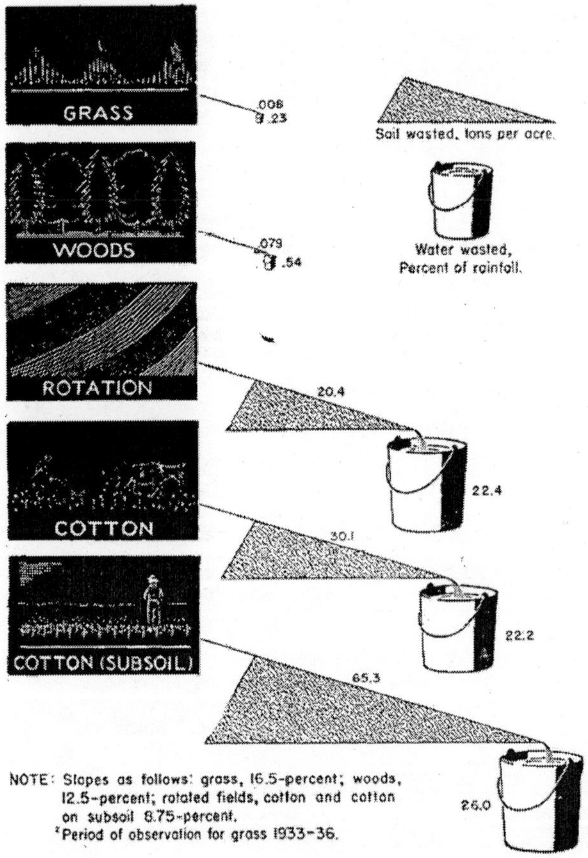


GRAPH 13.—Annual losses of soil and water from same kind of land (Kirvin fine sandy loam, various slopes), with same rainfall, from grass, woods, rotated fields, cotton, cotton on subsoil. (Soil and water conservation experiment station, Tyler, Tex., 1932–1936.) (*Soil Conservation Service.*)

mately representative, of one of the most important types of erodible land in the 48 million acres comprised in the predominantly sandy land area stretching from west of the Brazos River in Texas to the vicinity of Little Rock in Arkansas and to the south of the Red River in Louisiana. From the standpoint of costly erosion, this large region—the Interior West Gulf Coastal Plain—represents an important problem area within the Cotton Belt (see *Problem Area* map).

Graph 13 illustrates the comparative rates of erosion and runoff from comparable areas of Kirvin fine sandy loam devoted to forest, grass, cotton, a cotton-corn-lespedeza rotation, and cotton grown on erosion-exposed subsoil.

Below are shown the lengths of time indicated as necessary to strip off 7 inches of soil material from Kirvin fine sandy loam under: grass, forest, a 3-year crop rotation, cotton on topsoil and on erosion-exposed subsoil:

INDICATED TIME NECESSARY TO ERODE 7 INCHES OF SOIL MATERIAL FROM KIRVIN FINE SANDY LOAM

| | *Years* |
|---|---|
| Grass (16.5 per cent slope) | 171,500 |
| Forest (12.5 per cent slope) | 27,400 |
| Rotation (8.75 per cent slope) | 67 |
| Cotton (8.75 per cent slope) | 46 |
| Cotton on subsoil (8.75 per cent slope) | 21 |

## Losses from Representative (Erodible) Land in the Eastern Part of the Red Plains (Vernon Fine Sandy Loam)

Soil and water losses from the most representative type of erodible farm land in the eastern part of the Red Plains of Oklahoma—Vernon fine sandy loam of 7.7 per cent slope—are shown in Graph 14. Under various

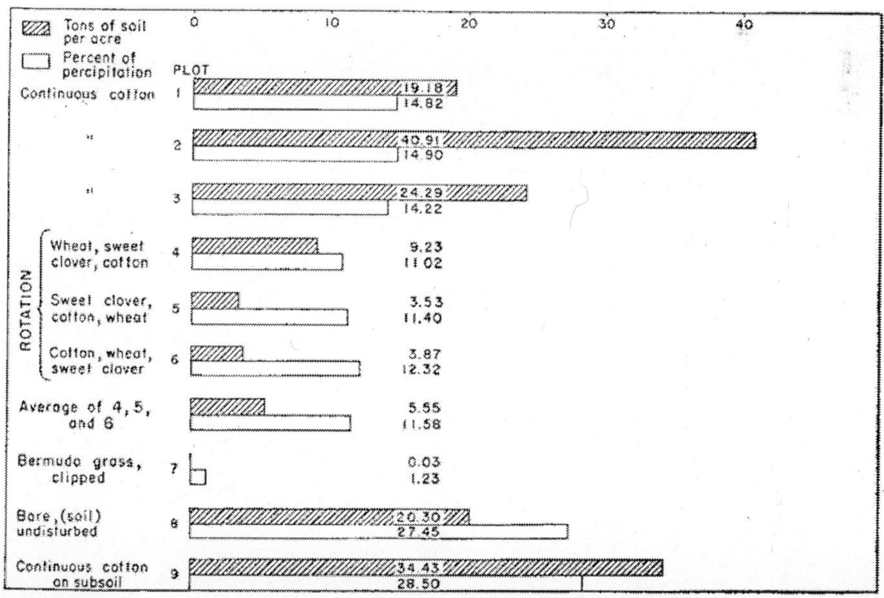


GRAPH 14.—Average annual soil and water losses, 1930–1935, inclusive, from 7.7 per cent slope of Vernon fine sandy loam, at the soil and water conservation experiment station, Guthrie, Okla. Average annual precipitation, 33.12 inches. (Plot 1 = 6 by 36.3 feet; plot 2 = 6 by 145.2 feet; plots 3 to 9, inclusive, = 6 by 72.6 feet.) (*Soil Conservation Service.*)

treatments, the measurements were made on comparable areas at the soil and water conservation experiment station near Guthrie, Okla. The Red Plains problem area comprises approximately 37 million acres, situated principally in Oklahoma and Texas (see *Problem Area* map). It is a highly important agricultural section; and although it is comparatively new with respect to settlement and general farming, erosion already is widespread and very serious. The principal type of erodible farm land is losing soil in the runoff 759 times faster under clean tillage (cotton grown continuously) than where the same type of land is stabilized with a good cover of

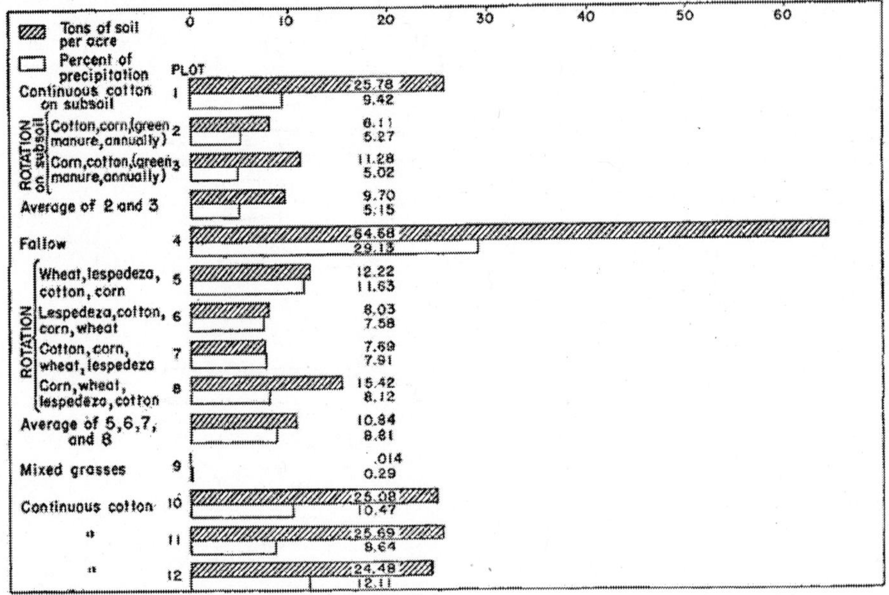


GRAPH 15.—Average annual soil and water losses, 1932–1936, inclusive, from 10 per cent slope of Cecil sandy clay loam, at the soil and water conservation experiment station, Statesville, N. C. Average annual precipitation, 48.42 inches. (Plots 1 to 10, inclusive, = 6 by 72.6 feet; plot 11 = 6 by 145.2 feet; plot 12 = 6 by 36.3 feet.) (*Soil Conservation Service.*)

grass (24.29 tons per acre annually, as against only 0.032 ton per acre annually). Corresponding losses of water as immediate runoff have been 14.22 per cent of the total precipitation from land in cotton, as compared with only 1.23 per cent from the same kind of land, receiving the same rainfall, covered with grass. In other words, $11\frac{1}{2}$ times more of the rainfall has been lost from the cultivated land.

The measured losses (Graph 14) show that a crop rotation of wheat-sweetclover-cotton greatly reduces the wastage of both soil and water. Continuing erosion very largely increases the rate of soil loss and doubles the rate of water loss; that is, after the topsoil is removed, the subsoil erodes at the rate of 34.43 tons per acre annually and sheds water at the rate of 28.5 per cent of the total rainfall.

## Losses from Representative Erodible Land of the Southern Piedmont (Cecil Sandy Clay Loam)

The details of soil and water losses from Cecil sandy clay loam of 10 per cent slope, as measured under various types of cover and tillage treatment at the Statesville, N. C., soil and water conservation experiment station, are

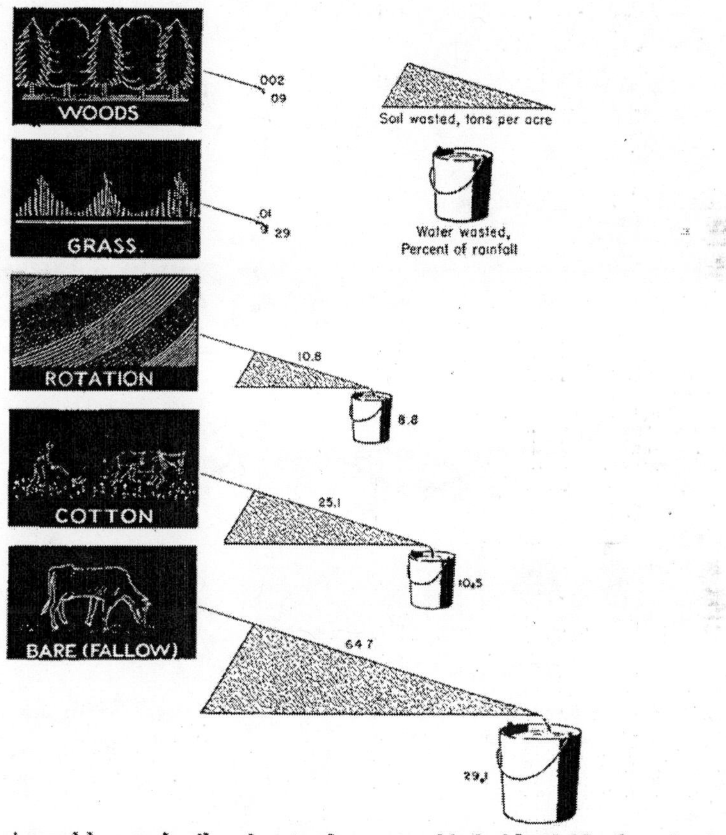


GRAPH 16.—Annual losses of soil and water from same kind of land (Cecil sandy clay loam, 10 per cent slope), with same rainfall, from woods, grass, rotated fields, continuous cotton, bare land. (Soil and water conservation experiment station, Statesville, N. C., 1932–1936.) (*Soil Conservation Service.*)

shown in Graph 15. This is one of the most important and representative types of erodible farm land of the approximately 41 million acres in the southern Piedmont region (foothill sector of the Southern Appalachian region). Graph 16 illustrates the comparative rates of runoff and of soil lost in the runoff from comparable areas of Cecil sandy clay loam of 10 per cent slope, in virgin timber, grass, continuous cotton, rotation, and bare ground. In this connection, it is interesting to note the tremendous differences in the time indicated as necessary to remove a 7-inch layer of soil (which represents about the average depth of the humus layer, or topsoil, of this type of land) by erosion, under different conditions of cover and treatment:

Indicated Time Necessary to Remove 7 Inches of Topsoil from Cecil Sandy Clay Loam, 10 Per Cent Slope

| | *Years* |
|---|---|
| Virgin timber | 575,000 |
| Grass | 82,150 |
| Rotation | 110 |
| Cotton | 46 |
| Bare ground | 18 |

Quantitative data presented in Graph 15 show that under a rotation of cotton-corn-wheat-lespedeza, erosion is only about one-third as severe as on the same kind of land, affected by the same rainfall, where cotton is grown continuously. The same type of land with no vegetative cover

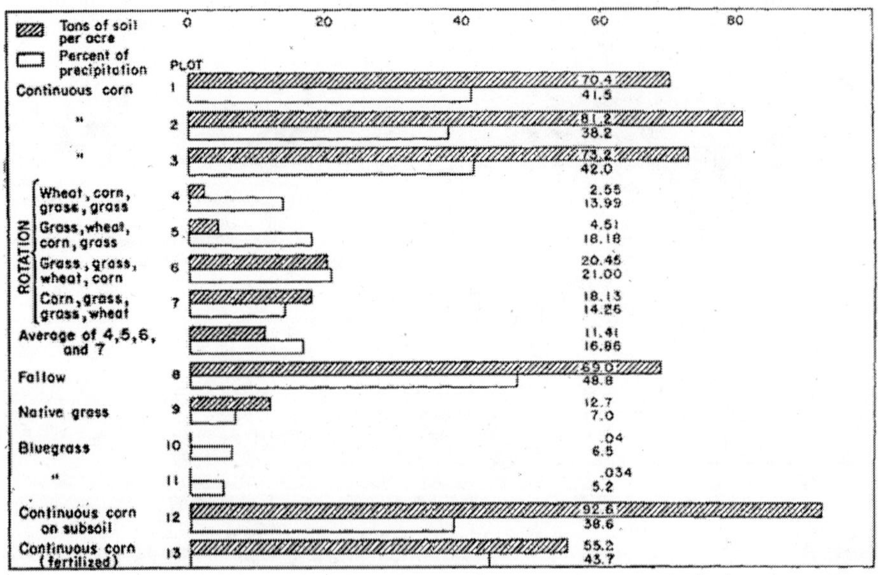


Graph 17.—Average annual soil and water losses, 1934–1936, inclusive, from 12 per cent slope of Muskingum silt loam, at the soil and water conservation experiment station, Zanesville, Ohio. Average annual precipitation, 36.46 inches. (Plot 1 = 6 by 36.3 feet; plot 2 = 6 by 145.2 feet; plots 3 to 13, inclusive, = 6 by 72.6 feet.) (*Soil Conservation Service.*)

erodes more than two and one-half times as rapidly as land in cotton and more than four thousand times as fast as that in grass. Also, from bare ground nearly three times as much water runs to waste as from the same kind of land in cotton, and one hundred times as much as from the same kind of land in grass.

## Losses from Representative Erodible Land of the Southwestern Part of Northern Appalachian Region (Allegheny-Ohio Section) (Muskingum Silt Loam)

Graph 17 presents the details of soil and water losses from one of the most representative types of erodible farm land in the southwestern sector of the Northern Appalachian region—Muskingum silt loam of 12 per cent

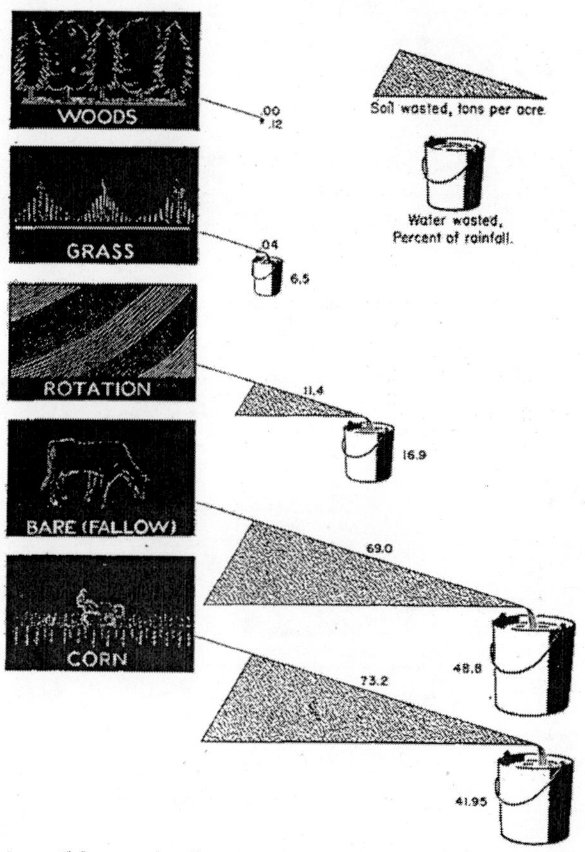


GRAPH 18.—Annual losses of soil and water from same kind of land (Muskingum silt loam, 12 per cent slope), with same rainfall, from woods, grass, rotated fields, bare land, corn. (Soil and water conservation experiment station, Zanesville, Ohio, 1934–1936.) (*Soil Conservation Service.*)

slope. The measurements were made at the Zanesville, Ohio, soil and water conservation experiment station, under a variety of treatments of comparable areas. This problem area—the Allegheny-Ohio section—comprises some 31 million acres.

Graph 18 illustrates the comparative rates of erosion and runoff from similar areas of Muskingum silt loam of 12 per cent slope under forest, grass, a rotation of corn-wheat-grass, bare fallow, and corn.

Below are shown the indicated periods that would be required to erode 7 inches of soil from comparable areas of Muskingum silt loam under forest, grass, a crop rotation, bare fallow, and corn:

INDICATED TIME NECESSARY TO ERODE 7 INCHES OF SOIL FROM MUSKINGUM SILT LOAM, 12 PER CENT SLOPE

| | *Years* |
|---|---|
| Forest | 173,700 |
| Grass | 28,900 |
| Rotation | 100 |
| Bare ground | 17 |
| Corn | 16 |

## Losses from Representative Erodible Land of the Texas Black Belt (Houston Black Clay)

Graph 19 shows the losses of soil and water from the most representative type of erodible cropland in the main Black Belt area of Texas, comprising some 8 million acres, not including the outlying areas (see *Problem Area* map). This type of land—Houston black clay of 4 per cent slope—has been subjected to quantitative erosion and runoff studies under various treatments at the soil and water conservation experiment station near Temple, Tex.

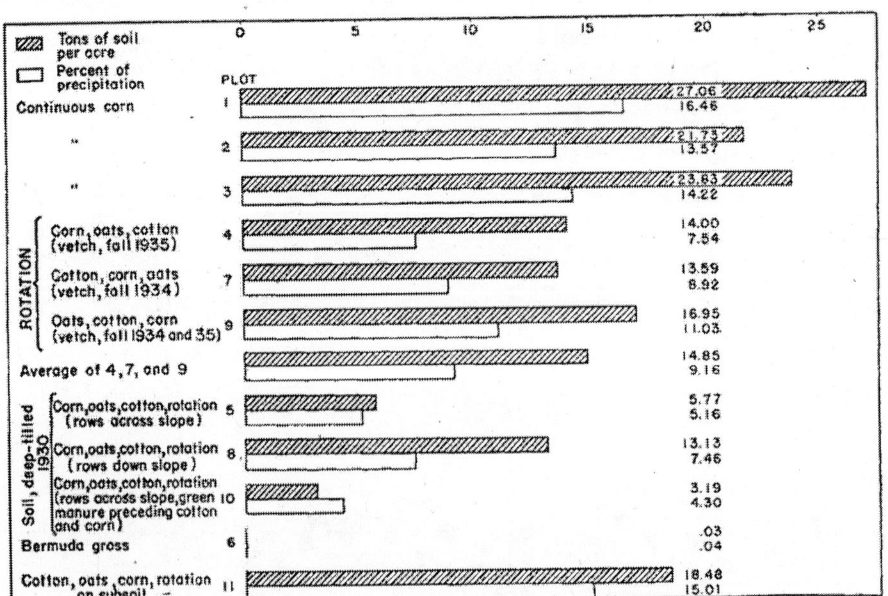


GRAPH 19.—Average annual soil and water losses, 1931–1936, inclusive, from 4 per cent slope of Houston black clay, at the soil and water conservation experiment station, Temple, Tex. Average annual precipitation, 32.76 inches. (Plot 1 = 6 by 36.3 feet; plot 2 = 6 by 145.2 feet; plots 3 to 11, inclusive, = 6 by 72.6 feet.) (*Soil Conservation Service.*)

## Losses from Representative Erodible Land in the Palouse Region of the Pacific Northwest (Palouse Silt Loam)

Graph 20 shows the soil and water losses from comparable areas of one of the most representative types of erodible cropland in the 8 million acres comprising the Palouse wheat area in southeastern Washington and adjacent Idaho and Oregon (see *Problem Area* map)—Palouse silt loam, 30 per cent slope. These losses were measured under a variety of tillage and cover treatments at the soil and water conservation experiment station near Pullman, Wash. The Palouse silt loam in its virgin condition is one of the most productive wheat soils of the world.

| | Plot | Tons of soil per acre | Percent of precipitation |
|---|---|---|---|
| Winter wheat, summer fallow | 1 | .12 | .83 |
| " | 2 | .37 | 43 |
| " | 3 | .42 | 3.59 |
| Grass | 6 | 1.45 | 3.71 |
| Spring wheat (fertilized) | 7 | 1.65 | 3.23 |
| Summer fallow, winter wheat | 8 | 16.61 | 15.22 |
| ROTATION: Sweet clover, winter wheat, spring wheat, peas and sweet clover | 9 | 6.99 | 8.27 |
| ROTATION: Winter wheat, spring wheat, peas and sweet clover, sweet clover | 10 | 10.24 | 11.88 |
| ROTATION: Spring wheat, peas and sweet clover, sweet clover, winter wheat | 11 | 6.01 | 7.93 |
| ROTATION: Peas and sweet clover, sweet clover, winter wheat, spring wheat | 12 | 7.69 | 12.01 |
| Average of 9, 10, 11, and 12 | | 7.73 | 10.02 |
| Fallow | 13 | 27.82 | 25.03 |

GRAPH 20.—Average annual soil and water losses, 1932–1935, inclusive, from 30 per cent slope of Palouse silt loam, at the soil and water conservation experiment station, Pullman, Wash. Average annual precipitation, 21.74 inches. (Plot 1 = 6 by 36.3 feet; plot 2 = 6 by 145.2 feet; plots 3 and 6 to 13, inclusive, = 6 by 72.6 feet.) (*Soil Conservation Service.*)

## Effect of Slope on Soil and Water Losses under Clean Tillage

The importance of slope in relation to soil and water losses from cultivated land is revealed in the measurements of rates of erosion and runoff, under comparable conditions of soil, cover, and rainfall, presented in Table 8.

In the instance of Houston black clay, at Temple, Tex., the average annual soil loss under corn has been at a rate nearly three times that from the same kind of land having a slope only half as steep, that is, 30.4 tons an acre from an area of 4 per cent slope, as against 10.6 tons from a similar area having a slope of 2 per cent. From Kirvin fine sandy loam, near Tyler, Tex., the average annual soil loss has been at a rate more than twice as rapid as that from a slope only about half as steep but otherwise comparable; and from Shelby silt loam, at Bethany, Mo., more than three times that from practically the same type of soil at Columbia, Mo., where the slope is not quite half so steep.

TABLE 8.—EFFECT OF SLOPE ON ANNUAL SOIL AND WATER LOSSES UNDER CLEAN TILLAGE[1]

| Soil and location | Period, inclusive | Rainfall, inches | Length of slope, feet | Slope, per cent | Crop | Soil loss, tons per acre | Water loss, per cent precipitation |
|---|---|---|---|---|---|---|---|
| Miles clay loam, Texas | 1926 to 1937 | 20.73 | 97 | 0 | Cotton | 2.2 | 6.4 |
| | | | | 1 | | 5.2 | 15.1 |
| | | | | 2 | | 7.0 | 15.5 |
| Muskingum silt loam, Ohio | 1934 to 1936 | 36.46 | 72.6 | 8 | Corn | 60.0 | 30.4 |
| | | | | 12 | | 73.2 | 42.0 |
| Houston black clay, Texas | 1933 to 1936 | 34.90[2] | 72.6 | 2 | Corn | 10.6 | 13.4 |
| | 1933 to 1936 | 35.47[2] | | 4 | | 30.4 | 16.6 |
| Kirvin fine sandy loam, Texas | 1931 to 1936 | 40.82 | 72.6 | 8.7 | Cotton | 27.9 | 20.9 |
| | 1933 to 1936 | 43.00 | | 16.5 | | 72.0 | 14.6 |
| Shelby loams, Missouri | 1918 to 1931[3] | 40.37 | 90.75 | 3.7 | Corn | 19.7 | 29.4 |
| | 1931 to 1935[4] | 34.79 | 72.6 | 8 | | 68.8 | 28.3 |

[1] Measurements at soil and water conservation experiment stations, Soil Conservation Service.

[2] Rainfall difference due to location.

[3] Columbia, Mo.

[4] Bethany, Mo.

It has not been determined, however, that the rates of increase in soil loss conform with increase of declivity in accordance with any definite degree of regularity. For example, the corresponding soil losses from adjacent areas of Miles clay loam at Spur, Tex., have been at the rates of 5 tons an acre annually from cotton grown on a 1 per cent slope and 7 tons annually from cotton grown on a 2 per cent slope. In this connection, it is important to observe that from level land of the same soil (Miles clay loam), there has been an average annual soil loss of 2 tons an acre over a period of 11 years. Erosion on level land is probably due to the piling-up effect of rainfall.

Under cultivation, water losses are usually greater from the steeper slopes, as the results presented in Table 8 show, but they frequently do not increase with increase of slope, under conditions otherwise comparable, nearly so rapidly as do the losses of soil. As a matter of fact, runoff sometimes is no greater, occasionally even less, from steeper slopes, otherwise comparable, from which the rate of soil loss is greater. This point is illustrated in the results obtained on Shelby loam in Missouri, where the runoff from an 8 per cent slope was less (28 per cent of total precipitation) than from a 3.7 per cent slope (29 per cent of total precipitation), regardless of the fact that the soil loss from the steeper area was more than three times as great. Such differences probably are involved with complex interrelations of rainfall intensity, infiltration rates, soil character and condition, and amount of material carried in suspension by the surface water.

## Effect of Slope Length (or Cross Section) on Erosion and Runoff

Comparative losses of soil from areas of different slope length are of great interest not only because of the complex hydrological problems involved but because of the practical application that can be made of such information. Measurements of rates of erosion from various types of soil in regions of different rainfall characteristics show that, with few exceptions, the process of erosion is accelerated with increase in length of slope. As well as can be determined with the information available, the intensification is due to the usual greater volume of water accumulating on long slopes and the consequent increased velocity of runoff. Only one exception is noted in the measurements recorded for eight widely separated types (Table 9), that of the Houston black clay, a soil of decidedly unique characteristics.

Measurements of erosion on Houston black clay, near Temple, Tex., show that from a slope 36.3 feet long, the average annual soil loss, over a six-year period, was 27 tons per acre whereas the corresponding loss from a slope 72.6 feet long was 24 tons; and from another 145.2 feet long, only 22 tons an acre. In like manner, the water lost as runoff decreased with increase in length of the slope cross section.

TABLE 9.—EFFECT OF LENGTH OF SLOPE ON SOIL AND WATER LOSSES FROM EIGHT REPRESENTATIVE TYPES OF ERODIBLE FARM LAND, UNDER CLEAN TILLAGE[1]

| Soil and location | Slope, per cent | Period, inclusive | Rain-fall, inches | Crop | Length of slope, feet | Annual | |
|---|---|---|---|---|---|---|---|
| | | | | | | Soil loss, tons per acre | Water loss, per cent precipitation |
| Clinton silt loam, La Crosse, Wis. | 16 | 1933 to 1936 | 32.28 | Corn | 145.2<br>72.6<br>36.3 | 116.1<br>100.5<br>64.3 | 21.6<br>25.7<br>23.9 |
| Muskingum silt loam, Zanesville, Ohio | 12 | 1934 to 1936 | 36.46 | Corn | 145.2<br>72.6<br>36.3 | 81.2<br>73.2<br>70.4 | 38.2<br>42.0<br>41.5 |
| Cecil sandy clay loam, Statesville, N. C. | 10 | 1931 to 1936 | 45.22 | Cotton | 145.2<br>72.6<br>36.3 | 26.2<br>22.5<br>22.6 | 8.8<br>10.2<br>11.7 |
| Kirvin fine sandy loam, Tyler, Tex. | 8.7 | 1931 to 1936 | 40.82 | Cotton | 145.2<br>72.6<br>36.3 | 43.6<br>27.9<br>18.6 | 19.4<br>20.9<br>18.3 |
| Houston black clay, Temple, Tex. | 4 | 1931 to 1936 | 32.76 | Corn | 145.2<br>72.6<br>36.3 | 21.7<br>23.8<br>27.0 | 13.6<br>14.2<br>16.5 |
| Vernon fine sandy loam, Guthrie, Okla. | 7.7 | 1931 to 1936 | 31.45 | Cotton | 145.2<br>72.6<br>36.3 | 38.6<br>22.5<br>17.2 | 14.8<br>14.1<br>14.6 |
| Shelby silt loam, Bethany, Mo. | 10 | 1934 to 1935 | 33.52 | Corn | 270<br>180<br>90 | 66.4<br>54.1<br>23.7 | 23.4<br>20.8<br>19.2 |
| Marshall silt loam, Clarinda, Iowa | 8 | 1933 to 1935 | 26.94 | Corn | 630<br>315<br>157.5 | 21.3<br>16.3<br>11.7 | 6.2<br>7.9<br>10.3 |

[1] Measurements at soil and water conservation experiment stations, Soil Conservation Service.

Runoff appears to be much more variable than erosion in its relation to length of slope. Average rates of water loss are less from the longer slopes of the Clinton, Cecil, Marshall, and Houston soils, all of which are highly permeable to water when dry (the Houston by reason, particularly, of its tendency to crack deeply on drying). This appears to be due to the increased opportunity for infiltration afforded by the longer time involved with the flow of water across the longer slopes, especially in the instance of rains of low intensity. Such slope-length effect tends to diminish or even disappear on soils of low permeability, especially during heavy rains. In this connection, it should be observed that the results shown in Table 9 involve considerably more rains of low intensity than of high intensity as well as more that fell on relatively dry soil than on less absorptive wet soil.[1]

## Effect of Rainfall Intensity on Soil and Water Losses

Investigations relative to the effect of rainfall intensity on the permeable Marshall silt loam, 9 per cent slope, near Clarinda, Iowa, show[2] that, whereas runoff from long slopes has exceeded that from short slopes for rains of high intensity, the reverse has been true with respect to rains of low intensity. Soil losses for the corresponding period of study have followed in the same course. Analysis of 18 runoff records over the period from Aug. 15, 1932, to the end of 1934, including 5 rains of high intensity (rains considerably exceeding immediate infiltration capacity of the soil) and 13 of low intensity (rains approximately twice the duration and only slightly above the infiltration capacity of the soil), clearly shows this distinctly different direction of trend for this particular type of land:

LOSS OF WATER AS RUNOFF FROM SLOPES OF DIFFERENT LENGTH
(Marshall silt loam)

| *Slope Length, Feet* | *High-intensity Rains, Per Cent Precipitation* | *Low-intensity Rains, Per Cent Precipitation* |
|---|---|---|
| 630 | 20.3 | 13.5 |
| 315 | 18.0 | 16.5 |
| 157.5 | 10.8 | 28.0 |

The entire precipitation of the five intense rains, amounting to 8.2 inches, fell over a period totaling nearly 32 hours (average $6\frac{1}{2}$ hours), whereas the thirteen low-intensity rains, totalling 17.4 inches fell over a period of 133 hours (average $10\frac{1}{3}$ hours).

The rate of erosion proceeded in much the same order:

LOSS OF SOIL FROM SLOPES OF DIFFERENT LENGTH
(Marshall Silt Loam)

| *Slope Length, Feet* | *High-intensity Rains, Tons per Acre* | *Low-intensity Rains, Tons per Acre* |
|---|---|---|
| 630 | 32.86 | 5.70 |
| 315 | 18.26 | 6.35 |
| 157.5 | 8.64 | 7.68 |

[1] Musgrave, G. W. Some relations between Slope-length, Surface-runoff, and the Silt-load of Surface-runoff, Am. Geophysical Union *Trans.*, 1936.

[2] *Ibid.*

Here it is seen that, although rains of high intensity have removed much more soil from the longer slopes, the trend is reversed in respect to the effect of precipitation of less intensity, although the actual losses have not been greatly different. The average amount of soil carried away in the off flowage of all the rains has averaged 5.29 pounds per cubic foot of runoff from the long slopes, 3.18 pounds from the slopes of intermediate lengths, and 1.56 pounds from the short slopes.

These results indicate the need for providing special protection for long slopes, such as strip cropping and terracing. The need of such protection for relatively permeable soils, such as the Marshall silt loam, probably is not so great as it is for impermeable soils. For example, distance between terraces (terrace spacing) and contour strips of protective vegetation probably can be greater, generally, on long slopes of the more permeable soils.

The effect of slope length appears to be further complicated on some soils by a little understood process of loading and unloading on the part of thick and thin sheets of soil-transporting water flowing across the slopes.

Information gained from these studies of the influence of slope length on erosion and runoff points to the need for modifying this effect by breaking up the length of slope with such practical measures for diversion or retention of rainfall as field terraces, contour furrowing, contour ridging (or listing), and strip cropping.

## Effect of Season on Soil and Water Losses

Water erosion generally is most severe during summer and spring, especially in the humid East. In the Pacific Northwest, specifically in the Palouse Wheat Belt of southeastern Washington and adjacent Idaho and Oregon, much serious erosion is caused by snow melting under the impact of warm winds from the Pacific (*chinooks*), usually from January to March, in fields of winter wheat seeded on summer-fallowed ground. August and September commonly are the months of most excessive soil erosion over much of the semiarid and arid grazing areas of the Southwest. As a rule, erosion is most serious over the greater part of the Pacific Coast region in winter and spring, because this is the season of heavy precipitation.

In general, the period of heaviest erosion by water is that season represented by a combination of unstable ground condition and intensive rainfall. This obviously is dependent on both climate and type of land use. In most humid sections, heaviest soil washing takes place in the spring or early summer on freshly plowed land or land supporting a sparse or weak cover of vegetation or crop stubble, provided the heaviest rains fall at that time. In the same locality, the season of most disastrous washing for the following year may be delayed until summer or late summer, if the heaviest and most intense rains chance to be delayed that long.

Wind erosion is influenced by seasonal conditions in much the same manner as water erosion. It is most severe when the ground is highly desiccated, loose, and bare of anchoring vegetation and, of course, when the wind is most constant and violent. In the Great Plains, it is generally most disastrous in winter and early spring, because of the frequency of soil-transporting wind and, during drought, the excessive instability of the soil at that time of the year. However, wind erosion may be serious in time of protracted drought at any time of the year, not only in the Plains but in many dry-land areas west of the Plains, especially on the less stable soils, as the more sandy lands, bared of vegetation by cultivation, overgrazing, or crop failure.

Average seasonal variations in erosion and runoff can be accurately ascertained for different regions only through long-time measurements, since seasonal rainfall varies markedly from year to year. The data contained in Table 10 indicate the wide variations that have occurred from season to season at three of the soil and water conservation experiment stations, all of which are in the humid region. Erosion from clean-tilled crops has been greater generally during the period of greatest precipitation at the sites referred to—usually in summer. Runoff from the

TABLE 10.—AVERAGE SEASONAL SOIL AND WATER LOSSES UNDER: (1) CULTIVATION AND (2) PROTECTIVE COVER OF VEGETATION FOR PERIOD 1932–1936[1]

| Soil and location | Slope, per cent | Season | Average precipitation, inches | Crop | Clean tilled | | Crop | Dense cover | |
|---|---|---|---|---|---|---|---|---|---|
| | | | | | Soil loss, tons per acre | Water loss, per cent precipitation | | Soil loss, tons per acre | Water loss, per cent precipitation |
| Marshall silt loam, Clarinda, Iowa | 9 | Winter | 1.99 | Corn | 1.07 | 47.80 | Grass | 0.002 | 35.00 |
| | | Spring | 6.32 | | 6.80 | 24.50 | | 1.510 | 14.07 |
| | | Summer | 10.10 | | 15.84 | 16.15 | | 0.260 | 5.91 |
| | | Fall | 7.85 | | 3.38 | 12.85 | | 0.004 | 1.58 |
| Shelby silt loam, Bethany, Mo. | 8 | Winter | 2.94 | Corn | 0.15 | 25.15 | Grass | 0.003 | 16.05 |
| | | Spring | 8.05 | | 15.68 | 21.05 | | 0.031 | 1.11 |
| | | Summer | 10.14 | | 25.15 | 29.78 | | 0.020 | 11.20 |
| | | Fall | 10.30 | | 13.35 | 31.30 | | 0.059 | 0.55 |
| Cecil sandy clay loam, Statesville, N. C. | 10 | Winter | 12.58 | Cotton | 0.34 | 3.54 | Grass | 0.000 | 0.11 |
| | | Spring | 12.28 | | 4.74 | 5.10 | | 0.024 | 0.64 |
| | | Summer | 12.80 | | 16.87 | 15.92 | | 0.001 | 0.18 |
| | | Fall | 11.07 | | 1.67 | 8.01 | | 0.001 | 0.04 |

[1] Measurements at soil and water conservation experiment stations, Soil Conservation Service.

highly permeable Marshall silt loam has proceeded at the highest rate during the winter period when the soil was frozen, whereas at the station located near Bethany, Mo., on the much less permeable Shelby silt loam, it has been most rapid in summer and fall. Rainfall loss from Cecil sandy clay loam at the Statesville, N. C., station has been most rapid during summer. From comparable areas protected with a dense cover of grass, soil losses have been relatively low at all seasons, with some tendency toward greatest severity in spring and summer, except from Shelby loam in north-central Missouri, where the higher rate of erosion, thus far, has been in the fall. Runoff losses from grassed areas, on the other hand, have been highest on the Marshall and Shelby types in winter and on the Cecil in spring.

Similar studies at the University of Missouri[1] covering a longer period of investigation (1918–1931) show that the greatest losses of soil under fallow or sparse cover conditions have occurred during April, June, August, and September. The average rainfall at this site was highest in September, with June, May, and April following in order. It is interesting to note that in spite of the higher rainfall of May as compared with April, erosion has averaged lighter for the former month, probably because of a somewhat greater vulnerability of the soil at the earlier season. Runoff at the Columbia, Mo., station has been highest, commonly, during April, June, and September from the fallow or sparsely vegetated areas, as well as from the area continuously covered with grass.

Much additional information is needed with respect to the effect of season on rates of erosion and runoff, as a means of assisting in the perfection of the best combination of vegetative and mechanical controls and in proper adjustment of such measures to the land (see Chap. VII, "Climate and Soil Erosion").

## Effect of Crop Rotation on Erosion and Runoff

Since a good crop rotation provides more cover of a protective nature for a given field over a period of years than the production of clean-tilled crops continuously and improves the physical condition of the soil (and probably also the biologic condition) by building up the supply of organic matter in the soil, it is not surprising to find that such protective practice generally results in marked reduction of erosion and runoff, as shown in Table 11, summarizing the losses from comparable areas and conditions under (1) continuous clean tillage and (2) crop rotation.

[1] Miller, M. F., and Krusekopf, H. H. The Influence of Systems of Cropping and Methods of Culture on Surface Runoff and Soil Erosion, Univ. Missouri Research *Bull.* 177, 1932.

TABLE 11.—ANNUAL SOIL AND WATER LOSSES FROM SEVERAL IMPORTANT TYPES OF FARM LAND, UNDER CROP ROTATIONS AND CONTINUOUS CULTIVATION[1]

| Soil and location | Slope, per cent | Period, inclusive | Rain-fall, inches | Crop | Soil loss, tons per acre | Water loss, per cent precip-itation |
|---|---|---|---|---|---|---|
| Vernon fine sandy loam, Guthrie, Okla. | 7.7 | 1930 to 1935 | 33.12 | Cotton in rotation | 14.33 | 12.9 |
| | | | | Cotton continuously | 24.29 | 14.2 |
| Houston black clay, Temple, Tex. | 4.0 | 1931 to 1935 | 31.34 | Corn in rotation | 18.46 | 12.1 |
| | | | | Corn continuously | 20.89 | 13.7 |
| Colby silty clay loam, Hays, Kans. | 5.0 | 1930 to 1935 | 20.36 | Kafir in rotation | 11.96 | 18.6 |
| | | | | Kafir continuously | 11.74 | 16.2 |
| | | | | Wheat in rotation | 2.13 | 10.8 |
| | | | | Wheat continuously | 2.25 | 10.7 |
| Shelby silt loam, Bethany, Mo. | 8.0 | 1931 to 1935 | 34.79 | Corn in rotation | 27.33 | 20.6 |
| | | | | Corn continuously | 68.78 | 28.3 |
| Cecil sandy clay loam, Statesville, N. C. | 10.0 | 1931 to 1935 | 45.22 | Cotton in rotation | 13.87 | 8.8 |
| | | | | Cotton continuously | 22.58 | 10.2 |
| Marshall silt loam, Clarinda, Iowa | 9.6 | 1933 to 1935 | 26.82 | Corn in rotation | 13.41 | 8.5 |
| | | | | Corn continuously | 18.82 | 8.6 |
| Clinton silt loam, La Crosse, Wis. | 16.0 | 1933 to 1935 | 34.12 | Corn in rotation | 53.82 | 16.4 |
| | | | | Corn continuously | 88.35 | 20.6 |
| Muskingum silt loam, Zanesville, Ohio | 12.0 | 1934 to 1935 | 34.51 | Corn in rotation | 27.57 | 24.0 |
| | | | | Corn continuously | 59.62 | 35.2 |
| | | | | Grass in rotation | 0.20 | 14.1 |
| | | | | Grass continuously | 0.05 | 4.5 |

[1] Measurements at soil and water conservation experiment stations, Soil Conservation Service.

The rotations cited in Table 11 for eight important types of farm land, within as many problem areas, include in the cycle of crops grown one or more of dense cover such as provide greater stability to the soil and produce heavier residues of vegetable matter in the form of roots and stubbles for increasing the organic-matter content of the soil. Some of them provide for the turning under of the entire crop; others include legumes that add not only humus but nitrogen obtained from the vast

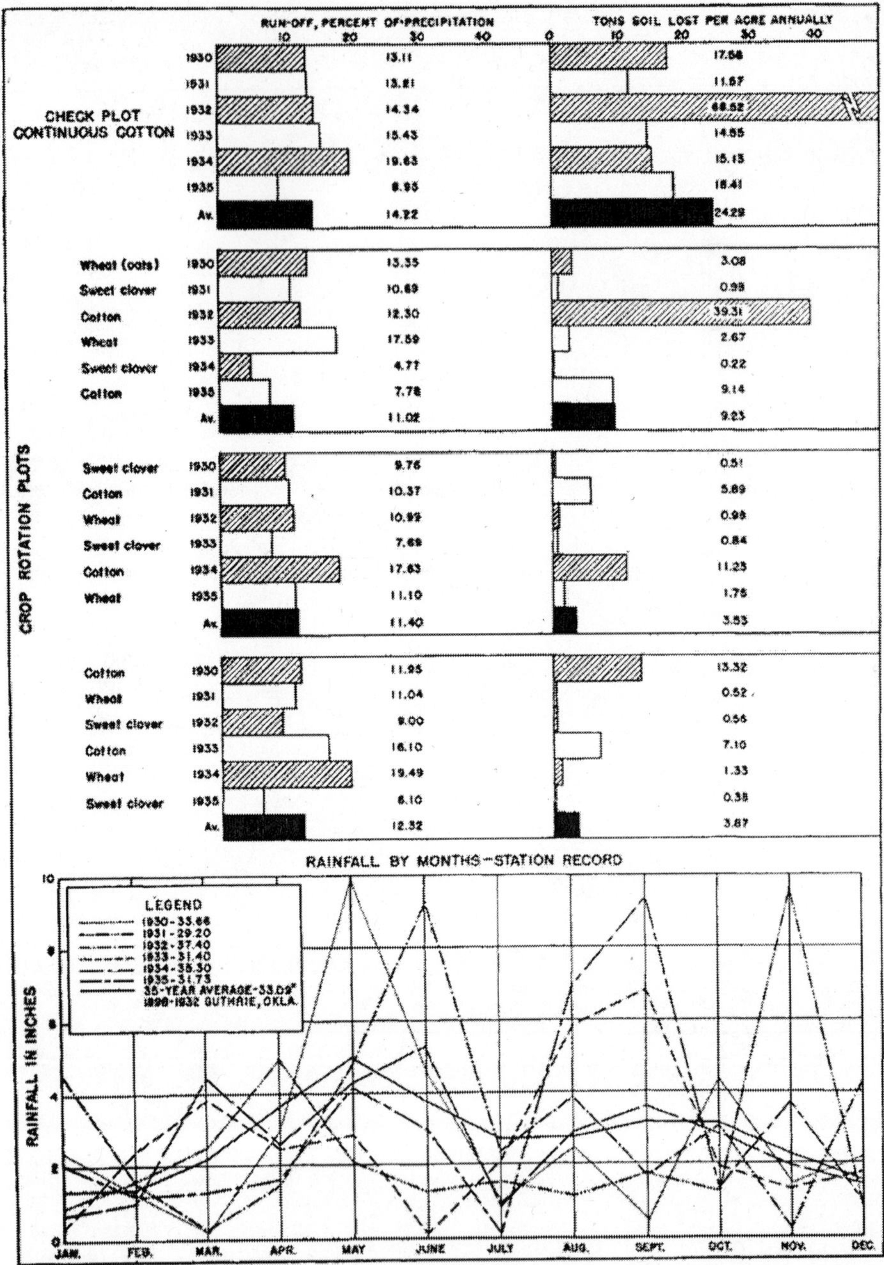


GRAPH 21.—Effect of crop rotation on soil and water losses, 1930–1935, inclusive. Plots $\frac{1}{100}$ acre on Vernon fine sandy loam, Guthrie, Okla., 7.7 per cent slope. Three-year rotation: cotton, wheat, sweetclover (wheat seeded in cotton in fall, and sweetclover ensuing winter.) (*Soil Conservation Service.*)

supply of that important element of plant food contained in the atmosphere. With respect to the protective efficiency of a given rotation, much depends on the kinds of plants included in the succession of crops. For example, grass in rotation usually is not so effective in restraining runoff and erosion as where it is grown continuously.

The best crop rotation for reducing the loss of soil and water is one that not only provides for a reduction of the area devoted to clean-tilled crops but also provides for dense cover of protective vegetation, in the form of the growing crop or its stubble, during the season of intense rains

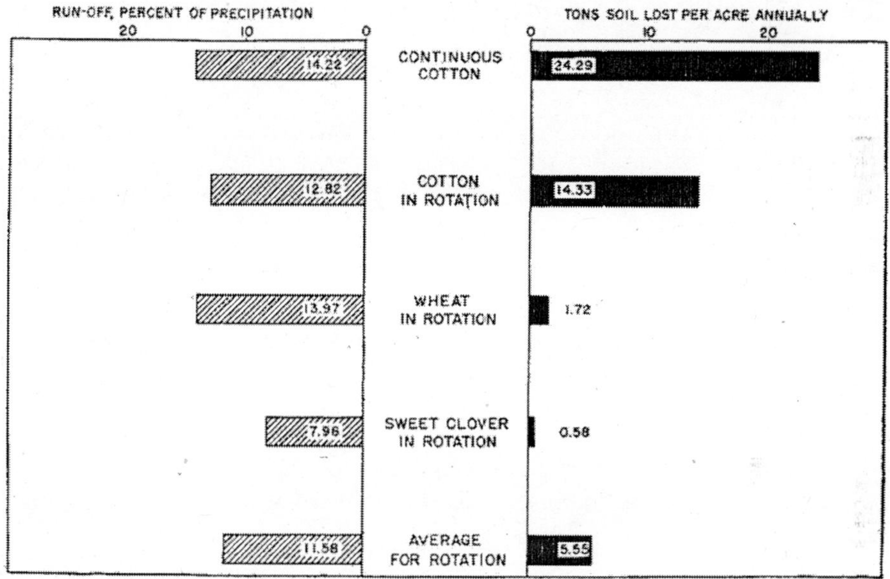


GRAPH 22.—Effect of crop rotation on soil and water losses, 1930–1935, inclusive. Average for continuous cotton, for each crop in the rotation, and for the rotation. Plots $\frac{1}{100}$ acre on Vernon fine sandy loam, Guthrie, Okla., 7.7 per cent slope. Three-year rotation: cotton, wheat, and sweetclover (wheat seeded in cotton in fall, and sweetclover in wheat ensuing winter.) (*Soil Conservation Service.*)

as well as the season following harvest. Thus, small grain, as wheat, barley, or oats, gives considerable protection, especially when the crop has attained sufficient growth to provide good coverage of the land. In Ohio, for example, intense rains ordinarily do not occur so frequently in spring (time for seeding oats) as during autumn (time for seeding wheat); therefore, the protective effect of oats under the particular conditions of this locality may be expected, generally, to prove superior to that of wheat. Thus, within limits of practicability, it is important to adjust the crops included in rotations, to average seasonal characteristics, in accordance with their growing habits. It is also important to preserve the protective stubble of dense-growing crops fitted into rotation as long as possible, and to plow under no more than is immediately necessary of seasonal cover crops, especially during seasons characteristic of intense rains or winds, in the preparation of the field for the succeeding cultivated crop.

Graph 21 shows annual and average loss of soil and water under a 3-year rotation of wheat, sweetclover, and cotton, together with the annual and average precipitation, for Vernon fine sandy loam, near Guthrie, Okla. Graph 22 compares the respective annual losses of soil and water while in use for the three crops included in the 3-year rotation with the losses sustained under continuous use for cotton.

## Effect of Cropping Practices and Direction of Cultivation on Erosion and Runoff

Cultivation of intertilled, or row, crops on the contour instead of in the direction of slope (up and down the slope) is usually effective in reducing both soil and water losses. Examples of control effects of this kind are shown in Table 12 for four widely different, important types of farm land: Marshall silt loam of 8 per cent slope, Houston black clay of 4 per cent slope, Colby silty clay loam of 5 per cent slope, and Vernon fine sandy loam of 6.8 per cent slope.

Also, the use of strips of close vegetation, such as alfalfa in conjunction with cotton as cited in the results from Guthrie, Okla., is effective in the control of erosion. Here the practice reduced the soil loss below that of continuous cotton, even though the land slope was somewhat greater where the strips were used. At Tyler, Tex., the growing of cotton with vetch cover on the contour was not so effective as the production of cotton with alternate strips of sorghum or oats also grown on the contour. Likewise, on the Houston black clay, continuous cotton lost soil at the rate of 224 tons per acre in contrast to cotton and oats in alternation where the losses were 32 tons per acre.

## Effect of Organic Matter on Erosion and Runoff

Incorporation of vegetable matter with the soil, whether in the form of barnyard or green manures, is usually highly effective in reducing erosion and runoff. For example, soil losses from Marshall silt loam of 9 per cent slope (at the Clarinda, Iowa, station), used for corn (1) without manure, (2) manured with 8 tons an acre, and (3) manured with 16 tons an acre have been at the rates of 22, 9, and 4 tons per acre, respectively. Even bare soil has shown marked reductions in soil losses as the results of manure applied annually, showing the direct beneficial effect of organic-matter addition to the soil. Similar results have followed the manuring of Clinton silt loam in Wisconsin (Table 13). The principles involved with these effects are discussed in "Principles, Processes, and Types" (Chap. IV, Part 1). Aside from improvement of the infiltration capacity of soils, and consequent reduction of runoff and erosion, through the addition of organic matter, other benefits accrue. By improving the structural condition, organic matter increases aeration of the soil. It improves soil conditions favoring the growth of beneficial microorganisms. It aids in processing the inorganic constituents of the soil, changing unavailable materials into forms available as plant nutrients. It aids in conserving the easily soluble constituents of the soil as well as those

TABLE 12.—EFFECT OF VARIOUS CROPPING PRACTICES AND SLOPE DIRECTION OF CULTIVATION ON EROSION AND RUNOFF FROM AGRICULTURAL SOILS[1]

| Soil and location | Slope, per cent | Period, inclusive | Annual rain-fall, inches | Treatment | Row direction | Soil loss, tons per acre | Water loss, per cent of precip-itation |
|---|---|---|---|---|---|---|---|
| Marshall silt loam, Iowa.... | 8.0 | 1933 to 1935 | 26.94 | Corn continuously | Contour | 0.00 | 0.1 |
| | | | | | Up and down | 11.75 | 10.3 |
| Houston black clay, Texas.. | 4.0 | 1931 to 1936 | 32.76 | Rotation: corn, oats, cotton | Contour | 5.77 | 5.2 |
| | | | | | Up and down | 13.13 | 7.5 |
| Colby silty clay loam, Kansas | 5.0 | 1934 to 1935 | 17.25 | Wheat continuously | Contour | 1.50 | 12.4 |
| | | | | | Up and down | 2.04 | 15.8 |
| Vernon fine sandy loam, Oklahoma.............. | 6.8 | 1932 to 1935 | 33.96 | Cotton (wheat, winter cover) | Contour | 24.65 | 9.9 |
| | | | | | Up and down | 55.19 | 11.1 |
| Vernon fine sandy loam, Oklahoma.............. | 4.5 | 1935 to 1936 | 26.24 | Strip cropping: cotton, wheat, alfalfa | Contour | 0.23 | 2.8 |
| Vernon fine sandy loam, Oklahoma.............. | 4.0 | 1935 to 1936 | 26.24 | Strip cropping: cotton, oats, sudan | Contour | 1.50 | 10.3 |
| Vernon fine sandy loam, Oklahoma.............. | 3.0 | 1935 to 1936 | 26.24 | Strip cropping: oats, cotton | Contour | 1.33 | 6.9 |
| Vernon fine sandy loam, Oklahoma.............. | 3.5 | 1935 to 1936 | 26.24 | Cotton continuously | Contour | 2.21 | 9.7 |
| Bowie fine sandy loam, Texas | 5.5 | 1935 to 1936 | 41.33 | Strip cropping: cotton, sorghum | Contour | 10.46 | 16.0 |
| Bowie fine sandy loam, Texas | 5.5 | 1935 to 1936 | 41.33 | Cotton vetch continuously | Contour | 31.84 | 14.1 |
| Houston black clay, Texas.. | 3.5 | 1934 to 1935 | 38.16 | Strip cropping: oats, cotton | Contour | .036 | 0.3 |
| Houston black clay, Texas.. | 3.5 | 1934 to 1935 | 38.16 | Cotton continuously | Contour | 7.38 | 5.1 |
| Houston black clay, Texas.. | 3.5 | 1934 to 1935 | 38.16 | Cotton | Up and down | 18.63 | 16.7 |
| Houston black clay, Texas.. | 3.5 | 1935 to 1936 | 43.25 | Strip cropping: vetch, cotton | Up and down | 0.70 | 2.6 |
| Houston black clay, Texas.. | 3.5 | 1935 to 1936 | 43.25 | Cotton continuously | Contour | 16.65 | 9.2 |
| Houston black clay, Texas.. | 3.5 | 1935 to 1936 | 43.25 | Cotton continuously | Up and down | 27.78 | 18.0 |

[1] Measurements at soil and water conservation experiment stations, Soil Conservation Service.

contained in fertilizers applied to the soil. It develops a buffering capacity against toxic effects of materials developed within or added to the soil.

All of these benefits are reflected in increased vegetation, which prevents soil and water losses through (1) the binding effect of the more vigorous and widely spreading root systems produced and (2) increased protection of the ground surface from direct impact of rains provided by increased vegetative canopy. And the relatively abundant amount of organic matter supplied by the roots of grasses or legumes grown in rotation is of great value, for a period, in binding soil particles together when the land is subsequently used for those crops which expose it to wind or water erosion.

## Effect of Grass and Forest on Erosion and Runoff

Erosion proceeds so slowly under a good cover of grass, or woods growth old enough to have accumulated a layer of litter or mold over the entire forest floor, that the profound effect is restated here as a matter of emphasis. Table 14 shows the almost insignificant rates of soil loss and the usual very low rates of runoff from four extensive and important types of farm land. In only a single instance has the soil loss exceeded one ton per acre in one year. In some instances, only traces of soil have been lost; and in no instance has the water loss been as much as 10 per cent of the total precipitation, averaging for the 19 average measurements recorded 2.53 per cent of the total precipitation.

TABLE 13.—ANNUAL SOIL AND WATER LOSSES FROM TWO IMPORTANT TYPES OF FARM LAND TREATED WITH ORGANIC MATTER[1]

| Soil, slope, and location | Period, inclusive | Rainfall, inches | Manure, applied annually, tons per acre | Treatment | Soil loss, tons per acre | Water loss, per cent precipitation |
|---|---|---|---|---|---|---|
| Marshall silt loam, 9 per cent slope, Clarinda, Iowa | 1933 to 1935 | 26.87 | 16 | Corn | 4.73 | 4.0 |
| | | | 8 | | 9.24 | 7.4 |
| | | | 0 | | 22.07 | 10.4 |
| | | | 16 | Fallow | 36.60 | 13.7 |
| | | | 8 | | 46.07 | 17.8 |
| | | | 0 | | 58.47 | 20.2 |
| Clinton silt loam, 16 per cent slope, La Crosse, Wis. | 1933 to 1936 | 32.28 | 5 | Fallow | 145.77 | 24.9 |
| | | | 0 | | 166.02 | 25.3 |

[1] Measurements at soil and water conservation experiment stations, Soil Conservation Service.

TABLE 14.—ANNUAL SOIL AND WATER LOSSES FROM FORESTED AND GRASSED AREAS FOR FOUR IMPORTANT TYPES OF FARM LAND[1]

| Soil and location | Slope, per cent | Drainage area, acres | Rainfall, inches | Period, inclusive | Cover | Soil loss, tons per acre | Water loss, per cent precipitation |
|---|---|---|---|---|---|---|---|
| Kirvin fine sandy loam, Tyler, Tex. | 7.5 | 7.94 | 45.08 | 1933 to 1935 | Woods | 0.050 | 3.6 |
| | 9.0 | 0.89 | 42.12 | | Grass | 0.054 | 5.5 |
| | 12.5 | 0.01 | 41.89 | 1933 to 1936 | Woods | 0.079 | 0.5 |
| | 12.5 | 0.01 | 41.89 | | Woods, burned | 0.291 | 2.7 |
| | 8.7 | 0.01 | 40.53 | | Grass | 0.036 | 0.6 |
| | 16.5 | 0.01 | 43.00 | | Grass | 0.008 | 0.2 |
| Vernon fine sandy loam, Guthrie, Okla. | 4.8 | 5.62 | 31.62 | 1933 to 1936 | Woods | 1.57 | 3.6 |
| | 5.6 | 2.50 | 31.62 | | Grass | 0.04 | 1.7 |
| | 5.17 | 0.01 | 33.42 | 1931 to 1935 | Woods | 0.017 | 0.2 |
| | 5.17 | 0.01 | 33.42 | | Woods, burned | 0.195 | 6.1 |
| | 7.7 | 0.01 | 33.01 | | Grass | 0.029 | 0.9 |
| Cecil sandy clay loam, Statesville, N. C. | 10.0 | 0.01 | 45.26 | 1932 to 1935 | Woods | 0.002 | 0.1 |
| | | | | | Woods, burned | 0.627 | 3.6 |
| | | | | | Grass | 0.012 | 0.3 |
| Muskingum silt loam, Zanesville, Ohio | 12.0 | 2.23 | 36.46 | 1934 to 1936 | Woods | 0.01 | 1.2 |
| | | 3.57 | | | Grass | 0.13 | 9.8 |
| | | 0.01 | | | Woods | trace | 0.1 |
| | | 0.01 | | | Woods, burned | trace | 0.3 |
| | | 0.01 | | | Grass | 0.031 | 7.0 |

[1] Measurements at soil and water conservation experiment stations, Soil Conservation Service.

On Cecil sandy clay loam of 10 per cent slope, near Statesville, N. C., the loss of soil from virgin woodland has averaged only $\frac{1}{500}$ ton per acre annually over a period of five years; whereas the corresponding loss from grass has been at the rate of only $\frac{6}{500}$ ton per acre annually. The corresponding losses of water have been only one-tenth and three-tenths of 1 per cent of the total precipitation, respectively. These losses are so small that soil probably builds from beneath as rapidly as it is removed from the surface.

In comparison with cultivated land, it is interesting that on the same farm the same kind of land (Cecil sandy clay loam, 10 per cent slope) has lost, for the corresponding period, 11,290 times more soil under cotton and 102 times more water than was lost from the comparable forested area and 1,881 times more soil and 31 times more water than was lost from the grassed area. At the Tyler, Tex., station, the loss of soil from cultivated Kirvin fine sandy loam of 16.5 per cent slope (cotton) has exceeded the loss from comparable land well-grassed by 9,005 times, whereas the corresponding loss from the same cultivated area has exceeded the soil loss from a forested area, similar except for its milder slope (12.5 per cent), by 792 times.

## Effect of Erosion on Rates of Soil and Water Losses from Ten Important Types of Farm Land

Usually, erosion speeds up as the more absorptive, humus-charged topsoil is washed off, to expose less absorptive, less stable sublayers. Table 15 shows comparative losses of soil and water from (1) areas still retaining the topsoil (or part of it) and (2) comparable areas from which the topsoil has been completely removed to the depth of subsoil (*B* horizon). In the instance of each of these 10 important types of land, representing collectively many millions of acres of farm land, the losses of both soil and water have been greater where the subsoil was exposed under similar conditions of slope, treatment, and rainfall. For example, in the instance of Palouse silt loam, the loss of material from exposed subsoil exceeded that from the corresponding surface soil, both areas having been originally identical, by six times. Loss of water as immediate runoff was more than three times greater from the eroded land. The average loss of soil material from the exposed subsoil exceeded the corresponding loss from surface soil by 47 per cent. The average water loss from the severely eroded areas was one-third greater than from the uneroded or only slightly eroded areas.

It is interesting to note that the loss of both soil and water from the exposed subsoil of Cecil sandy clay loam, in the North Carolina Piedmont, was nearly the same as from the area with its topsoil intact.

Weaver and Noll[1] found wide differences in the rate of runoff and erosion from unplowed prairie land and overgrazed pasture in Nebraska. With respect to this, they say: "Runoff on a 10° slope from 26.88 inches of rainfall during 15 months was 2.5 per cent from prairie, 9.1 from overgrazed pasture, and 15.1 per cent from a pasture entirely bared of cover by close grazing. The soil was Carrington silt loam. No measurable amount of soil eroded from the prairie, only a small amount from the pasture, but 5.08 tons per acre were lost from the bare area."

Various factors other than those discussed above materially influence rates of soil and water losses. As a matter of fact, every obstacle in the path of wind or running water retards the rate of movement and con-

[1] Weaver, J. E., and Noll, W. C. Comparison of Runoff and Erosion in Prairie, Pasture, and Cultivated Land. Univ. Nebraska *Bull.* 11 (contribution from Department of Botany No. 96) pp. 34–35, November, 1935.

sequently the cutting effect of these most powerful agencies of erosion. Every plant, every patch of stubble, every stone and clod, every inequality of the ground surface—hummock, depression, basin, or change in declivity—each contour ridge or furrow and every hedge or fence following the contour deflects wind or retards runoff, to the end of reducing their power to pick up and transport soil.

TABLE 15.—EFFECT OF EROSION ON RATES OF SOIL AND WATER LOSSES FROM 10 IMPORTANT TYPES OF FARM LAND ORIGINALLY THE SAME[1]

| Soil, location, and years of measurement | Slope, per cent | Annual rainfall, inches | Crop | Uneroded | | Eroded | |
|---|---|---|---|---|---|---|---|
| | | | | Soil loss, tons per acre | Water loss, per cent precipitation | Soil loss, tons per acre | Water loss, per cent precipitation |
| Cecil sandy clay loam, North Carolina 1932–1936 | 10.0 | 48.42 | Cotton | 25.1 | 10.5 | 25.8 | 9.4 |
| Nacogdoches fine sandy loam, Texas 1932–1936 | 10.0 | 42.34 | Cotton | 8.2 | 16.5 | 31.9 | 21.9 |
| Kirvin fine sandy loam, Texas 1932–1936 | 8.7 | 40.52 | Cotton | 30.1 | 22.2 | 65.3 | 26.0 |
| Muskingum silt loam, Ohio 1934–1936 | 12.0 | 36.46 | Corn | 73.2 | 42.0 | 92.6 | 38.6 |
| Clinton silt loam, Wisconsin 1933–1935 | 16.0 | 34.12 | Corn | 88.7 | 20.8 | 108.7 | 25.2 |
| Vernon fine sandy loam, Oklahoma 1930–1935 | 7.7 | 33.12 | Cotton | 24.3 | 14.2 | 34.4 | 28.5 |
| Houston black clay, Texas 1931–1936 | 4.0 | 32.76 | Rotation | 14.8 | 9.2 | 18.5 | 15.0 |
| Marshall silt loam, Iowa 1933–1935 | 9.0 | 26.82 | Corn | 18.8 | 8.6 | 30.3 | 12.8 |
| Palouse silt loam, Washington 1932–1935 | 30.0 | 21.74 | Wheat, fallow | 0.7 | 3.4 | 4.2 | 11.8 |
| Colby silty clay loam, Kansas 1930–1936 | 5.0 | 20.36 | Wheat | 2.2 | 10.7 | 7.8 | 14.5 |
| Average losses..... | .... | ..... | ........ | 28.6 | 15.8 | 42.0 | 20.4 |

[1] Measurements at soil and water conservation experiment stations, Soil Conservation Service.

Weeds, for example, have reduced runoff, as compared with corn on the same kind of land, by more than a third and erosion by more than half, according to actual measurements.

Removal of stones larger than 2 inches in diameter from an important type of stony soil resulted in the loss of twice as much soil as from a comparable area with the stones left in place and increased the tendency of the land to gully. In the mountains of eastern Cuba, west of Cape Cruz, a stringent rule on sugar cane plantations is that neither stone nor tree stump can be removed in clearing land for cane. Experience has taught the growers that heavy rains accompanying the tropical storms of that region strip off the productive soil in a very short time when clean clearing and plow cultivation are practiced.

Table 16 shows the powerful effect of a vegetative mulch in reducing erosion on one of the most important types of land in the southern Piedmont. The mulch was applied at the rate of approximately 23 tons an acre and was renewed in September or October each year.

The effectiveness of various practical erosion measures in retarding runoff and holding soil against water and wind is discussed in other chapters dealing with control and prevention of erosion.

TABLE 16.—SOIL AND WATER LOSSES FROM FRESHLY CLEARED CECIL SANDY CLAY LOAM, 10 PER CENT SLOPE,[1] TREATED WITH FOREST LITTER[2]

| Treatment | Year | Rain-fall, inches | Soil loss, tons per acre | Water loss, per cent precipi-tation |
|---|---|---|---|---|
| Covered with 2-inch layer of undecomposed pine needles | 1934 | 49.34 | 0.27 | 10.98 |
| | 1935 | 42.80 | 0.18 | 11.33 |
| | 1936 | 60.00 | 0.11 | 8.98 |
| | Average 1934 to 1936 | | 0.19 | 10.43[3] |
| Covered with 2-inch layer of undecomposed hardwoods forest litter | 1934 | 49.34 | 0.25 | 10.87 |
| | 1935 | 42.80 | 0.12 | 8.57 |
| | 1936 | 60.00 | 0.09 | 6.08 |
| | Average 1934 to 1936 | | 0.15 | 8.51[3] |
| Continuous cotton, no treatment | 1934 | 49.34 | 15.15 | 8.60 |
| | 1935 | 42.80 | 40.72 | 22.38 |
| | 1936 | 60.00 | 40.92 | 17.41 |
| | Average 1934 to 1936 | | 32.26 | 16.13[3] |

[1] Average annual precipitation 50.71 inches.

[2] Measurements at soil and water conservation experiment station, Soil Conservation Service, Statesville, North Carolina.

[3] Average of yearly percentages.

## Effect of Biological Forms on Soil and Water Conservation

To a degree much more profound than is generally realized, erosion is influenced by the millions of organisms that inhabit the soil. This influence may be direct, as in the case of such organisms as earthworms, various insects, and rodents; or indirect, as in the case of those forms grouped under the general heading of microscopic organisms.

As long ago as 1881,[1] Darwin called attention to the fact that earthworms aid in the denudation of English soils. As a result of careful observation and weighings, he calculated that a large quantity of earth passes through their bodies and is deposited on the surface of every acre (about 10 tons annually). Castings left in this way on pasture land are frequently washed downslope with heavy rains, so that some soil is lost even where the ground is protected by a dense sod. Although quantitative data pertaining to this point are not available for the United States, there is evidence that similar activities take place in the humid parts of the country. On the other hand, earthworm burrows often extend to a depth of 3 to 4 feet or more and serve as water channels to increase the infiltration capacity of the soil. Earthworms also aid materially in carrying fragments of organic matter below the surface. In working over the soil with its content of organic matter, earthworms improve its physical condition, especially with respect to the development of a crumb structure, thus aiding in making soil more permeable to water.

In the arid and semiarid Western regions particularly, rodents play some part in erosion damage by bringing loose earth to the surface, where it is subject to wind and water erosion. Especially in overgrazed areas, the influence of such forms as the gopher, the prairie dog, and the kangaroo rat may sometimes be considerable. The chief effect of these mammals and of various insect forms, however, lies in their destruction of vegetative cover. The grasshopper has been responsible for exposing many areas to the beat and wash of the rain by stripping off the coverage of vegetation.

The biological effects on the soil of microorganisms, such as protozoa, bacteria, fungi, and algae, are enormously important, even essential to productivity. As Selman A. Waksman, microbiologist of the New Jersey Agricultural Experiment Station, points out:[2] "For every plant or animal living on the surface of the earth, there are hundreds of kinds of lower forms of life which lead their entire existence in the soil and which give life to the soil itself." The influence of these organisms depends on their ability, especially marked in the case of bacteria and fungi, to break down fresh organic matter.

When organic material, such as barnyard manure, green manure, and crop residues, is incorporated with the soil, it is attacked by fungi

[1] Darwin, Charles. "Vegetable Mould and Earthworms." Appleton ed. 1890.

[2] Waksman, Selman A. The Living Soil, *Soil Cons.*, Vol. 3, pp. 173–177, January, 1938.

and bacteria and is broken down into various organic compounds. Among the end products are ammonia, nitrate nitrogen, carbon dioxide, and various organic acids which tend to enrich the soil solution by converting unavailable mineral elements of the soil into available nutritive forms. When residues containing small amounts of nitrogenous and large amounts of carbonaceous materials are added to the soil, temporarily available forms of nitrogen tend to disappear, and, as a result, higher forms of plant life may suffer from a lack of this element. This is one reason why leguminous plant residues produce better results under some conditions than do nonleguminous plants. The residue left in the soil after decay of the more readily decomposable organic matter is complex colloidal organic material which greatly increases the base exchange capacity of a soil. Both the quality and the quantity of these organic colloids materially affect the stability of the structural aggregates of the soil. These aggregates largely govern the maintenance of favorable soil granularity, such as has much to do with resistance to erosion and capacity for absorption of rainfall.

From the standpoint of soil conservation, the activities of soil microorganisms may be considered both helpful and harmful. The breakdown of organic matter with the production of nitrates is essential to crop production. But the continued destruction of soil organic matter under cultural practices tends to create conditions favoring erosion. On the other hand, the production of the organic colloids tends toward the development of a desirable erosion-resistant granular structure, as noted above. It is apparent, then, that it usually is necessary in farm practice to restore organic matter to the soil by the use of manures, green manures, and crop residues as well as by growing in rotation those plants which supply vegetable matter through the decomposition of their roots. When such restoration is made, the action of soil microorganisms in the production of more humus is, in the end, favorable to soil conservation.

The organisms that inhabit the soil are of the utmost importance to soil productivity. Much or all of the nitrogen that crops use is derived from organic matter through the action of ammonifying, nitrite- and nitrate-forming organisms. An unknown amount is fixed by such independent organisms as Azotobacter which obtain their energy from decaying organic material. The Rhizobium microbes fix nitrogen in the soil in the nodules of legume roots that they invade. There are injurious organisms as well as beneficial ones, but, on the whole, the biological activities in a soil well supplied with organic matter make for soil productivity.

# Chapter VI. Relation of Physical and Chemical Properties of Soils to the Erosion Problem

When the seriousness of the problem of soil erosion was finally recognized, fundamental studies of the causes began, along with efforts to develop new and better methods of control and prevention. It soon became apparent that, aside from the effect of declivity or length of slope, some types of soil are much more susceptible to erosion than others. This further complicates the problem of control, since methods that may be adequate for one type of land may be entirely inadequate for another, under identical conditions of rainfall and use.

Investigations have shown that under a tropical climate, certain types of highly weathered, nonplastic clays are almost immune from erosion, whereas slightly weathered, plastic clays are subject to severe erosion on cultivated slopes.[1] Analyses of representative samples have revealed marked differences in the chemical composition of these groups. It has been shown that the molecular ratio of silica to iron oxide plus alumina in various clay types of high colloid content (particles having diameter of 0.002 millimeter or less) covers a wide range and that opposite ratios (low and high) are associated with markedly different soil characteristics, including degree of erodibility.

Average analyses of samples representing three groups from Cuba, each having distinctive characteristics in this respect, show the following relation of contained silica to iron and alumina:

| *Soil Group* | *Consistence* | *Ratio of Silica to Iron and Alumina*[1] |
|---|---|---|
| Nipe clay | Extremely friable | 0.17 |
| Matanzas clay | Intermediate friability | 1.86 |
| Bayamo clay | Extremely plastic | 4.13 |

[1] This silica-sesquioxide ratio is expressed as $\frac{SiO_2}{Fe_2O_3 + Al_2O_3}$

[1] Bennett, H. H. Soil Erosion and Flood Control, lectures before the Graduate School, U. S. Dept. Agr. (mimeographed), 1928.

It was found[1] in all the tropical soils examined that those most resistant to erosion have a molecular ratio of silica to iron and alumina of less than 2. Also, marked differences in the base-exchange capacities of these soils were revealed. The highly weathered material of Nipe clay has a very low base-exchange capacity, whereas the little-weathered plastic material of Yaguajay clay (Bayamo group) has a high base-exchange capacity. Table 17 shows the base-exchange capacities of the three groups.[2]

These preliminary investigations led to more intensive studies of the properties of soils that affect their erodibility. Some of these properties (all closely related and interdependent) are:

1. Texture
2. Structure
    *a.* Ease of dispersion
    *b.* Degree of aggregation
    *c.* Porosity
    *d.* Permeability
3. Organic-matter content
4. Chemical composition

## Effect of Texture

The direct effect of texture on erodibility can be stated in approximate terms only. In general, coarse texture indicates high permeability and the necessity of comparatively high velocities on the part of runoff water to move the individual particles of soil. On the other hand, coarse texture is usually associated with single-grain structure (particles separated, that is, not clustered in granules or aggregates); and for this reason, unprotected coarse-grained soils usually are less resistant, within certain limits (probably up to the texture of coarse sand), to concentrated flows of high velocity than fine-textured soils.[3]

The effect of structure, coupled with the influence of those properties associated with chemical composition, is so pronounced in most fine-grained soils, except those of very high silt content, that texture alone has little significance. Some types of high silt-content soils are extremely

[1] Bennett, H. H. Some Comparisons of the Properties of Humid-tropical and Humid-temperate American Soils, with Special Reference to Indicated Relations between Chemical Composition and Physical Properties, *Soil Sci.*, Vol. 21, pp. 349–375, 1926.

[2] Bennett, H. H., and Allison, R. V. "The Soils of Cuba." Tropical Plant Research Foundation. 1928.

[3] See Gerdel, R. W. Reciprocal Relationships of Texture, Structure, and Erosion, Soil Sci. Soc. America *Proc.*, Vol. 2, pp. 537–545, 1937.

susceptible to erosion, especially where the silt particles are not associated with sufficient clay to produce an aggregated or granular structure.[1]

TABLE 17.—TOTAL EXCHANGEABLE BASES PER 100 GRAMS OF SOIL AND PER 100 GRAMS OF COLLOID OF NIPE, MATANZAS, AND YAGUAJAY CLAYS EXPRESSED AS MILLIEQUIVALENTS[1]

| *Soil Group* | *Exchangeable Bases per 100 Grams of Soil, Milliequivalents* | *Approximate Percentage of Colloid* | *Approximate Total Exchangeable Bases per 100 Grams of Contained Colloid, Milliequivalents* |
|---|---|---|---|
| Nipe | 1.1 | 50 | 2.2 |
| Matanzas | 17.0 | 70 | 24.0 |
| Bayamo (Yaguajay clay) | 79.7 | 70 | 114.0 |

[1] The equivalent weight of an element in grams is the atomic weight of the element divided by its valence. A milliequivalent is one-thousandth of an equivalent.

## Effect of Organic-matter Content

The quantity of organic matter contained in soils is important from many standpoints. With respect to erodibility, its greatest effect is on structure.[2] The organic fraction of the soil has a greater capacity, proportionately, for absorbing and storing water than the mineral fraction, but its most important effect is in forming water-stable aggregates that produce increased porosity and permeability of the soil. Almost without exception, soils of low organic-matter content are subject to comparatively rapid erosion; usually, they are less retentive of moisture and less able to deliver to plants that which is retained.

## Effect of Chemical Composition

Chemical composition of soil, particularly of the colloidal fraction, affects erodibility directly through its effect on physical properties and

[1] Bouyoucos, G. J. The Clay Ratio as a Criterion of Susceptibility of Soils to Erosion. *Jour. Am. Soc. of Agronomy*, Vol. 27, pp. 738–741, 1935.

Slater, C. S., and Byers, H. G. A Laboratory Study of the Field Percolation Rates of Soils. U. S. Dept. Agr. *Tech. Bull.* 232, 1931.

[2] A few of the more important articles dealing with the effect of organic matter on soil structure are as follows: Baver, L. D. Factors Contributing to the Genesis of Soil Microstructure, Am. Soil Survey Assoc. *Bull.*, Vol. 16, pp. 55–56, 1935. Baver, L. D., and Rhoades, H. F. Aggregate Analysis as an Aid in the Study of Soil Structure Relationships, *Jour. Am. Soc. of Agronomy*, Vol. 24, pp. 920–930, 1932. Browning, G. M. Changes in the Erodibility of Soils Brought about by the Addition of Organic Matter, Soil Sci. Soc. America *Proc.*, Vol. 2, 1938. Burr. W. W., and Russell, J. C. The Importance of Organic Matter in Soil Structure and Tilth, 1st Internat. Cong. Soil Sci. *Abstract Proc.*, Vol. 1, p. 68, 1927. Chapman, J. E. The Effect of Organic Matter on the Tillage of a Clay Soil, 1st Internat. Cong. Soil Sci. *Proc.*, Vol. 1, pp. 443–445, 1928.

Fig. 64.—Fracturing of plastic, impervious clay subsoil of Iredell silt loam as result of desiccation on exposure. (*Photograph by Soil Conservation Service.*)

Fig. 65.—Practically no fracturing of porous, granular clay subsoil, Davidson clay loam, as result of desiccation on exposure. (*Photograph by Soil Conservation Service.*)

indirectly by reason of its influence on the growth of plants. The silica-sesquioxide ratio appears to be the most significant index of erodibility obtainable by chemical analysis. It is an index of the degree of weathering and, as a result, of erodibility. General observation, together with some quantitative data, indicates that highly weathered soils are less susceptible to erosion than young or slightly weathered soils. This ratio is of particular value in comparing soils of the same general locality.

Although no such striking differences have been found in the chemical composition of the soils of the United States as in those of the Tropics, the same general trend exists. The Davidson and Iredell soils of the Piedmont region of the Southeast are among the more outstanding examples of contrasts in susceptibility to erosion of the many types that have been studied from this standpoint. The Davidson, a highly weathered, relatively nonplastic soil, is exceptionally resistant to erosion, whereas the Iredell, an immature soil with an exceedingly plastic subsoil (*B* horizon), is subject to severe erosion on sloping areas. The properties of these soils have been studied intensively.[1] Figure 64, showing the clay subsoil of Iredell silt loam, affords some conception of the plastic nature of the material in the severe fracturing produced by the shrinkage on drying. The profile of Davidson clay loam (Fig. 65), on the other hand, shows no shrinkage or cracking as the result of drying. Table 18 shows the sesquioxide ratios of the colloids contained in the *A*, *B*, and *C* horizons of these contrasting types:

TABLE 18.—SILICA-SESQUIOXIDE RATIOS OF COLLOIDS OF IREDELL AND DAVIDSON SOILS

| | *Silica-sesquioxide Ratio* | |
|---|---|---|
| *Horizon* | *Iredell* | *Davidson* |
| *A* | 1.66 | 1.33 |
| *B* | 1.78 | 1.44 |
| *C* | 1.76 | 1.28 |

The important Nacogdoches and Kirvin soils of east Texas and adjacent parts of Louisiana and Arkansas, which are notably different in their susceptibility to erosion (see Table 7), also show considerable contrast in their silica-sesquioxide ratios.[2]

[1] Lutz, J. F. The Physico-chemical Properties of Soils Affecting Soil Erosion, Missouri Agr. Exper. Sta. *Research Bull.* 212, 1934.

Middleton, H. E. Properties of Soils Which Influence Erosion, U. S. Dept. Agr. *Tech. Bull.* 178, 1930.

[2] Middleton, H. E., Slater, C. S., and Byers, H. G. The Physical and Chemical Characteristics of the Soils from the Erosion Experiment Stations (2d rept.), U. S. Dept. Agr. *Tech. Bull.* 430, 1934.

## Effect of Structure

Structure is of primary importance in determining erodibility of a soil. Since it is the result, in large degree, of complex interactions of texture, contained organic matter, and chemical composition, it is difficult to devise satisfactory methods for its direct measurement, although various indirect methods have been used. Determination of the ease and degree of

Fig. 66.—Characteristic erosion of the exposed subsoil of Helena sandy loam. The fine material of the sandy clay subsoil goes readily into suspension and flows away, thus developing a highly unstable or skeletonized condition. (*Photograph by Soil Conservation Service.*)

dispersion of the soil particles, together with the degree of aggregation, probably provides the most useful information on the subject. Measurements of dispersion and aggregation are complementary in that particles of soil not dispersed by water remain aggregated or clustered in granules, which are favorable to infiltration of rainwater. Various methods have been described for measuring the degree of dispersion and of aggregation.[1]

[1] Some of the methods for determining degree of dispersion and aggregation are described in: Baver, L. D., and Rhoades, H. F. Aggregate Analysis as an Aid in the Study of Soil Structure Relationships, *Jour. Am. Soc. of Agronomy*, Vol. 24, pp. 920–930, 1932. Lutz, J. F. The Physico-chemical Properties of Soils Affecting Soil Erosion, Missouri Agr. Exper. Sta. *Research Bull.* 212, 1934. Middleton, H. E. Properties of Soils which Influence Erosion, U. S. Dept. Agr. *Tech. Bull.* 178, 1930. Yoder, R. E. A Direct Method of Aggregate Analysis of Soils and a Study of the Physical Nature of Erosion Losses, *Jour. Am. Soc. of Agronomy*, Vol. 28, pp. 337–351, 1936.

Soils containing a large proportion of water-stable aggregates generally are more resistant to erosion than dense or compact soils characterized by dispersed particles. Higher velocities of flow are necessary to suspend and move the aggregates than for transporting the lighter individual particles (of which aggregates are composed). Figure 66 shows the type of erosion that may be expected on nonaggregated or easily dispersed soils.

Figure 66, showing the subsoil of Helena sandy loam (an important type of the southern Piedmont region), illustrates the readiness with

FIG. 67.—Gully erosion in Cecil clay loam. Note undercutting of the *B* horizon, where water has attacked the unconsolidated *C* material, and resultant caving. (*Photograph by Soil Conservation Service.*)

which the fine colloidal material goes into suspension and flows away to leave a honeycombed or skeleton section of unconsolidated sandy material which offers very little resistance to erosion. In contrast, the clay subsoil of the associated Cecil soils is more granular and resistant to concentrated flow of water. However, it is underlain usually by a third layer (*C* horizon) of soft weathered rock (gneiss or granite), which erodes very rapidly where exposed, undermining the overlying clay (the *B* horizon) and resulting in caving, as illustrated in Fig. 67.

As previously explained elsewhere, when muddy water penetrates a soil, the suspended particles are strained out near the surface and lodge in the pore spaces of the soil. Thus, the openings are choked, and infiltration is reduced. The greater the proportion of rainfall that may be induced to enter the soil the smaller the quantity of water remaining to run off the surface and cause erosion. For this reason, soils of open structure that do not shrink or crack very much with drying or wetting and are not dispersed readily have a greater capacity for intake of water and, accordingly, are less susceptible to erosion.

Porosity of soil as ordinarily measured may be of little value in estimating susceptibility to erosion. If the difference between the capillary and noncapillary porosity of a soil[1] is known, however, susceptibility to erosion can be measured more accurately. Unfortunately, few data showing the relation of noncapillary pore space to erodibility[2] and infiltration are available. It is evident from the available information, however, that soil with a large volume of noncapillary pore space has a high rate of infiltration and a low susceptibility to erosion.

The rate of percolation of water through soil profiles has been the subject of much study. In general, no significant relation has been found between results obtained with disturbed soil samples in the laboratory and soils of undisturbed natural structure in the field.[3] Comparative results obtained in the laboratory are of value in a qualitative estimation of erodibility.

Methods and apparatus have been developed with which it is possible to measure directly in the field the rate at which water enters a soil profile.[4] Such measurements furnish direct comparisons of the infiltration rates of soils with undisturbed structures. They are also of great value in comparing the erodibility of various soil types. Since the apparatus is readily portable, these fundamental data may be obtained from a large number of soil types in a comparatively short time.

Structure is not always constant throughout the extent of a soil; at any one location, it may vary with seasonal changes and cropping procedures. For this reason, it is difficult to evaluate the erodibility of a soil on the basis of any one index. Plot experiments have shown that erosion from a soil type at a given location may vary from a fraction of a ton to 150 tons or more per acre annually, depending on the cultural treatment of the soil. For this reason, an index is of little value in comparing soil types that

[1] Baver, L. D. Soil Porosity as an Index of Structure, Am. Soil Survey Assoc. *Bull.*, Vol. 14, pp. 83–85, 1933.

[2] Baver, L. D. Soil Characteristics Influencing the Movement and Balance of Soil Moisture, Soil Sci. Soc. America *Proc.*, Vol. 1, pp. 431–437, 1937.

[3] Slater, C. S., and Byers, H. G. A Laboratory Study of the Field Percolation Rates of Soils, U. S. Dept. Agr. *Tech. Bull.* 232, 1931.

[4] Musgrave, G. W. The Infiltration Capacity of Soils in Relation to the Control of Surface Runoff and Erosion, *Jour. Am. Soc. of Agronomy*, Vol. 27, pp. 336–345, 1935.

Musgrave, G. W., and Free, G. R. Preliminary Report on a Determination of Comparative Infiltration Rates on Some Major Soil Types, Am. Geophys. Union, *Trans.*, Part II, pp. 345–349, 1937.

are not immediately adjacent or, at least, in the same general locality where the factors that affect erosion are constant or nearly so.

Measurements of soil and water losses from two soil types under nearly identical conditions have been made at the soil and water conservation experiment station near Tyler, Tex. Some of the field and laboratory data pertaining to these contrasting soils are shown in Table 19.

The Nacogdoches fine sandy loam was less than half a mile from the Kirvin area. The principal difference, other than soil character, was the slightly greater slope of the former (10 per cent, as against 8.75 per cent). Nevertheless, where the topsoil remained, erosion from Kirvin fine sandy loam under cotton culture was three times greater than that from Nacogdoches fine sandy loam; under grass, it was approximately nine times greater. Where subsoil was exposed, erosion from the Kirvin was more than twice that from the Nacogdoches. The differences in the water losses from the plots were not so striking, but the losses from both surface soil and subsoil, respectively, were higher from the Kirvin fine sandy loam.

TABLE 19.—COMPARISON OF SOIL AND WATER LOSSES AND OF VARIOUS PHYSICAL AND CHEMICAL PROPERTIES[1] OF KIRVIN AND NACOGDOCHES FINE SANDY LOAMS AT TYLER, TEX., 1932–1934.[2]

| Soil | Crop | Run-off, per cent | Average annual erosion, tons per acre | Average annual yield, pounds per acre | Dispersion ratio | Pore space, per cent by volume | Texture, per cent | | | Silica-sesquioxide ratio |
|---|---|---|---|---|---|---|---|---|---|---|
| | | | | | | | Sand | Silt | Clay | |
| Kirvin fine sandy loam......... | Cotton | 21.3 | 19.7 | 323 | 37.7 | 35.7 | 75.1 | 15.5 | 8.5 | 2.02 |
| Nacogdoches fine sandy loam.. | Cotton | 16.8 | 6.5 | 338 | 28.2 | 43.4 | 69.7 | 9.8 | 18.0 | 1.07 |
| Kirvin fine sandy loam......... | Grass | 1.0 | 0.06 | | | | | | | |
| Nacogdoches fine sandy loam... | Grass | 0.76 | 0.007 | | | | | | | |
| Kirvin subsoil................. | Cotton | 26.2 | 66.7 | 60 | 13.4 | 53.4 | 30.9 | 7.3 | 60.9 | 1.74 |
| Nacogdoches, subsoil........... | Cotton | 21.6 | 31.7 | 10 | 13.8 | 48.3 | 45.7 | 4.9 | 48.4 | 1.21 |

[1] Data from *Tech. Bulls.* 316 and 430. U. S. Dept. Agr.

[2] Average annual rainfall 42.06 inches on Kirvin; 42.6 inches on Nacogdoches. Respective slopes, 8.75 and 10 per cent.

Laboratory data indicate a much greater susceptibility to erosion on the part of the surface soil of the Kirvin fine sandy loam. Its soil material is dispersed more readily than that of the Nacogdoches and has less porosity, indicating lower infiltration capacity (as actually shown by the runoff data). The higher content of clay in the Nacogdoches, together with its low silica-sesquioxide ratio and higher content of organic matter (1.89

as against 0.54 per cent), is indicative of a much more erosion-resistant structure than that of the Kirvin.

With respect to the subsoil of these types, laboratory findings are not so conclusive as for the surface soils. It is difficult to determine, however, whether the distinction is due to differences in the characteristics of the soil and subsoil or to differences in the amount of cover contributed by the crops grown. The relatively low silica-sesquioxide ratio of the Nacogdoches subsoil indicates that it has a more favorable structure in relation to susceptibility to erosion than the subsoil of the Kirvin.

# Chapter VII. Climate and Soil Erosion

## The Development of Climatology as an Applied Science

Climate has always directly and indirectly influenced man's mode of living. The farmer, the navigator, the craftsman, the laborer, all men in all cultures have always been acutely aware of the bearing of "the weather" on their lives. It was not, however, until men began to travel widely and to write accounts of their journeys that it was recognized that climate varies from place to place and that along with this variation a corresponding differentiation in plant and animal life occurs, as well as in the specific techniques and devices that man employs in obtaining food, clothing, and shelter. The Greeks were well aware of these differences. Thus, Ptolemy divided the world as he knew it into "climate zones," pointing out that as one went southward it became warmer, whereas as one moved northward from Alexandria the winters became longer and colder until finally a land would be reached where there was no summer.

With the period of exploration and migration, these early ideas were revised greatly, but the relationships between climatic and other elements were still expressed only in a general way. Not until the eighteenth century, when such meteorologic instruments as the thermometer and barometer were invented, did scientists begin to express these relationships in quantitative terms; and it was not until the end of the nineteenth century, by which time weather records had begun to accumulate, that generalizations were made. Even today, we are seriously hampered by the paucity of instrumental data: The period of observation is short, and there are too few stations providing climatic records. Nevertheless, certain significant principles have been developed, and it has become more and more apparent that an understanding of climate, as it varies from region to region, is extremely important.

The early ecologic work of such investigators as Von Humboldt[1] and Wallace[2] has been revitalized and strengthened through the application of

[1] Humboldt, Alexander von. "Personal Narrative of Travels to Equinoctial Regions in the New World, 1799–1804." London. 1822. "Views of Nature." London. 1849.

[2] Wallace, Alfred Russel. "Man's Place in the Universe." New York. 1903. "The Geo-

climatic data to problems of plant distribution. Similarly, owing largely to the work of such Russian pedologists as Dokutschajeff[1] and Glinka,[2] the significance of climate in soil genesis and distribution has been recognized and given quantitative expression. Finally, geomorphologists, dealing with the evolution of land forms and the physiographic processes that mold them, have been led to conclude that different climatic regimes find expression in correspondingly different minor land forms and in different erosion complexes.

Soil erosion is intimately related to all of these factors: vegetation, soil, land form, and physiographic processes. And, more directly, it is related to the climatic factors themselves: to wind velocity, rainfall intensity, temperature, and many other such elements. Thus, whether it is natural or man-induced, erosion finds different expression from climate to climate, as to both the forms that it assumes and the rate at which it occurs. Soil conservation problems must, therefore, be approached with a clear knowledge of the climatic environment.

## The Climatic Complex

Everyone knows what "climate" is, but it is still not an easy word to define. Perhaps the most satisfactory definition is that complex of meteorologic conditions which exists in any given area and imparts an individuality to the landscape of that area. This is not a rigorous definition, but it conveys the important idea that climate is not an abstract term but a set of physical conditions directly related to the distribution of natural phenomena.

Among the numerous meteorologic elements are precipitation, temperature, relative humidity, evaporation, fog, wind, and air pressure. The meteorologist is interested in all of these, for as a physicist he is concerned with all the dynamic relationships that exist in the atmosphere. But the climatologist is principally interested in those climatic factors which play the most important role in influencing plant distribution, animal distribution, erosional processes, and the activities of man. Air density, for example, is a significant element to the meteorologist; but the climatologist is mainly concerned with such factors as rainfall and temperature which are of more immediate significance.

Precipitation, evaporation, temperature, wind, and relative humidity are the most important climatic elements. Comparing conditions in the southeastern United States with those of the southern Great Plains, the

graphical Distribution of Animals, with a Study of Living and Extinct Faunas as Elucidating the Past Changes of the Earth's Surface." London. 1876.

[1] Dokutschajeff, V. V. Discussion *in* Nat. Hist. Soc. St. Petersburg *Proc.*, 10, 1879.

[2] Glinka, K. D. "The Great Soil Groups of the World." Trans. from the German by C. F. Marbut. Ann Arbor, Mich. 1927.

significance of these climatic factors becomes immediately apparent. In the Southeast, summers are hot and long, winters mild and short; rainfall is moderate to heavy throughout the year; relative humidity is high, particularly in the summer when it is "sticky" or "muggy"; and wind velocities are abnormally high during the hurricane season. Correspondingly, forest constitutes the natural vegetation; soils are deep and slightly acid; streams are numerous; and hills are gently rounded. Also, agriculture thrived in the region during Indian time as well as during the period following white settlement. On the southern Plains, summers are long and hot, winters moderately cold; rainfall is low and mainly concentrated in the summer; evaporation is high, especially in summer; and wind velocities are high throughout the year. Short-grass is the natural vegetative type; soils are only moderately deep and are usually neutral to slightly alkaline; streams are few; and where hills occur at all, they tend to display sharp angles. Until the advent of mechanized agriculture, grazing was the basis of the regional economy.

In these contrasting climatic environments, soil erosion progresses at different rates and in different ways. In the Southeast, gullying and sheet washing cause the greatest damage; in the southern Plains, wind erosion is the principal hazard.

This comparison emphasizes the fact that the climatic elements are real and significant quantities. By determining the quantitative values of these factors from region to region and their relationship to biotic and geomorphologic conditions, it should be possible to develop a number of principles that will have practical meaning in soil conservation.

## Climatic Classification

Designation of the climate of any given region as a certain type implies more than mere reference to the amount of rainfall, temperature characteristics, relative humidity, etc. For example, the wet, hot regions of the tropics are associated with that whole assemblage of plant-, animal-, soil-, and land-form characteristics typical of this kind of climate. The relationship of these natural elements to climate is both immediate and once removed: Climate acts directly on them and, since they are interdependent, acts on them indirectly, also.

In the instance of plants, it is most convenient to think of the different species, genera, and families as competing with one another to survive under the existing climatic and soil conditions. Certain plants—called xerophytes—can stand drought much better than others. Therefore, xerophytes are found in areas where rainfall is very low or where evaporation is so high that what rain does fall largely passes rapidly back into the air. Plants also have a variety of responses to temperature. For example, the coffee plant cannot live where temperatures fall below the freezing point. Thus it is that in any given climatic region, certain groups of plants will thrive better than any other kind and will become or tend to become dominant. Such a plant complex is called *climax* vegetation. Every type of climate has its own particular climax vegetation.

In addition to determining the optimum conditions for any particular kind of vegetation, climate sets certain limits beyond which plants cannot penetrate even as subclimax types. Such trees as the southern cypress are limited to warm, moist climates. Red oak, on the other hand, is restricted to cool regions.

Climate also has a profound influence on the distribution of soils. This can best be understood by considering the way in which soil is formed.

FIG. 68.—Rounded topography characteristics of humid climate land forms. Limestone Valley Country, Maryland. (*Photograph by Soil Conservation Service.*)

Through chemical and physical weathering, the rock is broken up to form an unconsolidated mantle. Climatic conditions determine to a large extent how rapidly and in what ways this primary weathering process takes place. As soon as a few inches of fragmented material have accumulated, the climate begins to participate in actual soil genesis. If the rainfall is high and the evaporation low, water percolates down through the soil and carries material in suspension and solution from the top downward into the lower portions of the soil. With high evaporation rate and low precipitation, this leaching process does not take place. Instead, another kind of differentiation occurs in which evaporation has the effect, apparently, of drawing the basic salts upward in the soil to contribute to the formation of a horizon of carbonate accumulation.

Temperature is also important; for if the temperatures are high, then chemical action will proceed rapidly, and certain changes occur that could not take place at lower temperatures. As plants begin to grow on the soil cover, they will also influence the soil, since they remove certain constituents from it, and since on decaying they add humus to the soil. But the type of plant that will grow depends to a large extent on the climate; so the climate also indirectly influences the soil type that is formed. When soils have been acted upon by the same climatic conditions for a long time, they are said to be *mature*. A mature soil, therefore, reflects the climatic condition just as does a climax type of vegetation.

Fig. 69.—Angular, vertical-walled land-form characteristics of much dry climate country. Arizona. (*Photograph by Soil Conservation Service.*)

In surface configuration is found another example of the influence of climate on the landscape. Gentle, rounded slopes (Fig. 68) are characteristic of humid areas where the vegetative cover protects the land against violent erosion by running water and wind. Sharp angles—deeply entrenched arroyos, cliff-bounded mesas, "pedestal" rocks—are typical of arid regions (Fig. 69) where the water runs off rapidly because of the lack of natural vegetation and where the wind, carrying sand particles as tools, abrades the bare rock surfaces and scours the basin floors. Between these two extremes exists a variety of intermediate conditions, all of which are, in some measure at least, related to climate.

Finally, types of soil erosion vary from climate to climate. Gullying and sheet wash are the mechanisms of accelerated erosion in humid lands.

Wind is effective chiefly in the arid and semiarid climates. Frost action is at a maximum where the temperature crosses the freezing point the greatest number of times annually. Ice action is mainly limited to *polar, tundra, taiga,* and *microthermal* climates.[1]

A climatic classification is valuable in so far as it defines climatic types in such a way that the relationships and correspondences are emphasized and distinctly brought out. Two such classifications, Köppen's and Thornthwaite's, are useful in this connection. The first, that of Köppen,[2] was presented in 1900 and has been revised many times since. The second, Thornthwaite's[3] classification, follows Köppen's general scheme but introduces certain refinements. Both illustrate the regional relationship of climatic factors to plant, soil, and erosion characteristics. Thornthwaite's classification lays greater emphasis on moisture relationships and is, therefore, more applicable to erosion studies.

It is recognized that although two areas may receive the same amount of precipitation annually, or even monthly, one may still be considerably drier than the other. Thus, the *effective precipitation* (that part of precipitation remaining available for plant use) is greater at Fargo, N. D., than at Amarillo, Tex., although the actual annual precipitation is the same at both places. At Amarillo, the amount of evaporation is much greater than that at Fargo, so that the same amount of precipitation is not so effective at Amarillo as it is at Fargo. Thornthwaite has devised an expression involving monthly temperature and rainfall figures that serves as an index of effective precipitation. This he terms the *PE* index; and since this index is a much more sensitive indicator of variations in plant and soil distribution than are simple precipitation figures, it is used as the quantity denoting usable moisture.

The other index concerns temperature and is abbreviated *TE.* This was arrived at by summing the monthly temperature indices, the total indices thus obtained being arranged and evaluated on the basis of plant distribution. The index is of practical application, since the rate of chemical reactions that take place in plants and in the soil varies with the temperature. Such variations accord with definite types of vegetation and soil.

[1] Polar climate is characterized by perpetual snow and ice. Under tundra climate, there is sufficient warmth to produce such low vegetation as reindeer moss and stunted willow. Taiga climate is typified by cold winters, cool summers, and vegetation consisting of forests of spruce and fir or their equivalent. The term microthermal applies to climatic regions having cold winters and mild or warm summers, for example, New England.

[2] Köppen, Wladimir. Versuch einer Klassifikation der Klimate, vorzugsweise nach ihren Beziehungen zur Pflanzenwelt, *Geogr. Zeitschr.*, Vol. 6, pp. 593–611, 657–679, 1900. Köppen, Wladimir, and Geiger, R. Klimakarte der Erde (Scale: 1:20,000,000), Gotha, 1928.

[3] Thornthwaite, C. Warren. The Climates of North America According to a New Classification, *Geog. Rev.*, Vol. 21, pp. 633–655, 1931.

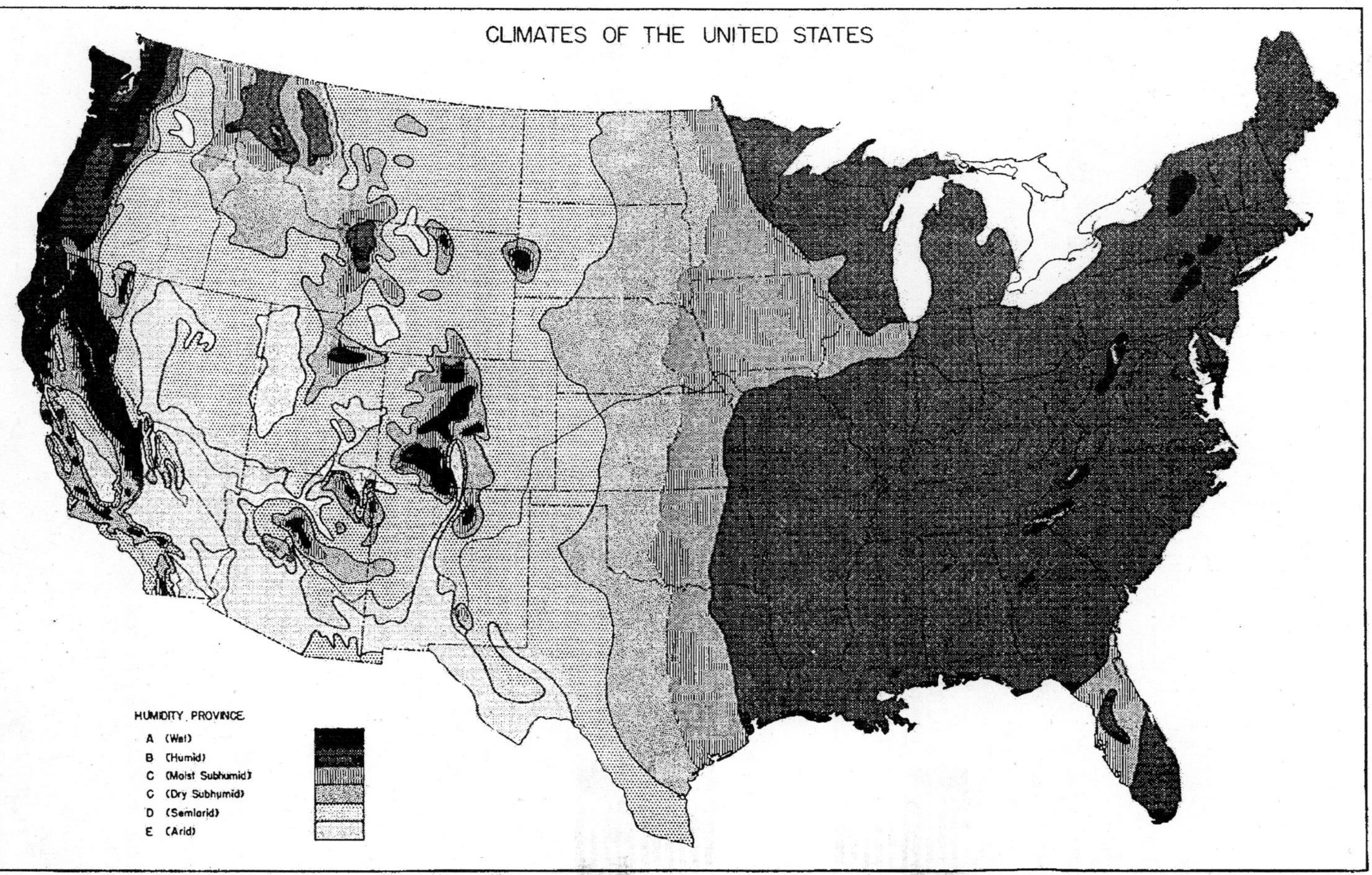


FIG. 70.—Distribution of climates of the United States. (*After Thornthwaite.*)

On the basis of these two indices, and with constant reference to plant and soil conditions, five moisture provinces have been determined: wet, humid, subhumid, semiarid, and arid. Similarly, six temperature provinces have been recognized: tropical, mesothermal (hot summers and mild winters, as in southeastern United States), microthermal, taiga, tundra, and polar. Another distinction is made on the basis of seasonal variations in effective precipitation. The notation *r* is used to indicate climates in which a surplus of moisture is available for plant growth throughout the year; *d* designates those climates characterized by moisture deficiency throughout the year; and *s* refers to summer-dry and *w* to winter-dry climates. In this way, climates are identified with reference to three factors: *PE*, *TE*, and seasonal rainfall distribution. For example, there is a subhumid mesothermal summer-dry climate, such as is found in central California; a subhumid microthermal climate deficient in moisture throughout the year, such as exists in the northern Great Plains; and a humid tropical climate with an adequate moisture supply throughout the year, which in this country is found only in the southernmost part of Florida. In the case of taiga, tundra, and perpetual frost climates, no moisture term is used, since the temperatures are so low that they alone are of chief significance. The distribution of climates in the United States is shown in Fig. 70.

The applications of this classification are numerous and of practical importance. This system, for instance, makes it possible to study the climatic fluctuations from year to year in any area. Thornthwaite[1] made such a study with reference to the problem of drought in the Great Plains. The Plains region lies mainly along the semiarid and subhumid boundary, and therefore even slight variations in climate are critical as regards agriculture. The climatic maps for 1910 and 1923 illustrate this variability (Fig. 71). In 1910, the Texas Panhandle experienced an arid climate, whereas in 1923 it was moist subhumid. Fluctuations of like magnitude in the southeastern United States would not be nearly so critical, for even in the drier years the *PE* index would be sufficiently high to permit a fair growth of crops.

Bodman[2] has applied the Thornthwaite classification to the study of soils, and Halliday[3] has shown its application to the distribution of vegetation in Canada. In numerous instances, the classification has proved

[1] Thornthwaite, C. Warren. "The Great Plains." Carter Goodrich *et al.*, "Migration and Economic Opportunity." pp. 202–250. Philadelphia. 1936.

[2] Bodman, G. B. The Forest Floor Developed under Conditions of Summer Rainfall Deficiency in a Californian Pine and Fir Forest, Am. Soil Survey Assoc. *Bull.* 16, pp. 97–101, 1935.

[3] Halliday, W. E. D. A Forest Classification for Canada, Dominion of Canada Forest Service *Bull.* 89, Ottawa, 1937.

directly applicable to agricultural, ecologic, and pedologic problems; and because of these relationships, it is a useful tool in solving soil erosion and water conservation problems.

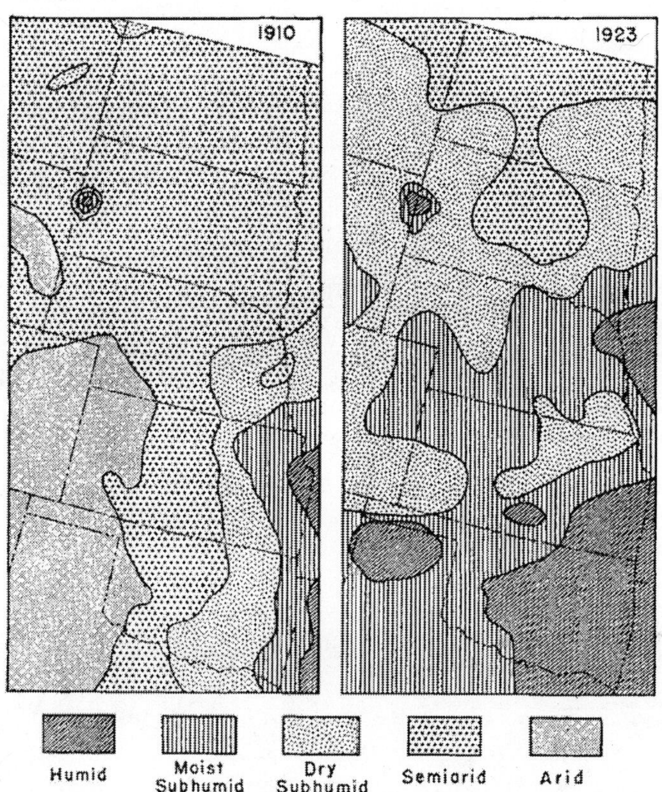


Fig. 71.—Climatic patterns for 1910 and 1923 illustrating the variability in climate in the Plains region. (*After Thornthwaite.*)

## The Relation of Individual Climatic Factors to Erosion

Precipitation. Through climatic classification, it is possible to establish the general setting within which particular soil erosion problems must be placed. It is also useful to consider certain specific climatic factors in their relation to erosion and conservation. Foremost of these climatic elements is precipitation. Since all mechanisms of denudation except wind and gravity movements involve water in either the liquid or the solid state, there is a very direct relationship between precipitation and erosion. The amount of water carried by streams or gullies that cut headward or laterally to erode the land is dependent on the amount of precipitation. Similarly, the amount of precipitation determines the amount of water that can seep into the soil to aid in mass movement or the amount that will flow over the surface to erode the land. The kind of erosion that occurs and the amount of soil removed depend largely on the characteristics of the rainfall.

Unfortunately, climatologists have concerned themselves mainly with precipitation in terms of monthly and annual means. This has been due partly to the fact that data have been collected in such a form that more refined statistics could not be readily isolated and partly to the fact that these mean quantities have been considered more significant than they really are. Actually, such averages tend to mask the true rainfall characteristics of a particular area. It is the quantitative characteristics of individual storms that are related to soil erosion and floods, not average quantities. The latter are really only a composite of these individual events. For example, two stations may each record 5 inches of precipitation in a given month; but if one station received this amount in one 5-hour storm, whereas the other experienced 25 storms of 0.20 inch each, the two situations would be actually quite different. In the area where only one storm occurred, the erosion damage will have been much greater than in the area where twenty-five 0.20-inch storms occurred.

Since the storm is the unit with which it is necessary to work, it is essential to have detailed information about rainstorm types. Such investigations as those now being carried out at the Muskingum Climatic Research Center[1] will provide the primary material needed. Thus, the data from closely spaced rain gages maintained at the Oklahoma Climatic Research Center have formed the basis for the detailed study of rainstorm morphology.[2] The material collected was used in the preparation of precipitation maps, both discrete and cumulative, plotted at 15-minute intervals from simultaneous observations. For the first time, concrete evidence was produced as to the actual degree of "spottiness" of rainfall (Fig. 72). It was strikingly shown that in any given storm, marked intensity differences may appear within short distances. Two types of storm pattern sequences were recognizable: General rains of light to moderate intensities seemed to invade an area rather uniformly, whereas more intense local storms were characterized by a much more spotty distribution, which frequently resulted in high precipitation at two stations only 10 to 20 miles apart but left the intermediate area without even a trace of rain.

The recognition of the nature of storms throws a great deal of light on the actual storm conditions that produce different flood and erosion situations. Storm centers of maximum intensity are relatively small in

[1] Thornthwaite, C. Warren. Microclimatic Studies in Oklahoma and Ohio. *Science*, Vol. 86, No. 2222, pp. 100–101, July 20, 1937.

[2] Thornthwaite, C. Warren. The Life History of Rainstorms, progress report from the Oklahoma Climatic Research Center, *Geog. Rev.*, Vol. 27, pp. 92–111, 1937.

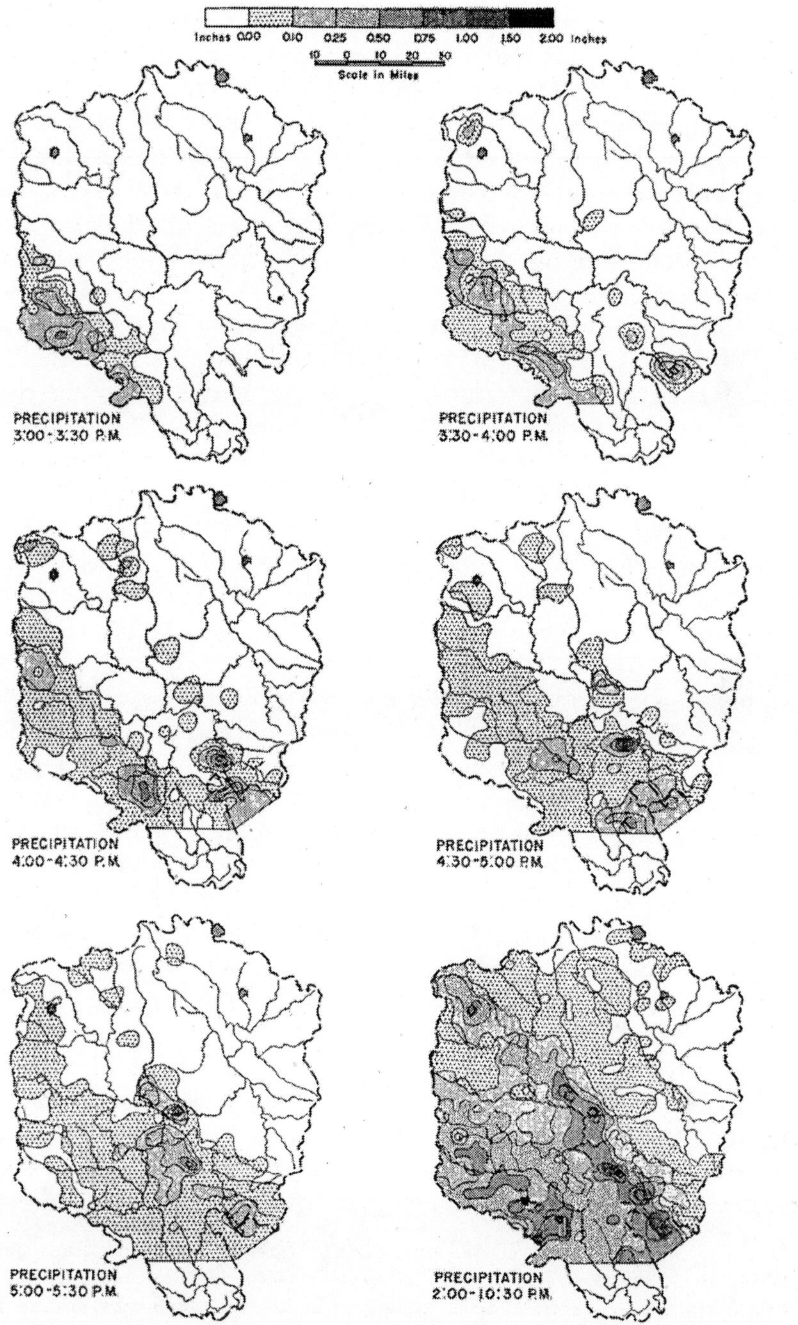


Fig. 72.—Life history of a rainstorm on the watershed of the Muskingum River, Ohio.

area. Their path is irregular; and where they strike, more damage may be done by a single storm than has been caused in several previous years of rainfall. For example, the storm of May 22–24, 1908, in eastern Oklahoma, produced 20 per cent of that year's precipitation, and those areas affected by the storm center suffered the most serious erosion ever experienced.[1] In this particular year, the rainfall in eastern Oklahoma was almost twice the mean annual precipitation; but if the amounts recorded in three storms were subtracted, there would have been no excess above the average. This instance provides a striking illustration of the statement that the individual storms are the true natural units which must be considered in any treatment of floods and erosion.

Serious floods on large watersheds are invariably caused either by general storms or by a combination of general and local precipitation. The Miami flood of 1913[2]; the Pennsylvania floods of March, 1936[3]; the Brazos River flood of 1899[4]—all were produced as the result of extensive rainfall of long duration. Small basins, on the other hand, are more likely to suffer from intense local showers which happen to be centered over the drainage area. In one such case near Pullman, Wash., a thundershower lasting less than an hour and a half removed an average of 0.57 ton of soil per acre from the watershed, and the loss was "many times this amount on the bare, fallow fields."[5] It is estimated that such local storms produce, in the aggregate, more damage than the widespread, more spectacular storms.[2]

Although only detailed rainfall maps can yield a true picture of time-area-intensity relationships of any particular storm, much can be gained through careful analysis of precipitation records from stations having recording rain gages. As a recent study[6] has shown, it is possible to determine the expectancy of various rainfall intensities and durations and the expectancy of periods without precipitation of any given length. Such specific climatic elements, critical in soil erosion, can be studied on an actuarial basis, and the climatic risk can then be quantitatively expressed.

[1] McDonald, Angus. Erosion and Its Control in Oklahoma Territory, U. S. Dept. Agr. *Misc. Pub.* 301, 1938.

[2] Houk, Ivan E. Rainfall and Runoff in the Miami Valley, Miami Conservancy District *Tech. Repts.*, Part VIII, Dayton, Ohio, 1921.

[3] The Floods of March, 1936, in Pennsylvania, Pennsylvania Dept. Forests and Waters, Div. Hydrography, Harrisburg, 1936.

[4] Hathaway, Gail E. "Special Report on Hydrologic Studies for the Possum Kingdom Project, Brazos River Basin, to the District Engineer." U. S. Engineer's Office, Mineral Wells, Tex. 1937.

[5] McGrew, P. C., and Horner, G. M. Soil and Water Conservation Investigations, Pacific Northwest Soil Cons. Exper. Sta. *Progress Rept.*, 1931–1935, Pullman, Wash., 1937.

[6] Blumenstock, David I. Rainfall Characteristics as Related to Soil Erosion, U. S. Dept. Agr. *Tech. Bull.* 698, in process.

Rainfall intensity is closely related to the amount and type of erosion. The amount of soil removed by sheet wash and the rate of headward gully cutting are largely dependent on the rate of rainfall. Gentle precipitation tends to cause mass movement; intense rainfall results in more rapid soil removal through the operation of running water.[1] Precipitation intensities are highest in summer and lowest in winter over most of the United States. Since, however, certain parts of the West Coast receive mostly winter precipitation, the generalization does not hold in such areas. Also, in the mesothermal subhumid climate of the southern prairies region, spring precipitation intensities are of the same order of magnitude as those experienced in the summer. It is necessary to know the rainfall regime in terms of seasonal intensity variations in order to adjust conservation and cropping practices to provide adequate protection for the land during those months which are most likely to experience high-intensity precipitation.

Rainfall duration is the complement of rainfall intensity. Together, duration and intensity determine rainfall amounts. Since the most intense showers are usually local and of short duration, whereas the lighter falls are associated with prolonged warm-front general rains, there is a general inverse relationship between rainfall intensity and rainfall duration. Correspondingly, although intensities are usually highest during summer and lowest during winter, rainfall durations are generally shortest in summer and longest in winter. Important quantitative differences obtain, however, from region to region; and these are particularly important in their bearing on the flood hazard. Floods on large watersheds generally are associated with rains of long duration, and, in those areas where snow melt may contribute to the stream flow, it is essential to know the spring rainfall characteristics not only with respect to intensities that may be expected but also with reference to rainfall duration.

Since drought leads to crop failures and consequent soil exposure, and since it is impossible to effect erosion-prevention measures without including land-use adjustment in general, the frequency of rainless periods of varying length and at different seasons comprises an essential climatic index in soil conservation work. The drought problem is, of course, most acute in the subhumid and semiarid parts of the country. Dust storms and crop failure have called attention particularly to the Great Plains region. The eastern portion of the Plains is less subject to drought than that part farther west, but this is not the only comparison to be made. There are seasonal contrasts also. Along the eastern border, the length of rainless intervals is less in spring than in summer, rainfall being moder-

[1] Ireland, H. Andrew, Sharpe, C. F. Stewart, and Eargle, D. Hoye. Principles of Gully Erosion in the Piedmont of South Carolina, U. S. Dept. Agr. *Tech. Bull.* 633, 1938.

ately reliable during both seasons. Farther west, however, as in the Texas Panhandle, spring becomes a particularly hazardous season, whereas summer is the most reliable season for precipitation. Such seasonal shifts in drought hazard from season to season must be taken into consideration in adjusting land-use practices in order to decrease soil erosion.

WIND. That wind is an active agent of erosion is well-known. However, it is customary to think of wind action as being limited largely to semiarid and arid regions. Although it is true that the most spectacular and widespread evidences of wind scouring and deposition are found in these regions, the wet and humid areas also are subject to wind erosion. Throughout the world, land without a protective cover is subject to some degree of alteration by wind. Soil particles are mixed by shifting or swirling air currents; the coarser grains and fragments or clusters of soil are driven across open country by a process of rolling or short leaps (saltation); the finer material is transported in suspension over long distances. Coming to rest, the wind-driven coarser soil particles and fragments develop into sand dunes, clay dunes,[1] and vast expanses of sandy waste. Even solid rock is sculptured through the action of sand blast.

The effectiveness of wind in causing erosion is dependent on the velocity and amount of turbulence and on the degree to which the soil is protected. If the land is bare, even winds of low velocity and without any appreciable vertical component may remove a great deal of material. If the vegetation is profuse, wind velocities must be of hurricane magnitudes to produce soil loss. In the Great Plains, wind erosion has been especially active and spectacular, because velocities are high, and much soil has been exposed through plowing. Drought, particularly in the spring, has added to the wind erosion hazard. Under such circumstances, crops have failed to mature after planting, and the soil has remained exposed at that particular season when velocities are the highest and when turbulent movement in the atmosphere is most common.

Arid regions suffer excessively from wind erosion, because the climax vegetative types do not offer sufficient protection to the soil. Here, however, the process sometimes operates even under natural conditions, whereas in such semiarid regions as the Moenkopi Plateau, in northern Arizona, the damage is largely the result of the activities of man—in this case, overgrazing.

Wind, then, is potentially an agent of erosion in all climates. Ordinarily, it acts in humid areas only in soil mixing through the moving of soil particles short distances along the ground. If the land is bared, however, excessive wind erosion may develop. In arid areas, the difficulties

[1] Soil Survey, Cameron County, Texas. Bur. Chemistry and Soils, U. S. Dept. of Agr., (Series 1923) No. 17, pp. 552, 566.

of combating wind erosion are very great, since it is often impossible and always hard to establish an effective, protective, vegetative cover. Although the problem is most acute in such dry lands, it is everywhere potentially present. In any climate, the wind when provided with a bare dry soil on which to operate can rapidly impoverish or ruin what ordinarily would be "good land."

TEMPERATURE. Temperature acts in two ways to influence erosion. Through its effect on weathering processes, it participates in the preparation of material for ready transportation by water, wind, or glaciers. More directly, it has a bearing on the rate of movement of soil material downslope through mass movement.

Both chemical and physical weathering involve the temperature factor. Rate of solution and of chemical reactions is directly related to temperature, usually being faster at higher temperatures. In the tropics, this is evidenced by the profound alterations that the soil undergoes and by the depth to which these changes are effective. Deeply weathered soil and complex chemical transformations are characteristic of hot, moist regions; under similar *PE* conditions in cold lands, the weathered soil mantle is relatively thin because of retardation of solution and chemical action.

Physically, temperature is active in disrupting rock surfaces through alternate heating and cooling and through the wedge action of ice formed from water in the interstices and joints of the rock. Similarly, with the freezing of water in the surface soil, *frost heaving* occurs and serves to loosen the soil so that the particles and aggregates can be more easily carried away.

Temperature operates directly in the erosional process through its bearing on mass movement. "In all heating and cooling of material on slopes there is a tendency to greater expansion *toward* the downhill side and greater contraction *from* the uphill side, this producing a gradual downward migration."[1] This action takes place more rapidly in the instance of frost heaving on slopes; and although under all circumstances the process is a slow one, in the aggregate an appreciable and significant amount of material is moved downslope through the influence of temperature.

## Climate and Normal Erosion

It is neither essential nor feasible to check natural geologic erosion, but an understanding of what the natural erosion processes are and how they operate with varying degrees of effectiveness from one climatic area

[1] Sharpe, C. F. Stewart. "Landslides and Related Phenomena." pp. 28–29. New York. 1938.

to another is necessary in order to recognize man-induced erosion and to plan soil conservation measures intelligently. The same mechanisms operate in producing both normal and accelerated erosion; certainly an understanding of the natural operation of these forces is necessary for an appraisal of abnormal situations. "Under natural conditions a partial balance exists. The ground is protected by the vegetative cover. Erosion is retarded, and topsoil develops, very slowly it is true, but faster than normal erosion can carry it away."[1]

The natural erosion balance varies with the climate. In the wet and humid regions where rainfall intensities are high, the ground over large areas is well protected by forest. Tree crowns shield the soil from the direct impact of falling rain; litter on the forest floor impedes the overland flow of water; soils are deep and are capable of holding a large amount of moisture. In subhumid lands, tall prairie grasses replace the forests. Such cover also serves naturally to conserve the soil through the binding action of its roots and through the protection afforded by its foliage. Finally, in semiarid areas, short grasses are highly effective in protecting the soil against excessive erosion by wind or water. Under natural conditions, it is only in desert areas that the climatic elements are able to function with real effectiveness in causing erosion; and even then, desert shrubs are partially able, through their elaborate root systems, to protect the soil.

The climatic zone of greatest natural weakness as regards erosion, therefore, lies on the drier margins of semiarid climate and within arid regions. For although other climatic regions experience rains of higher intensities and longer duration, the vegetative cover is correspondingly dense and partially checks these potentially effective climatic factors.

## Climate and Accelerated Erosion

Under natural conditions, the vegetation and certain properties of the soil itself tend to hold the erosive power of the elements in check. With the stripping of the vegetative cover through the activities of man, however, the most effective natural safeguard is removed, and accelerated erosion sets in. In humid climates, this results in the introduction of erosion processes similar to those characteristic of arid lands. The gently rounded hills normally present in areas of high precipitation tend to give way to steeper slopes and sharper angles. Steep-walled gullies are introduced, and wind becomes effective for the first time.

Although features such as these are formed very slowly in arid regions, in humid areas characterized by heavy precipitation these abnormal erosion forms develop rapidly once the vegetation has been removed—that is, where adequate preventive measures are not taken.

[1] Sharpe, C. F. Stewart. What is Soil Erosion? U. S. Dept. Agr. *Misc. Pub.* 286, p. 1, 1938.

Such activities as lumbering, farming, and grazing may thus shift the zone of maximum erosion into humid climates. Potentially, the climatic characteristics of humid areas are such that erosion progresses at a maximum rate where the land has been bared. Evidence of this is found in southeastern United States, where gullying and sheet washing have completely removed the topsoil from large areas. This soil took centuries to develop under natural vegetative conditions which permitted the accumulation of organic matter. A few decades of land misuse have not only made further accumulation of this kind impossible but resulted in the sluicing away of the exposed soil.

Since baring the soil allows the climatic elements to operate to their fullest capacity, the climatic differential from region to region becomes extremely important in considering accelerated erosion. Where rainfall intensities are highest and rainstorm durations are at a maximum, gullying and sheet washing cause the greatest damage. Similarly, areas experiencing high wind velocities are subject to the maximum wind erosion once natural vegetative checks have been removed. Temperature contrasts from region to region will likewise be reflected more sharply in the abnormal erosion produced. Where temperatures are high, weathering proceeds faster; and under certain conditions, the material may be removed with greater rapidity.

Accelerated erosion is, therefore, an expression of what are normally partially latent erosive forces which vary from region to region and are expressed in the climatic characteristics of those regions. Maltreatment of the land has permitted the forces to operate with abnormal effectiveness. The balance can be reestablished only by understanding the nature of the climatic forces that must be combated and by appreciating the fact that these elements vary so much from area to area that no solution of general applicability can ever be found. Each climatic complex offers a different problem which must be solved separately.

## Climate and Floods

The overflow of streams is a natural phenomenon which always has occurred and which is responsible for the alluvial soils found along their courses. The waterways that nature has provided have led people to settle in valleys occasionally subject to flood damage. Thus, man has imposed upon himself the problem of controlling the flow of streams by building dams and levees.

In order to carry out flood-protection measures, it is necessary to understand the climatic conditions that give rise to floods. Every flood that occurs—whether within a large watershed such as the Ohio or within one of less than 100 square miles in area—is the result of a particular meteorological situation. Floods may result from prolonged rains of moderate intensity or from shorter rains in which the intensity is very high. They may or may not involve the melting of snow that has accumulated throughout the winter months. In certain instances, ice jams may be important in causing local flooding through the backing up of water.

In all situations, the rate of runoff is of critical importance. It depends on the rate and duration of precipitation, the moisture content of the soil, the infiltration capacity of the soil, the height of the ground-water table, and the amount of surface detention by vegetation or irregularities in the ground surface. If the rate of precipitation exceeds the rate of infiltration, runoff will occur whether the soil is saturated or not. Similarly, if the soil is frozen, surface runoff will most likely take place.

The flood hazard in an area should be determined not only with reference to the amounts, intensities, and durations of precipitation to be expected but also with respect to the numerous factors indicated above. Soil moisture deficiency depends not only on rainfall but also on the rate of removal of moisture from the soil through transpiration and evaporation. The rate of accumulation of surface waters is associated not only with the precipitation rate but also with the soil, slope, and vegetation characteristics of the particular basin under consideration. It is important to know when the ground is frozen and to consider temperature relationships, particularly as they affect the melting of snow.

With all these factors interacting in their influence on floods, it is frequently difficult to analyze the flood problem with sufficient precision to know with certainty all the steps that may be needed to afford immediate protection. The problem primarily rests, however, on the fact that climatic conditions are the principal variables and must always receive due consideration in any particular area. The soil and slope characteristics of a basin remain fairly constant. After determining the runoff characteristics of a watershed, the next step, therefore, is to find out what magnitudes of rainfall intensity and duration are to be expected. This involves the consultation of past records and an attempt to predict what future frequencies will be. Maps of the variations in precipitation intensities and amounts to be expected within various periods must be prepared for use in planning flood-control measures. Furthermore, specific storms that have resulted in floods, such as that in the Miami basin in 1913, must be studied carefully with a view toward determining under just what climatic conditions the greatest flood hazard exists.

Through the accumulation of climatic records and increased knowledge of rainstorm morphology, the climatic problems relating to floods will, as time passes, become more and more clear. Moreover, since all flood conditions are intimately associated with climatic situations, the problem of floods and flood control should receive corresponding clarification.

## Climate and Its Relation to Conservation Practices

A conservationist confronted with the problem of planning suitable soil-conserving practices for any area must know what accelerated erosion has already occurred, what constitute further erosion hazards, how the actual and potential conditions can best be combated, and to what degree such problems as moisture conservation are also involved. All these phases of the general problem must be analyzed from the climatic point of view if an adequate solution is to be worked out.

Although in virtually every area in the United States climatic factors have found expression in accelerated erosion forms, it is not to be assumed that they have acted to their fullest extent in any climatic region. The *total climatic hazard* must be comprehended so that preventive as well as curative measures may be inaugurated. In this connection, it is important to appreciate the relationships existing between agricultural practices and incipient erosion damage; that is, the climatic risks associated with a given type of agriculture must be recognized. Seasonal variations in climatic characteristics have a particularly significant bearing on this point. In the Great Plains, for example, although wind velocities are highest in spring, wind is sufficiently high throughout most of the year for soil exposure to constitute a dangerous practice at any time.

Similarly, in the foothills of the Wasatch Range, although 95 per cent of the stream runoff occurs in the spring from melting snow, 85 per cent of the erosion damage results from summer thundershowers.[1] Here it is particularly necessary to remedy those conditions which permit such a rapid summer discharge of rainwater.

In every region, then, the precise nature of the climatic hazard must be taken into consideration. The climatic risk must be ascertained by studying the agricultural practices and the crop calendar with reference to the seasonal variations in this hazard. If the practices are such that they permit the climatic elements to operate with maximum efficiency, the agricultural program should, where possible, be altered. In any case, climatic factors must always receive careful consideration in determining what preventive measures must be adopted.

[1] Forsling, C. L. A Study of the Influence of Herbaceous Plant Cover on Surface Run-off and Soil Erosion in Relation to Grazing on the Wasatch Plateau in Utah, U. S. Dept. Agr. *Tech. Bull.* 220, 1931.

# Chapter VIII. Infiltration in Relation to Runoff, the Erosion Process, and the Utilization of Rainfall

When water comes in contact with sloping land in quantity exceeding immediate surface evaporation, part or all of it normally enters the soil. If the amount exceeds the immediate intake capacity of the soil, the excess flows along the surface as runoff, provided impounding obstacles are not present. It is the runoff that produces soil erosion.

Where liberal amounts of water enter the soil, part proceeds downward to the ground-water level (water table), frequently beyond the reach of ordinary plants; another part, retained within the upper horizons of the soil, is partly utilized by plants and finally lost in the form of transpiration. Still another fraction is lost as direct evaporation from the ground surface. The proportion of the total precipitation that is finally disposed of in one or more of these forms (runoff, evaporation, transpiration, percolation, absorption) is greatly affected by the quantity and intensity of rainfall, by the soil type, slope, kind and density of vegetative cover, surface roughness, and such artificial barriers to off flowage as contour ridges, terraces, and water traps, as well as several other factors, such as soil and air temperatures and wind velocity.

Obviously, if it were possible to cause all precipitational water to enter the soil somewhere near its point of contact with the land, there would be no runoff and, therefore, no water erosion problem. Although this is commonly impracticable on steep land in regions of high rainfall, nevertheless one of the first objectives in any comprehensive soil and water conservation program is to manage the land so that the largest possible portion of the rainfall enters into the body of the soil (the profile of soil, subsoil, and substratum, as illustrated by Fig. 43, Chap. IV, Part 1).

An indication of the large practical possibilities of soil intake of rainfall is seen in the vast storage space for water that is contained in the soil itself. If the water-holding capacity of all soils is assumed to be 33⅓ per cent of the volume (soils not uncommonly have a 50 per cent or greater capacity), a 3-foot soil section would contain, if filled, 12 surface inches of water. Even a surface foot of soil would have a theoretical storage capacity of around 4 inches. Achievement of water intake up to the practical water-holding capacity of a soil will depend primarily on surface and subsurface conditions of the land favorable to rapid infiltration; the capacity for retention of such infiltered water will depend principally on subsurface conditions unfavorable to excessive percolation.

## Water Intake

The movement of water from the surface into the soil profile by way of openings (natural pores, cracks, root and animal holes, and cavities introduced by tillage) is known as *infiltration*. That part of such intake which moves down through the soil and sublayers to the water table or to the subterranean water level[1] or onward by way of porous beds or underground openings to surface outlets, as springs or seeps, is defined as *water of percolation*, or *gravitational water*. In the practical sense, percolation very largely has to do with the water that moves through openings large enough to favor relatively free flowage by overcoming surface tension under the pull of gravity. For the most part, water thus passing into and through a soil moves as through a filter, leaving behind the suspended material.

*Absorption*, or *sorption*, has to do with that part of the water intake which is retained through the direct and indirect effects of capillary or molecular forces. As commonly understood, capillary water is that which, overcoming gravity, collects, through movement in all directions, in the small interstices between soil particles and granules. Much of this is available to plants (*available soil moisture*) where the soil and subsoil are sufficiently tractable for the roots to penetrate the material in search of moisture. Part of the soil water, apparently, is held as exceedingly thin films around or in soil particles with such force (a force on the order of molecular attraction) as to be unavailable to plants and incapable of moving from particle to particle by capillarity (*hygroscopic or unavailable soil moisture*). For example, a "bone dry" soil in which most plants will not live over any considerable period still contains uncombined, yet unavailable, water. The dividing line between available and unavailable soil water is usually referred to as the "*wilting point*."[2]

[1] See the Occurrence of Ground Water in the United States, U. S. Geol. Survey *Water Supply Paper* 489. Tolman, C. F. "Ground Water." McGraw-Hill Book Company, Inc. New York. 1937.

[2] Briggs, L. J., and Shantz, H. L. The Wilting Coefficient for Different Plants and Its Indirect Determination, U. S. Dept. Agr. Bur. Plant Industry *Bull.* 230, 1912. Veihmeyer, F. J., and Hendrickson, A. H. Soil Moisture at Permanent Wilting Point of Plants, *Plant Physiology*, Vol. 3, pp. 355–357, July, 1928.

Recent studies have shown that the infiltration rate for different soils differs widely. For example, tests (Table 20) carried out on Ruston sandy loam, a friable, porous soil of the southeastern coastal plain, revealed that the infiltration rate, averaged 2.06 inches of water per hour for 3 continuous hours, whereas the corresponding infiltration rate for the regionally associated Susquehanna clay loam, a stiff, relatively impervious soil, was less than 0.10 inch per hour. Similar wide differences in rates of water intake have been found between Davidson clay loam, a friable, relatively pervious soil of the southeastern Piedmont (derived from basic igneous rocks) and the regionally associated Iredell loam (also derived from basic igneous rocks) which has a dense and much less permeable clay subsoil. The average infiltration for the former was 0.82 inch an hour over a 3-hour period, whereas for the latter the corresponding rate was only 0.01 inch an hour.[1] These measurements were all made on firm, bare ground, under comparable initial conditions of soil moisture. Similar

TABLE 20.—INFILTRATION CAPACITY OF EIGHT IMPORTANT SOILS OF THE PIEDMONT AND COASTAL PLAIN SECTIONS OF GEORGIA, OVER A 3-HOUR PERIOD OF CONTINUOUS WATERING FOLLOWING PREVIOUS WETTING[1]

| Soil | Depth, surface soil, inches | Location | Amount of infiltration during successive time intervals, inches | | | | | |
|---|---|---|---|---|---|---|---|---|
| | | | 0 to 15 minutes | 15 to 30 minutes | 30 to 60 minutes | 60 to 120 minutes | 120 to 180 minutes | Total for 3 hours |
| Cecil sandy loam | 11 | Watkinsville | 0.42 | 0.14 | 0.23 | 0.43 | 0.42 | 1.65 |
| Cecil sandy clay loam | 3 | Watkinsville | 0.27 | 0.10 | 0.19 | 0.37 | 0.33 | 1.26 |
| Cecil clay loam | 1 | Watkinsville | 0.07 | 0.04 | 0.03 | 0.06 | 0.09 | 0.28 |
| Davidson clay loam | 6 | Monticello | 0.37 | 0.18 | 0.38 | 0.74 | 0.79 | 2.47 |
| Iredell loam | 6 | Lexington | 0.01 | 0.00 | 0.01 | 0.00 | 0.01 | 0.04 |
| Ruston sandy loam | 8 | Americus | 0.94 | 0.46 | 0.91 | 1.88 | 1.99 | 6.18 |
| Greenville sandy clay loam | 3 | Ellaville | 0.20 | 0.04 | 0.09 | 0.12 | 0.15 | 0.60 |
| Susquehanna clay loam | 4 | Ellaville | 0.04 | 0.03 | 0.05 | 0.07 | 0.10 | 0.29 |

[1] Each measurement followed a similar 3-hour continuous wetting of the soil 24 hours previously.

[1] Musgrave, G. W., and Free, G. R. Preliminary Report on the Determination of Comparative Infiltration Rates on Some Major Soil Types, Am. Geophys. Union *Trans.* Part II, pp. 345–349, 1937

differences have been found for many other important types of agricultural land in various parts of the country.

The importance of these differences in infiltration characteristics is seen when it is considered that a rain falling at an intensity of 1 inch an hour would be taken up entirely over a 3-hour period by Ruston sandy loam under conditions like those obtaining where the foregoing measurements were made, whereas the same rain falling on Susquehanna clay loam under like conditions would produce an average runoff amounting to more than 0.90 inch an hour and would normally result in severe erosion on unprotected slopes. Such a rain on Davidson clay loam would, under the same conditions, cause a runoff of only about 0.2 inch an hour, whereas from Iredell loam the water lost as runoff would amount to 0.99 inch an hour. In considering these differences, it is important to observe that the high rates of runoff (meaning low rates of infiltration) were from the two relatively impermeable types (Susquehanna and Iredell), whereas the low rates of runoff meaning high rates of infiltration were from the two comparatively permeable soils (Ruston and Davidson).

These variations in rates of infiltration for different soil types are closely correlated with (1) total amount of pore space and (2) average size of the individual openings. Other things being equal, soils having a large combined pore space (cavity), as well as large individual spaces between the soil particles or aggregates of particles, have infiltration rates much higher than soils whose individual openings are small. By the same token, soils of large total pore space have greater capacity for potential water storage, provided the individual orifices are not so large as to favor excessive percolation. On the other hand, fine-textured soils with large potential storage capacity may not be able to avail themselves of such capacity because of slow intake due to the small diameter of the individual openings, especially where the slope and other surface conditions favor rapid runoff. With respect to available moisture,[1] however, those dense clay soils having exceedingly small pores may contain a small effective supply regardless of their high water-retaining capacity, because small interstices usually are associated with a soil density that tends to impede penetration by plant roots in search of moisture. Such soils are characteristically difficult to drain, and in dry weather they tend to crack into resistant blocks which become so hard on the surface as to trap the contained moisture and so hold it unavailable to plants.[2]

[1] For a discussion of available moisture see Edlefsen, N. E. Effect of Soil Moisture Characteristics on Irrigation Requirements, *Agr. Eng.*, Vol. 18, pp. 247–250, 1937.

[2] The behavior of the Bayamo clay of Cuba (Bennett and Allison, "Soils of Cuba"), a noncalcareous, exceedingly dense clay, is illustrative of the tendency of very fine-textured, plastic soils to hold their moisture unavailable to plants. Bayamo clay contains more than 75 per cent of clay particles (85 per cent of which is colloid). When wet, it is intensely plastic

The results of the infiltration tests presented in Table 20 show that the term *saturated soil*[1] is not easily defined. It has been used freely in appraising ground moisture conditions in relation to runoff where actually it had little meaning beyond a general qualitative expression. For example, Ruston sandy loam (to which 14.41 inches of water had been applied over a 3-hour period, 24 hours previously) was still taking in water at a rapid rate following a steady 3-hour application of water in amount sufficient to keep the surface covered. The rate of water intake by this permeable soil was nearly as rapid toward the close of the test as in the beginning: 1.99 inches during the third hour as against 2.31 inches during the first hour.

These measurements of water intake, together with others made in other parts of the country, indicate that most soils continue to take in water after prolonged periods of rainfall. True, the point of saturation is approached in the instance of some soils, but usually infiltration continues in some degree under almost all conditions. Precise saturation points have not been determined for a sufficient number of soil types (if for any) to define the exact conditions under which soils, in place, cease to take in water. It appears that hidden openings leading into the reservoir of the substrata continue to conduct water into the body of the soil under conditions that would seem sufficient to prevent any such movement. Frequently, perhaps usually, soils that are frozen or covered with what appears to be an impermeable sheet of ice still take in some water. This probably means that openings still exist—holes and cracks in the ground as well as in the blanket of ice.

---

and sticky; but on drying, it cracks widely and deeply, forming hard, resistant blocks which, as the result of horizontal fracturing, can be lifted out of position intact. These, with their desiccated rind, encase considerable moisture, cut off from adjacent parts of the soil body and so firmly held as to be almost completely unavailable to plants (sugar cane suffering severely for lack of moisture, under such conditions, even where the soil contains more than 25 per cent of uncombined moisture). The ultimate result of such cracking and recracking is to develop, with continuing dry weather, a coarsely fragmental condition, resembling somewhat that of the "buckshot" structure of the extensive Sharkey clay of the Mississippi alluvial plain from Cape Girardeau to New Orleans. These fragments frequently become so hard as to represent, in that condition, the essential equivalent of gravel. On wetting, the soil swells, the clods melt, the cracks close, and the entire soil mass runs back to its original condition of density.

Another Cuban clay, the polvillo (powdery) phase of Truffin clay, which contains 60 per cent or more of clay in the soil and 85 per cent (80 per cent colloid) in the subsoil, forms, on drying, a surface layer of loose material of small, moderately hard pellets with dustlike soil between. Such land becomes so exceedingly desiccated with the arrival of the dry season of winter (even when containing 20 per cent of moisture at 110°C.) that sugar cane parches rapidly, necessitating early harvesting.

[1] Lowdermilk, W. C. Water Intake of Saturated Soils, Am. Geophys. Union *Trans.*, Part II, pp. 355–361, 1937.

Effect of Tillage on Infiltration. Soil management practices have much to do with rates of water intake. Plowing, cultivation, subsoiling, or other stirring normally increases, temporarily at any rate, the openness (combined cavity space) of a soil—the disturbed part—chiefly by increasing the size of the cavities. Measurements of infiltration rates of cultivated and uncultivated soil of the same kind have shown, for the types investigated, that the rate of intake on land plowed 4 inches deep averaged 0.99 inch an hour, and 1.20 inches where plowed 6 inches deep; as against an intake of only 0.77 inch where there was no cultivation.[1] The effects of cultivation on soil porosity and infiltration, however, are generally transitory, lasting only until the soil settles back, under the effect of subsequent rains, to its former condition of density or even to a condition of increased density or until the openings are closed with fine particles filtered from inflow of muddy surface water. Bradfield says, in this connection:

> "In many cases from 25 to 30 per cent more soil is crammed into a cubic foot than was present in the virgin soil. This has reduced porosity, especially the volume of larger pores through which water penetrated readily and through which the soil received the necessary ventilation. As a result of these changes in structure, root development is hampered, the storage capacity of the soil for water is reduced, flood hazard is increased, and the damage from . . . droughts is magnified."[2]

Beneficial effects of cultivation are likely to be more lasting, within certain textural limitations, in those soils which have a relatively low content of very fine material (clay and colloid), especially where such material is of a plastic or waxy consistence.[3] On the other hand, loose, porous sands obviously are not likely to be especially benefited by artificial disturbances, since they generally are too open in the first place, do not fracture very much, and tend to settle back rapidly to their former condition, except under the drifting effect of wind.

It is pointed out elsewhere that cultivation frequently tends to reduce infiltration by closing up the natural surface openings that lead into the body of a normal soil.

Effect of Organic Matter on Infiltration. When coarse organic matter is incorporated with a soil, the porosity remains high for a com-

[1] Musgrave, G. W., and Free, G. R. Some Factors Which Modify the Rate and Total Amount of Infiltration of Field Soils, *Jour. Am. Soc. of Agronomy*, Vol. 28, pp. 727–739, 1936.

[2] Bradfield, Richard. Soil Conservation from the Viewpoint of Soil Physics, *Jour. Am. Soc. of Agronomy*, Vol. 29, February, 1937.

[3] Bennett, H. H., and Allison, R. V. "The Soils of Cuba." Tropical Plant Research Foundation (Boyce Thompson Institute, Yonkers, N. Y.), pp. 73–74, 152, 315, 318–321, 1928.

paratively long period, depending largely on the rate of decomposition of the material. For example, measurements of the infiltration rate of Clarion loam, at the Iowa Agricultural Experiment Station,[1] have shown that where no organic matter was used the intake of rainfall was at a rate less than 2 inches in two hours, whereas more than 3 inches of rainfall entered the soil in the same time, under comparable conditions, where 8 tons of manure per acre was plowed in; and 4½ inches, or more than twice as much, where 16 tons per acre was applied.

Similar effects on soil porosity are produced as the result of crop rotations, turning under crop residues, growing grasses and legumes, and using other methods that increase the soil supply of organic matter.[2] Work at the West Virginia Agricultural Experiment Station[3] has indicated that soils receiving such treatments ordinarily are characterized by a higher content of relatively large aggregates of soil particles (soil granules) than of small aggregates. This means essentially that soils of coarsely aggregated structure have larger pores and, consequently, greater water intake capacity, as well as greater absorption capacity, as a rule. As compared with land under continuous cultivation to clean-tilled crops, infiltration measurements have shown generally a higher rate of intake for those areas which have been efficiently managed with respect to maintenance of the organic supply, as under good pasture practice or good crop rotations.

Effect of Turbidity of Surface Water on Infiltration. The infiltration rate of a soil is profoundly affected by the turbidity of surface water. Clear water penetrates a soil much more rapidly than turbid (muddy) water, as shown by Lowdermilk.[4] This is because the suspended material of muddy water tends to filter out as surface water enters the soil openings, closing them and impeding downward movement. Those soils having a considerable proportion of their contained colloid (ultrafine material) in a flocculated or aggregated condition are much less subject to the sealing effect of muddy infiltration than those whose particles are deflocculated or separated as individual grains. This is because such granular material, as well as the material of soil in good condition of tilth or of favorable organic matter content, generally does not enter so readily into suspension to produce highly turbid surface water as nongranular soil or soil deficient in humus. Turbidity is also of much less importance,

[1] Smith, F. B., Brown, P. E., and Russell, J. A. The Effect of Organic Matter on the Infiltration Capacity of Clarion Loam, *Jour. Am. Soc. of Agronomy,* Vol. 29, July, 1937.

[2] Peele, T. C. The Effect of Lime and Organic Matter on the Erodibility of Cecil Clay, Soil Sci. Soc. America *Proc.*, Vol. 2, pp. 79–84, 1937.

[3] Browning, G. M. Changes in the Erodibility of Soils Brought about by the Application of Organic Matter, Soil Sci. Soc. America *Proc.*, Vol. 2, pp. 85–96, 1937.

[4] Lowdermilk, W. C. Influence of Forest Litter on Runoff, Percolation, and Erosion, *Jour. Forestry*, Vol. 28, No. 4, 1930.

or altogether lacking, on areas having a dense cover of vegetation. Water flowing from thickly grassed areas, for example, is frequently essentially clear following heavy rains, whereas that from adjacent cultivated land is always muddy in some degree.

Lowdermilk found in the Berkeley Experiment[1] that a muddy suspension percolates through soil columns at about one-tenth the rate of clear water. Four 10-inch soil lysimeters, 10, 11, 12, and 13, were uniformly filled

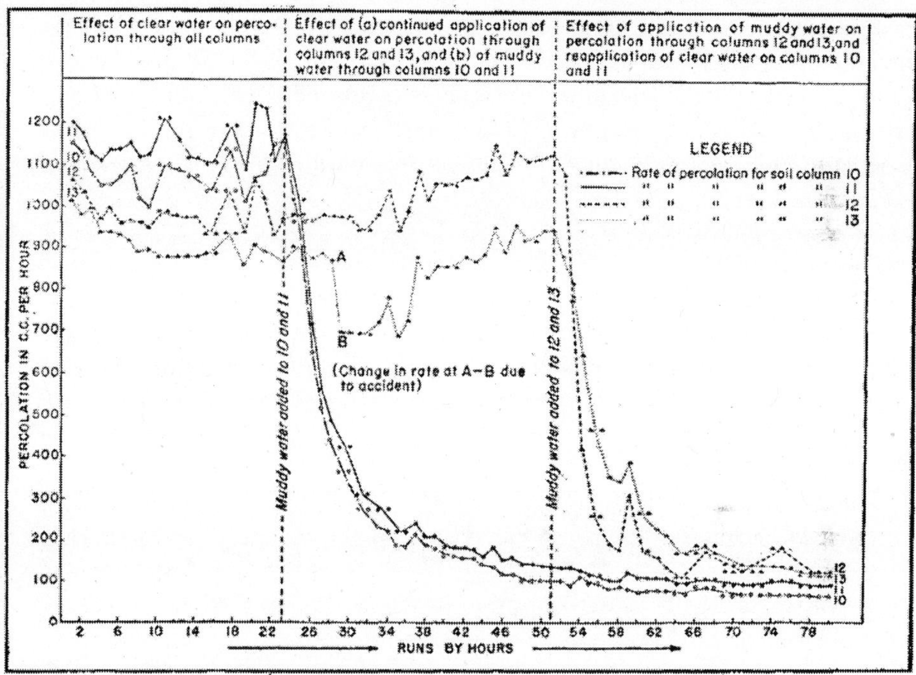


GRAPH 23.—Effect of muddy suspensions on rate of infiltration and percolation of storm waters.

with samples of a fine sandy loam (Holland series, California). After running clear water through the soil columns for one week, a muddy suspension ranging from 1.7 to 1.9 per cent silt content was added to lysimeters containing soil columns 10 and 11. Immediately, the rate of percolation through the columns dropped off, as shown in Graph 23, to about one-tenth the rate for clear water. Then, after two weeks of clear-water application, the muddy suspension was added to soil columns 12 and 13. The rates of percolation through these dropped off in much the same manner as with the other two columns. But the rate of percolation for soil columns 10 and 11 did not increase under the reapplication of clear water at the end of two weeks.

[1] Lowdermilk, W. C. Further Studies of Factors Affecting Surficial Runoff and Erosion, International Cong. Forestry *Proc.*, 1929.

After the experiment, it was found that the soil in suspension had filtered out at the soil surface to form a thin layer of fine-textured material, silt and clay, which determined the rate of intake or infiltration of water into the soil column irrespective of the capacity of the material below the surface for percolation.

This experiment establishes one of the most fundamental facts in land use. To a considerable degree, it explains why clearing land of its natural cover of vegetation and its cultivation creates a special hazard in the acceleration of soil erosion under agricultural use. The natural cover prevents the formation of soil suspensions by rainwater at the surface; the water remains clear, and the rate of intake is maintained at the natural capacity of the soil profile. But bare soil surfaces produce muddy suspensions under the impact of rain, and such suspensions, as already pointed out, tend to seal off the surface, thereby reducing the rate of intake of water into a soil.

The turbidity of runoff from bluegrass and alfalfa grown on dark prairie loessial soil of good organic content (Marshall silt loam), at the soil and water conservation experiment station near Clarinda, Iowa, averaged (over a period of three and one-half years) less than 0.2 pound of suspended soil per cubic foot as against a turbidity of more than 4.0 pounds per cubic foot, or twenty times as much, in runoff from the same kind of land on which corn was grown continuously, under the same rainfall. The corresponding turbidity of water flowing from areas of the same soil type, on which crop rotations were practiced, ranged from 1 to 3.8 pounds of soil to the cubic foot, or from five to nineteen times the muddiness of water flowing from the densely vegetated lands. From exposed, untreated subsoil of the same type, the runoff from continuous corn averaged more than 4.5 pounds of soil per cubic foot; whereas from comparable areas, also used continuously for corn, the corresponding densities were 2.8 and 1.6 pounds of suspended material per cubic foot, respectively, where (1) green vegetation and (2) barnyard manure were incorporated with the subsoil.[1]

Measuring the runoff and erosion from small plots (3 by 10 feet) of Kirvin fine sandy loam of 8¾ per cent slope and in dry, loose fallow, to which application of (1) clear water and light suspensions of (2) fine sandy loam (topsoil of Kirvin fine sandy loam) and (3) moderately sandy clay (subsoil of Kirvin fine sandy loam) was made, Hendrickson[2] obtained the following results:

[1] Musgrave, G. W., and Norton, R. A. Soil and Water Conservation Investigations, Clarinda, Iowa, Progress Report, 1931–1935, U. S. Dept. Agr. *Tech. Bull.* 558, 1937.

[2] Hendrickson, B. H. The Choking of Pore-space in the Soil and Its Relation to Runoff and Erosion, Am. Geophys. Union *Trans.*, 15th annual meeting, pp. 500–505, 1934.

| Application | Runoff, Per Cent | Erosion, Tons per Acre |
|---|---|---|
| Clear water | 4.97 | 0.001 |
| Water plus fine sandy loam | 6.79 | 0.003 |
| Water plus sandy clay | 8.91 | 0.705 |

In the instances of the fine sandy loam and sandy clay suspension, the contained solids filtered out to form thin surface sheets. The clay material developed a more impermeable covering which not only increased imperviousness by nearly 100 per cent, as compared with infiltration under the clear-water application, but greatly increased erosion (by more than seven hundred times in this instance).

Thus it is indicated that by sealing surface openings, muddy water not only reduces infiltration, especially where the suspended material is fine textured, but results in marked acceleration of erosion.

Other measurements made in connection with these studies of the effects of muddy water on runoff and erosion on Kirvin fine sandy loam, where clean sand was used as the soil medium, showed that fine material (silt and clay separately) by sealing over the surface from applications of muddy water not only impeded infiltration greatly but reduced, to an almost negligible degree, transposition of fine particles to lower depths by surface-applied water.

These and other experiments have shown that the turbidity of surface runoff can be greatly reduced by practical farm treatments. Such diminution means also, of course, a reduction of soil loss by erosion.

EFFECT OF EROSION ON RATE OF INFILTRATION. The infiltration rate of soils frequently is reduced, also, by the erosion process. Thus, slightly eroded Cecil sandy loam (a granite-derived soil with moderately stiff red clay subsoil), at Watkinsville, Ga., was found to have an infiltration rate of 0.55 inch an hour, as compared with a rate of 0.42 inch an hour for an adjacent area of Cecil sandy clay loam, which was more severely eroded but otherwise similar. The infiltration rate of near-by comparably treated Cecil clay loam, excessively eroded, was found to be only about one-fifth as fast, or 0.09 inch an hour.

As erosion progresses, more impermeable material (dense clay subsoil) is eventually uncovered in the instance of many types of soil. This is one of the reasons why erosion so frequently is accelerated once the process has started. Rainfall usually runs off eroded land more rapidly and in larger quantity than from uneroded land and for that reason is more destructive.

## Methods for Increasing Infiltration

Infiltration of rainfall into the body of the soil can be increased in a variety of practical ways. For example, some soils peculiarly subject to surface crusting following rains shed water at relatively rapid rates if the crust remains unbroken until the next rain. Accordingly, breaking of the hardened or compact surface by cultivation is likely to have considerable effect, temporarily at any rate, toward increased intake of water.[1] Subsoiling under certain conditions, particularly in soils having relatively impermeable sublayers susceptible to fracturing, also speeds up the rate of infiltration. The breaking up of such material has the effect, as noted above, of enlarging the openings and, because of this, of increasing the capacity of the soil to take in more water—that is, as long as the condition of artificially increased porosity obtains.

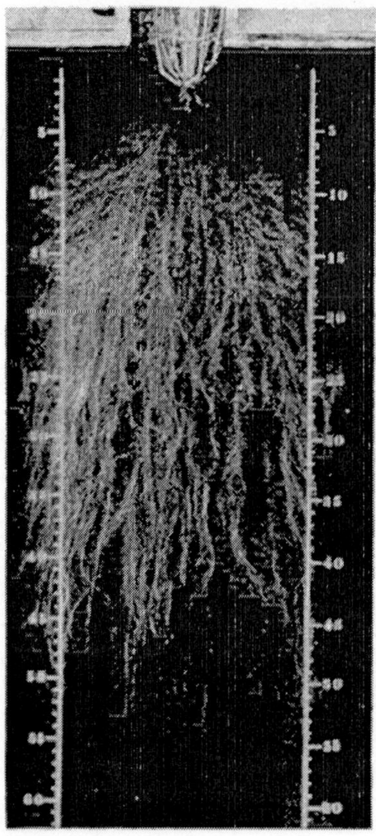

FIG. 78.—The root system of a single wild-oat plant, grown free from competition and excavated 80 days after emergence. Total length of roots, 54 miles. On decaying, each root leaves, at least temporarily, a soil tunnel. Some of the tunnels extend 5 feet into the ground. (*Courtesy T. K. Pavlychenko.*)

The total amount of water entering a soil is governed to a considerable degree by the time available for infiltration. If water can be made to move slowly rather than rapidly across the land, or if it is impounded on the surface, the amount of intake is thereby greatly increased. A variety of feasible methods are available for retarding or, on moderately sloping land, even preventing runoff. One of the most satisfactory is the use of vegetation in complete or partial coverage of the ground. As already pointed out, both runoff and erosion from sloping land densely covered with such vegetation as grass, clover, or alfalfa are much less than from comparable land under cultivation. Measurements show that surface discharge of water from a cultivated crop is frequently more than two and one-half times, and in some instances several hundred times, that from land under such cover as clover, alfalfa, bluegrass, or Bermuda grass.[2]

[1] Harper, Horace J. Soil Structure and Moisture Movement, Soil Sci. Soc. America *Proc.*, Vol. 2, pp. 15–20, 1937.

[2] Norton, R. A., and Smith, D. D. Effect of Density of Vegetation on Rate of Runoff of Surface Water, paper presented at annual meeting of Amer. Soc. of Agronomy, Chicago, Nov. 30, 1937.

In addition to the surface effect of vegetation in restraining runoff and increasing infiltration, other benefits result from underground effects, such as increased organic supply and channels opened by ramifying root penetration. Pavlychenko finds, for example, that a single wild-oat plant 80 days old had a root system totaling 54 miles in length (Figs. 73 and 74).[1]

Fig. 74.—One crown root from the main root system shown in Fig. 73. Total length of this crown root, 4.05 miles. The ground tunnel formed by the main stem of this crown root probably is an effective water channel after decay of the root material. (*Courtesy T. K. Pavlychenko.*)

Forest with a good ground cover of litter or vegetal mold is also highly effective in retarding runoff as well as surface movement of soil.

As noted above, however, another very important effect of both grass and forest litter is the low turbidity of runoff and good porosity of the soil screened by such effective cover.

[1] Pavlychenko, T. K. The Soil-block Washing Method in Quantitative Root Study, *Canadian Jour. Research*, Vol. 15, 1937.

Other effective methods of retarding surface runoff and increasing infiltration, such as incorporation of organic matter with the soil, contour furrowing and cultivation, terracing, and strip cropping, are described elsewhere.

## Need for Increased Water Conservation in Humid Regions

Increased infiltration, aside from its powerful effect in reducing runoff and erosion, is important throughout the country generally because of the dependence of plants on soil water reserves in times of drought. Even in the humid regions, deficiencies in soil moisture are the cause of seasonal crop deterioration nearly every year on many upland areas. The Atlantic seaboard has a rainfall of 40 to 50 inches, yet, during July and August, pastures frequently deteriorate or give out, and field crops are sometimes seriously damaged because of moisture shortage. Since it is ordinarily impracticable to reduce evaporation, such deficiency must be made up through water conservation practices—storage in the reservoir of the soil during periods of rain. Practical methods for conserving runoff or waste water are given elsewhere.

Relative losses of rainfall, as runoff, percolation, evaporation, and transpiration, have been measured carefully for cropped and uncropped areas of the important Marshall silt-loam soil of the Corn Belt, at the Clarinda, Iowa, soil and water conservation experiment station. Comparisons were made between untreated bare ground and similar areas devoted to corn. One comparison was with corn untreated, and another with corn that had received an application of manure. The results, covering $3\frac{3}{8}$ years, are summarized in Table 21.[1]

In the instance of bare ground (no crop), not manured, 14.16 inches of a total precipitation of 102.41 inches (14 per cent) for the period passed below the root zone (3 feet below surface), 58.78 inches (57.4 per cent) evaporated, and 29.47 inches (28.8 per cent) was lost as runoff. At the same time, only 5.77 inches passed below the root zone in the untreated corn area; 19.16 inches was lost immediately as runoff, and 77.48 inches disappeared as evaporation and transpiration.

From the corresponding manured areas, the runoff from bare ground was 8 inches less than from the unmanured bare area, and approximately 8 inches more passed below the root zone. The loss by evaporation remained substantially the same. At the same time, the manured corn land lost only 13.2 inches as runoff, and only 10.63 inches passed below the root zone. Evaporation and transpiration were about the same from both the manured and the unmanured corn areas.

[1] Musgrave, G. W., and Neal, O. R. Rainfall and Relative Losses in Various Forms, Am. Geophys. Union *Trans.*, Part II, pp. 349–355, 1937.

TABLE 21.—SUMMARY OF RUNOFF, AMOUNT OF RAINFALL PASSING BELOW ROOT ZONE, AND VAPOR LOSS ON A PERMEABLE SOIL (MARSHALL SILT LOAM), INCHES[1]

| Type of loss | Not manured | | Manured | |
|---|---|---|---|---|
| | No crop | Cropped to corn | No crop | Cropped to corn |
| Runoff | 29.47 | 19.16 | 21.27 | 13.20 |
| Amount passing below root zone | 14.16 | 5.77 | 23.51 | 10.63 |
| Vapor loss | 58.78 | 77.48 | 57.63 | 78.58 |

[1] Under cropped and uncropped conditions, with and without manurial treatments, at Clarinda, Iowa. (Total precipitation 102.41 inches for period of three and two-thirds years.)

These excessive losses of water from a soil of such good permeability as the Marshall silt loam, coupled with an annual precipitation of only about 30 inches, show the importance of making the greatest possible use of all feasible measures for increasing the penetration of rainfall. The loss by runoff and vapor loss of 88.2 per cent of the total precipitation from bare ground is a striking indication of what is taking place, probably rather generally, on unplowed fallow; and what happens on bare ground is an indication that the water losses from cultivated fields are also entirely too excessive. If it is assumed that evaporation accounts for half the vapor loss from the untreated corn plot of Table 21, then runoff and evaporation are responsible for the loss of 58 per cent of the total precipitation.

Clearly, the practical farmer cannot afford to lose much water as runoff. The value of manure, aside from its fertilizing effect, in reducing runoff by 8 inches of water becomes obvious when it is considered that this amount of rainfall theoretically is more than sufficient to produce 40 bushels of corn per acre.

It is important to understand that those practical soil treatments which increase penetration of water into the ground do not cause increased losses by evaporation. In bare soils of high permeability, whether natural or artificially introduced, most of the water moves downward to the water table. But where a crop is grown on such areas, a considerable part of such percolating water is intercepted by the plant roots and utilized in their growth. On the other hand, in a relatively impermeable soil, whether bare or cropped, a much smaller proportion of the water proceeds in the direction of the water table, and losses by both runoff and evaporation are greater. Thus, improvement in the degree of permeability of a soil has been shown to reduce runoff and increase water intake without increasing evaporation. In effect, such conservation of rainfall is equivalent to increasing the rainfall, in so far as plants are concerned.

In general, then, it is important to increase the penetration of rainfall into the body of the soil, by whatever practicable means available, for two reasons: (1) consequent reduction of runoff and erosion; (2) increase of available water for plant growth.

Although different soils differ markedly in respect to the amount and rate at which they take in water, it has been shown that even refractory and relatively impermeable soils can be improved vastly in this respect by such treatments as the incorporation of organic matter, good cropping practices, and good tillage.

# Chapter IX. Relation of Erosion to Crop Yields

"Most of the worn-out lands of the world are in their present condition because much of the surface soil has washed away, and not because they have been worn out by cropping. Productive soils can be maintained through centuries of farming if serious erosion is prevented."[1]

In spite of the success achieved in breeding better varieties of corn and the increased employment of improved farm machinery designed for more efficient tillage, and regardless of the more general use of improved crop rotations, lime, and various soil amendments; and, furthermore, in spite of all the education provided in the schools of every state and in books, bulletins, the press, corn clubs, and farmers' meetings, with frequent prizes for the best producers, the nationwide acreage yield of corn has not increased to any important degree in the United States—at least it had not prior to the inauguration of the recent nationwide programs of farm adjustment and soil and water conservation. When it is considered that corn growing has not extended on a very large scale into the less favorable environment of semiarid regions and has not suffered from any far-reaching, devastating insect or disease scourge, but one conclusion can be reached: that, for the nation, erosion has thwarted much of the stupendous efforts to increase the acreage yield of the crop in the United States. Actually, the per acre yield has increased on the better lands; the decrease, or failure to increase, has been restricted principally to soil worn lean by erosion. But for the technical and educational assistance provided, coupled with replacement of impoverished areas by land still retaining a cover of topsoil and by land reclaimed through drainage and irrigation, the acreage yield doubtless would have declined much more than it has.

The average yield of corn in the United States for the ten-year period 1871–1880, inclusive, was 26.4 bushels an acre, as against 25.9 bushels for the period 1921–1930, inclusive (Table 22). That the maximum and minimum annual yields through the former period were larger, respectively, than the corresponding yields for the latter decade seems to

[1] Controlling Surface Erosion of Farm Land, Missouri Agr. Exper. Sta. *Bull.* 211, p. 1, 1924.

indicate that the two periods are reasonably comparable. Wheat yields increased from a 10-year average of 12.6 bushels an acre for the period 1871–1880 to 14.5 bushels for the period 1901–1910; but for the decade 1921–1930, there was a slight decline to 14 bushels. The average yield of cotton for the 10-year period 1871–1880 was 173.5 pounds an acre, as against 156.6 pounds an acre for the period 1921–1930.[1]

In numerous localities, the yields of these crops have declined much more than these nationwide averages, even to the extent of necessitating abandonment of many thousands of fields severely impoverished or worn out by erosion, many of them entirely stripped of the more productive topsoil, and others gutted with deep erosion channels impassable to farm machinery.

In Illinois, the average yield of corn for the period 1896–1905 was 36.7 bushels; for 1906–1915, 36.6 bushels; and for 1916–1925, 36.2 bushels. The average corn yield since 1925, for the five-year period up to and including 1930, was 33.5 bushels. This last figure is hardly comparable, because the one year of 1930 turned out only 25.5 bushels an acre. During these same ten-year periods, the average yield of oats varied as

TABLE 22.—ANNUAL AND AVERAGE ACRE YIELDS OF CORN AND COTTON FOR THE DECADES 1871–1880 AND 1921–1930

| Year | Corn,[1] bushels per acre | Cotton,[2] pounds lint per acre | Year | Corn, bushels per acre | Cotton, pounds lint per acre |
|---|---|---|---|---|---|
| 1871 | 27.2 | 159 | 1921 | 28.4 | 132.5 |
| 1872 | 29.4 | 182.3 | 1922 | 27.0 | 148.8 |
| 1873 | 22.9 | 168.3 | 1923 | 28.4 | 136.4 |
| 1874 | 22.2 | 157 | 1924 | 22.1 | 165.0 |
| 1875 | 27.7 | 181.2 | 1925 | 27.6 | 173.5 |
| 1876 | 26.7 | 167.6 | 1926 | 25.6 | 192.9 |
| 1877 | 25.8 | 170.4 | 1927 | 26.6 | 161.7 |
| 1878 | 26.2 | 167.5 | 1928 | 26.6 | 163.3 |
| 1879 | 28.2 | 180.5 | 1929 | 25.8 | 164.2 |
| 1880 | 27.3 | 190.9 | 1930 | 20.5 | 157.1 |
| Average acre yield........ | 26.4 | 172.5 | .... | 25.9 | 159.5 |
| Weighted average acre yield | 26.4 | 173.5 | .... | 25.9 | 156.6 |

[1] Compiled from data published in *Agr. Statistics*, U. S. Dept. Agr., 1937, pp. 39–40.

[2] Compiled from statistical data, Cotton Revision, Acreage, Yield and Production, 1866–1935, Bur. Agr. Econ., 1936.

[1] Agricultural Statistics, U. S. Dept. Agr., 1937.

follows: 1896–1905, 33.2 bushels; 1906–1915, 32.2 bushels; 1916–1925, 35.4 bushels. For the five years following 1925, the yield of oats was 31.3 bushels. In McLean County, the corn yields in five-year averages since 1911 have been: 1911–1915, 40.8 bushels; 1916–1920, 40.1 bushels, 1921–1925, 38.4 bushels; and 1926–1930, 35.4 bushels. It is readily seen from these figures of acreage production of corn in Illinois that for this period something was reducing the yields faster than improved agricultural practices were raising them. On those better lands of the state which are not subject to erosion and where cropping systems involve the application of liberal amounts of limestone and superphosphate, or both, yields are being maintained or increased. But on land where sheet erosion is active, it will be difficult to maintain the yield, even with the best of farming systems, unless those types of farming are employed which consistently prevent or decrease loss of soil by erosion.[1]

## Measured Yields from Topsoil and Subsoil

Highly significant quantitative information pertaining to the effect of erosion on yield has been acquired at the soil and water conservation experiment stations, where, in order to determine the precise effect on important types of farm land, the same crop was grown on adjacent areas, one with its topsoil, or a considerable part of it, remaining, the other stripped of topsoil down to clay subsoil (the *B* horizon). The same variety of crop was grown on each set of contrasting plots, with the same number of plants in each and with identical cultural methods applied at the same time. Since the slope was identical, and rainfall the same on both areas, it appears reasonable to conclude that, for the periods involved, the results probably are as accurate as scientific technique could make them. Moreover, the measurements are considered significant, particularly for humid and subhumid conditions, because they were made, for the most part, on widely separated types of land carefully selected for their representativeness with respect to the erosion hazard within problem areas comprising some 200 million acres of land, much of it cropland.

The results of these determinations are presented in Table 23. Where the soil has been stripped down to the subsoil level, as erosion has removed it from more than 100 million acres of what formerly was fair to good cropland, the average production for the 10 types investigated has been 77 per cent below that of the corresponding areas still retaining a good cover of topsoil. With respect to individual types, the maximum decline of productivity, as between soil and subsoil, has been from 35.3 to 1.1

[1] "The Cost of Soil Erosion, with Control Suggestion." Illinois Farmers' Institute, Springfield, Ill. 1934.

TABLE 23.—AVERAGE ACRE YIELDS FROM TOPSOIL AND CORRESPONDING SUBSOIL OF 10 REPRESENTATIVE TYPES OF FARM LAND UNDER COMPARABLE CONDITIONS OF SLOPE, RAINFALL, AND CULTURAL TREATMENT[1]

| Type of land | Soil amendment | Rainfall, inches | Period, inclusive | Crop | Average yield | | Decline in yield, per cent |
|---|---|---|---|---|---|---|---|
| | | | | | Topsoil | Subsoil | |
| Houston black clay, 4% slope, Tex. | None | 26.56 | 1931, 1934 | Corn | 26.8 bu. | 2.9 bu. | 89 |
| | | 38.95 | 1932, 1935 | Oats | 60.6 bu. | 22.5 bu. | 63 |
| | | 32.76 | 1933, 1936 | Cotton[2] | 288 lb. | 102 lb. | 64 |
| Marshall silt loam, 9% slope, Iowa | None | 27.3 | 1932 to 1935 | Corn | 30.7 bu. | 6.5 bu. | 79 |
| Clinton silt loam, 16% slope, Wis. | None | 32.6 | 1933 to 1934 | Corn | 49.3 bu. | 21.2 bu. | 57 |
| Muskingum silt loam, 12% slope, Ohio | None | 39.5 | 1933, 1935, 1936 | Corn | 35.3 bu. | 1.1 bu. | 97 |
| | | | | Stover | 4,258 lb. | 510 lb. | 88 |
| Palouse silt loam, 30% slope, Wash. | None | 21.7 | 1932 to 1935 | Winter wheat | 23.9 bu. | 7.2 bu. | 70 |
| Colby silty clay loam, 5% slope, Kans. | None | 19.9 | 1931, 1935 | Winter wheat | 12.5 bu. | 5.3 bu. | 58 |
| Kirvin fine sandy loam, 8.75% slope, Tex. | None | | | Cotton | 365 lb. | 50 lb. | 86 |
| | 400 lb. 4-8-4 fertilizer | 40.6 | 1931 to 1934 | Cotton | 580 lb. | 206 lb. | 65 |
| | None | 41.0 | 1931 to 1934 | Cotton | 308 lb. | 13 lb. | 96 |
| Nacogdoches fine sandy loam, 10% slope, Tex. | 400 lb. 4-8-4 fertilizer and green manure | 34.4 | 1936 | Cotton | 450 lb. | 130 lb. | 71 |
| Cecil sandy clay loam, 10% slope, N. C. | None | 46.1 | 1932 to 1934 | Cotton | 950 lb. | 290 lb. | 69 |
| Vernon fine sandy loam, 7.7% slope, Okla. | None | 30.9 | 1929 to 1936 | Cotton[2] | 159 lb. | 96 lb. | 39 |
| | Virgin soil | 33.1 | 1930 to 1935 | Cotton | 495 lb. | 313 lb. | 37 |
| Average decline | ........... | ...... | ......... | ........ | ........ | ........ | 77 |

[1] Measurements at soil and water conservation experiment stations, Soil Conservation Service.

[2] Lint cotton; other cotton yields relate to the unginned product ("seed cotton").

bushels of corn per acre (97 per cent reduction), in the instance of Muskingum silt loam, and the minimum decline has been from 495 to 313 pounds of seed cotton an acre (37 per cent reduction) on Vernon fine sandy loam.

Table 24 shows the comparative annual yields for the contrasting conditions of topsoil and exposed subsoil for three types of land used, respectively, for corn, cotton, wheat, and summer fallow.

EFFECT OF FERTILIZATION ON PRODUCTIVITY OF ERODED LAND. In the humid parts of the country, all other things being equal, yields from land treated with fertilizer will run generally higher than from areas not so treated. In this connection, however, it is instructive to observe (Table 25) the effects on cotton yields produced by the application of 400 pounds per acre of 4-8-4 fertilizer on two plots of Kirvin fine sandy loam (one with a cover of topsoil and the other with the subsoil exposed), near Tyler, in east Texas. Although the average per acre yield was increased by the application of fertilizer in both instances, it will be noted that the average subsoil yield, even after fertilization, was still 43 per cent less than the yield from unfertilized topsoil. When cotton was grown on Cecil sandy clay loam, one of the most important soils of the southern Piedmont, the yield from fertilized subsoil was 20 per cent below that from the corresponding unfertilized topsoil. The yield of corn on unfertilized topsoil of Muskingum silt loam, in southeastern Ohio, was twenty times the yield from the corresponding subsoil, even where it was fertilized.

Similarly, the subsoil of the Nacogdoches fine sandy loam, on the same Texas farm, produced cotton at an average per acre yield 71 per cent

TABLE 24.—ANNUAL YIELDS FROM TOPSOIL AND CORRESPONDING SUBSOIL OF THREE TYPES OF FARM LAND[1]

| Type and location | Rainfall, inches | Year | Crop | Yield per acre | | Decline in yield, per cent |
|---|---|---|---|---|---|---|
| | | | | Topsoil | Subsoil | |
| Muskingum silt loam, 12 per cent slope, Ohio | 18.4[2] | 1933 | Corn | 60.9 bu. | 0.9 bu. | 98 |
| | 25.0 | 1934 | Corn | 26.3 bu. | Very small | 99* |
| | 41.3 | 1935 | Corn | 28.0 bu. | 2.3 bu. | 92 |
| | 39.1 | 1936 | Corn | 17.0 bu. | 0 bu. | 100 |
| Kirvin fine sandy loam, 8.75 per cent slope, Texas | 36.1 | 1931 | Cotton | 490 lb. | 20 lb. | 96 |
| | 46.7 | 1932 | Cotton | 220 lb. | 50 lb. | 77 |
| | 44.3 | 1933 | Cotton | 510 lb. | 90 lb. | 82 |
| | 35.2 | 1934 | Cotton | 240 lb. | 40 lb. | 83 |
| Palouse silt loam, 30 per cent slope, Washington | 30.0 | 1933 | Wheat | 44.4 bu. | 8.9 bu. | 80 |
| | 19.7 | 1934 | Fallow | | | |
| | 14.6 | 1935 | Wheat | 51.3 bu. | 20.0 bu. | 61 |
| | 14.5 | 1936 | Fallow | | | |

[1] Measurements at soil and water conservation experiment stations, Soil Conservation Service.

[2] For July to December, inclusive, only.

* Estimated.

below the yield obtained from the corresponding topsoil, where both soil and subsoil were treated with 400 pounds of 4-8-4 fertilizer per acre.

TABLE 25.—AVERAGE ACRE YIELDS FROM FERTILIZED AND UNFERTILIZED TOPSOIL AND CORRESPONDING SUBSOIL OF THREE IMPORTANT TYPES OF FARM LAND[1]

| Type and location | Rainfall, inches | Years, inclusive | Crop | Average yield, topsoil | | Average yield, subsoil | |
|---|---|---|---|---|---|---|---|
| | | | | Not fertilized, pounds | Fertilized, pounds | Not fertilized, pounds | Fertilized, pounds |
| Kirvin fine sandy loam, 8.75% slope, Texas | 40.52 | 1931 to 1934 | Cotton | 365 | 580 | 50 | 206 |
| | 40.82 | 1931 to 1936 | Cotton | .......... | 507 | ....... | 176 |
| Cecil sandy clay loam, 10% slope, North Carolina | 46.44 | 1932 to 1934 | Cotton[2] | 950 | 1,123 | 290 | 759 |
| | 51.40 | 1935 to 1936 | Cotton[2] | .......... | 923 | 50 | 774 |
| Muskingum silt loam, 12% slope, Ohio | 41.40 | 1935 to 1936 | Corn | 22.5 bu. | 28.6 bu. | ....... | 1.1 bu. |
| | | | Stover | 2,487 lb. | 2,750 lb. | ....... | 525 lb. |

[1] Measurements at soil and water conservation experiment stations, Soil Conservation Service.

[2] Nonfertilized cotton was continuous; the fertilized topsoil received green manure.

These results indicate that, for the soils tested, erosion drives down the productivity of land violently, reducing it, at the stage marking complete removal of topsoil, to a condition where ordinary methods of soil building generally fail by a wide margin to restore the productiveness of the depleted land. It should be observed further in this connection that erosion does not stop at the stage of topsoil removal; on the contrary, it usually speeds up at that stage of planation, cutting away the subsoil to expose eventually the deep-lying raw parent material (*C* horizon), which usually is less productive than any of the layers above:

COMPARATIVE PRODUCTIVITY OF TOPSOIL, SUBSOIL, AND SUBSTRATUM MATERIAL OF CECIL SANDY LOAM, NEAR SPARTANBURG, S. C., 1936

| *Degree of Erosion* | *Average Yield per Acre, Pounds Seed Cotton* | *Reduction in Yield, Per Cent* |
|---|---|---|
| Surface soil (*A* horizon)............ | 1,609 | |
| Subsoil (*B* horizon)................ | 234 | 85.4 |
| Substratum (*C* horizon)............ | 74 | 95.4 |

It is true that the subsoil of some types of land is much more productive, relatively, than that of other types, as shown in Table 23; but in most instances, if not all, the exposed subsoil is less productive than the surface soil that once covered it. Lowering of the productivity level as the result of severe erosion seems to be more pronounced in the humid region than in arid and semiarid areas. Unfortunately, the subject has not been investigated adequately for the low rainfall areas,[1] but field observations have shown that grass will return on erosion-exposed subsoil of some types of dry land soil at a surprising rate where the land is protected from heavy grazing. This may be due in part to less severe soil leaching in regions of light precipitation and the consequent more favorable supply of plant nutrients; it may be partly due, also, to differences in the effects of the activities of microorganic life in some of the soils of low annual rainfall areas, as compared with soils of more humid areas.

Although it is true that the productivity level of erosion-exposed subsoil can be raised very materially through the use of fertilizer, manure, green manure, and crop rotations, perhaps up to or, in some instances, even above the productivity of the original topsoil (especially in case of some of the more leached sandy soils), field observations throughout the nation indicate that, for the more important types of farm land, subsoil yields generally cannot be made to equal those of the virgin soil by any practical type of farming. Similar treatment of comparable areas of exposed subsoil and of land retaining a reasonable depth of topsoil will, with very few exceptions, maintain the productivity level of the latter well above that of the severely eroded areas, as indicated by the results shown in Table 25.

In 1931, comparative acreage yields of oats produced in adjacent areas of Shelby loam, one with the soil intact, the other with subsoil exposed (at the Bethany, Mo., soil erosion experiment station), were as follows:

| | *Yield* | |
|---|---|---|
| *Treatment* | *On Topsoil, Bushels* | *On Subsoil Bushels* |
| None | 37.0 | 9.8 |
| Lime, superphosphate, and manure | 41.2 | 19.8 |

In other words, the exposed subsoil, although limed and liberally supplied with superphosphate and manure, produced under identical conditions of slope, rainfall, and cultural treatment only a little more than half as much as the untreated topsoil.

A further important consideration from the standpoint of economy is that erosion-exposed, raw subsoil clay is more intractable than the humus-

[1] Forsling, C. L., A Study of the Influence of Herbaceous Plant Cover on Surface Runoff and Soil Erosion in Relation to Grazing on the Wasatch Plateau in Utah, U. S. Dept. Agr. *Tech. Bull.* 220, 1931.

enriched, mellow layer that once overlaid it and for this reason is more difficult and expensive to till. The cost of fertilizer, manure, and green manure used to replenish the productivity of severely eroded land is a matter of no small economic importance to those who cultivate erosion-impoverished land.

## Cause of Reduced Productivity of Severely Eroded Land

Among the more important apparent causes of the relatively low yields obtained from erosion-exposed subsoil are: deficiency in organic matter, impaired structural efficiency, and reduced availability of moisture and plant nutrients.

Table 26 shows the low content of organic matter and nitrogen in erosion-exposed subsoil of a number of important types of farm and grazing land, as compared with the topsoil of corresponding types collected from near-by virgin areas or land only slightly eroded. It will be observed that the content of organic matter is higher in all the samples of topsoil and much higher in many of them, the average being 4.3 per cent for the uneroded and 1.0 per cent for the eroded. Similarly, the content of nitrogen in the samples for which this constituent was determined averages 0.24 per cent for the soil as against only 0.09 per cent for the subsoil.

There is considerable evidence that available plant nutrients, other than nitrogen, also run higher in the layer of naturally conditioned topsoil than in the unconditioned (less weathered) raw material of the subsoil. Beyond this, the availability of both plant nutrients and moisture contained in mellow topsoil, easily penetrated by rootlets and root-hairs, exceeds that in freshly exposed subsoil material, especially clay subsoil too tough for easy penetration by roots.

Eroded and uneroded Kirvin fine sandy loam used for cotton in adjacent areas at the Tyler, Tex., erosion station, carried practically the same amount of moisture as an average throughout the year—24.5 and 24.2 per cent, respectively, for the 6-foot profile. During a late summer drought, however, the area with topsoil retained carried 25.6 per cent as against 20.8 per cent for the exposed subsoil. That the yield of cotton on topsoil was at the rate of 490 pounds an acre, as compared with only 20 pounds an acre on subsoil, indicates that under the conditions of more favorable structure, organic supply, and probably other factors, the uneroded soil had a much better reserve of available moisture for the exigencies of drought.

At the beginning of the 1931 growing season on May 13, at the Red Plains soil and water conservation experiment station near Guthrie, in central Oklahoma, an area of virgin soil—Vernon fine sandy loam—planted to cotton, contained approximately the same amount of moisture as an adjoining area from which the topsoil had been removed, that is,

15.08 and 15.38 per cent, respectively; but at harvest time, Sept. 1, the virgin soil contained 5.91 per cent of moisture, and the exposed clay

TABLE 26.—EFFECT OF EROSION ON ORGANIC MATTER CONTENT OF 12 IMPORTANT AGRICULTURAL SOILS ORIGINALLY IDENTICAL OR NEARLY SO

| Soil and location | Uneroded[1] | | | | Eroded | | | |
|---|---|---|---|---|---|---|---|---|
| | Depth, inches | Condition | Organic matter, per cent | Nitrogen, per cent | Depth, inches | Condition | Organic matter, per cent | Nitrogen, per cent |
| Carrington gravelly loam, Rice Co., Minn. | 1½ to 9 | Forest,[2] gr. loam | 8.0 | 0.39 | 0 to 9 | Cultivated, gravelly sandy loam | 2.8 | 0.13 |
| Carrington loam, Rice Co., Minn. | 0 to 4 | Grass, loam | 5.0 | 0.21 | 0 to 8 | Cultivated, clay loam | 2.7 | 0.08 |
| Palouse silt loam, Whitman Co., Wash. | 0 to 16 | Cultivated, silt loam | 3.4 | 0.18 | 0 to 5 | Cultivated, silty clay loam | 1.4 | 0.10 |
| Shelby loam,[3] Nodaway Co., Mo. | 0 to 7 | Virgin hardwoods | 4.3 | 0.19 | 0 to 5 | Cultivated, clay | 1.3 | 0.06 |
| Manor silt loam, York Co., Pa. | 0 to 9 | Hardwoods | 5.3 | .... | 0 to 4 | Cultivated, clay loam | 0.2 | |
| Dekalb silt loam, Meigs Co., Tenn. | 0 to 6 | Virgin pine-hardwoods | 5.7 | .... | 0 to 7 | Second-growth pine, clay loam | 1.1 | |
| Dutchess gravelly loam, Dutchess Co., N. Y. | 0 to 7 | Virgin hardwoods | 6.2 | .... | 0 to 4 | Abandoned clay | 0.2 | |
| Frederick silt loam, Washington Co., Ind. | 0 to 3 | Virgin hardwoods | 3.0 | .... | 0 to 3 | Pasture, clay | 0.2 | |
| Susquehanna silt loam, Marengo Co., Ala. | 0 to 4 | Virgin pine and hardwoods | 2.6 | .... | 0 to 4 | Cultivated, clay | 0.4 | |
| Mountain gravelly sandy loam, Garfield Co., Utah | 0 to 10 | Brush, grass, grazed, gravelly sandy loam | 1.4 | .... | 0 to 10 | Brush, overgrazed, gr. sandy loam | 0.4 | |
| Boone loamy fine sand, Jackson Co., Wis. | 0 to 10 | Forest, loamy fine sand | 2.3 | .... | 0 to 10 | Cultivated, loamy fine sand | 0.6 | |
| Hays silt loam, Rooks Co., Kans. | 0 to 6 | Grass, silt loam | 4.1 | .... | 0 to 6 | Cultivated, clay loam | 0.6 | |
| Average.................... | ..... | ............ | 4.3 | 0.24 | ..... | ............ | 1.0 | 0.09 |

[1] Uneroded or only slightly eroded.
[2] 0–1½ inches consists of forest mold containing 23.7 per cent organic matter and 1.15 per cent nitrogen.
[3] Average of two samples.

subsoil 7.12 per cent. The fact that the former produced 162 pounds of cotton per acre, as against only 98 pounds from the subsoil, indicates that moisture consumption by the much heavier vegetative growth on the virgin soil was considerably greater than on the area representing severely eroded soil.

This kind of plant behavior under the sharply contrasting environments of eroded and uneroded soil has been observed on numerous comparable areas at the soil and water conservation experiment stations throughout the country. It is also an obvious characteristic that can be seen over the country by comparing plant growth on vast areas and numerous kinds of eroded and uneroded (or slightly eroded) land, originally the same (Fig. 75). That plant growth is almost universally

Fig. 75.—Good cotton on slightly eroded flat limestone soil and almost complete failure of cotton on eroded slope (foreground) of soil that originally was the same. Floyd County, Georgia, October, 1936. (*Photograph by Soil Conservation Service.*)

inferior on severely eroded land, where clay subsoil is exposed, is undoubtedly due, in a large measure, to the fact previously pointed out that a smaller proportion of the moisture rigidly fixed in the smaller interstices of clay is available to plants than of that contained in the larger cavities of loamlike, easily penetrable topsoil. In other words, the *moisture efficiency* of a soil, which obviously has so much to do with productivity, depends, to a large degree, on the *structural efficiency* of that soil. Structural efficiency, as pointed out in "Effect of Structure" (Chap. VI, Part 1), depends on such soil characteristics as texture, content of organic matter, character of colloids, and porosity.

Another interesting example of the plant-moisture relationship under the contrasting conditions of topsoil and subsoil is that of measurements made on Houston black clay in central Texas, at the Temple soil and water conservation experiment station. In 1931, with a light rainfall, the uneroded black clay surface soil at the beginning of the crop season (corn), on Apr. 10, contained 23.5 per cent of moisture as compared with only 16.8 per cent in the adjacent area of exposed subsoil (humus-free, light-colored chalky limestone and clay). The moisture in the uneroded soil continued to exceed that contained by the exposed subsoil through the growing season until about the time the corn began to mature, when the moisture supply of the former area dropped below that of the subsoil in the proportion of 13.8 to 15.9 per cent. The fact that the uneroded soil produced several times as much corn as the subsoil area probably largely explains the reversal of the moisture situation toward the end of the season, where again the heavier vegetative growth drew more heavily on the supply of available moisture. Here, too, the higher structural efficiency of the topsoil apparently had much to do with the greater availability of water when needed.[1]

## Shift from Land Impoverished by Erosion to More Productive Land

The yield of a given tract of cultivated land at a particular time is a more accurate measure of its productivity than average per acre yields of all the land within an area comprising important variations of soil and soil conditions. As already noted, acreage yields may be maintained or increased through abandonment of fields worn lean by continuing erosion or by shifting crops from less to more productive land.

Maintenance of average yields per acre through removal of poor land from production is illustrated by the results of investigations carried out in South Carolina.[2]

There was a decrease in the cotton acreage for that state from an average of 2,728,000 acres for 1912–1916 to an average of 1,415,000 acres for the period 1932–1936. In view of an increase in acreage yield from 215 pounds for the former period to 251 pounds for the latter, the indication is that much of the land removed from cultivation was of inferior productivity.

Several million acres of land in the rolling parts of the Southeast that formerly produced good cotton have been forced out of production because of severe erosion; most of it is either so infertile as to be practically worthless for further cultivation or so badly dissected with gullies that

[1] Soil Erosion, a National Problem, unpublished paper presented before the 3d Southwest Soil and Water Cons. Conference. Fayettesville, Ark., 1932, by H. H. Bennett.

[2] Fulmer, J. L. A Statistical Study of Agricultural and Related Trends in South Carolina, South Carolina Agr. Exper. Sta. *Bull.* 312, 1937.

plowing is a physical impossibility. Yields per acre, however, have been fairly well maintained for the whole section. Retirement of the depleted areas in favor of land still retaining a part or all of its topsoil has made this possible.

In Ohio, the average acreage yield of corn for the period 1920–1929 was practically the same as that for the decade 1870–1879. The acreage of corn declined by 20 per cent during this period in five counties in the southeastern section of relatively low-producing shale and sandstone land, but the area used for corn increased by 393 per cent in five counties of the rich glacial land in the northwestern part of the state.

It is therefore necessary to know what shifts in land use have been made in order to evaluate correctly acreage yields for large areas.

## Influence of Improved Cultural Technique on Crop Yields

Many factors other than the level of soil productivity have a bearing on maintenance of yields. As observed above, improvement in varieties

Fig. 76.—Soil-improving rotations are inducive to increased crop yields. Maryland. (*Photograph by Soil Conservation Service.*)

and strains of crops has been an oustanding scientific achievement of the past 50 years. Spread of soil-improving rotations (Fig. 76) and increased use of seasonal crops and soil-building legumes have contributed much to the stability of nation-wide yields, as have improved techniques and increased activity in the control of insect pests and plant diseases.

In 1879, with 223 million acres of *improved land tilled* (*including fallow and grass in rotation*), farmers of the nation spent more than $28,-

000,000 for about a million tons of commercial fertilizer; fifty years later, with 413 million acres of cropland, expenditures for the 7,500,000 tons of fertilizer used amounted to $271,000,000. North Carolina increased its expenditures for fertilizer seventeen times and its tonnage nearly sixteen times from 1879 to 1929 (from $2,000,000 to $34,000,000; from approximately 71,000 to 1,107,000 tons), but the cropland area increased by only 18 per cent (from 5,926,000 to 7,010,000 acres).[1]

In many localities, artificial drainage of land formerly too wet for cultivation by reason of swampiness or overflow has to a considerable degree relieved the stress on eroding upland fields. Large areas of rich alluvial land have been brought into use in many parts of the country, notably along the lower Mississippi alluvial plain and numerous stream bottoms throughout the humid sections of the Mississippi Valley. According to the Census (1930), irrigation has turned some 19 million acres of formerly unproductive dry land to productive use, with yields in many instances exceeding those of virgin areas within the humid belt.

Marked improvement in farm machinery designed for more efficient tillage, seeding, placement of fertilizer, and distribution of manure, together with increased use of such machinery, have contributed to more efficient yields, as have various other cultural practices. The effects of these advances are not easily measurable, but there is no doubt that the combined effect of all elements of technical agricultural progress has aided greatly in the maintenance of yields over the last half century.

Obviously, yields should have increased much more than they have, if erosion had not cut so deeply into the productivity level of a vast aggregate area. In this connection, R. M. Salter, R. D. Lewis, and J. A. Slipher say[2]:

> "Certainly . . . all of these changes and improvements should have raised acre-yields considerably—how much, it is difficult to say exactly, but we believe an increase of 40 to 60 per cent would have been conservative. . . . There can be but one explanation for the stubbornness with which acre-yields have resisted the farmer's efforts to improve them. The natural productive capacity of the land has been deteriorating at a rate almost fast enough to offset all these improvements in soil and crop management. With every step ahead we have slipped back almost if not quite as far."

[1] All these data, except the fertilizer tonnage figures for 1879, are from the Tenth Census of the United States, 1880, and the Fifteenth Census of the United States, 1930. The fertilizer tonnage figures for 1879 are based on estimates made by the National Fertilizer Association of Washington, D. C.

[2] Our Heritage, Ohio Agr. Exper. Service *Bull.* 175, p. 5, 1936.

# Chapter X. Relation of Erosion to Vegetative Changes

One of the most conspicuous effects of the violent soil changes produced by accelerated erosion is expressed in the altered responsiveness of the land to plant growth.

Bodily, progressive soil washing strips off the most productive part of the land—the humus-charged, spongelike layer, with its host of microorganisms and relatively high content of readily available plant nutrients. In place of the original topsoil characterized by optimum conditions of humus supply, mellowness, granulation, and openness or permeability, such as favor infiltration of rainfall, conservation of moisture, and ready penetration by fibrous plant roots seeking moisture and food, raw subsoil of opposite, or nearly opposite, characteristics is exposed at the surface. Such freshly exposed material is of much lower organic-matter content; usually it contains more clay and is stiffer (dense and tenaceous), less absorptive of rainfall, and much more subject to running together, baking (hardening), cracking, and desiccation. Erosion thus replaces favorable soil with unfavorable subsoil, to create a new plant environment. As a result, the original type of vegetation is markedly or completely altered over millions of acres of grazing land and abandoned or idle farm land.[1]

Such change in the vegetative responsiveness of the soil is not only of interest as a geographic and ecologic problem but of great economic importance in relation to the productiveness of forest and range land and to hazards of floods and siltation.

## Relation of Soil to Vegetation

Soil type exerts marked influence on vegetative characteristics, because of the varying adaptive factors involved. For example, blueberries, slash pine, and some varieties of azalea do best on moist sandy soils of acid reaction, whereas black walnut, redbud, and alfalfa prefer well-drained soils of neutral or alkaline reaction. Such edaphic characteristics pertain not only to plants in their natural environment but to many of

[1] Bennett, H. H. Cultural Changes in Soils from the Standpoint of Erosion, *Jour. Amer. Soc. of Agronomy*, Vol. 23, June, 1931.

the ordinary farm crops as well. For instance, certain varieties of sugar cane grown in Cuba that thrive on deep moist soils give poor yields on dry shallow upland (as POJ 2878), whereas other varieties, giving but indifferent yields on the same moist soils, get along nicely under the environment of shallow, droughty soil (as POJ 2727).

This profound effect of soil character on plant growth is strikingly revealed in many of the well-drained cane fields of the Caribbean Islands. On certain upland areas, the plants remain green and continue to grow throughout the dry season, normally of five to six months' duration, whereas those of the same variety on adjacent areas of markedly different soil parch, stop growing, or die. Examination of these areas invariably reveals that the green fields are on mellow soil, usually rich in lime and humus, such as retain favorable supplies of moisture throughout the dry period; the fields of dormant or dead plants, on the other hand, invariably are restricted to soils which, though occupying the same degree of slope and receiving the same rainfall and cultural treatment, lose their available moisture soon after cessation of the rains. Here, soil character is sufficiently potent in its relation to growth of sugar cane (and a number of other crops) to maintain the equivalent of a humid climate within a temporary desert area.[1]

Exactly the same behavior may be witnessed in the corn, wheat, and cotton fields of America, especially where severely washed, soil-stripped land is associated with land still retaining its topsoil or a considerable part of it. Almost invariably, crops growing on erosion-exposed subsoil suffer much more in dry weather than those growing on uneroded soil. This is particularly true of the more succulent plants, as corn, vegetables, and melons. Unquestionably, the difference is the result of (1) deficiency of *effective moisture* (moisture available to plants) and (2) deficiency of readily available plant nutrients in the areas of exposed raw clay.

If soils differ in their virgin condition sufficiently to affect plant growth to the extent of counterbalancing an environmental factor equivalent to the difference between a humid climate and a desert climate, such as can be seen in the sugar-cane fields of Cuba every year, even on freshly cleared forest land, what, then, is to be expected when the original differences between two such soils are still further increased by the complete removal of that part of one of them, the humus-charged topsoil, which (of all the soil layers from the surface deep into the subsoil) most nearly matches the corresponding layer of the other type, and which (of all the layers) normally has the greatest capacity for manufacturing available plant food? In other words, soils whose responsiveness to plant growth varies widely in the virgin condition frequently show even greater varia-

[1] Bennett, H. H. "Some Cuban Soils," a supplement to the "Soils of Cuba," *Sci. Contr.* 22, Tropical Plant Research Foundation, Yonkers, N. Y., 1932.

tion following removal of the surface layer from one of them. As between two inherently different soils, the normal tendency of erosion is not merely to minimize or eliminate the soil-equalizing effect of the organic matter and the population of microorganisms, contained chiefly in the uneroded topsoil, by removing this layer; but it is to impoverish land, also, by bringing quickly to the surface unweathered, unconditioned (in relation to normal soil plant growth) material entirely unlike and inferior to that which has been bodily swept away. If it can be called soil, then this freshly exposed material obviously is of a very different type, one constituting an entirely changed plant environment.

## Conditions in Oklahoma

In 1930, an erosion survey[1] of Oklahoma was completed that showed that of approximately 16,000,000 acres in cultivation in the state, 13,000,000 were subject to accelerated erosion, nearly 6,000,000 acres having reached the stage of gullying, with 374,000 acres, still included in fields, so badly gullied that good farm machinery could not cross the devastated areas. The survey revealed further that during the "last three or four years" 1,359,000 acres of formerly tilled land had been abandoned because of erosion.

The vegetative changes from the original virgin post oak forest and "bluestem prairies" of many localities were found to have been so great over the rolling part of the state that this feature was helpfully employed as a soil-condition indicator in connection with the classification of certain types of idle eroded land. In the Red Plains region, the original valuable prairie vegetation (particularly that on the very extensive type of farm land, the Vernon fine sandy loam) composed chiefly of Andropogons (as little bluestem and big bluestem), together with some grama grasses, has been largely replaced on the erosion-exposed, droughty red-clay subsoil by an essentially worthless sod composed of weeds and such inferior forage plants as the needle grasses.

This sweeping vegetative change has accompanied an almost hopeless decline in the economic value of a tremendous aggregate area of farm land within the Red Plains of both Oklahoma and Texas.

## Hardwoods Replaced by Persimmon and Sassafras

In the east Texas sandy lands region, where numerous fields have been abandoned because of erosion, thickets of small persimmon trees are very common, especially on the red soils of the Kirvin group.[2] Almost

[1] Oklahoma Agr. and Mechanical Coll. *Circ.* 79, 1931.

[2] For description of the Kirvin soils, see Soil Survey, Nacogdoches County, Texas. Field Operations, Bur. Chemistry and Soils, U. S. Dept. Agr., 1925.

without exception, examination of these sites shows that they consist of stiff red clay or a very thin covering of comparatively infertile grayish sandy loam over red clay. Remnants of the regional virgin hardwoods, including principally oaks and hickory, reveal that all these soils once had a covering of about 6 to 8 or 10 inches of light-brown to yellow fine sandy loam over red clay, the true topsoil or humus-charged layer averaging about 4 or 5 inches in depth. This surface soil has been partly or completely removed, and the lighter colored subsurface layer also, over thousands of acres. Following such denudation, persimmon trees and smilax have sprung up abundantly, along with broom sedge, poverty grass, and a variety of worthless weeds. Wherever seed trees occur, there usually has been, also, a rather wide dispersal of pine.

In the rolling areas of the formerly rich limestone soils of the Highland Rim section of southern Kentucky, the thick clumps of sassafras almost universally present in abandoned fields usually mean simply one thing: that the former covering of mellow silt loam, richly charged with organic matter, has been washed off down to much stiffer clay material which dries out and hardens much more quickly. The vegetation coming in on the abandoned fields is entirely different from that of the hardwoods originally covering these productive uplands, and it is not known what length of time would be required for reestablishment of anything like the virgin stand of timber.

In 1930, a soil survey was completed of Washington County, Indiana, where the base map was prepared from air photographs made during the very dry season of that year. Much land has been completely or partly abandoned here, because of excessive soil washing—something like 25 to 30 per cent of the entire county as indicated by the air map and soil survey. The vegetation on the critically eroded areas is so unlike that where all or part of the original topsoil still remains, it was an easy matter to pick out the denuded areas on the photographic sheets. Abandoned fields which had washed down to clay were essentially bare of vegetation during the excessively dry period referred to, and those old fields which had lost all but a thin covering of the original silt loam supported principally broom sedge. These features are all plainly legible on the aerial map (Fig. 77).

Detailed studies of a number of the virgin profiles of the limestone soils of this southern Indiana county, which originally supported a splendid stand of oak, hickory, maple, elm, basswood, cherry, buckeye, and walnut, show that the immediate topsoil of the most extensive types is considerably more alkaline than the sublayers. The deeper lying substratum material, however, is more alkaline than either the topsoil or the intermediate layers. Apparently, the plant roots, bringing up soluble basic constituents from the alkaline substratum and storing them in the leaves of forest trees, have in this indirect way transferred alkaline material to the surface soil, to make it more productive than the layers immediately beneath. It was found that the immediate topsoil, that is, the mineral soil beneath the forest litter, of the Frederick and Canton silt-loam soils, is nearly neutral, whereas the various layers of the subsoil proper are of acid reaction, some of them very acid. Also, the material of the substratum was found to be acid in the upper part and decidedly

Fig. 77.—Air photograph of about 317 acres, Washington County, Indiana. Twenty-five per cent timbered (dark clumpy areas), with one-half representing rather scrubby second growth on land which was abandoned because of erosion; 11 per cent plowed in 1930, though partly eroded (light-colored geometric patterns); 64 per cent abandoned because of erosion (light-colored areas of irregular outline, deeply eroded and supporting little or no vegetation; and gray areas representing abandoned fields, now covered largely with broom sedge). Total abandonment within a rather large surrounding similar area of formerly highly productive limestone soil, because of impoverishment by erosion, amounts to approximately 75 per cent of the area. Approximately this condition obtains over a very large aggregate area in the rolling limestone regions of southern Indiana, southern Ohio, Kentucky, and Tennessee. (*Courtesy, Purdue Univ. Agr. Exper. Sta.*)

nonacid (alkaline) beneath, where freshly decomposed limestone material is present. It was shown, too, that the humus cover, to depths of about 1, $1\frac{1}{2}$, or 2 inches, ranges from approximately neutral to alkaline in every instance.

These characteristics of the successive layers through the vertical soil section indicate what happens to the productivity of such lands under those unwise farm practices which permit the topsoil to wash off. Actually, bluegrass growing abundantly on the comparatively "sweet" (alkaline) virgin soil has been displaced, on the exposed acid, droughty subsoil, principally by persimmon, broom sedge, and worthless weeds, with numerous areas supporting no vegetation or only scattered weeds and grass of dwarfed growth.

## Effect of Erosion in the Southern Appalachians

While a reconnaissance survey of erosion conditions in the Southern Appalachian region was being made during 1931, it was observed that practically all the broad-leaved forests had been cleared from the slopes of numerous valleys within the Blue Ridge Mountains (where the prin-

FIG. 78.—The original granite-derived soil (Cecil sandy clay loam) of this Southern Piedmont area was first cultivated until the soil was completely removed by sheet erosion, as well as part of the subsoil, and then abandoned. Volunteer pines covered the area, but failed to stop the washing, especially in the gullies. Continuing erosion removed the more productive part of the subsoil down to comparatively unproductive deep subsoil, and now the pines not only are making no headway, but many of them are dying. In other words, this eroded area continued to wash after abandonment toward a still more impoverished condition in spite of the second growth.

cipal soils are derived from granitic rocks and micaceous schists) for a distance ranging from about ½ to ¾ mile or more up the slopes rising from the alluvial plains bordering the streams. About one-third to more than three-fourths of these formerly cultivated areas have been abandoned because of the severity of erosion. The character of the vegetation coming in after the stripping off of the surface layer (as well as the subsoil in many instances) contrasts sharply with the former heavy forests of oak, chestnut, hickory, maple, sourwood, cucumber tree, tulip poplar, rhododendron, laurel, azalea, etc. Most of the abandoned land is covered with pine, although some areas support dense stands of tulip poplar.

Out in the Piedmont country, east of the Blue Ridge, where erosion has devastated numerous areas, a variety of vegetative changes have taken place on the extensive Cecil soils (granite-derived soils having brittle, brick-red clay subsoils), following the gradual washing off first of the topsoil and then of the subsoil, on down into the lighter colored substratum, consisting of intermingled clay and particles of disintegrated rock. The exposed subsoil (that is, the upper subsoil) usually is farmed for awhile with the aid of fertilizers, although all crops suffer on it during droughts, corn frequently failing completely at the same time when fairly good crops are produced on the associated less severely washed soils of the same original type. Continuing erosion lowers the surface of the ground until finally the poor, intractable deep-subsoil material is reached, at which stage the land is very nearly worthless for crop production. Even the pines that grew well on the upper subsoil now fail to make much headway. They are characteristically stunted; some of them actually die as the result of continuing washing (Fig. 78).

Many of these deeply eroded areas are bare of vegetation, save for an occasional dwarfed blackberry or smilax. The broom sedge, blackberries, and weeds that flourished on the upper subsoil of the fields that were thrown out of cultivation immediately following the stripping off of the surface soil usually grow stuntedly and sparsely, or not at all, on this essentially sterile material, representing the last stages of erosion before exposure of bedrock.

W. D. Lee has the following to say with respect to vegetation on the Cecil clay loam of Burke County, North Carolina,[1] an erodible soil (usually representing exposed subsoil of areas that formerly consisted of Cecil sandy loam and loam):

> "Abandoned fields first grow up in broom sedge and brambles, followed the second year by sassafras, sumac bushes, and yellow pine, and in a few years, except on badly eroded areas, there is a good stand of pine. Forested areas support a fair or good growth of white, red, black, post, scarlet, and chestnut oaks, shortleaf or yellow pine, spruce, pitch and white pines, hickory, black gum, yellow poplar, dogwood, and a few persimmon, locust, sourwood, black walnut, white elm, sweet gum, red cedar, and hemlock trees."

This represents the natural secondary succession in old fields, which in many localities tends to remain more or less permanently in the stage of a pine subclimax as a result of frequent fires. But where serious erosion continues, pine woods cannot develop effectively, and even the broom-sedge stage is much interrupted and dwarfed.

[1] Soil Survey, Burke County, North Carolina. Bur. Chemistry and Soils, U. S. Dept. Agr., 1926.

The following observations[1] relate to the far-reaching vegetative changes that have taken place over millions of acres throughout the southern Piedmont region, mainly on the Cecil soils:

"The original forest growth was quite different from the forests of the present time. On the highlands the oak, hickory, and chestnut were of large growth and stood far apart. There was no underbrush and the woods were carpeted with grass and the wild pea vine. Along the streams and in the valleys the distinctive growth was willow, beech, birch, black walnut, ash, poplar, and gum. The cane also flourished best here, although it often grew upon the higher ground. The cane growth was the standard by which the early settlers estimated the value of the land. If it grew only to the height of a man's head, the land was esteemed ordinary, while a growth of from 20 to 30 feet indicated the highest fertility.

"Not only the forests, but the cultivated fields as well, present a very different aspect now from what they did after the country was first opened up. It was then new and beautiful and as remarkable for the luxuriant richness of its landscape as it is now for the striking features of its rolling hills and long, narrow valleys. The original forest has disappeared almost entirely, and has been replaced by scrubby oaks, by underbrush, and by shortleaf pines of the abandoned fields. The chestnut and chestnut oaks have been dying out for the past sixty years, and the cane has likewise almost disappeared."

Second-growth pine in the Piedmont region usually means eroded land wherever found in essentially pure stands on sloping areas. Much of it grows where the original soil has completely disappeared. A rough picture of what has taken place can be formulated from conditions in Spartanburg County, South Carolina. In this one county alone, 297,216 acres have been classed and mapped as clay loam and sandy clay loam.[2] Examination of the soil profile as found in remnants of the virgin stands of mixed hardwoods and shortleaf pine ("forest pine") shows the original soil to consist of some 4 to 8 inches of brownish or yellowish, mellow sandy loam and loam. This top layer is gone, or largely gone, from 297,000 acres of clay loam and sandy clay loam. These soils (classed as Cecil clay loam, sandy clay loam, and gravelly sandy clay loam and as Louisa clay loam and sandy clay loam) are, in a sense, new soils, representing products of the excessive erosion that has taken place over the cultivated slopes.[3] They are markedly different from the virgin soils not only texturally but structurally as well. On the basis of studies pertaining to disturbed forest

[1] Soil Survey, Abbeville area, South Carolina. Field Operations, Bur. Soils, U. S. Dept. Agr., 1902.

[2] Soil Survey, Spartanburg County, South Carolina. Field Operations, Bur. Soils, U. S. Dept. Agr., 1921.

[3] Soil Survey, Union County, South Carolina. Field Operations, Bur. Soils, U. S. Dept. Agr., 1914.

conditions at the erosion experiment stations, on the same or similar soils, the most important structural change due to excessive erosion, from the standpoint of plant response, is a marked diminution of the effective pore space and an enormously increased tendency of the exposed clay subsoil to run together while saturated and to bake and crack during dry periods, with consequent serious losses of available moisture. Most of the original forest of hardwoods was removed long ago from the 209,152 acres of Cecil sandy clay loam of Spartanburg County. Now about 60 per cent of this land is under cultivation, the remainder largely covered with old-field pine. Occasionally, a few patches of second-growth white oak, hickory, or dogwood are seen, and here and there thickets of blackberry, plum, sassafras, and smilax.

## In the Northern Appalachians

On Nov. 6, 1931, a forester, an agronomist, and a soil specialist of the State of Ohio, an agricultural engineer from the United States Department of Agriculture, and the author walked over the erosion-denuded slopes of a large section about 3 miles northeast of Roseville, in the hills of the Appalachian border, southeastern Ohio. From one point, the view took in an abandoned area of approximately 2 square miles. All this had been cultivated and once was considered good soil. A fine stand of hardwoods was cut from the area long ago. Remnants of such woodlands attested the fact. Those lands were abandoned. The soil over the greater part had been worn away, down to a mixture of clay and shale, very deficient in calcium, phosphorus, and nitrogen. On this we found, principally, goldenrod, poverty grass (*Danthonia spicata*), dewberry, and blackberry. Patches here and there were bare of vegetation. The land was very acid; nutritious grasses were so scarce that a large proportion of the land was not even used for pasturage. The prevailing type of cover was of a kind that permits continuing erosion although at a slower rate than from bare ground.

In remaining clumps of the original forest type was a good ground cover of forest litter; below that, a half-inch layer of duff (mixed vegetable matter and mineral soil); and beneath the latter, a 7- to 9-inch layer of yellow-brown mellow silt loam of good organic-matter content. This, in turn, was underlain by 5 to 11 inches of yellow or buff-colored silty clay and finally by a mixture of pale yellow, acid clay and partly decayed shale fragments (representing profile of Muskingum silt loam). The lowest section corresponded to the material exposed over the greater part of the 2 square miles of eroded slopes as well as over many thousands of acres elsewhere throughout the county.

The question was asked: "What can be done with land of this kind?" The soils and crop specialists asserted that, without reseeding, a fair

Fig. 79.—Above, virgin hardwoods, Piedmont area near Flemington, N. J. Soil is brown mellow loam about 8 inches deep overlying reddish clay loam containing some rounded gravel (Washington loam). At depths of about 36 to 40 inches, red clay, such as that shown in the middle picture, is encountered. The middle picture shows a large abandoned area formerly cultivated from which 3 feet or more of soil and subsoil like that in the upper picture has been washed off. This area is now exceedingly poor and roughly gullied; about 40 per cent of it is bare or practically bare, and the remainder covered chiefly with poverty grass shown in the lower photograph.

stand of bluegrass could be established with the addition of lime and superphosphate, at a cost of about $10 an acre, the grass coming in voluntarily following such treatment. It was unanimously agreed that the average farmer could not afford such an expenditure on such denuded land. The forester was of the opinion that, although it would be exceedingly difficult and entirely impractical to reestablish hardwoods forest on such infertile, acid soil (subsoil), conifers probably could be established. Again, it was agreed by all that farmers probably would not, on their own resources, do much in this direction. It was further agreed that it was too late to accomplish very much with soil-saving measures, since so little soil was left to conserve. In other words, the soil poverty brought about by cultivating these steep slopes without provision for erosion control, resulting in stripping off from 8 to 20 inches of soil and subsoil, was so extreme that this group of specialists was unable to decide what remedy, if any, might be resorted to in order to restore those submarginal lands to any considerable degree of productivity beyond that which nature might accomplish through a long process of partial control by whatever vegetation should come in spontaneously. The magnitude of the problem unfolds itself when it is considered that in the one county where this worn-out land was found there was, at that time, more than 100,000 acres of practically idle land that once was cultivated, over which the condition of erosion and degraded vegetation described, or an approximation of it, obtained. This condition is quite general over many parts of the Allegheny Plateau in Pennsylvania, West Virginia, and Ohio.

Much the same thing is taking place in numerous other parts of the Northern Appalachian region. Figure 79 shows characteristic views of west-central New Jersey, in the Piedmont section, where erosion has seriously affected a large area.

Summing up the relation of erosion to vegetative changes over immense areas formerly occupied by hardwoods and mixed hardwoods-pine forests in eastern United States, these few observations and measurements give some conception of the widespread changes, such as have been, in almost countless localities, about as complete as could be imagined. The vast areas of erosion-exposed subsurface, subsoil, and substratum materials represent newly formed soils—new plant habitats or environments, having scarcely a semblance to former conditions and in comparison with which they are tremendously inferior.

## Changes in Texas Black Belt

Much the same thing has happened in the more rolling parts of the prairies and plains west of the eastern forests. When it is considered that one rain, on May 10, 1930, removed 23 tons per acre of black soil (actual measurement) from a 4 per cent slope of the central Texas Black Belt, near Temple, Tex.; and, furthermore, that every rain causing runoff from those areas devoted to clean-tilled crops carries away a load of topsoil,

FIG. 80.—Upper, virgin grassland ("bluestem meadow") in the Black Belt near Temple, Texas. The dense vegetation is composed of 82 per cent little bluestem, 7 per cent big bluestem, 5 per cent feather bluestem, 4 per cent dropseed, and 2 per cent of other grasses. Soil is Houston black clay, slope 3 per cent; no erosion. Lower, cultivated soil originally like that above. Black topsoil and most of the lighter colored subsoil washed off, then abandoned. After 4 years, more than 75 per cent of the ground surface was still bare of vegetation. Eroded land like this produces only about $\frac{1}{10}$ bale of cotton per year as an average.

it is not at all difficult to interpret the changes now rapidly taking place in this great area of originally highly productive prairie soil—the transmutation from a genuine "black belt" to a patchy composite of (1) rich black clay soil on the more nearly level areas, (2) less productive yellowish clay subsoil on numerous eroded slopes, and (3) whitish chalk (or marl) of relatively low productivity on erosion-exposed deep substratum.

The virgin soil was densely matted with nutritious prairie grasses (predominantly bluestem), which in spring and early summer were

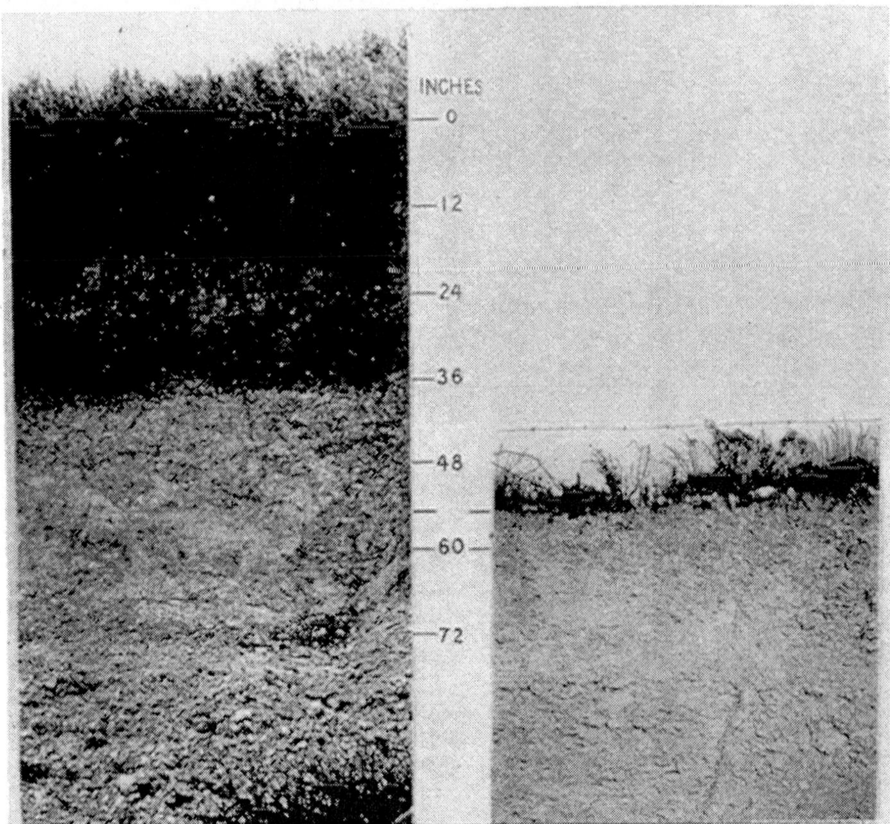


Fig. 81.—Virgin (left) and severely eroded (right) Houston black clay, Texas. Both areas originally were the same, occupying a $3\frac{1}{2}$-per cent slope. About 54 inches of soil and subsoil have been removed from the eroded area, following about 40 years of cultivation. The vegetation on the virgin soil is largely composed of grasses, with 96 per cent of the surface area covered; while that on the erosion-exposed chalk to right (corresponding to the substratum of the virgin area, as indicated on the scale) supports principally weeds, 60 per cent of the ground being bare.

beautifully sprinkled with numerous species of flowers. On the other hand, the exposed whitish chalk, resulting from 40 to 50 years' cultivation, now supports sparse stands of vegetation, consisting largely of unpalatable weeds (Fig. 80), whereas the erosion-bared subsoil proper, the intervening layer, supports chiefly weeds and stunted grasses, with a relatively high proportion of unoccupied ground surface. The virgin soil produced upward of a bale of cotton an acre; the exposed chalk produces at the rate of about $\frac{1}{10}$ bale in years of favorable rainfall and at the rate of about $\frac{1}{20}$ to $\frac{1}{40}$ bale in dry years.

The results of vegetative surveys on the principal soils of the region (the Houston black clay and its exposed sublayers of Sumter clay and chalk),[1] employing small plots of equal size (surveys carried out on same date, summer of 1931, in Bell County, Texas),[2] are given in Table 27. These surveys were made to determine as accurately as possible the vegetative changes that have taken place as the direct result of soil washing on the principal soil type of the region (Fig. 81). The conditions revealed are duplicated, or essentially duplicated, in numerous instances throughout the extensive rolling parts of the Texas Black Belt. In connection with the changes shown in this table, it should be observed that chalk was not present at the surface of the areas studied. Where chalk is actually exposed, the percentage of bare ground is relatively larger than where only the yellowish-brown clay subsoil (Sumter clay) is exposed; the composition of the cover is more largely of weedy growth; and recovery with respect to reestablishment of grasses, following abandonment, has been, generally, much less.

## Changes in the Corn Belt

The Shelby loam, a glacial drift soil,[3] of northern Missouri, southern Iowa, northeastern Kansas, and southeastern Nebraska, originally was densely covered with true prairie grasses, such as the bluestems and side-oats grama, more or less invaded by the introduced bluegrass of the East (Fig. 82), growing on dark-colored, mellow loam, highly charged with organic matter. Under continuous cropping to corn, this rich topsoil has been swept from innumerable areas by erosion, down to yellow clay subsoil, within a farming period of about fifty to sixty years on 4 per cent slopes and in about ten to twenty years on 8 per cent slopes. The exposed stiff, yellow clay produces little grass of any value and only about 20 bushels of corn per acre (no corn is produced in dry years), as against 60 bushels or more for the best years on uneroded or only slightly eroded soil (Fig. 83).

The vegetative changes resulting from erosion on this extensive prairie soil have been extremely violent—a change from almost exclusive stands of introduced bluegrass, in density of nearly 100 per cent ground cover, to scattering stands of weeds and dwarfed grasses of very low grazing value.

[1] Soil Survey, Milam County, Texas. Bur. Chemistry and Soils, U. S. Dept. Agr. (Series 1925), No. 25.

[2] Surveys by H. V. Geib, U. S. Dept. Agr., Bur. Chemistry and Soils, Blackland Station, Temple, Tex.; and Simon E. Wolff, Texas Agr. Exper. Sta.

[3] Soil Survey, DeKalb County, Missouri. Field Operations, Bur. Soils, U. S. Dept. Agr., 1914

Fig. 82.—Virgin Shelby loam (northwestern Missouri) with dense turf of bluegrass. Recent gullying due to excessive runoff from cultivated land along ridge crest.

Fig. 83.—Corn failure on severely eroded Shelby loam (photograph made in August); same farm as shown in Fig. 82.

In his studies of the effect of plant roots on erosion under Nebraska prairie conditions, Weaver says:[1]

"Under continued grazing the true prairie grasses are gradually replaced by bluegrass, by the short grama grass and buffalo grass, or by a mixture of the

[1] Weaver, J. E. Effects of Roots on Vegetation in Erosion Control. Paper presented before Am. Soc. of Agronomy (mimeographed). December, 1937.

TABLE 27.—COMPOSITION AND GROUND COVER OF VEGETATION ON VIRGIN AND SEVERELY ERODED HOUSTON BLACK CLAY, BELL COUNTY, TEXAS[1]

| Ground condition | No. 1: virgin grassland, 2 to 3 per cent slope | | | No. 2: severely sheet eroded, with some gullies, 3 to 8 per cent slope[2] | | | No. 3: 11–18 in. of soil and subsoil removed, 6 to 7 per cent slope[3] | | |
|---|---|---|---|---|---|---|---|---|---|
| | Area of ground cover, per cent | Common name | Botanical name | Area of ground cover, per cent | Common name | Botanical name | Area of ground cover, per cent | Common name | Botanical name |
| Bare | 6 | | | 35 | | | 60 | | |
| Grass cover | 93 | | | 6 | Buffalo grass | *Bulbilis dactyloides* | 1 | Silver beard grass | *Andropogon saccharoides* |
| | 65 | Indian grass | *Sorghastrum nutans* | | | | | | |
| | 14 | Little bluestem | *Andropogon scoparius* | | | | | | |
| | 9 | Drummond dropseed | *Sporobolus drummondii* | | | | | | |
| | 4 | Silver beard grass | *Andropogon saccharoides* | | | | | | |
| | 1 | Three-awned grass | *Aristida sp.* | | | | | | |
| | | Needle grass | *Stipa leucotricha* | | | | | | |
| | | Big bluestem | *Andropogon furcatus* | | | | | | |
| | | Buffalo grass | *Bulbilis dactyloides* | | | | | | |
| Weed cover | 1 | | | 59 | | | 39 | | |
| | | Sensitive briar | *Morongia angustata* | 34 | Snakeweed (Texas broomweed) | *Gutierrezia texana* | 18 | Wreath aster | *Aster multiflorus* |
| | | Kuhnia | *Kuhnia glutinosa* | | Snakeweed (broomweed) | *Gutierrezia dracunculoides* | 13 | One-seed croton | *Croton monanthyogynus* |
| | | Ruellia | *Ruellia tuberosa* | 20 | ...... | *Hartmannia sp.* | 4 | Kuhnia | *Kuhnia glutinosa* |
| | | Prairie clover | *Petalostemum multiflorum* | 3 | One-seed croton | *Croton monanthyogynus* | 2.75 | Texas lupine (blue bonnet) | *Lupinus subcarnosus (L. texensis)* |
| | | Gum elastic | *Bumelia lanuginosa* | 2 | Bee balm | *Monarda sp.* | 1 | Annual sunflower | *Helianthus annuus* |
| | | Great azure sage | *Salvia pitcheri* | | Annual sunflower | *Helianthus annuus* | .25 | Spurge | *Euphorbia bicolor* |
| | | ...... | *Vincetoxicum biflorum* | | Hoary bindweed | *Convolvulus incanus* | | | |
| | | | | | Spurge | *Euphorbia malaca* | | | |
| | | | | | Spurge | *Euphorbia serpens* | | | |
| | | | | | Nodding spurge | *Euphorbia nutans* | | | |
| | | | | | Ironweed | *Vernonia interior* | | | |
| | | | | | Hoarhound | *Marrubium vulgare* | | | |

[1] Eroded areas, though originally Houston black clay, now represent Sumter clay.
[2] Practically worthless, sometimes affording a little spring grazing.
[3] Formerly cultivated, now abandoned. Not plowed or grazed for 4 years.

three grasses. Not only is the forage yield greatly decreased, but the underground plant parts are reduced one-third or more. Despite this loss in weight, due to the finer fiber in the root system, the soil is still anchored quite as firmly as before. But under long continued grazing and trampling the bluegrass and short grass sod is broken, and weeds become abundant. A final stage in deterioration occurs where only fragments of these grasses are found and the dominant grasses are annual weeds such as wire grass, poverty grass, crab grass, and other weeds as woolly plantain and fetid marigold. These are relatively poorly rooted, much soil is only partly occupied by roots. . . . "

He points out that the soil under this final stage of grass deterioration erodes very rapidly.

## Erosion and Vegetation in Western Kansas

The effects of erosion with respect to soil removal in the western Kansas Wheat Belt are illustrated by an erosion survey of a 159-acre farm in Township 13 S, Trego County, Kansas, completed in 1931.[1]

The virgin grassland or mixed prairie of this region, as represented by the upland part of Class 7, Map 4, consists of dark-brown, mellow silt loam (Colby silt loam), averaging about 12 inches deep. The material beneath this top layer consists of brown silty clay, which at approximately 22 inches below the surface passes into light-brown, silty clay of a limy nature. By the time of the first cultivation, the mixed prairie type of cover had been reduced by grazing to a dense cover of short grasses; and hence this tract, representing a typical area of highly productive western Kansas wheat soil, was broken out of plains "short grass" in 1922, just 10 years prior to the making of the erosion survey. All the eroded areas were washing at a considerably accelerated rate, as compared with the rate measured at the Hays, Kans., erosion experiment station on practically virgin soil of the same original type occupying about the same slope. All the eroded areas on this Trego County farm had suffered marked reductions in productivity. The most severely washed portion was excessively droughty and had very low value for the regional crops. The composition of the vegetation on the typical virgin land, on moderately eroded land, and on severely eroded land (formerly cultivated but then untilled), all of which originally was nearly the same with respect to soil and slope, is shown, respectively, under sites 1, 5, and 6 in Table 28.

With respect to ground cover, the change from virgin grassland condition (site 1) to deeply eroded, unproductive land of the class representing site 6, came within 5 per cent of amounting to a complete reversal of the virgin condition—a switch from a condition of essentially 100 per cent vegetated surface to 95 per cent nonvegetated surface. The vegetation on

[1] Davis, R. H., U. S. Dept. Agr. at soil and water conservation experiment station, Hays, Kans.

both of the eroded sites, Nos. 5 and 6, contained not a single plant that was present in the native sod (all three plots of equal area).

Similar conditions of vegetative change on former grassland areas are to be seen in other localities within the Kansas plains as well as in numerous places in the other Plains states. For several years the attention of the nation has been focused on the Great Plains, primarily because of the great damage that this region has suffered from excessive wind erosion.

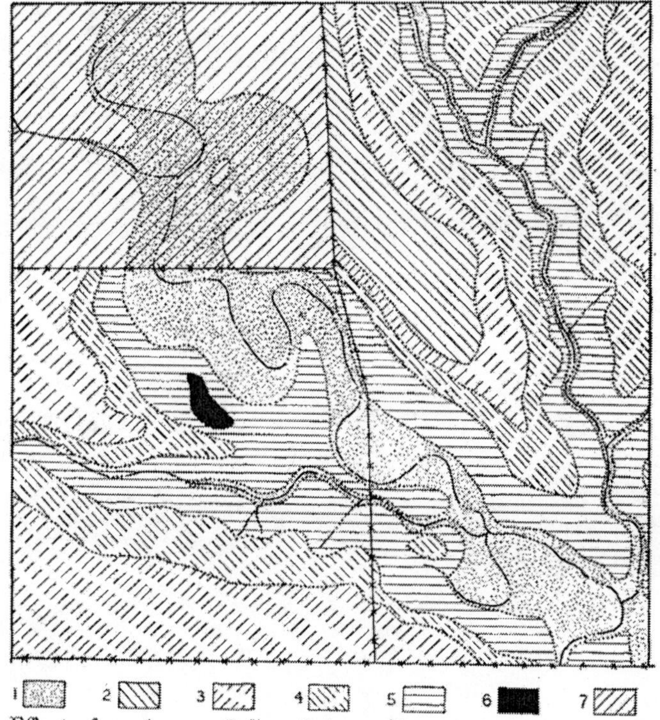


MAP 4.—Effect of erosion on Colby silt loam, Trego County, Kansas. 1, stream alluvium—no erosion but considerable deposition of eroded material (35.08 acres); 2, level, cultivated—very little erosion (7.7 acres); 3, gently sloping, cultivated—3 inches of soil removed by erosion (31.04 acres); 4, moderately steep, cultivated—3 to 6 inches of soil removed (30.6 acres); 5, moderately steep, cultivated—6 to 10 inches of soil removed (35.84 acres); 6, moderately steep, cultivated—all of soil and part of subsoil removed, down to 24 inches below the surface (0.5 acre); 7, level, and sloping upland and level alluvium—virgin grassland showing no erosion (18.04 acres).

Local regions in all the 10 states—Montana, Wyoming, North Dakota, South Dakota, Nebraska, Kansas, Colorado, Oklahoma, New Mexico and Texas—lying within the boundaries of the Great Plains have suffered from an increase in the number and intensity of dust storms. Probably the hardest hit and more widely known area is the so-called "dust bowl" of the central and southern Great Plains, comprising a large area in southeastern Colorado, southwestern Kansas, the Panhandles of Oklahoma and Texas, and northeastern New Mexico.

TABLE 28.—COMPOSITION AND GROUND COVER OF VEGETATION ON VIRGIN AND ERODED SOIL, TREGO COUNTY, KANSAS

| Ground condition | No. 1: Virgin grassland | | | No. 5: 6 to 10 inches topsoil removed | | | No. 6: All of topsoil and part of subsoil removed[1] | | |
|---|---|---|---|---|---|---|---|---|---|
| | Area of ground cover, per cent | Common name | Botanical name | Area of ground cover, per cent | Common name | Botanical name | Area of ground cover, per cent | Common name | Botanical name |
| Bare | 0 | ........ | ........ | 85 | | | 95 | | |
| Grass cover | 85 | 80 Buffalo grass | *Bulbilis dactyloides* | | | | | | |
| | | 10 Little bluestem | *Andropogon scoparius* | | | | | | |
| | | 10 Blue grama | *Bouteloua gracilis* | | | | | | |
| Weed cover | 15 | 60 Common ragweed | *Ambrosia elatior* | 8 | 85 Redroot amaranth (pigweed) | *Amaranthus retroflexus* | 2 | 80 Redroot amaranth (pigweed) | *Amaranthus retroflexus* |
| | | 30 Little barley | *Hordeum pusillum* | | 10 Witch grass | *Panicum capillare* | | 20 Witch grass | *Panicum capillare* |
| | | 10 Daisy fleabane | *Erigeron annuus* | | 5 Russian thistle | *Salsola pestifer* | | | |
| Old kafir stubble | | | | 7 | | | 3 | | |

[1] Abandoned several years.

In its natural undisturbed state, the vegetation of the Great Plains was grassland characterized for the most part by a short grass cover with mid and tall grasses scattered throughout. In the past 50 years, a great deal of this native grassland has been placed under cultivation, and the remainder has been used for grazing purposes. During the 6 years of the recent unprecedented drought, many of the cultivated fields failed to produce either any crops at all or sufficient vegetative growth to protect the soil and prevent it from being blown or washed away, and in many places the vegetation on the range and pasture lands has been depleted seriously, undergoing striking changes in composition and density.

Soil shifting has occurred throughout the entire Great Plains. Vast quantities of soil materials have been removed from one locality to accumulate in another. As a result, much of the present vegetative cover is far different from that which was formerly to be found in this region. Many cultivated fields, now abandoned, originally dominated with dense stands of blue grama, buffalo grass, and Western wheat grass are either denuded or characterized by a scattered growth of annual weeds and grasses such as Russian thistle and false buffalo grass. The rapidity with which these areas recover depends on a number of factors, namely, the length of time the field has been broken, the degree of erosion, the proximity of grassland to furnish a seed supply, the amount of protection received, and the amount of precipitation. In the past, the process has required from 5 to 50 years depending on the intensity of the foregoing factors. However, in severely wind-eroded areas where soil has been removed to plow depth in some places and piled into dunes 25 feet in height in others, recovery may take a much longer period.

## Succession of Vegetation in Central America

Evidence of long-continuing erosional effect on vegetative succession is found in the forests of western Honduras and eastern Guatemala. Here, forest composed almost exclusively of pine frequently adjoins tropical forest, where there is no important difference in climate, topography, or character of underlying rock. With respect to soil depth, however, the pine land usually is very different from that occupied by the characteristic tropical growth. Along the southerly slopes of the Espiritu Santo Range, west of the Chamelecon River, in Honduras, pine covers many small and large areas that are surrounded or bordered by tropical growth. Much of the soil under pine is a stony red clay, only a few inches deep (rarely more than 2 feet deep), overlying crystalline schist (closely resembling the Talladega soils of the Blue Ridge area of southeastern United States), whereas that of the near-by tropical woods is of the same character and origin but richer in humus, more friable, much deeper to bedrock (rarely less than 5 feet), and usually stone-free.[1] In the areas of pine highlands, bedrock is exposed in many places.

The indications are that the pine lands were cultivated, severely eroded, and abandoned generations ago. Most of them show no evidence of recent habitation, but there is abundant evidence of former occupation, including even the remains of ancient buildings in places.

Cook has the following to say with respect to past and present conditions of vegetation in this region[2]:

"That the ancient occupations of the humid mountain regions of eastern Guatemala by agricultural civilizations were very prolonged or were repeated in several prehistoric ages is indicated by the very severe erosion which this region has suffered. It is not likely that such deeply dissected contours would have been formed if the country had not been kept in a denuded condition for long periods of time.

"There is practically no erosion at all on slopes covered with dense tropical forest. The soil is never loosened by frost, but is held in place by the matted roots of trees. . . .

"The water of the torrents that gather from the forested slopes is still clean, even after heavy rains, or is made only slightly milky by the washings of the leaves. Only the erosion of the streams loosens any solid matter. After the forest has been cut and the dry brush has been burned off, the loose surface soil is left fully exposed. More erosion can then take place in a single season than would be possible in many centuries of undisturbed forest growth.

"The driest and most sterile localities, those too forbidding for human occupation, have retained their forest growth, scarcely excepting sheer precipices and exposures of bare rocks. The regions which are now treeless and barren or covered only with grass are those naturally well suited for the growth of forests, for the formation of fertile soil, and for human occupation. Indications of prehistoric agriculture are found in all the denuded areas, as well as in other regions now covered with forests.

"The abundance of rubber and other temporary types of trees and the absence of humus-inhabiting arthropods and forest palms enable regions of recent reforestation to be distinguished from forests of older growth.

"Facts of several different kinds thus support the conclusion that the Central American region had a continuous forest covering before the advent of agricultural man. If human interference were withdrawn the normal growth of the vegetation would again cover the Central American region with dense and continuous forests.

" . . . Though the effects of human activities may not be so direct or so obvious in temperate climates as in tropical countries, it is well to be aware of the fact that natural conditions can be profoundly altered by human agencies.

[1] Treadwell, J. C., Hill, C. R., and Bennett, H. H. Possibilities of Para Rubber Production in Tropical America, *Trade Promotion Ser.* 40, Bur. Foreign and Domestic Commerce, U. S. Dept. Commerce, 1926.

[2] Cook, O. F. Vegetation Affected by Agriculture in Central America, U. S. Dept. Agr. Bur. Plant Industry *Bull.* 145, 1909.

"Apart from dangers of war or pestilence to which the ancient communities of Central America may have been exposed, their existence was definitely limited by methods of agriculture which denuded the country of its forests and destroyed the fertility of the soil. Civilization is at an end when an agricultural country ceases to be adapted to agriculture. To recognize these natural limitations of the primitive civilizations of Central America should make us more careful to appreciate and to correct the harmful tendencies of some of our own systems of agriculture."

## Changes in the Southwest

In the vast domain of grazing country represented by the arid and semiarid Southwest, soil erosion following overgrazing has caused severe denudation of an enormous area of uplands, with consequent vegetative changes, such as have seriously reduced the grazing capacity of millions of acres of formerly fair to good range country, even to the extent of essentially 100 per cent damage in some localities. So tremendous has been the vegetative change in consequence of this abuse that it is impossible to undertake any comprehensive statement of the details here. There are areas in southwestern Texas and various parts of New Mexico, Arizona, Utah, California, and other Western states over which the vegetation has been completely destroyed, with little evidence of any important recovery following years of comparative freedom from invasion by livestock. Usually, however, protection from further grazing is followed by slow or rapid recovery, even from a practically bare initial condition, to a stage where there has been sufficient recuperation for the plants to give considerable or even complete protection from erosion. Some areas that have been swept almost bare of plant cover make a remarkable comeback under protection, depending on the soil, slope, underground seepage from higher areas, and other local conditions. Such second-growth vegetation, however, is often markedly different from the original cover, particularly with respect to various nutritious grasses characterizing the original condition.

On Aug. 16, 1929, a heavy downpour of rain 6 miles southeast of Panguitch, in southern Utah, swept from a 50 per cent slope of sheep range country great quantities of soil and practically all the grass. In many places the soil, a brown gravelly sandy loam possessing considerable mellowness, was gouged out to depths of 10 inches or more (18 inches in places). Samples collected the day following this erosive rain showed in the uneroded soil, which had been conserved during that particular rain by protective clumps of sage, an organic matter content of 1.4 per cent, as against only 0.4 per cent in the exposed lighter colored subsurface material (from which 10 inches of surface soil had been washed off). Such eroded areas were locally appraised as having little further value for grazing. It was claimed that grass would come back with exceeding slowness on slopes thus denuded.[1]

Depletion of the upland vegetation has brought about a very serious problem in the alluvial plains of almost countless valleys throughout this great Western grazing region. Excessive runoff from the denuded slopes has had the effect of trenching many valley plains, formerly well-vegetated, with broad channelways and lateral arroyos (gullies), where, 40 or 50 years ago, there was either no channel or only a diminutive runway. In many places these channels have undercut and widened to such degree that the entire valley floor has been riddled or bodily swept out, from the

FIG. 84.—The disappearing alluvial plain of Rio Chaco, New Mexico, looking downstream from the east side, at Pueblo del Arroyo, 1933.

foot of the uplands on one side to the corresponding position on the other (Fig. 84). Soil, whose accumulation has taken millennia under the slow natural process of valley filling, has been torn away completely or stripped of its productive surface material to the extent of wholly changing the type of vegetation for an undeterminable period. Creosote bushes and a few other woody shrubs scattered in an ephemeral cover of annual grasses and forbs are all that can be seen over some of the denuded areas which formerly were densely vegetated with nutritious grasses and herbaceous growth. This process of partial and essentially complete devastation of soil and vegetation is continuing steadily. There still remains, in some localities, enough land of approximately virgin char-

[1] Bennett, H. H. Cultural Changes in Soils from the Standpoint of Erosion, *Jour. Am. Soc. of Agronomy*, Vol. 23, June, 1931.

acteristics to verify, unmistakably, what the changes have been; in other localities, where the original conditions have been essentially obliterated, there is much historic evidence as to what the alterations have been.[1] The student of soil conditions needs no historic proof, however, to measure the violent changes that have taken place in such areas as that of the Puerco River Valley in New Mexico, for example. Here the valley plain is fast being gouged out, and there are large areas of adjacent slopes, now nearly bare of vegetation, where the soil has been largely swept away (as shown by what is left about the protected base of an occasional brushy tree).

## Changes Due to Silting

In the fall of 1929, a very destructive flood swept down the Rio Grande. The Bureau of Chemistry and Soils had completed a detailed soil survey of the alluvial plain from about the head of water in Elephant Butte Reservoir to about 11 miles above the confluence of the Rio Grande with the Rio Puerco. This flood covered so much of the surveyed area with clay and sand that most of it had to be resurveyed in 1930.[2] Areas that had been mapped as clay were changed to loose sand, and sandy lands were buried deeply with clay (Fig. 85). In some places, the depth of freshly deposited sand was as much as 7 feet, and clay was laid down in strata exceeding 2 feet. Thus, droughty soils were changed to

[1] The following letter to the author from Thomas A. Field, of Field, N. M., July 1, 1930, is an example of the type of local information that bears on the vegetative changes that have taken place since the coming of white man into the region: "I came here Aug. 10, 1886, saw the Rio Puerco first about the 10th, in fact, crossed it with a large herd of cattle where now is the station known as Puerco. The river then was not very wide, nor was the channel deep, and there were vegas and marshes all along the valley. Now the river has washed deep and wide and the marshes have disappeared. I know more about the Alamosa or Rio Salado, which is about 40 miles south of the Puerco by road, and in Socorro and Catron Counties, is about 150 miles long and drains a very large and mountainous region. When I first came here in 1886 and went into the stock ranching business, there was no channel to speak of in these valleys. The valley was covered with grass, shrubs, and in many places groves of cottonwoods, and whenever it rained in the mountains the flats would be flooded and grass and everything grew fine. Now there is a wide, deep channel, cut from the source to the mouth of this river, which has not only taken away nearly all the cottonwood trees, but in some places has washed the houses, farms, and orchards of some of the old-time Mexican people away. In some places it is from 10 to 30 feet deep and from 200 feet to one-fourth of a mile wide. Water runs a little at different places for from one-half to two miles, very shallow, sinking into the sand between the runs; and in some places there are dry stretches of from five to fifteen miles in length. Our rainfall here is about the same as always, but owing to the big washes, the water does not penetrate the ground, only runs off down the arroyos, taking more of the soil every year. So, the stock-grazing capacity per acre here is only about 50 percent of what it was when I first came to the country."

[2] Poulson, E. N., and Fitzpatrick, E. G. Soil Survey, Socorro and Rio Puerco Areas, New Mexico. Bur. Chemistry and Soils, U. S. Dept. Agr. (Series 1929), No. 2.

soils of good moisture retentiveness, even to a condition of swampiness, and vice versa. And the vegetation has undergone accordant changes.

Subsequent floods have raised the level of the alluvial plain, piled sand into the river bed, and developed considerable areas of waterlogged land, much of it with a salinity too high for crop production. Tamarisk and

Fig. 85.—This field of Anthony sandy loam was changed to Gila clay by the 1929 flood along the Rio Grande, New Mexico. Here 12 inches of dense, sticky clay was deposited over the loose sandy soil.

other plants have invaded many areas once covered with nutritious grasses and other plants of an entirely different order.

## Effect of Lowering the Water Table

Another phase of the effect of the erosion process on vegetation is represented by the lowering of the water table following the gouging out of deep channels. A striking example is to be seen in Canada de los Alamos, north of Los Angeles, Calif. Here an old arroyo that became active a few years ago, as the result of increased runoff from the overgrazed watershed, rapidly increased its depth into the substratum of porous materials of alluvial and colluvial origin. The violent lowering of the water table resulting from this channeling so changed the moisture condition that bordering groves of cottonwoods of large size died forthwith. Similar effect on vegetation is clearly observable along the Puerco (New Mexico) and numerous other Western streams.

## Effect of Overwash

A very large proportion of the alluvial plains in the Piedmont region (as well as of many other regions) has been so altered by overwash derived from eroding slopes that the soil (and the vegetation on it) has been changed markedly. Soil that formerly consisted of loam has been buried deeply by sands of varying texture, clay deposits, and silt, all so variable, both on the surface and through the vertical section, that it has been impossible to classify most of the areas within the Piedmont as uniform soil types. Moreover, such land (classed as Meadow in soil surveys) has been converted into swamp or semiswamp by the increased overflows resulting from the clogging of stream channels with erosion debris. Formerly, these alluvial areas were heavily timbered with ash, sycamore, oak, sweet gum, and pine. Much of the land was put into cultivation and considered the best of the region. Since abandonment, following these changes, most of it has grown up to thickets of willow, alder, birch, blackberry, pine, rushes, cattail, and coarse grasses.

# Chapter XI. Sedimentation

## The Problem

Direct damage by soil erosion is not confined to the farms or ranges from which soil is lost but extends to those places where the debris of erosion is deposited as sediment. Where the sediment accumulates,

Fig. 86.—Sediment deposited by flood of 1935, farmstead on Republican River, near Guide Rock, Neb. The depth of the deposit can be judged by the partially buried farm machinery. (*Photograph by Soil Conservation Service.*)

reservoirs are filled, irrigation and drainage ditches are clogged, floodways lose capacity, navigable channels and harbors are shoaled, farm lands on lower slopes and flood plains are damaged (Fig. 86) or ruined. Although some of these effects have been recognized by a few critical observers for at least half a century, and notwithstanding the growing magnitude of the problem, no comprehensive investigations were undertaken until the Soil Conservation Service developed specialized sedimentation studies as part of its research program. These studies are designed to evaluate the downstream public interests in the erosion problem and to provide specific information as a basis for their protection.

Public and private investments in storage reservoirs alone amount to more than 2 billion dollars.[1] The aggregate investments in or dependent on all forms of water utilization must be counted in terms of tens of billions. Throughout the country, every year, damage and maintenance costs chargeable solely to the effects of soil erosion on water utilization facilities run into tens of millions of dollars. It is, therefore, self-evident that a large public interest demands the complete determination of causes of, and the rapid development of adequate remedies for, the progressive impairment or destruction of the resources and enormous investments involved in valley agriculture, water utilization, and flood control.

## Significance of the Assorting Action of Water

When soil or subsoil materials are removed by running water from cultivated fields, pastures, ranges, or other upland sources, they usually begin an intermittent journey toward a resting place in the sea, a journey that commonly includes many halts during which the eroded material is deposited temporarily as colluvial, alluvial (or fluvial), or lacustrine sediment at favorable resting places. This process of intermittent transportation involves a progressive sorting of the original soil material, and the nature and completeness of this sorting depend on both the character and the amount of sediment and the distance and rate of transportation.

Ordinarily, the coarser fractions of the soil material move more slowly and by shorter jumps and thus tend to be concentrated as sediment in the upper parts of the valleys and in the beds of the stream channels. On the other hand, the lighter, finer materials—silt, clay, and very fine sand for the most part—are concentrated more largely in the lower valleys (see Chap. IV, Part 1, "The Assorting Effect of the Erosion Process,") particularly as overbank deposits from flood waters, which from time to time cover the flood plains of streams. It is these finer sediments that initially have made the alluvial plains exceptionally fertile. Moreover, the occasional additions of such fine sediment, often abounding in plant nutrients, are important fertilizing factors in maintaining the productiveness of valley lands. Most notable and historic of such instances is the alluvial plain of the Nile, cultivated continuously since very ancient time yet maintaining its productivity because of seasonal flooding.

When, however, as has been the case in many parts of the United States, soil erosion by surface wash is greatly accelerated on the sloping lands of a watershed, such heavy sediment loads result that the normal stream regimen is upset; in consequence, a readjustment must take place

[1] Brown, Carl B. Rates of Silting in Representative Reservoirs throughout the United States, Am. Geophys. Union *Trans.*, 18th ann. meeting, p. 554, 1937.

in the orderly process of seaward transportation and assortment of erosional debris. The results of such readjustment may be of grave concern to inhabitants of the valleys (Fig. 87) or to people dependent on the use of the rivers or valley lands. In many places, misuse of land not only has accelerated erosion on upland slopes but has caused previous accumulations of coarser material in the upper parts of valleys to become unstable. As a result, larger loads of both coarse and fine material are moved with greatly increased rapidity, filling stream channels that were devel-

Fig. 87.—Cleaning streets of Newport, Ky., of silt deposited by 1937 flood on the Ohio. Note high-water mark on building is at second-story window sills. (*Photograph by Soil Conservation Service.*)

oped in adjustment to the transportation of finer materials. Less productive sand and gravel are carried out of the channels during flood periods and distributed over the surface of flood plains composed of fine-textured, rich alluvium.

As sediment loads are increased, there is correspondingly more rapid filling of reservoir basins, which on many streams serve as traps catching increased fractions of the sediment, until within relatively short periods storage capacity is depleted so largely that the reservoirs no longer serve a useful purpose. Also, the enlarged burdens of river-borne sediment add to the cost of maintaining harbor facilities and navigable channels.

## Historical Background

No integrated nation-wide study of the downstream effects of accelerated soil erosion as related to specific cause and cure had been undertaken prior to the present decade. The general nature of the problem was, however, recognized long ago. In the Southern States, it was described in part as early as the beginning of the nineteenth century.[1] The process as observed in northern Mississippi was described in 1860 by Hilgard,[2] who particularly noted the damage caused by deposition of sand on flood-plain soils. Similar damage had been noted in soil reports covering portions of the southern Piedmont.

A situation developed in California about 1860 that focused attention on the dangers of sedimentation. Hydraulic gold mining in the Sierra foothills resulted in delivery of vast quantities of sand and gravel to the stream channels, particularly the Yuba River. Increased overflows resulted from the choking of stream channels with this mining debris, and fields were ruined by overwash of sand and gravel. In 1884, hydraulic mining was prohibited by state law in order to protect the lowlands. Attempts were made to stabilize the material in the upper valleys by construction of barriers across the streams. A large part of the sediment continued to move, however, resulting in what has been described as a wave of aggradation moving downstream. Classical studies of the processes of stream transportation by Gilbert[3] grew out of this situation.

Silting has been recognized as an important factor in reservoir and irrigation problems along the Rio Grande in Texas and New Mexico almost since the beginning of extensive reclamation activities. Special studies of the sediment load of the Rio Grande were carried out as early as 1897.[4] Since the construction of Elephant Butte Reservoir on the Rio Grande, in New Mexico, in 1915, several surveys have been made to determine its rate of silting. More recently, attention has been devoted to the accumulation of sediment in the channel and on the flood plain of the Rio Grande above the head of Elephant Butte Reservoir, where aggradation has contributed to flood damage, obstruction of drainage for irrigated lands, and waterlogging.

Glenn[5] investigated the effects of deforestation and erosion on the principal southern Appalachian rivers and concluded that floods had increased markedly in severity and frequency as a result of cultural changes in their drainage basins. He stated that much damage was being

[1] Moore, T. "The Great Error in American Agriculture Exposed; and Hints for Improvement Suggested." Baltimore. 1801.

[2] Hilgard, E. W. Report on the Geology and Agriculture of the State of Mississippi. Jackson, Miss. 1860.

[3] Gilbert, G. K. The Transportation of Debris by Running Water, U. S. Geol. Survey *Prof. Paper* 86, 1914.

[4] Follett, W. W. Silt in the Rio Grande. U. S. Dept. State Internat. Boundary Comm. for the Equitable Distribution of the Waters of the Rio Grande. 1913.

[5] Glenn, L. C. Denudation and Erosion in the Southern Appalachian Region and the Monongahela Basin, U. S. Geol. Survey *Prof. Paper* 72, 1911.

done to valley lands by the resulting increased bank erosion and deposition of sand, gravel, and boulders. These damages extended on downstream into the Piedmont, but no detailed study was made then to determine how far downstream the sedimentation was serious. In the mountain valleys, however, the damage was reported to be severe at least locally; and because of the scarcity of arable land sufficiently smooth to be reasonably safe from excessive erosion, the damage to the bottomlands was much more serious than might be supposed merely on the basis of the acreage involved. Glenn also mentioned, as a common form of sedimentation damage, the filling of many millponds with sand.

Silting surveys have been made of 39 reservoirs in the United States over a period of 40 years by various persons and agencies concerned about the menace to storage capacities of specific reservoirs or reservoirs in certain problem areas. A wide variety of methods have been employed in these surveys, which have ranged from scarcely more than estimates to rather elaborate measurements. As a group, they do not offer sufficiently comparable standards of accuracy or adequate geographical distribution to serve for nation-wide evaluation of causes, effects, and remedies for reservoir storage depletion by silting.

A fairly extensive foreign literature exists on many phases of the sediment problem, but little or no application has been made in this country of important principles of sediment control discovered abroad.

Scattered measurements of suspended loads in streams have been made, mostly by the United States Geological Survey. Some have been made on Western streams by the Bureau of Reclamation and by the Bureau of Agricultural Engineering. With the exception of the Colorado River, the measurements covered short periods and, being taken at varying intervals, are not highly significant. They constitute, nevertheless, useful indices of relative silt loads. Measurements of material carried as bed loads have been little more than rough estimates. For instance, by comparing the rate of deposition on bars in the Gulf of Mexico with suspended load measurements, Humphreys and Abbot[1] estimated in 1876 that the bed load at the mouth of the Mississippi amounted to 10 per cent of the suspended load. Recently, engineers of the Mississippi River Commission by similar calculations have estimated that the bed-load movement in some localities might amount to 200 per cent of the normal suspended load.

## Sedimentation Studies by the Soil Conservation Service

Recognizing the fact that previous studies were neither sufficiently extensive nor intensive enough to answer the problems involved in such

[1] Humphreys, A. A., and Abbot, H. L. Report upon the Physics and Hydraulics of the Mississippi River. Government Printing Office. 1876.

varied programs as soil and water conservation, water facilities, flood control, drainage, irrigation, and adjustments in land use, the Soil Conservation Service developed a research section for sedimentation studies. The need for specific investigations, which became apparent as action programs on the land evolved, led to the initiation of four distinct research projects in this field.

## Reservoir Silting Investigations

Investigations of reservoir silting, with which the first sedimentation project started by the Soil Conservation Service had to do, are designed to establish the true relation between soil erosion and silting damage to the concentrated investments represented by storage reservoirs. Work under these investigations includes detailed sedimentation surveys of representative reservoirs, following uniform survey standards set by Eakin[1]; reconnaissance examination of silting in all other important reservoirs; compilation of a comprehensive inventory of existing reservoir resources of the country; and intensive field studies of the factors involved with reservoir silting and their relation to watershed conditions.

It is highly important that the nature, extent, and rates of this silting damage be determined accurately, in order that erosion-control measures, water conservation, and land-use readjustments in the drainage basins may be commensurate, in costs, scope, and benefits, to the reservoir values involved as well as to agricultural interests. Correspondingly, an essential phase of such studies is the development and adaptation of economical methods designed specifically to reduce reservoir silting and supplementary to the widely used soil-conservation practices.

## Stream and Valley Sedimentation Studies

Investigations of stream and valley sedimentation are directed toward the determination of the nature, extent, and probable future trend of damages in local areas where excessive sedimentation is under way.

Since the recent sediments are for the most part in direct contact with older sediments deposited by essentially the same geologic processes (but not necessarily at the same rates), the approach in these studies has been largely by application of pedologic and geologic methods and criteria, modified where necessary to fit the specific problems at hand. Detailed studies are made of representative parts of regions in which conditions are generally similar, supported by briefer inspection of the entire area and laboratory sediment analyses where required. This procedure provides a basis for isolating the factors in the problem and for developing general principles to guide plans for the practical application of control or remedial measures.

[1] Eakin, H. M. Silting of Reservoirs, U. S. Dept. Agr. *Tech. Bull.* 524, 1936.

## Investigations of the Total Sediment Load of Natural Streams

One of the most important unknown variables in the whole sediment problem is the percentage of sediment moving as bed load in a natural stream and its rate of transportation. It is a common statement that the volume of material moved along the stream bed can be estimated only roughly or is a wholly unknown factor. As late as 1934, Stevens[1] summarized the situation with the unequivocal statement that "as far as known, no successful attempt has yet been made to measure the material transported in a river as bed load."

Accurate knowledge of the total sediment load is essential, however, to the most economical construction not only of large reservoirs and flood-control works but also of the smaller developments involved in upstream engineering or soil-conservation work on watershed areas.

By a new type of control installed on representative southern Piedmont streams near Statesville, N. C.; Greenville, S. C.; and Dadeville, Ala.; the Soil Conservation Service is undertaking to make exact measurements of bed loads, suspended loads, and various related hydraulic functions of the streams. At these stations, the bed load is removed hydraulically from subdivisions into which the bottom of the stream is divided, and simultaneous measurements are made of suspended load and stream velocities. Data from these stations will be correlated with the topography, land use, and hydrologic conditions of the watersheds.

## Laboratory Investigations of Sediment-laden Flows

Many field problems call for more knowledge of the fundamental mechanics underlying the entrainment, transportation, and deposition of erosional debris by flowing water. To provide for such investigations, a specially designed laboratory was set up in cooperation with the California Institute of Technology at Pasadena, Calif., early in 1935. Here, the individual factors involved with these complex natural phenomena are being isolated and studied under controlled conditions.

Transportation of suspended load has been selected as a primary point of attack on the whole problem. The importance of this particular study lies in the influence that the character of sediment in a stream has on its capacity for bed and bank erosion and for harmful deposition of sediment.

Another problem being studied is the cause and rate of abrasion of sediments, that is, reduction in size of particles in the course of stream transportation. This study has an important bearing on the problems of

[1] Stevens, J. C. The Silt Problem. Am. Soc. Civil Eng. *Trans.*, Vol. 101, pp. 207–288, 1936.

stream-bank erosion, formation of bars, transportation of stream loads, and generally with regard to the effect of change in the character of erosional debris on the regimen of streams.

The apparatus for this study consists of a closed circuit flume in which water and bed load are continuously circulated. Rate of wear is measured by periodic analysis of samples of the sediment. Simultaneous studies of abrasion of the same material in jar mills are being made as a means for more rapid future determination of the abrasion coefficient.

Other problems under investigation are (1) *localized scour* caused by high-intensity disturbances of a local nature, such as occur in drops over certain types of dams or along curves of short radius on stream meanders; and (2) *underflow* of sediment-laden water in reservoirs. These underflows, which are of wide occurrence in nature, influence the location at which the stream load of debris is deposited in the reservoir. A thorough knowledge of underflows may lead to practical methods for reduction of reservoir silting.

## Additional Studies Needed

Much work remains to be done along the present lines of investigation, for only a promising beginning has been made so far. The true importance of the sedimentation problem from a national viewpoint has not yet been appraised.

Nevertheless, certain additional lines of investigation beyond the scope of the present program should be started in the near future. As a basis for developing erosion and flood-control measures in the watersheds above reservoirs or areas of channel and valley damage, systematic sampling of suspended load should be carried out on many streams.

The relative importance of stream-bank erosion and of mass movement of soil in contributing sediment to stream channels should be determined by careful measurements and field study. Studies and experiments are needed also to develop the most practicable control measures where bank erosion is found to be an important source of damaging sediment.

A satisfactory program of watershed protection should, in many instances, include supplemental sediment-control methods in addition to erosion control on farm and range lands. In many places, relatively large eroded areas of low value are contributing much harmful sediment to reservoirs, channels, and valley soils, such as sand, gravel, and unproductive subsoil material. Whenever the topographic and other physical conditions are favorable, it may be economical to provide for trapping the sediment from such areas, while soil-conservation measures are beginning to take effect on the watershed. Comprehensive studies of the physical and economic possibilities along these lines are needed.

Development of methods for promoting spread of vegetation on deltas and flood plains is a promising line of research for protection of reservoirs.

In many foreign countries, silt has been used extensively for building up and improving worthless land. The method of reclamation is known as *colmatage* when applied to the utilization of sediments on inland areas and *warping* when referred to reclamation of tidal marshes. This method of land reclamation and improvement was first used in Italy. Documents of the twelfth century mention reclamation by deposition of stream-borne silt in Tuscany, and such reclamation probably was carried on before the fall of the Roman Empire. From Italy, the practice spread to France and England and subsequently to Russia, India, China, Switzerland, Scotland, Ireland, Canada, and other countries. The possibilities of making practical application of such practices in the United States should be looked into carefully.

The development of suitable methods for reclamation of valley lands already damaged by sediment necessarily will require considerable experimental work, as well as a critical study of past or existing reclamation efforts. The form of treatment will, of course, vary considerably. Particular attention should be given to the possibility of filling up swamped areas by simple diversion and detention methods and to protecting drainage channels necessary for reduction of flood hazards. In many instances of severe valley sanding, an adequate reclamation program probably should be directed toward deposition of a finer textured, more productive silt blanket over the sand. It may also be found possible to develop special crops, either for continuing use in the valleys as they are or selected primarily to aid in improving the agricultural character of the sediments for future use after other measures have been applied to reduce the severity of existing harmful conditions.

The entire field of the relation of sediment to flood-control problems needs extensive and intensive investigation. Loss of capacity in detention reservoirs is, of course, a factor of major importance in considering the use of reservoirs for flood control, but very little is known as to the expectancy of silting under probable operating conditions in the more severely eroding sections of the country.

Only a small beginning has been made in accurate measurement of the rates of accumulation of *lag sediments* in natural stream channels throughout the country. This is an important factor in planning a long-range program of flood protection and may deserve special attention in certain watersheds where relatively coarse erosional debris is being fed to the streams at excessively rapid rates.

## Findings and Examples

From time to time, references have appeared in literature citing examples of streams that formerly ran clear except during spring freshets and that now are turbid the year round; of formerly cultivated bottom-lands that are now sandy wastes (Fig. 88); and of coastal inland streams that were formerly navigable but now are difficult to traverse even by canoe. Some studies have been made in recent years with respect to the clogging of drainage and irrigation channels, the filling of reservoirs, and reduction in capacity of floodways; but it is apparent that the full significance of these isolated studies was not generally recognized until

FIG. 88.—Sand deposited over a large area of Ohio River alluvial plain by flood of January, 1937. In 1936, this field produced about 100 bushels of corn per acre and was valued at $100 an acre. It is now practically worthless. (*Photograph by Soil Conservation Service.*)

very recently. In the following pages, results of the more important previous studies are cited as well as representative findings of the Soil Conservation Service research.

### Sediment Load of Streams

During the past 80 odd years, many measurements have been made of the material carried in suspension and solution by streams. Unfortunately, these records are inadequate for accurate estimates of any increased amounts of water-borne material resulting from agricultural use of the land. Various authors have presented estimates of the rates of denudation, based in some instances on both the suspended and the dissolved load of streams; in others, on total solid load as revealed in reservoir deposits, with or without regard to dissolved load. These figures generally do not reflect the extent of soil wastage, since they represent only

TABLE 29.—SUSPENDED SOLIDS AND DISSOLVED MATTER CARRIED BY REPRESENTATIVE STREAMS

| Stream and location | Drainage area, square miles | Period of measurements, years | Average annual runoff during period, second-feet per square mile | Average dissolved material by weight, parts per million | Average suspended silt by weight, parts per million |
|---|---|---|---|---|---|
| Kennebec, Waterville, Me. | 4,270 | .... | 1.76 | 48 | 4 |
| Hudson, Hudson, N. Y. | 9,530 | 1.0 | 1.82 | 108 | 16 |
| Raritan, Bound Brook, N. J. | 800 | 1.0 | 2.14 | 85 | 36 |
| Susquehanna, Williamsport, Pa. | 5,640 | 1.1 | 1.68 | 74 | 18 |
| Potomac, Great Falls, Md. | 11,400 | .... | 1.12 | 115 | 85 |
| James, Richmond, Va. | 6,820 | 1.0 | 1.33 | 89 | 71 |
| Dan, South Boston, Va. | 2,750 | 0.7 | 1.34 | 71 | 264 |
| Wateree, Camden, S. C. | 4,500 | 1.0 | 1.44 | 73 | 214 |
| Chattahoochee, West Point, Ga. | 3,300 | 1.0 | 1.98 | 52 | 136 |
| Alabama, Selma, Ala. | 15,400 | 0.9 | 1.81 | 82 | 100 |
| Pearl, Jackson, Miss. | 3,120 | 1.0 | 1.29 | 59 | 46 |
| Tennessee, Gilbertsville, Ky. | 38,400 | 1.0 | 1.60 | 101 | 127 |
| Kentucky, Frankfort, Ky. | 5,140 | 1.1 | 1.60 | 104 | 142 |
| Monongahela, Elizabeth, Pa. | 5,580 | 1.1 | 1.80 | 81 | 84 |
| Muskingum, Zanesville, Ohio | 5,830 | 1.0 | 0.96 | 244 | 70 |
| Great Miami, Dayton, Ohio | 2,450 | 1.0 | 0.73 | 289 | 94 |
| Wabash, Azabia, Ind. | 2,100 | 1.1 | 0.90 | 279 | 48 |
| Mississippi, Minneapolis, Minn. | 19,600 | 1.0 | 0.81 | 200 | 8 |
| Mississippi, Quincy, Ill. | 135,500 | 1.0 | 0.53 | 203 | 119 |
| Illinois, Kampsville, Ill. | 26,730 | 1.0 | 0.83 | 267 | 145 |
| Chippewa, Eau Claire, Wis. | 6,740 | 1.0 | 1.31 | 90 | 4 |
| Minnesota, Shakopee, Minn. | 15,100 | 1.1 | .... | 480 | 142 |
| Cedar, Cedar Rapids, Iowa | 6,320 | 1.0 | 0.65 | 228 | 61 |
| Missouri, Howard Bend, Mo. | 529,000 | 1.9 | 0.24 | ..... | 2,800 |
| Grand, Sumner, Mo. | 7,900 | 1.9 | 0.67 | ..... | 2,910 |
| Kansas, Bonner Spring, Kans. | 61,300 | 1.9 | 0.19 | ..... | 3,065 |
| Platee, Plattsmouth, Neb. | 90,200 | 1.9 | 0.19 | ..... | 2,040 |
| Big Sioux, Akron, Iowa | 9,420 | 1.9 | 0.87 | ..... | 625 |
| James, Scotland, S. D. | 21,500 | 1.9 | 0.01 | ..... | 153 |
| Bad, Fort Pierre, S. D. | 3,110 | 1.9 | 0.11 | ..... | 38,610 |
| Yellowstone, Glendive, Mont. | 66,900 | 1.9 | 0.42 | ..... | 2,840 |
| Arkansas, Little Rock, Ark. | 148,000 | 1.9 | 0.19 | 630 | 748 |
| Red, Shreveport, La. | 56,900 | 1.0 | 0.28 | 561 | 870 |
| Mississippi, New Orleans, La. | 1,261,000 | 1.0 | .... | 190 | 600 |
| Brazos, Waco, Tex. | 28,500 | 2.9 | 0.20 | ..... | 8,600 |
| Brazos, Waco, Tex. | 28,500 | 0.9 | 0.08 | 1,136 | 1,188 |
| Colorado, Austin, Tex. | 38,000 | 1.0 | 0.07 | 321 | 351 |
| Rio Grande, San Marcial, N. M. | 30,000 | 34.9 | 0.14 | ..... | 14,200 |
| Colorado, Yuma, Ariz. | 242,000 | 20.9 | 0.23 | ..... | 9,400 |
| Colorado, Grand Canyon, Ariz. | 138,700 | 6.9 | 0.38 | ..... | 13,400 |
| Gila, Florence, Ariz. | 17,850 | 1.0 | 0.01 | 944 | 14,130 |
| Santa Maria, Santa Maria, Calif. | 1,000 | 0.9 | .... | 2,412 | 1,302 |
| San Joaquin, Lathrop, Calif. | 12,500 | 0.8 | 1.02 | 161 | 60 |

the net removal of material in suspension in streams from large watersheds in which much nonagricultural land may be contributing little or no erosional debris. It is noteworthy, however, that even these estimates are being constantly revised upward as additional data, longer term records, and more efficient sampling devices and technique become available. For another reason, stream-load measurements do not give full evaluation of the magnitude of sedimentation problems. They disregard entirely the immense quantities of debris temporarily stored behind artificial barriers, such as reservoirs, and in colluvial and alluvial deposits in the watershed above the points of stream sampling.

Although suspended- and dissolved-load measurements alone do not give a true measure of land wastage, they are of value in comparing rates of debris production in the various sections of the country and in individual streams. Table 29 shows the suspended and dissolved material of a number of representative streams of the United States.

The dissolved load of streams is governed mainly by three watershed characteristics: (1) location with respect to climatic conditions, (2) chemical constitution of rocks and soil, and (3) contamination by municipal and industrial wastes. To illustrate, the rivers of the Atlantic slope, draining areas underlain by crystalline rocks of low solubility and occupied by soils that have been subjected to a relatively high degree of leaching, show uniformly low salinity. In contrast, the rivers of the dry Southwest, particularly those draining semiarid and arid regions in which the soil contains large quantities of soluble salts, show high concentration of dissolved material. However, it is evident that the most important factors in individual instances are not too easily isolated, since the other factors may tend to obscure the true relationship. The effect of contamination or pollution by industrial wastes prevalent in the East is particularly difficult to isolate, except by reference to the specific analysis of the samples themselves.

The suspended load of streams, on the other hand, tends to show striking correlations with soil, climate, vegetative cover, and, to some extent, type of agriculture. Although available measurements of suspended solids are distributed too unequally to allow broad regional generalizations, it is evident that sediment loads are generally higher in the dry regions of the Southwest and least in the Eastern coastal regions.

When records of individual streams are compared, some important relationships become evident. For example, the Bad River of South Dakota, draining country with large areas of badlands, carries much more silt than the James River of the Dakotas, draining a region comprising much permeable glacial soil. Still another striking contrast is revealed by records from the Mississippi River at Minneapolis and the Illinois River at Kampsville. These two streams, although having comparable watershed areas and similar climatic conditions, show strikingly different sediment values. It appears that the large extent of erodible loessial soil and extensive cultivation in the watershed of the Illinois is associated with the higher sediment load of this river as compared with that of the upper Mississippi, which drains a region including much permeable glacial soil, and numerous lakes, peat bogs, and forested areas.

The highest recorded sediment concentrations are found in streams draining arid and semiarid regions of the Southwest. A maximum concentration of 26 per cent by weight (42 per cent by volume) was measured on the Rio Puerco, near its junction with the Rio Grande in central New Mexico, and a concentration of 21 per cent by weight (25 per cent by volume) has been recorded on the Rio Grande, at San Marcial, N. M.[1] Some high values are recorded also for the Colorado River system. Pierce[2] records a silt concentration of 12 per cent by volume and 9 per cent by weight on the San Juan River, one of the upper Colorado tributaries. Recent measurements by the United States Geological Survey have shown a maximum concentration on the Colorado River, at Grand Canyon, of 14 per cent by weight.[3] Fortier and Blaney[4] report concentrations of 3 per cent by weight for the entire Colorado River discharge at Yuma, Ariz., for short periods. Forbes[5] records a maximum silt concentration of 9.4 per cent by weight (22 per cent by volume after one month settling) for the Gila River near Yuma. The only other streams showing like silt concentrations are those which drain the badlands of South Dakota. According to the chief of engineers, United States Army,[6] the Bad River, South Dakota, which drains much badlands country, may during times of high precipitation carry a silt load in excess of 10 per cent by weight.

### Reservoir Silting

An incomplete inventory of water-storage development reveals that at least 8,400 dams and reservoirs, exclusive of farm ponds, have been constructed in the United States. No state is lacking in reservoir developments; the minimum value in any state probably exceeds $100,000, and some have investments aggregating more than $200,000,000, notably California, where more than 775 reservoirs have been built. Water-

[1] Follansbee, R., and Dean, H. J. Water Resources of the Rio Grande Basin, U. S. Geol. Survey *Water-supply Paper* 358, p. 712, 1915.

[2] Pierce, R. C. Measurement of Silt-laden Streams, U. S. Geol. *Water-supply Paper* 400-C, p. 41, 1917.

[3] Howard, C. S. Suspended Matter in the Colorado River in 1925–1928, U. S. Geol. Survey *Water-supply Paper* 636, p. 32, 1929.

[4] Fortier, S., and Blaney, H. F. Silt in the Colorado River and Its Relation to Irrigation, U. S. Dept. Agr. *Tech. Bull.* 67, 1928.

[5] Forbes, R. H. The River-irrigating Waters of Arizona—Their Character and Effects, Arizona Exper. Sta. *Bull.* 44.

[6] United States Congress. Missouri River. Letter from the Secretary of War transmitting report from Chief of Engineers. 73d Cong., 2d Sess. *House Doc.* 238, pp. 1032–1236.

storage reservoirs for New York City alone are reported to have cost in excess of $80,000,000, and many other major cities throughout the country have proportional investments and are likewise dependent on these water-storage resources.

Data on rates of silting are far from adequate to give an accurate record of the total annual losses in reservoir storage attributable to accelerated soil erosion. Altogether, reliable silting investigations have been completed on only 98 reservoirs, unequally distributed throughout the country. Of this number, 34 surveys have been made by agencies other than the Soil Conservation Service, since 1895. The Soil Conservation Service has completed surveys of 64 reservoirs since 1935. A summary

FIG. 89.—The reservoir behind the original dam (Barrett Dam, East Sandy Creek, near Athens, Ga.) was filled with sediment; the dam was raised, and again the reservoir is filled. (*Photograph by Soil Conservation Service.*)

of representative data from these investigations is given in Table 30. If it is assumed that data on 98 reservoirs represent a fair sampling of the total number in the United States, that rates of silting will remain generally uniform, and that when the average reservoir has lost 80 per cent of its storage capacity its usefulness will be at an end, the existing data indicate that about 39 per cent of these reservoirs will be useful for less than 50 years from the date of their construction, 25 per cent will be useful for 50 to 100 years, 21 per cent for 100 to 200 years, and only 15 per cent for more than 200 years.

Much significant information is developed when conditions of reservoir silting are appraised on a regional basis. In the southern Piedmont region, where farming has been important for 200 years, widespread damage has already occurred (Fig. 89). The reservoirs of this section may be classified in three groups: (1) channel reservoirs without appreciable overbank or flood-plain storage, used mainly for power by grist mills and cotton mills; (2) basin-type reservoirs impounding generally from 1,000 to 20,000 acre-feet of water, used principally for municipal purposes but occasionally for power and recreation; and (3) large storage reservoirs impounding

TABLE 30.—RESULTS OF SELECTED RESERVOIR SURVEYS BY THE SOIL CONSERVATION SERVICE

| Name of reservoir[1] | Location | Age at time of survey, years | Area of watershed, square miles | Original capacity, acre-feet | Sediment volume, acre-feet | Annual capacity loss, per cent | Total capacity loss, per cent |
|---|---|---|---|---|---|---|---|
| Burnt Mills (ws) | Silver Spring, Md. | 7.8 | 27.0 | 170 | 79 | 5.96 | 46.47 |
| Byllesby (p) | Byllesby, Va. | 23.7 | 1,310 | 8,892 | 5,354 | 2.54 | 60.21 |
| Washington Mills (p) | Fries, Va. | 33.5 | 1,140 | 2,954 | 2,443 | 2.47 | 82.69 |
| Burlington (ws) | Burlington, N. C. | 10 | 105.2 | 1,488 | 163 | 1.10 | 10.95 |
| High Point (ws) | High Point, N. C. | 10.25 | 62.8 | 4,354 | 316 | 0.71 | 7.26 |
| Lake Lee (ws) | Monroe, N. C. | 11.1 | 50.5 | 821 | 169 | 1.85 | 20.58 |
| Lancaster (ws) | Lancaster, S. C. | 13.4 | 9.4 | 242 | 52 | 1.60 | 21.49 |
| Spartanburg (ws) | Spartanburg, S. C. | 8.2 | 91.0 | 2,700 | 463 | 2.09 | 17.15 |
| Lloyd Shoals (p) | Jackson, Ga. | 24.3 | 1,414 | 112,538 | 13,960 | 0.51 | 12.40 |
| Bayview (ws) | Birmingham, Ala. | 24.6 | 72.3 | 11,866 | 2,352 | 0.81 | 19.82 |
| Lake Decatur (ws) | Decatur, Ill. | 14.2 | 906 | 19,738 | 2,808 | 1.00 | 14.23 |
| Lake Calhoun (r.) | Galva, Ill. | 11.9 | 13.1 | 286 | 149 | 4.37 | 52.10 |
| Lake Taneycomo (p) | Branson, Mo. | 22.4 | 4,610 | 43,980 | 20,266 | 2.06 | 46.08 |
| Clinton (ws) | Clinton, Okla. | 7.4 | 23 | 4,415 | 434 | 1.33 | 9.83 |
| Lake Waco (ws) | Waco, Tex. | 5.9 | 1,662 | 39,378 | 7,790 | 3.34 | 19.78 |
| Rogers (ws) | Rogers, Tex. | 12.0 | 0.55 | 164 | 37.5 | 1.91 | 22.87 |
| White Rock Lake (ws) | Dallas, Tex. | 25.0 | 99.1 | 18,158 | 3,882 | 0.86 | 21.38 |
| Mission Lake (ws) | Horton, Kans. | 13.0 | 11.4 | 1,852 | 289 | 1.20 | 15.60 |
| Wellfleet (r) | Wellfleet, Neb. | 5.6 | 43.0 | 519 | 55 | 1.89 | 10.60 |
| Hayes Lake (r) | Hayes, S. D. | 4.2 | 40.0 | 629 | 49 | 1.86 | 7.79 |
| Baker (r) | Baker, Mont. | 29.1 | 5.2 | 756 | 254 | 1.15 | 33.60 |
| Black Canyon (i) | Emmett, Idaho | 12.0 | 2,540 | 37,659 | 4,037 | 0.89 | 10.72 |
| Elephant Butte (i) | Hot Springs, N. M. | 20.25 | 26,312 | 2,638,860 | 365,186 | 0.68 | 13.84 |
| San Carlos (i) | Coolidge Dam, Ariz. | 6.33 | 13,540 | 1,247,999 | 36,896 | 0.47 | 2.96 |
| Gibraltar (ws) | Santa Barbara, Calif. | 16.25 | 215.4 | 14,500 | 4,314 | 1.83 | 29.75 |
| Santa Anita (fc) | Monrovia, Calif. | 10.2 | 10.8 | 1,043 | 360 | .... | 34.52[2] |

[1] ws = water supply; p = power; r = recreation; i = irrigation; fc = flood control.

[2] Total capacity loss prior to 1938, 3.17 per cent. Loss during storm of Mar. 2, 1938, 31.35 per cent.

generally over 100,000 acre-feet of water, used almost entirely for hydroelectric power development. The reservoirs of the first class have long since lost all storage capacity beyond that represented in the channel required to carry the normal flow of the stream. Thirteen representative reservoirs of this type, with dams averaging 30 feet in height, were completely silted at an average age of 29½ years. Striking examples of this class are Gaston Shoals on Broad River, South Carolina, which had an estimated original capacity of 10,000 acre-feet; Holiday Reservoir on Saluda River, South Carolina, with an estimated original capacity of

3,500 acre-feet; Lockhart Reservoir on Broad River, South Carolina, with an estimated original capacity of 2,300 acre-feet; and Parr Shoals Reservoir also on Broad River, with an estimated original capacity of 36,000 acre-feet.

The excessively rapid filling of channel-type reservoirs is not confined to the Piedmont but also has occurred in the mountainous regions of the Southeast. Of four reservoirs on New River,[1] in southwestern Virginia, Washington Mills Reservoir, with an original capacity of 2,954 acre-feet, lost 82.7 per cent of its capacity in 33.5 years, of which more than 80 per cent was lost in the first 18 years. Byllesby Reservoir, just below on the same stream, lost 60.2 per cent of its original capacity of 8,892 acre-feet in 23.6 years. Buck Reservoir, just below Byllesby, lost 23 per cent of its capacity of 1,225 acre-feet in 23.6 years.

These and similar reservoirs have not been rendered entirely useless by silting, since many of them still produce some power when there is ample flow in the river. Because of their tie-in with power generation and distribution networks, this secondary power (or residue of reservoir power) may be used advantageously at times. If these reservoirs were single units, however, the secondary power generated could not be sold at sufficient price to justify their continued operation, and the reservoirs would, therefore, be totally worthless.

Schoolfield dam on the Dan River just above Danville, Va., completed in 1904, had an original capacity of approximately 4,000 acre-feet. A survey by the United States Army Engineers showed that by 1915 this reservoir had lost 80 per cent of its original capacity. The annual production of power in kilowatt-hours has been reduced materially by this loss of storage. Islands of silt, formed within the reservoir, now cover about 400 acres out of the original surface area of 540 acres and are partly used for farming.

Three large storage reservoirs built for hydroelectric power development, High Rock Reservoir near Salisbury, N. C.,[2] Lloyd Shoals Reservoir near Jackson, Ga.,[3] and Lay Reservoir near Clanton, Ala.,[4] are, respectively, silting at annual average rates of 0.62, 0.51, and 0.52 per cent of their capacity. This uniformity appears significant; supported by reconnaissance measurements in other reservoirs of the same class in the region, it suggests that projects of this type in the Piedmont generally may

[1] Brown, C. B., and Barnes, F. F. Advance report on the sedimentation investigations of reservoirs and navigation improvements on the New River, Virginia and West Virginia. Soil Conservation Service (mimeographed), SCS-SS-6, 1936.

[2] Eargle, D. H. Advance report on the sedimentation survey of High Rock Reservoir, Salisbury, N. C. Soil Conservation Service (mimeographed), SCS-SS-10, 1937.

[3] Eakin, H. M. Silting of Reservoirs, U. S. Dept. Agr. *Tech. Bull.* 524, 1936.

[4] Barnes, F. F. Advance report on the sedimentation survey of Lay Reservoir, Clanton, Ala., Soil Conservation Service (mimeographed), SCS-SS-13, 1937.

be expected, under existing conditions of watershed utilization, to silt at rates of one-half to three-quarters of 1 per cent a year. Silting rates under present land-use conditions indicate a useful life not much in excess of 100 years for these costly and irreplaceable developments.

There are notable contrasts within the Southeast between reservoirs in headwater areas with almost completely forested watersheds and those located in the middle or lower Piedmont where a large part of the tributary drainage comes from agricultural lands. For example, the enormous Lake James in the eastern part of the Blue Ridge Mountains, on the Catawba River near Marion, N. C., with an original capacity of 288,000 acre-feet, is estimated on the basis of a few silt measurements to have filled less than one-half of 1 per cent in the 12 years after its completion. By contrast, Mountain Island Reservoir, also on the Catawba River but situated in the middle Piedmont region, is estimated to be filling at an average rate of more than one-half of 1 per cent a year.

A few reconnaissance measurements in larger reservoirs of New England have indicated almost negligible rates of filling, a fact considered significant because most of the watersheds are 90 per cent or more in forest cover. Brief examination of some of the major reservoirs in the headwater tributaries of the Ohio River, north and east of Pittsburgh, have likewise indicated low rates of silting. Most of the larger reservoirs north of Virginia and east of Ohio are estimated to be losing only 0.1 to 0.2 per cent of their capacity annually. Large parts of this region, particularly the reservoir watersheds, are forest-covered and are thus generally well protected against the ravages of soil erosion. However, a few reservoirs situated on large streams receive heavy loads of sediment.

The Western Gulf drainage systems, covering most of Texas, are carrying particularly heavy loads of sediment and many reservoirs built on these streams are suffering high, often disastrous, rates of silting. An outstanding instance of severe silting revealed by surveys of the Soil Conservation Service is Lake Waco on the Bosque River at Waco, Tex. Built for municipal water supply in 1930 at an initial cost of more than $2,000,000 and storing originally 39,378 acre-feet of water, this reservoir had lost over 14 per cent of its original capacity by 1935. A resurvey in 1936, when the reservoir was still less than 6 years old, showed a storage loss of 19.78 per cent, representing an average loss of 3.34 per cent a year and an actual loss of over 5 per cent of storage space, or 693,000,000 gallons, in 11 months. If the average 6-year rate of silting continues, the reservoir capacity will be exhausted by about 1960.

White Rock Reservoir at Dallas, Tex., costing $765,000, lost 21.38 per cent of its original capacity of 18,158 acre-feet in 25 years, from 1910 to 1935. A survey of Lake Worth on the West Fork of Trinity River near Fort Worth, Tex., made in 1928, showed that during the previous 13-year period the reservoir had lost storage capacity at the rate of 2.3 per cent annually.[1] This rate was materially reduced, of course, when large water-supply and flood-control reservoirs at Eagle Mountain and Bridgeport were constructed above Lake Worth, in 1932 and 1933. Even so, it is estimated that Lake Worth had lost 39 per cent of its capacity by 1937, over a period of 22½ years.

The prodigious rates of silting in the Old and New Austin Reservoirs on the Colorado River at Austin, Tex., are historically noteworthy. Old Lake Austin was built in 1893 at a cost of $1,400,000[2] and failed in April, 1900, during one of the heaviest floods of record on the Colorado. A survey made in January of that year showed that the reservoir had lost 47 per cent of its original capacity of 49,300 acre-feet in a period of only 6¾ years. The depletion of storage took place at the rate of slightly more than 7 per cent a year. In 1911, a new dam was built on the same site to replace the old structure. In 9 years, the new reservoir, with 32,029 acre-feet original capacity, had lost 83 per cent of its storage and after 13 years had lost more than 95 per cent of its storage.

The small municipal reservoir at Rogers, Tex., impounding 164 acre-feet of water and draining 350 acres of Black Belt land, entirely in cultivation, lost 23.2 per cent of its capacity in 12 years. The accumulation represented an average loss of more than 19 tons of soil per acre from the entire watershed during this period.

Lake Taneycomo on White River, Missouri, the second largest reservoir and hydroelectric development in the state, with an original storage of 43,980 acre-feet, lost 46.08 per cent of its capacity in 22.4 years, representing an annual loss of 2.06 per cent of its original storage volume.[3]

Lake Decatur on the Sangamon River, Illinois, municipal water-supply reservoir for Decatur and second largest in that state, storing originally 19,738 acre-feet, lost 14.2 per cent of its capacity in 14.2 years.[4] Of the smaller storage reservoirs in Illinois, Lake Bracken at Galesburg, has filled nearly 8 per cent in 13 years; West Frankfort Reservoir, 8 per cent in 10 years[5]; and Lake Calhoun, a recreational center at Galva, 52 per cent in 12 years.[6]

[1] Taylor, T. U. Silting of Reservoirs, Texas Univ. *Bull.* 3025, pp. 79–86, 1930.

[2] Schuyler, J. D. "Reservoirs for Irrigation, Water-power, and Domestic Water Supply," p. 245, John Wiley & Sons, Inc. 1901.

[3] Kesler, T. L. Advance report on the sedimentation survey of Lake Taneycomo, Taney County, Missouri, Soil Conservation Service (mimeographed), SCS-SS-8, 1936.

[4] Glymph, L. M., Jr., and Jones, V. H. Advance report on the sedimentation survey of Lake Decatur, Decatur, Ill. Soil Conservation Service (mimeographed), SCS-SS-12, 1937.

[5] Jones, V. H. Advance report on the sedimentation survey of Lake Bracken, Galesburg, Ill. Soil Conservation Service (mimeographed), SCS-SS-14, 1937.

[6] Glymph, L. M., Jr., and Jones, V. H. Advance report on the sedimentation survey of Lake Calhoun, Galva, Ill. Soil Conservation Service (mimeographed), SCS-SS-16, 1937.

The Keokuk Dam on the Mississippi River at Keokuk, Iowa, one of the first large power and navigation projects on the upper Mississippi River, built in 1913 with a capacity of 370,300 acre-feet, had lost 30.2 per cent of its capacity in 15 years, according to a survey of the Army Engineers.[1]

Surveys of 10 reservoirs in the Great Plains and central Kansas revealed an average silting rate of 1.20 per cent a year. All but two of these reservoirs were built within the last decade and had original storage capacities ranging from 519 to 3,213 acre-feet.

In the arid Southwest, reservoir storage is perhaps of more critical importance than in any part of the country, since the very existence of the irrigation civilization is dependent on a continuous supply of stored water. Available information on rates of storage loss by silting in this region indicate that the concern so often expressed is fully warranted. The great Elephant Butte Reservoir on the Rio Grande, with an original storage capacity of 2,638,860 acre-feet, silted at the rate of 18,034 acre-feet, or 0.68 per cent annually, for the $20\frac{1}{4}$ years from the beginning of storage in 1915 to 1935. Comparison with previous surveys indicates that the annual rate of silting declined during the period 1925–1935 from a previous average of approximately 20,000 acre-feet per year. If the present rate of filling is maintained, only about 80 years more will be required to reduce the capacity for storage of water to the normal annual draft required for present irrigation purposes; thereafter, a progressively increasing deficiency in water supply may be expected, unless in the meantime the serious erosion problem over the contributing watershed can be controlled.

McMillan Reservoir on the Pecos River, supplying irrigation water for the Carlsbad project in New Mexico, had lost $55\frac{1}{2}$ per cent of its capacity of 90,000 acre-feet by 1932, when it was 38 years old. It is significant, however, that the rate of silting in this reservoir has been continually declining since 1912 as a result of the natural spread of dense thickets of tamarisk above the reservoir, which have the effect of holding back silt. The initial rate of filling of 2.14 per cent a year had declined to 0.31 per cent during the period 1925–1932. Zuni Reservoir, supplying water for the irrigation project on the Zuni Indian Reservation, in New Mexico, lost 76 per cent of its original capacity of 14,800 acre-feet in $21\frac{1}{2}$ years from the date of its construction in 1906.

Surveys in southern California made after the exceptionally severe flood of Mar. 2, 1938, showed great damage to orchards by deposition of silt and boulders (Fig. 90) and disastrous rates of reservoir silting.[2] Santa Anita Reservoir in Los Angeles County, built in 1928 for flood

[1] Stevens, J. C. The Silt Problem, Am. Soc. Civil Eng. *Trans.*, Vol. 101, p. 210.

[2] Burke, M. F. Flood of March 2, 1938. Los Angeles County Flood Control District (mimeographed), 1938.

control with an initial capacity of 1,043 acre-feet, lost 31.35 per cent of its capacity during this one storm. During the preceding 10 years, it lost altogether only 3.17 per cent of its capacity.

Most of the California reservoirs are located in mountainous regions where watershed protection is good as long as the original forest and chaparral covers are maintained. When the cover is removed by fire or cutting, a very serious erosion and runoff hazard is established. In 1926, the chaparral cover was burned off the watershed of Lake Harding, in southern California. Following the fire, in February, 1927, a series of

Fig. 90.—Citrus orchard damaged in 1938 by deposition of silt and rocks on delta cone of mountain canyon. Note size of boulders transported by the flood. (*Photograph by Soil Conservation Service.*)

heavy rains occurred which caused severe erosion of the denuded loose granitic soils and practically filled the reservoir within a period of a single month. Previously, the reservoir had silted to only a negligible degree.

## Source of Material

The common belief that bank erosion along the shoreline and soil washing along the immediate slopes leading down to the shore are responsible for much of the filling of reservoirs has been shown to be incorrect. An investigation of Black Canyon Reservoir at Emmett, Idaho, shows that only 4 acre-feet of sediment has come into the lake from approximately 11,000 feet of wave-cut shoreline. This amount was only 0.1 per cent of the total accumulation of 4,037 acre-feet in the lake, as of 1936.[1] It was found in 1936 that of the total accumulation of 2,808

[1] Hough, J. L., and Flaxman, E. M. Advance report on the sedimentation survey of Black Canyon Reservoir, Emmett, Idaho. Soil Conservation Service (mimeographed), SCS-SS-19, 1937.

acre-feet of sediment in Lake Decatur, Illinois, only 4 acre-feet came from the rather widespread shoreline erosion.

No reservoir has been found where more than 1 per cent of the total sediment has been derived from wave cutting or from soil wash on the immediate slopes above the reservoir. The bulk of the material obviously is derived from the intake of silt-laden stream water.

## Sedimentation in Channels

Harmful sedimentation occurs not only in reservoirs but commonly also in channels where the coarser parts of stream-borne materials accumulate faster than the stream can move them. Such deposits tend to accumulate chiefly in those places where the velocity of flow is decreased for any reason, as immediately above the heads of deltas or at the mouths of relatively steep tributaries that enter a master stream of markedly lower gradient. Such accumulations contribute to increased overflow; higher water-table levels, with consequent waterlogging or swamping of adjacent valley lands; and increased overbank deposition of material that may be harmful to the alluvial soils.

One of the most striking instances of the relation of channel sedimentation to flood control and land-drainage problems is furnished by the conditions in the Middle Río Grande Valley, in central New Mexico. Here, a population of about 55,000 has depended chiefly on irrigation with water taken directly from the river.

Much of the drainage area tributary to the Rio Grande has been so heavily used for grazing that severe erosion is contributing greatly increased quantities of silt to the converging flood waters. As a result, the channel is filling at an alarming rate (Fig. 91), at least in the lower part of the Middle Valley, where the only accurate records are available. Immediately above the head of Elephant Butte Reservoir, the river bed is steadily rising with the inpouring products of erosion; flood damage is increasing; and waterlogging is ruining the former productive alluvial plain. The town of San Marcial, which had a population of over a thousand prior to 1929, is now almost abandoned. Since 1914, when Elephant Butte was built, the bed of the Río Grande a short distance above the head of water in the reservoir had risen an average of about 13 feet. Two miles upstream, just above San Marcial, the increase in level was about 7 feet during the same period. In view of these conditions, it is not surprising that dikes built to protect the town and adjacent irrigated lands have been overtopped repeatedly and broken by flood waters, with disastrous results. The former irrigated fields west of the river were almost completely converted into swamp as a result of a severe flood in 1929. In the spring of 1937 the Rio Grande broke out of the floodway to which it had been confined by dikes and took a new course directly across the remaining irrigated lands lying east of the river.

Channel filling has also been rapid where the Rio Grande is joined by one of its principal tributaries, the Rio Puerco, some 50 miles above San Marcial. The Puerco, draining a large area of eroding range land, is delivering silt to the Rio Grande more rapidly than that stream can remove it. As a result, flood waters in the spring of 1937 overtopped and washed out the levee and portions of a paved highway and converted hundreds of acres of pasture land into a sandy waste. Much formerly irrigated land near the village of Contreras has become swampy, because the water cannot be drained into the river now flowing at a level above that of the fields.

FIG. 91.—Complete filling of channel under San Antonio bridge across the Rio Grande in central New Mexico. (*Photograph by the author.*)

The filling of drainage canals and ditches is a serious form of channel sedimentation in many localities, as in a number of drainage districts of northern Mississippi (Fig. 92) and western Tennessee. Many of the ditches have been largely or completely filled with sand derived from the eroding uplands. In the Wells Drainage District of Lafayette County, Mississippi, three-fourths of the canals, dug at a cost of $71,000 in 1920, were completely filled with sand by 1936,[1] although special taxes to pay off the bond issue continue at least until 1943. The rate of filling of the other canals has been estimated to amount to an average monthly damage of $500, on the basis of the original investment, without including the damage to adjoining valley lands.

[1] Happ, S. C. Fertile Valleys Laid Waste by Upland Erosion, *Soil Cons.*, Vol. 2, No. 9, 1937.

In the Obion and Forked Deer Drainage basins in western Tennessee, according to a report of the State Planning Commission,[1] drainage canals dug at a cost of nearly 6 million dollars during the period 1910–1930, are largely ineffective, partly because of sedimentation. The damage here is reported to include development of serious malarial conditions adjacent to the swamped valleys and ponded streams as well as destruction of standing timber and abandonment of cultivated fields due to waterlogging and overspreading of sand.

On the upper Mississippi, channel sedimentation involves heavy expenditures for dredging operations to maintain a navigable channel

FIG. 92.—Three generations of fence posts that illustrate the progressive filling of Hurricane Creek Valley, Lafayette County, Mississippi. (*Photograph by Soil Conservation Service.*)

and also contributes to flood damages in tributary valleys. Although accurate data are not available to show the rates of channel filling, numerous accounts of early settlers and old residents testify to a large increase in turbidity of the streams that drain the adjoining uplands, where soil erosion has become very active since the clearing of forests and cultivation of the sloping lands. These streams now deliver large quantities of sediment to aggravate the problem of channel maintenance. In many instances the tributaries bring down the products of erosion faster than the material can be removed by the Mississippi and hence are building up the lower parts of the side valleys. In these areas of tributary aggrada-

[1] Tennessee State Planning Comm. The Obion River and the Forked Deer River Watersheds. 1936.

tion, channels are inadequate to carry the higher flows, so that flooding is causing much damage. As the filling progresses upstream, flood damage is increasing, and considerable areas of valley lands are being swamped by the rising ground-water table.

In a compilation of characteristics of navigable streams, the Preliminary Report of the Inland Waterways Commission in 1908[1] stated concerning the Pearl River of Mississippi:

"Pearl River has completely changed its character in the last half century; from a slow, clear stream it has become a swift, muddy one, and from a good channel with a depth of five or six feet it is now shallow and much obstructed with drifts and logs. This rapid filling is due to the washing of the sandy uplands that have been opened up along the stream, with consequent choking of the river channel."

Concerning the Tallahatchie, a tributary of the Mississippi River, Lowe, then state geologist,[2] wrote:

"The Tallahatchie was formerly a navigable stream. Even as late as 1900 a small steamer drawing four feet of water plied on the Tallahatchie from Batesville downstream. Now the stream is choked with sand bars, and can be easily waded at almost any place. . . . "

A condition of silt debris common to the uplands of the loessial region along the Missouri River in western Iowa is cited by Towl[3] as follows:

"Silt deposits in ditches result in great damage to drainage districts and eventually affect railroads and other interests. For instance, in Fremont County, Iowa, the Burlington Railroad raised its main line 7 feet at a cost of $36,000 to permit the construction of the Plum Creek floodway ditch to carry silt directly to the Missouri River. In Woodbury County, Iowa, the C. M. & St. P. Railway has recently raised its bridge over the west fork of the Little Sioux River, and is planning other works, on account of drainage ditches which have been reduced in capacity by silt deposits."

In southwestern Kansas, the bed of the Arkansas River has risen as much as 5 feet in a period of forty-five years, according to Meeker,[4] who cites this as "indicative of heavy sand and silt burden resulting from erosion induced by settlement."

[1] United States Congress. Preliminary Report of the Inland Waterways Comm. 60th Cong., 1st Sess., *Senate Doc.* 325, p. 65, 1908.

[2] Lowe, E. N. Reforestation, Soil Erosion and Flood Control in the Yazoo Drainage Basin, 4th Southern Forestry Cong. *Proc.*, Vol. 4, pp. 10–11, 1922.

[3] Towl, R. N. Silt and Silting Basins on Drainage Works, abs., *Eng. News*, Vol. 92, pp. 488–489, 1924.

[4] Meeker, R. I. Discussion of Silting of the lake at Austin, Texas, Am. Soc. Civil Eng. *Trans.*, Vol. 93, pp. 1681–1735, 1929.

## Sedimentation on Bottomlands

In parts of the headwater section of the Yazoo Basin of northern Mississippi, sand deposition in minor valleys as a result of accelerated soil erosion was described as seriously harmful by State Geologist E. W. Hilgard[1] as early as 1860, only about twenty years after the area was

Fig. 93.—Twenty inches of cross-bedded, flood-deposited sand over productive alluvium (Congaree fine sandy loam), North Tyger River, South Carolina. The original soil now buried was formerly cultivated. (*Photograph by Soil Conservation Service.*)

settled. Recent investigations in the same area show severe damage to bottomlands in a north-south belt in which the sandy Holly Springs formation outcrops or has been exposed by gullying. In many places the original thin mantle of loess has been removed by erosion since the

[1] Hilgard, E. W. Report on the Geology and Agriculture of the State of Mississippi. Jackson, Miss. 1860.

agricultural occupation of the area, but most of the sand debris is derived from the many gullies that have incised the formation.

In the southeastern Piedmont, especially through the Carolinas and Georgia, accelerated sedimentation has reduced the capacity of the stream channels and thus caused more frequent flooding of the bottomlands. Also, a large part of the first bottoms, particularly of the narrower bottoms which formerly were highly productive, fine-textured soils, has now been covered with comparatively infertile sand (Fig. 93) and gravel. Much of this formerly cultivated land has been changed to a condition of swampiness. These conditions are recorded in numerous soil survey reports, from one of which the following typical description is taken[1]:

"Some of the areas mapped as Meadow were originally good areas of Congaree silt loam or Congaree fine sandy loam over which a mixed covering of sandy material has been deposited following the clearing of adjacent hillsides. The bottom land of Mill Shoal Creek about a mile from the county line is typical of this condition. It was reported that all the bottom was cultivated and the soil was Congaree fine sandy loam 12 years ago. At present the bottom land is a sandy wash with poor drainage and is covered with alders and willows. . . ."

In and adjacent to the unglaciated portion of the Upper Mississippi Valley, comprising a large area in southwestern Wisconsin and adjacent sections of Minnesota, Iowa, and Illinois, damage by sedimentation on valley lands frequently takes the form of accumulation of alluvial fans about those points where gullies and streams debouch upon larger valleys of lesser gradient. More than a thousand acres of bottomland along the Zumbro River, southeastern Minnesota, have been covered with alluvial-fan deposits to such depths as to reduce the productivity by an estimated average of 80 per cent. In one instance, as the result of a single storm which produced excessive erosion within the basin of a tributary, an area of 600 acres in the Zumbro Valley was covered with sand and silt to depths up to 5 ft.

Damage by overwash and development of alluvial fans is not confined to the land but frequently necessitates considerable repair and maintenance work on highways. In Winona County, Minnesota, for example, it was reported recently that the average cost of removing such material from the county and township highways amounted to approximately $30,000 annually.

Deposition of fine material from irrigation water is troublesome to the water users of many irrigation districts, because fine silt and clay deposited on the land by irrigation water reduces permeability by clogging

[1] Fuller, G. L., and Hendrickson, B. H. Soil Survey of Elbert County, Georgia, U. S. Dept. Agr. Bur. Chem. and Soils (Series 1928) No. 15.

of the soil pores. Fortier and Blaney[1] quote M. J. Dowd, general superintendent of the Imperial Irrigation District in California, to the effect that:

"The silt in the water increases the difficulties of raising certain kinds of crops. In the case of alfalfa grown on the harder type of soil with little fall to the land, the silt depositing a thin film seals the ground surface, thus increasing the length of time the water stands on the land after an irrigation and during the hot summer this results often in the scalding of the alfalfa. In the case of lettuce and cantaloupes which are furrow-irrigated, it is a common occurrence for a farmer to run water in a furrow for several days trying to 'sub' the moisture to the top of the hill and often before this is accomplished it is necessary to shut off the water, break up the film of deposited silt on the bottom and sides of the furrow, and then turn the water on again."

In the same report, they state further that:

"It is estimated that the annual expense to the farmers of Imperial Valley caused by silt averages $2 an acre. Applying this cost to the acreage irrigated by 1924, and adding thereto the cost of canal cleaning, brings that year's cost of silt disposal and control in its various forms above $1,333,000."

Silt-detention Basins. Detention of the finer fractions of eroded soil in silting basins has resulted in the building of excellent farm land in some localities. The Bennett desilting basin on Wilson Creek,[2] in east-central Washington, provides for both irrigation and fertilization of the formerly poor soil of a bottomland farm. A dam 30 feet high across this valley has an outlet gate at its base which permits the basin to be completely drained. The mean annual precipitation in the watershed is less than 8 inches, and the creek is usually dry except during a few weeks in the spring when melting snow frequently produces a flood. The flood water is held in the basin for about two weeks, thus saturating the soil sufficiently to provide moisture for the production of an excellent crop of winter wheat and two or three crops of alfalfa during summer. Erosional material derived from the silty uplands is deposited in considerable quantity. A survey of the basin showed that 470 acre-feet of silty material has been deposited on the 345 acres within the basin in 18 years. Fertility tests made by the pot-culture method at the State Agricultural Experiment Station indicate that the newly deposited sediment is more productive than either the original valley soil or the upland soil from which the material is derived.

[1] Fortier, S., and Blaney, H. F. Silt in the Colorado River and Its Relation to Irrigation, U. S. Dept. Agr. *Tech. Bull.* 67, 1928.

[2] Hough, J. L., and Flaxman, E. M. Advance report on the sedimentation survey of the Bennett irrigation and silting basin, Wilson Creek, Washington. Soil Conservation Service (mimeographed), SCS-SS-27, 1938.

Another type of silt-detention basin, designed primarily to prevent sedimentation damage to valley lands, is represented by a 26.6-acre debris basin on the Missouri River flood plain in Doniphan County, Kansas.[1] Here an area of 4,300 acres of productive bottomland was being damaged by local flooding caused by filling of the drainageways with soil washed from the adjacent uplands. The principal stream carrying material down over this area is Chase Creek, with an upland drainage area of 1,011 acres. In 1920, a debris basin was formed by constructing a dike approximately 6 feet high around an area of 26.6 acres along the foot of the bluff where the creek issues from the highlands. The creek was diverted into this area and provisions were made for drainage of surplus water along the edge of the bluff into the river. This produced a stilling basin for flood waters coming from the small watershed of Chase Creek, and permitted the deposition of the sediment over the diked area. In ten years, from 1920 to 1930, the basin accumulated 135.6 acre-feet of soil, representing the equivalent of a loss of 0.13 foot of topsoil from the entire watershed. The material deposited in the debris basin ranged from 2.8 to 8.2 feet deep and proved to be exceedingly productive. It has produced from 65 to 80 bushels of corn per acre and correspondingly high yields of vegetables.

## Correlation of Sedimentation Data with Watershed Conditions

Although sedimentation surveys indicate the places and extent of damage by erosional debris, they do not necessarily serve as a basis for planning control programs for the contributing watersheds, unless correlated with the silt-producing areas revealed by surveys of the entire drainage basin.

Within any part of a major watershed, the contribution of a unit area to sedimentation downstream may be quite different from what would be indicated solely by the extent of upland erosion as delineated on a conservation survey map.

The greatest contribution to reservoir silting may be from erosion of stream banks and roadside ditches or other nonagricultural areas. Hence, a conservation program planned solely for protection of farm land may not at the same time accomplish maximum reduction in the sediment load of streams or in sediment damage. Likewise, the degree of past erosion, as shown by conservation surveys, may not be a complete index to the most prolific agricultural sources of sediment at the present time. Yet downstream water-storage values may justify and demand the control of soil erosion in critical areas of sediment production to an even greater extent than the agricultural interests of the watershed will economically justify treatment solely on the basis of land values.

[1] Brown, C. B. Protecting Bottom Lands from Erosional Debris: A Case History, *Soil Cons.*, Vol. 3, No. 4, 1937.

In some watersheds, mineralogical analyses of sediment samples collected from the delta areas of a reservoir and from stream beds are helpful in indicating the primary sources of bed-load material. Examination of the material of recent flood-plain deposition may lead to identification of the particular parts of a watershed that, because of their heavy contribution of silt, may need special control treatment.

Much sediment is transported so slowly from its point of origin in some watersheds that it has little significance with regard to deposition in reservoirs. Large quantities of silt, produced by slope wash in some watersheds, are deposited on lower slopes and alluvial plains far above the reservoir and there stabilized indefinitely by voluntary vegetation. In other instances, the main inflowing streams, eroding unconsolidated terrace deposits, carry the debris immediately into the lake basin. Such erosion may be prevented by installation of rip-rap or a protective vegetative cover on the actively eroding areas. In some places, the establishment of a barrier of trees, grass, or shrubs across the flood area near the head of water or delta area has served to reduce greatly the amount of sediment entering the reservoir.

The extent of sediment damage to downstream resources is, of course, an important factor in determining total costs that are justified for erosion control, especially for determining the extent of the public interest in such control programs. This, however, is not the only relation between the two problems. Valley alluvial lands are generally not subject to severe erosion and thus offer relatively advantageous sites for development of a permanent type of agriculture. They may be potentially the best lands of a locality, provided the streams that they border can be kept under reasonable control and irrigation water can be stored, where needed. The reclamation of alluvial lands, therefore, offers one potential outlet for some of those farmers at present cultivating unprofitably poor hill lands.

In some foreign countries, infertile or swampy bottomlands have been brought into productive use by covering them with silt deposited from streams directed over the areas so as to lay down the more fertile finer fractions of the stream load (see Chap. IV, Part 1, "The Assorting Effect of the Erosion Process"). In some instances, reservoir silt has been used for crop production or as a top dressing for improvement of less fertile soils located relatively nearby. In some parts of the irrigated orchard districts of southern California, the rich silt that collects in small irrigation reservoirs is regularly removed during the period when the reservoir is not in use, hauled to the orchards, and spread around the trees. However, such utilization of silt is exceptional in the United States; but with decreasing soil resources, some localities may be forced to make use of this waste product of erosion.

# Chapter XII. Mass Movement an Important Process of Soil Wastage

Mass movement of soil is not only an important factor in the denudation of land under natural conditions, but it also plays an important role in the acceleration of soil removal as the result of man's activities. Such movement involves a group of interrelated processes, many of which are intimately associated with soil erosion.[1]

Under natural conditions, the part played by mass movement and its involved processes varies greatly with the climate, slope, soil, rock type, and physiographic characteristics of the land. In areas where large landslides have occurred, the power of mass movement is easily recognized. Where slumping and earthflows or mudflows are common phenomena, the effect of mass movement in shaping the landscape is clearly identified. But downhill movement of soil by the slower and less obvious process of *soil creep* is likely to escape notice unless it is given special attention. Such inconspicuous movement, however, may equal or exceed in importance the spectacular displacements of large masses.

A clear understanding of all types of land movement is essential to a proper interpretation of soil removal under natural conditions. Such information is needed by those who are engaged in the control of accelerated soil displacement as the result of man induced conditions. What effect conservation measures may have on mass movement must be considered. Water conservation by terracing, for example, probably should not be attempted generally on hillsides subject to serious slippage. In other words, measures employed for soil and water conservation must be adjusted to local hazards of mass displacement wherever such hazards are known to be serious.

Mass movements may be divided into those that take place by *flowage* and those that take place by *slippage*. The distinction is not always sharply defined, but it is a useful one, closely related to the causes of the displacements.

[1] Sharpe, C. F. S. "Landslides and Related Phenomena." New York. 1938; What Is Soil Erosion? U. S. Dept. Agr. *Misc. Pub.* 286, 1938.

## Soil Creep

The most nearly universal but least spectacular type of mass movement is soil creep, a form of slow flowage. It has been defined by Sharpe as the "slow downslope movement of superficial soil and rock debris, usually imperceptible except to observations of long duration."[1] Creep is effective with large fragments of rock as well as the finest textured soils. It is essentially a surface phenomenon, extending to a depth of only a few feet, but it transports vegetation and also man-made structures. Its rate of movement depends on slope, character, and water content of the material, climatic conditions, and vegetative cover. It operates on forested land as well as on bare or grass-covered areas and is one of the processes that keeps soil profiles on slopes thinner than on level land. Soil creep beneath a cover of vegetation is believed to be one of the major factors in producing the smoothly rounded topography characteristic of humid lands under natural conditions. Unlike stream erosion and gullying, this process of soil removal often proceeds over the entire interstream areas. Sheet washing also operates over the whole surface but is effective only where there is little or no protective cover of vegetation.

Cultivation of the land by man modifies soil creep in several ways. Removal of natural vegetation increases the rate of runoff and reduces the amount of surface water to enter and "lubricate" the soil. This tends locally to retard the rate of creep. With the cutting of gullies and the deepening of stream channels, however, slopes are steepened, and the rate of creep near the channels is increased.[2]

Evidences of creep are commonly overlooked. One of the most obvious is the curvature of trees whose trunks are concave on the uphill side, owing to the contest between downslope tilting of their bases and the tendency of their crowns to grow vertically. Fence posts are often tilted downslope and sometimes are so moved out of alignment that one fence line will be abandoned altogether and a new set of posts installed a few feet farther up the hillside. Movements of telegraph and telephone poles, gravestones, and other monuments and the tilting and fracturing of retaining walls often give a key to the activity of soil creep.

The movement, however, often is obscured by vegetation, or it may take place largely beneath the surface. Creep is believed to play an important part in the formation of the shallow crescentic scars found in parts of the Great Plains from Texas to Montana and sometimes spoken of as "stepped crescents" (Fig. 94). These scars are commonly attributed

[1] Sharpe, C. F. S. "Landslides and Related Phenomena." New York. 1938; What Is Soil Erosion? U. S. Dept. Agr. *Misc. Pub.* 286, 1938.

[2] Lowdermilk, W. C. Acceleration of Erosion above Geologic Norms, Am. Geophys. Union, *Trans.*, pp. 505–509, 1934.

to the effects of sheet wash and overgrazing. The latter undoubtedly is an important factor in their formation, but sheet wash seems to play little part. The process, so far as it is now known, is a combination of downslope flowage of the soil beneath the sod and a vertical or almost vertical downsinking of the surface immediately within the crescent.

The importance of creep in other areas has been pointed out by Sauer,[1] who believes it to be the major factor in forming the "dales" of parts of California. These small depressions, normally lacking channels and characterized by rounded floors, he says, are the normal small valleys

FIG. 94.—Low crescentic breaks—"stepped crescents"—on a 7 per cent slope in Chouteau County, Montana. (*Photograph by Soil Conservation Service.*)

of California and of large sections of the Southwest. The dale is gently concave, where typically developed, and its surface passes smoothly into the flanking slopes. It is in no sense stream-cut but is both formed and maintained by mass movement. The weathered products from the dale slopes are evacuated entirely by gravity in the form of a soil flowage down the flanks. Sauer[2] states that:

"By continuation of weathering the soil particles are broken down more and more. By growing finer they become more easily moved and thus make more unstable the slopes on which they are forming. Growth in thickness of weathered cover, of course, also increases instability of slope. Lubrication by wetting facilitates movement in mass down grade. Continuing, though not continuous, down-

[1] Sauer, Carl. Land Forms in the Peninsular Range of California as Developed about Warner's Hot Springs and Mesa Grande, Univ. California, *Geog. Jour.*, Vol. 3, pp. 210, 212, 1929.

[2] Sauer, C. O. Land Resource and Land Use in Relation to Public Policy, Sci. Advisory Board *Rept.*, p. 184, Washington, D. C., 1934.

grade migration of soil appears to be characteristic of all slopes under normal processes of degradation, and such a slope is a momentary curve of balance, a surface of slow transport. . . . Observations now under way in California have found measurable displacement of every point staked out in the lapse of a year, with large differences in relative movement and apparently consistent variation in mobility of surface."

It is the opinion of Penck[1] and others that the character of vegetation has little influence on the rate of soil creep and that colloidal materials and other fine soil particles can pass through the network of roots with ease, even on forested slopes. According to his observations, coarser materials move less readily; but as the root mat is shallow and has its greatest effect within 10 centimeters of the surface, creep is not greatly retarded.

The causes of soil creep vary in different regions, and the character of movement differs accordingly. Normally, it progresses by a multitude of minute movements of the individual particles or aggregates. Frost is one of the most active agents, both by its wedging action and by its upward heaving. Where there is sufficient soil moisture, a freeze will produce layers of *needle ice*, or *spew frost*,[2] which will lift the overlying soil and vegetation as much as several inches. Melting of the ice during the succeeding thaw allows the heaved soil to settle downward again. Although this process is very common, it may escape notice on lands having a good turf. The spongy feeling underfoot after a thaw is a clue to what has been going on beneath the surface. On bare ground, as on the walls of road cuts and gullies, the formation of this needle ice, or spew frost, can be observed readily. When the temperature drops below freezing, the ice crystals can be seen growing perpendicular to the slope to a height of 1, 2, or even 3 or more inches, carrying up with them a layer or crust of drier soil particles, stones, twigs, and miscellaneous debris (Fig. 95). Successive freezes unaccompanied by complete thawing may cause additional stages of growth of the ice crystals. As the ice melts and the layer of raised material is let down, the place where it will come to rest will depend in part on the slope of the ground. On perfectly level surfaces, the material may return to approximately the point from which it was raised. On a slope, however, the action of gravity tends to deposit the particles farther down the slope than they were originally. Sharpe[3] reports that it

[1] Penck, Walther. Die morphologische Analyse, Geographische Abh. (2d Reihe), Heft 2, pp. 65–68, Stuttgart, 1924.

[2] Bennett, H. H. Soil Erosion and Flood Control. Lectures before U. S. Dept. Agr. Graduate School. Soil Conservation Service (mimeographed), SCS-TP-7, 1928. (See Lecture 2, pp. 1–2.)

[3] Sharpe, C. F. S. "Landslides and Related Phenomena." New York, 1938; What Is Soil Erosion? U. S. Dept. Agr. *Misc. Pub.* 286, 1938.

is common for the ice crystal to melt first at its base so that, instead of allowing the particle that it had lifted to return vertically downward, the crystal falls downslope and deposits the solid matter at least the length of the crystal farther down the incline (Fig. 96).

This process of soil transportation is particularly effective under a climate characterized by frequent alternation of freeze and thaw. Its effect in loosening the surface of clayey soils is commonly seen in the

Fig. 95.—Needle ice, or frost crystals, 3 inches high on a steep slope in South Carolina. Melting at the bases of the crystals allows the ice and the thin layer of soil and debris it carries to fall downslope. (*Photograph by Soil Conservation Service.*)

Southeastern States. The walls of gullies and road cuts often exhibit a distinct line of separation. Below the line, where more moisture is present, an abundant coating of spew frost is formed; but on the relatively dry soil above, the process is not effective. On slopes of sufficient steepness, where frost crystals and their load of soil and debris slide or roll to the bottom of the incline, the amount of material accumulated is often considerable. Depending on its frequency of repetition, this process of mass movement by frost may be highly effective in causing the removal of soil and the recession of the walls of ditches and gullies.

Many other processes take part in the downhill creep of the soil. Wedging by the growth of plant roots, prying by the swaying of trees in the wind, the upturn of soil by the roots of falling trees, compaction by the walking of animals, filling of holes made by animals and of cavities left by decaying vegetation, expansion and contraction caused by heating and cooling or by absorption and desiccation, all tend to move material slowly down the slope. Even the disintegration of rocks and the crumbling of clods and other soil aggregates contribute to the downhill mass

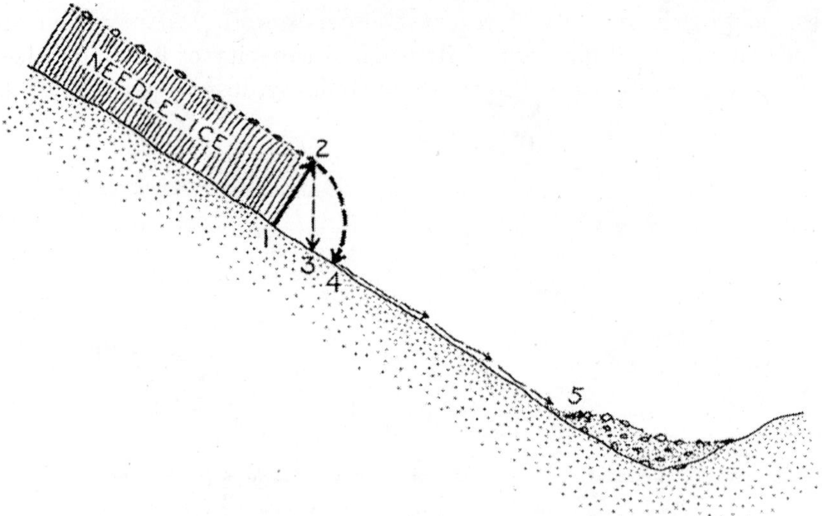


FIG. 96.—A particle at position 1 is lifted perpendicular to the slope to position 2 by the formation beneath it of needle ice, or spew frost. Gravity would let the particle fall to approximately position 3, but because the crystal melts first at the base and falls downhill as a unit the particle reaches the ground at position 4. On a steep slope it is likely to roll or slide to position 5.

migration of the soil. Movement brought about by man through cultivation is a particularly important factor. Where the ridge is thrown always to the downhill side, each plowing may move the surface soil several inches or a foot or more down the slope. Movement produced in this way far exceeds the normal rate of creep for surface soil under natural vegetation.

## Solifluction

In the colder climates at high altitudes or high latitudes, where the soil is frozen to considerable depths for long periods, solifluction rather than the common soil creep is the dominant type of mass movement. Owing to its climatic restrictions, this powerful process, with its associated polygonal soils, stone rings, stone stripes, and other land forms, is of only secondary importance to agriculture and will not be discussed here.

## Other Forms of Soil Flowage

The more rapid forms of soil flowage common to temperate climates may be described as earthflows, debris avalanches, and mudflows. These differ from creep and other types of slow flowage in that their movement usually is perceptible to casual observation. In some parts of the country, rapid flowage movements are the outstanding processes in the sculpturing of the landscape. Such movements are active both in humid and in semiarid climates but usually can be correlated closely with precipitation and therefore are most common at particular seasons of the year.

## Earthflows

Earthflows are especially abundant in a large area of hilly to mountainous country in the southeastern Ohio, West Virginia, and south-

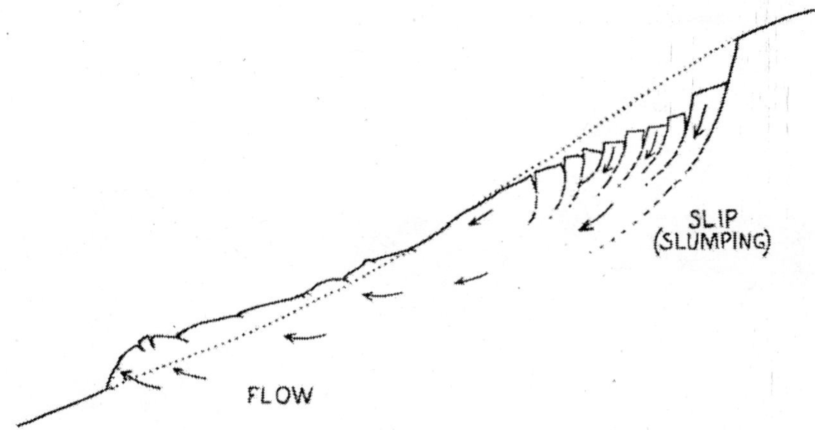


FIG. 97.—Longitudinal section through a typical hillside earthflow, showing the association of flowage in the lower part with slippage of slump type higher on the slope.

western Pennsylvania sections of the Appalachian Plateau. The principal soils of this region, such as the Muskingum, Westmoreland, Meigs, and Upshur, are derived largely from shales and sandstones interbedded with coal seams and thin limestones. Movement of ground water through the basal rock appears to be controlled largely by the impervious fire clays that occur beneath the coals, as indicated by the seepage commonly occurring where these beds outcrop on hillsides. Following protracted rains, masses of locally saturated soil often move downhill as earthflows, commonly known as *slides*, or *slips*.

In the movement most typical of this region, flowage is dominant in the lower part of the disturbed area; slumping is the major process along the upper side; and the central part shows characteristics of both (Figs. 97 and 98). At the downhill terminus of the movement, surface soil and vegetation are forced upward in one or more ridges. This bulging toe may

Fig. 98.—An earthflow which resulted from the heavy rains of January, 1937. Tuscarawas County, Ohio. (*Photograph by Soil Conservation Service.*)

Fig. 99.—Slumping at the head of the earthflow shown in Fig. 98, Tuscarawas County, Ohio.

rise 6 feet or more above the ground level and, depending on the material and degree of wetness, may be smoothly rounded or broken into irregular blocks and ridges. If the moving material is very wet, water and mud frequently break through the bulge and flow downhill. The area of slumping at the head of the disturbance is broken into a series of blocks bounded by bare scarps (Fig. 99). Individual blocks sometimes move downward and rotate toward the slope, so that they tilt backward or at least have a lower angle of inclination than in their original position. Between the bulging toe and the area of slumping, the ground often appears to have been little disturbed, or it may be badly broken and disordered, with some blocks protruding upward and others sunk beneath the general level. In earthflows complicated by more than one stage of movement, there are commonly some duplication and overlapping of the three zones.

Few of the earthflows in this Eastern area are more than an acre in size, but a single hillside may show traces of many flows. Some slopes have a peculiar choppy or benched configuration as the result of repeated flows over a long period.

Earthflows are most frequently seen in pasture land, but they are also found in wooded areas. For the reason explained above, they seldom occur on cultivated land. Cultivation favors soil removal by surface washing rather than by mass movement.

In various parts of the country, earthflows not only cause considerable inconvenience to agricultural operations but result in the actual loss of tillable land. On the basis of present knowledge, it appears that the best method of handling such displacements is to smooth over the areas and seed them to grass. If the fractures along the slip planes are allowed to remain open, they serve as channels by which water enters the ground to promote further movement. Unfortunately, the best treatments for the prevention of mass movement do not always accord with the best methods for promoting plant growth or preventing surface erosion. Probably the best solution generally would be to install stone or tile drains on those slopes known to be particularly susceptible to mass flowage of soil, but such procedure seldom is practicable.

In the Palouse Wheat Belt of Washington, Idaho, and Oregon, earthflows are of common occurrence on steep northward slopes following rapid melting of snowdrifts formed along the brink of such slopes. Strips of trees and shrubs planted over these critical areas have had the highly beneficial effect of distributing the snow so as to prevent deep accumulations. This has reduced erosion markedly on the slopes below and also appears to have affected favorably the hazard of slides. Earthflows, or "flow slides," as they are sometimes called by engineers, have been particularly destructive to highways, railroads, and other engineering works in the Ohio-West Virginia-Pennsylvania region. Various measures have been employed in attempts to retard or prevent such slides. Rows of wooden piles driven near the base of the disturbed area generally have failed after short periods, either by being moved out of alignment or by being sheared off underground. A more successful modification of that procedure has been the use of well casings set 6 or 8 feet into solid rock and filled with reinforced concrete. One or more horizontal rows with the piles spaced several feet apart at the base of active areas have been successful in retarding and sometimes in stopping these movements. A somewhat simpler and fundamentally more sound procedure is the use of intercepting trenches or drains which prevent water from reaching the sliding area by removing whatever water falls directly above.[1]

Earthflows are of importance in many other areas. One of the most striking of these is the Gros Ventre River Valley in Wyoming where an area extending more than a mile up the hillside began moving some 30 years ago. Portions of it are still unstable.

In addition to these hillside earthflows, similar movements are very common on the clay terraces of glaciated regions. Heavy rains, especially those following long dry spells, have penetrated and so lubricated the clays that hundreds of acres of terrace land have sometimes moved outward and downward into the river or to lower positions in the valley. A striking feature of many of the occurrences is that the clay along the bank appears to have been firmer and more resistant than that farther back in the terrace. In such instances, although the break in the firm material at the river bank has been relatively narrow, very large areas of the terrace have flowed outward through the bottleneck within a few hours. At other localities where the opening has been wider, flowage has been combined with slumping of the river banks, and as much as 25 million cubic yards of material have moved outward at one time.

Many earthflows of this sort have taken place from the terraces which, for a distance of 200 miles or more, border the tributaries on the north side of the St. Lawrence River. In the area around Portland, Me., several such movements have been recorded, and there are traces of many unrecorded flows. In the upper Hudson Valley, movements of similar type have occurred; but in most instances, the fluid material was some distance below the surface, and outflowage of the liquid mass allowed an almost vertical downsinking or subsidence of the more coherent material. Other somewhat similar examples in California have been reported.

## Debris Avalanches

On steep slopes in humid areas having a good covering of vegetation, very rapid mass movements, called *debris avalanches*, are common. In such areas as the White Mountains of New England, the long, narrow, bare scars produced by these avalanches are abundant. Movements of this type usually take place during or following heavy rains, when the in-

[1] Ladd, G. E. Landslides, Subsidences and Rock-falls, Am. Ry. Eng. Assoc. *Proc.*, Vol. 36, pp. 1128–1150, 1935.

creased weight caused by saturation of the soil and the lubrication furnished by the water surmount the forces that previously held the soil on the slope. Once loosened from position, the mass flows rapidly downhill.

Movements of this type usually destroy part of the vegetative cover on the slope and often strip off the soil down to bedrock. Damage to agriculture usually results from flooding produced by choking the channels of streams flowing through the more nearly level land of the lower valleys or from the deposition of the avalanche debris over tillable land. The history of New England records many avalanches of this kind. Such movements, however, are not restricted to glaciated regions. Debris avalanches have taken place in the Blue Ridge Mountains of North Carolina at several different times, although there, owing probably to the more gradual transition between soil and rock, vegetation is more successful in obscuring the scars.

## Mudflows

The two preceding types of rapid flowage are most common in temperate humid climates. In semiarid to arid lands, their place is taken by the *mudflow*. Although mudflows are not exclusively the result of accelerated erosion, their frequency and violence have been increased by man's misuse of the land. Such flows are common where the topography is rugged and where fire, overgrazing, or cultivation has left the ground poorly protected by vegetation. During heavy rains, soil and loose rock are picked up and carried along by the runoff, usually in a canyon or stream channel. The muddy flow is a pasty mass of high viscosity and can transport blocks and boulders of large size.

In the usual progress of such a flow, the mass moves along until enough water has been lost by absorption or evaporation for frictional resistance to overcome the movement. The debris then temporarily dams the canyon or depression in which it stops until additional water accumulating behind furnishes the lubrication necessary for further movement. Thus, by successive stages of flowage and damming, the mudflow moves down the canyon. Where it emerges from the canyon mouth, it spreads out and deposits its material in a broad fan consisting of a heterogeneous or poorly sorted mass of silt, sand, gravel, cobbles, and rock masses, some of which may weigh more than 100 tons.

Mudflows differ from earthflows in character of material, in that they are found most commonly in canyons and stream channels, and in that they tend to recur in the same places just as do floods. Earthflows, on the contrary, seldom occupy stream channels but are found on valley sides and terraces and even on convex portions of slopes. They do not have such a marked tendency to recur in the same area.

Mudflows are common in many of the Western States. Along the front of the Wasatch Mountains north of Salt Lake City, Utah, almost every canyon has its mudflow fan. Much of the area along the base of these mountains formerly was valuable and productive land used for nurseries and truck gardens. It was part of the floor of old Lake Bonneville, which once occupied this basin. In recent years, heavy grazing and fires on the mountains have depleted the vegetal cover and accelerated runoff. Mudflows also have been common. During severe rains in 1923, and several times in 1930, flows coming from some of the canyons did serious damage,

Fig. 100.—The coarse blocks and heterogeneous debris of this mudflow fan at the mouth of Parrish Canyon, north of Salt Lake City, Utah, covered productive farm lands. (*Photograph by C. F. S. Sharpe.*)

blocking highways, damaging or destroying farm buildings, and covering agricultural land to depths of 5 to 10 feet or more (Fig. 100).

Mudflows are not restricted to Utah or the Wasatch Mountains. They may occur wherever steep slopes of unconsolidated or deeply weathered material are unprotected by a substantial cover of vegetation and where intermittent rainfall is interspersed with heavy downpours. They are not limited to semiarid climates but occur under alpine conditions, where vegetation is sparse and slopes are steep. They also occur in areas of unconsolidated volcanic debris.

Much of the damage caused by mudflows can be avoided. Prevention or control of grazing where such flows are common will do much to keep the runoff clear and prevent the carrying away of valuable soil. Protection of valley lands from damage by deposition can be accomplished in part by the construction of settling basins and overflow sluices. Many of these, as well as training walls for the mudflows, have been built by communities along the mountain front north of Salt Lake City. Permanent control measures, however, must aim to prevent the flows, not merely to control them after they have formed.

## Slippage Forms

*Landslides*, or mass movements by slippage, are an important accessory to the soil-erosion process. Large slides, such as those at Frank, Alberta, in 1903, and in the Gros Ventre River Valley of Wyoming in 1925, have displaced 30 to 50 million cubic yards of rock and have covered or ruined several square miles of land. Even larger slides have taken place in foreign countries. Fortunately, landslides of such magnitude are rare in agricultural regions and therefore do not play a major role in the destruction of tillable soil. Smaller movements by slippage, however, are very common; they not only transport soil but prepare the way for erosion by water and wind.

*Slumping*, one of the most common types of slippage, has been defined by Sharpe[1] as "the downward slipping of a mass of rock or unconsolidated material of any size, moving as a unit or as several subsidiary units, usually with backward rotation on a more or less horizontal axis parallel to the cliff or slope from which it descends."

Most slumping takes place slowly and intermittently over long periods, but some of the movements may be completed in a single rapid slip. Displacement is small in comparison with the size of the individual masses or blocks of material moved. The slip surface usually is spoon shaped and is seen at the ground level as a crescent the points of which are directed down the slope.

Slumping is an important process in the recession of shores and in the widening of stream channels and gullies. Undercutting is the major cause of these slumps (Fig. 101), but seepage of ground water from the base of the bank aids the movement.

Large slumps are found on the north shore of Long Island and along the bluffs and floodplains of rivers such as the Mississippi and many of the streams in the Southwest. Slumping is common along man-made channels where dredging has produced an oversteepened bank. Examples of this may be seen along some of the cutoffs recently dredged across the necks of meander spurs of the Mississippi River (Fig. 102).

[1] Sharpe, C. F. S. "Landslides and Related Phenomena." New York, 1938; What Is Soil Erosion? U. S. Dept. Agr. *Misc. Pub.* 286, 1938.

Man has directly or indirectly caused slumping in many other situations. Hillside excavations for foundations, roads, and railroads or for

FIG. 101.—This large slump in Ventura County, California, settled downward an average depth of about 10 feet during the night of June 17, 1938, and continued moving at a slower rate. There was also a sliding movement in the direction of the deep barranca at the bottom of the valley—enough to shove some of the material across the bottom of the ravine in places. (*Photograph by Soil Conservation Service.*)

FIG. 102.—This slump extends for almost one-quarter mile along the east bank of the Giles Bend Cut-off of the Mississippi River, near Natchez, Mississippi. (*Photograph by Soil Conservation Service.*)

gravel, clay, and other materials commonly induce movement of the soil mantle higher up (Fig. 103). Interruption of the natural slope of the land by works of this kind introduces a major hazard, the full significance of which often is overlooked. Even if a relatively small amount of material is removed from the base of a slope, the break may migrate rapidly uphill and destroy fields and woodlands.

Not all slumping is produced by undercutting. Large natural movements of this kind are common in the Cascade Mountains of Washington and in the Columbia Plateau, where segments of the capping basalt flows move slowly downward as a result of weathering of the underlying less resistant layers. Backward rotation of successive large slump units has

FIG. 103.—Slumping on this hillside in New York State was started by excavation of material for a road fill. It already has eaten its way 50 feet into the field and may go much farther. (*Photograph by Soil Conservation Service.*)

produced a kind of terraced slope. On the steps of such slopes, lakes or swamps sometimes develop. Slumping often is associated with flowage, as in some of the earthflows previously described.

*Terracettes.* Slumping of masses only a few feet wide is the cause of many of the hillside steps, variously known as "cattle terraces," "sheep paths," or "catsteps," found on sloping lands in many parts of the country. Because these forms are most abundant and most obvious in pastures, they generally have been attributed entirely to the trampling of cattle and sheep. Many of them are true stock paths, the result of slow wearage (attrition), such as has contributed comparatively little to actual mass movement of the land. Some of the small irregular steps, however, have developed as the result of slippage, apparently with little or no aid of animals. Such steps are particularly well-developed in loess, two characteristics of which are a tendency to vertical cleaving and the ability to stand in vertical walls for long periods. These characteristics favor slippage along more or less vertical planes.

In the loess areas of the Missouri River Valley, many hillsides are covered from bottom to top with these small slippage steps or terracettes (Fig. 104). Near the base of hills where the slopes are gentle, such steps are indistinct. On steeper angles, however, especially between 20 and 40 degrees, they are better developed, and the scarps between them are much higher. Where hillsides are very steep, the more or less flat steps tend to give way to typical crescentic slumps.

Debris Slides and Caving. On steep slopes, debris slides sometimes cause the entire removal of the surface soil and even portions of the sub-

Fig. 104.—A Nebraska hillside with characteristic distribution of terracettes. (*Photograph by Soil Conservation Service.*)

soil. Some slides of this kind aid in the widening of gullies. Still more active in gully widening is the caving or falling of masses of soil from the walls, a process that takes place largely during seasons of slow saturating rains. Wetting of the soil causes slight changes in volume, and tension cracks develop parallel to the gully walls. If there is not sufficient earth or fallen material in the gully to buttress the banks, the tension accumulates, and in time the blocks of material break away and fall to the gully floor. This process may take only a few hours or days or may require many months, but the cumulative effect is of considerable importance in the enlargement of caving gullies, such as those characteristic of the Cecil and related soils of the Southeastern States.

Caving of gully walls is of itself not a highly destructive process, because if the caving material were allowed to remain where it fell, slopes would be built up which would support the walls and aid in stabilization of the gullies. It is the alternation of periods of gully caving with periods of clearing out of the gully channel by erosion that has the most destructive effect.[1] Caving is also an effective agency of soil removal in road cuts and along stream banks and other steep slopes.

## Subsidence

Although agricultural land is frequently damaged by subsidence, this type of mass movement contributes very little to actual soil removal.

Gradual lowering of the surface has resulted in many places from mining operations. Subsidence of land in such areas produces shallow depressions in which water sometimes accumulates and drowns the crops. Deeper and sharper depressions, such as open cracks and pitholes, also

FIG. 105.—This hillside field in West Virginia is badly pitted by sinks or cave-ins over old coal mine workings. (*Photograph by Soil Conservation Service.*)

result from subsidence and leave the surface too severely broken to be suitable for agriculture. In mining areas, such as some of the coal fields of the Appalachian Plateau, where the workings have been close to the surface, pastures and fields frequently are badly pitted (Fig. 105), and surface soil is lost by sliding and washing into the depressions.

## Conclusion

Although mass movement, unlike wind or water, does not transport soil or rock out of the region in which it was formed, it is in many different ways, nevertheless, an important factor in soil removal. Mass movement causes widespread downhill migration of sloping soil and, in many locali-

[1] Ireland, H. A., Sharpe, C. F. S., and Eargle, D. H. Principles of Gully Erosion in the Piedmont of South Carolina, U. S. Dept. Agr. *Tech. Bull.* 633, 1938.

ties, prevents the formation of deep profiles. Bare scars produced by such movement prepare the way for erosion. Slumping and caving aid in the widening and headward extension of gullies and contribute large quantities of soil for subsequent removal by water. Creep, earthflow, slumping, caving, and other mass-movement processes are constantly feeding soil and rock material to the streams.

Mass movement, therefore, is a process that must be carefully considered in investigations of the causes and mechanics of accelerated soil erosion.

# Chapter XIII. Geology and Soil Erosion

The distinctive features of the earth's surface are formed through the reactions between physical forces which elevate land masses by deep-seated earth movements to a position within the influence of the atmosphere and those mechanical, chemical, and biological activities which disintegrate, corrode, erode, remove, and redeposit the uplifted materials. Thus, land masses are formed, modified, and removed, and soils are developed and removed as a step in the process. Geological erosion creates, removes, and recreates on the world-wide scale throughout long periods of time.

These physical, chemical, and biological forces operate under all climatic conditions from polar to tropic, from excessively humid to excessively arid. They operate upon original rock masses of great diversity of mineral composition, hardness, permeability to atmospheric waters, and elevation and declivity. The topography of the earth's surface is developed by these forces; surficial conditions, including the nature and distribution of soils, are the results of the reactions between the atmosphere with its included water, the existing biological forms, and the parent rock.

The natural forces act through long cycles of elevation, weathering, plant occupation, erosion, transportation, and deposition. Periods of reconstruction intervene between those characterized dominantly by processes of destruction. Active geological erosion is slowed through the reduction of elevations and slopes, through the creation of surfaces increasingly capable of absorbing meteoric water, and through the natural covering of land surfaces by protective vegetation. Thus, nature employs in erosion control the engineering method of slope reduction and the vegetative method of surface protection by appropriate plant cover.

Accelerated erosion is caused by man's interference with the natural cycle. Wherever groups of men have passed the hunting and herding stages of development and have come to depend upon the annual production of crops to sustain themselves and their livestock, natural vegetation is removed and replaced by domesticated plants. Slopes are cleared that do not permit the retention of the surface soil under continued or repeated tillage operations. Desiccation beyond that experienced under natural cover of trees or grass ensues. Both wind and rain gain easier access to the exposed surface, and the removal of soil material far outstrips in pace its renewal under natural conditions. Accelerated erosion, caused by man's interference with the natural cycle, results in the impairment and destruction of the thin surface layer of soil and in the ultimate eviction of the human operator, until through natural processes the underlying materials can be reclothed with a mantle of soil economically useful to man.

Theoretically, at least, a mass of mineral material of complex chemical composition and of heterogeneous structure ultimately would be weathered into a mass of silicates, chiefly of aluminum and iron; of silica; and of the remains of original minerals. With reference to advanced weathering, Merrill says: "The ultimate product of the weathering of rocks of any but the purely siliceous types is a more or less ferruginous clay, which may be contaminated or admixed with coarser foreign particles. It is the extent of the decomposition, more than its lithological derivation, that determines both the chemical composition and physical characteristics of any soil."[1]

Because of geological erosion and climatic influence, such extreme weathering is rarely approached very closely, except under conditions of long-continued, undisturbed occupation of land surfaces by vegetation in humid, tropical localities.

Under arid conditions, solution and leaching out of soluble materials are of much less importance as instrumentalities of rock weathering and soil formation; here, mechanical processes—extreme changes in temperature and abrasion by wind—are relatively more important. Periods of precipitation intervening between periods of desiccation occasion some solution. The dissolved material commonly is transported only short distances either vertically or laterally; deposition takes place through the concentration of the solution by evaporation. This may occur at shallow depths within the mass, or the dissolved material may be concentrated at the surface as the result of evaporation. Desert soils commonly are marked, in consequence, by an excess of the readily soluble salts over humid-region soils and by the presence of minerals in relatively fresh, unaltered condition.[2]

The fundamental materials from which soils are derived vary with such complexity over the surface of the land, and the atmospheric and

[1] Merrill, G. P. "A Treatise on Rocks, Rock Weathering, and Soils," p. 378. The Macmillan Company. New York, 1906.

[2] See results of analyses of 466 soils from the humid region of southeastern United States and of 313 soils from the arid areas of California, Washington, and Montana. The Data of Geochemistry, U. S. Geol. Survey *Bull.* 770, 5th ed; pp. 493–494, 1924 .

biologic agencies that operate upon them are so divergent, that soils are formed, transformed, and even removed to create a surface pattern almost as complex and varied as that produced by the passage of clouds across a grassy plain.

In the natural process of continental degradation, geological erosion probably has served to prevent the development, in numerous localities, of such unfavorable types of land as excessively leached and infertile areas, imperfectly drained claypan areas, and soils of high salinity. By contrast, accelerated erosion is the result of human interference with the natural order. It effects within a brief period the removal of a soil created throughout long ages by weathering. The time element involved in the natural progress of geological erosion is so shortened that impairment or destruction ensues without intervals for restoration. Natural forces distribute periods of removal and transportation between long rest periods of recreation, but human eagerness accelerates the removal process by failing to allow intervening periods for recovery of eroded surfaces under the protection of vegetation.

It is fairly obvious that no type of erosion, even the most recent forms of accelerated erosion, may be separated completely from the geological process. The character of the soil, the depth to decomposed or solid rock material, the surface configuration formed by geologic processes, the degree of slope, the gradients of streams, all influence the rate and type of soil erosion as it is experienced on arable, pasture, and wooded lands.

Directly or indirectly, the speed at which soil erosion progresses and the form that it takes are influenced profoundly by the lithologic and structural characteristics of the basic materials from which soils are formed. Even after the surface material, the soil and subsoil, has been washed away, exposing the zone of parent material, the rate and pattern of further erosion are determined, quite naturally, by the character and structural form of the exposed geologic materials. It should be pointed out, however, that slope of the land and the structural features of exposed geologic formations exert much less influence on the rates and patterns of wind erosion than they do on water erosion.

There are many and varied examples of the influence of geologic materials on the formation and removal of soils. Sandstones, decaying and disintegrating under the impact of weathering, usually give rise to coarse-textured soils, such as sands, sandy loams, and sandy clay loams. Fine-grained shales, on the other hand, generally give rise to fine-textured soils, such as clays, clay loams, and silt loams. All soils, however, do not always exhibit such close similarity to their parent materials. Even sandstone may be ground to the consistency of flour by glaciers or flowing water and eventually become fine-grained soil. A soil may, or may not, retain the color of the rock from which it is derived. Much depends on the intensity and type of weathering processes involved in the transition from parent material to soil. Processes of chemical change and solution (leaching) may alter or remove the original coloring constituents. Moreover, the color of the surface zone of any soil is dependent to a considerable degree on the character of the vegetative cover and the degree to which the decaying accumulation of organic matter has become incorporated with the underlying earthy material.

Lateritic soils are particularly illustrative of profound variance from the parent rock. For example, some of the old red soils derived from limestone under the warm, humid conditions of Cuba are very similar to the old red soils derived from serpentine. As previously indicated, this similarity occurs, however, only in areas where the topography is sufficiently smooth to prevent rapid erosion. Moreover, in both instances, prolonged leaching has deprived the material derived from these strongly contrasting parent rocks of a large portion of their original identifying constituents. The residuum, representing the present soils, contains much larger proportions of highly insoluble compounds of iron and aluminum than the parent rocks. These residuary constituents dominate the character of such lateritic soils and often make them so much alike that differentiation, in the absence of chemical analyses or differences in vegetative response, is difficult except through identification of the respective parent materials. Conceivably, with continuing weathering, the derivative soils could come to have identical compositions.[1] Such precise conformity, however, seldom is found in nature; some characteristic of the parent material usually is expressed in some property of the derivative soil.

Soils derived from limestone frequently contain little or no lime carbonate, the latter having been removed in solution during the change from rock to soil.

Soils formed from rocks of high salinity, like some of the extensive beds of shale in the northern Great Plains and in the Colorado Basin (as the Chinlee and Mancos formations), usually contain much easily dissolved salt and are especially susceptible to erosion. Such soils support only a sparse cover of protective vegetation, and the surface material frequently fractures or scales on drying, forming detached fragments which are moved easily by water or wind. Soils of this type are also peculiarly susceptible to the development of V-shaped rills and gullies.

Granitic rocks tend to weather to great depths, especially in humid regions, as in southeastern United States. Far beneath the zone of true soil, the soft "rotten" rock succumbs to rapid incision by gullies that have been permitted to cut through the soil layer (Fig. 106). Because of this tendency, steeply sloping land with soil derived from granite or gneiss

[1] Bennett, H. H., and Allison, R. V. "The Soils of Cuba." Boyce-Thompson Institute. Tropical Research Foundation. New York, 1929.

generally yields, when it is stripped of vegetation, to rapid destruction by deep gullies of the undercutting, U-shaped type.

In rolling areas where highly erodible, silty soils have developed, as over large portions of the Alleghenies and the Appalachian Plateau, the soil is generally shallow, even under natural conditions. On the steeper slopes, bedrock commonly is encountered at depths of less than 3 or 4 feet.

FIG. 106.—A crystalline schist weathered to great depth in the Southern Piedmont. (*Photograph by Soil Conservation Service.*)

Apparently, this shallowness is due to a combination of slow rock decay and relatively rapid soil removal. Whatever the cause, the relatively shallow depth to bedrock prohibits the formation of deep gullies. Instead, broad and shallow gullies, or washes, are likely to develop (Fig. 107) when the soil is exposed.

Where shales, slates, and schistose rocks are so tilted and fissured that rainfall percolates rapidly into the substrata, the overlying soils are often relatively shallow (Fig. 108) and susceptible to such marked desiccation that they support no more than scanty vegetation. The indirect effect of such structural character of the parent rocks—the effect on moisture—is to favor rapid erosion. Under conditions of this kind,

FIG. 107.—On steep slopes bedrock is encountered frequently at depths less than 3 feet. On this area of formerly cultivated Lordstown silt loam, Steuben County, New York, erosion has cut away the soil to bedrock so that gullying spreads laterally instead of down. (*Photograph by Soil Conservation Service.*)

FIG. 108.—Where shale, slate, and schistose rocks are so tilted and fissured that rainfall percolates rapidly into the substrata, derivative soils often consist of little more than disintegrated rock material and support scanty vegetative growth. (*Photograph by Soil Conservation Service.*)

erosion on steep slopes is so rapid that the derivative soils sometimes consist of little more than disintegrated rock material. Under cultivation, such soil and rock material is sometimes swept away, and the bedrock exposed within three or four years.

Numerous examples of the effect of rock character on land forms, soil depth, and rates of erosion could be cited for nearly all parts of the world. For example, the monadnocks standing as conspicuous hills hundreds of feet or more above the general level of the southeastern Piedmont, such as Pine Mountain in west-central Georgia[1] and the Kings Mountain group in the Carolinas, mark the location of rocks, like quartzite, that

Fig. 109.—Unequal solution of limestone formations gave rise to this sinkhole or Karst topography. Kentucky. (*Photograph by Soil Conservation Service.*)

have worn away much more slowly than the surrounding rocks, simply because of their superior resistance to weathering and erosion. Here the parent rock has not only exerted a definite influence on the local topography but has given rise to a distinctive shallow soil—Herndon stony sandy loam.

Limestone, because of its tendency to form underground cavities by a process of unequal solution, permits a type of erosion that not only presents a difficult problem of control but influences the topography (sinkhole or Karst topography) to a marked degree (Fig. 109). Extensive areas of this kind occur in many parts of the world.[2]

[1] See Soil Survey, Talbot County, Georgia. Field Operations, Bur. Soils, U. S. Dept. Agr. 1913.

[2] Dicken, S. N., and Brown, H. B., Jr. Soil Erosion in the Karst Lands of Kentucky, U. S. Dept. Agr. *Circ.* 490, 1939.

In its effect on major land forms, differential erosion is largely a geologic process, the results of which are determined chiefly by rock character and the inclination of the exposed strata (in areas of sedi-

FIG. 110.—The Badlands of the Dakotas owe their outlines principally to abrasion by water and wind-driven sand. (*Courtesy of N. H. Darton.*)

FIG. 111.—Badlands of New Mexico. (*Photograph by Soil Conservation Service.*)

mentary rocks). Fantastic land forms in the Badlands of the Dakotas and other Western States owe their outlines principally to the varying degrees of resistance offered by the exposed material to abrasion by water and wind-driven sand (Figs. 110, 111).

The direct and indirect relationships of rock character, faulting, folding, river action, wind action, glaciation, and the stratigraphic and structural features of sedimentary beds to erosion rates and processes and to underground water supplies are so numerous and varied that an exposition here seems undesirable. It is sufficient to state that the relationships do exist and are of great importance to the soil-erosion scientist.

Geologists have emphasized the powerful effects of erosion in altering the character and form of the earth's surface. Their studies and findings relating to the subject fill a great many pages in the literature of geology. In turn, erosion scientists recognize the erosion process, whether normal or accelerated, as a geologic process. However, erosion science seeks to ascertain all the factors involved in the mechanics of the erosion process and the effects of the process on the soil resource when accelerated through disturbance of normal conditions. This science stresses, in other words, the processes and effects of accelerated, abnormal erosion.

In a somewhat similar sense, soil science is related to the science of geology. Whereas soil scientists are concerned with the nature of soils, their distribution, and the processes involved with the development of soils, geologists emphasize the character, origin, and distribution of the parent materials of soils (geologic materials).

# Chapter XIV. Relation of Entomology to Erosion

Excessive populations of insects may be an important indirect cause of erosion by destroying the vegetative cover so essential to soil conservation. Grasshoppers, beetles, caterpillars, ants, and other insect pests sometimes remove the ground cover completely or defoliate plants to such extent that the soil is left bare and susceptible to erosion by wind or water (Fig. 112). Some insects, feeding on the roots of plants, damage or kill vegetation. Unfavorable weather conditions or improper cultivation may so weaken plants that they are unable to recover from insect attacks.

White grubs, for example, sometimes contribute to serious erosion in closely grazed bluegrass pastures, particularly during periods of drought. Such damage may be reduced by liming and fertilizing, to encourage the development of a more vigorous turf; or by planting legumes, such as sweetclover and alfalfa, which the beetles avoid when depositing their eggs.

Grasshoppers have been the indirect cause, or partial cause, of considerable soil erosion, particularly in times of drought in the Great Plains area. In this connection, Bishopp says:[1]

"The part played by grasshoppers in the removal of soil cover is of first rank. In severe outbreaks not only is every green leaf removed but many of the stems, crowns, and dead protective vegetation as well. Even in the face of droughts or other unfavorable growing conditions, the vegetation would usually develop sufficiently to form a reasonably satisfactory cover if grasshoppers were not present. Hundreds of thousands of acres of land in Nebraska, Wyoming, Montana, and the Dakotas were stripped of vegetation by grasshoppers during the last few years, permitting the soil to be severely blown and clouds of dust to be formed.

"This condition is often attributed to the drought and overgrazing by livestock. The part which grasshoppers play in this denudation has been demonstrated in a striking manner by experiments conducted by J. R. Parker in Montana in 1936. Screen cages were placed over grazing land to exclude the grasshopper and similar areas were fenced to exclude livestock. At the end of the

[1] Bishopp, F. C. Entomology in Relation to Conservation, *Jour. Econ. Entomology,* Vol. 31, No. 1, 1938.

season, the fenced areas were practically devoid of grass, whereas those protected from grasshoppers were well covered."

Overgrazed range lands are especially susceptible to damage by grasshoppers. Such conditions may also encourage the growth of weeds that harbor and encourage the propagation of insects harmful to crops in adjacent irrigated fields. An outstanding example is the beet leafhopper

FIG. 112.—Left, Kansas corn crop destroyed by grasshoppers, 1936. Right, shows grasshoppers at work. (*Photographs by Soil Conservation Service.*)

which transmits the curly-top disease of sugar beets. Summer weeds, such as Russian thistle, growing on intermittently cultivated land; and winter weeds, as plantain, mustard, and peppergrass, favor the breeding of this insect. Hosts of leafhoppers, breeding on these weeds early in the season, later move on to inflict great damage to fields of sugar beets, beans, and tomatoes by transmitting the curly-top disease. Frequently, much of the damage can be prevented through the reestablishment of grasses for range purposes.

Insects may also be the indirect cause of erosion in woodlands. The destruction of trees by the bark borers often leaves a forest or farm woodlot in a highly inflammable condition. Fires occurring under such conditions bare the land to accelerated runoff and erosion.

Erosion, in turn, may encourage insect infestation. For example, areas retired from cultivation and planted to trees are often so seriously impoverished by erosion that the slow-growing trees lack the vigor to resist insect attacks. The locust borer, for example, causes more damage to stunted or slow-growing trees than to those of vigorous, healthy growth. Such damage to young plantings can be eliminated partially by mixed plantings and the use of fertilizers.

Practices important in the control of erosion may also affect the seriousness of insect damage. The effect of such practices on insect populations, breeding places, etc., must be taken into consideration in the formulation of erosion-control plans. For example, Bishopp has the following to say with respect to strip cropping and insect damage:

"The principles of strip planting are by no means new, for they have been widely practiced for many years in the older countries of the world. In general, insect depredations are less severe in these old countries than in our own. The planting of extensive acreages to one crop has led, in the past, to serious insect losses. There is, therefore, some practical as well as theoretical ground for the argument that strip planting may be helpful in combatting insects. It is logical to suppose that the spread of a deleterious insect which has become established locally is less hampered in a large field of one crop than in an interrupted planting. There are certainly instances in which natural control of insect pests is favored by strip planting. This results from provision of conditions suitable for the development of alternate hosts of the parasite or predatory enemies of the crop pest.

"Certain destructive insects may be favored by strip farming if the crops are not carefully selected and insect relation kept clearly in mind. The practice advocated by some soil conservationists of planting crops in strips, with permanent sod intervening, provides an ideal setup for the differential and two-striped grasshoppers which by this plan are enabled to pass directly into the succulent crops as the young hatch from eggs concentrated in the sod strips."

In Montana, poisoning of insects has been reported as especially effective by spreading the poison bait along the edge of the sod strips. The insects feed greedily upon the bait while passing to the more succulent intertilled crops. Strip cropping may intensify other insect pests, such as the pea aphid, when alfalfa and peas are grown together, as they are in the Palouse country of Washington and Idaho; and the chinch bug, when corn and grain are grown side by side. Consideration, therefore, should be given to the danger of insect damage when selecting crops to be grown in a strip-cropping system. Every precaution should be taken to prevent any increase in the pest population and any intensification of insect depredations.

The use of cover crops in orchards, which is so important in the conservation and improvement of orchard soil, may in some instances increase insect infestation. It has long been considered necessary to practice clean cultivation in orchards in order to aid in the control of plant pests and make the most efficient use of soil-stored water. Preliminary surveys[1] of a number of orchards in California indicate, however, that insect and plant-disease control cost no more where orchard soils are protected by a continuous cover. It seems logical to assume, therefore, that continuous cover may, under some conditions, favor the increase of parasitic and predacious life forms that aid in keeping the pests in control.

In some localities, the construction of farm ponds to provide water for livestock has increased the hazard of mosquitoes. Several methods of control are effective in checking mosquito infestation, such as introducing top minnows; applying larvicides, as pyrethrum extract and oils; deepening the water along the edges; removing debris; and manipulating the water level.

Other practices important in erosion control, such as spring plowing, the maintenance and use of plant residues, fire protection for fence rows and woods, and the establishment of wildlife areas, may make conditions more favorable for plant pests. However, agriculture in the United States has been combating dangerous insects for years, and it is recognized that these pests have cyclical periods of heavy population in rotation with periods of very small or at least greatly reduced population. It is possible that any increase in numbers of insects and extent of damage resulting from the introduction of soil conservation practices may be counterbalanced by the contribution of such measures to an increase of predatory birds and mammals as well as insect parasites and diseases that tend to reduce the total population of harmful forms.

Agricultural losses resulting from insect infestations must be contrasted with the enormous reductions in crop production caused by erosion of soil. Crop rotations and other soil-conserving practices tend to increase soil productivity and the production of more vigorous crops, which in turn help to reduce losses through insect damage.

There is need for additional information on the effect of various soil conservation practices on insect population. The sequence of crops in crop rotations, time of plowing, mosquito control in farm ponds, use of crop residues, and many other soil and water conservation practices should be studied in relation to insect control, so that the science of entomology may contribute most effectively to programs of soil defense.

[1] Ahlson, C. B., and Hutchinson, George. Permanent Cover in Irrigated Orchards, *Soil Cons.*, Vol. 4, No. 8, 1939.

Erosion scientists recognize the important favorable role of microorganism and insect life[1] in making soils more permeable to rainfall and thereby more resistant to erosion. They understand that the holes bored by crickets, earthworms, and other insects function as conduits to increase the infiltration capacity of the soil and that organic matter passing through the bodies of insects is converted into the kind of humus that contributes to the development of favorable soil granulation. They also appreciate the need for more information with respect to the precise interrelationships of cultural practice and biological activities and the effects of these on the agricultural economy.

[1] F. C. Bishopp says (*Jour. Econ. Entomology*, Vol. 31, 1938): "The number of living things in the soil is almost beyond our comprehension. It has been estimated that an ounce of soil may contain 30,000,000 microorganisms such as bacteria, algae and fungi. There are also tremendous numbers of animals . . . in soils. A large percentage of these are insects. W. L. McAtee has estimated that there were 1,216,880 animals, principally insects, in an acre of woodland and 13,654,710 in an acre of meadow, based upon an examination of 4 square feet of soil to a depth a bird could scratch. H. W. Morris made a census of the fauna of a permanent pasture in England, which indicated an insect population of 3,586,088 per acre. G. N. Wolcott has made an intensive study of limited areas in Illinois and New York. From his figures the average number of animals per acre is computed at 3,611,124 in the Illinois area and 2,980,810 in the northern New York area. In a tropical forest in Demerara, C. W. Beebe determined the animal population to be 10,640,000 per acre."

# PART TWO

# Soil Conservation

# Chapter XV. A National Program of Soil Conservation

Stated in simplest terms, conservation of the soil is a matter of using land as it should be used. Some land is too steep to be farmed; some soils are too erodible to be cultivated; some regions are climatically unsuited to intensive agriculture; and so on. Proper use of the land, in other words, is governed by certain physical factors. When it is used without regard for these factors, the processes of waste begin.

Certain factors of an economic and social character also influence the use of land. Men who farm the land must make a living. That is their most realistic and immediate concern. Economic need is a powerful factor in determining how they use the land, a factor that usually outweighs less urgent considerations of conservation. Social patterns likewise are a land-use factor. In some sections of the country, men are attempting a type of agriculture too extensive to be carried out on small farms; in other instances, farms are too large for sound operation with available facilities; elsewhere, the tenure system impedes good land-use practice. As they affect the interests of the individual operator, therefore economic and social influences of this kind are as much a part of the problem of soil conservation as the purely physical processes through which soil is washed or blown away.

National action to conserve the soil resource, consequently, must not only deal with soil erosion and related physical phenomena but also cope with the complex economic and social considerations that affect land use.

The ultimate objective of national action—a secure land resource, adequate permanently to sustain the national economy—will be reached only when the principles of good land use have been accepted and put into practice by agriculture as a whole.

But agriculture as a whole is based on the activities of individuals. The land is owned and used by many men. They enjoy the rights and privileges of private ownership under the American system. In this democracy, national action to conserve soil must be generated by these millions of land users. If they are active and willing participants in such a movement, it will endure; otherwise it will fail. Democracy does not give government the unbridled right to compel conservation; but it does recognize the responsibility of government to lead and assist its citizens in conservation, in the interest of the public welfare.

This recognition of the democratic process is essential. National conservation action must spring from people on the land and, to a large extent, be advanced by them as individuals, with the help of government.

In the last 5 years, Congress, through a series of acts, has erected the framework of a comprehensive action program looking to the conservation of the nation's soil resource. This program, carried on as an integral part of a broader program of basic land-use adjustment, is administered, for the most part, by the United States Department of Agriculture.

Late in 1938, the Secretary of Agriculture directed the Soil Conservation Service of the Department to assume responsibility for administering certain of these acts bearing on the physical aspects of the conservation problem. As a result, a number of governmental activities on the land, authorized by various legislative actions, were brought closely together to provide a unified program broad enough to help farmers make the physical adjustments in land use fundamental to the conservation of natural resources and the establishment of a permanent and balanced agriculture.

The current work of the Soil Conservation Service constitutes a comprehensive national action program to conserve and properly utilize soil and water resources. Other agencies of the Department, at the same time, are administering other acts of Congress directed toward the correction or alleviation of certain economic and social maladjustments that contribute to land abuse. The work of these agencies complements the work of the Soil Conservation Service to form a single unified program of action bearing on the physical, economic, and social aspects of the land-use problem.

The program of the Soil Conservation Service, primarily concerned with physical problems, follows several lines of action having a common, ultimate goal of better land use, conservation, and a more abundant life for the people of the country. The lines of action are as follows:

I. Soil Erosion Control. A program authorized by the Soil Conservation Act of 1935 (*Public* No. 46, 74th Congress) and designed to conserve basic soil and water resources by extending sound land-use practices to all land vulnerable to soil erosion. Action in this program involves (1) research to determine the character, causes, and effects of soil erosion and to develop practical measures for erosion control; (2) surveys to determine the facts needed in planning and prosecuting erosion control and moisture conservation work; (3) demonstrations to illustrate the practical effectiveness of soil-conserving land-use practices and to prove the techniques of erosion control through actual application on the land; (4) cooperation with local soil conservation groups formed under the authority of state law; (5) dissemination of information regarding erosion and its control through the normal media of public communication; (6) cooperation with Federal, state, and local agencies in the fields of conservation, land-use adjustment, and related fields.

II. Submarginal Land Purchase. A program authorized by Title III of the Bankhead-Jones Farm Tenant Act and designed to correct social and economic maladjustments in rural areas by changing the pattern of occupancy and consequent agricultural use of land. Although the program deals directly with the conservation of physical and human resources, it also involves consideration of those social and economic problems, such as land ownership and settlement, size of holdings, type of farming, and systems of local public finance, which frequently must be dealt with before proper land use and the welfare of landowners and operators can be established on a permanently satisfactory basis.

Action under this program involves (1) the purchase of submarginal land or land not primarily suited to cultivation; (2) the development and improvement of purchased areas for sustained use; (3) the management of purchased lands under uses to which they are adapted; and (4) cooperation with local groups of farmers and public agencies in furthering such programs as range management, rural zoning, water development, fire prevention, and more efficient government.

III. Flood Control. A program authorized by the Flood Control Act of 1935 and supplemental legislation and designed to reduce the hazard of floods to human life and property, by land treatment on the watersheds from which flood waters originate.

Action under this program, carried on through cooperation between the Soil Conservation Service, the Bureau of Agricultural Economics, and the Forest Service of the Department of Agriculture and the Corps of Engineers of the War Department, involves (1) preliminary investigations of watersheds to determine the existence and seriousness of flood problems and the practical possibility of alleviating flood hazards and siltation by adaptable land treatments; (2) detailed surveys of watersheds to obtain essential physical, social, and economic information and, on the basis of such information, to develop technically and economically sound plans for watershed treatment in the interests of flood control and prevention of siltation; and (3) actual watershed protection work in accordance with the programs developed through the surveys.

IV. Water Facilities. A program authorized by the Pope-Jones Water Facilities Act of 1937 and designed to assist farmers and ranchers in the improvement and development of farm and range water supplies in arid and semiarid areas, with a view to promoting better use of the land and advancing human welfare.

This program is limited to 17 Western States and is conducted cooperatively with the Farm Security Administration, Bureau of Agricultural Economics, and other agencies, through the Water Facilities Board. The Soil Conservation Service assists farmers and groups of farmers in (1) planning the development of water supplies that will facilitate improvements in land use; (2) constructing and installing, or assisting farmers to construct and install, such small water facilities as wells, ponds, small reservoirs, small dams, pumps, springs, stock water tanks, spreading systems for utilizing runoff water, and similar improvements.

V. Farm Forestry. A program authorized by the Norris-Doxey Cooperative Farm Forestry Act of 1937 and designed to foster the practice of farm forestry in agriculture, with a view to conserving soil and water resources and aiding in the establishment of sound and economical land-use methods.

This program is carried out jointly with the Forest Service, Agricultural Extension Service, State Experiment Stations, State Foresters, and the Bureau of Agricultural Economics. The Soil Conservation Service is responsible for the action phases of the program in (1) advising farmers regarding the establishment, protection, and management of farm forests; (2) investigating the economic and other benefits of farm woodland management; and (3) training personnel in methods of bringing about the use of proper farm forestry practices in agriculture. The Service cooperates in producing, procuring, and distributing forestry planting stock to farmers.

VI. Drainage and Irrigation. A program authorized by the Agricultural Appropriation Act of 1932 and designed to develop efficient and economical methods of draining and irrigating agricultural land with a view to promoting better land use.

Activities of the Soil Conservation Service under this program involve (1) the study of laws and regulations affecting the organization and administration of drainage and irrigation districts and companies; (2) development of basic hydraulic information involved in the design of drainage ditches, tile drains, and pumping plants; (3) investigation of methods of applying irrigation water to farm lands; (4) development of apparatus for accurate measurement of the quantity of irrigation water delivered to the farmer or rancher; (5) development of pumps and equipment for utilizing underground waters; (6) development of diversion dams and desilting works for diverting flood waters into irrigation canals; and (7) the making of snow surveys for the purpose of forecasting irrigation water supplies.

## Broad Objectives

For purposes of discussion, this program of soil and water conservation and land-use betterment may be divided broadly into three major phases, namely: (1) education, to develop widespread understanding of the problems of sound land use and their solution; (2) action by individuals through local mechanisms with the help of government; and (3) efforts to relieve economic and social maladjustments in the use of land that impede conservation.

## Education

Education is a prerequisite of conservation. It is essential to develop an intelligent public understanding of the value and importance of natural resources in terms of individual and national life. Only out of such an understanding will the impetus to act arise.

Many of the techniques through which the process of education is advanced are familiar. They include the diffusion of information through radio broadcasts; through the printed word—in official publications, newspaper and magazine articles, textbooks, and so on; through visual media, such as exhibits and motion pictures; through the integration of soil conservation subject matter in school curricula; through direct contact with individuals and groups interested in the movement; and so on. These well-known procedures are most effective in presenting the problem of conservation to the public as a whole.

But the people close to the land—the men and women who use the land day in and day out as a source of livelihood—want to know more than the need for conservation; they want to be shown *how* to conserve. The answer to this demand is the demonstration.

In the last 5 years, large-scale demonstrations of good land use and conservation have been remarkably effective in giving farmers an opportunity to examine, criticize, and learn modern conservation farming methods.

Soil and Water Conservation Demonstrations. Demonstrations in soil and water conservation have now been established in most of the important farming sections (see *Problem Area* map). To a large extent, each of them is a cross section of the agricultural lands for miles around. The soils, topography, climate, and cropping and grazing systems are closely similar to those found on outlying farms and ranges. The farming and grazing methods and land-use adjustments needed to control erosion and conserve rainfall are typical of those generally required over the whole problem area.

The method employed in these demonstrations is to help land operators with the actual job of controlling erosion, conserving rainfall, and making the necessary changes in the pattern of land use. Agronomists, soil specialists, agricultural engineers, foresters, wildlife experts, and other technicians combine their skills to attack the problem simultaneously along several fronts. They plan measures and practices to fit the needs of every different parcel of land, and coordinate their application so that one measure or practice supplements the others, and what is done on one field or on a given farm contributes to the protection of adjacent or downstream areas. Wherever possible the demonstration areas are staffed with technical men who are familiar with local conditions.

In setting up a demonstration area, the first step is to obtain a base map of the entire project area. If such a map is not already available, one is made with aerial photographs as the base. Then a farm-by-farm survey is made, showing soil types, erosion conditions, slopes, current land use, and other pertinent features directly on the base map.

With this map as a guide, the technicians are able to draw up individual farm plans for each farmer in the area who is interested in adopting a conservation system. Land-use adjustments, cropping changes, improved tillage methods, and structural devices are discussed with the farmer, usually on the ground. Every step is considered on the basis of need, adaptability, practicability, economic feasibility, and physical relationships with adjacent lands.

If the farmer decides to adopt the conservation plan finally worked out, he signs a cooperative working agreement with the Federal Government. He agrees to follow the recommended land-use practices over a 5-year period and to contribute as much as possible in the way of labor, power, seed, and materials toward the establishment of an effective conservation system. The Government agrees to lay out the work, draw up the necessary structural specifications, provide whatever labor and materials the farmer is unable to supply, and furnish suitable planting materials for lands taken out of cultivation because of their highly erodible character.

An example of the major land treatment and farm reorganization procedure involved in a representative demonstration area will partly illustrate the complexity of the soil conservation program on a single watershed:

The Green Creek Soil and Water Conservation Project in Erath County, Texas, comprises approximately 30,000 acres and is fairly representative of the West Cross Timbers problem area of approximately seven million acres in Texas and an additional large area in Oklahoma. The soils are predominantly sandy loam with moderately stiff clay subsoil; the topography ranges from undulating to moderately rolling. For the most part, farming has been carried on for only about 40 years.

The number of farms in the county decreased by 28 per cent between 1910 and 1935, according to the census. The value of farms in the county, including land and buildings, decreased by 54 per cent between 1920 and 1935, and farm population decreased by 33 per cent between 1910 and 1930.

Surveys in this watershed in 1935 disclosed that 13 per cent of the area had been essentially ruined for cultivation by erosion and that approximately 50 per cent of the topsoil of all the cultivated land had been washed or blown away. Many farmers were attempting to control erosion by terracing and by contour tillage of the terraced acres; some were doing nothing to defend their wasting fields. On 4,082 acres of terraced land on 103 farms in the project area, only 1.6 per cent of the terraces were satisfactory, and only 3 per cent could be rebuilt to function properly. Of the cultivated land terraced prior to 1935, a total of 29 per cent had been retired from cultivation or abandoned by the owners and operators because of erosion damage. An additional 18 per cent of these terraced acres was retired when the 103 terraced farms were placed under cooperative working agreements with the Soil Conservation Service.

Under these agreements, an integrated 17-point program was employed. It involved:

*A.* Detailed farm surveys of 217 farms, comprising 29,990 acres and including 132 farms actually cooperating in the watershed program.

*B.* Joint preparation of detailed working plans for each farm by project conservation specialists and cooperating farmers, on the basis of the physical findings of the individual farm surveys.

*C.* Application of the following soil and water conservation measures to meet the diverse needs of the various land and farming conditions of the cooperating land users.

1. Strip cropping.
2. Terracing of slopes ranging up to 5 per cent, largely in conjunction with strip cropping.
3. Crop rotation.
4. Cover crops for seasonal protection of fields.
5. Retirement of severely eroded cultivated land to the permanent protection of grass or trees.
6. Contour cultivation.
7. Planting critically erodible areas to feed and cover plants for rehabilitation of wildlife.
8. Control of gullies, chiefly with grass.
9. Control of terrace outlets with grass.
10. Construction of small grassed channelways to carry water discharged from terraces safely downhill.
11. Development of broad meadow strips for safe disposal of excess runoff from terraced and contoured fields.
12. Pasture development by contour furrowing and reseeding.
13. Pasture improvement through rotation grazing, reduced grazing, and mixed grazing.
14. Consolidation of small fields into larger fields adaptable to contour cultivation, stripping, terracing, and other treatment necessary for adequate protection of soil and proper conservation of rainfall.

15. Construction of small reservoirs for continuous supply of stock water.

16. Planting black locust to protect erodible slopes and to provide an adequate supply of fence posts.

17. Field stripping to check wind erosion.

The mere enumeration of these major steps reveals very little of the manner in which they were employed. It fails to indicate how one measure was used to support another or how protection of one field provided protection to an adjacent field. However, since the establishment of these measures within the watershed, observations made by farmers, local businessmen, and the project technical staff reveal that for the first time in many years, Green Creek did not overflow its banks during 1936, although 11 inches of rain fell in May and 10.9 in the last 14 days of September. Moreover, the creek ran during a greater number of days in 1936 than in 1935, although there was approximately 12 inches less rainfall, with unfavorable distribution.

Similar results have been obtained generally in most of the demonstration areas now in operation. The adoption of a planned system of land use, involving the integrated use of a variety of conservation measures, has in many instances effectively secured the soil against erosion. Simultaneously, flood crests have been lowered, and the silt content of the water reduced along many streams draining project areas. In the Great Plains, practically all the rain falling on properly treated fields and pastures has been held on the ground to be absorbed, furnishing moisture for a cover of vegetation adequate to protect the soil from both wind and water erosion. Moreover, in a number of instances, much or all of the rain that normally ran to waste through roadside ditches, gullies, and other drains has been directed to useful purposes by diversion into fields and pastures.

There is ample evidence that, on many of the demonstration areas, lands that a short time ago were declining steadily, along with the agriculture on them, are now well safeguarded from erosion. Furthermore, as a result of erosion-control work, many of these farms are moving steadily in the direction of increased yields per acre. In some instances, larger profits have been due in part to better prices; but to no inconsiderable degree, the upturn is the result of wiser land use and the reorganization of farm operations to achieve maximum efficiency in production. The conservation of rainfall alone has resulted in approximately a 25 per cent increase in productivity on both crop and grazing land in many of the demonstration areas.

Work of CCC Camps. Closely related to the series of watershed demonstration areas are a number of conservation projects worked by camp units of the Civilian Conservation Corps under the direction of the Soil Conservation Service. A CCC camp with its resources of man power and equipment is in itself an effective work unit. Many camps have been attached directly to demonstration projects in order to supplement the facilities of the project organization in executing its work program, and some are assigned to the larger Western watershed projects to provide labor for various types of work on public lands. The greater number, however, are located outside the boundaries of the regular project areas and work on privately owned farm and grazing lands.

These camps operate within specified work areas selected to serve as effective community erosion-control demonstrations. They work on farms within a radius of approximately 15 miles from the location of their headquarters.

Some camps have been used almost entirely for special operations, such as the construction of large dams, the quarrying and grinding of limestone, and the collection of planting materials for soil conservation nurseries and for immediate use on the land. Certain camps have been assigned to emergency tasks, when occasion demanded. They have fought forest fires, performed rescue work in connection with floods, and cleaned up the debris left by swollen waters.

The major portion of the work in erosion-control camps, however, involves the installation of soil-conserving measures on farm lands. This work includes such activities as gully control operations, terrace outlet construction and protection, planting of trees and shrubs, running contour lines for furrows and terraces, and building structures for streambank protection. Many of the enrollees, on completing their course of CCC enrollment, return to rural communities and homes, where their training and experience aid them in furthering the spread of soil conservation principles and practices.

CCC labor, organized to work together on complicated areas, requiring the application of many practical control measures, properly integrated to meet a diversity of land needs, is peculiarly well suited for the performance of soil and water conservation work.

Through these demonstration projects and the work of the CCC camps, farming people throughout the country have had an opportunity to observe the effects of conservation farming and to learn how it is done. Primarily, the function of the demonstrations has been educational, although no inconsiderable amount of land within the project and camp areas—some 33 million acres of private and public lands, all told—had been brought under the protection of conservation methods by the end of 1938. Their primary purpose is merely to introduce in a small cross-sectional area the conservation measures adapted to a large natural land-use region. Obviously, they do not attempt to solve the problem on all the land of the region but seek only to point the way toward the solution.

## Action

One of the most important developments of recent years in the field of soil conservation has been the emergence of the soil conservation district as a mechanism for cooperative conservation action on a local scale by the users of the land themselves. Through these districts, established under state law, the facilities of government, both state and Federal, can be brought most effectively into play in response to the direct demand of farmers and ranchers. The districts represent a significant development in bringing democratic processes to bear upon the problems of the land.

The rapid growth of soil conservation districts in the last two years is significant, moreover, of a trend toward the assumption of conservation responsibility by local people. Thirty-six states have enacted legislation authorizing the formation of these local cooperative land-use organizations. As of Sept. 15, 1939, 133 districts had been formed, covering an aggregate area of some 70 million acres of private farm and grazing land.

The function of the district is to develop and carry out a program of proper use for all the land within its boundaries. The district itself may undertake to carry out the work; or it may request the assistance of the Soil Conservation Service and other agencies of the Government. Actually, the latter course usually is followed, since districts, in the beginning at least, seldom have the facilities for extensive reorganization of land use.

The program drawn up and proposed by a new district, embracing five counties in north-central Georgia, will indicate the part that these local agencies are now taking in the national land-use picture.

The declared objectives of the program of this Georgia district are, in general, to "enable farmers to raise and maintain a suitable standard of living and to perpetuate agricultural resources within the district." More specifically, they are (1) to bring about the adoption of necessary practices for the conservation of soil resources; (2) to make the adjustments necessary to a wise land-use program, such adjustments being directed toward an increased income for individual farmers; and (3) to develop necessary land-management practices such as would provide for the efficient utilization of extra feed and pasture resources resulting from the realignment of farm cropping systems.

In approaching these objectives, the district proposes a number of land-use readjustments, including an increase in total cropland of approximately 12,500 acres; an increase in pasture land of approximately 7,500 acres; a decrease in woodland pasture of approximately 8,000 acres; a decrease in idle land of about 60,000 acres; the reforestation of some 30,000 acres of eroding land currently in crops; the purchase and development of approximately 200,000 acres of submarginal land; the rehabilitation of some 2,900 farmers through loans by the Farm Security Administration (FSA); the protection and improvement of 26,000 acres of farm woodlands; the development of food and cover for wildlife; and the installation of sound conservation practices, such as strip cropping, terracing, cover cropping, and planned rotations on all cultivated land vulnerable to erosion.

Into this district, at the specific request of the district supervisors, the Soil Conservation Service is bringing the facilities of a major public action program. It has the authority to assist in planning and applying soil-conservation measures on individual farms; in establishing good woodland and pasture management practices; in reforesting or regrassing croplands retired from cultivation; in purchasing the areas of submarginal land recommended for permanent retirement by the district; and so on. Through alliance with the district, the Service is enabled not only to readjust land-use practices on individual farms but to fit these single farm realignments to the larger pattern of desirable land use in the district as a whole.

Wherever possible, activities of the Soil Conservation Service are now being carried out in direct cooperation with soil conservation districts. As rapidly as new districts are formed in the various states, the Service endeavors to cooperate with them to the extent of available resources, where the districts request such cooperation. In this way the Service will be able to assist most effectively the large groups of farmers who are cooperatively engaged in the prevention of soil wastage on a large scale, and whose holdings and operations represent all land within a complete watershed or natural land-use area.

This does not mean that the demonstration program is to be abandoned or that the principle of demonstration is to be ignored as the Service prepares to assist districts. The projects and camp areas already in operation, and individual demonstration farms established in cooperation with the State Extension Services, together with new demonstration units to be established in the future, will be necessary to provide a background of substantial experience in planning and applying control measures and to serve as proving grounds where measures and practices can be applied and studied under conditions peculiar to each natural area.

In accordance with the policy of the Department of Agriculture, however, new watershed demonstration projects are being established at present only in those states which have passed adequate soil conservation district laws. Exceptions to this policy are found, of course, in those states in which the legislatures have not had an opportunity to consider such legislation and in those in which the Service has not previously established watershed demonstration projects.

Where demonstration areas are necessary in the future, to provide proving grounds for proper control measures, they will be established, as a general rule, in cooperation with soil conservation districts organized under state acts. Likewise, the services of CCC camp enrollees working under the technical supervision of the Service will be provided in the future in accordance with this same principle, in so far as is feasible.

The operations carried on by the Soil Conservation Service in districts will follow the same principles of soil and water conservation, through the application of the coordinated method of land treatment, according to adaptability and economic feasibility, as have been employed on the demonstration projects.

Direct Action on Federally Controlled Land. Several direct action projects have been established on areas consisting largely or entirely of Federal lands. One outstanding example is that of the Navajo Indian Reservation, with an area of about 16 million acres; another comprises the Gila River watershed above the Coolidge Dam in Arizona and New Mexico, with an area of 8,200,000 acres; a third covers the watershed of the Rio Grande River above the Elephant Butte Reservoir, with an area of 11,500,000 acres; and still another is located in the Shoshone Indian Reservation in Wyoming. The Navajo project includes lands that contribute vast quantities of silt to the reservoir above Boulder Dam and involves the preparation and extensive application of comprehensive erosion control, land use, range control, and water-conservation measures and devices. It also involves the reorientation and stabilization of the entire agricultural economic system of the 45,000 Navajo Indians. The Gila project involves seriously eroded and eroding lands constituting the basis of a widespread livestock industry. From these overused lands, great quantities of silt are washed into Coolidge Reservoir. The Rio Grande project covers the most densely populated and productive area in New Mexico, now threatened with abandonment because of serious erosion and sedimentation resulting from overgrazing and other malpractices with respect to land use over the watershed. The Elephant Butte Reservoir is also subject to rapid silting. The problem of land depreciation in the flood basin of the Rio Grande and its tributaries is an exceedingly serious one, involving impoverishment or outright ruin of the range by erosion, troublesome sedimentation, waterlogging, and increasing flood damage.

On these Western lands, the task of cutting down soil and water wastage is complicated by a number of physical and economic factors peculiar to the region.

In the first place, the pattern of land ownership over much of the range country is quite different from that which prevails in the Eastern and Central sections of the nation. An ownership map of many sections will show Indian reservations, unallotted tracts of public domain, school lands, state and national forest reserves, Federal grazing districts, railroad lands, and small and large private holdings, frequently intermingled in such complexity as to make it difficult to achieve adequate coordination in the application of needed soil- and water-conservation measures.

Any sound approach to the twin problems of controlling erosion and conserving rainfall must recognize the value of the land involved. Obviously, elaborate and costly measures can scarcely be justified when used to protect land of inherently low economic worth, except in those situations where the work provides worth-while protection to other more valuable areas or to important reservoirs. On the large Western watershed projects, where the Soil Conservation Service is operating, much of the area falls in the category of relatively cheap land. Except for the developed holdings within irrigation districts, the great bulk of these areas consists of large tracts of low-priced land used for grazing. Expenditures for erosion control must therefore be carefully limited and measured against damage to reservoirs, irrigated areas downstream, etc. Every effort must be made to extend the benefits from a few inexpensive measures and devices over the largest possible area. Numerous scattered areas of exceptionally favorable soil and situation, however, warrant more intensive treatment. Certain areas of valley land, especially of alluvial or valley-fill soil of low salinity, can be made to produce excellent stands of grass and even good crops of corn, beans, etc., by controlling runoff in such a way as to spread the water over the surface and so conserve it within the soil.

The low annual rainfall on these project areas, typical of vast sections in the West, is a factor that seriously limits the production of an adequate vegetative cover. Moreover, the torrential quality of the infrequent rains compounds the problem of controlling runoff. In addition, many areas requiring treatment are difficult of access, and some have developed, within both historic and prehistoric times, desert and near-desert conditions.

In spite of these difficulties, however, the general method of attack on the soil erosion problem in these areas is similar to that employed on the smaller demonstration projects throughout the country. The chief difference lies in a greater emphasis on the problem of restoring vegetation to misused ranges. Also, a consistent attempt is made to select work areas scattered throughout these extensive projects at the most advantageous locations. Aerial surveys furnish the base for the necessary mapping and planning by the technicians. Work plans are developed, as they are on the smaller projects, under the terms of cooperative agreements that specify in some detail the respective contributions of the Service and of individuals and cooperating agencies.

Since the major portion of the land in these four large Western projects consists of open range, the program of the Soil Conservation Service is based largely on the scientific management of livestock and improved treatment of grazing lands. Grazing is controlled in such a way as to encourage the reestablishment and continued healthy growth of protective vegetation useful for forage. This aim is realized through a many-sided program, including the use of permanent and temporary fences, the development of wells and stock ponds, and the limitation of livestock numbers to the actual carrying capacity of the range. Supplementary mechanical aids, such as contour furrows and water-spreading dams or dikes, are employed to hasten revegetation. In addition to protecting the land surface on these watersheds, all this work, including the range management, is helping to reduce silting in the reservoirs that receive the runoff from the areas under treatment.

The work on this federally controlled land is largely in the nature of land rehabilitation. It is aimed primarily at the improvement and stabilization of extensive tracts, most of which have suffered from improper use. The present work areas, however, also have a decided demonstrational value. As the effectiveness of the measures employed becomes evident, it is expected that similar work will be undertaken through local initiative on large and small adjacent or neighboring holdings. With this probable spread of local effort and with the continued assistance of the Federal Government, the satisfactory stabilization of vast areas of these extensive Western lands seems a distinct possibility within the next 20 or 30 years.

## Purchase of Submarginal Land

An important phase of the broad program of soil and water conservation involves the establishment of practical methods for the utilization of marginal and submarginal lands, since such lands must be retired from cultivation if erosion is to be controlled effectively. When this type of land occurs with relative infrequency between productive fields, it is planted to grass, legumes, shrubs, vines, or trees in such a manner as to control erosion and provide, wherever possible, the supplementary income needed to replace that which the owners or operators formerly obtained through cropping and other land-use practices that caused exhaustion of their capital land assets. Where practicable, those types of vegetation are utilized which furnish a food supply and coverts for the development of game resources—resources that the farmers can turn to profitable use by disposing of hunting privileges. Betterment of farm habitability is also given consideration in the operations aimed at the establishment of physical security for the land and economic safeguards for those living by the land.

A much more difficult land utilization problem is presented in those regions where marginal and submarginal lands exist in such large blocks that they cannot be operated profitably under individual ownership. Where such areas occur, they may be purchased by the Federal Government, the states, or the local communities, in order to establish thereon Government, state, county, municipal, or community forests, recreational facilities, or some other public welfare interest.

For example, in the Great Plains, there are many areas where both public and private lands are broken up into tracts that are unsuited to cultivation and yet too small for grazing units. By buying out the private lands, the Government can include them, together with the public lands, in large grazing areas that can be managed efficiently. Near-by stockmen can then run cattle on these grazing areas, through the operation of some leasing system, and the grass can be protected from overgrazing and wind blowing as it could not be before.

In a similar way, farms on unproductive land scattered through forest areas can be bought and turned over to forest use. Wasteful land practices will thereby be stopped, and the heavy costs for schools and roads to serve these unproductive farms will be brought to an end.

These land purchases are being concentrated in areas where other measures are being combined with public acquisition to bring about a more productive use of land. For example, purchase preference will be given to areas where rural zoning regulations are in effect or where soil conservation districts have been organized.

Moreover, lands are selected for purchase where public ownership will help achieve as many different aspects of good land use as possible. For example, land will be bought in an area where public purchase will help control floods, restore needed range resources, and eliminate heavy costs for local government, in preference to an area where the public acquisition of land would further only one of these objectives.

Before lands are purchased, arrangements are made for the management of the lands by some public agency, of either the state or the Federal Government, that is best qualified for that work. Most of the development of the land will be done by the agency in the care of which the land will be placed. No land purchased under this program will be resold to private individuals. It will be kept in public ownership, although grazing lands may be leased to private operators.

When an area has been selected for purchase in the light of the land-use plans mentioned above, individual landowners are notified of the opportunity to sell their land to the Government at a fair price, based on its appraised value. Whether or not a purchase project is consummated in any such area depends on the willingness of landowners to sell.

## Flood Control

Activities under the flood control program of the Soil Conservation Service involve the survey of watersheds and treatment of lands contributing to the flood hazard within those watersheds approved by the Congress for flood control activities, including such lands as may be purchased under the acquisition program described on pages 326 and 327.

For many years, workers in the fields of agriculture and forestry have contended that proper watershed-protection practices or measures for upstream control are needed on most watersheds to supplement downstream works, if the flood problem is to be solved adequately. This contention has resulted from research work indicating the tremendous possibilities for watershed absorption of rainwater by utilizing the vast storage capacity of the soil. The reservoir capacity of the land is especially significant where there is adequate vegetal cover or when the land is provided with proper mechanical checks against rapid runoff.

It has not been contended that upstream control measures generally could be relied on as the sole solution of the flood problem but rather that such measures would represent a first line of defense in any comprehensive flood control and water utilization program. Many who have studied the situation are convinced that man, through his activities with axe, plow, herds, and fire, has created conditions that increase the possibilities of greater floods. To the extent to which this has occurred, watershed-protection practices should be used in any program of flood control.

The principle of protecting the entire watershed was first recognized as a part of a national flood control policy in the Omnibus Flood Control Act of 1936. This act and subsequent amendments have authorized the Secretary of Agriculture to make preliminary examinations and surveys for runoff and waterflow retardation and soil erosion prevention on watersheds involving more than 90 per cent of the land area of the country. Moreover, the Secretary has been authorized to prosecute measures of improvement on the watersheds of all waterways for which the War Department has been authorized to carry on flood control works. An appropriation was made by Congress for initiating the application of watershed-protection measures during the fiscal year 1938–1939.

The objective of the Flood Control Act of 1936 is to protect existing values from flood damage through the control of destructive flood waters and flood-borne sediment. The act does not limit or retard any agency of the department in carrying out similar and related activities authorized under other legislation. Neither is it intended that the related activities of departmental agencies are to be expanded through the use of flood control funds, except as they constitute an integral part of a specific flood control project.

Upstream flood control measures are being distinguished from the land-use adjustment and conservation programs of the department, although many of the techniques employed are similar. Upstream flood control measures are directed to the control of flood flows and the stabilization of soils in headwater areas for the protection of existing values downstream. The land conservation programs are directed to the better use of land and water resources for the maintenance of a permanent agriculture. Each class of work creates a distinct kind of Federal and local relationship. Each requires particular standards of procedure, justification of costs, and obligations of maintenance. Each can be integrated in the case of many watersheds to further a common program for the conservation and proper utilization of the soil and water resources of the nation.

General direction of the flood control work of the department is vested in a Flood Control Coordinating Committee consisting of one representative each from the Bureau of Agricultural Economics, the Forest Service, and the Soil Conservation Service. This committee formulates policies, establishes priorities of watersheds for preliminary examinations and surveys, allocates funds, and assigns leadership for surveys of the several watersheds to one of the participating bureaus. The operations phases of the program are to be directed by either the Forest Service or the Soil Conservation Service, on the basis of the predominance of forest or farm land in the watershed.

## Water Facilities

During the successive dry years of 1934, 1935, and 1936, rural distress in the arid and semiarid sections became so acute that the President appointed a Great Plains Committee to study the problem and recommend a program that would lead to a sound and permanent agriculture in low-rainfall areas of the West. In its report, this committee points out that an irregular rainfall is the most striking general characteristic and one of the greatest natural handicaps to much of the Plains country west of the 100th meridian. The committee states further that if farming and ranching in many parts of the Great Plains are to survive, and if the population of village and urban trade centers is to be sustained, all available water resources will have to be conserved carefully and utilized wisely. The committee's report recommends Federal aid in the development of small water facilities in arid and semiarid areas.[1]

[1] Cooke, Morris L., REA, Chairman; Bennett, H. H., SCS; Gray, L. C., RA; Harrington, F. C., WPA; Moore, R. C., Corps of Engineers; Page, J. C., Bur. Reclamation; and Person, H. S., REA, The Future of the Great Plains, Great Plains Comm. *Rept.*, Washington, 1936.

In line with this recommendation, the 75th Congress passed the Water Facilities Act (*Public No.* 399, 75th Congress, sometimes referred to as the Pope-Jones Act) directing the Secretary of Agriculture, through existing agencies of the Department, to assist farmers in constructing and installing small water facilities in the arid and semiarid areas of the United States.

In accordance with this Act, the Department has initiated in the 17 Western States a water facilities program under which funds are being expended or loaned to farmers for the development of carefully planned water facilities. The program aims to encourage wiser use of agricultural lands over a period of years and, through this better land use, to promote the rehabilitation and welfare of the people who live on the land.

Under this program, small facilities are being provided to effect the storage or utilization of water for livestock and for irrigation of lands suitable for farming. Farmers and ranchers are receiving direct Federal assistance in developing ponds, small reservoirs, wells, springs, pumping facilities, water spreaders, stock water tanks, small irrigation facilities for individual families or groups of families, and retention and diversion dams. This assistance is available to needy landowners and to renters with long-term leases or annual renewable rental contracts. It may be in the form of technical aid, or it may be in the form of a long-term loan at 3 per cent interest to pay for construction or installation of the facilities needed. The Department is not requiring reimbursement for its technical services or other general administrative costs. However, the benefiting farmer or rancher must agree to make the maximum possible contribution toward the cost of the facility that his economic condition permits, by furnishing labor, teams, materials, and other equipment. The amount to be repaid on the loan is adjusted to the individual borrower's ability to repay, as determined by the Department.

After an area has been selected for water facilities work, the first step is to formulate a general plan. All types of land within the area are classified according to the uses for which they are best suited, and recommendations are made for the development of suitable water facilities. Detailed plans are then drawn up for each farm or ranch where the work is to be carried on.

These detailed plans include a water facilities plan that indicates the type and location of the facilities needed to promote the best possible use of the land and water resources; a plan of conservation operations, which is simply a field-by-field description of necessary erosion control measures and practices; and a farm and home management plan setting forth a detailed cropping system that will provide living expenses, operating costs, and repayments on the loan (if one has been granted), as well as adequate supplies of food and feed for family and livestock.

## Work of Other Agencies

In addition to the Soil Conservation Service, of course, other agencies of the Federal Government and the states are engaged in programs of physical land use looking to the conservation of soil and water. Among these allied units are:

I. THE FOREST SERVICE. Among the major functions of the Forest Service are the protection, development, and administration in the public interest of the federally owned national forest system and its resources, products, values, and services; and the prosecution of a program of research in problems involving protection, development, management, renewal, and continuous use of these resources, products, and values.

The national forests are located in various parts of the Appalachians from New Hampshire to Georgia, around the Great Lakes, in the Southern Pine Belt, and on the slopes of the Rockies, Cascades, the Sierra Nevada, and the Coast Ranges between Canada and Mexico.

> "Including slightly more than 1,000,000 acres approved for purchase . . . the national-forest system . . . as of June 30, 1938, comprised 175,238,168 acres of federally owned lands. Established irrespective of State boundaries, these public properties are distinct from national parks in that they are administered on a multiple-use basis, with renewal of all their resources and values.
>
> "Today's national forests offer striking contrasts between civilization and wilderness, industrial activities and pastoral, material values and spiritual ones. More than 1,280 million feet of timber was harvested from them, under provisions that assure continuity of the forest stand, in the fiscal year 1938. They already provide a living for almost a million people and recreation for 30 million each year; are home and refuge for most of our remaining big game; include some 70,000 miles of fishing streams and more than 3,500 developed public campgrounds; furnish forage for more than 6,857,000 domestic livestock; help prevent floods and erosion; and provide domestic water for 6 million city people. They have a public transportation system that includes more than 138,000 miles of highways and roads and 153,000 miles of trails. Gross receipts from the national forests in the fiscal year 1938 were $4,671,133, with $1,178,883 returned to the States."[1]

Increasing numbers of stockmen, who use portions of the national forests for grazing, recognize the fact that conservative management means a lower death loss (especially in times of drought), increased lamb and calf crops, better animals, better wool, and a higher net return. Irrigationists now recognize their interest in the national forest lands, from which they derive their water supply. Townspeople, at least indirectly dependent on the welfare of forest, range, and farming industries, are becoming increasingly aware of their responsibilities in the manage-

[1] Report of the Chief of the Forest Service for 1938.

ment of public forests. These and other benefits are a part of the record of the Forest Service—a national achievement in conservation.

II. Biological Survey. The Bureau of Biological Survey had its inception in 1885 as the Branch of Economic Ornithology in the Division of Entomology. In the half century since C. Hart Merriam gathered together a handful of workers to investigate the relations of birds to agriculture, it has evolved from a purely research institution into an organization of considerable complexity, reflecting the profound changes that have taken place in the Department of Agriculture. Whereas the first tasks essayed had to do largely with the effects of birds and mammals on human interests, the migrations of birds, and the distribution and classification of the native fauna, the functions of the Survey now encompass such diverse activities as the large-scale acquisition of land for wildlife refuges, the enforcement of certain laws bearing on the welfare of wildlife, and the allotment of Federal funds for state aid and the management of huge tracts of land as well as a greatly expanded research program. The embryonic bureau, composed of a few naturalists with vision, was of scarcely more than academic interest to the agriculturist; today the Biological Survey is one of the major land-use agencies of the nation, and the results of its research influence the management of millions of acres of land not actually under its jurisdiction.

The latest report (as of Aug. 31, 1938) shows that the Biological Survey administers 248 wildlife refuges with a combined area of more than 11 million acres. Several million of these acres are lakes, marshes, and swamps which contribute directly to the conservation of the water resources of the country and which, by retarding the heavy discharge of runoff water into stream channels, play an important part in reducing flood hazards on a number of drainages. More than 5 million acres of the land under the jurisdiction of the Biological Survey are managed for the benefit of herds of large game species that have been threatened with extinction. The latter areas, consisting chiefly of upland, play a part in the general scheme of soil and water conservation, since overgrazing is not permitted, and definite efforts are made to encourage the reestablishment of as nearly natural conditions as possible. This means the return of normal vegetative protection to the land, which in turn means a minimum of runoff and erosion. In fact, the establishment of these wildlife refuges has insured safe use for many thousands of acres of submarginal land retired from crop production and consequent deterioration, through Federal purchase under the land-use adjustment program.

The influence of the research work of the Biological Survey extends far beyond the confines of the lands that it actually controls. Facts concerning food habits and life histories of birds and mammals, developed by this research, are used regularly in the development and administration of millions of acres of land under the jurisdiction of the Forest Service, the Soil Conservation Service, the states, and even private owners. The Soil Conservation Service, particularly, has relied on the plant-food habit studies of the Biological Survey as a guide in the selection of plants to control critically eroding areas. Thus, plantings for both wildlife and erosion control are being demonstrated to thousands of farmers and ranchmen throughout the country.

In other words, although the program of the Biological Survey is aimed directly at the conservation of wildlife, it makes an important contribution to the conservation of the basic resources of soil and water.

The Biological Survey was transferred to the Department of the Interior July 1, 1939.

III. Division of Grazing. The Division of Grazing of the Department of the Interior, established under the Taylor Grazing Act of 1934, is actively engaged in improving the grazing use of some 120 million acres of public domain. This activity represents the most important step toward the control of erosion and excess runoff on these lands. The primary objectives of this agency are: "To stop injury to the public grazing lands by preventing overgrazing and soil deterioration, to provide for their orderly use, improvement, and development, and to stabilize the livestock industry dependent upon the public range. . . . "

In cooperation with local associations of stockmen, state land officials, and various other agencies, the Division of Grazing conducts range surveys to determine the carrying capacity and proper seasonal use of the land; prescribes rules and regulations for range use; issues permits for such use through district advisory boards made up of representatives elected by the stockmen; and engages in operations designed to rehabilitate, improve, and preserve the public range land included within established districts.

IV. State Agencies. State agricultural colleges; experiment stations; extension services; vocational training services; planning boards; and conservation, forestry, and agricultural departments are actively engaged in educational, research, planning, or action programs bearing directly or indirectly on the problem of soil and water conservation.

In one or more phases of the conservation problem, the Soil Conservation Service, Forest Service, Agricultural Adjustment Administration, Farm Security Administration, Bureau of Agricultural Economics, Farm Credit Administration, Bureau of Plant Industry, and other agencies of the Department of Agriculture are cooperating with the state agencies. In addition, other Federal agencies, such as the Bureau of Reclamation, Division of Grazing, Office of Indian Affairs, Biological Survey, and Geological Survey of the Department of the Interior; the Corps of Engineers of the War Department; the Civilian Conservation Corps, National Resources Committee, Tennessee Valley Authority, Works Progress Administration, and Public Works Administration are cooperating with one or more of those state agencies whose activities are related to the land and its welfare, the welfare of people living on the land, conservation of rainfall, flood control, prevention of silting, or conservation of forests and wildlife.

In a number of states, the highway departments, the Soil Conservation Service, and the Bureau of Public Roads are cooperating for the control of erosion along public highways. This work is contributing also to highway beautification and traffic safety.

As soil conservation work progresses, the role of state extension services becomes increasingly important. With their staffs of county agricultural agents, they are in an excellent position to stimulate and assist in the widespread adoption of conservation measures. In some parts of the country, state extension services have long recommended certain individual practices, such as terracing, for erosion control. Currently, however, more and more states are advocating erosion control through the coordinated use of all adaptable conservation measures. To present their recommendations more effectively, a number of states, through their extension departments, working with the Soil Conservation Service, have established cooperative demonstrations for erosion control and water conservation in the farming sections.

Staff members of state agricultural colleges and other educational and research institutions are giving more time to the study of different phases of the erosion problem, including its economic and sociological aspects. Information of this type, essential to the practical adoption of conservation methods, is being developed on both a localized and statewide basis. Wherever possible, the Soil Conservation Service is cooperating in this work.

State experiment stations also are playing an important part in the furtherance of soil conservation. In cooperation with the Soil Conservation Service, they are participating in the development of fundamental facts bearing on the erosion process, on the improvement of present methods of control, on discovery of new methods, and on many other aspects of the erosion problem requiring research. The relation of erosion to such factors as soil type, soil condition, topography, and the intensity of precipitation is being given particular attention.

## Relief of Economic and Social Pressures

The economic and social ramifications of the land-use problem are both extensive and complex. Economic and social maladjustments both cause and stem from physical land ills; almost any action affecting the physical use of land will have immediate or ultimate consequences of an economic and social nature. The realignment of fields, adoption of rotations, use of moisture conservation devices, and other similar purely physical land-use practices eventually bring about, in most instances, a larger return per acre in terms of crops or income. For this reason, virtually all the conservation activities heretofore outlined exert some influence in either correcting or mitigating the economic pressures that cause men to abuse the land.

There are, however, a number of agencies directly concerned with economic and social aspects of the land-use problem. They form an integral part of the broad-scale national program that has developed in recent years. Notably, these agencies are:

I. The Agricultural Adjustment Administration. This Federal agency seeks to bring farm income into its proper relation to industrial income through payments to farmers for the adoption of conservation land-use practice. The purposes of these payments, in other words, are to promote the spread of conservation practices and to relieve farmers of the economic stress caused by crop surpluses and the consequent low prices for farm products.

In 1936, following the Hoosac Mills case in which the United States Supreme Court invalidated the crop control and processing tax aspects of the Agricultural Adjustment Act, the making of benefit payments to farmers in return for the adoption of soil-conserving practices on their farms became the principal basis of operation in the Agricultural Conservation Program.[1] Schedules of payments were offered for reduction in the acreage of soil-depleting crops—generally those cash crops of which there was a national surplus. In addition, payments were authorized for soil-building practices, such as the use of legumes and grass, and, to a limited extent, for terracing and contour furrowing and the application of fertilizer and crushed limestone.

The practices for which benefit payments were made varied through the several agricultural regions of the United States. In actual operation, during the first year of the Agricultural Conservation Program, the bulk of the payments were made for reductions in the acreage of the soil-depleting crops, such as cotton, wheat, corn, and tobacco. On the average, only about 20 per cent of the total payments were made for such soil-building practices as the seeding of grass and legumes, the use of fertilizer, terracing, and contour furrowing.

In 1937, the program shifted its emphasis more toward payments for definitely soil-building practices, although it continued to make most of the payments for reductions in soil-depleting acreages. In the Agricultural Adjustment Act of 1938 (*Public No.* 430, 75th Congress), the soil-

[1] Agricultural Adjustment Administration *Ann. Repts.*, 1936, 1937.

conservation objectives were preserved as major provisions of the farm program.

II. THE FARM SECURITY ADMINISTRATION. This Federal agency has been assigned the task of helping needy and low-income farm families to become self-supporting. A majority of these families probably are victims of ill-fated attempts to farm on submarginal land, on farm units of inadequate size, on eroded soil, or under unstable tenure arrangements and one-crop systems. Perhaps a combination of all these adverse factors is involved.

To several thousand families who formerly lived on the submarginal lands purchased by the Resettlement Administration, the Bureau of Agricultural Economics, and the Soil Conservation Service, the FSA has offered a fresh start on good land in its Rural Homestead Projects. Other thousands in such areas have been helped to find and rent farms better adapted to agriculture.

A larger segment of the farm population, however, has been helped to adjust its land-use practices through the FSA's rehabilitation program. Under this program, low-income families unable to obtain adequate credit from any other source are given assistance in working out sound farm-management plans and are loaned the funds necessary to put the plans into effect.

The FSA has helped more than 650,000 families through rural rehabilitation loans. A recent sample survey[1] of the progress made by 231,661 of these families indicates some of the important shifts in land use achieved through the system of planning and credit that is the heart of the FSA's work. Since joining in the rehabilitation program, these families have increased their cultivated acreages by more than 27 per cent—often by renting additional land. They have diversified their cropping systems, nearly doubled their ownership of work animals and cattle, increased feed acreages by 50 per cent, and increased garden plots enough to double the amount of food canned and stored for personal use during the winter.

The farm plans worked out by each family, in cooperation with FSA management specialists, usually call for terracing and contouring, planting of cover crops, and other improvements less measurable in terms of statistics but nevertheless important.

The FSA has encouraged better leasing arrangements among its tenant borrowers and their landlords. Between the time of entry into the rehabilitation program and Jan. 1, 1938, 98,065 of the 231,661 families mentioned above obtained longer renewable leases, 65,480 changed from

[1] Summary of Rural Rehabilitation Supervisors Progress Reports. Farm Security Administration (mimeographed), Dec. 31, 1937.

sharecropper to tenant status, and 66,566 obtained tenure of farms better adapted to their needs.

Since July 1, 1937, the Farm Security Administration has administered a program authorized by the Bankhead-Jones Farm Tenant Act, under which approximately 5,000 farm tenant families will be loaned funds for purchase of farms of their own during the fiscal year 1939. Thus, incentives to good land use, which come with ownership and stability, are added to the farm-management training and assistance offered under the rehabilitation program.

In summary, by offering to needy farm families instruction in sound farming practice, credit to finance needed adjustments, and new incentives to conservation, the FSA has done much to encourage conservation of the land itself.

## The Future

Unless the United States goes ahead vigorously, persistently, and speedily to defend and conserve the soil and to make far-reaching adjustments in its complex land economy, national decadence lies ahead. The nation must continue to capitalize on experience and to advance through research. From the standpoint of the country's soil resource alone, the need for action is now clear enough. Failure to act in the past already has caused the essential ruin or serious impoverishment of some 280 million acres of farm and grazing land. Now erosion is active on an additional 775 million acres. The average citizen does not yet fully understand the deep significance of this waste or realize the hardships that it has caused by lowering thousands of land users virtually to the level of bankrupt farming, with its attendant discouragement and inertia. To date, the social and economic implications of this single facet of the erosion problem have not been fully probed and appraised.

*There is no longer a question of the need for coping with these evils. There is no longer a question as to whether the nation can cope with them. We know that we can. Millions of acres already have been effectively anchored against erosion. New and practical conservation measures are being developed continually through research and experience on the land. Many of our economic and social difficulties on the land are beginning to be solved. We are moving constantly ahead though not yet with sufficient speed.*

It should be remembered that today's necessity for public action is the outgrowth of yesterday's failure to look more carefully to our land. Hindsight is easy; but foresight during the last century, when our present land-use picture was in the making, would have produced a different result. Today, we are simply retracing our steps across the land in an effort to correct past mistakes in the interest of the future.

# Chapter XVI. Agronomic Practices in Soil and Water Conservation

Soil completely covered with vegetation is in an ideal condition to absorb moisture and resist the inroads of erosion, provided the cover is continuous and the soil is well permeated with roots. Before the advent of white man, the soils of North America nearly everywhere were covered with forest, prairie, or plains vegetation, bush growth, or desert vegetation. Under such conditions, erosion was limited to a normal, harmless rate. The accelerated erosion of comparatively recent years is, as already shown, a product of wholesale land denudation. It is definitely connected with the destruction of forests, the plowing of prairies and plains, and the overgrazing of pastures and ranges. While it is not possible to restore original vegetative conditions in the United States and at the same time maintain the nation's agricultural economy, measures taken to conserve soil and water must be patterned, as far as possible, on nature's own methods of soil defense. This means the use of close-growing vegetation in connection with regular farm operations and for special purposes.

There are several reasons why close-growing vegetation functions so effectively in checking soil and water losses. In the first place, such vegetation serves to reduce or eliminate entirely the direct impact of rain on the soil surface. Wherever a drop of rain strikes a bare soil, particles are agitated and loosened, and the finer ones are taken into suspension. By sheltering the ground surface with dense top growth, this first step in soil wastage is controlled. Then, too, the speed of water flowing over a surface is checked by the numerous stems of a close vegetative cover and by the vegetal debris left on the surface. This reduction in the speed of runoff decreases its power to pick up and carry soil particles, while allowing more time for its penetration into the earth. Finally, an increase in soil organic matter by root growth, especially in the instance of a permanent cover, or by the incorporation of vegetation as green manure, improves the water-holding capacity of the soil and favors infiltration of water. Even vegetative debris, such as crop residues and straw, has an important place in checking both wind and water erosion and increasing the water-holding capacity of soil.

These principles are basic; they underlie any attempt to conserve soil and water by the use of close-growing vegetation. The particular vegetative methods used in any given situation, however, will depend on local farm needs, slope, soil type, and physical condition of the soil, on local adaptability of crop plants, and on the need of vegetation to support or complement engineering methods.

Some vegetative methods of erosion control are applicable to all problem areas, while others, such as the use of winter cover crops and range revegetation, are more serviceable in some parts of the country than in others. In no region, however, should one method be depended on to the exclusion of all others. All are interrelated, and even on a single farm two or more methods may need to be used. For instance, crop rotation alone may be used where this will assure control, while other parts of the same farm may require strip cropping or a combination of strip cropping and crop rotation. All these methods may be used with or without terraces or contour cultivation, depending on slope and physical character of the soil involved.

The principal methods by which vegetation is used to insure conservation of soil and water are: crop rotation, strip cropping, pasture and meadow improvement, and the growing of crops for seasonal cover and green manure. Vegetation is also used in gully control, in the stabilization of field waterways and diversion ditches, in windbreaks, in buffer strips, in field "edges" for protection of game, and in stabilizing range land to control wind and water erosion while increasing the grazing capacity.

Frequently two or more of these methods are used in combination, as, for example, in the instance of crop rotation and strip cropping (strip cropping on a rotation basis). Simple crop rotation involves the growing of only one crop on one field in any one year, the crops being rotated by whole fields. In strip cropping, different crops are grown in a single field by planting them in alternate strips, not necessarily rotated. The rotation may, however, be by strips as well as by fields.

## Crop Rotation

Crop rotation is generally defined as a more or less regularly recurrent succession of different crops on a single piece of land. The crops used are commonly a cultivated crop; a small grain; and a grass, legume, or legume-grass mixture. Of these, the cultivated crop exposes the soil to the maximum erosion; the small grain allows less erosion; and the grass or legume-grass crop effectively controls erosion during the period of its life. The fact that the cultivated crop follows the sod crop also tends to reduce soil loss below what it would have been had one cultivated crop succeeded another. The net result for the rotation period is almost invariably a reduced soil and water loss.

HISTORY. The rotation of crops on tilled fields is a very old practice, referred to by Roman writers before the Christian era and in the first century A.D. During the Middle Ages, however, all agriculture fell to such a low state that the only rotation for many centuries was the three-field system in which one field was in winter grain, one in spring grains or beans, and one fallow. No true soil-saving rotation was possible until the advent of cultivated grasses and clovers in the seventeenth century.[1]

The lack of systematic crop rotation in the agricultural practices of the American colonists was roundly condemned in 1859 by Baron von Liebig,[2] who said:

> "The deplorable effects of the spoliation system of farming are nowhere more strikingly evident than in America, where the early colonists in Canada, in the State of New York, in Pennsylvania, in Virginia, Maryland, etc., found tracts of land, which for many years, by simply plowing and sowing, yielded a succession of abundant wheat and tobacco harvests. . . . In less than two generations, [these fields] though originally teeming with fertility were turned into deserts . . . . "

Sometime after the establishment of an independent nation, speculative farming became the general practice in the United States, and all possible effort was made to produce "cash crops." Over long periods, continuous production of cotton in the South, corn in the Ohio and Upper Mississippi Valleys, and wheat in the central and northern Great Plains and in the Palouse country of Oregon, Washington, and Idaho, to the exclusion of all other crops in each of these sections, has resulted in an enormous loss of valuable land through erosion. For many years Federal and state agricultural agencies have advocated crop rotation, but the spread of the practice has been slow. Generally, rotations have been adopted in the past only as a result of immediate economic needs rather than through any real appreciation of their value in reducing soil wastage. With increased understanding of the need for saving soil, however, crop rotations are now assuming a greater degree of importance in American agricultural practice.

PURPOSE AND ADVANTAGES. Crops are rotated in order that soil productivity may be preserved and crop yields maintained. These objectives are gained because the practice of crop rotation systematizes farming; saves labor; helps to control weeds, insects, and plant diseases; aids in maintaining soil organic matter and nitrogen; and lessens soil loss through erosion.

TYPES OF ROTATIONS. Crop rotations vary with the soils, economic conditions, and cropping systems of different farming regions. The 3-year rotation of cultivated crops, small grain and grass or legume-grass, how-

[1] Simkhovitch, Vladimir G. Hay and History, *Political Sci. Quar.*, Vol. 28, pp. 358–403, 1913.

[2] Hopkins, Cyril G. "Soil Fertility and Permanent Agriculture." 1910.

ever, may be considered as basic. This can be modified in various ways by dropping one crop to make a 2-year rotation or by having one of these crops on the same field for more than 1 year, thus lengthening the rotation to 4, 5, or even more years. When grass or a legume, or a mixture of the two, forms part of the rotation, the cultivated crop usually follows immediately, and the small grain serves as a companion crop for the next stand of grass or legume.

Of the various rotations recommended by state agricultural experiment stations, the following will serve as examples: In the extreme northeast of the Northern Appalachian and New England area (see Map 1, *Problem Areas*, "Results of Erosion," Chap. III), approval is given to a potato, crimson clover or potato, oats, crimson clover rotation. Farther south in the same area, a corn, wheat, clover rotation is common. In the Southern Appalachian region, a rotation of corn with cowpeas, small grain, lespedeza, and cotton with a winter legume provides for three legume crops in 4 years. Where cotton is not grown, a 3-year rotation of corn, corn, meadow may be used. In the Central Prairie and Eastern Timbered Border region, in parts of which the practice of crop rotation is well established, an approved rotation includes corn, small grain, and meadow, with the last sometimes being left for 1 or 2 years as pasture. In the Ozark Highlands, corn interplanted with cowpeas is followed by cotton, and this by small grain followed by cowpeas.

In parts of the Great Plains region, corn is replaced by sorghum, and wheat occupies the ground for a greater period, so that a typical rotation includes grain sorghums, oats and sweetclover, sweetclover alone, and finally wheat for 2 years. A variation is wheat 3 years, followed by 1 year of oats and sweetclover and 1 year of sweetclover alone. In the Palouse section of eastern Washington, a common rotation is seed peas planted with sweetclover, sweetclover plowed under the following June, spring wheat, seed peas, and spring wheat. In this system, the ground is left rough after plowing the sweetclover so that damage by erosion over winter is minimized. Another rotation that is used on land where peas are not grown is sweetclover and grass, usually tall meadow oat grass[1] for 2 years, wheat 2 years, summer fallow, and wheat. If crop residues are utilized properly, and a method of "trashy" fallow is employed, this rotation can be used with very little soil loss on slopes that are not too steep.

[1] Other grasses well suited to mixtures with sweetclover in this region are blue wild rye (*Elymus glaucus*), mountain brome (*Bromus marginatus*), slender wheat grass (*Agropyron pauciflorum*), Canada wild rye (*Elymus canadensis*), crested wheat grass (*Agropyron cristatum*), smooth brome grass (*Bromus inermis*), tall fescue (*Festuca elatior var. arundinacea*), and orchard grass (*Dactylis glomerata*). The seed of some of these are not yet commercially available.

The principles to be followed in a soil-saving rotation are to reduce the time the land is occupied by a cultivated crop as much as the economic situation of the farm will permit, increase as much as possible the time it is covered by a legume or a grass, and reduce soil tillage. Furthermore, the same principle is applicable when some other type of crop is used for soil protection. For example, in western Kentucky a rotation of tobacco 1 year with strawberries 2 years may meet all soil-saving requirements, even though a legume or grass is not used. In this rotation, the ground is protected by the strawberries. Such rotations may cover several fields, with each crop occupying a full field each year; or they may be combined with strip cropping, the crops being rotated from strip to strip in a single field. The best practice to use will depend on the slope, soil type, and the management requirements of the individual farm.

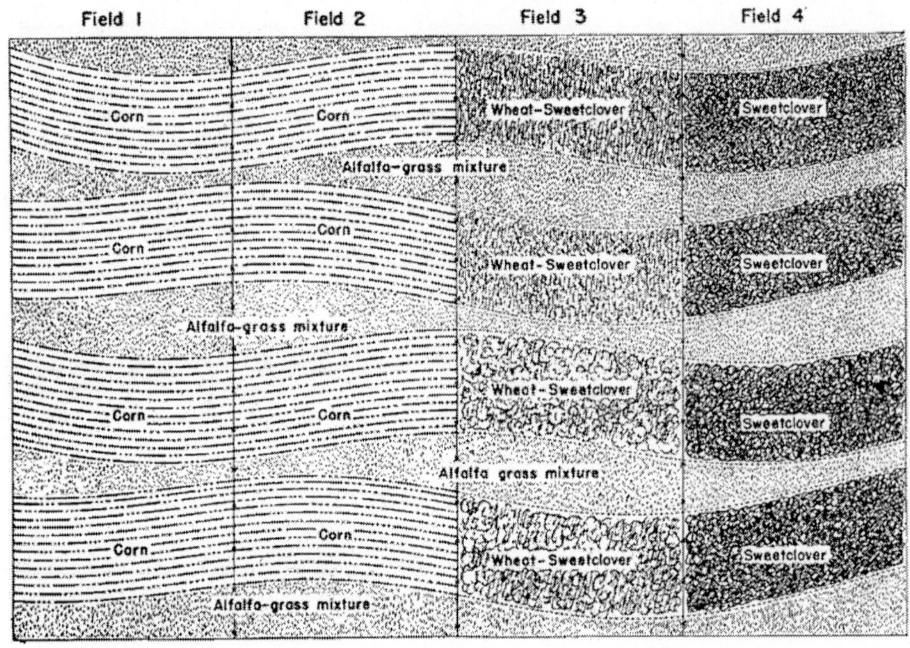


Graph 24.—Adaptation of contour strip cropping to livestock farming. In this instance, a four-year rotation of corn, corn, wheat-sweetclover, sweetclover, permits grazing without extra fencing. The fields may be adjacent, as indicated here, or located on different parts of the farm. The permanent strips of alfalfa-grass can be grazed or cut for hay, depending upon the crop in the field. (*Soil Conservation Service.*)

Rotations on Contour. On sloping land other measures will be needed to supplement crop rotations if soil and water losses are to be reduced to a minimum. Of these measures, contour cultivation, terracing, contour strip cropping, and the use of seasonal cover crops are the most important.

The flexibility of a strip-cropping system makes it possible to meet practically any crop or livestock problem through proper arrangement of the strips. For example, permanent close-growing contour strips of legumes, legume-grass mixtures, or grass may be used in those instances where it is undesirable to have all crops of a rotation in one field at one time. These permanent soil-conserving strips permit the use of a rotation of tilled crops, small grains, or legumes on a field basis. Such an arrangement meets the requirements of the livestock farmer who finds it necessary to graze stubble fields after the grain is cut, without building temporary fences to protect the corn or other crops that might interfere with grazing in the same field. As an illustration, in a rotation of corn, corn, wheat, and sweetclover, both fields 1 and 2 (Graph 24), when used for corn, would also produce alfalfa-grass hay from the permanent strips. Field 3, which at the same time would be in wheat seeded with sweetclover, could be grazed the remainder of the season following the harvest of the wheat and the cutting of a crop of hay from the alfalfa-grass strips. Field 4, which for the corresponding year would be in sweetclover, would be available for grazing throughout the season or as long as the sweetclover lasts. The permanent strips, in addition to checking erosion, take up field irregularities and allow contour cultivation without involving point rows.

Rotation of Crops Saves Soil. From the standpoint of soil and water conservation, the value of crop rotations must be measured by comparative soil and water losses, as from plots with and without rotation. A number of experimental studies show that when crops are rotated, the soil loss from a given field is less for the rotation than it would have been had the field remained continuously in a cultivated crop.

At Guthrie, Okla., records from a 3-year rotation of cotton, wheat, and sweetclover, covering two rotations or a 6-year period, show the following average annual losses of rainfall and soil:[1]

| | *Runoff, Per Cent of Precipitation* | *Average Annual Soil Loss, Tons per Acre* |
|---|---|---|
| Continuous cotton | 14.22 | 24.29 |
| Cotton in rotation | 12.72 | 14.33 |
| Wheat in rotation | 13.93 | 1.72 |
| Sweetclover in rotation | 8.00 | 0.58 |
| Average of rotation | 11.5 | 5.5 |

The plots in this experiment were on virgin Vernon fine sandy loam of 7.7 per cent slope. The results are shown in detail in Graph 21 (Chap. V, Part 1).

Similar results were obtained at La Crosse, Wis. The records from a 3-year rotation of corn, barley, and clover, expressed as average annual runoff in percentage of total rainfall and tons of soil lost annually per

[1] Soil and water conservation investigations. Red Plains Erosion Experiment Station, Guthrie, Okla., July, 1937. Soil Conservation Service (mimeographed), SCS-ESR-3.

acre, are given below. The soil was Clinton silt loam of 16 per cent slope.[1] The record covers 1933, 1934, and 1935:

| | *Runoff, Per Cent of Precipitation* | *Average Annual Soil Loss, Tons per Acre* |
|---|---|---|
| Continuous corn | 21.00 | 89.00 |
| Corn in rotation | 16.37 | 53.83 |
| Barley in rotation | 13.10 | 21.44 |
| Clover in rotation | 7.26 | 0.90 |
| Average for rotation | 12.00 | 25.00 |

It is evident that the soil loss from corn following clover in the rotation is less than where corn is grown continuously; also, that the soil loss during the entire rotation occurs mainly while the field is in a cultivated crop. This emphasizes the need for additional protective measures during the year when a cultivated crop occupies the field.

Crop Rotation Maintains Yields. Not only does crop rotation serve to reduce soil and water loss, but it also maintains crop yields. Hence, this practice is a profitable one for the farmer to adopt regardless of its value in saving soil and moisture. In the Palouse country, a rotation including sweetclover and grass produced 37 bushels of wheat per acre, as compared with 22 bushels per acre on similar land not under rotation.

At the Illinois Agricultural Experiment Station,[2] where a rotation experiment has been carried on for 38 years, the yields from corn grown continuously from a 2-year rotation of corn and oats, and from a 3-year rotation of corn, oats, and clover are given below:

Yield of Corn (in Bushels) Grown Continuously, as Compared with 2- and 3-year Rotations

| Years | Corn continuously | 2-year rotation | | 3-year rotation | | |
|---|---|---|---|---|---|---|
| | | Corn | Oats | Corn | Oats | Clover |
| 1888–1903 (16-year average) | 39.7 | 41.0 | 44.0 | 48.0 | 47.6 | 2.03 T. |
| 1915–1926 (12-year average) | 24.0 | 34.0 | 33.0 | 43.0 | 55.0 | 1.90 T. |

Using the 1915–1926 average yields given in the preceding table for calculations, 30 acres of land continuously planted to corn would produce

[1] Soil and water conservation investigations. Upper Mississippi Valley soil and water conservation experiment station, La Crosse, Wis., September, 1937. Soil Conservation Service (mimeographed), SCS-ESR-1.

[2] Hughes, H. D., and Henson, E. R. "Crop Production," p. 750. The Macmillan Company. New York. 1935.

720 bushels annually. The 2-year rotation, with 15 acres in corn and 15 acres in oats, would produce 758 bushels of corn equivalent, at a lower cost per bushel. The 3-year rotation, with 10 acres in corn, 10 acres in oats, and 10 acres in clover, would produce 1,065 bushels of corn equivalent, at a still lower cost per bushel, and would add materially to the soil-conserving value of the whole cropping system.

CROP ROTATION HELPS TO MAINTAIN THE ORGANIC MATTER AND NITROGEN OF THE SOIL. Reference already has been made to the fact that when the cultivated crop in a rotation follows the grass or legume crop, soil loss is less than when the cultivated crop is grown continuously. This fact may be related to the maintenance of soil organic matter by the sod crop. A study in Ohio[1] showed the effect of rotations on the organic matter and nitrogen in soils in comparison with continuous crop production on land limed and drained but not fertilized. The results are given below:

SOIL NITROGEN AND ORGANIC MATTER CONTENT AS INFLUENCED BY ROTATION[1]

| Cropping system | Duration of test, years | Pounds per acre in surface soil | |
|---|---|---|---|
| | | Organic matter | Nitrogen |
| Corn, continuous | 32 | 12,516[2] | 820 |
| Oats, continuous | 32 | 21,722 | 1,300 |
| Wheat, continuous | 32 | 21,826 | 1,320 |
| Corn, oats, wheat, clover, timothy rotation | 32 | 26,515 | 1,540 |
| Corn, wheat, clover rotation | 29 | 29,549 | 1,760 |
| Original soil (approximate) | .. | 36,825 | 2,240 |

[1] Ohio Experiment Station.
[2] Organic carbon times 1.724.

Incidental benefits also may accrue from crop rotation. In the section of Texas where cotton root rot is common, this disease was less prevalent where a rotation of cotton and grain was practiced than where cotton was grown continuously. Hubam clover also can be used in rotation with cotton, as it completes growth early enough to escape the root rot, gives excellent erosion control, and improves the soil.

PLANTS USED IN CROP ROTATIONS. Except in parts of the Great Plains, in the Palouse country, and occasionally elsewhere, a crop rotation ordinarily includes at least one cultivated crop. This may be corn, cotton, tobacco, or potatoes. Another will be a small grain crop, the choice of which will depend on local adaptability and economic needs. Wheat,

[1] *Ibid.*, p. 753.

barley, rye, oats, or sorghum may be used, depending on the severity of winter moisture conditions, and the individual farmer's requirements. A third crop should, wherever possible, be a biennial or perennial legume, a grass, or a mixture of grass and legume.

Under some conditions, the small grain crop may be omitted, as in northern Maine, where a common rotation is potatoes followed by crimson clover. In the Palouse, the cultivated crop is dropped, and the rotation may begin with 2 years of sweetclover and grass followed by several years of small grain. Also, in parts of Kansas the rotation may be 3 years of wheat followed by oats and sweetclover. Two-year rotations commonly consist of corn followed by small grain, with the grass or legume omitted. This obviously is not a good soil-saving rotation but may be used on level land of good productivity. Ordinarily, one crop in the rotation should be a legume or grass-legume mixture, and the longer this can be left on the land, especially on sloping areas, the better the rotation will be for preventing soil wastage.

## Strip Cropping

Water flowing down an unprotected slope gathers both volume and speed as it moves; consequently, the erosive power of runoff increases with the length of the slope over which it flows, especially where the surface is of even declivity. The practice of strip cropping, in effect, divides a long slope into a series of shorter ones. It is based on the principle that anything that checks the downhill flow of water will reduce its capacity both to pick up soil particles and to carry them in suspension. Strips of close-growing crops planted across the slope will not only decrease runoff and soil loss on that part of the field which they occupy but will act to desilt the water flowing over them from cultivated parts of the field above (Fig. 113).

Strip cropping, then, is a system under which ordinary farm crops are planted in relatively narrow strips, across the slope of the land, and so arranged that the strips of nonerosion-resistant crops are always separated by strips of close-growing, erosion-resistant crops.

In *contour strip cropping*, the strips are planted on the contour, at right angles to the natural direction of the slope. In *field strip cropping*, the strips are laid out parallel across the general slope, not necessarily on the exact contour. *Wind strip cropping* involves the planting of regular farm crops in straight parallel strips at right angles to the direction of the prevailing winds, without regard to the contour.

The most important of the strip-cropping practices in controlling erosion and conserving water is contour strip cropping.

History. Although strip cropping is a relatively new farming practice, it has been used for years in a few localities. The practice probably was brought to this country from Europe by a small group of farmers who settled in western Pennsylvania, West Virginia, and eastern Ohio. More recently, it has been widely used on demonstration projects of the Soil Conservation Service.

A study of 65 farms in Pennsylvania, West Virginia, and Ohio[1] showed that in several instances the practice had been used for 20 to 30 years, and one farmer was found who had strip-cropped his fields over a 43-year period. In some fields with slopes ranging from 5 to 40 per cent, crop yields are larger now than they were 20 years ago. Strip cropping alone

Fig. 113.—Strip-cropped field in Dallas County, Texas. Contour strips of oats 36 feet wide (light-colored bands) separate strips of cotton and corn (dark bands) 108 feet wide. (*Photograph by Soil Conservation Service.*)

has not been entirely responsible for these increased yields; other practices, such as rotations, use of fertilizers and lime, and the growing of legumes, were contributing factors. Strip cropping of these slopes was, however, the foundation practice that made continuous cultivation possible and permitted the use of other good farming methods.

Contour Strip Cropping. Contour farming is conducted on the level across the slope instead of up and down the hill in order to interrupt the flow of water downhill so that it will penetrate the soil or travel slowly across the slope along a grade approximately on the contour. Strip cropping on the contour, combined with terracing, where necessary, has

[1] Strip Cropping in Southwestern Pennsylvania. Pennsylvania Agr. Exper. Sta., Soil Cons. Service, and Bur. Agr. Econ., 1937. (Mimeographed.)

been proved the most practical means for conserving soil and water on a vast area of cultivated land (Fig. 114).

The value of strip cropping as a measure for soil conservation has been shown by results obtained between 1932 and 1936 at the Temple, Tex., soil and water conservation experiment station—on Houston clay of 3.5 per cent slope. In this test, a field contour strip-cropped with cotton and an erosion-resistant crop lost an annual average of less than 4 tons of soil per acre. From an adjoining field, cotton planted on the contour but not strip-cropped lost an annual average of nearly ten times as much soil, amounting to 39 tons per acre.

FIG. 114.—Contour strip cropping on a 37 per cent slope, Appalachian region, West Virginia. On steep slopes the clean-tilled strips should be on the exact contour and usually not more than 50 feet wide. In this instance the corn strips are protected by the intervening strips of hay crops, which effectively check erosion. (*Photograph by Soil Conservation Service.*)

Table 31 shows the soil losses from cotton grown alone on the contour and from cotton strip-cropped with soybeans on comparable areas having slopes of 5, 10, 15, and 20 per cent—at the experiment station of the Alabama Polytechnic Institute. In every instance, the loss was less from the strip-cropped areas than from the plots simply cultivated on the contour, and the benefit from strip cropping was greatest on the steepest slope.

In Table 32, the results of strip cropping obtained on an area slightly larger than an acre, at the Temple, Tex., soil and water conservation experiment station, are compared with results obtained, in one instance, from a similar adjoining area planted on the contour continuously to cotton, and from another much smaller area planted up and down the slope according to common farm practice. The erosion losses, amounting to 10.6 tons per acre annually from the strip-cropped area, have been much lower than those from either the contour-cultivated area or the one cultivated up and down the slope. The respective losses from the latter areas have been approximately 72 and 78 tons per acre annually.

TABLE 31.—SEASONAL SOIL LOSS FROM CECIL CLAY LOAM PLANTED TO COTTON AND TO COTTON STRIP-CROPPED WITH SOYBEANS[1]

| | Rain-fall, inches | Cropping system | Slope of land, per cent | | | |
|---|---|---|---|---|---|---|
| | | | 5 | 10 | 15 | 20 |
| | | | Soil losses, pounds per acre | | | |
| Early summer (growing period of soybeans) | 13.95 | Cotton alone | 6,975 | 14,139 | 51,193 | 47,348 |
| | | Cotton, soybean strips | 5,517 | 7,528 | 11,354 | 14,473 |
| Late summer and fall | 8.52 | Cotton alone | 2,480 | 2,824 | 12,105 | 14,773 |
| | | Cotton, soybean-stubble strips | 2,097 | 2,205 | 3,665 | 3,939 |
| Entire growing season | 22.47 | Cotton alone | 9,455 | 16,963 | 63,298 | 62,121 |
| | | Cotton, soybean strips | 7,614 | 9,733 | 15,019 | 18,412 |

[1] Agricultural Experiment Station, Auburn, Ala., 1932.

TABLE 32.—COMPARISON OF CROPPING PRACTICES ON HOUSTON BLACK CLAY, 4 TO 6 PER CENT SLOPE[1]

| Treatment | Size of area, acres | Period | Water loss, per cent of precipitation | Annual soil loss, tons per acre |
|---|---|---|---|---|
| Not stripped: | | | | |
| Cotton on contour | 1.39 | 1934 to 1936 | ... | 72.23 |
| Cotton up and down slope | 0.137 | 1934 to 1936 | ... | 77.86 |
| Stripped: | | | | |
| Rotation of cotton and oats, on contour (vetch preceding cotton) | 1.38 | 1934 to 1936 | 9.9 | 10.6 |
| Average not stripped | .... | ........... | ... | 75.0 |
| Average, stripped | .... | ........... | 9.9 | 10.6 |

[1] Soil and water conservation experiment station near Temple, Tex. Average annual rainfall 38.7 inches.

On the strip-cropped area, a contour rotation of cotton and oats was followed with vetch seeded as a green manure crop preceding cotton. In 1934 and 1936, cotton was grown on the lower strip; in 1935, oats were grown on this lower strip. Table 33 shows comparative results, with respect to erosion losses, as between strip cropping and the other methods of cultivation, for these two sets of years. In 1935, the strip-cropped area lost an average of 2.7 tons of soil an acre, as compared with 104.6 tons lost from the area cultivated up and down the slope. In 1934 and 1936, the corresponding losses were 14.5 and 64.5 tons per acre. The cotton on the lowest strip accounted for a large part of the loss occurring in 1934 and 1936.

TABLE 33.—COMPARISON OF CROPPING PRACTICES ON HOUSTON BLACK CLAY, 4 TO 6 PER CENT SLOPE[1]

| Treatment | Size of area, acres | Annual rainfall, inches | Period | Water loss, per cent of precipitation | Annual soil loss, tons per acre |
|---|---|---|---|---|---|
| Not stripped | .... | 34.76 | 1934 and 1936 | ... | 64.5 |
| Not stripped | .... | 46.65 | 1935 | ... | 104.6 |
| Stripped:[2] | | | | | |
| Bottom strip in cotton | 1.38 | 34.76 | 1934 and 1936 | 9.9[3] | 14.5 |
| Stripped: | | | | | |
| Bottom strip in oats | 1.38 | 46.65 | 1935 | 9.9 | 2.7 |
| Average, not stripped | .... | ..... | ............. | ... | 84.5 |
| Average, stripped | .... | ..... | ............. | 9.9 | 8.6 |

[1] Soil and water conservation experiment station near Temple, Tex. Average rainfall 38.7 inches.

[2] Same plot as in Table 32.

[3] Average for 1934–1935; no record for 1936.

Strip cropping is employed with or without terraces, depending chiefly on the slope and erodibility of the land. Where used without terraces, care must be taken to follow the contour as closely as possible. Irregularities of relief, however, sometimes make it necessary, in order to avoid too many "point" rows (short rows, which frequently involve difficulties of cultivation), to keep the rows parallel, even though this may call for some deviation from the contour. If the soil is fairly permeable and the deviation is slight and continues for only a short distance (100 feet or less), such departure from the contour is not likely to cause serious damage. No great deviation should be permitted, however, unless provision is made for grassed waterways to conduct the accumulated water through the succeeding strips to the foot of the slope. Where conservation of rainfall is an important objective, contour operations should be maintained as nearly on the contour as may be practicable.

Laying out a strip-cropping system in fields that have been level-terraced is relatively simple because the contours are permanently marked by the terraces. The width of the strips can be adjusted easily to the terrace interval.

Strip cropping usually involves the growing of at least one row crop in every field, this in alternation with at least one erosion-resistant or semi-erosion-resistant, close-growing crop, such as a small grain, an annual

FIG. 115.—Even-width strips of corn planted on the terrace intervals, with small grain or hay in the irregular intervening terrace strips. The next time corn is planted in the rotation it will occupy the terrace ridges with the irregular strips of small grain or hay on the terrace interval. (*Photograph by Soil Conservation Service.*)

legume, or meadow. Two or more such crops may be used in one field at the same time, as corn, grain, and clover or meadow. All these crops may be rotated in the same field, thus combining the advantages of contour cultivation, strip cropping, and crop rotation.

TECHNIQUE. The establishment of a strip-cropping system for an entire farm may or may not necessitate complete rearrangement of the farm plan. Where this is unavoidable, proper adjustments may call for such changes as relocation of fences, retirement of the steeper lands to trees or permanent grass, the bringing into cultivation of level pasture land, the use of shorter rotations on land least subject to erosion and longer rotations on the steeper, more erodible lands. In planning such a strip-cropping system, careful consideration should be given to such physical conditions as variation in relief, runoff concentration points, size and nature of the drainage area above each field, length of slope, soil

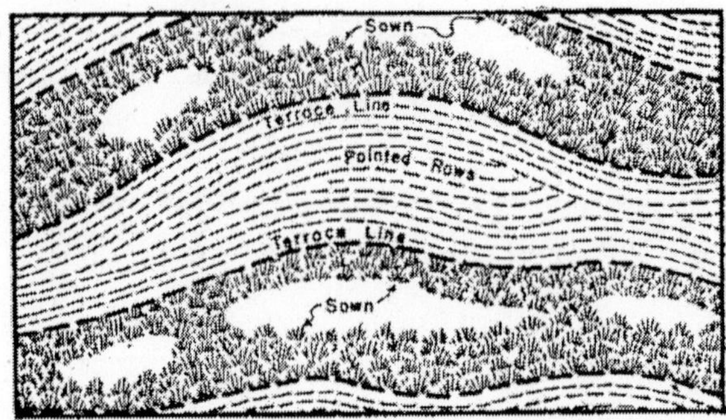


GRAPH 25.—Strips laid out on the terrace intervals with terrace ridges forming the boundaries of the strips. (*Soil Conservation Service.*)

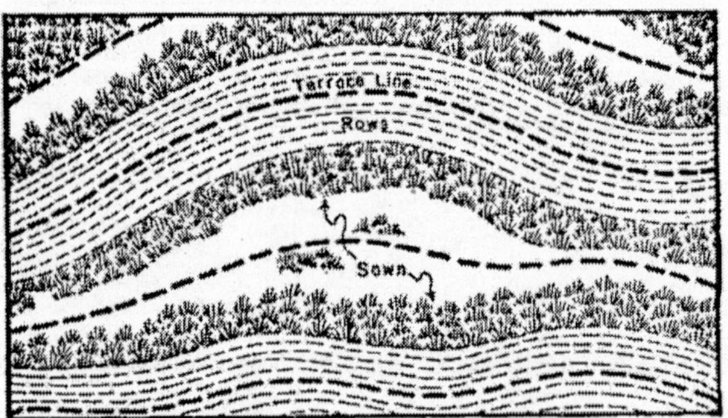


GRAPH 26.—Strips laid out to straddle the terrace ridges. This is a very practical method and is easy to follow with a rotation of crops. (*Soil Conservation Service.*)

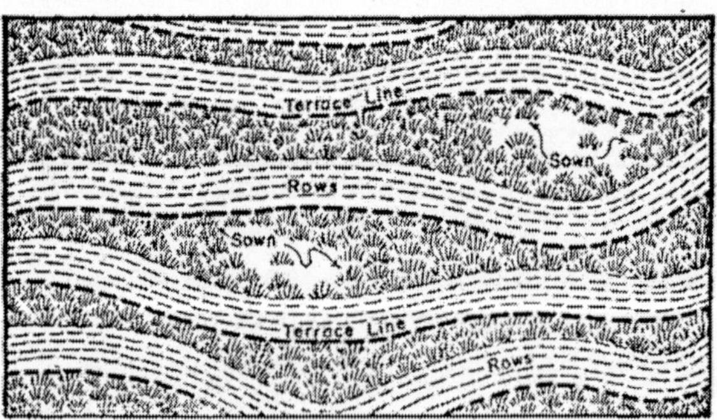


GRAPH 27.—Strips laid out on either side of the terrace ridge, occupying only part of the terrace interval. (*Soil Conservation Service.*)

type, condition of the soil, crops grown in the rotations, and the kind of farm machinery employed.

The line to be used as a base in laying off strips should be on the contour and so located that the maximum number of strips can be measured from it both up and down the slope. Such base line may mark either the upper or the lower boundary of one of the strips, or it may be used for the center line. If the surface is very uneven, a single base line usually will serve for the laying of only two or three strips. Under such conditions, additional base lines will be necessary where more strips are needed. This often will necessitate the use of a *correction strip* between the last two even-width or *regular cropping strips*. Such strips are of irregular width and so placed between contour plantings of cultivated crops as to permit the cultivated strips to be of uniform width (Fig. 115).

In establishing a cultivated strip upward from the base line, the rows are laid out parallel to the base line until the point of maximum allowable

FIG. 116.—Small grain seeded on terrace ridges with corn on the terrace intervals. Close-growing, erosion-resistant crops are frequently seeded to reinforce newly constructed terraces. After the terraces settle the strips can be made wider to facilitate rotating the crops. (*Photograph by Soil Conservation Service.*)

departure from the true contour is reached. From this point, it becomes necessary to go farther up the slope to the next established base line and lay off strips parallel to the downward side until the maximum deviation from the true contour is again reached.

LOCATING STRIPS ON TERRACED FIELDS. Several methods are used in locating strips on terraced fields. They can be placed on the terrace interval, extending from terrace ridge to terrace ridge (Graph 25); or they can be located so as to straddle the terraces with approximately half of the strip above and half below the ridge (Graph 26). These basic procedures may be modified to suit special soil conditions or unusual crop requirements. For example, a portion of the terrace interval may be seeded to erosion-resistant crops and part to row crops (Graph 27); or the terrace ridge may be seeded to a close-growing crop which will serve to strengthen the structure (Fig. 116). Where the variety of crops grown in rotation permits, row crops can be planted to straddle alternate terrace ridges, and the rows extended above and below the ridge to a uniform distance so that all rows will be parallel. The intervening terraces and adjacent areas can then be planted to close-growing crops in such ways that all of the irregular areas will be covered with a protective cover. This makes it possible to eliminate all point rows. Where these are not objectionable, and the row crops are planted over the terrace interval, rows on each side of the strip should be parallel to the terraces so that resultant point rows will be in the middle of the terrace interval.

Width of Strips. In terraced fields, the width of strips is usually adjusted to the terrace interval; but on unterraced land, the strips frequently should be narrower than the standard terrace interval. Here, the width will be governed by the length and steepness of the slope and by the permeability of the soil type, since all these factors influence the amount of runoff. As a general rule, the steeper the slope the narrower the strips. On long slopes, strips nearer the lower part of the slope should be narrower than those at the top, because of runoff accumulation from above.

Strips should seldom be wider than 200 feet to meet the requirements of soil conservation; but for practical farming, they can seldom be narrower than 50 feet. The width frequently can be adjusted advantageously to an even number of rows or to an even multiple of the width of the machinery that will be used in farming the strips. In determining the optimum width of strips, good judgment should be used to the extent of making slight adjustments, where possible, so that the width of the strips will fit the farming implements, particularly planters, drills, and harvesters.[1]

Rotation of Strips. Crop rotations can be used as effectively on strips as on large fields. It is practicable to have approximately the same area of tilled crops, small grains, hay crops, and grass under a strip-cropping system as under the regular field system. The only rearrangement is within the field (Fig. 117).

Maintenance of a good rotation is one of the principal features of the best type of strip-cropping system. The crop sequence within or between the strips themselves is especially important. No two cultivated strips, as well as no two strips having the same planting or harvesting dates, should be adjacent. By rotating the strips, the clean-tilled row crops usually can be made to follow crops like grass or legumes that have a dense-fibrous root system, which tends to hold the soil and retard erosion

[1] For discussion of formulas for determining the width of strips, see Kell, Walter V., Strip Cropping for Soil Conservation, U. S. Dept. Agr. *Farmers' Bull.* 1776, 1937.

until the roots are broken down by decay. In this way, the organic-matter content of the soil is increased or maintained, and its physical condition improved, so that the land absorbs water more readily and has increased capacity to resist erosion. The method of rotating strips is illustrated in the diagram (Graph 28).

FIG. 117.—Upper, terraced field, farmed with 5-year rotation on contour strip-cropped basis—alfalfa, oats, alfalfa, corn, alfalfa—1936; lower, same field 1937, with strips moved up slope. Upper barley strip follows oats because of failure of alfalfa seeding previous year. Note narrow strip of alfalfa was left below corn as buffer strip. Clinton silt loam, slope range about 6 to 12 per cent. Vernon County, Wisconsin. (*Photograph by Soil Conservation Service.*)

PERMANENT OR ANNUAL STRIPS NOT ROTATED. In some parts of the country, it is occasionally advisable to establish either permanent or temporary strips which are not a part of the rotation. These are made use of generally in the form of buffer strips of perennial legumes, shrubs, grass, a legume-grass mixture, Sudan grass, grain sorghums, or, in some instances, small grain. The row crop or cultivated strips under this system may or may not be rotated. In many localities, it is entirely practicable to establish permanent strips in connection with either field stripping or contour stripping on irregular areas, in the correction strips, or on critical slopes or other odd corners or portions of the field that are difficult to maintain in the regular rotation strip-cropping program. Frequently, these permanent strips can be devoted to the production of black locusts for fence posts or planted to kudzu or perennial lespedeza for forage. Incidentally, all these plants provide food and cover for wildlife.

Buffer or Spreader Strips. A buffer strip is a more or less permanent contour strip, usually of variable width, planted to grass or other erosion-resistant vegetation which is not a part of the regular farm rota-

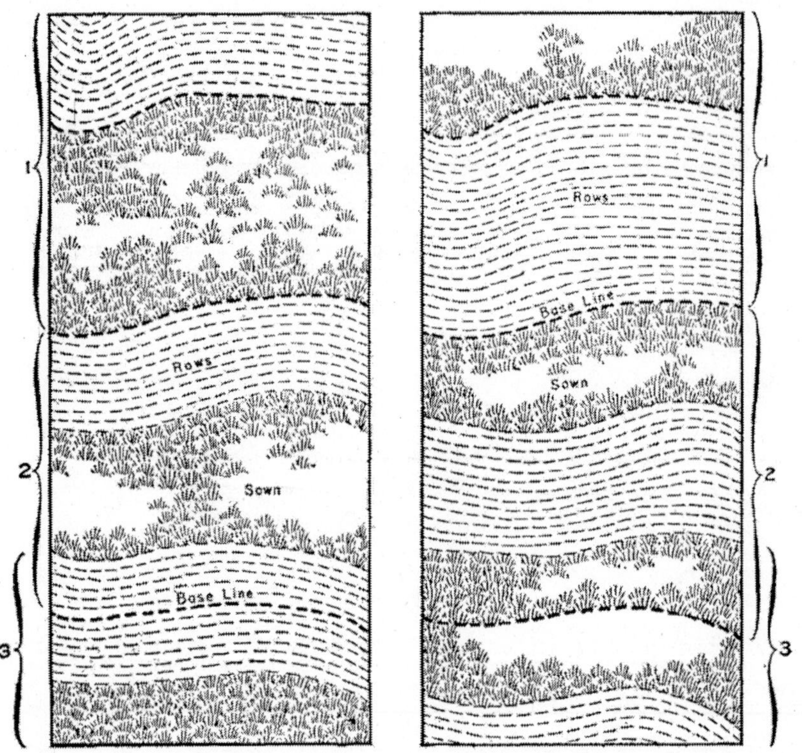


Graph 28.—This diagram illustrates how crops can be rotated in a strip-crop system.

tion, and which may or may not be harvested. The area planted to erosion-resistant crops under a buffer-strip system is much smaller than under a contour strip-cropping system; the buffer strips usually vary between 8 and 20 feet in width.

Buffer strips often are used on truck farms where land is valuable and where erosion-resistant crops have no sale value and are not needed for farm use or in orchards either in the tree rows or in the alternate spaces between rows. If the tree row itself is left in sod, and one-way cultivation is practiced, in time a benchlike terrace will be built up favoring level cultivation.

In rotated cultivated fields, a buffer system has little practical value. However, where erosion-resisting crops do not occupy a sufficient part of the field to allow strip cropping, buffer stripping offers the next best method of vegetative erosion control.

The most effective use of buffer strips is on critical slopes and other vulnerable parts of cultivated fields, which cannot be controlled by annual crops or by ordinary strip-cropping methods.

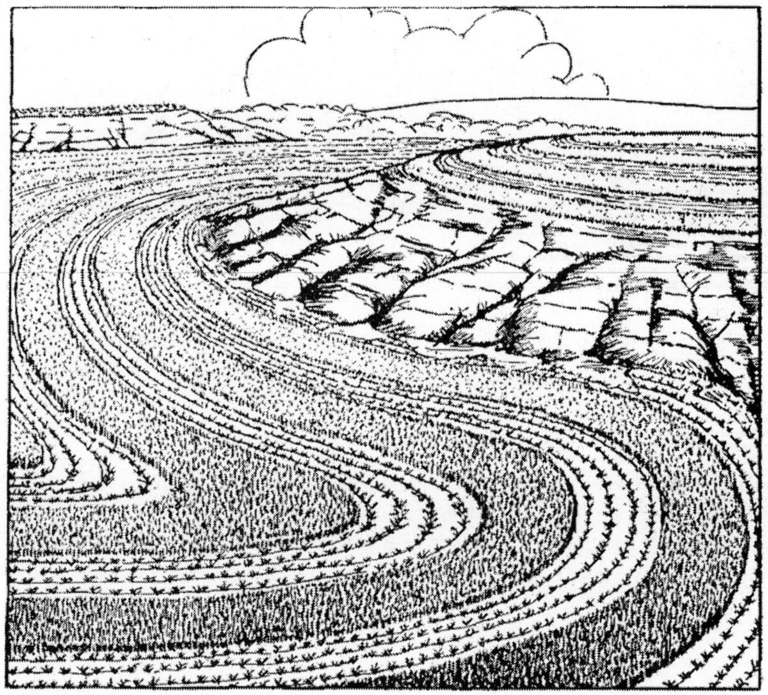

GRAPH 29.—Critical slopes, where erosion progresses more rapidly than on lesser slopes, should be taken out of cultivation and seeded to perennial forage crops before erosion destroys the soil above the critical point or damages by sedimentary deposits the productive land below.

TURN AND BORDER STRIPS. In some fields, it is necessary to run strips around points or narrow ridges. Here it will help to keep the rows an equal distance apart if at the apex of such an area a small section across the strip is left in sod the year the strip is in a cultivated crop. When such a strip is in small grain or hay, this turning area can be plowed with the rest of the strip. Moreover, a *headland* or *border strip* of this kind will make an excellent roadway from one strip to another, as well as a good terrace or diversion-ditch outlet or grassed waterway.

CRITICAL SLOPES. Where a field is broken by areas having slopes of 15 to 20 per cent or more, strip cropping, either alone or in combination with other soil-conserving practices, must be supplemented by the use of special types of "heavy-duty" perennial vegetation on these steep portions of the field. *Critical slopes* (Graph 29) may extend as fairly regular bands across the entire field, or they may extend only for a short distance along some surface irregularity. Where such slopes occur in cultivated fields, the vegetation generally used should be a perennial forage crop rather than trees or other woody vegetation. Such plantings of heavy-duty perennial vegetation will provide year-round protection for those field inclusions where erosion is most critical, and when well established they may be utilized for hay.

If, however, these slopes are continued under a system of row-crop farming, it becomes almost impossible to control erosion anywhere in the field. Soil washed from such critical areas clogs terrace channels and contoured rows below, causing breaks and permitting the formation of gullies. Poor subsoil material, sand, and gravel derived from these areas often damage lower lands over which they are spread.

Unfortunately, adaptable perennial forage plants are not available for all parts of the country. Over much of the Southeast, perennial lespedeza and kudzu give good results in the treatment of these critical areas. In the Northeast and in the Corn Belt, a combination of grass and legume mixtures, such as alfalfa with smooth brome grass, timothy, orchard grass, or bluegrass, makes good permanent cover for land of this nature. Grass or legume-grass mixtures may be used effectively in many parts of the West.

Where it is necessary to extend terraces across the acutely erodible areas, it may sometimes be advisable to plant only part of one or more terrace intervals with a permanent cover crop, particularly where the slope breaks sharply. The remainder of the field could then be used for annual crops.

Field Strips. Field strip cropping is a modified form of contour strip cropping, under which the strips are laid out parallel and across the general slope but not exactly on the contour. This system is best suited to land of uniform slopes and to gentle slopes with minor surface irregularities that make accurate contour stripping impracticable. Where field strips cross depressions, the rows depart from the contour to such an extent that water usually accumulates in the low places. To avoid such dangerous accumulation, grassed waterways should be established and maintained in the depressions. If the strips deviate from the contour for only short distances—not more than 100 feet—little damage may result, especially on soils of low erodibility. However, if the deviation should be more than 3 per cent, and the distance of such deviation greater than 100 feet, field stripping should give way to contour stripping.

### Adaptation of Strip Cropping to Major Erosion Regions

The principle of strip cropping has wide application. It can be used in one form or another in the Corn Belt, the Cotton Belt, and the wheat lands of the Great Plains and in many specialized crop areas of the country. Its practical application necessarily will vary according to the nature of the problem and the types of vegetation applicable and available to the region.

The Corn Belt. The Corn Belt is especially adapted to strip cropping by reason of prevailing conditions of soil and topography and the large number of crops besides corn that can be grown successfully. Small grains, such as rye, wheat, oats, and barley, are all available; hay crops, such as red clover, sweetclover, alfalfa, alsike, timothy, brome grass, and mixtures of legumes and grass, as well as many of the pasture grasses, are generally adaptable to the region. With a 4-year rotation of corn, small grain one or two years, and hay one or two years, all the cultivated fields can be laid out in strips to provide the necessary acreage of corn while other portions of the same field are being devoted to the production of small grain or hay. Here both the small grains and the hay crops would constitute the protection strips for the corn. Usually, it is an advantage to have only two crops (for example, corn and hay or small grain and hay) in one field, so that after harvest the crop residue can be pastured if necessary. Fields containing strips of small grain can be pastured after the grain and hay are harvested, whereas fields containing corn strips can be pastured later in the season after the corn is gathered. Thus, the fields can be pastured in rotation.

In the Corn Belt, Kentucky bluegrass, Canada bluegrass, alfalfa or a mixture of alfalfa with timothy or smooth brome grass, sweetclover, red clover, and clover-grass mixtures are among the most erosion-resistant types of vegetation readily available for use. Reasonably early fall seedings of rye and wheat are also very good for erosion control in winter months. The spring seeded crops, such as oats and barley, are less desirable but much better than soybeans, cowpeas, and buckwheat, which have a loosening effect on the soil and usually should be classed as erosion-permitting crops.

The Cotton Belt. The Cotton Belt presents a somewhat more complex problem than the Corn Belt in that much of this area is older farming country, and erosion has taken a greater toll of the productive topsoil, especially with respect to relative area.

Cotton is likely to continue as the major crop around which the strip-cropping system must develop, although corn will need to be dealt with on most farms, and tobacco on many of those within certain districts, as in the Bright Tobacco sections of the Carolinas and Georgia.

The effectiveness of strip cropping on cotton farms will depend to some extent on a reduction of the cotton acreage and on an increase in the acreage of erosion-resistant crops, such as grass, legumes, small grains, and green manure crops, grown in rotation with cotton. Some of the crops that can be used, where adaptable, in rotation with cotton and in alternate strips are: hairy vetch, rye, a mixture of rye and vetch, fall-seeded oats, Austrian winter peas, crimson clover, corn and soybeans, or corn and velvet beans, sweetclover, annual lespedeza, crotalaria, rye grass, sericea lespedeza, and kudzu. For the maintenance of organic matter in the soil, perennials are most desirable. Another consideration is to select plants that mature early enough to be followed by a hay, seasonal cover, or green-manuring crop, thus making it possible to produce two crops in one year on the protection strips and so provide maximum protection.

On land of irregular relief, annual vegetation that provides ample protection for gentle slopes is sometimes inadequate to hold the soil on the steeper areas, particularly those that have lost all topsoil and now present to the lash of rain a more erodible subsoil. Where these steeper lands occur in fields, it usually is desirable to protect them with permanent vegetation, such as sericea lespedeza, kudzu, or mixtures of grass and legumes, all of which can be maintained as permanent hay strips and made a part of the strip-cropping system. In some instances, these critical areas are so rough and badly eroded as to be unfit for hay production. They may, however, be planted to low-growing shrubs or other plants that will provide food and shelter for wildlife. This type of plant growth may have a slightly unfavorable border effect (competition for moisture and sunlight), but generally it will not be so objectionable with respect to adjacent cropland as trees.

Because of the long seasons and open winters in the Cotton Belt, all clean-tilled strips should, where possible, be protected with winter cover crops to retard erosion, prevent leaching, and add organic matter to the soil. Where this cannot be done, the stubble or residues of summer crops should be preserved to the fullest possible extent.

THE GREAT PLAINS. Strip cropping in the Great Plains is practical for controlling or retarding both water and wind erosion and for the conservation of moisture. Here the problem differs from that in humid farming areas in that very little runoff occurs during the summer growing season on some of the more permeable cultivated soils. Runoff may occur, however, during winter and spring thaws, which frequently melt the snow while the ground remains frozen and unabsorptive. Some runoff occurs, of course, even on permeable soils during exceptional downpours.

To prevent the loss of needed moisture, mechanical barriers, such as terraces, contour furrows, and contour strips of close-growing vegetation, are necessary to hold the water. Two types of strip cropping are commonly used: contour strip cropping and *wind strip cropping.* Where runoff occurs, contour strips are preferable on soils not particularly subject to drifting. Wind strips are sometimes used in level or nearly level fields susceptible to drifting, the strips being placed crosswise the direction of the prevailing wind, regardless of the contour. The function of the wind-resisting strips is to minimize soil drifting by retarding and deflecting ground air currents. For this purpose, a type of vegetation is needed that makes an erect growth, withstands air movement, and resists deterioration for as long a time as possible. Wind-resisting strips may be used in the form of narrow buffers composed of 4 to 10 rows placed at 5- to 10-rod

FIG. 118.—Wind buffer strips of Sudan grass, placed on the contour in a cotton field in Lamb County, Texas. (*Photograph by Soil Conservation Service.*)

intervals across fields, or they may be used as wide rotation strips. The effectiveness will depend on the density, width, and distribution over the exposed areas.

Since strip cropping reduces the expanse of individual clean-tilled areas, it is one of the most effective controls for wind erosion now commonly employed. It has been extensively used in western Canada as well as in the United States. Because of better moisture conservation, contour strip cropping is rapidly replacing the older method of wind stripping.

Crops adapted to wind stripping in the Great Plains are limited. Sudan grass is probably the best, because of its rapid and dense growth and its wide adaptability (Fig. 118). The grain sorghums also are very effective and, except in the northern part of the Plains, are widely adaptable. Broomcorn gives good results over a wide area.

One of the grain sorghums is usually the first crop grown to stabilize land that is out of control and drifting badly. It is generally seeded over entire fields for a year or two, until the soil is stabilized. Subsequently, it is seeded in alternate strips to protect strips of other crops, such as wheat, corn, beans, and potatoes, or strips of fallow land. In some of the extensive wheat-producing areas of those regions where rainfall is fairly dependable, alternate strips of wheat and fallow are combined with proper utilization of crop residue (leaving the trash, or part of it, on the surface) to control wind erosion. This practice has been demonstrated as practical and effective; it is much safer than plowing and cultivating, at one time, large unprotected areas subject to drifting.

Both wind and contour strip cropping catch and hold snow which would otherwise drift into large accumulations, where the concentrated moisture would be of little economic value. By deflecting ground air currents, the tall-growing strips also probably help reduce both evaporation and transpiration from the intervening areas.

Furthermore, strip cropping affords an opportunity to maintain a protective crop stubble and residue on a portion of the land until it is discovered whether or not soil moisture conditions are favorable for the establishment of a subsequent crop. In emergencies, when crops fail because of inadequate moisture, the clean-tilled barren areas are still protected by the undisturbed strips of residual vegetation. By means of this constant protection, the soil is held in fields that otherwise become serious blow hazards.

THE NORTHWEST WHEAT AREA. Strip cropping is showing practical possibilities for effective erosion control on the long steep slopes typical of numerous wheat farms in the Pacific Northwest. Because of frequent steep declivity, the heavy farm machinery used in wheat production must be operated across, rather than up and down, the slope; but until recently, little or nothing was done to keep these operations exactly on the contour. Where large fields of wheat were to be harvested with combines, they usually were laid out in "lands" across the slope. This practice necessitated the disadvantageous procedure of driving through the wheat in opening these lands. Contour strip cropping has proved an aid to these mechanical operations and has provided needed protection against concentration of runoff in low areas and consequent cutting of gullies impassable to farm machinery.

Sweetclover, or sweetclover and grass, in alternate rotation strips with wheat has given good results. Crop rotations, rough tillage, and proper utilization of crop residues are also dependable allies of strip cropping.

Grass buffer strips may be used effectively to define the contour lines permanently and to reinforce the rotation strips by functioning as water spreaders. On irregular slopes where the strips deviate from the contour and cause water to concentrate in the lower positions, permanent grassed waterways offer a good method for conducting the runoff safely across the strips to the foot of the slope. Some of the better plants for this purpose are brome grass, crested wheat grass, and slender wheat grass.

The large acreage of spring wheat produced in the Pacific Northwest necessitates fall plowing, which leaves the soil bare over winter. Protection strips at frequent intervals down the slopes will prevent enormous losses of productive topsoil, which otherwise would occur on these freshly plowed slopes. Peas and beans are other crops that mature early and leave the soil highly susceptible to erosion unless the field is protected with thick-growing contour strips.

## Vegetative Control of Gullies

Gullies tend to form wherever runoff is concentrated in unprotected depression channelways. Once started, they increase in size and, unless checked, ultimately develop into deep, spreading chasms which ruin the land. The best method of gully prevention is by providing permanent cover for critical slopes, by strip cropping, contour cultivation, crop rotation, water diversion, and other methods that check runoff and prevent the concentration of water.

Where gullies have developed to considerable size, special methods must be used to control their growth and, where possible, to reclaim the gullied area for agricultural use. For this purpose, vegetation is most economical and satisfactory, although in some instances, especially in large gullies, the use of close-growing plants may need to be supplemented with mechanical measures. Even in small gullies, brush, straw, log, loose rock, or sod dams may be helpful in checking runoff and collecting silt favorable to plant growth while the protective vegetation is becoming established.

CLASSES OF GULLIES. Gullies may be conveniently classified according to size as follows: small gullies, or those that can still be readily crossed with farm implements; medium-sized gullies, or those that cannot be readily crossed with farm implements but have not yet attained the size of ravines; and large gullies or ravines, barrancas, and broad, but not necessarily deep, "washes."

*Small Gullies.* The growth of small gullies may be readily checked by plowing them in and smoothing off the surface of the field, followed by the installation of a proper contour strip-cropping rotation where the slope is gentle, or by retiring the field to grass or other permanent vegetation, such as alfalfa and lespedeza. Kudzu may be used to advantage for control of somewhat larger gullies (Fig. 119). Seeding such areas to orchard grass with annual lespedeza or to brome grass will also check the erosion and aid in reclaiming gullied fields. It is especially important to control gullies while still small, since at this stage the field can be saved without material interference with regular crop production.

FIG. 119.—Upper, a typical *critical slope* (steep and highly erodible) which was cultivated until completely stripped of its productive topsoil;—it has been terraced and plowed for planting to a permanent protective cover. Below, a similar slope adequately protected with one year's growth of kudzu, which will produce excellent hay. Coastal plain of southern Alabama. (*Photographs by Soil Conservation Service.*)

*Medium-sized Gullies.* When gullies have attained medium size, more drastic measures are usually necessary to check their growth. The first step generally will be to control runoff and erosion higher up the slope and immediately about the gully heads. For this purpose, diversion waterways frequently can be used to great advantage. Adequate erosion-control measures, however, must be used to protect the entire field in which the gullies occur or the entire watershed feeding into them. If the gullying has spread over a large area, terracing is frequently necessary, since other contour measures are not so likely to overcome the difficulties. In some instances, it may be desirable partly to fill a gully by plowing along its sides so as to turn soil into the bottom or by grading down the sides with a bulldozer. This results in a gentler slope on which grass or other protective plants may be seeded. Where such work is done in summer, sorghum or

Fig. 120.—A diversion ditch around the head of a gully. The gully is being fenced and planted to trees and grass. The ditch drains both ways. (*Photograph by Soil Conservation Service.*)

Sudan grass may be seeded as a temporary cover. When the work is done in the fall, a grain should be used, both to protect the soil and to serve as a nurse crop for the grass seeding. Sometimes, temporary dams may be needed in gully channels to reduce the erosive force of running water until vegetation is established (Fig. 120).

Where bluegrass is adaptable, it is particularly effective in the stabilization of the banks of fairly deep and troublesome gullies. Sod of this grass may be used also to control overfalls in gullies, if the drainage area and overfall are not too great. It is especially adapted to pasture gullies where the overfall is less than 4 feet and the drainage area not more than 4 or 5 acres.[1] In the South, Bermuda grass can be used in this same way. Often old feed or fertilizer bags filled with strips of sod can be

[1] For full discussion of bluegrass sod in gully control, see U. S. Dept. Agr. *Farmers' Bull.* 1760.

used to establish grass dams or barriers across small washes. If the bags are filled with good soil, and a handful of seed placed in the upper part

Fig. 121.—Upper, low wire check dams were used to prevent this gully from increasing in depth. Lower, same gully as above after check dam had silted full and bank had been sloped. Bermuda grass has been sodded to keep this waterway stable and prevent further erosion. (*Photographs by Soil Conservation Service.*)

of the bags before placing them across the bottom of the gully, the grass will grow through the meshes of the bag and soon make a firm vegetative dam. Soil deposited behind this obstruction will tend to fill up the gully and reduce the channel grade.

For full control, the sides of the gully must be covered completely with vegetation. This must be protected from livestock, especially during the early stages of growth, by fencing or other means. In the South, kudzu or Bermuda grass, and in the North, bluegrass or grass and legume mixtures, are recommended for this purpose. If the slope is not too steep, the lespedezas also will give control in those sections where they grow well. Where gullies have attained considerable size, vegetation should be started in the bottoms. If the drainage area is large, mechanical checks, such as woven wire, brush, or rock, may be needed to prevent further cutting and to flatten out the bottom by collection of silt. In these silt deposits, kudzu can be planted later.

Gullies of medium size often can be utilized for vegetated waterways. Water must first be diverted, however, the sides smoothed by plowing, and suitable grasses or grass-legume mixtures seeded (Fig. 121).

Although it is desirable to treat small and medium-sized gullies wherever practicable to make them productive for meadow or pasture, many gullies of less than ravine size can be stabilized only by planting trees, shrubs, or vines.

*Large Gullies.* Large gullies can be defined roughly as those which have gone beyond the stage where reclamation for cropping or meadow is practicable. In many instances, the soil has been so deeply eroded that the sides of the gullies and intervening ridges consist of essentially sterile material. For the control of such ravines, tree planting is often the only recourse, although vines, such as kudzu and Japanese honeysuckle, are useful. The tree most commonly used for gully control in eastern United States is the black locust. Another tree useful in the South is the shortleaf pine (*Pinus echinata, Mill*), which is often the only tree that will grow on the raw subsoil material. The loblolly pine (*P. taeda, L*) can also be used, but it is more exacting in its soil requirements.

Cottonwood and willow cuttings can be planted advantageously in closely spaced rows at right angles to the direction of flow. Helpful brush dams may be anchored by pegging them down with stakes of green cottonwood or willow. These will take root and withstand considerable covering with moving soil material while aiding to increase the effective height of the dam.[1]

On the more gentle slopes, plantings of black locust should ultimately be followed by grasses that will grow in light shade and benefit from the nitrogen fixed by the nodules on the black-locust roots. Among the various native shrubs and small trees that can be used effectively are: wild plum, snowberry or buckbrush, sassafras, persimmon, dogwood, bearberry, and

[1] For more detailed discussion, see U. S. Dept. Agr. *Farmers' Bull.* 1697.

blackberry. Such plants also serve to convert large gullies into excellent wildlife refuges. Vines like kudzu and honeysuckle generally should not be used where trees are wanted, since they tend to smother or impede the growth of young trees.

In the West, many deep gullies may be stabilized with vegetation. Usually, some mechanical assistance in the way of check dams will be needed. Several plants have been found to grow well in the bottoms and on the sides of such gullies. Among them are tobosa grass, bearberry, and Australian salt bush. One of the most useful plants to establish at the base of gullies and above check dams is the sand-bar willow (*Salix exigua*). This tree is especially suited for planting along washes subject to severe floods, since it sprouts from the roots and forms dense thickets. Another small tree is the wild olive (*Forestiera neomexicana*) which, because of its layering habit, does well on gully banks. The desert willow (*Chiolepsis linearis*), which sprouts from the stump, is also useful for gully control in the Western States.

A study of Western plants has brought to light a number of species that are promising for stabilizing gullies. Some are suited to one section; some to another. Before large gullies can be effectively stabilized, however, measures must be taken to control runoff and erosion over the entire drainage area. So long as recurring torrents are permitted to rush down these arroyos, almost any control measures will remain ineffectual; with adequate protection over the drainage area, however, many may be stabilized by relatively simple vegetative means.

## Vegetated Waterways

Waterways are natural or artificial channels used to carry runoff from fields or pastures to lower levels of land. They consist of relatively narrow channels laid out to meet special conditions and broad natural depressions. In any case, they must be protected by vegetation or mechanical measures, or both, to prevent conversion into gullies. Natural drainage depressions, seeded with an adaptable grass or grass-legume mixture, often can be utilized as satisfactory water channels and meadows at the same time. Such areas frequently supply especially good crops of hay because of the increased water supply (Fig. 122). The narrower waterways should be 20 to 100 feet wide, depending on the amount of water to be carried and the steepness of the slopes. Seedings should not only cover the channel but extend well back on the shoulders to prevent scouring and undercutting (Fig. 123).

Some artificial waterways must be protected with concrete or wooden baffles and strips of sod. For this purpose, Bermuda grass is well adapted to Southern conditions, and bluegrass to the North. Where no baffles are used, sod must be established solidly over the bed of the channel, particularly where the slope exceeds 15 or 20 per cent. Where the slope is gentler, the sod may be planted in strips of varying width, depending on the slope

FIG. 122.—A vegetated waterway that not only conducts surplus water from the field safely but pays its own maintenance cost by producing a good yield of hay. (*Photograph by Soil Conservation Service.*)

FIG. 123.—Grassed waterway in corn strips. Strip cropping in the background. (*Photograph by Soil Conservation Service.*)

and erodibility of the soil. Solid sodding is more expensive than strip sodding but is much more satisfactory, especially if the waterway must be used immediately or carries large amounts of water.

Terrace outlets call for special stabilization care. For best results, they should be constructed and seeded at least a year before they are to carry water. If the outlet is to be used immediately, the seeding should be reinforced by sod strips and sometimes by mechanical means. Control by vegetation alone should be limited to slopes of not more than 10 to 12 per cent, with a maximum drainage area of 5 or 6 acres. Western wheat grass (*Agropyron smithii*) and range mesquite (*Panicum obtusum*) have given excellent results in stabilizing terrace-outlet channels and waterways in many parts of the West. In the South, Bermuda grass is used especially where the channel is relatively narrow and occasionally carries large quantities of water. In the wider meadow outlets, Bermuda grass, grass-legume mixtures, kudzu, or sericea lespedeza may be used. In the North and Northeast, Kentucky bluegrass is recommended for the narrower channels; in pastures and for meadow outlets, a standard grass-legume mixture is generally most desirable.[1] In the drier areas of the North, awnless brome grass gives good results.

## Seasonal Cover Crops

Seasonal cover crops are very important in protecting fields from erosion. They also serve to absorb plant nutrients that otherwise would be lost by leaching and thus aid in their preservation for subsequent crop use.

The use of the term *cover crops* has become almost synonymous with *green manure crops;* but from the standpoint of soil conservation, there is a significant difference. Crops may, and do, serve as cover crops that are not used as green manure crops. Green manure crops, however, are always of necessity cover crops.

If it were economically possible to grow only soil-protecting crops, other measures for preventing erosion would be less necessary. As it is, however, cultivated crops must be grown, so it is important to make the widest possible use of cover crops—grass, where its use is feasible; green manure crops; grains for winter and spring cover; and *catch crops* of various kinds—in order to leave the soil exposed to erosion for a minimum period during the year. The purpose in growing a seasonal cover crop is, therefore, to protect the soil when the regular cultivated crops are off the ground.

Examples of the soil-saving effect of a dense cover have been given in preceding pages. Although specific data on the effect of crops grown especially for seasonal cover are not available, there is no doubt that their effect in reducing runoff and soil loss is of the same order as for any other comparable dense cover.

[1] For seed mixtures for different localities, see Pasture Handbook, U. S. Dept. Agr. *Misc. Pub.* 174.

Cover crops, as noted before, are helpful in conserving those soluble plant nutrients subject to loss by leaching. This process goes on wherever water percolates through a soil containing soluble plant food, especially nitrates. Under mild temperatures, nitrates are being continually formed by the decomposition of organic matter. This process goes on day and night, even in winter, in localities where the ground does not freeze. Unless plants are ready to absorb these nitrates, they are carried down by percolating water, beyond the reach of roots. Lysimeter experiments have shown that seventeen times as much nitrogen may be lost by leaching from bare ground as from cropped ground. In Texas, these losses have been shown to range from 67 to 168 pounds of nitrogen per acre per year, which is equivalent to an application of 400 to 1,000 pounds of 16 per cent nitrate of soda per acre.

Barring the application of nitrogenous fertilizers, the organic matter of soils is the source of nearly all the nitrates used by nonleguminous crop plants. The loss of this organic matter by decay cannot be avoided, but it is important that the products of such decay should not be allowed to go to waste. The nitrates produced can be largely absorbed by plant roots and removed from the path of percolating waters if the soil is covered at all times by a growing crop capable of curbing erosion.

Protection of the soil from erosion and the absorption of nitrates that would otherwise be wasted are not the only roles played by cover crops. They also play an important part in soil economy by adding organic matter and, in the instance of leguminous cover crops, soil nitrogen. This is true in part, whether the cover crop is finally turned under or not. Most of the organic matter in soils is derived from the decay of roots.

Experiments at the New Jersey Agricultural Experiment Station have shown that when land, of the kind tested, is left in grass for two years, the amount of organic matter in the soil is actually increased by about 1,300 pounds per acre. Grass is not only the best possible cover from the erosion standpoint but one of the best crops where an increase of soil organic matter is desired. Not only is organic matter a storehouse of nitrogen, but its mechanical effect on the soil is highly favorable. A soil rich in organic matter absorbs rainwater more readily than one with a deficiency of organic matter.

*Crops Used.* The choice of a cover crop will depend on local conditions, costs, use to be made of the crop, and other factors. These crops may be used for green manuring purposes or merely for cover and to absorb nitrates or as catch crops to fill in between the harvesting of one principal crop and the planting of another. Such crops may be used for winter or for summer cover. In some instances, the protective effect can be extended by plowing only part of the field, when preparing land for a subsequent crop. For example, crimson or bur clover preceding cotton may be plowed on the balk system, only enough furrows being run to seed the cotton. Subsequently, the intervening strips can be plowed out as cultivation progresses. This not only extends the time during which the cover crop furnishes protection but provides for reseeding of the clover, particularly bur clover.

Some summer cover crops may be left on the ground through winter, thus providing, even though dead, a measure of soil protection. For example, crotalaria, soybeans, and cowpeas, grown to improve the soil, should not be plowed under in the fall unless a winter cover is to be seeded. If turned under in fall without a winter crop following, decay will result in the loss of much of the nitrogen, and the bare soil will be exposed to erosion.

Technique. The seeding and use of cover crops should be managed as a part of the regular farm operations. Grass or legume-grass mixtures are seeded in the regular farm rotation in accordance with local requirements. Temporary cover or catch crops are seeded as soon as possible after the main crop is harvested. For instance, a grain field in which no clover or grass-clover mixture has been seeded is plowed as soon as conditions permit, so that a catch crop, such as Sudan or soybeans, can be planted. On poor Northern soils, buckwheat is commonly used. Such seeding can be successful only if moisture conditions are favorable. If not, it is better to leave the stubble and such weeds as may grow for soil protection.

In the Great Plains, cover crops for protection against wind erosion are best seeded on the contour if there is any slope, since contour cultivation aids in the conservation of moisture needed for the protective crops. If the season is favorable for a good growth, the upper portion of such cover as Sudan grass or sorghum may be cut for fodder, leaving a stubble about 18 inches for protection purposes. When only a small growth is made, the entire stand should be left, since it is worth much more for soil protection than for forage. Where a field has been severely blown, it may be advisable to seed the entire area to a protective crop until moisture conditions permit better growth. In other instances, the protective crop may occupy strips of varying width, placed at suitable intervals.

Winter cover crops in the South are used both for soil protection and for soil improvement. They should be seeded early enough to establish some growth before cold weather, but extremely early seeding is generally undesirable. The seeding date will depend on latitude. For example, in Virginia, North Carolina, Kentucky, and Tennessee, seeding in corn is not commonly satisfactory, as the corn furnishes too much shade. Farther south, velvet beans, soybeans, and the early-maturing strain of *Crotalaria spectabilis* are used extensively with corn as summer legumes. Crab grass, Florida pursley, and beggar weed, all of which generally make a heavy growth after corn is laid by, also furnish satisfactory cover. In the

North, seeding must be earlier, late August or early September being the best time for the latitude of Washington, D. C. In the coastal plain area of Georgia, Alabama, Mississippi, and Florida, October seeding gives best results.

APPLICABILITY OF CROPS TO PROBLEM AREAS. Winter cover crops are not extensively grown in the Northern Appalachian and New England area. Hairy vetch is the only legume used to any considerable extent, although rye and other winter grains are planted by many farmers. A grass-clover mixture in a regular rotation gives excellent protection. Alfalfa and biennial sweetclover also afford good protection on suitable soils. For summer cover, crimson clover is used in the extreme north, whereas Sudan grass, millet, and buckwheat succeed in nearly every part of the region. On productive soils, rape, cowhorn turnip, and mustard may be used as catch crops, and soybeans may be used generally throughout the region.

Within the Southern Appalachians, crimson clover, hairy vetch, Austrian winter peas, and bur clover are used successfully in South Carolina, Georgia, Alabama, southern Tennessee, and the less elevated parts of North Carolina. Elsewhere in the region, red clover, seeded alone or with a grass mixture, together with alfalfa and sweetclover on adaptable soils, are among the more commonly used winter cover crops. Hairy vetch, seeded with rye or oats, or winter grains seeded alone can be grown as beneficial cover crops throughout the region. In the southern part, rye grass, either alone or with crimson clover, has proved very successful. For summer cover, soybeans and cowpeas with Sudan grass, millet, or sorghum are important wherever a temporary cover is needed. Annual lespedeza is perhaps the chief leguminous summer cover for the Southern Appalachians, and perennial lespedeza is taking an important place both as summer and as winter cover on land not suited to alfalfa or red clover.

The chief winter legumes now grown in the Atlantic and Gulf Coastal Plains region are Austrian winter peas, hairy vetch, crimson clover, and bur clover. Local experience should be the guide in choosing from this test.

In the Florida peninsula, winter legumes are not extensively planted. Here winter grains may be used generally, as may also rye grass and Natal grass, on the better soils. On the limestone soils of Alabama and Mississippi, biennial sweetclover, Hubam clover, and sour clover are grown successfully. Hubam clover is promising for the area affected by cotton root rot in Texas.

For summer cover, a wide choice is available for the Atlantic Coastal plain. Soybeans, and cowpeas, annual lespedeza, velvet beans, crotalaria, and sericea lespedeza may all be used, depending on soil and other conditions. Biennial sweetclover and alfalfa also make both winter and summer cover where soil conditions are suitable. For permanent cover, kudzu and sericea lespedeza are recommended.

In the Central Prairie and Eastern Timbered Border region, red clover, clover and grass mixed, alfalfa, and biennial sweetclover constitute the principal winter cover. Hairy vetch is grown to a limited extent in Michigan, usually with rye. Except in the extreme northwestern part of the region, winter grains make a satisfactory cover. In parts of Oklahoma and Texas, Italian or domestic rye grass has given promising results for winter cover. For summer, soybeans can be used nearly everywhere, and cowpeas and annual lespedeza southward from northern Missouri. In some localities, canning peas and Canada field peas are grown, but they do not afford protection in late summer because of their early harvest season—unless interseeded with clover. Clover and grass or clover alone, and alfalfa or biennial sweetclover seeded in grain, also furnish excellent summer cover throughout this region. Sudan grass and millet can be used also.

For the Ozark Highlands, small grains are regarded as the most reliable cover for soil protection in winter. Soybeans, cowpeas, and lespedeza, sorghum, Sudan grass, and spring oats are effective at other seasons.

In the southern part of the Great Plains, winter grain is the most effective seasonal cover crop for winter. For summer, the sorghums and Sudan grass are most reliable. In the northern Plains, spring grain affords protection for spring and early summer, and the perennial grasses, especially crested wheat grass and western wheat grass, for winter. Sweetclover does well over parts of the area, particularly the eastern part, where alfalfa also succeeds.

In the drier western areas, only irrigated crops can be relied on generally for seasonal cover. Aside from irrigated lands, chief reliance for winter cover must be placed on the native grasses, although winter grains can be used in some localities.

In the Pacific Southwest, special cover cropping is confined to the orchard areas. On nonirrigated land, winter grain, wild oats, bur clover, and alfilaria provide fair to good cover. Cover cropping is a common practice in most of the citrus groves and deciduous orchards. Vetches, bur clover, sour clover, and mustard are effective for winter. Some use is made of domestic rye grass, alfilaria, fenugreek, and small grain, especially wild oats. In the orchards of the Placerville section of California, permanent cover of domestic rye grass, redtop, or legumes is used, and sometimes Johnson grass. Satisfactory cover crops have not been found for the olive groves on dry land and vineyards on steep slopes. This is equally true of land annually planted to grain, hay, and beans.

Sesbania and other legumes are grown during summer in the Imperial Valley on land intensively cropped in winter, but their primary function is for soil improvement.

Cover crops are used regularly in the orchards of the Pacific Northwest. In the Palouse, wheat and, in some cases, sweetclover provide some protection during winter, whereas peas and spring wheat give considerable protection to the less erodible lands. Where sweetclover is seeded in peas, the cover may be extended through summer and the following winter and spring.

In the Snake River Valley, special cover crops are not used; but on irrigated land, alfalfa and clover, grown as a regular practice on many farms, provide excellent cover. In the Willamette Valley of Oregon, alfalfa, red clover, alsike clover, and sweetclover cover thousands of acres every year.

## Green Manuring

Green manuring is the practice of turning under plant material to improve the soil. While any thick crop will provide seasonal cover, special crops are grown for green manuring. Red clover, sweetclover, and alfalfa make excellent green manuring crops when finally turned under, but they are rarely seeded for this particular purpose.

Crimson clover, hairy vetch, and Austrian winter peas are the most important plants used for green manuring in the South. Bur clover, sour clover, and Hubam clover also are used but on a much smaller acreage. Crotalaria and cowpeas are the chief summer legumes used for this purpose. Lespedeza is not often seeded for green manure, but the stubble of the grazed crop serves the same purpose in the South as red clover stubble does in the North. In the Corn Belt, sweetclover is often seeded especially to be turned under for soil improvement. A native legume volunteering in wheat fields of western Iowa, known as Wood's clover (*Dalea alopecuroides*), makes an excellent green manure crop. Rye is used rather extensively, especially in the potato section of Delaware. It is turned under early and serves to maintain the organic supply of the soil at a satisfactory level.

Green manuring in the intermountain and Pacific coast regions is confined to irrigated orchard or truck croplands. The crops most generally used are vetch, sour clover, bur clover, and mustard. Recently, a new system of mowing the cover crop and leaving it on the surface has been introduced in orchards and citrus groves of California. Italian rye grass is commonly used. It reseeds enough to maintain the stand. With repeated cuttings, a large amount of plant material accumulates on the surface. How successful this practice, which is a form of mulching, will be remains to be seen, but results so far have been encouraging.

In the Pacific Northwest, especially in Washington, where alfalfa is used as a permanent cover in irrigated orchards, the crop is mowed several times a year, and the material allowed to accumulate on the ground. This is disturbed only by the furrowing necessary for irrigation. In the Willamette Valley, vetch is frequently used as a regular green manuring crop. In the Palouse, sweetclover seeded with peas is plowed under in June the following year in preparation for grain, thus serving the purpose of green manuring.

## Use of Crop Residues

Crop residues consist of those parts of crops left in fields after harvest. Grain straw baled and sold is not considered a residue; but if it is scattered in the field by a combine, it is so considered. Corn stalks, cotton stalks,

Fig. 124.—Burning wheat stubble to facilitate plowing for next crop. (*Photograph by Soil Conservation Service.*)

tobacco stalks, small grain stubble, scattered wheat straw, and the refuse from potato plants are illustrations of crop residues.

In the Palouse region, the practice of burning wheat straw and stubble was found to be conducive to severe erosion. On the other hand, where this residue was partly turned under and the ground left in a cloddy condition, erosion, both by wind and by water, was greatly reduced. The method is also applicable to the Great Plains (Figs. 124 and 125).

Where wheat is combined in this region, the stubble and straw left in the field frequently amounts to about 1,000 pounds per acre. This can be incorporated with the soil without any marked effect on the yield of the next crop. But heavier amounts of such residue may depress the succeeding yield. Insufficient experimental data are available on the subject, but experience has shown that soil losses may be decreased or prevented by plowing under part of the material and leaving part protruding from the soil.

Where wind erosion is serious, the stubble of such crops as sorghum, Sudan grass, or even the entire plant when there has been little growth, will catch and hold snow and retard or prevent soil blowing for periods of a year or more. In parts of Kansas, stubble of Sudan or sorghum has been found favorable for seeding sweetclover without any seed-bed preparation. Oats and sometimes rye are successfully seeded in cotton fields

FIG. 125.—Disking wheat stubble to incorporate part of the residue with the soil for the control of wind erosion. (*Photograph by Soil Conservation Service.*)

without plowing up the stalks. Some farmers report that small grain seeded on lespedeza stubble without preliminary plowing produces just as well as where the entire field is broken out. In Missouri, especially good results have been obtained with lespedeza-barley pasture where the barley is seeded on the lespedeza with only moderate disking.

There seems to be little doubt that many mistakes have been made in general agricultural practice by plowing under every vestige of a crop in order to plant the succeeding crop. Much improvement is needed in this phase of farm procedure, in view of the fact that it usually amounts to unnecessary baring of the soil to the full erosive effects of wind and water.

# Chapter XVII. Farm and Range Plants Useful for Erosion Control and Water Conservation

It has been explained that a dense cover of vegetation is the ideal implement for soil conservation. On farm and range lands, such perfect protection is most nearly approached, ordinarily, under conditions of a good pasture, meadow, or range sod; it is approximated through the use of various temporary and long-term protective cover crops.

Plants that are useful for soil conservation may be divided conveniently into two groups: common farm and range plants and hitherto overlooked plants that offer promise for controlling erosion. Almost without exception, the former are introduced plants from Europe, Asia, and Africa; the latter group consists largely of native American species whose utility was not recognized until the need for plants suitable for controlling erosion under various environmental conditions called attention to their value.

Good stands of trees, maintained in proper locations and protected from fire and grazing, also give adequate protection against erosion. The profitable use of this crop on the commonly cultivated and better lands, however, is limited because of the relatively long time required for its establishment. On range lands, its use is also limited because of climatic restrictions.

## Common Plants Useful in Erosion Control

The more common farm crops useful in erosion control are those grown in pastures and meadows, as soil-building crops in rotations, and as seasonal cover crops. These plants have a dual value: they provide a satisfactory soil-protecting cover, and at the same time they furnish hay or pasturage or serve as soil-building crops.

### Pasture Grasses

Permanent pasture vegetation consists mainly of those plants which make an effective cover for soil and water conservation. This is particularly true of sod-forming grasses when care is taken to establish and maintain a good ground cover. Where a good cover of forest occupies the summit and a good sod the adjacent slopes, even steep lands are safe from serious erosion.

Since numerous plants can be used for pasture, adaptable species generally can be found wherever it is economically feasible to devote the land to permanent grass. Any classification of grasses as pasture or meadow plants is more or less arbitrary. When utilized for both purposes, they will be discussed here under the most common usage.

KENTUCKY BLUEGRASS (*Poa pratensis*) AND CANADA BLUEGRASS (*P. compressa*). In those sections where Kentucky bluegrass makes a

FIG. 126.—This Virginia pasture is adequately protected from erosion by a good stand of bluegrass. (*Photograph by Soil Conservation Service.*)

good growth, this plant usually occupies a place of first importance in erosion-control procedures by vegetative means. No other grass will make a more dense sod nor more effectively hold the soil on which it grows (Fig. 126). Ordinarily, Kentucky bluegrass requires a good soil; but even where soil conditions are not naturally favorable, much can be done to develop a good sod by appropriate fertilization and careful pasture management.

This grass, which is highly palatable when young, will withstand a considerable amount of trampling. During the heat of summer, however, its grazing value is reduced, since it is not especially drought-resistant and prefers a cool rather than a hot climate. During hot, dry weather, Kentucky bluegrass pastures should be rested or grazed lightly to prevent dying and resultant loss of protective cover.

In the North, Kentucky bluegrass is valuable for use on terrace outlets, in grassed waterways, and for sodding gullies. It occurs quite generally through the Northern States, usually accompanied by white clover. The growth of both is commonly improved by an application of phosphatic fertilizer. Such treatment is often essential if white clover is to be maintained in the pasture. Kentucky bluegrass is also valuable for permanent cover in many of the Rocky Mountain valleys that have an adequate supply of rainfall and a moderately cool climate.

Canada bluegrass is less satisfactory for erosion control than Kentucky, since it makes a thinner sod. On the other hand, it will grow on poorer land and more droughty soil.

REDTOP (*Agrostis alba*) AND THE BENT GRASSES (*A. spp.*). These species of Agrostis, with their good sod-forming habits, are useful on land too acid for Kentucky bluegrass. Redtop not only flourishes on wet land but does fairly well on poor land. Although the density is not equal to that of good bluegrass on productive land, redtop has much the same habits of growth and can be recommended for many locations where bluegrass fails to thrive.

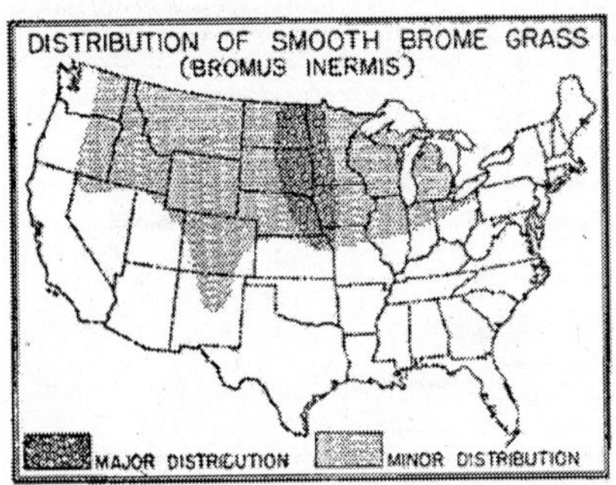


FIG. 127.—Distribution of smooth brome grass. (*Soil Conservation Service.*)

The bent grasses, although grown chiefly for lawns and golf courses, are used frequently in the pastures of New England and the northern Pacific coast region. The numerous runners of some species and the dense growth of others form a heavy mat which acts as a cushion against rainfall and runoff. For this reason, the bent grasses are particularly useful in grassed waterways.

SMOOTH BROME GRASS (*Bromus inermis*). Because of its drought-resistant character, this grass is peculiarly suited to semiarid conditions (Fig. 127). Its dense sod forms an ideal cover for any slope on which it can be established, and the sod resists trampling. It may be utilized for grazing or cut for hay. But for the fact that it becomes "sod bound" and declines in productivity, it might be regarded as a permanent cover. A stand will remain reasonably productive for about 5 years; at this stage, a good disking that will not turn up the soil too much may extend productive life for a year or two longer. Eventually, any smooth bromegrass pasture must be plowed and turned into some rotation. Even so, the soil will have been enriched through the accumulation of organic matter, and its resistance to erosion materially increased.

BERMUDA GRASS (*Cynodon dactylon*). This is the most common grass used for hay and pasturage throughout the Southern States. It produces underground rootstocks and creeping stems that readily take root at the nodes, which results in a dense sod of outstanding value for erosion control (Fig. 128). Notwithstanding its habit of spreading, Bermuda grass permits a fair growth intermixed with lespedeza. This combination adds to the grazing value of the pasture. Bermuda makes a good permanent pasture, even on erodible sloping land. It is especially effective as a protective lining for terrace outlets and diversion-ditch channels and gives good results in the stabilization of gullies. It is readily established from seed, stolons, or rootstocks.

Objection is sometimes made to Bermuda grass on the ground that it may become a weed in cultivated fields. Although every effort should be

FIG. 128.—Bermuda grass makes a good cover and good pasture. (*Photograph by Soil Conservation Service.*)

made to keep it out of fields in the regular rotation, its value in the program of soil conservation on the upland areas of the South is so great that it is not possible to dispense with its use. No grass of proved adaptability can take its place in this part of the country.

In parts of Arkansas, Louisiana, and east Texas, Bermuda grass is being grown extensively for pasturage. In getting it started, many farmers are cultivating and fertilizing the crop in row plantings.

CARPET GRASS (*Axonopus compressus*). This Southern grass occupies but a minor place in the catalogue of erosion-control pasture plants. Its habitat is the moist sandy soils of the southern Atlantic coastal plain, where the land is seldom subject to serious erosion, aside from gullying sometimes started by runoff from eroding uplands. After it has become well established, carpet grass makes a tough sod, so dense as practically to exclude all other vegetation. Wherever the location to which the grass is suited is subject to erosion, its use can be recommended.

DALLIS GRASS (*Paspalum dilatatum*). This grass, while not ideally suited to erosion control because of its slowness in spreading, is helpful, nevertheless, especially in gullies and waterways subject to gullying. Since the seed are always rather poor, even heavy seedings seldom produce a dense stand.

Dallis grass often is used in Southern pastures, where it may be of considerable value. It generally should be used, however, in such mixtures as will insure a complete ground cover promptly. These mixtures may include Bermuda grass, lespedeza, or some kind of clover, the particular selection varying with local conditions. Dallis does especially

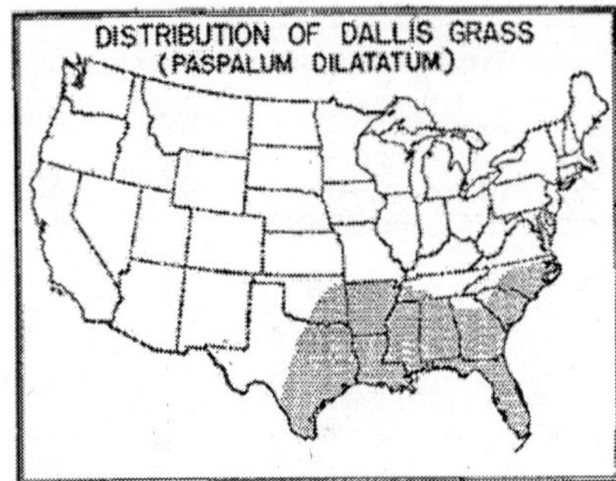


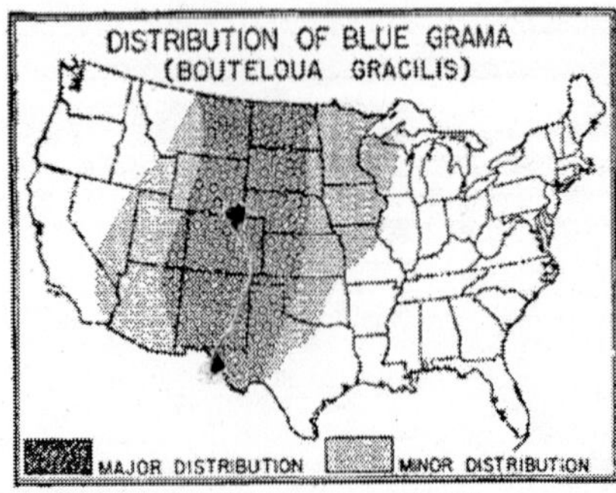


FIG. 129.—Distribution of Dallis grass (left) and blue grama (right).

well on the calcareous lands of the Mississippi-Alabama Black Belt. It is also useful on good lands elsewhere in the South (Fig. 129).

BLUE GRAMA (*Bouteloua gracilis*). Although seed of blue grama are not yet on the market, the plant is so common and occurs in pure stands over such a wide range that a good seed crop can be counted on almost every year. Blue grama occurs generally throughout the Great Plains and farther west in New Mexico and Arizona (Fig. 129). It is highly palatable and is one of the chief grazing plants on much of the Western range. Commonly associated with buffalo grass, it is sod-forming in habit, with a relatively shallow but very dense root system which makes it especially desirable for use in erosion control. Quick germination (with proper moisture, it will emerge within 48 hours after planting) is a distinct asset under semiarid conditions.

Since the seedlings are shallow-rooted, good survival is closely related to the care exercised in preparation of the seedbed and the protection of the young seedlings against wind damage. A rather common practice in parts of the Great Plains is to seed in stubble, which gives both protection and the desired seedbed. The seed must be planted shallow. Conservation of all available moisture by methods of contouring has proved desirable in the Plains country. Seedings on loose sandy soil have resulted in many failures. Some good results have been attained, however, by seeding on snow, immediately after a rain, and by contour seeding on contour-furrowed land.

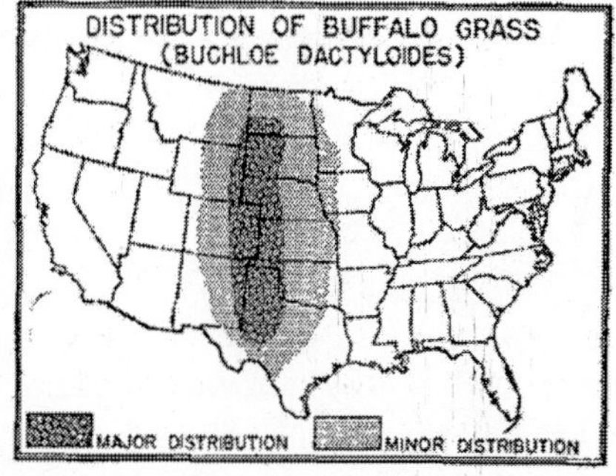


Fig. 130.—Distribution of buffalo grass.

There appear to be distinct strains of blue grama, with the Southern varieties tending to be later in maturing and more vigorous than the Northern strains. This fact provides a starting point for selection of the best strains for both erosion control and range revegetation. Seed can be harvested easily by the use of bluegrass strippers and special power strippers developed by the Soil Conservation Service.

Buffalo Grass (*Buchloe dactyloides*). This is one of the most characteristic grasses of the semiarid Great Plains (Fig. 130). It makes a close, even turf, produces a large amount of nutritious forage, and forms

Fig. 131.—Buffalo grass: development from sodding after one season's growth.

a most effective erosion-resistant sod. In the Southern Plains, it ranks first in importance among the native grasses for control of wind erosion, possessing all the characteristics desired for this purpose, except that of producing heavy seed crops. Being very drought-resistant, it does well throughout the region (except on loose sandy soil), spreading rapidly by means of stolons and withstanding heavy grazing. Besides its primary use for grazing and soil binding over extensive areas, buffalo grass is useful in vegetative linings for terrace channels and outlets, for gully strip sodding, and for the protection of earth fills around ponds and reservoirs.

Despite its excellent qualities for erosion control, the use of buffalo grass is somewhat limited by lack of seed. The seeds are borne in burs which are carried on stalks so short as to preclude the use of regular harvesting machinery. In the experimental work of the Soil Conservation Service, some seed have been gathered by hand and with specially developed suction machines which collect fair quantities. Seed crops are light, however, and the expense of collecting may always interfere with extensive seeding.

Buffalo-grass seed remain in good condition long after shattering; old seed have been found more readily germinable than fresh seed. An ordinary corn planter can be used for seeding; in Texas, cotton planters have been used. Another method of establishment is the use of small pieces of sod. This method (Fig. 131) offers the opportunity for extending the area of this valuable grass steadily although not rapidly.[1]

## Leguminous Pasture Plants

Several legumes suitable for pastures have an important role in the control of soil erosion, either alone or in mixtures with grass.

**White Clover** (*Trifolium repens*). This plant, probably the most widely distributed of the true clovers, occurs in most humid areas wherever soil conditions are at all favorable. Except for seed purposes, it is never used alone but always in mixtures with grasses. The creeping stems root at the nodes and thus help to bind the soil. Since white clover will not endure much shade, it is a typical pasture rather than a meadow plant. Where grass is allowed to grow high, white clover eventually disappears. Although it demands less lime in the soil than red clover, it is especially responsive to treatment with phosphatic fertilizers. On many Northern hillsides, a good growth of white clover often can be obtained without seeding, following an application of about 500 pounds of superphosphate per acre. This vigorous growth of white clover will, in turn, stimulate the grass and make for a denser sod (Fig. 132).

**The Hop Clovers.** Two low-growing winter annual clovers (*Trifolium dubium* and *T. procumbens*) are useful for late winter and early spring cover in the South. They are used chiefly in association with summer grasses. *Trifolium dubium*, the *least hop clover*, is most abundant in the South and on the Pacific Coast; whereas *T. procumbens*, the *low*

[1] Savage, D. A. Methods of Reestablishing Buffalo Grass on Cultivated Land in the Great Plains. U. S. Dept. Agr. *Circ.* 328, 1935.

*hop clover*, is the dominant species farther north, particularly in Tennessee and North Carolina.

Although both species occur scatteringly throughout the humid parts of the United States, they are of importance as cover and for grazing only in the South and Far West. Growth is so rapid in early spring that by June the plants are mostly mature and so disappear for the remainder of the season.

SWEETCLOVER. Although sweetclover often is grown for hay, its main uses are for grazing and soil improvement. However, since the most commonly used species is a biennial, it cannot serve for a long-term cover. Requiring an abundance of lime in the soil, sweetclover will thrive on rundown land only after artificial liming. Seeded with small grain, it

FIG. 132.—White clover is very desirable in a mixed-pasture. North Carolina. (*Photograph by Soil Conservation Service.*)

covers the soil for the remainder of the summer and begins a vigorous growth early the following spring. Then it may be heavily grazed until midsummer, at which time the seed crop is matured. Sufficient seed for new plantings may be harvested readily from any pasture or field of second-year sweetclover.

Since it is winter-hardy and drought-resistant, sweetclover has a wide range. If lime is available, this legume will grow almost anywhere. It is a valuable plant to combat water loss after grain harvest and to add organic matter to the soil. From the standpoint of soil and water conservation, sweetclover is perhaps as important in adaptable portions of the Great Plains as any legume.

Sweetclover requires a firm seedbed and may be drilled in Sudan grass or sorghum stubble without disturbing the land. This stubble protects the ground until the sweetclover has attained a sheltering growth. In the drier localities, seeding should follow a period of fallow. Seeding in contoured furrows has given fair success in eastern Colorado and other dry areas of the Great Plains.

THE ANNUAL LESPEDEZAS. Few plants except grasses are more satisfactory for the conservation of soil and water than the annual lespedezas (Fig. 133): *common lespedeza, Tennessee* 76, *Kobe* (*Lespedeza striata*), and *Korean* (*L. stipulacea*). Although not usually thought of as permanent cover, they can be made to serve practically the same purpose over a term of years if allowed to volunteer.

Adapted to both good soil and soil of relatively low productivity, these plants make a dense, low cover of innumerable stems and leaves which

FIG. 133.—The annual lespedezas make ideal summer pasture and a dense soil cover. Tennessee.

prevent direct impact of rain on the soil and retard runoff. As a result, the soil loss from a field carrying a good stand is small. At Statesville, N. C., the average annual soil loss per acre for the period 1931–1935 from an area (on a 10 per cent slope of the extensive Cecil sandy clay loam) in annual lespedeza was only 0.63 ton, compared to a loss of 20.19 tons from a similar plot in corn and 13.87 tons from another similar area in cotton.

Annual lespedezas warrant extensive use. They may be established at low cost and will maintain themselves for several years at no cost whatever. Meanwhile, they yield a return in hay or grazing and protect the soil at the same time. Even if grown only as a measure of soil conservation, their use is justified by the improved condition of the soil when it is returned to other crops. Although the lespedezas do well on soil of only moderate fertility, they respond to treatment with phosphatic fertilizers on poor, badly eroded land. It generally will pay to use some lime and phosphorus on eroded acid land, since a better growth of lespedeza will mean greater protection from erosion, more hay or grazing, and more rapid soil improvement.

Annual lespedezas may be seeded alone or with a grass like Bermuda. In many localities, they lend themselves to a dual use of the land, which at the same time makes for more perfect erosion control. In Missouri, for example, a year-round pasture system has been devised in which small grain is disked into a growth of annual lespedeza after the seed have ripened. The grain covers the field during winter, and the volunteer lespedeza occupies the ground during summer. At present, it is impossible to state how generally such a combination can be used or the number of years that it can be continued. But the system is proving of such great value in the conservation of soil and water that it deserves careful consideration wherever it may be applicable. Seed are so readily harvested that any farmer can secure from his own fields an adequate supply at almost no cost.

The varieties of *L. striata* are generally best suited to the territory from northern Tennessee to the Gulf. Korean, on the other hand, because of its early maturity, will give better results for the area from Tennessee to about central Illinois.

## Meadow Grasses

Meadow plants are not commonly left on the land so long as pasture crops; they are used sometimes merely as one member of a 3-year rotation. In some instances, however, especially after alfalfa, a stand of some meadow grasses may be left for many years.

Timothy (*Phleum pratense*). Timothy is the most widely used grass in the Northeastern States. Its range extends as far south as the highlands of western North Carolina. It also is used in mountain meadows and on irrigated land in the Northwest. Although its lack of drought resistance makes it unavailable for semiarid regions, in the more humid sections of its habitat timothy is used in most farm rotations. Seeded alone or with clover, it makes a splendid cover for 2 years or more. On erodible land, it is desirable to leave timothy as long as it remains productive and in many instances even after its period of optimum productivity has passed. A field may be used for pasture after it has lost its usefulness for hay. Because of the well-understood use of timothy, coupled with the abundance and usually low price of seed, this is one of the best grasses for semipermanent cover in the Northeast.

Orchard Grass (*Dactylis glomerata*). Since orchard grass forms clumps or tussocks, it makes a less satisfactory cover than sod-forming grasses. When seeded thickly, however, or when interseeded with lespedeza, it may give good results. It grows on poorer soils and will endure more shade than the better sod-forming grasses. Orchard grass has proved especially useful in the southeastern Piedmont and Ozark Highlands and is grown also in other sections. If not grazed too heavily, it can be used as a permanent cover for moderate slopes, in permanent strip crops in cultivated fields, and in orchards where a permanent cover is needed. Seedings should be heavy enough to insure a thick stand. Annual lespedezas, particularly Kobe and Korean, interseeded in thin stands will help complete the cover. The crop may be utilized for hay or for grazing.

Reed Canary Grass (*Phalaris arundinacea*). Although primarily a swamp plant, Reed canary grass succeeds also on some upland soils. With its interlacing rootstocks and heavy growth, it makes a dense protective cover (Fig. 134). Ordinarily it flourishes in the Northern States on seepage

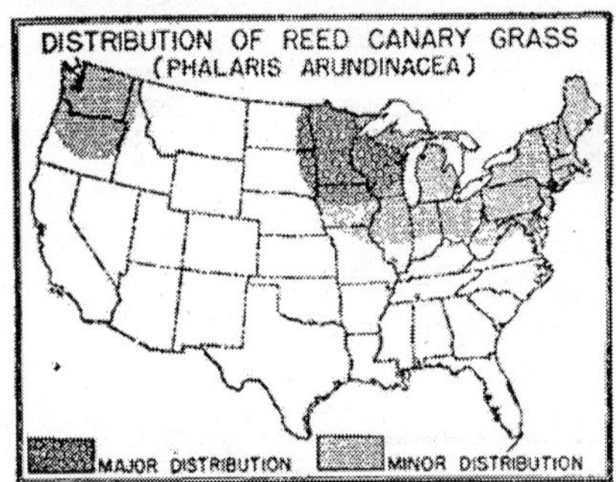


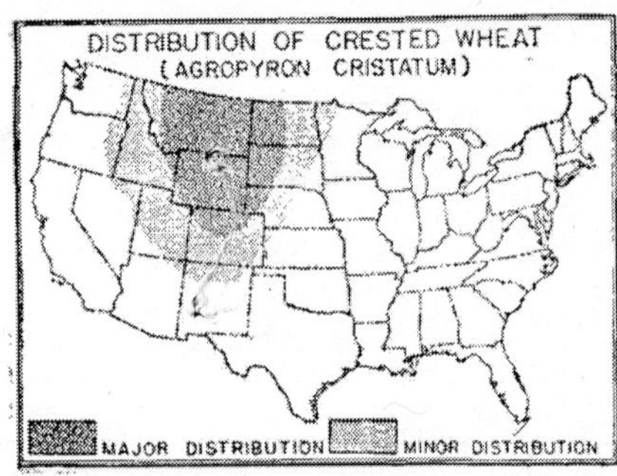


Fig. 134.—Distribution of Reed Canary grass (left) and crested wheat grass (right)

slopes and is helpful in the stabilization of terrace outlet channels. Reed canary makes a permanent cover and may be used for hay or grazing. Although the seed shatter readily, farmers generally can gather their own supply.[1]

Crested Wheat Grass (*Agropyron cristatum*). Introduced from Russia, this perennial bunch grass (Fig. 135) has been found particularly well adapted to the soils and rigorous climatic conditions of the northern Great Plains (Fig. 134). Its marked resistance to drought, its deep, extensive root system, and its early spring growth and excellent reseeding capacity make it a most useful plant for erosion control over an enormous area in subhumid and semiarid regions. On farms where the cost of seeding is an important item, crested wheat grass is an especially suitable species. About 2 to 3 pounds of seed per acre will produce a stand. Such a planting, if allowed to seed, will in a few years cover the entire area with a good stand. A quicker cover can be obtained by the use of more seed, sown in drills or broadcast; but on many tracts of abandoned

[1] Arny, A. E., Hodgson, R. E., and Nesom, G. H. Reed Canary Grass for Meadows and Pastures, Univ. Minnesota, Agr. Exper. Sta. *Bull.* 263, 1930.

land, the lower cost seeding plan will give satisfactory results. Late fall planting has been found advantageous, as it obviates the necessity of loosening the soil in spring when it is more susceptible to blowing. Since seed are easily harvested, the crop may produce an income from seed, hay, or grazing.

SUDAN GRASS (*Sorghum vulgare var. sudanense*). This standard forage grass can be grown nearly everywhere in the United States. In the humid East it is frequently used as an emergency hay or pasture crop and also

FIG. 135.—Plants of crested wheat grass. North Dakota. (*Photograph by Soil Conservation Service.*)

for temporary cover. Its most important place, however, is for checking wind erosion in the Great Plains. The seed germinate quickly, and the plants make rapid, sturdy growth. One of the most common treatments for eroded fields in the Southern Plains is blank listing and the seeding of Sudan grass in rows, preferably on the contour. In harvesting, the stalks should be cut high; where checked by drought, the entire crop should be left as a protective cover.

## LEGUMES FOR MEADOWS OR PERMANENT COVER

In some ways, legumes are preferable to grasses for conservation purposes, since they produce hay with a higher protein content and increase productivity by adding nitrogen to the soil. On the other hand, many of the most valuable legumes do not grow on badly worn soils, whereas some of the grasses thrive on all but the most sterile land.

Alfalfa (*Medicago sativa*). In those areas where it is adaptable, alfalfa is almost invariably the leading hay crop. However, it thrives only on land that is at least fairly productive and even then requires plenty of lime and phosphorus. It is grown principally in the country north of the Ohio and Potomac Rivers, thence westward and southwestward. South of this line, it is grown locally in the upper Piedmont; the Appalachian limestone sections; the Black Belts of Alabama, Mississippi, and Texas; the Mississippi delta; and other favorable localities. In the drier sections west of the 100th meridian, successful growth depends largely on irrigation.

Fig. 136.—Alfalfa is an excellent plant for permanent strip cropping. North Carolina. (*Photograph by Soil Conservation Service.*)

Alfalfa is well suited for the development of permanent or semipermanent cover on sloping land where the abundant growth effectively controls erosion and runoff. It is also well suited for permanent strips between cultivated crops (Fig. 136).

Red Clover (*Trifolium pratense*) and Alsike Clover (*T. hybridum*). Used in standard rotations, these crops make excellent cover. Commonly seeded on or with small grain, the first year's stand covers the ground after harvest, during the following winter, and for at least part of the next summer.

Red clover usually is seeded with timothy. This mixture may be left for hay or grazing for more than one year, making excellent cover. Plowed under, red clover improves the soil, making it more permeable and resistant to erosion. Although it requires a productive, well-drained soil, its lime requirement is not so high as that of alfalfa or sweetclover. Although it is a northern plant, it can be grown in the upper part of the southern Piedmont where lime and phosphatic fertilizers are applied. In the South, however, it does not last through summer, nor will it grow under semiarid conditions.

*Alsike clover* is a good substitute for red clover and has the added advantage of succeeding on poorer soils—particularly those less well supplied with lime and of somewhat poorer drainage.

LESPEDEZA SERICEA. In the South, the principal legumes available for permanent or semipermanent cover are *L. sericea* and kudzu. The former is a perennial well suited to a wide variety of soil conditions, including eroded land (Fig. 137). Seeded on land where gullies are forming,

FIG. 137.—*Lespedeza sericea* is a good soil conserver and builder, and is valuable for wildlife. North Carolina. (*Photograph by Soil Conservation Service.*)

a good stand of sericea will not only check erosion but help smooth out the gullies that already have started. It is also suitable for strip cropping, buffer strips, meadow waterways, and marginal field and odd-corner plantings. Although it makes slow growth the first season, subsequent progress is characteristically good, even under conditions of heavy weed infestation and poor soil. The crowns give rise to an increased number of stems in successive years, and accumulation of decaying leaves and branches builds up a protective ground mulch.

*Lespedeza sericea* already has proved to be a valuable plant for erosion control throughout the region to which it is adapted. This, according to present experience, appears to be generally south of the Ohio and Potomac Rivers and as far west as eastern Kansas and Oklahoma. It has succeeded in some places north of the Ohio but is subject to winter killing where the temperatures go much below zero.

This crop has an important economic place on the farm for the dual purpose of erosion control and hay production. It is also an excellent plant for use in the development of wildlife resources. The dense growth offers splendid refuge, and the seed are relished by quail. The plant produces an abundance of seed.

**Lespedeza Juncea.** This is another perennial of similar habit to Lespedeza sericea, which gives promise of equal, or perhaps even greater, usefulness in erosion control. Since its habit of growth is more bushy than

Fig. 138.—Kudzu taking hold on badly eroded slope. Clark County, Georgia. (*Photograph by Soil Conservation Service.*)

that of sericea, it may make a better cover. However, information is not yet available as to its value for hay or grazing.

**Kudzu** (*Pueraria thunbergiana*). Kudzu is a perennial vine of great value both for control of sheet washing and for the reclamation of gullies (Figs. 138 and 139). Established plants produce each year a number of runners from 30 to 50 feet in length. These, together with the broad leaves, almost completely cover the ground. The vines root at the nodes wherever they are in contact with moist soil and thus establish new plants from which runners spread over a wider area the following year. The result in three or four years is the establishment of a dense cover which effectively checks runoff and erosion.

Planted on the margins or on the sides of gullies, the vines eventually fill even large chasms by catching silt and gradually filling in the depression. Meanwhile, such a gully may be used for reserve summer grazing long before it is filled. Although kudzu plants are killed by severe frosts, such a mass of vines and dead leaves is left on the ground as to aid materially in the control of winter erosion.

Kudzu should not be depended on alone for control of gully heads. Since water may cut the soil from under the plants at such points, more close-growing vegetation or mechanical structures frequently are needed to supplement the kudzu.

FIG. 139.—Severely eroded land completely stabilized with kudzu, which is producing excellent hay. Alabama. (*Photograph by Soil Conservation Service.*)

VELVET BEANS. These are strong-growing viny plants very commonly planted with corn in the Cotton Belt. Of the several varieties, the later maturing ones are more vigorous growers than the earlier varieties but are restricted in their range to Florida and the extreme southern parts of Georgia, Alabama, and Mississippi. The earlier varieties can be grown throughout the Gulf States and as far north as eastern North Carolina.

Planted with corn, velvet beans make an excellent late summer and early winter cover. They make good growth even on poor land and add liberally to the organic supply of the soil. They are immune to nematodes and wilt and carry numerous large nodules on their roots.

## WINTER COVER CROPS

The use of cover crops has been discussed in a previous chapter. Those seeded in the fall for the purpose of erosion control and soil improvement are called *winter cover* or *green manuring* crops. With the exception of rye and other small grains occasionally used in this way, such crops consist of legumes, and the list is relatively small.

Crimson Clover (*Trifolium incarnatum*). This clover is best suited to the northern part of the Cotton Belt and the Atlantic coastal plain as far north as New Jersey. In northern Georgia and other parts of the southern Piedmont, it makes a splendid winter cover and adds much organic matter when turned under (Fig. 140). It also may be used for winter grazing and for hay. Once a stand has been established, seed may be collected easily by a farmer for his own use. Seed in the hull are somewhat more difficult to handle than hulled seed, but a stand is more certain when unhulled seed are used.

Fig. 140.—Crimson clover is a good winter cover and green manure crop. South Carolina. (*Photograph by Soil Conservation Service.*)

Vetches (*Vicia spp.*). Two of the more important vetches are the *hairy* and the *smooth* forms of *V. villosa*. The latter makes better growth in the average winter of the South; it is not so well suited to the more northerly parts of the United States. Hairy vetch, which is winter-hardy from the Gulf of Mexico to Michigan, is the most commonly used winter legume in the United States. Both are weak-stemmed, viny plants which ascend only with support (Fig. 141).

*Monantha vetch* (*V. monantha*) resembles hairy vetch in the growth habit but has smaller stems and leaves. It is valuable in Florida and the southern parts of Georgia, Alabama, and other similar regions. *Common vetch* (*V. sativa*), although not reliably hardy in the Southern States, is used extensively in western Oregon and Washington for hay, pasturage, seed, and soil improvement. *Hungarian vetch* (*V. pannonica*) is especially adapted to heavy, wet soils and is as winter-hardy as the smooth form of *V. villosa*. The stems are stronger than those of hairy vetch. Its present use is confined largely to the Pacific Northwest. *Purple vetch* (*V. atropurpurea*), although winter-hardy, is most frequently used for green manure in the citrus groves of southern California. It is not adapted to the Cotton Belt.

AUSTRIAN WINTER PEA (*Pisum arvense*). Next to hairy vetch, this is the most widely used winter cover crop in the Cotton Belt. It makes good growth in winter and spring, producing considerable quantities of organic matter.

FIG. 141.—Hairy vetch and rye make a good combination for protective cover and soil improvement, even in the North. (*Photograph by Soil Conservation Service.*)

BUR CLOVER (*Medicago spp.*). The *southern bur clover* (*M. arabica*) is adapted to most of the Cotton Belt, preferring productive soil. The seed are sown in the bur and can be readily collected by sweeping up the burs in the field. On well-drained, good land, the plant makes excellent cover; with proper handling, it can be made almost permanent. When bur clover land is prepared for cotton or corn, unplowed strips or balks may be left between the cotton rows. On these, the clover will ripen seed. The residue can be worked down as the cotton is cultivated.

*California bur clover* (*M. hispida*) is one of the chief winter covers on millions of acres of hill land of the Pacific Coast, from southern California to Oregon. It furnishes excellent grazing and volunteers with the advent of fall rains.

SOUR CLOVER (*Melilotus indica*). This is one of the minor winter legumes of the Cotton Belt. It is a yellow-flowered annual related to biennial white sweetclover and, like that species, demands an adequate supply of lime and phosphatic fertilizer.

HUBAM CLOVER (*Melilotus alba annua*). This annual variety of the biennial white sweetclover is being used effectively in parts of Texas affected by cotton root rot. Making its principal growth in spring and early summer, it escapes most of the ravages of the disease (which affects all legumes).

Hubam clover is also used in some parts of the North as a temporary late summer cover.

## Overlooked Vegetation Useful for Erosion Control

Although for the present and immediate future, control of erosion by vegetation must depend largely on the use of those established crops for which a seed supply is available, it is recognized that many native species, as well as some newly introduced, hold promise for wide areas and for special purposes or local use.

As the work at Soil Conservation Service nurseries has progressed, many species and strains that heretofore had received very little consideration have been found to have exceptional value for various erosion-control practices. Overlooked until this particular need developed, many such plants because of especially desirable characteristics are now being brought into the soil and moisture conservation program.

### GRASSES

THE GRAMAS (*Bouteloua spp.*). Among the most valuable grasses of the Great Plains and various parts of the West are the gramas. They stand high in the scale both for forage and for erosion control. The species so far tested are all perennials, although Rothrock grama (*B. rothrockii*)[1] is rather short-lived. *Blue grama* (*B. gracilis*) was referred to in the first part of this chapter. Of the others, *side-oats grama* (*B. curtipendula*) has a wide distribution, develops a bunch habit of growth, and is found usually in scattered stands associated with other gramas. Because of its vigorous growth, leafiness, heavy stems, strongly branching root system, and the ease with which it is seeded and harvested, combined with a wide soil and climatic adaptation, this plant promises to become one of the most important native grasses for general revegetation purposes (Fig. 142). Seed production under cultivation has been good, with yields running as high as 400 pounds an acre.

[1] The nomenclature followed is that of Hitchcock, A. S. Manual of the Grasses of the United States, U. S. Dept. Agr. *Misc. Pub.* 200, 1935.

*Black grama* (*B. eriopoda*) is distributed more or less generally through southern New Mexico, Arizona, and western Texas. Spreading by tillers and stolons, it has good soil-binding properties. Since it does not produce seed consistently under natural conditions, development of improved methods of seed production must be sought.

*Hairy grama* (*B. hirsuta*) is a perennial grass which occurs in scattered stands throughout most of the Central and Rocky Mountain states but reaches its maximum usefulness as a range plant in Arizona and New Mexico. Its adaptation to soil and climatic conditions is similar to that

Fig. 142.—Native stand of side-oats grama. Hartley County, Texas, 1936. (*Photograph by Soil Conservation Service.*)

of blue grama. Like the latter, it tends to form a sod in the North and to assume a bunch habit of growth in the South. It appears to succeed on land unfavorable to various other grasses, but the scattered stands have made it necessary thus far to harvest the seed by hand.

*Rothrock grama* is limited to the southern half of Arizona and southwestern New Mexico, where it occurs extensively in pure stands. Although unable to withstand too close grazing, it seeds readily in favorable seasons and produces an excellent cover on relatively poor soils where few other grasses will grow.

Wheat Grasses (*Agropyron spp.*). The native wheat grasses of the United States have been found to be of great importance. *Western wheat grass* (*A. smithii*) (Fig. 143), a perennial that spreads by creeping root stocks, is an important hay grass. It produces a quick ground cover and a heavy sod. Its tenacious soil-binding habit makes it especially useful for sodding terrace outlets and channels, stabilizing gullies, and protecting stream and reservoir banks against wave action. It is useful in the reclamation of abandoned farm lands and some types of depleted range.

Fig. 143.—Western wheat grass is of great assistance in the control of erosion. Wyoming. (*Photograph by Soil Conservation Service.*)

It is quite generally distributed through the West and has been introduced in some localities east of the Plains (Fig. 144).

Seed are readily harvested and cleaned with ordinary small grain equipment. Seeding can be deep enough to insure germination and establishment, even under low rainfall conditions. Best success has been attained on fairly deep soil, particularly on alluvial and valley-fill land.

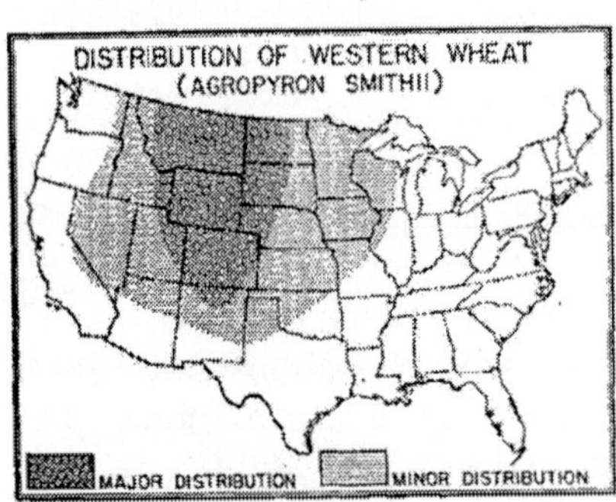


Fig. 144.—Distribution of western wheat grass.

*Blue-bunch wheat grass* (*A. spicatum*) and *beardless blue-bunch wheat grass* (*A. inerme*) are much alike, but the latter, being awnless, makes a more desirable grazing plant. The former occurs from Alaska to California and eastward into New Mexico, Colorado, Wyoming, and Montana, whereas *A. inerme* is found chiefly in the intermountain region and the Pacific Northwest. Both are perennial bunch grasses, characterized by tenacious root systems and marked drought resistance. Reproduction is by seed. Superior strains of *A. inerme*, showing exceptional qualities of vegetative vigor and seed production, are now under observation in Soil Conservation Service nurseries.

Another stolon-forming species is *thick-spike wheat grass* (*A. dasystachyum*), frequently found on drifting, sandy soil. It is a sod-forming grass but often produces rather sparse cover and a light seed crop. A promising strain has been found, however, which makes a more nearly complete cover and gives exceptional yields of seed. This has been established successfully in test plots under a wide variety of conditions and is believed to have special value for control of wind erosion.

Another species of the group is *quack grass* (*A. repens*), which long has been considered one of the most serious weeds of the North (extending as far west as the Plains). In spite of this bad reputation, however, its

FIG. 145.—Big bluestem in Soil Conservation Service nursery. North Dakota. (*Photograph by Soil Conservation Service.*)

very aggressiveness and habit of rapid spread by creeping rootstocks make it valuable as a soil binder. Wherever it is commonly established, its use on slopes that should be in permanent grass may well be given consideration, as it is a good grazing and a fair hay plant.

BLUESTEMS (*Andropogon spp.*). The big and little bluestems once covered wide areas in the prairies, particularly from Illinois to Oklahoma. These stands have given way to the plow, although scattered remnants are to be found. Recently, there has been a revival of interest in these species. *Big bluestem* (*A. furcatus*, Fig. 145) is a tall, leafy plant of bunch habit but with a deep and extensive root growth which furnishes good soil protection. It occurs in the eastern parts of Oklahoma, Kansas, Nebraska, and South Dakota. Seed may be harvested by power strippers or by combine harvesters. Higher yields of better quality seed have been obtained from cultivated plantings made where the plants have room to develop.

*Little bluestem* (*A. scoparius*) is also a bunch grass of wide distribution. It occurs in dense stands where the soil is good and moisture plentiful. Its leafiness and vigorous root system make it a very good plant for control of erosion. Several strains showing wide differences in vegetative characteristics have been isolated by the United States Department of Agriculture and state experiment stations. Excellent yields of high-quality seed have been obtained under cultivation (Fig. 146).

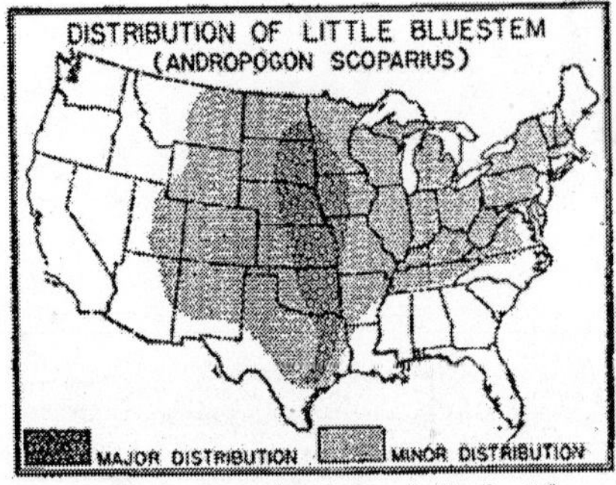


Fig. 146.—Distribution of little bluestem grass.

*Sand bluestem* (*A. hallii*), widely distributed throughout the Great Plains, is especially suited to sandy soils. It has an extensive creeping root system and has given good results in stabilizing certain types of "blow soil" in the central Great Plains. Extensive use for reseeding purposes is somewhat restricted by present scarcity of seed, but this difficulty is being overcome gradually by producing seed under cultivation. Matured hay of sand bluestem spread on eroded land serves as a protective cover until the seedings become established.

*Broom sedge* (*A. virginicus*), a bunch grass that resembles little bluestem, is widely distributed on abandoned land from Pennsylvania to the Gulf. Establishing itself readily in dense stands, it has served a highly useful purpose in conservation of soil, especially on temporarily idle and abandoned land. When young, it is of value for grazing and produces fair hay when cut at a height of about 10 or 12 inches. It might well be encouraged until better grasses can be substituted.

Reed Grasses (*Calamovilfa spp.*). Two of the reed grasses—*Giant reed grass* (*C. gigantea*) and *longleaved reed grass* (*C. longifolia*)—of wide distribution on light sandy soil through the Great Plains have been found useful in erosion-control practices. Although not very palatable, their vigorous horizontal rootstocks are highly effective in binding the soil, especially on "blowout" sandy areas. Their use for this purpose is facilitated by their extreme resistance to drought; by their tall, coarse, tough growth characteristics; and by their responsiveness to planting either vegetatively or with seed. An effective method of establishing these grasses on sand dunes has been by scattering the full-ripe hay over the surface. This temporarily stills the sand and furnishes protection until the young plants are established.

Blowout Grass. Another grass useful in controlling "sand blows" is *blowout grass* (*Redfieldia flexuosa*), which grows naturally on the deep, loose sands of the Great Plains, particularly the dunes of the Texas

Panhandle section and the Nebraska sandhills. It has a tenacious, strongly rhizomatous root system which often extends 20 feet or more from the parent plant. This feature, together with the toughness of stem and foliage, enables the plant to withstand drifting sand and makes it especially useful for controlling such areas. Although seeding habits are poor, fairly good supplies have been produced under cultivation.

BEACH GRASSES. The grass most widely used in the control of dunes along the seacoast and shores of the Great Lakes is *beach grass* (*Ammophila spp.*). The European species, *A. arenaria* (Holland grass), has become established along the Pacific Coast north of San Francisco,

FIG. 147.—Canada wild rye in Soil Conservation Service nursery. Washington. (*Photograph by Soil Conservation Service.*)

whereas the similar American species, *A. breviligulata*, is native to the Atlantic Coast and the region around the Great Lakes. Both species are large, coarse, perennial grasses which grow rapidly from creeping rootstocks. They have been used extensively on both the Pacific and the Atlantic seaboards for stilling coastal dunes while other permanent vegetation is being established. Since the plants produce little or no seed, increase must be by vegetative propagation.

WILD RYE GRASSES (*Elymus spp.*). Several species of wild rye grasses occur from the Great Lakes to the Pacific Coast. Although generally inferior as forage plants, they have a definite place in the conservation program because of their ability to establish quick ground cover. *Canada wild rye* (*E. canadensis*, Fig. 147) and *blue wild rye* (*E. glaucus*) are coarse, vigorously growing perennials which produce an abundance of seed. Both grow rapidly and produce good ground cover. Seed must be produced under cultivation, since they do not occur in pure stands. Blue wild rye is of greatest importance in the Pacific Northwest, where it is useful for grazing as well as for erosion control. *Beardless wild rye* (*E. triticoides*) is a sod-forming grass well suited to the Pacific Northwest. Because of its extensive creeping rootstocks and colonizing habit of growth, it has excellent erosion-control value. *Elymus mollis* is being used effectively to supplement beach grass in the stabilization of sand dunes on the Pacific Coast. A species of lime grass recently introduced from

Fig. 148.—Nevada bluegrass, four selections taken from the native vegetation in the Pacific Northwest and growing in the Soil Conservation Service nursery at Pullman, Washington. (*Photograph by Soil Conservation Service.*)

South Africa (*E. sabulosa*) appears to offer possibilities in the control of wind erosion because of its underground stoloniferous habit, coarse tough foliage, drought resistance, and adaptation to sandy soils.

Blue Grasses (*Poas.*). Two of our native species of Poa deserve special attention: *Nevada bluegrass* (*P. nevadensis*, Fig. 148) and *Sandberg bluegrass* (*P. secunda*). Both are tufted perennial bunch grasses, extremely drought resistant and highly valuable for forage and erosion control. Since reproduction is by seed, range-management practices should be such that natural seeding can take place. Superior strains of Nevada bluegrass have been found through selection. Sandberg bluegrass comes up and persists on semidesert and rocky land where other native grasses have disappeared. It withstands both trampling and vegetative competition. Because of early seed production, the plant is particularly useful for erosion control. Seeding trials, however, indicate difficulty in establishment from seed.

FIG. 149.—Curley mesquite (*Hilaria belangeri*), showing tufted, stoloniferous habit which makes it of particular value in erosion-control work. New Mexico. (*Photograph by Soil Conservation Service.*)

*Rough-stalked meadow grass* (*P. trivialis*) is less resistant to heat and drought than Kentucky bluegrass, which it resembles. It is especially suited to moist, shady places and appears to be a good grass in northerly areas for seeding in diversion ditches and runoff-disposal channels where some moisture is constantly present.

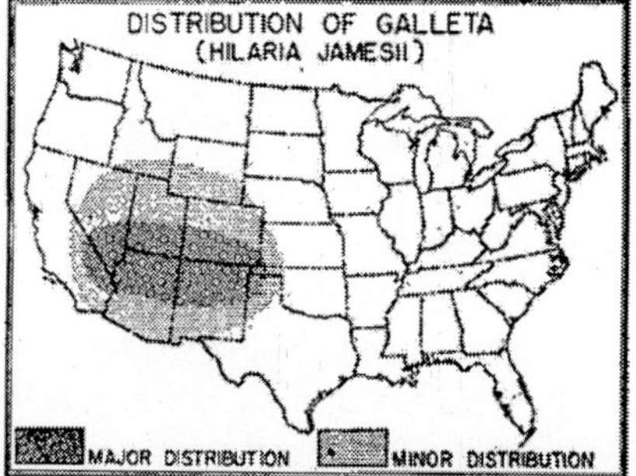


FIG. 150.—Distribution of Galleta grass.

HILARIA SPP. The genus Hilaria furnishes three perennial species, all of which occur in parts of Texas, New Mexico, and Arizona. All are valuable for control of erosion. *Curly mesquite* (*H. belangeri*, Fig. 149) has a tufted, stoloniferous growth habit and is highly resistant to drought and heavy grazing. Although it reproduces both by seed and by stolons, seed production is sparing except under cultivation.

*Galleta grass* (*H. jamesii*) (Fig. 150) has scaly rootstocks and a deep, tenacious root system which makes it important as a soil binder. It is resistant to heavy grazing, recovering rapidly under protection. Harvesting of seed is difficult because of the tendency to shatter. It germinates readily under range conditions, however, and the young seedlings are very drought-resistant and winter-hardy. It is one of the best forage grasses throughout its range.

*Tobosa grass* (*H. mutica*) (Fig. 151) is a strong, tufted perennial confined chiefly to heavy soils. It does especially well in swales subject to occasional flooding. The new growth supplies good grazing, but it soon becomes woody and unpalatable. Accumulation of dead foliage furnishes

Fig. 151.—Characteristic growth of tobosa grass (*Hilaria mutica*) in swale near San Simon, Arizona. (*Photograph by Soil Conservation Service.*)

an excellent protective cover. Seed can be harvested from native stands although shattering readily.

Bush Muhly (*Muhlenbergia porteri*). This native perennial, formerly one of the dominant species on the dry mesas and foothills of the Southwest, has been almost eliminated from the range because of the excessive grazing induced by its high palatability at all seasons. Its extreme drought resistance and deep, wiry root system give it distinct advantages as an erosion-resistant plant. Although somewhat difficult to harvest and handle, the seed germinate readily, and the young seedlings are established without difficulty. A woody base is produced, and the plants develop wiry procumbent stems which form offsets at the nodes. These often take root on contact with the ground.

Indian Rice Grass (*Oryzopsis hymenoides*). This tufted perennial grass (Fig. 152) is useful for stabilizing sandy land subject to blowing.

Seed are produced abundantly and are harvested easily, but they are hard-coated and remain dormant for a long time after planting. A strain is being propagated that has less pronounced seed dormancy as well as a larger and more vigorous growth.

*Oryzopsis miliacea* is a perennial bunch grass introduced from Europe and Asia. It produces thick top growth and a deep tenacious root system; it is resistant to drought and adapted to a variety of soils. It produces abundance of seed and may prove useful for soil conservation in the southern half of the United States.

Vine Mesquite (*Panicum obtusum*). This grass, found generally in western Texas, Arizona, New Mexico, and southern Colorado, makes a

Fig. 152.—Indian rice grass for seed production on the Soil Conservation Service nursery, Shiprock, New Mexico. This grass is helpful for stabilizing active sand dunes of low-rainfall areas. (*Photograph by Soil Conservation Service.*)

viny growth and occurs more commonly on good soils that contain more than the normal amount of moisture for the region. Growing rapidly, the creeping runners take root at the nodes, anchor themselves to the ground, and produce new plants to form quickly a dense soil-binding mat. Vine mesquite reproduces by creeping runners and rootstocks which transplant readily, as well as by seed. Its greatest value is for lining waterways and protecting their banks. Successful establishment for such purposes has been achieved by transplanting rootstocks. Seed may be harvested by power strippers or threshed from the hay. They usually are of low quality and germinate slowly.

*Switch grass* (*P. virgatum*) is a widely distributed perennial which reaches its greatest usefulness in the Great Plains. It develops extensive rootstocks and a leafy foliage, which, if cut at the right time, makes fair hay. Under grazing, the excess foliage becomes tough and unpalatable.

If left on the ground during winter, the dead material offers protection against the wind and forms a natural barrier to snow and soil drifting. For this reason, the plant is being utilized advantageously in parts of the Northern Plains.

Strains that show marked differences in resistance to insects (grasshoppers) and disease (rust) are being propagated at the Soil Conservation Service nurseries. Seed, frequently mixed with big bluestem, Indian grass, and other tall grasses, may be collected by power strippers. Excellent yields of high-quality seed have been obtained from cultivated stands.

*Panicum antidotale*, a native of the southern hemisphere, appears to be quite at home in the Southwest. It is drought-resistant, has spreading rootstocks, makes a large, rapid growth, and produces heavy crops of seed suitable for birds. It promises to become useful for several conservation purposes, including planting for windbreaks.

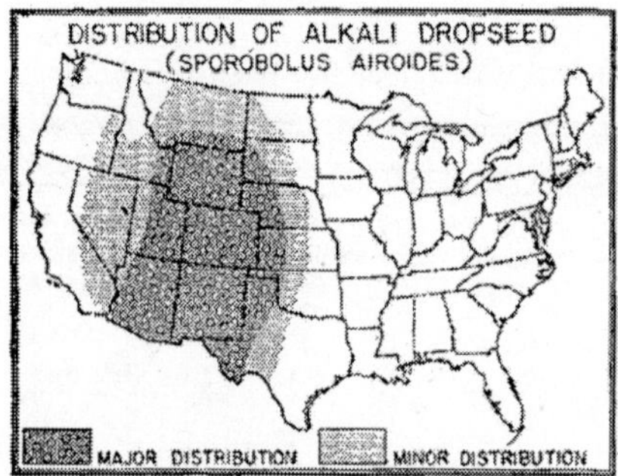


FIG. 153.—Distribution of alkali dropseed grass.

DROPSEED GRASSES. *Alkali Dropseed* (*Sporobolus airoides*), because of its heavy foliage, deep root system, drought resistance, ease of harvesting and seeding, and wide distribution, is very desirable for general revegetation purposes, particularly on highly alkaline soils. If allowed to mature, the plant becomes woody and unpalatable but retains its excellent soil-holding characteristic. Seed are easily collected from native stands or may be threshed from the hay (Fig. 153).

*Sand dropseed* (*S. cryptandrus*), an important native plant of the Great Plains, is commonly found on sandy soils. It is among the first grasses to appear naturally on unused farm land. Although not highly palatable, it provides effective ground cover. Seed harvested from native stands with strippers are often clean enough to be planted directly with farm equipment. A seeding rate of 2 pounds an acre has given a satisfactory cover. Scarification of the seed is necessary for best germination.

THE NEEDLE GRASSES (*Stipa spp.*). The needle grasses constitute a large group, among the more important of which are *needle-and-thread grass* (*S. comata*), *purple needle grass* (*S. pulchra*), and *green needle grass* (*S. viridula*). They occur generally throughout the Western States, particularly on dry, sandy plains and foothills, where they begin growth early and remain green over a long period. The needle grasses find their greatest use for conservation purposes by virtue of their persistent pioneering qualities, resistance to grazing, deep fibrous root system, drought resistance, and ability to produce good cover under severe extremes of both soil and climate. Excellent yields of seed have been produced under cultivation.

THE FINE-LEAVED FESCUES (*Festuca ovina and F. rubra*). In northeastern United States, these fine-leaved fescues will grow under conditions of poor soil, drought, and shade that are too severe for bluegrass. Their use is recommended on slopes of this character or for seeding ski trails and other places where summer erosion may start serious gullying.

KIKUYU GRASS (*Pennisetum clandestinum*). This South African grass spreads by runners that root at the nodes. It has been used to a limited extent in southern California in gully-control work and has proved valuable as a protective cover for small dams. Where adaptable, it is a vigorous grower and an excellent grazing grass. It will not endure frost.

WOOLLY FINGER GRASS (*Digitaria eriantha var. stolonifera*). This is another South African grass introduced into the Southern States as a pasture plant by the Bureau of Plant Industry. Its stoloniferous habit and drought resistance give it distinct erosion-control value. Its use has been limited by the fact that it seeds very sparingly. Plants have been found, however, in the Soil Conservation Service nurseries at Tucson, Ariz., that set viable seed. This latter strain is being propagated.

Other South African grasses that may have value are *love grass* (*Eragrostis curvula*) and (*E. lehmanniana*). The first is a large bunch grass; the latter, a semicreeper. Both produce an abundance of cover and heavy crops of seed. Although apparently adapted to the southern half of the United States, they may also have, because of ease of seeding and rapidity of development, an important place as annuals in control of erosion farther north.

CENTIPEDE GRASS (*Eremochloa ophiuroides*). This creeping perennial grass, introduced from China, has been found well suited to dry sandy soils in the South. Although commonly thought of as a grass for dry sands, it has proved useful in stabilizing waterways on better soils. The stems lie close to the ground and root at the nodes to form a dense mat. Producing no seed, centipede grass must be propagated by cuttings. These root readily. It is limited in range by its susceptibility to cold.

HARDING GRASS (*Phalaris tuberosa var. stenoptera*). This grass, introduced from Australia, where it is more commonly known as *Toowoomba grass*, promises to become a useful plant for controlling erosion in the warmer parts of the Southwest and along the Pacific Coast. It forms large, dense clumps with an extensive root system and produces an abundance of leafy foliage which remains green and suitable for grazing during most of the winter. It is quite resistant to drought. The large, plump seeds are much relished by birds.

## Shrubs, Vines, Succulents, and Other Plants of Possible Value

Although shrubs, vines, and herbaceous perennials do not, like grasses, find a place in extensive revegetation planting, many of them are useful in bank and gully control; some produce food acceptable to wildlife; and others are valuable browse plants which grow under difficult conditions. Some of the species studied and propagated are described below.

Bearberry, or Kinnikinnick, (*Arctostaphylos uva-ursi*). This plant grows on a variety of soils and in exposures ranging from dense shade to open slopes in the Northern States and throughout the Western mountains and foothills southward to Mexico. It forms a low-spreading bush 3 to 6 feet in diameter from which the branches take root and become firmly anchored to the soil. This property makes it especially useful for stabilizing the soil on steep, exposed banks and gully sides. The heavy crops of red berries hang on all winter and provide good food for game. Propagation is by layering.

Fig. 154.—Australian saltbush is useful for control of erosion on steep slopes and gully banks of southwestern United States. California. (*Photograph by Soil Conservation Service.*)

Southernwood (*Artemisia abrotanum*). This introduced shrub, which grows 3 to 5 feet high, is drought-resistant, winter-hardy, and easy to propagate and establish by cuttings (in the same manner as willow or cottonwood). Under the severe conditions of the northern Great Plains, it may provide an initial snowbreak or a component of more permanent windbreak plantings.

Saltbushes (*Atriplex spp.*). The native, as well as a number of introduced, saltbushes are markedly resistant to drought and high soil salinity. They develop as low, compact bushes 3 to 6 feet high and often spread over several feet. Some species are valuable for browse. One of the best, particularly useful for revegetation purposes, is *chamise* (*A. canescens*), a native, whose seed remain viable for long periods, awaiting favorable conditions for germination. *Atriplex nummullaria*, the "*old-man*" *saltbush* introduced from Australia, has a layering habit and is especially promising in the protection of dry banks and control of sandy washes. Under such conditions, the procumbent branches take root and form a compact vegetative mass.

The procumbent *Australian saltbush* (*A. semibaccata*, Fig. 154) has been established in parts of Arizona and along the Pacific Coast, where it has proved useful on steep slopes and gully banks. It is propagated from seed and makes a compact mass 3 to 6 inches thick and 3 to 4 feet in diameter the first season, indicating that it may be useful as an annual beyond its normal southern range. Besides having forage value, it produces heavy crops of sweet, fleshy fruits which are greatly relished by birds.

Desert Hackberry (*Celtis pallida*). The desert hackberry is a large, almost evergreen shrub generally found along dry washes in the Southwest. It is propagated by seed and has the habit of sprouting from the ground and forming dense thickets, which, together with its deep and extensive lateral root systems, makes it very effective in preventing erosion along washes. The plant also provides heavy crops of berries, as well as cover, for birds.

Mountain Mahogany (*Cercocarpus spp.*). Several species of this plant occur in the mountain regions of the West at elevations ranging from about 4,000 to 6,000 feet, frequently forming almost solid stands on mountainsides. The different species grow from 3 to 12 feet high, make excellent browse for cattle and deer, and propagate by self-planted seed. The sharp, pointed seed have long spiral awns which, when alternately moistened and dried, bore into the ground. This characteristic makes mountain mahogany especially valuable for erosion control, since it often is difficult to establish tree and shrub transplants under the adverse conditions obtaining in the drier parts of the West.

Desert Willow (*Chilopsis linearis*). This large, easily propagated and transplanted shrub or small tree, occurring mainly along washes throughout the warmer parts of the Southwest, succeeds under conditions of extreme drought and unfavorable soil. The plant has a deep taproot with well-branched laterals, sprouts vigorously from the stump after cutting, and tends to colonize. The latter characteristic makes it especially effective for use along desert washes and draws and in sandblow areas. In addition, the wood is very durable and is prized for fence posts.

DOGWOODS (*Cornus spp.*). The dogwoods, varying in size from shrubs to small trees, succeed on almost any type of land except dry or alkali soil. They produce berries relished by birds, and they have a definite place in soil conservation. Being especially adapted to shady situations, a number of species, including *C. florida*, *C. paniculata*, *C. asperifolia*, *C. fomina*, and *C. baileyi*, occur in the wild throughout the East and South. One of the best of the dogwoods for erosion-control purposes is the *red-osier dogwood* (*C. stolonifera*), commonly found in favorable locations throughout the North and westward from the Mississippi. Its prostrate, stoloniferous-tip layering habit and its adaptation to exposed situations make it peculiarly suitable for gully and streambank plantings. The introduced dogwood, or *cornelian cherry* (*C. mascula*), because of its dense growth and large, edible fruit, is also a valuable plant for use in erosion control and wildlife plantings. All of the dogwoods propagate from seed, either planted or stratified in the early fall before drying out. The red-osier dogwood also reproduces readily from tip layers.

EPHEDRAS. Both the introduced *Mahuang* (*E. sinica*, Fig. 155) and the native *E. viridis* have deep tenacious root systems, thrive under conditions of severe drought, and form a low-spreading, matted growth. They are suitable for sandy lands and sand-blow areas. *Ephedra sinica* has withstood temperatures of 30 degrees below zero. It contains ephedrine and produces heavy crops of berries that are edible by birds. *Ephedra viridis* has proved useful for control of soil blowing, particularly on the Navajo Indian reservation. It is effective in holding the soil, even where closely browsed. Although both species spread and propagate by natural layering, the simplest means of propagation is by seed.

WINTER FAT (*Eurotia lanata*). This small, woody perennial, distributed from North Dakota to Arizona and California and resembling white sage in general appearance, is an excellent forage plant. Extremely hardy and drought resistant, it is one of the best plants for use in range revegetation, since it can be seeded directly without the necessity of soil preparation. The seed come up very quickly without covering. The heavy production of seed is easily harvested with strippers. The forage is so palatable that care must be exercised to avoid destructive overgrazing.

APACHE PLUME (*Fallugia paradoxa*). This drought-resistant, near-evergreen shrub, growing 3 to 6 feet high, is widely distributed in the Southwest, mostly at elevations of 4,000 to 6,000 feet or more, and on a great variety of soils. In washes, the plants bend downstream during floods and often become buried only to sprout up from the tops. Underground stems resembling rhizomes or horizontal roots often sprout and grow into plants, thus making a tangled mass of roots and tops which is very effective in holding the soil. Propagation is by seed. The seedlings transplant readily.

Wild Olive (*Forestiera neomexicana*). A pendulous, heavily branched form of this deciduous shrub, native to southwestern United States, has been discovered that promises to be especially valuable for stabilizing gully slopes. It layers naturally, making dense, widely spreading clumps 3 to 6 feet high. Wild olive tolerates a wide variety of soils and occurs

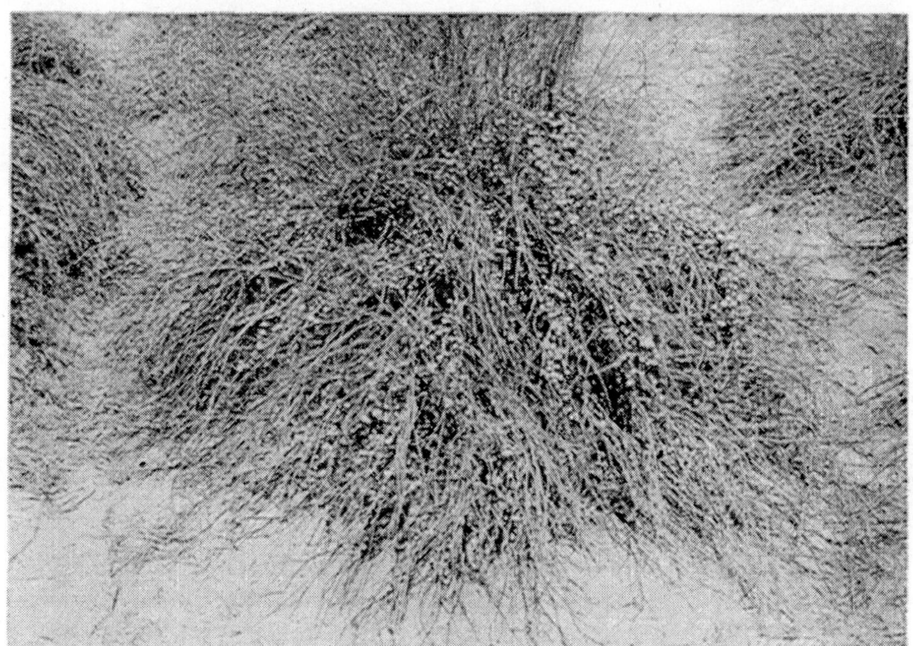

Fig. 155.—Mahuang (*Ephedra sinica*), a compact stoloniferous plant valuable for stabilizing sandy soil in the Southwest. (*Photograph by Soil Conservation Service.*)

both on steep hillsides and along the banks of washes and gullies. It is found at elevations of about 7,000 feet in the Southwest and seems to be hardy much farther north. It is easily propagated from seed and cuttings. The seedlings are easy to transplant. Heavy crops of berries good for bird food are produced.

Ozark White Cedar (*Juniperus ashei*). The Ozark white cedar is found principally in the Ozark Mountains, extending from northwestern Arkansas into southwestern Missouri, where it grows on extremely rocky, exposed sites. It is hardy as far north as the Dakotas. Its distinctive value over the other junipers is its apparent immunity to cedar-apple rust. In addition, it produces heavy crops of fleshy fruits which make good bird food and attains a size sufficient to make fence posts. Propagation is by seed sown or stratified early in the fall, before thoroughly dried.

BADLAND JUNIPER (*Juniperus horizontalis*). This most distinctly creeping of all the native junipers derives its special soil-binding value from the fact that it forms a very dense mat of tenaciously rooted branches. In nature, the plant grows on steep slopes under the most adverse soil conditions. As the name signifies, it is a truly "badland" plant. Although it occurs naturally mainly in the foothills of the Black Hills and westward into Montana, nursery trials show that it has possibilities of succeeding over a much wider range. Like most junipers, it is not a fast grower; but individual plants eventually attain a spread of 10 to 20 feet in diameter, effectively covering and holding the soil. This splendid erosion-control characteristic, combined with extreme drought resistance and hardiness, suggests its special use as a permanent cover for steep, badly eroding banks. Although this is one of the few junipers that propagates readily by layers taken from older plants, for general use the growing of young plants from seed planted or stratified in the early fall before drying is preferable.

DUNE BROOM (*Parryella filifolia*). This small shrub, native to the loose sands of northern Arizona and New Mexico, has succeeded as far north as North Dakota and Washington. It has a deep, widely branching root system and spreads by layers, making broad, dense clumps. The fact that the branches root as the sand piles around them enhances the plant's excellent qualities for wind resistance. It propagates readily from seed, and the seedlings transplant easily.

Another plant of similar habit useful on sandy lands is *bitterbush* (*Purshia tridentata*), which is distributed through the mountainous region from Oregon and Montana to Arizona and California. It has a special place for stabilizing sandy hillsides and gully banks.

SHEEP BUSH (*Pentzia incana*). This extremely drought-resistant native of the South African desert plateau survives and reseeds under the semiarid range conditions of the Southwest. It also has withstood temperatures as low as 17 degrees below zero, which indicates its possible value for much of the country where prolonged droughts occur and where erosion is severe. It is a distinctly layering shrub, has a well-branched root system, and forms a dense ground cover. It becomes dormant during the dry season but revives with the first rain. It is one of the few useful woody perennial shrubs able to compete with such unpalatable species as snakeweed and rabbit brush.

It sets heavy crops of seed. The most generally satisfactory method of propagation has been to seed directly during seasons of favorable moisture, either in the fall or in the early spring.

Bushmint (*Poliomintha incana*). The *bushmint* is another small shrub that has proved useful for holding sand dunes. Thriving on deep, sandy soils of the higher, colder elevations of the Southwest, the low, spreading plant layers readily and forms a dense surface root system which is very effective in stabilizing drifting sand. Its extreme drought resistance and winter hardiness give it a wide range of usefulness throughout the semiarid West, particularly since it is also valuable as a browse plant. Bushmint produces rather sparse quantities of seed which shatter shortly after ripening, thus making quantity collection rather difficult. This difficulty has been overcome, however, by producing the seed under cultivation. As the seed germinate quickly, direct seeding is an effective way for establishing bushmint. Nursery plants are produced easily and transplanted with a high percentage of survival.

Plums and Cherries (*Prunus spp.*). In some localities, especially throughout the Central States and westward, some of the low-growing native plums and cherries are finding a special place in soil-conservation practices. Combined with the value of the fruit for game and human food, the main characteristics that make these plants useful for erosion control are ability to succeed under adverse soil and drought conditions and capacity for holding the soil, especially in sand-blow areas. The low, heavily branching growth habit and the readiness with which the plants are established, either by layering or by sprouting, make them peculiarly adaptable for holding loose sand.

Among the more erosion-resistant of these stone fruits are: the *beach plum* (*P. maritima*), common along the sandy seashores of the North Atlantic Coast and adaptable to the Central States and farther west; the *sand plum* (*P. angustifolia var. watsoni*), conspicuous in the dry sandy areas of northwest Texas and New Mexico; the *sand cherry* (*P. pumila*), native among the sand dunes about the Great Lakes and adaptable to the country much farther west; and the *Western sand cherry* (*P. besseyi*). The Western sand cherry is especially valuable in that it is distinctly prostrate, layering in habit, and succeeds well under the trying conditions of the Great Plains. Here also the *Rocky Mountain chokecherry* (*P. melanocarpa*), because of its drought resistance, hardiness, and profuse sprouting habit, is very useful in preventing sand blowing.

Sumacs (*Rhus spp.*). Sumacs occur generally throughout the United States, the different species growing into small or large shrubs, or, as with *R. copallina*, *R. typhina*, *R. ovata*, and *R. sempervirens*, even to small trees. Some of the forms are found in shady or moist situations, whereas others, more especially *R. trilobata*, *R. laurina*, and *R. microphylla*, succeed under extremely dry conditions. The wide soil and climatic adaptations of the sumacs, together with their more or less thicket-forming habit, ramifying root system, and ability to produce heavy crops of berries for birds, make them particularly suitable for use in erosion-control practices. The *smooth sumac* (*R. glabra*) and *mountain sumac*

Fig. 156.—Japanese trailing raspberry (*Rubus parvifolius*). Washington. (*Photograph by Soil Conservation Service.*)

(*R. copallina*) are most commonly used throughout the Eastern States, and the more drought-resistant *lemonade berry* (*R. trilobata*) is better adapted in the West. The latter is widely distributed under very adverse conditions of soil and exposure, being equally at home on sand dunes, gravelly hillsides, and dry washes. Spreading by means of suckers, it frequently forms dense clumps 3 to 6 feet high and 10 to 20 feet in diameter. Most of the sumacs can be propagated from suckers, but the simplest method is from seed, cleaned and planted soon after collection or stratified in the fall and planted the following spring.

BLACK LOCUST (*Robinia pseudoacacia*). The black locust is used extensively for erosion control, in gullies as well as on eroding slopes. An especially superior strain is the "*shipmast*" *locust*. The latter's upright habit of growth, durability of wood, and apparent resistance to the locust borer make it especially desirable. It is seedless and must be propagated by root cuttings obtained from the younger wood.

RASPBERRY (*Rubus spp.*). Blackberries, dewberries, and raspberries are useful for gully control, bank protection, and hillside planting. Most of them are subject to disease and insect pests, but an immune form has been found in the *Japanese trailing raspberry* (*R. parvifolius*, Fig. 156), brought to the United States by P. H. Dorsett of the Bureau of Plant Industry in 1929. Selections of this layering trailer have proved resistant to all known insect and disease pests. The fruit is an excellent bird food and is fairly good for human consumption. These selections have good soil-binding qualities, wide climatic and soil adaptations, and are easily propagated by seed, tip layering, and root cuttings.

SAND-BAR WILLOW (*Salix exigua*). *Sand-bar willow* has proved useful in the Southwest for control of gullies and washes subject to occasional heavy flooding. It is a drought-resistant, fast-growing willow of brush type, forming dense thickets from a widely branching root system which sprouts prolifically. Planted at the base of gullies and above check dams, cuttings of this species very quickly form effective erosion barriers.

BUFFALO BERRY (*Shepherdia spp.*). Two of the native buffalo berries, *S. canadensis* and *S. argentea*, by reason of their resistance to extreme cold and drought, colonizing habit, and fruit production, have proved helpful in soil conservation work, particularly in the colder parts of the West. The fruit is useful for birds and human beings. *Shepherdia canadensis*, which forms a spreading, twiggy bush 3 to 8 feet tall, and *S. argentea* (*bullberry* or *silver leaf*), which attains a height of 10 to 15 feet, are both useful in soil conservation. These plants are propagated readily from seed.

JOJOBA, GOAT-NUT (*Simmondsia chinensis*). This compact evergreen shrub and good browse plant is found in the foothills of southern Arizona and California at elevations between 2,000 and 4,000 feet. The seed (or "nuts"), containing almost 50 per cent oil, are relished by sheep, goats, mule deer, and squirrels. The plant is drought-resistant but grows slowly and does not transplant readily. It has a place for erosion control along gullies and on steep rocky slopes within its range of adaptability.

SNOWBERRY (*Symphoricarpos spp.*). The *snowberry*, locally known as *buckbrush*, *waxberry*, or *wolfberry*, is native throughout the United States. It is a small shrub, more or less procumbent in habit of growth; it suckers freely, layers readily, develops a compact root system, and colonizes to form dense masses of vegetation. This combination of favorable characteristics gives it high rank as an erosion-control plant. The common Eastern form (*S. orbiculatus*) occurs usually on shaded hill-sides and along small drainages where moisture conditions are favorable. The Western species, widely represented by *S. racemosus*, *S. occidentalis*, *S. oreophilus*, and *S. utahensis*, are much more drought-resistant. An added quality of both types is the fact that they produce heavy crops of berries edible by birds, the Eastern form red and the Western white. Based on their natural characteristics and adaptations, all the species are especially useful for gully and hillside plantings.

LILAC (*Syringa vulgaris*). This hardy plant withstands extremes of heat, cold, and drought and for this reason has a special place in the composition of windbreaks.

LONICERA ALBERTI. This native of the dry, colder parts of Turkestan is a small, drought-resistant spreading shrub. Its arching, procumbent branches layer readily and form a compact mass. It has promise for the northern Great Plains.

BUCKTHORN (*Sageretia wrightii*). This is a low, spreading shrub of procumbent, layering habit, native to the foothills of Arizona and the country eastward into western Texas. It makes a tangled mass of vegetation on dry, rocky situations and may be useful for stabilization of gullies and steep hillsides. It produces an abundance of sweetly flavored berries edible to birds.

CEDAR OR THUJA. This is a special type of the *Chinese arborvitae* (*Thuja orientalis*) which is fast-growing, drought-hardy, and capable of maintaining itself in a windbreak planting under very adverse conditions. It is easy to propagate and more promising than other species of Thuja.

ASPARAGUS (*Asparagus officinalis*). This plant, with its persistent woody crown, tenacious root system, and ability to endure drought and alkali, has proved an effective soil binder of particular value in gully plantings and for protection of banks.

MORTON'S MILK VETCH (*Astragalus mortoni*). A native of the Pacific Northwest, this vetch spreads by underground rootstocks and grows on a variety of soils under conditions of scanty moisture. It may fit into cropping systems planned for control of erosion in low-rainfall areas. *A. rubyi*, a native of western Montana, produces good ground cover on alkali land. The growth is prostrate, and volunteer seedlings serve to thicken the stand. The plant has a high forage value.

Crown Vetch (*Coronilla varia*). A perennial legume from Europe, *crown vetch* has a trailing, compact habit of growth and a tenacious root system. It has escaped from cultivation in the northern and eastern parts of the United States and is well suited for protection of gullies and steep embankments.

*Euryops multifidus.* This herbaceous perennial composite from South Africa offers promise for erosion-control purposes in the warmer, drier parts of the Southwest. Producing woody crowns and spreading procumbent stems, 2 to 3 feet high, it survives severe drought and reseeds naturally. It may have special value for range revegetation purposes. It makes good browse, and the flowers attract bees in large numbers.

Ice Plant (*Mesembryanthemum roseum*). A native of South Africa, *ice plant* is a fast-growing, trailing succulent, which furnishes good ground cover for dry, exposed sites of the Pacific Coast where it is difficult to establish other vegetation. It is especially useful on road banks and terraces. Various strains or varieties are used in California for protection of roadside banks.

Bird's-foot Trefoil (*Lotus corniculatus*). A strain of this species, found in New York State, appears particularly promising for control of erosion in pastures. The plants are low-growing, have a heavy root system, and produce good cover. The species is a good seed producer and apparently is adapted to poor, eroded soils where clovers do not succeed.

Alfilaria (*Erodium cicutarium*). This winter annual from the Mediterranean region is well distributed in the Southwest and is useful even in southern Oregon and Washington. The seed germinate with the first fall rains, and the plants make an effective ground cover. Because the seeds bore into the ground, it is possible to establish the plant on rough, badly eroded land without disturbing the soil. Seed may be swept up in early spring after the plant debris has been cleared away. It should be planted in early summer without cleaning, since this would break off the awns and prevent the seed from boring into the soil.

## Weeds

Although weedy vegetation should be replaced wherever possible by more desirable plants, weeds often occupy a useful place in the control of erosion, especially as volunteer cover on abandoned and temporarily idle land. From the standpoint of soil conservation, it is better to have ground cover even with weeds than to have no cover at all. As a matter of fact, man has not had much choice about the spread of weeds over untilled, neglected land. Nature rather effectively attends to this.

Throughout the country, great areas of formerly used, erodible land are now covered with weeds. Some of this is well protected; other areas are partially protected; and most of the land is more stable than if no cover were present. Scores of plants, known as *weeds*, enter into this far-reaching cover of volunteer vegetation: goldenrod, ragweed, poverty grass, rabbit brush, snakeweed, wild oats, cheat, Russian thistle, crab grass, and hosts of others. Many of the weeds are objectionable of course; some are not particularly harmful; and others are of great service in their proper place. Russian thistle, for example, is of much value as a protective volunteer growth on land subject to wind erosion in the Great Plains and other Western areas, even though it spreads some hay-fever pollen and is a host to the beet leafhopper. Wild sunflowers are sometimes seeded with good effects in regions of low-rainfall as an aid in restoring grass to wind-whipped soil. Crab grass, although its control calls for hard labor in cotton, corn, and tobacco fields, provides considerable protection for the fields after the last cultivation and produces some fair hay in the Southeastern States. Other weeds tend to crowd out grass from pastures and range, especially where heavily grazed; and still others, following overgrazing, are poisonous to livestock. Experience indicates, however, that the use of sound conservation measures tends to favor the better grazing plants at the expense of those of less favorable characteristics, particularly in many parts of the West.

Ragweed, one of the major sources of hay-fever pollen, pioneers idle fields, fields in stubble, and abandoned areas in great profusion over a large portion of the United States, but it retreats rapidly under competition with other plants, including weeds. It is one of those weeds which, regardless of its ability to grow on very poor land and of any temporary conservation benefit performed, should be kept under control wherever possible, along with poisonous plants and other extra-noxious weeds.

One plant which has been extensively used in erosion-control operations, common Japanese honeysuckle, can cause much trouble if not handled with caution. Because of its spreading and twining characteristics, it can do much damage among other plants, even woodland growth, especially by smothering or deforming the smaller plants.

# Chapter XVIII. The Place of Forestry in Soil and Water Conservation

When Columbus discovered America, forests covered most of the eastern part of the continent. East of the Central Prairie region (see *Problem Area* map, "Results of Erosion," Chap. III, Part 1), with the exception of some relatively large bodies of prairie in what is now Indiana and Illinois and smaller areas in the sections now known as Alabama and Louisiana, all was comparatively unbroken forest. The first task of the pioneers was to convert some of this forest land to agricultural use. Each succeeding generation applied itself diligently to this task until the point was reached several decades ago when the process should have been reversed.

The report of the Chief of the Forest Service for the fiscal year 1937 gives the total forest area of the United States at 615 million acres (Fig. 157). This is divided as follows:

| | *Million Acres* | *Per Cent* |
|---|---|---|
| Publicly owned forests | 180 | 29.3 |
| Private forests | | |
| Farm woodlands | 185 | 30 |
| Other than farm woodlands | 250 | 40.7 |

The 185 million acres classified as farm woodlands is composed of many thousands of individual patches and relatively large bodies of woods. These little farm forests comprise almost a third of the entire forested area of the United States and represent more than 17 per cent of the total area of farm land. In the New England States very nearly half of the agricultural lands, an average of 48 acres per farm, is wooded. In the Ohio River Valley one-sixth of the farm land is forested. In the Southern States two-fifths of each farm is wooded.

Publicly owned forests generally are carefully guarded in order to provide protection from destruction by fire, and grazing is carefully controlled. Such protection and control safeguard the land against erosion. Forests, thus managed, contribute to the maintenance of stream flows and provide considerable protection from flood. From the standpoint of national welfare, it is important that these public forests be maintained and protected. The 250 million acres of private forests, other than farm woodlands, include most of what may be called the *commercial forests.* This vast area is only slightly less important than public forests in its relation to the stabilization of stream flow and protection against floods. It is relatively less important only because private forests generally occupy lower lands and gentler slopes. Such forests are, nevertheless, a major factor in relation to the general land and water economy of the nation. For this reason, it should not be too much to expect that in the course of the next 50 years the majority of these lands will be well cared for, either directly by the owners or indirectly through public regulation.

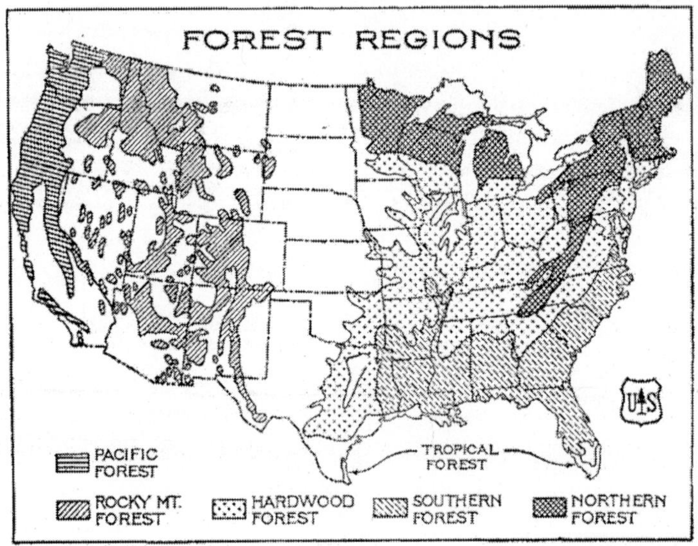


Fig. 157.—The total forested area of the United States amounts to approximately 32 per cent of the total land area. (*Forest Service, U. S. Department of Agriculture.*)

The 185 million acres of farm woodland are also a matter of particular concern, not only because the trees can and do play an important part in farm economy, but because farm woodland, properly handled, can make a substantial contribution to control of erosion and diminution of the flood hazard. Unfortunately, silviculture has not been fully recognized as a branch of agriculture; and partly for that reason, farm forestry generally has occupied a sort of marginal position in American agriculture. Too frequently, farm woodland has not been considered a productive unit of the farm enterprise, and income derived from it has been considered incidental and unusual. Any consideration of the fact that from 20 to 50 per cent of most farms in eastern United States is covered with woods of some sort should be sufficient to convince anyone of the importance of making this large area more productive than it is or has been in the agricultural past.

## Forests Control Soil Erosion

The capacity of trees to reclothe worn-out land is proved by the forests that have grown to maturity on what was once cropland (Fig. 158). Throughout many of the Eastern States, old stone fences can be found in the midst of newly grown forests. Once they marked the boundaries of cultivated fields. Now, corn rows still can be distinguished in the dense second growth covering some of these old fields. Even wild apple trees, descendants of farmers' orchards, sometimes are stumbled upon as one travels through such woodlands.

Fig. 158.—Good forests have grown to maturity on land formerly cropped and abandoned because of soil impoverishment, as on this Georgia hillside. (*Photograph by U. S. Forest Service.*)

In a good forest the treetops usually are close enough to touch and form a closed canopy; and frequently small trees, shrubs, and other forms of lesser vegetation make up a thin or thick undergrowth. Anything less is not first-class forest and is not likely to provide such good protection for the land beneath.

The leaves, twigs, branches, and stems of a forest expose innumerable little surfaces, aggregating, under good conditions, an area several times greater than that of the ground beneath. This loosely thatched roof, often 100 feet or more in thickness, is the first line of protection against soil erosion and excessive runoff. Driving rains beat upon this roof; the raindrops spatter, and the water slips gently down the stems or drips intermittently to the ground. As much as $\frac{1}{2}$ inch of rainfall may be completely intercepted by this intervening thatch. Part of this intercepted water is lost by evaporation and so never reaches the ground.

The main forest bulwarks against erosion and runoff, however, are still lower down. Covering the forest floor is a blanket of woods litter—a mass of leaves, twigs, and fragments of bark, in various stages of disintegration (Fig. 159). It is not always a smooth blanket, for beneath it are an endless series of little depressions. These catch part of the water that penetrates the thatched roof overhead and restrain much of it from running away. The blanket performs a double function—that of absorbing

Fig. 159.—The principal forest bulwark against erosion is the litter and mold covering the ground. (*Photograph by Soil Conservation Service*, 1936.)

part of the water and like a sieve directing the remainder downward to filter slowly to the soil beneath. This all-important blanket of vegetative material exerts a powerful influence on the soil in several ways, making it more permeable to water (Fig. 160). The surface of the soil is kept moist and absorbent, even in winter when exposed soil is deeply frozen. The litter and humus form the principal habitat for a vast population of organisms important to soil building and conditioning. From the standpoint of soil protection, this covering of the forest floor is the most effective element of the forest complex. Here is the gateway to water storage in soil and underground channels. Its function is to filter water, keep it clear, and so keep it moving downward into the soil. Muddy water, such as gathers on bare surfaces, clogs the channels, slows infiltration, and changes beneficial percolation to harmful runoff.

In any system of sound land use, it is essential that excessively steep slopes, such as usually prevail about the headwaters of numerous streams, be kept in some dense cover, preferably forests or grass. Trees make the

FIG. 160.—Virgin forest soils are generally characterized by highly permeable, mellow, or granular topsoil. Often in dry weather, when neighboring cultivated land is of almost stonelike hardness, forested soil is moist and soft enough to dig into with the bare hands. (*Photograph by U. S. Forest Service.*)

FIG. 161.—End view of a Nebraska windbreak. These shelters of hardy trees and shrubs protect adjacent land from soil blowing, protect livestock, and make homes of treeless areas more livable. (*Photograph by Soil Conservation Service*, 1937.)

most effective cover for vast areas of mountain and hill country, where it is not easy to establish and maintain an adequate cover of grass. The contrary is true, of course, with respect to the use of grass in those areas not climatically suited to rapid tree growth. Combinations of trees and grass, as well as combinations of trees and shrubs, have important places for special conditions in the control of runoff and erosion.

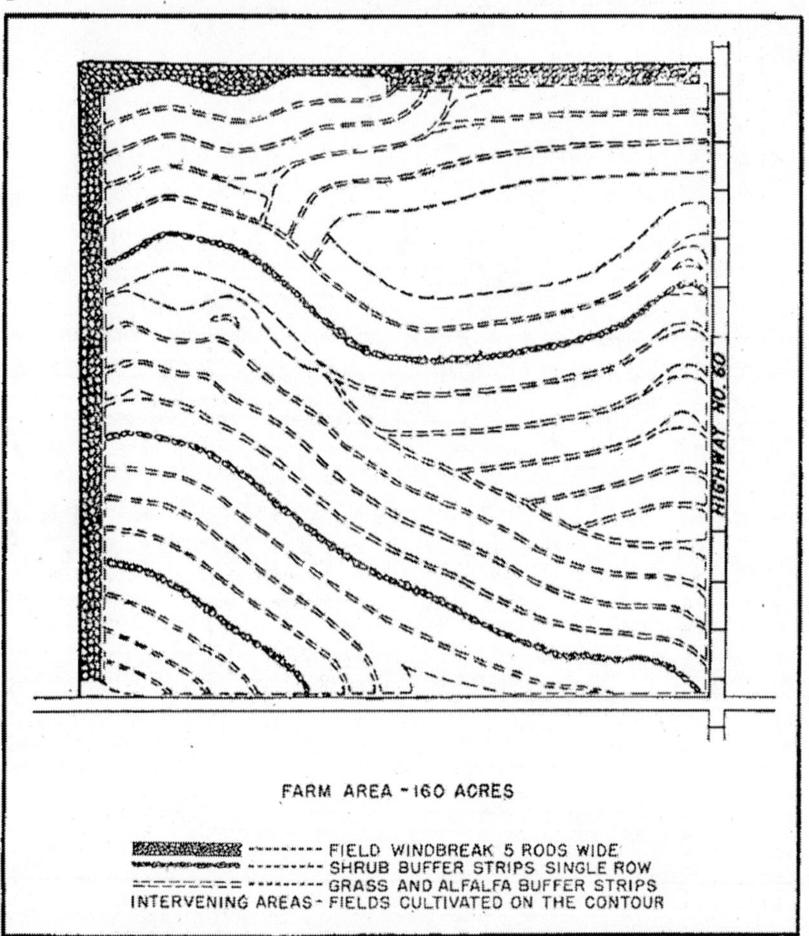


FIG. 162.—Soil-conservation plan of a North Dakota farm, showing how plantings of adaptable trees and shrubs are utilized. (*Soil Conservation Service.*)

Under conditions of low rainfall, a number of hardy trees and shrubs have proved successful in windbreak plantings. Such plantings are valuable not only from the standpoint of their utility but for their effect in making farm homes of treeless areas more livable and more comfortable in times of climatic extremes (Figs. 161, 162, 163).

## Tree Planting

Successful planting of trees, even on a small scale, is seldom a simple matter. The first, and probably most important, step is the selection of adaptable trees for particular sites. Correct determination of the species, size of stock, and cultural practices to be used is basic to successful

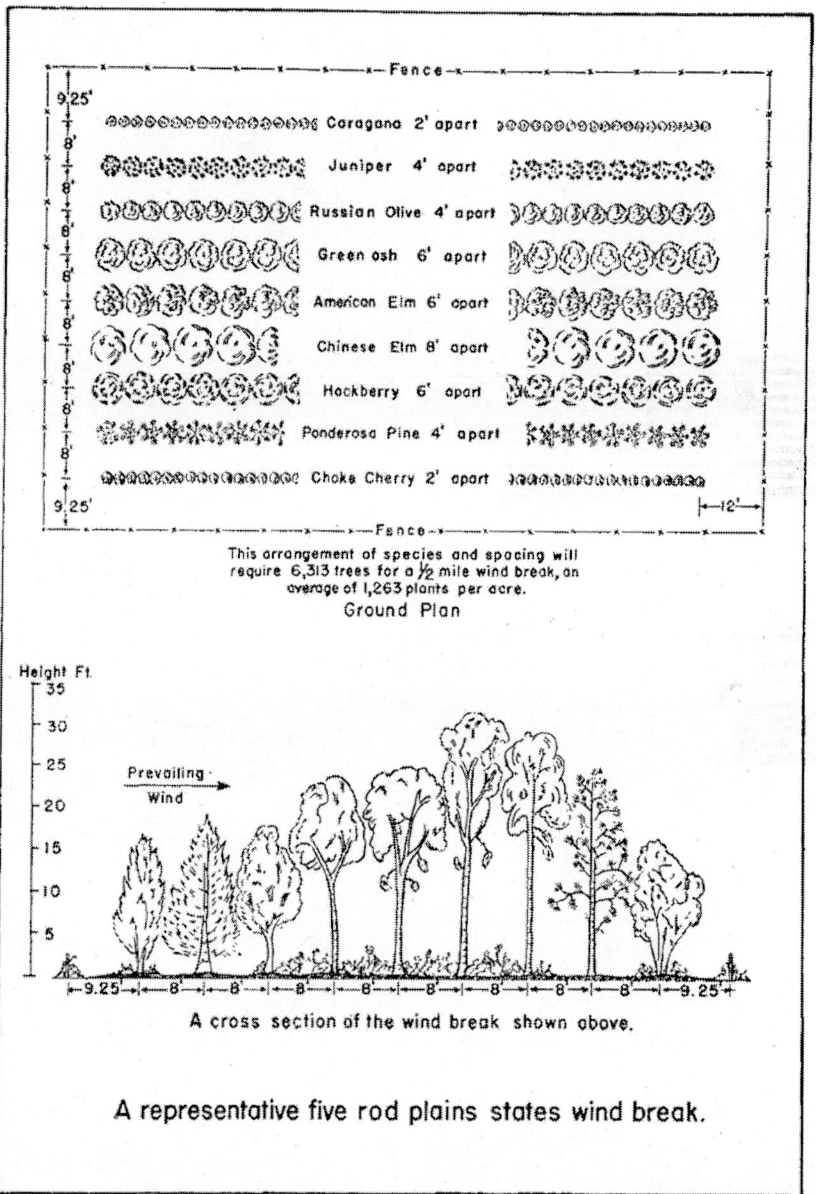


FIG. 163.—Successful windbreaks for the Great Plains need outside rows of shrubs to cut off ground wind, and taller trees in the center, to deflect above-ground wind. (*Soil Conservation Service.*)

afforestation. It frequently is necessary to plant, at first, species that may not ultimately occupy the areas to be protected. Unless cultural treatment promises to ameliorate quickly the adverse site conditions, a hardy species better suited to the prevailing conditions should be selected. Such local factors as elevation, exposure, quantity and distribution of rainfall, intensity and duration of sunlight, and intensity, constancy, and direction of the winds must be duly considered. Conditions of soil and erosion also must be given similar consideration. An understanding of the local indigenous flora frequently offers a useful guide. Where the site has been markedly disturbed by forestry or agricultural operations or by fire, remnants of the natural vegetation may be found in protected places, which afford visible evidence of the primitive plant association and often constitute the nucleus from which vegetation spreads to reclothe the affected area. This new vegetation, like the old, struggling against unfavorable conditions, frequently overcomes them and gradually reacquires the characteristic form of the original stand. Careful observation of these changes and conditions often affords an excellent basis for evaluating sites and selecting the trees best suited for plantings.

Such studies may, of course, show that the altered soil conditions caused by man's interference with the site have resulted in a marked degradation of the site, under which circumstances it would be a serious error to attempt to establish immediately the same species that once thrived in the locality.

Not only have the trees of such depleted areas been destroyed, but the shrubs and perennials as well. The problem of proper reforestation, therefore, may involve reestablishment of the entire flora, beginning with those plants that will lay down an essential ground cover of organic litter and mold. This would require the successful planting of those pioneer species which aid in preparing the land for a higher type of forest.

Every climatic zone has in its flora at least one species suitable as temporary or preliminary vegetative cover, even on poor land. Among the shrubs, the number of species is generally larger than among trees. Often the most adaptable species for conditioning the site have little commercial value.

As already indicated, it is frequently possible to overcome adverse site conditions by proper preparation of the soil. In some places where erosion has so impoverished the land that only worthless kinds of vegetation will grow, the landowner is neither satisfied to plant species without economic value nor willing to wait on the slow process of natural rehabilitation. It therefore becomes necessary to prepare the land for immediate planting by cultivation, mulching, fertilization, or other means.

Hardwoods or broadleaf trees generally must have either a better site or more careful preparation for planting and better care afterward than coniferous trees. The native pines of the South, for example, will grow on severely eroded land with little or no preparation of the soil and without subsequent care other than protection from fire and grazing. If black locust is desired for fence posts, however, preparation of the soil is generally needed, and even cultivation and fertilization of the trees during the first two or three years after planting.

As a rule, the more adverse the site the greater is the need for preparation for planting and frequently for attention after planting, especially with broadleaf trees. Adaptable trees, for example, can be grown in the more favorable situations of low-rainfall areas, such as the Great Plains, but real success depends on careful preparation of the land to be planted and later cultivation. Here the land generally should be plowed and contour-listed or basin-listed and carefully tilled to conserve rainfall and prevent weed growth for a year in advance of planting; and subsoiling may be needed in the tree rows. After planting, both trees and shrubs may require cultivation or weeding for several years or until the growth becomes too dense for the use of machinery.

On areas of erosion-exposed raw subsoil, it frequently is impossible to get trees or shrubs started satisfactorily without covering the surface with a mulch of such vegetative material as straw, forest litter, rotted hay, or barnyard manure. Where soil and topographic conditions permit, plowing under a legume cover crop is one of the best methods of soil preparation. On thin land, even the planting of pines is a good first-step measure of soil preparation for hardwoods. The mat of needles accumulating when fire is kept out gradually conditions the soil for the broadleaf trees. In the East, such hardwoods as oak, maple, beech, hickory, dogwood, and gum will gradually intermix with or even replace the pines.

PREPARATION FOR GULLY PLANTING. Preparation for planting gullies frequently requires the building of temporary dams in the bottoms to catch enough silt to get the trees started. Bank sloping is often advisable along steep-sided gullies; and in many instances, water must be diverted from the heads before any success can be attained with plantings in the gully itself.

NURSERIES. The Soil Conservation Service maintains a series of nurseries throughout the United States, with a productive capacity of between 150 and 200 million trees and shrubs. These seedlings are used in connection with the national program of soil and water conservation in which the Service is engaged. Additional millions of trees are being produced in Forest Service and state nurseries and in commercial nurseries.

The experience of the Soil Conservation Service has shown that much more farm land, both cultivated and uncultivated, should be restored to forest than previously had been suspected. On the basis of detailed farm surveys made in connection with the erosion-control program, it is estimated that approximately 50 million acres of land now in farms should be planted to trees and shrubs in order to protect it from erosion, conserve rainfall, produce wood supplies, and benefit wildlife. If this work is to be carried out, about two billion trees and shrubs will be required annually over a period of 25 years.

Because of the risks in handling planting stock—in taking the young trees from the soil, bundling, heeling in, and transportation—foresters agree that better results are had where the seedlings are obtained near the planting site. This reduces the rehandlings and the danger of losses from overheating, drying out, spread of disease, and other unfavorable results of handling. Farmers who can go into near-by woods and collect acorns or nuts or wild planting stock for immediate use on their lands will be able to meet part or all of their needs.

Planting and Maintenance. From the heel-in bed to planting site and the actual setting out of the seedlings, every care must be taken to move the work along according to a well-planned schedule, step by step, and with proper technique (Fig. 164) until the trees are in the ground. The difference between good planting and poor planting is often the difference between the success or failure of the operation. Large plantings, therefore, require the assembling of labor, the training of men, and adequate arrangements for transportation.

The time of planting is probably one of the most, if not the most, important factors in the survival of trees. Throughout most of the United States, the planting season is limited to a few weeks in the spring. It begins as soon as the frost is out of the ground and the danger of freezing is past. The sooner the trees are in the ground under these conditions the better the chance for success.

The spacing of trees ranges from 1-foot intervals (in both directions), as in the planting of willows along stream banks, to as much as 10 or 12 feet where natural growth may be expected to fill the gaps. Usually about 6 by 6 feet is considered standard spacing. Under conditions where the establishment of a quick cover is needed, the spacing frequently is 4 by 4 feet; and with shrubs, 2 by 2 feet. A certain amount of site preparation may be done at the time trees are planted, such as *scalping*, or removal, of the sod or other vegetation over a space usually about 18 to 24 inches square.

After planting, protection from fire and grazing must be provided. The safeguarding of large forest plantings from fire may require the construction of fire lines and patrol during dry seasons. Protection from stock may require either herding of the animals or their exclusion by fencing. Occasionally, artificial shading is necessary, as under the adverse conditions of semiarid climate or where the trees are of unusual value.

After the first year, some replanting is usually necessary because of the failure of some of the seedlings. Experience of the Soil Conservation

Service indicates that, on the average, failure of survival may be expected to require replanting about 15 per cent of the original number of trees set out.

In rare instances, watering of individual trees is necessary. Thinning of stands is a cultural operation which comes considerably later than other maintenance measures. Ordinarily, it is not considered practicable to thin stands until the trees are pole size or large enough to yield some fuel wood or fence posts. The need for thinning in young stands is usually indicated by a general slowing up of the rate of growth, particularly in diameter. Growth can be determined readily by cutting a few trees to determine the relative width between the annual growth rings. If the distance between

Fig. 164.—Men must be trained to plant trees properly. The difference between good planting and poor planting is often the difference between success and failure. West Virginia. (*Photograph by Soil Conservation Service.*)

the outer rings is about the same as or greater than that between the inner rings, the tree is still maintaining good growth. If it is found that the width of the inter-ring spaces is narrower toward the bark, then growth is slowing down, and the trees need more space.

It is sometimes necessary to remove the lower branches of growing trees in order that clear wood may be produced. Ordinarily, density of stand aids in natural pruning; but if the lower branches persist and are not removed artificially, many valuable species will not yield a high-grade product. Only those more promising trees should be pruned—those which are growing fastest and have the best form. Ordinarily, not more than 150 crop trees per acre should be selected for pruning. Trees over 8 inches in diameter generally should not be pruned.

For best results in pruning, the branches should be removed with a saw operated as close to the trunk as possible. New wood grows over such a cut very rapidly, with the result that more clear wood is produced. The height of pruning should not exceed the length of clear wood desired. Pruning up to 16 to 20 feet will give one clear high-grade log. The cost of such work must be justified by the increased income from logs of higher quality.

## Management of Farm Woodlands

The greatest enemy of forests is fire. Fire destroys the forest litter and young trees; it often kills even the larger trees. Frequently, if trees are not killed, they are scarred and made more susceptible to insect attack and

FIG. 165.—Forest rejuvenation as a result of fencing against grazing. Ohio. (*Photograph by Soil Conservation Service.*)

disease. The forest canopy is thinned, and the soil often exposed to erosion. Severe forest fires frequently destroy all the trees.

Continued grazing of farm woodland usually results in some degree of damage, frequently severe damage. A good fence to keep domestic animals out of the woods is one of the best forms of forest insurance (Fig. 165). Since livestock browse young trees, both hardwoods and conifers, grazing not only injures or destroys young trees and practically all the undergrowth but frequently leads to injury or destruction of many of the large trees. Grazing of woodland tends to impair or destroy forest litter, and this, in turn, reduces infiltration, speeds up runoff, and lowers the productivity of the forest. Investigations by the Ohio Agricultural Experiment Station[1] show that for the sites studied, open pasture produces two and one-half times more grass and that its nutritive value is 20 to 60 per cent greater than woods grass. Woodland grazing usually yields relatively poor returns in pounds of meat and at the same time damages the forest.

One of the important phases of the forestry job in connection with control of erosion is to convince the farmer of the value of his farm woods. When farmers come to see the value of protecting their woodlands from fire, grazing, and neglect, they usually set about the job of preventing abuse and building up the woods. Generally, they need some technical help; but once convinced of the benefits to be derived from protection, they are likely to accomplish a good deal on their own initiative.

VALUE OF FARM FORESTS. Millions of board feet of lumber are utilized every year in the construction and maintenance of farm buildings. The agricultural enterprise of the nation makes use of millions of posts annually in the maintenance of fences. Wood, the chief fuel on the farm, is used in nearly all parts of the country. Many other forest products are used, poles, stakes, braces, logs, fence rails, handles, and wood for curing tobacco and processing syrups.

Good farm woodlands create a favorable habitat for many kinds of wildlife. A burned or closely grazed woods offers little cover and almost no food for wildlife.

Where farm woodlands produce more than is required for the farm needs, the surplus often can be marketed to advantage. Moreover, many farms offer excellent opportunity for development of woodlands to the point of producing wood materials for sale, as cash crops are produced. On numerous farms, properly managed woodland is producing net annual returns in stumpage of $1 to $4 per acre, exclusive of labor returns.

By selective cutting, forests that are properly cared for generally can be maintained on a permanently productive basis. Under modern silvicultural practice, a variety of species and sizes of trees should be maintained. Farm needs for posts, poles, fuel, and rough lumber should be supplied by cutting defective, poorly formed, slow-growing trees and those species which make up more of the stand than is required in a desirable mixture. By constantly selecting the poorest trees for home use, cutting only mature trees of good quality for market, and by taking every precaution to hold the quantity removed equal to or less than the quantity grown in the same period, even a badly neglected woodland generally can be built up gradually to at least a fair condition of productivity. The growing stock in the woods must be built up in a way something like that of the

[1] Welton, F. A., and Morris, V. H. Woodland Pasture, Ohio Agr. Exper. Sta. *Bi-monthly Bull.*, Vol. 14, No. 1, 1929.

stockman who improves his herd by selective removals and replacements. In a forest, there is no objection, generally, to the presence of tall, straight trees of high value for saw logs, even if they are 80 to 100 years old, provided the undergrowth contains trees of all ages. By cutting only those trees that are ripe for the market, together with the wood needed on the farm, an annual or periodic income can be obtained. Cuttings at relatively short intervals are conducive to good forest growth.

Best results are obtained from farm woodlands where a good management plan for maintenance and marketing is followed. Development of such a plan requires, first, a determination of the amount of wood material that can be removed, based on rate of growth, and then a division of

FIG. 166.—Demonstration of selective cutting in a Pennsylvania farm woodlot. (*Photograph by Soil Conservation Service.*)

the woodland into blocks or parcels suitable for rotation cutting. If five blocks, or areas, are established, each would be selectively cut, under such a system, once every 5 years.

In its cooperative work with farmers, the Soil Conservation Service follows a simple procedure of education and the application of practicable silvicultural measures. Where practicable and needed, a stockproof fence is built. Applicable measures for protection against fire are provided, such as the establishment of fire lines and the most advantageous system of disposing of slash. In severely eroded localities, it may be entirely practicable to utilize the slash, or part of it, for gully-control purposes or for mulching raw subsoil areas as an aid to reestablishment of trees or other forms of vegetative cover.

In order to arouse farmers' interest and action, practical foresters must work with them—plan and help to install adaptable systems for planting, fire protection, cutting, disposal of slash, etc. Furnishing technical advice and aid has proved the most effective means for arousing interest and, in some localities, the only means of getting action (Fig. 166).

Unfortunately, some element of the pioneer farmer's philosophy that a piece of woods was something in the way of farm progress still persists in some localities. Even where cutting is not conducted with the idea of bringing the land immediately into cultivation or into use for grazing, it frequently involves some purpose connected with farming operations rather than with woodland improvement. These concepts must be changed before the farmer can understand the full advantages of forest management. Since agricultural practice has long been governed by a "groundline" philosophy which calls for the complete removal of crops above the ground surface and which relies for the next crop on seeding or reproduction from roots, it is not surprising to find that many farm woods consist of trees that have come from sprouts (*coppice woods*). Coppice woods are produced by clean cutting, just as a crop of asparagus springs from the roots, following previous cuttings. The difference is that the asparagus yields satisfactorily under the system, whereas sprout trees generally are of unsatisfactory quality. Fortunately, in the farm forestry demonstration projects, it has been possible to change this viewpoint along the lines indicated. The fundamental concept and policy of the Soil Conservation Service that, in so far as practicable and economically feasible, every acre of land must be treated in accordance with its needs and adaptabilities have aided greatly in educating farmers with respect to the value of good forest for protecting highly erodible land. Having gained this much, it is proving much easier to induce farmers to go ahead with good forestry practice. When the ugly scars of erosion are effaced with trees, when excessive runoff from critical slopes is stopped, when wildlife comes back as the result of forest plantings or forest upkeep, most farmers feel compensated for these advantages alone. Subsequently, when products for sale begin to mature in the wooded areas, and better crops are produced on land protected by the trees, it generally may be considered, according to the experience of the Service, that the average farmer is then definitely won over to a sound farm forestry program.

# Chapter XIX. Contouring

Contouring refers to any tillage practice or mechanical treatment of range, pasture, cultivated, or other class of land, applied across the slope on the level, that is, on the contour. Over most of the United States, it has been customary to apply the various tillage practices in straight lines approximately parallel to field boundaries, regardless of the direction of slope. In contrast, contour cultivation disregards field boundaries and straight lines completely, following curved lines whenever necessary to stay on the level.

In regions of low rainfall, the primary purpose of contour farming is to provide maximum conservation and distribution of rainfall; in humid regions, on the other hand, the primary purpose is to reduce soil loss by erosion. The objectives in both instances are obtained in precisely the same way. The furrows or ridges developed on the level by contour tillage operations catch and hold the water, or part of it, in such manner as to store it in the soil, thus reducing runoff and erosion and bringing about a more uniform distribution of moisture.

Contour farming has other advantages. Among the most important is the saving in power and time. Results of tests conducted at Manhattan and Hays, Kans., show that for the areas on which the measurements were made, tractor cultivation on the contour is 12.8 per cent faster and 9.4 per cent more efficient with respect to fuel consumption (Table 34). In general practice, these results would vary considerably, depending on the slope, kind of soil, and time consumed in turning the tractor because of the short rows usually caused by contour tillage.

Contour farming has been practiced in various parts of the world for hundreds of years. In the United States, contour cultivation was practiced in some localities more than a century ago (see Chap. XL, Part 2). It has been most extensively employed in the Southeastern States, where the use of hillside ditching and terracing probably encouraged its development. Outside this area, the practice still is not at all in common use. In southeastern Ohio and the West Virginia Panhandle, contouring has been practiced on some farms, in conjunction with strip cropping, for more than 50 years. It has been used extensively near La Crosse and in the Mormon Coulee Valley, Wisconsin, for several decades.

TABLE 34.—COMPARATIVE TIME AND FUEL CONSUMPTION FOR TRACTOR CULTIVATION ON THE CONTOUR AND ON 7.56 PER CENT SLOPE[1]

| | Slope, per cent | Average speed, miles per hour | Area covered per hour, acres | Fuel consumption, pounds per acre |
|---|---|---|---|---|
| Up- and downhill | 7.56 | 3.88 | 2.43 | 7.63 |
| Contour | 0 | 4.27 | 2.74 | 6.91 |

[1] Tests conducted at Manhattan and Hays, Kans., by the Soil Conservation Service and the state agricultural and engineering experiment stations, cooperating.

Since the establishment of the Soil Conservation Service, with its many demonstrational areas throughout the country, thousands of farmers have come to recognize contouring as a necessary part of effective erosion control on cultivated land. Ranchers also are finding it a very important tool in conserving rainfall for the replenishment of grass on overused ranges.

## Field Contouring

Contouring is a necessary adjunct to strip cropping of fields and should be used in connection with terracing as a means of providing additional protection for the interterrace area and for reduced maintenance costs. On some areas of gentle slopes, contouring is sometimes used alone as an efficient measure for control of erosion.

Contouring is employed extensively as a means of conserving water in dryland areas, especially those affected by wind erosion. Here, tillage operations generally are performed with some type of furrowing implement, principally the lister. The damming, or basin, lister, which develops a dam at intervals in the lister furrow, is coming into quite general use in some localities. The furrow produced is of sufficient size to hold a large amount of water which otherwise would be lost as runoff (Fig. 167), averaging 6 inches deep, 4 inches wide at the bottom, and 25 inches at the top. It has been estimated, for example, that contour furrows of this size, spaced at intervals of 42 inches, have a mechanical storage capacity of approximately 2 inches of rainfall, or about 56,000 gallons of water per acre. In areas where moisture is the limiting factor in crop production, and where a large part of the rainfall is of such intensity as to produce quick runoff, it is clear enough why such treatment gives outstanding response in vegetative growth.

In some wheat-growing areas, the one-way disk plow is being replaced to some extent as a tillage implement by the field cultivator. This is because the disk plow leaves a comparatively smooth surface which offers little mechanical resistance to water movement, whereas the field cultivator forms shallow furrows which hold and distribute rainfall. A similar effect results from drilling wheat on the contour with a deep or semideep furrow drill.

Unfortunately, lister furrows lose much of their effective water-holding capacity after the crops planted in them have been cultivated once or twice. Since it is undesirable in wind erosion areas to disturb crop residues by listing until after the blow season of late winter and early spring, this

FIG. 167.—Work of the damming lister on contour for conservation of rainfall. South Dakota. (*Photograph by Soil Conservation Service.*)

means that lister furrows on cultivated land retain their maximum water-holding capacity for a short period only. For this reason, level terraces have been used with contour tillage to obtain maximum conservation of water, except on the sandier soils.

Although contour furrows produced by wheat drills have comparatively little water-holding capacity, they retard the rate of runoff and so conserve some water. From the standpoint of water conservation, wheat land in the Great Plains generally is benefited by level terraces, even where the land has a slope of less than 1 per cent. Contour listing, both with and without basin attachments, is being used more and more in areas affected by wind erosion. This is especially true in the Southern Plains.

## Pasture Contouring

Contour furrowing of pasture land is a practice that was used to only a very limited extent, if at all, before the advent of the recent nation-wide soil conservation program. During the early days of the program of the Soil Conservation Service, terraces were used in pastures for moisture conservation as well as for control of erosion. It was found that on slopes

FIG. 168.—Contour furrows made with moldboard plow in a Minnesota pasture. For conservation of rainfall. (*Photograph by Soil Conservation Service.*)

FIG. 169.—Contour furrowing with machine that transposes a slice of sod to side of furrow. Kansas. (*Photograph by Soil Conservation Service.*)

ranging generally from about 5 to 25 per cent or more, pasture terraces gave very poor distribution of water. And so the terraces were made constantly smaller and brought closer together, until finally they became contour furrows or a combination of furrows and ridges.

Such structures are now made, generally, with ordinary moldboard plows (Fig. 168). Much good turf is ripped up when they are placed close enough together to provide optimum distribution and conservation of rainfall, especially in bluegrass pastures. A special contour furrowing machine recently developed in Kansas picks up the furrow slice and sets it to the side without overturning, thus saving much of the sod (Fig. 169). Another recent machine developed in Iowa lifts a slice of sod so as to destroy no sod and yet make a combination furrow and ridge 6 to 8 inches deep.

FIG. 170.—Eighty per cent increase in grass (buffalo and grama) following contour furrowing of this 240-acre tract in the Colorado Plains, at a cost of 46 cents per acre. This increased growth was due to conservation of rainfall that hitherto had run off the land as waste water. (*Photograph by Soil Conservation Service.*)

In contouring newly established pastures, a recent practice is to plow narrow strips 4 to 6 furrows wide. This produces a series of ridges and furrows which are not completely worked down in preparing the seedbed. The practice can be especially recommended because of its effectiveness and inexpensiveness. The seedbed is prepared at the same time the furrows are made.

## Range Contouring

The contouring of range lands has proved a very beneficial practice (Fig. 170). Observations indicate that on heavy land with a deep topsoil, furrows about 5 inches deep, spaced at intervals of 42 to 84 inches, give best results. Ridges have proved most effective on sandy land sloping not more than about 3 per cent.

Although various types of implements have been used to construct contour furrows, the three-row lister, with the middle bottom removed, has proved one of the most practical. It is highly desirable that a strip of undisturbed sod be left between the furrow slices. Accordingly, listers with the moldboards removed, and even with the shares clipped, as a means of reducing the area of grass to be covered, have given especially good results. This procedure should be followed, generally, where a two-row lister is used. A chisel fastened ahead of the lister point has proved helpful

Fig. 171.—Contour furrows in Colorado made with terracer equipped with furrowing attachment. The purpose is to conserve rainfall that runs off as so much waste. (*Photograph by Soil Conservation Service.*)

in the furrowing of land having impervious, compact subsoil. Moldboard and disk plows, graders, and various special-type implements (Fig. 171) also have been used in many parts of the country.

Contour furrows have proved more practical than contour ridges in those low-rainfall areas where smooth surface conditions prevail, except on the very gentle slopes. The furrow has a number of advantages: It revegetates much more quickly than the ridge and does not interfere with mowing operations. Cross sections of the various types of furrows are shown in Graph 30.

## Contouring for Flood Control

Since contour furrows store rainwater at or near the point where it falls, they obviously have a definite value for flood control. In the opera-

Standard lister bottom

Listers, without moldboards, with wings clipped

Special attachment for terracer blade

Iowa machine

Moldboard plow in preparing seed bed

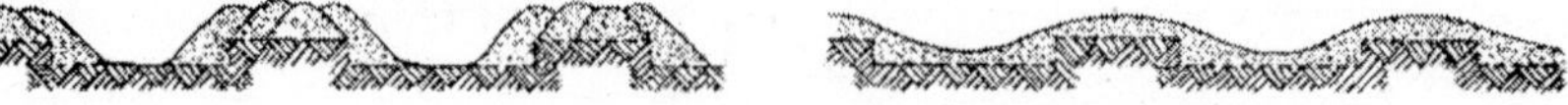

After plowing After harrowing

Graph 30.—Cross sections of various types of furrow. (*Soil Conservation Service.*)

tions of the Soil Conservation Service, floods apparently have been completely controlled on a number of small watersheds where all or most of the area was contour-furrowed. Table 35 shows the size of furrows and the spacing necessary to store various quantities of rainfall within watersheds susceptible of treatment.

TABLE 35.—SPACING AND SIZE OF CONTOUR FURROWS REQUIRED TO STORE VARIOUS AMOUNTS OF RAINFALL

| Storage capacity, inches of water | Cross section (square feet) and spacing (feet) | | | | | |
|---|---|---|---|---|---|---|
| | Square feet | Feet | Square feet | Feet | Square feet | Feet |
| 0.5 | 0.5 | 12 | 1 | 24 | 2 | 48 |
| 1.00 | 0.5 | 6 | 1 | 12 | 2 | 24 |
| 2.00 | 0.5 | 3 | 1 | 6 | 2 | 12 |
| 3.00 | 0.5 | 2 | 1 | 4 | 2 | 8 |
| 4.00 | 0.5 | 1.5 | 1 | 3 | 2 | 6 |
| 5.00 | 0.5 | 1.2 | 1 | 2.4 | 2 | 4.8 |

## Results

Contour furrowing of range land has caused marked improvement of both the quantity and quality of the forage in numerous instances through the western grazing area. For example, a representative area in the Hereford, Tex., soil and water conservation project treated in April, 1937, with furrows 4 inches deep and 8 inches wide, spaced 7 and 14 feet apart, produced 1,761 pounds of dry grass per acre, as against only 704 pounds produced on an untreated adjacent area of comparable land. At the Spur, Tex., experiment station, range land contour listed in 1932 produced 2,369 pounds of dry forage per acre, whereas an unlisted area of similar land produced only 725 pounds per acre.

At Dalhart, Tex., 21 contour-tilled fields, comprising 4,035 acres, produced a yield of 589 pounds of grain sorghum per acre. The average yield of seven fields of comparable land, totaling 882 acres, tilled according to the straight-row practice, gave only 461 pounds per acre.

Yields of wheat obtained from 56 fields contour-drilled and straight-row drilled, at the Vega, Tex., soil and water conservation project, are given in Table 36.

The cost of contour furrowing varies with the implements used and with the size and spacing of the furrows. A single-row lister, pulled by four horses, making furrows 3 inches deep by 12 inches wide, at intervals of 84 inches, has been used extensively at a cost of 27 cents an acre. Ridges ranging from 40 to 50 feet apart have been constructed with a turning plow at 10 cents per acre.

TABLE 36.—YIELDS OF WHEAT FROM 56 FIELDS CONTOUR-DRILLED AND STRAIGHT-DRILLED, SOIL AND WATER CONSERVATION PROJECT, VEGA, TEX.

| Contour-drilled | | Straight-row drilled | |
|---|---|---|---|
| Number of fields | Yield, bushels | Number of fields | Yield, bushels |
| *Following wheat, sorghum, or weeds* | | | |
| 5 | 0 to 2 | 12 | 0 to 2 |
| 6 | 3 to 6 | 9 | 3 to 6 |
| 1 | 11 to 14 | 0 | — |
| Average per acre per field..... | 2.8 | .. | 2.2 |
| *Following summer fallow* | | | |
| 0 | — | 1 | 0 to 2 |
| 3 | 3 to 6 | 2 | 3 to 6 |
| 7 | 7 to 10 | 2 | 7 to 10 |
| 4 | 11 to 14 | 1 | 11 to 14 |
| 2 | 15+ | 1 | 15+ |
| Average per acre per field..... | 9.25 | .. | 7.6 |

# Chapter XX. Terracing

Since Colonial times, American farmers have built various types of field terraces and hillside ditches to divert runoff and conserve soil on cultivated slopes. For undetermined centuries, agriculturists of older countries have used bench-terraces effectively to combat soil erosion and facilitate cultural practices on steep land. The field terrace, as used in the United States today, is essentially an artificial measure designed primarily for control of runoff in high-rainfall areas and for conservation of water in low-rainfall areas. Control of erosion is the ultimate objective in the more humid areas and a very important objective in the dry regions.

The basic mechanical feature of field-terracing and hillside ditching for surface drainage control is a channel, with a ridge below, constructed across the slope at vertical intervals to intercept and divert runoff before it gains sufficient volume and velocity to erode the soil. In the construction of the modern graded field terrace, spacing, size, and slope are so regulated that runoff can be collected and conveyed slowly along the channel to a protected outlet. Usually, both channel and ridge have a broad cross section, so designed as to permit cultivation over the entire structure. Terraces of the ungraded or level type are constructed somewhat along the same mechanical lines as the graded structure for retaining and distributing rainfall that otherwise would be lost as runoff. Outlets are not provided for, however, when the purpose is to hold all the rainfall.

The present development of terracing is the result of years of use, extensive field experimentation, and many modifications from early construction practices.

Properly located and constructed and adequately supported by cropping and tillage practices, they provide one of the most effective erosion-control measures applicable to cultivated land. If improperly constructed or not adjusted with proper land-use practices, they often accelerate rather than retard soil losses.

## Types of Terraces

As the result of attempts to improve the efficiency of terraces, several kinds of structures, and modifications thereof, have evolved.

Although there is fairly general acceptance of the broader functional classifications of terraces, techniques of construction are changing continually as the result of additional field experience and the accumulation of research data. Moreover, construction specifications vary from region to region, according to local conditions and requirements and, to a considerable degree, according to the opinion of the constructor.

Classified roughly, the three principal types are:

1. *Bench* type, with steplike, flattish surface between steep risers (steep slope to next terrace platform), designed (*a*) to control erosion by reducing the gradient of the cultivated area and (*b*) to make steep land available for cultivation.

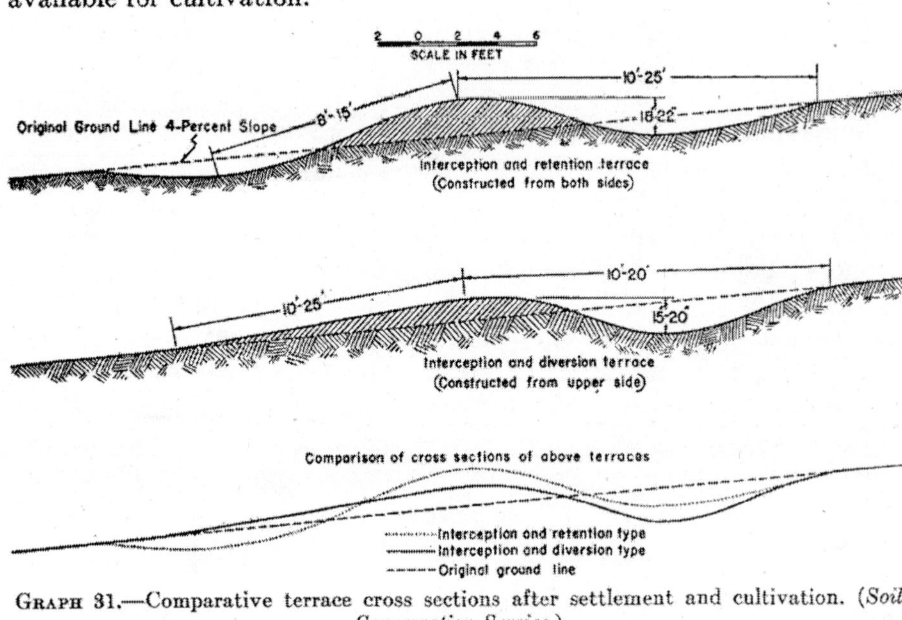


Graph 31.—Comparative terrace cross sections after settlement and cultivation. (*Soil Conservation Service.*)

2. *Graded-channel* type, without change of interterrace slope, designed primarily to intercept and divert runoff at nonerosive velocities. This *interception-and-diversion* type has been variously referred to as the *drainage-control* type, *runoff-control* type, *channel* type, and *American* type.

3. *Level* type, designed to impound runoff primarily for conservation of rainfall (Graph 31). This *interception-and-retention* type is sometimes designated the *absorption* or *retention* type.

## Bench Type

In various parts of the world, bench-terraces have been extensively constructed at great expenditure of labor, in order to make cultivable or to stabilize sloping land. Such structures consist of platforms, or nearly level benches, constructed in steplike series across slopes. The platforms are separated by steep to almost vertical risers or retaining walls supported by rock or vegetation.

In the strict sense, these bench structures represent the closest approach to a true terrace. Their use has been restricted largely to areas of dense population, where the supply of level land is insufficient for the needs of the people and where labor is cheap.

The practice of bench-terracing is almost as old as agriculture itself. Centuries before the Conquistadors ascended the Andes, the pre-Incas had attained great skill in the construction of essentially level terraces, supported by substantial artificial walls of stone. The structures were used extensively in the building of broad fields in the valleys and in the

FIG. 172.—Bench terraces on a 20 per cent slope, southern Piedmont, developed by gradual movement of soil toward vegetated terrace. Georgia. (*Photograph by Soil Conservation Service.*)

establishment of narrow contour fields on steep mountain slopes. Generally, soil was filled in behind the stone walls, and irrigation water, brought frequently from remote sources, was applied. Many of these ancient terraced fields are still in use.

In the Philippine Islands, the natives probably were building their present type of bench-terrace 2,000 years ago. Walled terraces have been used successfully for centuries in the Mediterranean Basin, the Orient, and other parts of the world. There is evidence that prehistoric Indians of Mexico and southwestern United States used bench-terraces to facilitate irrigation. The practice still is used to some extent by the Pueblos of New Mexico and the natives of Mexico.

The oldest examples of bench-terracing by American white men probably are those found in the rolling sections of the Southeastern States

(Fig. 172). They have been used frequently in landscaping. In the citrus and avocado districts of southern California, bench-terraces are coming into common use on steeply sloping land. In other states, they have been used occasionally and rather sparingly on truck farms and in vineyards and orchards.

Most of the old bench-terraces in the Southeastern States have been formed gradually through the combined effects of cultivation and erosion. Collecting in vegetation growing along the terrace banks (Graph 32), material moved downhill by tillage and erosion has had the effect, after years, of producing many steplike terraces which appear to have been originally constructed as such. Usually, however, the inner, or uphill, side

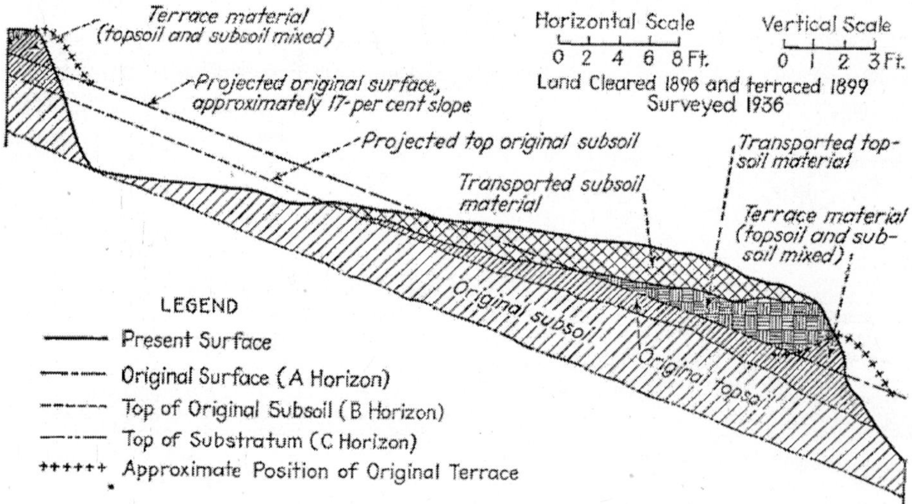


GRAPH 32.—Cross section of an old terrace in the South Carolina Piedmont. The inverted soil profile at the right has been developed by soil movement between terraces. Unless the topsoil is relatively deep in comparison with the vertical interval between terraces, subsoil is eventually exposed over the entire surface. (*Soil Conservation Service.*)

has been stripped of topsoil, and thousands of abandoned terraces have been cut to pieces by gullying. Most of these, if not all, originally were constructed as ordinary narrow-ridge field-terraces or as hillside ditches. Erosion between the ridges or ditches has resulted in filling and benching and finally in neglect and failure.

Graph 32, based on a detailed study of an old Mangum terrace and the terrace interval above it, in the Piedmont section of South Carolina, shows how erosion tends to develop a bench-terrace at heavy cost in soil wastage. Because of excessive slope (about 17 per cent), this terrace and its neighbors, together with the entire field, have been ruined by deep sheet washing and gullying (Fig. 173). Examination of the profile of the lower terrace shows what happens to the soil during the leveling process that goes on between most terraces on excessively steep land as well as on most terraces that are not supported by proper cropping systems.

The California type of bench-terrace, sometimes referred to as the *Reddick terrace* (because of H. E. Reddick's efforts in adapting it to citrus

FIG. 173.—Despite the use of terraces, this South Carolina Piedmont land is too badly eroded for agricultural uses. Same abandoned field as shown in Graph 32. (*Photograph by Soil Conservation Service.*)

FIG. 174.—California bench terrace. (*Photograph by Soil Conservation Service.*)

groves and orchards), generally is constructed in the following manner: Rows for planting trees are laid out on irrigation grade lines, varying with the soil and slope; and along these lines, a small ridge is thrown up with two or three furrows. On this ridge, the trees are planted. Irrigation water is applied along the upper side of the ridge. Subsequent cultivation is in the direction of irrigation and is conducted in such a way that a narrow strip of vegetation is left in the row of trees. By avoiding cross cultivation, a bench-terrace is formed gradually. When the bench is completed, after some years of cultivation under irrigation, the trees stand part way down the steep downhill slope, amply protected by vegetation (Fig. 174).

Before constructing bench-terraces in any area, thorough study should be made to determine whether or not justification exists for cropping steep slopes that require this type of protection. If suitable lands with gentle slopes are available, or if profitable returns cannot be expected, the steeper slopes should be utilized for purposes other than cultivation.

### Graded-channel Type

The graded-channel, or interception-and-diversion, type of terrace appears to have been developed in the United States and to have been used more extensively in this country than in any other. During the latter part of the eighteenth and the beginning of the nineteenth centuries, farmers in the Southeastern States began to use devices known as *hillside ditches.* These ditches were the real forerunners of the present-day field terrace.

Records indicate that the term *terrace* was applied to these erosion-control structures as early as 1847. From the Southeast, the use of terracing spread westward into Oklahoma and Texas. A number of farmers who had moved from the older Southern States were terracing in eastern Texas by 1903.[1] In Oklahoma, farmers of Payne and Carter counties were building terraces as early as 1907 and 1908.

The use of terracing also has spread northward to some extent from the Southern States. During the last few years, the American system of terracing land has been introduced into other countries. Reports from Australia, Rhodesia, and Kenya indicate a steady increase in the use of terracing in these countries since its introduction around 1930.

The hillside ditches used by early farmers in the Southeast were dug or plowed out across the slope to intercept and convey surface runoff from erodible fields. These ditches usually were placed from 200 to 400 feet apart, depending on the slope. Channel gradients frequently as steep as 2 or 3 per cent or more were used. Inadequate capacity caused overtopping; and water, accumulating in the steeply graded ditches, frequently flowed so swiftly that troublesome gullies were formed. Even with these failures, the structures were sufficiently successful to induce many farmers to continue their use year after year.

[1] In 1882, Duke Howell of Newton County, Texas, did some terracing.

Many of the early pioneers attempted to improve their hillside ditches, with varying results. One of the developments was a structure known as the *narrow-ridge terrace.* Generally, this also lacked adequate water-carrying capacity; the grades were steep; and the narrow ridges could not be cultivated. Soil movement down the slope between the ridges caused many of them to develop into bench-terraces.

Eventually, the need for such construction refinements as wide ridges and channels; variable grades; and spacing according to rainfall intensity, soil character, and slope was recognized and advocated by some of the early terrace builders. In general, however, farmers were slow to adopt such proposed improvements. One of the most important developments in the history of terrace construction came in 1885 as a result of the work of Priestley H. Mangum, a farmer living near Wake Forest, N. C. Mangum introduced the *broad-base* terrace, by widening the narrow ridges previously used, so that tillage operations could be conducted over the entire structure. Modifications of the *Mangum terrace* are used extensively today in many parts of the country.

It was not, however, until the practice of terracing received the critical attention of the agricultural colleges and the United States Department of Agriculture that extensive progress in terrace construction was made. After some preliminary study, an investigational survey of the use of terraces to combat soil erosion was begun in 1915 by Ramser, then drainage engineer of the Office of Experiment Stations. The results, based on studies of the methods of terracing used in the Southeastern States, the degree of success attained, and the factors affecting their success or failure, were published in 1917.[1] This report served as the basis for many later developments. Nichols later, while with the Alabama Polytechnic Institute, initiated studies relating to improvement in terrace design. His investigations led to certain modifications in construction procedure, particularly with respect to the terrace channel.[2] Additional information from the state agricultural experiment stations and from the Federal soil and water conservation experiment stations cooperating with the states and the field experience of the Soil Conservation Service have contributed to improvement in terrace construction. Since about 1915, county agents and others have devoted much time to educational work on the use of terracing.[3]

[1] Ramser, C. E. Prevention of Erosion of Farm Lands by Terracing, U. S. Dept. Agr. *Bull.* 512, 1917.

[2] Nichols, M. L. An Improved Channel-type Terrace for the Southeast, U. S. Dept. Agr. *Farmers' Bull.* 1790, 1937.

[3] Hamilton, C. L. Terracing for Soil and Water Conservation, U. S. Dept. Agr. *Farmers' Bull.* 1789, 1938.

## Technique of Construction

Grades. The effect of channel gradients on runoff and removal of soil from the terrace channel and the indicated effect on rate of soil loss from various soil types are shown in Graphs 33 and 34. These measurements, including similar measurements for a number of other important agricultural soils,[1] show that on most soils a reduction in channel gradient results in less soil loss, a lower rate of water loss at the terrace outlet, and, generally, less runoff. They indicate that the most satisfactory grade to use in the construction of terraces is the least grade that will provide adequate discharge of runoff.

This investigational work also shows that although the *variable-grade* terrace is slightly more efficient in conserving soil, it does not require so great a height or cross section toward the outlet as terraces of uniform grade, since the maximum rate of runoff is less for the structures of variable grades. The milder grades near the upper end of a variable-grade terrace tend to hold back water until that below has an opportunity to flow off. This prevents concentration near the outlet, thus obviating the need for an extra-large channel capacity. These advantages of the variable-grade terrace over the *uniform-grade* type far outweigh the objection that the former is somewhat more difficult to lay out.

Grades of more than 4 inches per 100 feet of length are seldom necessary or advisable. In field practice, a variable grade is established which increases uniformly every 300 to 500 feet in the direction of the outlet. The common grade for areas of relatively high-rainfall intensities and soils of average absorptive capacity is a fall of 1 inch per 100 feet for the first 400 feet of channel length; 2 inches per 100 feet for the second 400 feet; 3 inches per 100 feet for the third 400 feet; and 4 inches per 100 feet for the fourth length of 400 feet. Where the soil is of exceptionally good absorptive capacity, such as sandy loam, a common grade arrangement is level for the first 400 feet; a fall of 1 inch per 100 feet for the second 400 feet; 2 inches per 100 feet for the third 400 feet; and 3 inches per 100 feet for the fourth 400 feet. In areas of low rainfall and soils of average characteristics, a grade arrangement of level for the first 500 feet; a fall of 1 inch per 100 feet for the second 500 feet; and 2 inches per 100 feet for the third 500 feet may be advisable.

Spacing. Although a number of measurements indicate that soil erosion increases with increase in distance between terraces, other measurements indicate the contrary. Such inconsistent results probably are due largely to inadequate methods of measuring the soil movement

[1] Cecil sandy clay loam at Statesville, N. C.; Shelby loam at Bethany, Mo.; Houston black clay at Temple, Tex.; Muskingum silt loam at Zanesville, Ohio; and Clinton silt loam at La Crosse, Wis.

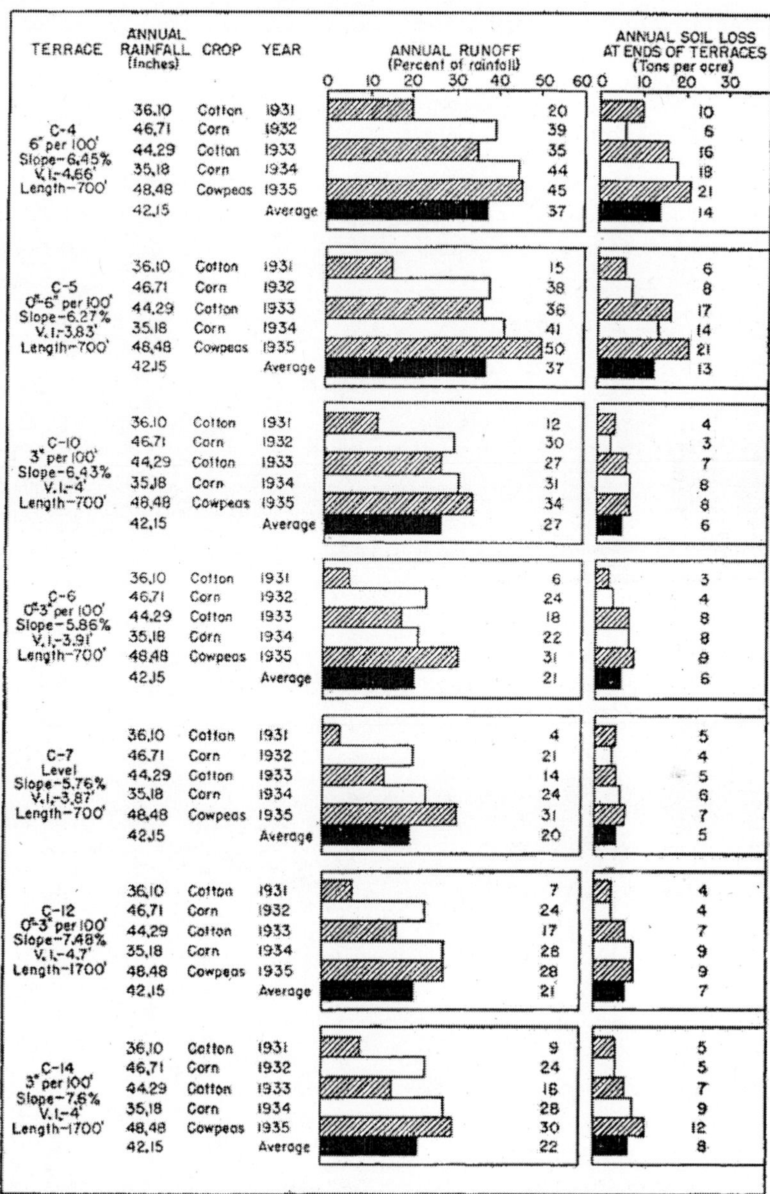


GRAPH 33.—Effect of terrace grade on runoff and soil removal from terrace channel, and indicated effect on rate of soil loss, from Kirvin, Nacogdoches, and Bowie fine sandy loams, as measured at the Tyler, Texas, soil- and water conservation experiment station, 1931 to 1935, inclusive. (*Soil Conservation Service.*)

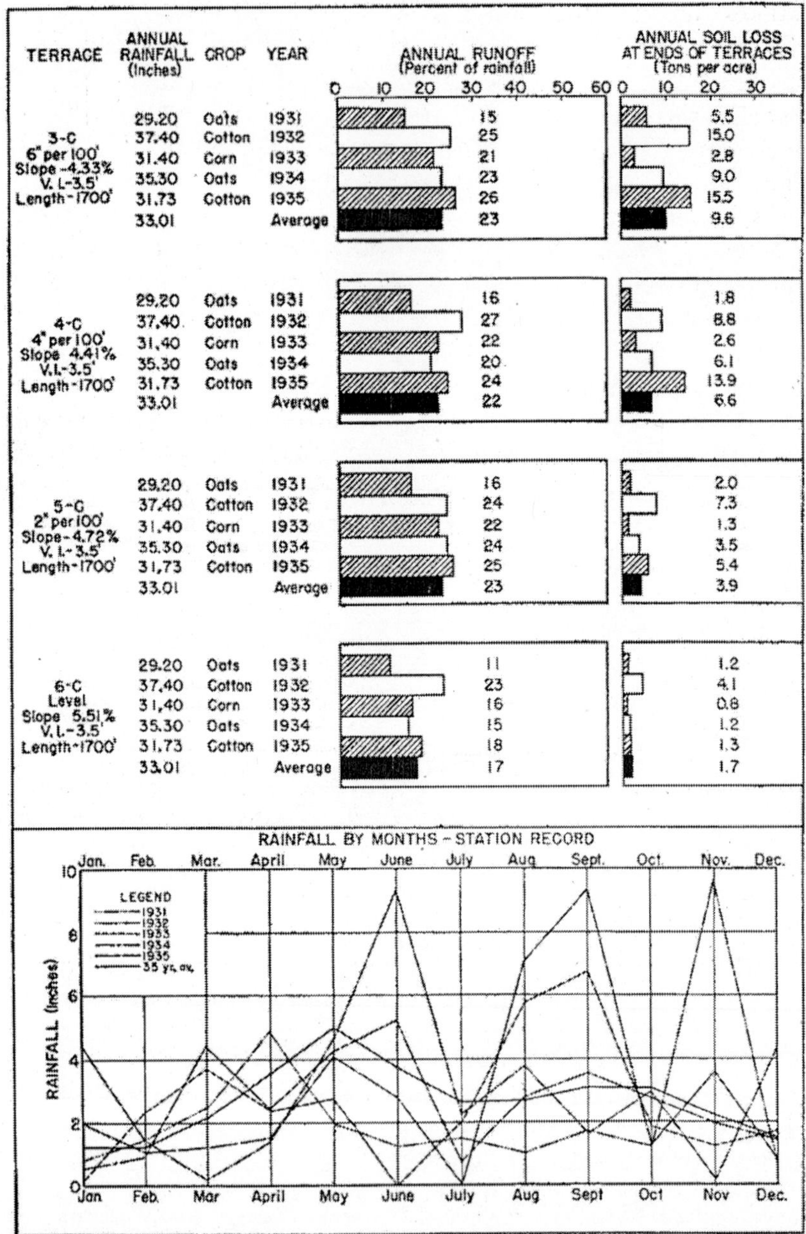


**Graph 34.**—Effect of terrace grade on runoff and soil removal from terrace channel, and indicated effect on rate of soil loss, from Vernon fine sandy loam, as measured at the soil and water conservation experiment station, Guthrie, Oklahoma, 1931 to 1935, inclusive. (*Soil Conservation Service.*)

that takes place between terraces. Soil losses measured at the ends of the terraces do not necessarily indicate the extent of soil movement between terraces.

Specifications for spacing terraces are based largely on field observations covering the results of many years of experience with terracing. A fundamental principle in effective terracing is to locate the structures so that they will intercept the runoff from the area above each terrace before its velocity has become great enough to cause severe erosion and its volume great enough to exceed the carrying capacity of the terrace channel. Also, the spacing must be wide enough to facilitate necessary tillage operations. Variations in soil, crops, and rainfall conditions have a marked influence on the extent of erosion that may occur on slopes of different lengths and so affect spacing requirements. Crop and soil conditions, however, usually cannot be made a major basis for adjustments in terrace spacing, since the structures must provide safe protection during the critical periods of maximum runoff and scarcity of vegetative cover. These are the periods when greatest soil losses occur and when terraces are most effective. The primary purpose of good cropping practices on terraced land is to improve the productivity of the soil and to increase its organic content and structure so as to increase its capacity for rainfall absorption rather than to permit major adjustments in terrace specifications. Such treatment provides for increased acre yields and minimizes soil movement between terraces.

In the North, terrace intervals may be somewhat greater on comparable soils and slopes than is permissible in the South, because of greater protection of the soil in winter by freezing and the frequent inclusion of more erosion-resistant crops in the rotations. Rainfall intensities also are generally lower in the Northern States.

The spacing specifications given in Table 37 are for the graded-channel type of terrace and are based on Ramser's recommendations. The minimum and maximum values shown vary from the average by 15 per cent. When good cropping practices are followed, and where the soil is exceptionally resistant to erosion, and low-rainfall intensities are characteristic of the area to be terraced, the spacing might be increased as much as 15 per cent with reasonable safety. Under opposite conditions, the spacing generally should be decreased as much as 15 per cent. With intermediate combinations of favorable or unfavorable factors, corresponding intermediate increases or reductions should be made in the spacing. In practice, a favorable factor frequently will be offset by an unfavorable one and so preclude any deviation from the recommended average for spacing. For example, the value of a good erosion-resistant rotation may be offset by a highly erodible soil or by high-rainfall intensity.

TABLE 37.—RECOMMENDED TERRACE SPACINGS[1] AND RELATED DATA FOR INTERCEPTION-AND-DIVERSION TYPE OF TERRACE

Southern States—$VI = 2 + S/4$

| Per cent slope ($S$) | Vertical interval ($VI$) | | | Horizontal distance | | | Acres per mile of terrace[2] | | | Acres per 100 feet of terrace[2] | | | Feet of terrace per acre[2] | | |
|---|---|---|---|---|---|---|---|---|---|---|---|---|---|---|---|
| | Minimum,[3] feet | Mean, feet | Maximum,[3] feet | Minimum,[3] feet | Mean, feet | Maximum,[3] feet | Minimum,[3] acres | Mean, acres | Maximum,[3] acres | Minimum,[3] acres | Mean, acres | Maximum,[3] acres | Minimum,[3] feet | Mean, feet | Maximum,[3] feet |
| 1 | 1.70 | 2.00 | 2.30 | 170.00 | 200.00 | 230.00 | 20.60 | 24.24 | 27.88 | 0.390 | 0.459 | 0.528 | 256.23 | 217.80 | 189.39 |
| 2 | 2.12 | 2.50 | 2.87 | 106.25 | 125.00 | 143.75 | 12.88 | 15.15 | 17.42 | 0.244 | 0.287 | 0.330 | 409.98 | 348.48 | 303.03 |
| 3 | 2.34 | 2.75 | 3.16 | 77.92 | 91.67 | 105.42 | 9.44 | 11.11 | 12.78 | 0.179 | 0.210 | 0.242 | 559.03 | 475.18 | 417.16 |
| 4 | 2.55 | 3.00 | 3.45 | 63.75 | 75.00 | 86.25 | 7.73 | 9.09 | 10.45 | 0.169 | 0.172 | 0.198 | 683.29 | 580.80 | 505.00 |
| 5 | 2.76 | 3.25 | 3.74 | 55.25 | 65.00 | 74.75 | 6.70 | 7.88 | 9.06 | 0.127 | 0.149 | 0.172 | 788.42 | 670.15 | 582.74 |
| 6 | 2.97 | 3.50 | 4.02 | 49.58 | 58.33 | 67.08 | 6.01 | 7.07 | 8.13 | 0.114 | 0.134 | 0.154 | 878.58 | 746.79 | 649.37 |
| 7 | 3.19 | 3.75 | 4.31 | 45.53 | 53.57 | 61.61 | 5.52 | 6.49 | 7.47 | 0.105 | 0.123 | 0.141 | 956.73 | 813.14 | 707.03 |
| 8 | 3.40 | 4.00 | 4.60 | 42.50 | 50.00 | 57.50 | 5.15 | 6.06 | 6.97 | 0.098 | 0.115 | 0.132 | 1,024.94 | 871.20 | 757.56 |
| 9 | 3.61 | 4.25 | 4.89 | 40.14 | 47.22 | 54.31 | 4.86 | 5.72 | 6.58 | 0.092 | 0.108 | 0.125 | 1,085.20 | 922.49 | 802.06 |
| 10 | 3.82 | 4.50 | 5.17 | 38.25 | 45.00 | 51.75 | 4.64 | 5.45 | 6.27 | 0.088 | 0.103 | 0.119 | 1,138.82 | 968.00 | 841.74 |
| 11 | 4.04 | 4.75 | 5.46 | 36.70 | 43.18 | 49.66 | 4.45 | 5.23 | 6.02 | 0.084 | 0.099 | 0.114 | 1,186.92 | 1,008.80 | 877.16 |
| 12 | 4.25 | 5.00 | 5.75 | 35.42 | 41.67 | 47.92 | 4.29 | 5.05 | 5.81 | 0.081 | 0.096 | 0.110 | 1,229.81 | 1,045.36 | 909.01 |
| Northern States—$VI = 2 + S/3$ | | | | | | | | | | | | | | | |
| 1 | 1.70 | 2.00 | 2.30 | 170.00 | 200.00 | 230.00 | 20.60 | 24.24 | 27.87 | 0.390 | 0.459 | 0.528 | 256.23 | 217.80 | 189.39 |
| 2 | 2.27 | 2.67 | 3.07 | 113.47 | 133.50 | 153.52 | 13.75 | 16.18 | 18.61 | 0.260 | 0.306 | 0.352 | 383.89 | 326.29 | 283.74 |
| 3 | 2.55 | 3.00 | 3.45 | 85.00 | 100.00 | 115.00 | 10.30 | 12.12 | 13.94 | 0.195 | 0.230 | 0.264 | 512.47 | 435.60 | 378.78 |
| 4 | 2.83 | 3.33 | 3.83 | 70.76 | 83.25 | 95.74 | 8.58 | 10.09 | 11.60 | 0.162 | 0.191 | 0.220 | 615.60 | 523.24 | 454.98 |
| 5 | 3.12 | 3.67 | 4.22 | 62.39 | 73.40 | 84.41 | 7.56 | 8.90 | 10.23 | 0.143 | 0.168 | 0.194 | 698.19 | 593.46 | 516.05 |
| 6 | 3.40 | 4.00 | 4.60 | 56.67 | 66.67 | 76.67 | 6.87 | 8.08 | 9.29 | 0.130 | 0.153 | 0.176 | 768.66 | 653.37 | 568.15 |
| 7 | 3.68 | 4.33 | 4.98 | 52.58 | 61.86 | 71.14 | 6.37 | 7.50 | 8.62 | 0.121 | 0.142 | 0.163 | 828.45 | 704.17 | 612.31 |
| 8 | 3.97 | 4.67 | 5.37 | 49.62 | 58.38 | 67.13 | 6.01 | 7.08 | 8.14 | 0.114 | 0.134 | 0.154 | 877.87 | 746.15 | 648.89 |
| 9 | 4.25 | 5.00 | 5.75 | 47 22 | 55.56 | 63.89 | 5.72 | 6.73 | 7.74 | 0.108 | 0.128 | 0.147 | 922.49 | 784.02 | 681.80 |
| 10 | 4.53 | 5.33 | 6.13 | 45.30 | 53.30 | 61.29 | 5.49 | 6.46 | 7.43 | 0.104 | 0.122 | 0.141 | 961.59 | 817.26 | 710.72 |
| 11 | 4.82 | 5.67 | 6.52 | 43.81 | 51.55 | 59.28 | 5.31 | 6.25 | 7.18 | 0.101 | 0.118 | 0.136 | 994.29 | 845.00 | 734.82 |
| 12 | 5.10 | 6.00 | 6.90 | 42.50 | 50.00 | 57.50 | 5.15 | 6.06 | 6.97 | 0.098 | 0.115 | 0.132 | 1,024.94 | 871.20 | 757.56 |

[1] The spacings are based on recommendations of C. E. Ramser, *Farmers' Bull.* 1669.
[2] Includes only the interval above each terrace.
[3] The minimum is 15 per cent below the mean, and the maximum, 15 per cent above the mean.

Cross Sections. The three main requirements of satisfactory terrace cross sections are (1) ample channel capacity, (2) channel and ridge side-slopes flat enough to permit operation of farm machinery along the terrace without undue tearing down of the ridge and without undue hindrance to operation of the machinery, and (3) economical construction.

Customary cross-section specifications for both the graded and the level runoff-control and absorption types of terraces are shown in Graph 31.

With respect to the terrace channel, the cross-sectional area that actually carries the water should seldom be less than 7 to 8 square feet, and this minimum should be increased somewhat along the lower part of long terraces. The water-carrying depth of a settled terrace should be from 15 to 20 inches, and the total width of the terrace from 15 to 40 feet, depending on the slope of the field and the type of farm machinery to be used. The side slopes of the channel or ridge should seldom be steeper than 4 to 1 (a fall of 1 foot in a horizontal distance of 4 feet), although 5 to 1 is preferable. On pasture or other noncultivated land, smaller cross sections may be satisfactory.

Length. Usually, 1,600 to 1,800 feet is the maximum distance that a terrace should carry water in one direction. Under favorable conditions, however, terraces half a mile long often will give satisfactory service if properly constructed and maintained. In order to eliminate the need for a second outlet ditch, structures of this length may sometimes be used to advantage in a system where the slopes are relatively uniform. But on gullied land a length of 1,500 feet rarely should be exceeded. Where an occasional terrace must exceed the maximum lengths recommended, it often can be handled most satisfactorily by draining the excess length to a convenient natural or vegetated outlet in the direction opposite to the outlet for the main part of the terrace. Or the entire terrace may be drained in one direction, if the cross section of the channel is increased toward the lower end to provide additional capacity.

Limiting Slopes. On slopes above 10 to 12 per cent, it is difficult to build and maintain graded terraces that have adequate drainage capacity or that can be satisfactorily farmed over with modern machinery. Under a soil conservation plan, steeper slopes are seldom recommended for the production of cultivated crops. In the majority of agricultural areas, the graded type of terrace generally can be used, if needed, on slopes considered suitable for cultivation under a good land-use program.

## Planning Terracing Work

In planning terracing operations, one of the important problems is to determine where and how the structures should be used. Terraces are most needed, generally, on sloping land that must be used for the production of farm crops and on which less expensive and applicable conservation measures will not provide adequate erosion control. Too often, terracing is considered as an alternative control measure to a permanent vegetative cover of grasses or trees. The practice should not be considered for land that can be retained under permanent vegetative cover, except where terraces may be required for moisture conservation or diversion of water for gully control or as an aid in establishing a satisfactory cover of vegetation. Neither can terracing be economically justified on land that can be protected adequately by such tillage and agronomic measures as contour cultivation, crop rotation, and strip cropping. These measures alone may provide sufficient protection under conditions of mild slopes, relatively low rainfall intensities, and high soil infiltration rates, especially where the rotations provide an erosion-resistant cover during a large part of the cycle of crop succession. But with highly erodible soils, long slopes, and high-rainfall intensities, and under conditions where a large percentage of erosion-permitting crops must be used in the rotations in order to provide a profitable farm income, the applicable agronomic control measures may give only partial control. Under such circumstances, the agronomic measures must be reinforced with terraces.

Under all conditions, terraces always should be supplemented by the best possible cropping practices, since the structures themselves do not add to the fertility of the soil but serve primarily as an aid to soil conservation and improvement. The use of proper rotations, contour strip cropping, and contour cultivation in conjunction with terracing, wherever terraces are needed, provides one of the most effective erosion-control combinations known for cultivated fields.

It is important in the interest of establishing, as quickly as possible, sound soil-conserving measures to all the land of the nation needing protection that no meaningless and deterring controversial issues be raised as to the relative merits of agronomic and mechanical control measures. Each has its useful place in a properly coordinated erosion-control program. It would be just as serious a mistake to employ only agronomic measures where they will not provide adequate control as it would be to use terraces without the support of necessary soil-improving crop practices.

It should be pointed out, however, that terracing has certain disadvantages. The practice may require abrupt changes in traditional farming practices and entail slightly higher construction, tillage, and maintenance costs. On thin soils, subsoil may be exposed in the terrace channels. Damage may also result from the diversion and concentration of runoff at uncontrolled points, unless due precaution is exercised. All these disadvantages, however, are not especially difficult to overcome where the land operator is really concerned about soil conservation. Terracing is an erosion-control measure that has been extensively tested and has been proved acceptable under actual farm conditions wherever it has been necessary, and its application has been in accord with proper land use.

FIELD RECONNAISSANCE. After it has been decided to install a terracing system, the first step in laying out the field work is to make a thorough physical inspection of the area under consideration. All such features as slope, ridges, hummocks, gullies, drainageways, field and property lines, roads, and buildings must be taken into consideration. Adjacent areas must be examined also to determine the size of the drainage area contributing runoff to land to be terraced, the vegetal cover, and available sites for disposal of excess water. Where the field to be terraced receives any considerable amount of runoff from adjacent land, it usually is good practice to divert this water or part of it by some form of diversion or interception ditch, in order to prevent overtopping and breaking the terraces.

In the preliminary planning, all necessary terracing for the entire farm should be considered at one time so that the system installed in one field can be fitted into the complete system for the entire farm without undue difficulty or expense. The practicability of rearranging fields, fences, and roads to conform with good terracing, land-use, and farm-management practices should always be carefully appraised.

Terracing usually is planned according to drainage units, that is, according to areas from which necessary water disposal can be satisfactorily handled through one outlet or system of outlets. Where fields on adjacent farms lie within the same drainage unit, a joint terracing system often can be worked out for both fields, provided a satisfactory agreement can be arranged between the landowners or operators for joint construction and maintenance of the terraces and outlets.

Location of the most desirable terrace outlet and determination of the necessary measures for outlet stabilization are important items to be considered in developing the field plans. After the outlets are located, the terracing system should be developed around them in such a way as to conform to the topography and at the same time make most advantageous use of the outlets. The most simple and economical type of outlet is an area having a well-established sod or other cover over which water can be discharged directly without danger of cutting. Caution must be exercised, however, in selecting such natural outlets, and they must be carefully watched for evidences of failure. Many terrace systems have failed because too much dependence was placed on a natural outlet that had far too thin a cover of grasses, shrubs, or forest litter to withstand the concentrated discharge from the terraces. The ideal natural outlet consists of a dense sod protected from overgrazing or other damage. On moderate slopes, woods with a good undergrowth and cover of ground litter make a very satisfactory outlet.

Where adequate natural outlets are not available, stabilized waterways (see Chap. XXI, Part 2) must be established.

STAKING TERRACE LINES. In laying out a terrace system, random stakes are first set out to mark the location and approximate width of the necessary outlet ditch or ditches. Then the upper terrace is staked, using the drainage divide as a starting point from which to measure the vertical interval for the first structure. An exception to this rule is sometimes made when it is desired to have a definite location for some particular terrace in the system. In that case, a particular terrace is located first, and a sufficient number of terraces are staked between it and the drainage divide to insure that the maximum vertical interval for any one structure will not be exceeded. Any necessary reduction in the spacing of terraces can then be divided proportionately. After the first terrace is staked, each of the succeeding structures is staked in turn.

It is not always possible, even for experienced terrace surveyors, to establish the most satisfactory layout in the first attempt. After a few lines have been staked, minor topographical features sometimes will necessitate changes in the terrace lines. If such changes are extensive, it is usually best to pull up all set stakes and start the job over.

In order to get proper alignment of terraces at the outlet ditch, it usually is most convenient to start staking at the outlet end. The selected terrace channel grades can then be used to locate all other stakes from that point to the other end of the terrace. Stakes should be set at intervals somewhere between 50 and 100 feet, except on curves and across depressions, where a 25-foot spacing is often necessary. It is customary and usually most convenient to have the stakes indicate the location of the center line of the terrace ridge.

After all the terrace lines have been staked, some realignment is usually necessary on each terrace in order to eliminate undesirable sharp curves, to obtain greater ease of construction, and to develop a terrace that will offer a minimum of inconvenience in later farming operations. Good field judgment must be exercised in order to provide the most satisfactory realignment of terrace lines. Graph 35 illustrates the principle of terrace realignment.

The realignment usually consists of moving certain stakes up or down the slope. Often such movement will, of course, be restricted by the drainage and construction features involved. Usually, the straightening should be limited in upward movement so that not more than 6-inch additional cuts will be necessary in the terrace channel. The straightening of terraces where they cross depressions should not be such as will introduce excessive ridge heights and ponding areas. The maximum ridge height generally should not exceed 3 feet. If a gully has formed, the settled height of the terrace ridge above the break or fill in the gully should seldom exceed 3 feet.

In order to facilitate terrace construction, continuous marking lines are usually preferred. If a final check of the terrace and outlet locations indicates that the entire layout will be satisfactory, the terrace lines are then marked with a plow furrow, since stakes are easily lost and are more difficult to follow with heavy terracing equipment.

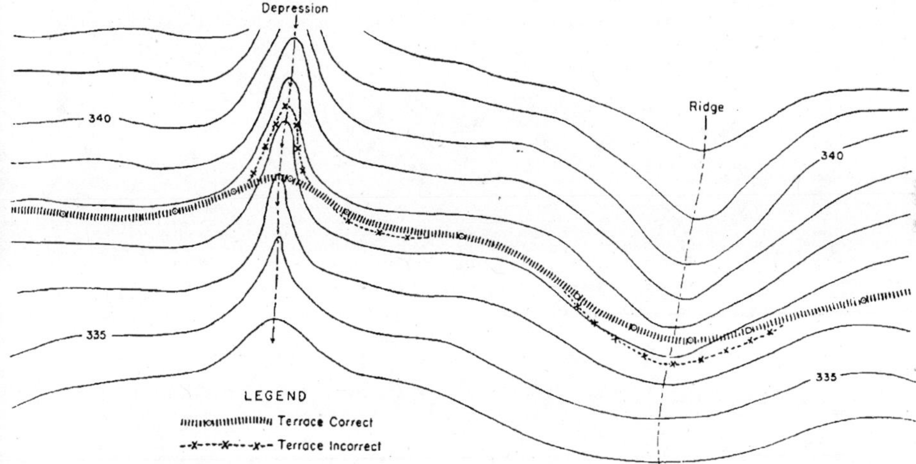


GRAPH 35.—Realignment of terraces across depressions and over ridges. (*Soil Conservation Service.*)

## Construction

Soil character has much to do with the technique and cost of terrace construction. Extensive experience has shown, for example, that the cost of building terraces on the sandy coastal plain soils of the Southeast, which have friable subsoils, is only a third or half the cost of construction on the principal types of soil in the Piedmont. In times of drought, the heavier soils of the latter region become hard and difficult to manipulate. When wet, the sticky Piedmont clay clogs terracing machines and so lowers their efficiency.

Generally, the cost of terracing increases with the degree of erosion, because of the toughness of the exposed or partially exposed clay subsoil and the frequency of gullies. In other words, it costs more to protect the less valuable land. If terraces are to be built eventually, construction should proceed as soon as possible after a field is put into cultivation.

In some areas, terraces may be impractical because of the unstable nature of the soil or the presence of rock or hardpan near the surface. Where shallow surface soil is underlain by relatively tough subsoil, difficult to penetrate, terracing usually is impractical because of excessive construction costs. Terracing of porous sandy land may be unsatisfactory because of the relative ease with which intercepted runoff penetrates the ridges to cause sloughing and breaks. Also, excessive interterrace erosion tends to discourage terracing on such soils because of the difficulty of keeping the channels open. Moderately loose soil frequently can be terraced satisfactorily by building wider ridges and placing the structures closer together. Fortunately, unfavorable soil conditions of this kind are of exceptional occurrence.

Shallow surface soil does not necessarily mean that terracing is inadvisable. If the subsoil is tough and difficult to plow, however, so much

FIG. 175.—Terrace construction with a 10-foot blade terracer and tractor. This type of equipment is used quite extensively for building terraces. (*Photograph by Soil Conservation Service.*)

topsoil may be required in developing the ridge that an excessive area of stubborn subsoil of low productivity is exposed, particularly in the terrace channel. But this objection does not always hold, because topsoil moving into the channel during the process of stabilizing the terrace interval often will gradually correct or improve the situation. Moreover, if the land must be cropped, decreased yields on the exposed areas may be more than made up for by increased yields on the other part of the treated land. And, of course, if the erosion is not checked, the entire area may quickly become unfit for cropping.

MACHINERY. In terrace construction, a wide variety of equipment is used, including the terracing plow, disk plow, blade grader (Fig. 175), V-drag (Fig. 176), scrapers, and other machines. Equipment designed specifically for terrace construction usually gives the most satisfactory results. Recently, marked improvement has been made in the design and adaptation of blade graders for terracing work, and two new types of terracing machines have been developed: the rotary-plow type and the elevating-grade type. Both offer considerable promise because of relatively low first cost and economical operation in comparison with their capacity.

Small-blade terracers, scrapers, V-drags, and plows pulled by farm tractors, horses, or mules have been used to build a large proportion of existing terraces. Use of such machinery, however, is generally laborious, except on the more tractable soils, and the tendency is to build structures out of conformity with standard specifications. Usually, the most eco-

FIG. 176.—Terrace construction with a plow and V-drag. It requires a large number of rounds to develop satisfactory terraces with this type of equipment. (*Photograph by Soil Conservation Service.*)

nomical and satisfactory terraces are constructed when the heavier tractors (40 to 50 horsepower) and terracing machines are used. Initial costs are so high, however, that many farmers cannot afford individual ownership. But through soil conservation districts and other farmer groups, this heavier type of terracing equipment can be made available.

In some states, counties are authorized to make road equipment available for terracing or to use county funds in purchasing terracing equipment. Farmers usually pay the operating costs and, in some instances, small additional amounts to cover any proportionate part of the equipment charges. Where state laws do not provide for such procedure, groups of local farmers can organize associations, purchase machinery, and rent it out to farmers on a cost basis. In some instances, a farmer or group of farmers have purchased a terracing outfit and, after completing their own terracing, have extended operations to neighboring farms. In

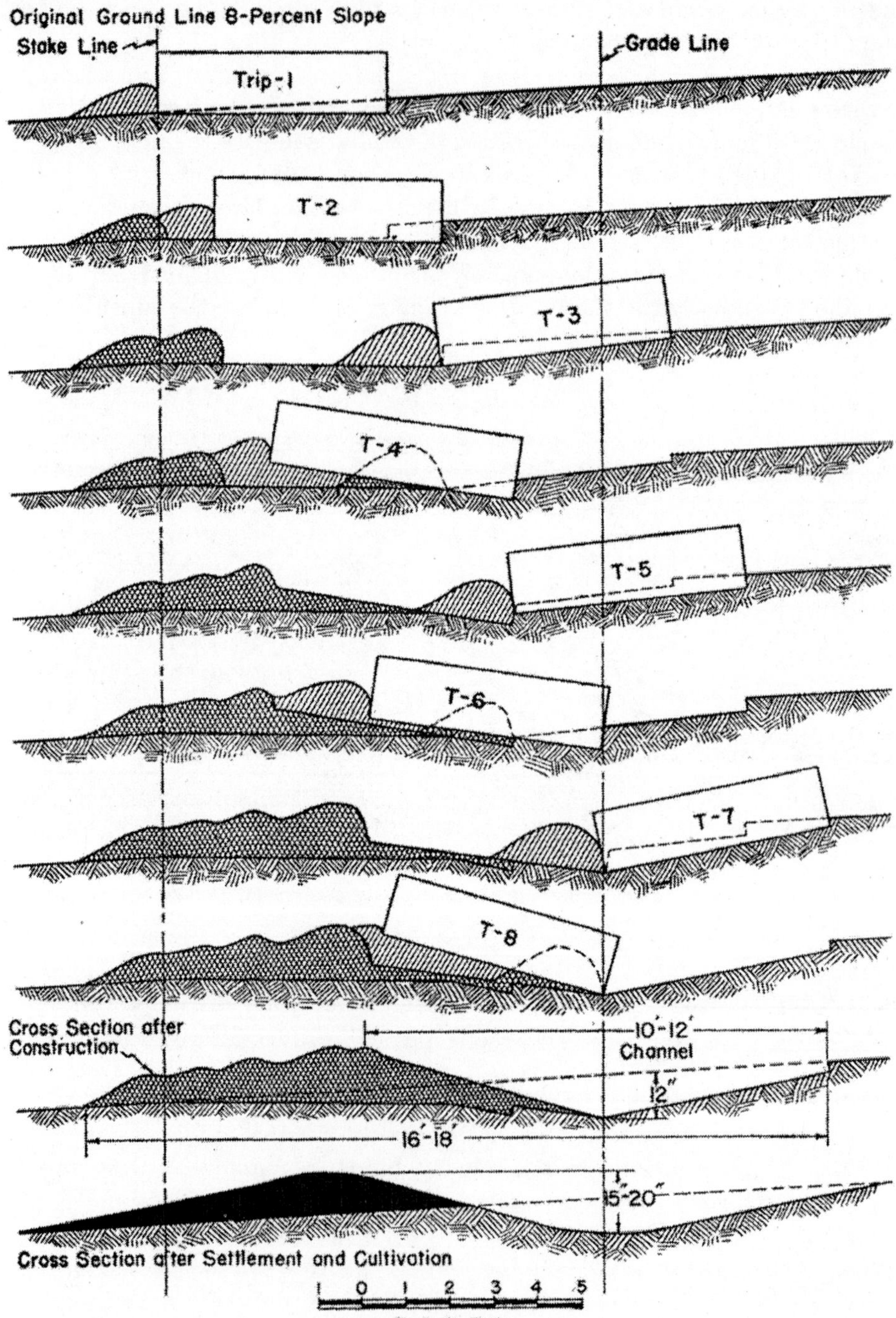


GRAPH 36.—Progressive steps in the construction of graded-channel terrace in areas where small tillage equipment is used, as in southeastern United States. Ten-foot blade used, working from upper side only. (*Soil Conservation Service.*)

some localities, private contractors who have entered this field seem to be doing the work in a satisfactory way.

Procedure. The upper terrace of a field should be constructed first; and then, in downslope succession, construction of the others should follow. If the lower terraces are constructed first, they are likely to be damaged if a heavy rain occurs before the upper ones have been completed. Not only should the top terrace be constructed first, but it should be especially well constructed because of the protection that it gives to the structures below. When an upper terrace breaks, the structures below are also likely to break under the stress of water thus concentrated against them.

A practice of delayed terracing, under which the construction of all structures in a field is extended over a period of several years, has been advocated in some sections in order to encourage terracing over a larger area without heavy initial cost. This frequently may be a dangerous procedure. Usually, the few rounds made on each terrace the first one or two years do not provide sufficient capacity to withstand runoff from torrential rains. Consequently, much overtopping and damage result. Generally, it is desirable to complete all terraces as quickly as possible. If the work must be distributed over a period of years, the practice of constructing a few of the upper terraces the first year and building additional ones each succeeding year is preferable to starting all structures at one time and doing only a little work on each.

Graphs 36 and 37 show the details by which cuts and fills can be made in constructing the graded-channel type of terrace with a 10-foot blade.

In building terraces, all the earth may be moved from the upper side, or a part from each side. A few years ago, the practice of moving the earth from both sides was common, but the tendency now is toward constructing more and more of the terrace from the upper side. Not only is this cheaper, but it also produces a more desirable cross section. In localities where a comparatively narrow terrace is suitable, it has been found most economical to build entirely from the upper side. In other places, particularly on slopes above 3 or 4 per cent, only a small part, if any, of the terrace is built from the lower side. This construction procedure necessitates reversible-type machines if the equipment is to be used in the most efficient way.

Where terraces cross gullies or even slight depressions, some extra fill work must be done to maintain proper location and ridge elevation. A slip scraper, fresno, or rotary scraper is used generally for such work (Fig. 177). Failure to construct fills properly is a common cause of trouble in terracing. Where even a small gully has developed, the fill should be made as impervious as an earth dam through proper placement of material and adequate compaction. A substantial bond between the fill and the gully

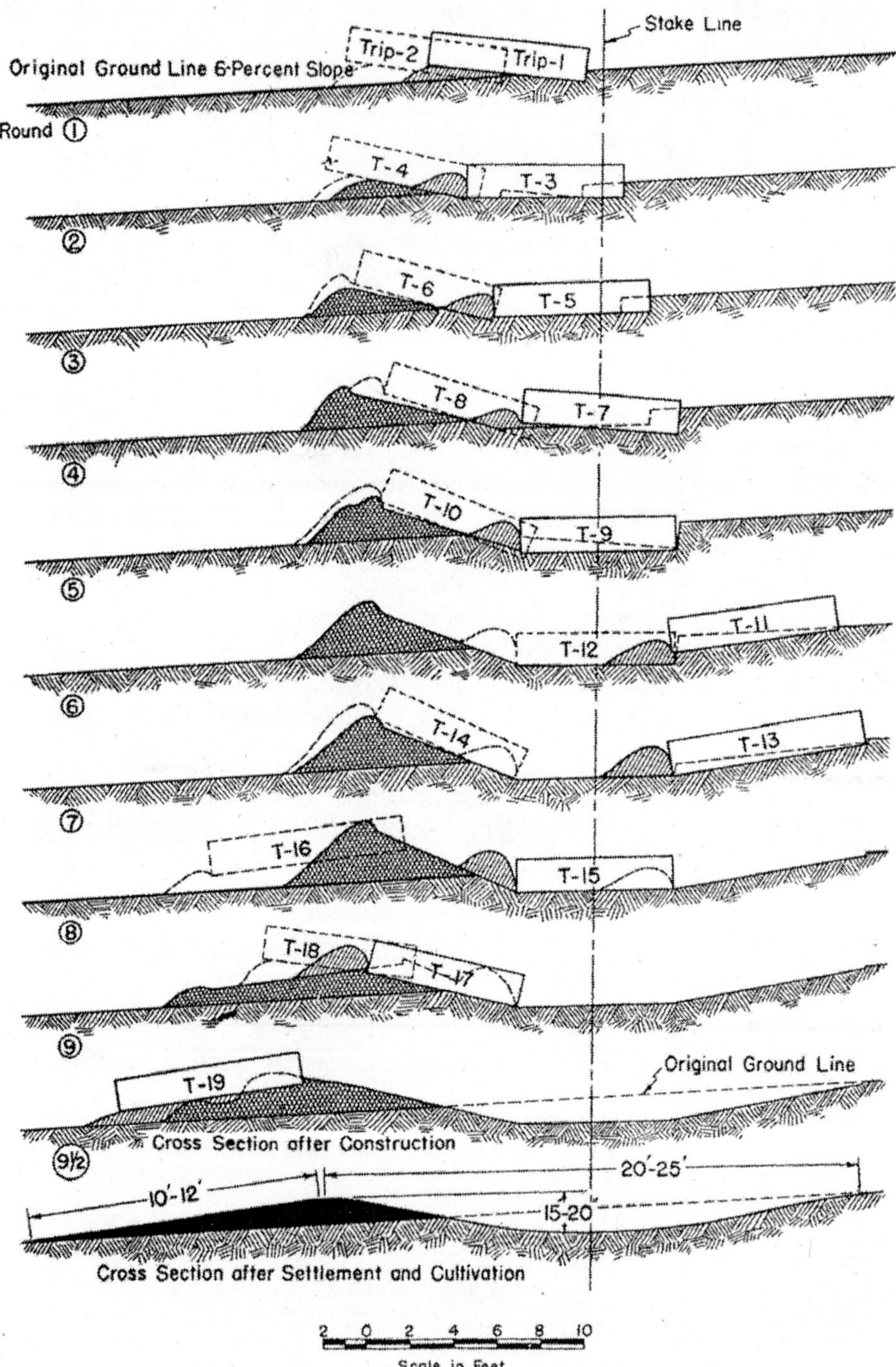


GRAPH 37.—Progressive steps in the construction of graded-channel terrace in areas where large tillage equipment is used, as in midwestern United States. Ten-foot blade used, working from upper side only. (*Soil Conservation Service.*)

sides always should be provided. The fill must be high and wide enough to prevent overtopping. Usually, the height of the terrace at points where fills are made should be increased by 15 to 20 per cent, to compensate for the extra settlement that will occur at such points.

Where terraces are to be continued from one field to another, it is necessary to build the intervening section by hand if a hedge or other obstacles prevent the use of regular terracing equipment. It is also frequently necessary to do some handwork on terraces where outlet ditches or field fences prevent regular equipment from covering the full length of the structure. It is highly important that all parts of the terrace be completed to full channel capacity, since one weak point will be enough to cause failure.

Fig. 177.—Slip scraper and team used to fill low places along terrace ridge. (*Photograph by Soil Conservation Service.*)

A terrace cannot be considered complete until it has been checked carefully for correct grade and height. To assure proper channel capacity and flow of water in the desired direction, low places along the ridge and high spots in the channel are usually marked and corrected before the equipment leaves the field. The level and rod are used in checking, and sufficient readings are taken to determine accurately where corrections are necessary. Elevations and grades are checked with special care around bends, across gullies, and at the outlets. A common fault in terrace construction is to provide too much grade near the outlet. If correctional work is required over an appreciable length of the terrace, it usually can be done most satisfactorily by using the regular terracing equipment.

COSTS. So many variables affect the cost of terracing that it is difficult in some localities to find two jobs that cost exactly the same. The more important features influencing terracing costs already have been discussed. When soil conditions are favorable, slopes uniform, and efficient terracing equipment available, the cost of terracing a field is often roughly equal to the cost of plowing the same field. Where conditions are unfavorable, however (as with heavy soil; an abundance of rocks, stumps, or gullies; and irregular topography), terracing may run as high as $5 or more per acre.

## Absorption-type Terrace

Although the absorption-type terrace is being used more extensively in the Great Plains, it is not entirely an American development. Some of the older countries, as Egypt and the Anglo-Egyptian Sudan, have used contour ridges or small dikes with closed ends for a great many years, in order to prevent runoff and increase moisture conservation for crop production. The principle probably originated many centuries ago. Construction procedure, however, has been improved by American effort. Special construction equipment has been developed, and the ridges have been widened out, so that the side slopes are flat enough to permit cropping with a minimum of inconvenience.

Erosion control, especially soil blowing, by the absorption-type terrace is accomplished to a large degree indirectly through water conservation. The increased water held by such terraces makes the soil more resistant to wind erosion, even while encouraging the growth of crops and other protective vegetative covers.

These terraces are particularly adapted to areas of low precipitation and to soil types that will absorb accumulated runoff fast enough to prevent damage to crops. Such areas are confined largely to the more absorptive soils and gentler slopes of the wind-erosion sections of the Great Plains and other low-rainfall localities of the West. The absorption-type terrace may be used with considerable success also on mildly sloping sandy land of some of the more humid regions, such as the coastal plain of the Southeast. Rainfall rates and infiltration capacity always should be known before this type of terrace is used. Figure 178 illustrates the absorption-type of terrace.

Before any mechanical measure for promoting absorption can be made fully effective, the soil itself must be in condition for optimum infiltration. The porous character of a soil, originally highly absorptive, may be seriously impaired by depletion of the organic matter, as the result of continuous cropping or the bodily removal of soil by erosion. Under such soil conditions, addition of organic matter is generally essential to any considerable improvement of the absorptive capacity.

In some localities, field tests of basin listing have shown 40 to 60 per cent more absorption where a legume had been turned under annually for a series of years than on comparable portions of the same field where no organic matter was returned to the soil. Avoidance of grazing or cultivation when the soil is wet enough to be miry or sticky also aids in the maintenance of a favorable condition conducive to infiltration.

Level terraces have not shown especially favorable results with respect to water conservation on cultivated land at any of the Eastern experiment stations. The ponded runoff is held so long above the terrace before it is absorbed that crops frequently are damaged. It is also difficult to construct economically terraces with sufficient capacity to handle the

Fig. 178.—Moisture conservation by level terracing and contour listing. The water from melting snow is well distributed and held by the furrows until it soaks into the ground to build up reserve moisture for the succeeding crop. Kansas. (*Photograph by Soil Conservation Service.*)

large amount of runoff. Closed-end level terraces even as far west as Guthrie, Okla., have not given favorable results. The use of such terraces does not appear to be practicable in localities where the annual rainfall is more than 25 or 30 inches, except on exceptionally pervious soils.

Beneficial results have been obtained, however, in western Oklahoma and Texas where moisture is a limiting factor in crop production. Experiments at the Oklahoma Agricultural Experiment Station at Goodwell, on nearly level Richfield silt loam soil, have shown that level terracing for moisture conservation has increased crop yields by 36.5 per cent over a ten-year period (1926–1935). All the rainfall was conserved on the terraced land, whereas 2.33 inches was wasted annually as runoff from comparable unterraced land. At the Texas Experiment Station near Spur, level terracing increased crop yields from 20 to 40 per cent over a three-year period. Fields receiving supplemental runoff from adjoining

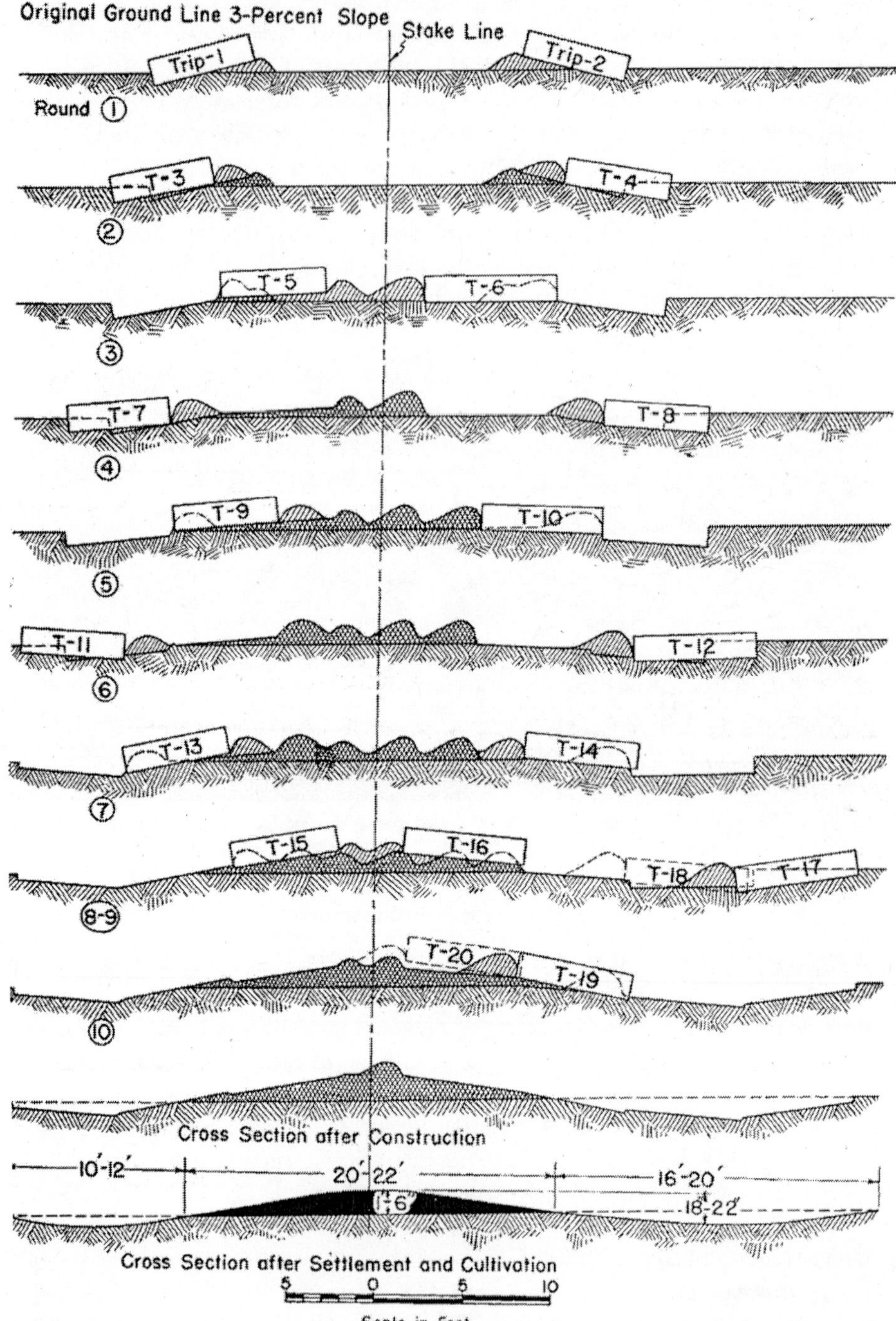


Graph 38.—Progressive steps in the construction of retention type of terrace in the Great Plains. Ten-foot blades used, working from both sides. (*Soil Conservation Service.*)

areas produced 16 pounds of lint cotton and 0.88 ton of alfalfa more per acre than comparable areas where all the rainfall was held but to which no additional water was diverted.

Construction. In the construction of level terraces, one of the most important objectives is to build them so that they will spread collected rainfall over the widest practicable area. Generally, they are built on slopes of less than 3 per cent; the ridge is made high enough to spread water over a relatively wide area (Fig. 179), and the earth required for the ridge is so excavated as to avoid concentration of the ponded runoff.

Construction of the terrace embankment from both sides (Graph 38) is usually advisable, not only because of the gentle slopes but because it is

Fig. 179.—Cross section of a level terrace in the Great Plains, western Kansas. The ridge projects well above the surface so as to spread the water over a wide area. The wide cross section is necessary to accommodate the large tillage machinery used in this region. (*Photograph by Soil Conservation Service.*)

easier to attain the necessary height in this way. Where maximum ponding effect is desired, the structure must be designed with ample cross section, and both ends must be closed. In general practice, however, the ends are often partly closed so that excess rainfall can escape before the terrace ridge overtops. In some sections, the ends are left open, and special outlets are used to dispose of excess water.

The most desirable spacing for absorption terraces has not been finally determined. The ideal spacing would be that which would give the best spread of intercepted runoff and minimum interference with tillage practices at lowest construction cost. Close spacing will provide more uniform distribution of water, but limitations are imposed by the cost and the factor of convenience of tillage. These difficulties can be met by spacing the terraces so that they will not interfere unduly with cultivation, depending on contour listing between the ridges as an aid in the effective distribution of the conserved water. In semiarid regions, contour listed furrows are capable of holding and distributing most or all of the surface water that ordinarily gathers on a field. The terraces provide the necessary safety factor for large rains and for those periods when the lister furrows are ineffective.

Reasonably good results can be expected where the terraces are spaced so that the ponded water can be flooded from one terrace to the base of the next one above (see Fig. 291, Chap. XXXV, Part 2). With an 18-inch terrace and 1 per cent slope, this would permit horizontal spacing at intervals of 150 feet. On steeper slopes, the water-storage capacity of level terraces with closed ends is often the limiting factor in determining spacings. This capacity should be sufficient to take care of the maximum accumulation of rainfall that ordinarily can be expected from the contributing drainage area. Such accumulation may be as high as 3 or 4 inches under semiarid conditions and 6 or 7 inches in humid areas. Maximum spacing for absorption terraces should seldom exceed those recommended for the runoff-control terraces, particularly where sheet erosion is a problem.

## Farming Terraced Land

Erosion-control efforts too often cease when a system of terraces has been installed. It is of greatest importance to recognize the fact that terraces may fail completely as the result of faulty cropping and tillage practices. Surveys have shown that a surprisingly high percentage of the terraces that have been in use for 5 years or more are no longer effective because of continuing erosion of the interterrace area as the result of one-crop farming and cultivation up and down the slopes. Under such poor practice, the channels have filled with eroded material, overtopped, and the land below has been cut to pieces. In other words, the structures, under these conditions, have aggravated rather than alleviated erosion.

One of the most desirable tillage practices for terraced land is contour farming. Operation of tillage equipment parallel to the terraces also minimizes the possibility of damage to the ridge and channel. By plowing parallel to the terrace embankment and regulating the location of dead furrows and backfurrows, terraces can be maintained. In this way, their cross sections can be changed to provide the most desirable slopes for particular fields (Figs. 180, 181, 182).

A particularly important point is the desirability of turning uphill as many of the interterrace furrows as possible.

Irregularity of slope and differences in the cross section of the terrace embankment and in equipment make it difficult to follow definite rules in plowing terraced land. The plowman has to exercise considerable ingenuity, since the starting and finishing points will vary not only from field to field but from year to year in the same field. He must keep in mind

FIG. 180.—Maintenance of terraces by plowing: *A*, the channel of the interception-and-diversion type of terrace can be enlarged by plowing it out. From the center of the channel it is important to turn as many of the furrows uphill as possible to offset the natural soil movement down slope. *B*, the ridge of the interception-and-retention type of terrace can be enlarged by backfurrowing to it. The location of the dead furrow should be changed from year to year to avoid excessive depression at any one point.

the most desirable terrace cross section and the principle of maintaining and developing the channel by plowing it out. He must look to adequate maintenance of the ridge by backfurrowing as well as to the development of the most desirable surface slopes over the terrace interval by proper regulation of dead furrows and backfurrows.

The two-way plow, although not commonly used for plowing unterraced land, appears to have distinct advantages on terraced fields. It eliminates the necessity of backfurrows or dead furrows in undesirable locations and makes it possible to turn the furrows up the slope. This

FIG. 181.—A terrace ridge can be made wider and the side slopes reduced by backfurrowing toward the ridge. (*Photograph by Soil Conservation Service.*)

FIG. 182.—Plowing a terrace to maintain proper cross section. Note how the channel is being plowed out and how the lower side of the terrace blends with the general slope of the land. (*Photograph by Soil Conservation Service.*)

gives the soil an uphill movement which partly counteracts the natural downhill movement caused by erosion and tillage. This type of plow has been used on terraced land at several of the experiment stations of the Soil Conservation Service with distinctly encouraging results.

Where it is not possible to maintain proper cross sections by the regular plowing operations, it will be necessary to use the blade or scraper on the terraces at more or less regular intervals. For this purpose, lighter terracing machines or home-made V-drags drawn by ordinary farm power generally will do a satisfactory job.

In many parts of the country, strip cropping is combined with terracing in a highly effective way. Several methods for arranging the alternate strips of close-growing and row crops on a terraced field have been worked out. The type of rotation, the particular crops, and the proportional area of each to be planted will determine, in part, the arrangement and width of strips. In combining the two control measures (Graph 39), it is generally advisable to (1) use strips of as nearly uniform width as possible, so that rotations may be properly balanced; (2) locate at least one boundary line of each strip between adjacent terraces, so that a portion of each terrace interval will be protected by a close-growing crop; (3) eliminate point rows in so far as practicable by including irregular areas in strips of close-growing crops; and (4) make use of the minimum number of strips as a means of simplifying field operations. It is important to stress the advantage of having the strips of close-growing vegetation to cover at least part of each terrace interval, as well as part or all of the terrace ridge, rather than merely to have alternate terrace intervals covered.

The latter procedure provides no better control of erosion than would be effected by terracing and rotating the same crops in the usual way.

Point rows can be arranged in various ways, choice of which will depend largely on the preference of the land operator. His choice may be influenced by such factors as past practice or the type of tillage equipment used. Graph 40 indicates three of the more common row arrangements, showing point rows in the terrace channel, at the base of the terrace ridge, and between the terraces.

A combination that appears to have some merit is to run the long rows parallel to alternate terraces so as to allow the short rows to terminate along the intervening terrace. By this method, the point rows terminate both in the channel and against the ridge of every other terrace. An objection to terminating point rows against the terraces is that they may be slightly off the contour and tend to concentrate the turning of tillage implements on the terraces. This objection can be offset to some extent by using parallel rows on the area occupied by the terrace and by ending the point rows just above or below this strip. Another arrangement (illustrated in Graph 40) nearly equalizes the digression of the point rows from the contour by placing them between the terraces.

Some farmers object to terracing because of the fear that the practice will interfere with their regular farming operations, forgetting that the gullies gradually forming in their fields eventually will cause far more serious interference with their farming operations. Farming terraced land

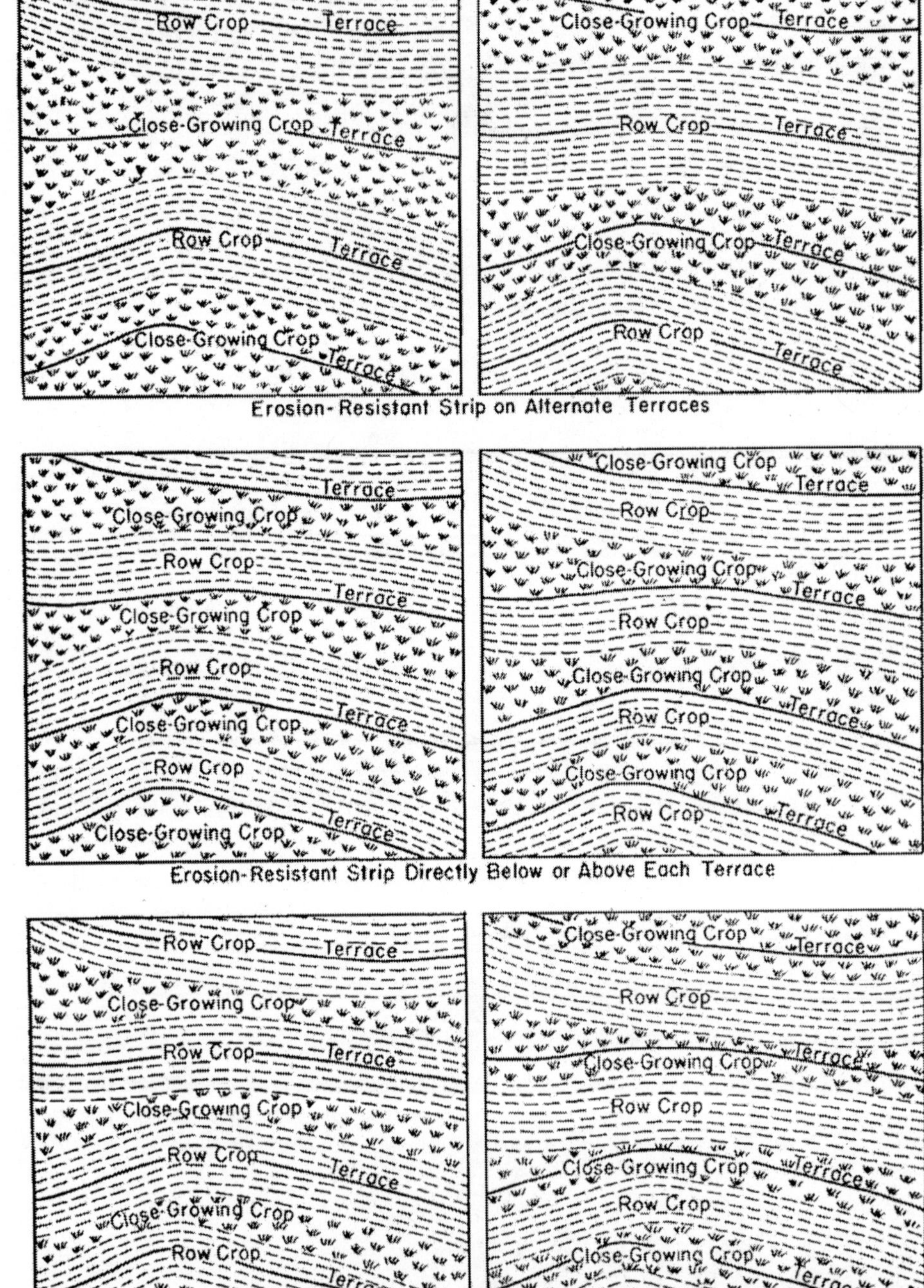


GRAPH 39.—Three methods for combining strip cropping and terracing. (*Soil Conservation Service.*)

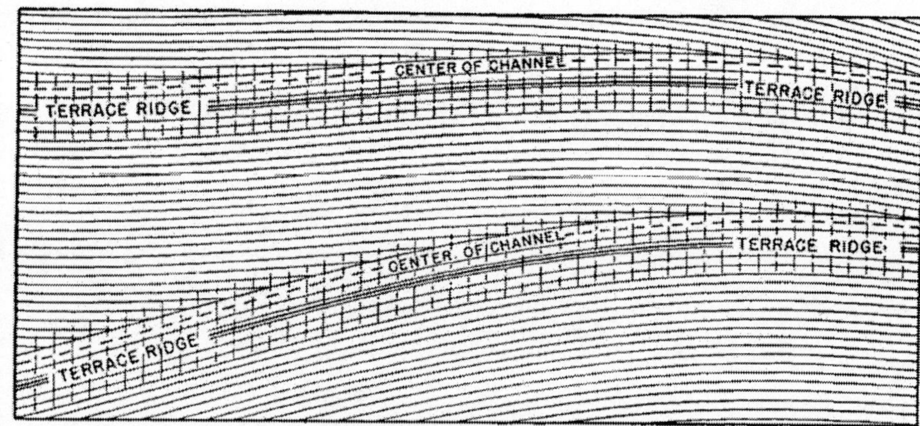


Point Rows in Terrace Channel

Point Rows at Base of Terrace Ridge

Point Rows Between Terraces

GRAPH 40.—Three methods for arranging crop rows in relation to terraces. (*Soil Conservation Service.*)

is not unduly difficult. What is required most is a willingness to give up notions about the nonexistent advantages of straight-row farming. Although contour farming sometimes introduces some minor inconveniences, the advantages far outweigh these unimportant difficulties. Many farmers have learned that even the turning of equipment necessitated by short rows is not nearly so difficult as expected.

# Chapter XXI. Runoff-disposal Channelways and Outlets

Under conditions of virgin vegetation, an intricate system of natural drainageways safely conducted excess rainfall from the earth's surface to the oceans. The rivers, with their branching tributaries, formed the main channelways of this natural runoff-disposal system. Millions of vegetated depressions without definite channels collected and conveyed runoff from the crests of watersheds down to the main watercourses. So dense was the protective cover, generally, that soil washing and gullying were inconsequential in these natural waterways. This widely ramified system of protected upstream waterways, together with the relatively stable main channels downstream, provided safe and adequate disposal of surface drainage for most of the North American continent.

Today the planning and establishment of adequate runoff-disposal systems to handle safely the unutilized rainfall running off cultivated lands of the United States are problems of extensive magnitude and they have been long neglected. As the virgin land was plowed up, the protective covering was destroyed in the natural depressions, as well as on the sloping fields, without much thought of the ultimate results. Millions of small and large gullies developed in the unprotected depressions and gradually branched out over the land. These land scars aided in the general acceleration of erosion that has accompanied the development of farming and grazing.

In the South, terracing has been used quite generally for many years to divert runoff from cultivated fields; now the practice is spreading rapidly to other parts of the country, not only to control runoff but to conserve rainfall. Although considerable care and experimentation have been devoted to the construction of the terraces, too little attention has been given to the final disposal of excess water at the ends of the structures. Increased concentration of water at this point has caused the development of numerous gullies that could have been avoided (Fig. 183). Remnants of terraces that originally were well constructed and ultimately failed because of the lack of suitable outlet protection can be found on many farms today. Progressing from the end of the terrace, inward along the channel, numerous terrace channels have thus grown into deep ravines, ruining fields and frequently damaging roadways.

Since 1935, the Soil Conservation Service has been studying the problem of establishing practical runoff-disposal systems for terraced land. The Service's opportunity for extensive field observations and trials, together with the concentration on the problem of men with special training in engineering, soils, and agronomy, has brought marked advancement in the technique of providing satisfactory terrace outlets and field or pasture drainageways. Because more information on the

FIG. 183.—Gully formed at end of terraces because of inadequate outlet protection. Kansas. (*Photograph by Soil Conservation Service.*)

design and construction of mechanically protected outlets was available at the beginning of the work, emphasis was placed on this type of stabilization. More recent developments, however, have shown that vegetation can be used effectively in the protection of terrace outlets and artificial channels in most agricultural areas. Where vegetation is properly used, its resistance to erosion from concentrated flowing water is proving much greater than originally was expected.

Natural drainageways, such as swales and streamlets, are the initial collecting places for runoff that eventually may reach the rivers and oceans. Such drainageways often serve admirably as outlets for terraces and other artificial waterways. More frequently, however, the outlets have to be either stabilized or constructed in their entirety. In order to provide for the completely safe disposal of runoff, drainageways are necessary in cultivated areas, even where terraces are not used. As a matter of fact, the establishment of satisfactory drainageways is one of the initial problems in developing farm conservation plans in all areas where runoff is involved.

A general classification of outlets and drainageways, now used in connection with the development of farm conservation plans, follows. The

- Nonchanneled drainageways
  - Natural
    - Grassed depressions and ravines
    - Timbered depressions and ravines
    - Shrub-covered depressions and ravines
    - Rocky depressions and ravines
    - Foot-slope seeps
    - Foot slopes of coarse absorptive materials
  - Artificial
    - Vegetated
      - Completely vegetated outlets
      - Stripped outlets
      - Completely vegetated depressions
    - Mechanical
      - Drop-checked outlets
      - Continuously lined, high-velocity outlets
    - Miscellaneous
      - Vegetated and mechanically controlled outlets
      - Absorption outlets
      - Accumulation outlets
- Channeled drainageways
  - Natural
    - Streamlet, creek, river
    - Partly vegetated washes
    - Ravine, gorge, canyon
  - Produced by accelerated erosion
    - Gullies, bare washes
  - Artificial
    - Graded terraces
    - Diversion ditches
    - Stripped, low-velocity channels
    - Completely vegetated, high-velocity channels
    - Drop-checked, low-velocity channels
    - Continuously lined, high-velocity channels (flumes)
    - Vegetated and mechanically controlled, high-velocity channels
    - Storm sewers
    - Aqueducts

two main types are *channeled* and *nonchanneled*. Both types are used in some form or other, under natural or controlled conditions. To the fullest extent possible, natural outlets and channels should be utilized in the disposal of excess runoff.

## Artificial Channels and Outlets

Since many of the lesser natural drainageways have now been damaged by accelerated erosion, it is necessary either to construct new ones or repair or rebuild old ones to provide adequate means for the safe disposal of runoff. In the repair or reconstruction of old drainageways, the most satisfactory method is to reproduce natural conditions as nearly as possible. Frequently, soil type, climatic conditions, or the artificial practices of agriculture make it advisable or necessary to modify the natural process of runoff disposal. Accordingly, a variety of artificial outlets and drainageways are now being used extensively to meet the requirements of such local conditions.

FIG. 184.—Stabilization of terrace discharge channelway with oat strip interplanted with grass. (*Photograph by Soil Conservation Service.*)

Vegetation, in the form of either strips or a dense cover of grass, or other plants, as well as plantings of trees, vines, and shrubs, is being extensively and effectively employed in the protection of runoff channelways and outlets (Fig. 184). Runoff from terraces, diversion ditches, and field or pasture water channels generally can be discharged safely over a firmly vegetated slope. Shallow, vegetated depressions usually are preferable to narrow, deep depressions or vegetated channels. Where grass is used as the stabilizer, the broad, shallow outlets or waterways are referred to as *meadow strips* or *pasture strips*, according to the use made of the forage.

Specially constructed vegetated channels are necessary in many fields and in some pastures where natural depressions suitable for meadow or pasture strips are not available. Because of increased velocity and depth of flow caused by the restricted cross section of such artificial channels, considerable precaution is necessary in their design, construction, and stabilization.

The use of mechanical outlets usually is limited to terracing systems and diversion ditches. Soil and climatic conditions unfavorable to the growth of suitable vegetation, however, may make it necessary to use them for practically all water-disposal purposes.

Low-velocity mechanical outlets are made by using a series of drop-check structures at intervals to reduce the channel grade. This slows the flow of water and prevents scouring. High-velocity channels are provided by the use of a continuous lining of some erosion-resistant material, such as masonry, asphalt, or metal.

A number of special types of outlets, designed to meet local conditions, are being used satisfactorily in various parts of the country. In *combination outlets*, both vegetative and mechanical protective measures are used in the same channel or waterway. With these, vegetative measures generally are used in the upper reaches and mechanical measures along the lower section, where a greater volume or concentration of water may be expected. The *absorption* type of outlet, such as the contour furrow, ditches, and embankment, is used sometimes in areas of low rainfall to collect and hold the runoff until it is absorbed by the soil. In limestone areas, where sinkholes are not uncommon, it is sometimes possible to dispose of runoff through the sinks. Occasionally, *accumulation outlets* are employed as a means of disposing of the discharge from terraces. These consist of ponds or dugouts and can be used to provide water for stock, recreation, or wildlife.

In the reorganization of farm plans for the establishment of proper land use and soil and water conservation measures, the planning of an adequate system for the control of surface water and the disposal of runoff is usually one of the first problems to be considered. Wherever there is any considerable slope, control of runoff on most cultivated land and on much grazing land is imperative, since some runoff generally is produced in nearly all areas used for these purposes.

In planning runoff-disposal systems, it is always necessary to consider the relation of the outlets to adjacent land on the same farm as

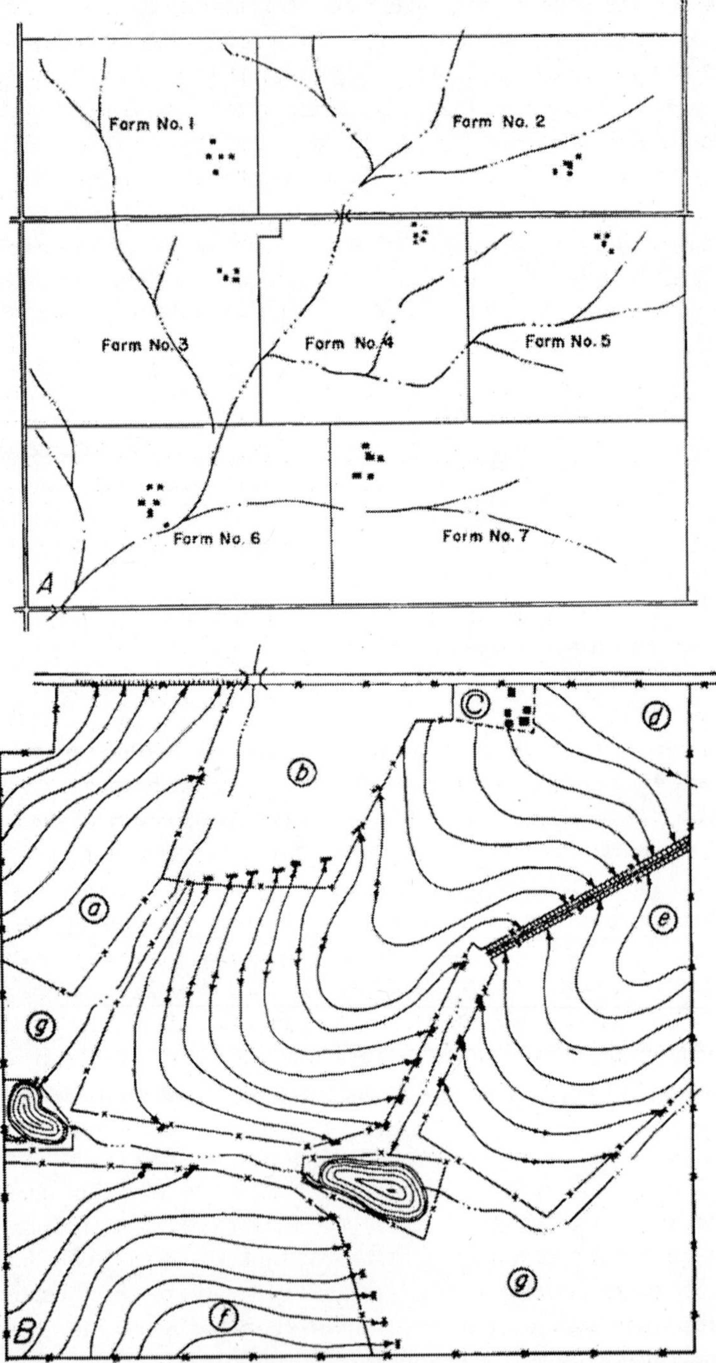


GRAPH 41.—*A* shows how natural drainageways cross farm boundary lines; *B* shows the details of a runoff-disposal plan for Farm No. 4. Fields *a*, *d*, *e*, and *f* are strip-cropped and terraced. *b* is woodland and *g* pasture. Note that all water channels and outlets are adjusted as nearly as possible to the natural drainage system, and that use is made of the advantages of woodland and pasture for safe disposal of runoff water. (*Soil Conservation Service.*)

well as to the land of adjacent and downstream farms. Where a complete water-disposal system is not planned at the outset, a series of unrelated and relatively costly channelways and outlets frequently have to be installed. Because of interference by farm roads, fences, buildings, field property lines, and topographic features, such installation usually develops many unforeseen difficulties and inefficiencies.

As a rule, natural depressions provide the most satisfactory and least expensive drainageways for disposal of excess rainfall. In so far as practical, therefore, it is generally advisable to base the entire runoff-control system on these natural drainage lines.

Graph 41 shows how natural drainageways frequently intersect farm boundary lines without relation to conservation measures or plans developed for individual farms. It also illustrates how a system of runoff controls for individual farms should be planned in relation to the natural drainage of adjacent farms. Properly planned, drainageways usually can be established so that integral parts on individual or adjacent farms can be fitted together without difficulty or extra expense when the final job is completed for the entire drainage unit, which may include several farms. The number of lateral drainageways required will depend not only on the topographic features but also on the type of soil conservation practices used. Where runoff-interception measures, such as terraces and diversion ditches, are used to divert the off flowage from one or more minor depressions, the number of drainageways that must be planned is naturally reduced. As land use and soil conservation plans are developed, the location of field boundaries, fences, meadow and pasture areas, and even farm roads often can be adjusted to facilitate economical establishment and maintenance of the selected drainageways. For example, pasture and meadow areas frequently can be arranged to include the main depressional drainageways, so that adequate grass cover will be provided and the principal maintenance requirements will be taken care of in the subsequent grazing and mowing.

WATER DISPOSAL ON UNTERRACED LAND. Where farm crops are produced on unterraced land, wide grassed waterways frequently are used for conducting the runoff down slopes that are steep enough to produce erosive velocities. Installations should provide for continuous protection of the main water-carrying depressions from field to field and from farm to farm until a stabilized watercourse is reached. If continuous protection is not provided, overfalls usually develop and undermine any inadequately protected points. Occasionally, where grassed waterways are not economically feasible, or where large amounts of runoff are involved, it is necessary to provide partial or even complete structural protection.

Protection of depressional drainageways on cultivated areas where no runoff-control measures are employed is sometimes more difficult than on terraced land. Where the topography is rolling, the upper reaches of watersheds usually branch out into numerous natural depressions, all of which carry runoff to the main drainageways. If these depressions are close together, efforts to establish adequate protection may lead to numerous grassed strips interlarding the field to such an extent that economical tillage is impossible. This situation, which is particularly troublesome in areas where large tillage machinery is used, can be corrected by retiring the entire field to a permanent cover or by using applicable runoff-diversion measures to reduce the number of vegetated drainageways.

WATER DISPOSAL ON TERRACED LAND. The problem of locating and establishing terrace outlets is inseparably associated with the planning of the entire terracing system. The cost of terrace construction and the ultimate success of the structures are dependent on proper outlet planning. On the other hand, a slight variation in the location of the first terrace or in the vertical interval between subsequent terraces may make it possible to discharge the runoff at a point where the whole problem of outlet protection is greatly simplified. Sometimes the direction of the terrace grade may be changed to advantage near the center, so as to distribute the runoff over established cover on adjoining areas. Where a special outlet strip or channel is required, savings often can be effected by directing the water flow from terraces toward it from two sides, so that the outlet will serve a larger area.

As a general rule, all terrace outlets should be planned so as to provide continuous protection from the upper terrace, on down the slope to a stabilized watercourse. All unnecessary bends in outlet channels should be avoided. Where runoff is discharged from higher points, it is necessary to provide outlets that will carry this additional water.

Unprotected roadside ditches and expanding gullies should be avoided as terrace outlets. In some instances, however, it may be desirable for the farmer and the highway maintenance agency to develop a joint outlet channel that serves the purpose of disposing of runoff from both the terraces and the highway. Similarly, it is frequently advantageous for two or more landowners having terraced fields within the same drainage unit to arrange for a common outlet system.

It is always important to determine the most satisfactory location for both terraces and outlets in advance of construction because, once established, relocation is a costly and difficult procedure. Improperly located and inadequately stabilized outlets may raise the cost of construction and maintenance so much that terracing becomes economically impractical.

Usually, the simplest and most economical type of outlet for terraces and diversion ditches is that where the water can be discharged directly on an established sod or other vegetative cover that will withstand the impact (Fig. 185). Although good pasture or meadow sod generally gives the greatest degree of protection, unburned, ungrazed woodland sometimes can be used satisfactorily, particularly on moderate slopes.

The development of natural depressions into meadows or pastures as discharge areas makes dual use of land that frequently lies unproductive. For such meadow-strip outlets, wide, shallow depressions, with slopes not exceeding 5 or 6 per cent, are preferable. The width required for efficient results will depend on the slope and the size of the area drained. Generally, a minimum width of 1 foot should be provided for each acre within the contributing watershed. Seldom, however, should

FIG. 185.—Nonchanneled, vegetated depressions of this kind are capable of disposing of large quantities of runoff without danger of gullying. They frequently make the best meadows on the farm. Iowa. (*Photograph by Soil Conservation Service.*)

a meadow strip be less than 100 feet wide, since narrower areas usually cannot be utilized satisfactorily for grazing or hay production.

Where suitable natural channelways or outlets are not available, vegetated channels are resorted to in those localities where an adequate stabilizing sod can be established and maintained. It is important that the cross section of such artificial waterways be designed to prevent flows of scouring velocity. The amount of water that can be carried safely by such a channel is limited by the gradient, the cross section, the permissible channel velocity (usually 5 to 8 feet per second), and the maximum practicable channel width (usually 15 to 20 feet). When larger amounts of runoff must be disposed of, two or more parallel channels are provided. Table 38 gives recommendations for depth and width of vegetated channels.

TABLE 38.—SPECIFICATIONS FOR DEPTH AND WIDTH OF SODDED TERRACE OUTLET CHANNELS WITH 4 TO 1 SIDE SLOPES TO DISCHARGE RUNOFF AS INDICATED[1]

| Channel slope | | Runoff, cubic feet per second | | | | | | | | | | | | | | | | | | | |
|---|---|---|---|---|---|---|---|---|---|---|---|---|---|---|---|---|---|---|---|---|---|
| Per cent | | 30 | 40 | 50 | 60 | 70 | 80 | 90 | 100 | 110 | 120 | 130 | 140 | 150 | 160 | 170 | 180 | 190 | 200 | 210 | |
| 10 | Depth........ | 0′8″ | 0′ 7″ | 0′6″ | 0′ 6″ | 0′ 6″ | 6′ 6″ | | | | | | | | | | | | | | |
| | Width........ | 2′6″ | 6′ 6″ | 10′4″ | 12′11″ | 16′ 5″ | 19′ 4″ | | | | | | | | | | | | | | |
| 9 | Depth........ | 1′0″ | 0′ 8″ | 0′7″ | 0′ 7″ | 0′ 7″ | 0′ 7″ | 0′ 6″ | | | | | | | | | | | | | |
| | Width........ | 2′0″ | 4′10″ | 8′7″ | 11′ 0″ | 13′ 8″ | 16′ 7″ | 20′ 0″ | | | | | | | | | | | | | |
| 8 | Depth........ | 1′2″ | 0′11″ | 0′8″ | 0′ 8″ | 0′ 7″ | 0′ 7″ | 0′ 7″ | 0′ 7″ | | | | | | | | | | | | |
| | Width........ | 1′0″ | 2′ 1″ | 6′2″ | 9′ 7″ | 11′ 5″ | 14′ 0″ | 16′11″ | 18′10″ | | | | | | | | | | | | |
| 7 | Depth........ | 1′2″ | 1′ 2″ | 1′0″ | 0′10″ | 0′ 9″ | 0′ 9″ | 0′ 9″ | 0′ 9″ | 0′ 8″ | 0′ 8″ | | | | | | | | | | |
| | Width........ | 1′0″ | 1′ 0″ | 2′1″ | 5′ 7″ | 8′ 2″ | 10′ 4″ | 12′ 4″ | 14′ 7″ | 17′ 2″ | 19′ 6″ | | | | | | | | | | |
| 6 | Depth........ | 1′2″ | 1′ 2″ | 1′2″ | 1′ 1″ | 1′ 0″ | 0′11″ | 0′11″ | 0′11″ | 0′11″ | 0′10″ | 0′10″ | 0′ 10″ | | | | | | | | |
| | Width........ | 1′0″ | 1′ 0″ | 1′0″ | 2′ 9″ | 4′10″ | 7′ 0″ | 9′ 1″ | 10′10″ | 12′11″ | 14′11″ | 17′ 0″ | 19′ 6″ | | | | | | | | |
| 5 | Depth........ | 1′2″ | 1′ 2″ | 1′2″ | 1′ 2″ | 1′ 0″ | 1′ 0″ | 0′11″ | 0′11″ | 0′11″ | 0′11″ | 0′11″ | 0′11″ | 0′11″ | 0′11″ | 0′11″ | | | | | |
| | Width........ | 1′0″ | 1′ 0″ | 1′0″ | 1′ 5″ | 4′ 4″ | 7′ 0″ | 9′ 2″ | 10′ 8″ | 13′ 0″ | 13′ 2″ | 13′ 7″ | 15′ 7″ | 17′ 2″ | 18′ 6″ | 19′ 4″ | | | | | |
| 4 | Depth........ | 1′2″ | 1′ 2″ | 1′2″ | 1′ 2″ | 1′ 2″ | 1′ 2″ | 1′ 2″ | 1′ 2″ | 1′ 2″ | 1′ 2″ | 1′ 2″ | 1′ 2″ | 1′ 2″ | 1′ 2″ | 1′ 2″ | 1′2″ | 1′ 2″ | 1′ 2″ | 1′ 1″ | 1′1″ |
| | Width........ | 1′0″ | 1′ 0″ | 1′0″ | 1′ 0″ | 2′ 0″ | 3′ 0″ | 4′ 0″ | 5′ 0″ | 6′ 0″ | 7′ 0″ | 8′ 0″ | 9′ 2″ | 10′ 6″ | 12′ 0″ | 13′ 4″ | 15′2″ | 16′ 4″ | 17′ 7″ | 18′ 0″ | 19′6″ |
| 3 | Depth........ | 1′2″ | 1′ 2″ | 1′2″ | 1′ 2″ | 1′ 2″ | 1′ 2″ | 1′ 2″ | 1′ 2″ | 1′ 2″ | 1′ 2″ | 1′ 2″ | 1′ 2″ | 1′ 2″ | 1′ 2″ | 1′ 2″ | 1′2″ | 1′ 2″ | 1′ 2″ | 1′ 2″ | 1′2″ |
| | Width........ | 1′0″ | 1′ 0″ | 1′0″ | 1′ 0″ | 2′ 0″ | 3′ 0″ | 4′ 0″ | 4′ 6″ | 5′ 4″ | 8′ 6″ | 10′ 4″ | 12′ 2″ | 13′ 6″ | 15′ 1″ | 16′11″ | 18′1″ | 1 8′11″ | 19′ 6″ | 19′10″ | 20′0″ |
| 2 | Depth........ | 2′0″ | 2′ 0″ | 2′0″ | 2′ 0″ | 2′ 0″ | 2′ 0″ | 2′ 0″ | 2′ 0″ | 2′ 0″ | 1′ 9″ | 1′ 8″ | 1′ 7″ | 1′ 6″ | 1′ 6″ | 1′ 5″ | 1′5″ | 1′ 4″ | 1′ 4″ | 1′ 4″ | 1′4″ |
| | Width........ | 1′0″ | 1′ 0″ | 1′0″ | 1′ 0″ | 1′ 0″ | 1′ 0″ | 1′ 6″ | 2′ 0″ | 2′ 6″ | 5′ 7″ | 7′10″ | 9′ 7″ | 11′ 4″ | 12′11″ | 14′ 6″ | 16′0″ | 17′ 6″ | 18′11″ | 19′ 8″ | 20′6″ |

[1] 3 to 1 side slopes are used for the channels on 2 per cent slopes. The velocities designed for this table do not exceed 8 feet per second or fall below 5 feet per second, except in some of the smaller channels carrying small amounts of runoff. As a factor of safety, the channel should be made 6 inches deeper than specified in the table. With steeper side slopes or with less stable sod, wider bottoms usually are necessary to reduce the velocity.

Mechanically stabilized channelways and outlets are necessary in those localities where soil and climatic conditions prevent the establishment of adequate vegetative protection and also where excessively large volumes of runoff water must be disposed of.

Mechanical protection is most commonly used as a supplement to vegetated water-disposal structures. Weak points in vegetated structures, particularly toward the lower ends, often call for some measure of mechanical support.

## Construction of Runoff Channels and Outlets

Individual terrace outlets should be so placed and shaped as to permit the minimum concentration of discharged runoff on vegetated slopes

Fig. 186.—Terrace outlet properly shaped for spreading runoff discharge thinly over a grass-protected slope. (*Photograph by Soil Conservation Service.*)

(Fig. 186). In order to discharge the runoff in a broad, shallow sheet of low erosiveness, the lower end of the terrace channel must be made broad and flat. Protective vegetation should extend up into the terrace channel (Fig. 187), so the flat cross section can be maintained and the development of overfalls prevented. Where the natural slope below the terrace outlet is at right angles to the terrace flow, it is good practice to extend the upper terrace a considerable distance into the grassed discharge slope. Such extension should be reduced progressively for each successive terrace down the slope, in order that the possibilities of concentration of the discharge water may be minimized.

To insure proper discharge, the bottom of the outlet channel should be several inches lower than the discharge end of the terrace channel. The overfall here should be sloped at a grade of at least 3 to 1 and well sodded to prevent cutting back into the terrace channel.

Where outlet vegetation is established before a terrace is constructed, it is generally advisable, if regular terracing machinery is used, to terminate the terrace construction far enough away to allow ample room for turning without damaging the outlet vegetation. The gap can be completed with scrapers.

Fig. 187.—Vegetation for protection of terrace outlets should extend some distance into the terrace channel to prevent overfall erosion. (*Photograph by Soil Conservation Service.*)

Excavation and filling should be reduced to a minimum. By careful selection, depressions frequently can be found that will serve as satisfactory disposal outlets without protective treatment. But where small gullies, hummocks, ridges, or shrubs are present, it usually will be necessary to level off the area and remove the shrubs. The presence of trees or shrubs in water channels of any type is generally undesirable because of their tendency to obstruct, divert, or concentrate the flow and produce troublesome meandering currents.

Where vegetated channels are to be used as outlets, some excavation is often required to provide the necessary channel cross section (Figs. 188 and 189). This generally can be done most effectively with a grader, terracing machine, or scraper. All outlet channels should be shaped with as gentle side slopes as possible (usually 4 to 1 or less, that is, a fall of 1 foot in a horizontal distance of 4 feet), so that mowing and maintenance work will not be hampered. Where the topsoil is shallow, the subsoil unproductive, and grass difficult to establish, it has been found desirable sometimes to restrict channel excavation to a minimum and to build up the sides, or channel berms, as much as possible from the outside.

FIG. 188.—Narrow terrace outlets with insufficient vegetative protection soon develop into gullies. Texas. (*Photograph by Soil Conservation Service.*)

FIG. 189.—Development of proper channel cross section for outlet shown in Fig. 188. (*Photograph by Soil Conservation Service.*)

VEGETATED OUTLETS. It is both hazardous and expensive to undertake establishment of vegetated outlets at the same time that they are being used for water disposal. Seed, fertilizers, and young plants are washed out readily, unless special precautions are taken. Ordinary seeding methods, along with necessary fertilization, usually will provide the most economical means of establishing an effective cover (Figs. 190, 191). Sometimes the placing of sod over the entire channel or in strips across it will provide adequate protection, but this is more costly than ordinary seeding. Moreover, it is difficult to anchor newly placed sod in some channels.

As already indicated, best results in the stabilization of terrace outlets with vegetation can be attained generally by doing the work before the terraces are completed. On some farms, where several types of outlets may be necessary and where all necessary terracing work cannot be done at once, those areas for which natural outlets are available, or for which channels with solid sodding or mechanical protection are required, can be terraced while seeded vegetation is becoming established in other outlets. Still another plan that may facilitate the use of seeding and reduce the cost of establishing vegetated outlets is to divert runoff from the terraces through a temporary outlet until plantings in the permanent outlet are well established.

FIG. 190.—With proper care this broad grassed channel not only will dispose of the runoff from the contributing watershed without gullying, but will produce considerable hay. A masonry structure at the lower end safely delivers the water to a stabilized outlet. Kentucky. (*Photograph by Soil Conservation Service.*)

In some localities, sod has been established successfully on seeded outlets with the aid of a mulch of straw, brush, or old hay, applied at the rate of 2 or 3 tons per acre and anchored with wire. A good mulch not only protects the seedbed from erosion but aids in the conservation of enough moisture to encourage germination and growth of the grass. Such mulching material is usually rotted sufficiently by the time the grass has become established to make its removal unnecessary.

The use of strip sodding generally is restricted to moderate slopes (5 per cent or less) and to relatively small drainage areas (15 to 20 acres or less) where the volume of runoff is not too large. Even with these limitations, however, there have been some failures. In other instances, some repair has been necessary because of the tendency, particularly on friable sandy soil, for overfalls to develop along the strips. The sod

FIG. 191.—Well-grassed, broad, disposal channels effectively handle large volumes of runoff. No erosion took place in this adequately protected channel, near Hartford, Conn., during the heavy downpour accompanying the September, 1938, hurricane. (*Photograph by Soil Conservation Service.*)

strips are usually cut 12 inches wide and 1 to 3 inches thick. Sometimes they are placed in trenches cut to the dimensions of the sod across the bottom of the outlet channel. In other instances, they are laid on the surface of the channel, while the spaces between them are filled with good soil and seeded. Regardless of which method is used, the upper surface of all strips should be flush with the bottom of the completed channel in order to reduce turbulence and the consequent danger of washing out during heavy rains.

Sod strips are spaced anywhere from 3 to 6 feet apart, according to the slope of the channel, but a 4-foot interval is most commonly used. Usually, the vertical interval between strips probably should not exceed the thickness of the sod. The strips should be extended up the sides of the channel to at least the maximum depth of runoff that may reasonably be expected.

The cost of strip sodding is almost as great as for solid sodding. Since the latter generally gives better results, strip sodding is used chiefly in those localities where not enough sod is available for solid channel lining.

Spot and broadcast sodding are used to establish Bermuda grass from root stolons. Spot sodding is accomplished by planting root sprigs in handmade holes, spaced on 12- to 24-inch centers, with 2 to 3 inches of soil packed around the roots. Another method of spot seeding is to drop the sprigs at more or less regular intervals and then plow or disk them in. The latter procedure is most economical where such equipment can be used and where large areas are to be covered.

For broadcast sodding, the sod source area is usually disked so as to cut up the roots and mix them with the topsoil. This mixture is then removed with shovels, scrapers, or fresnos and spread over the area to be sodded in a continuous layer. In some localities, manure spreaders have been used for this purpose. After spreading, the land usually is disked and packed. This has the advantage of adding considerable topsoil, as well as roots, to unproductive areas. The resulting mulch offers some resistance to erosion, although it is not nearly so effective in this respect as solid sodding. The economy of broadcast sodding is largely due to the fact that the gathering and planting procedure can be handled almost entirely by machinery.

In bluegrass areas where seeding has been impractical, solid sodding has given the most satisfactory results. Even in the Bermuda-grass sections, solid sodding frequently has been necessary where an immediate channel cover is required or where slopes are steep. With this method, the sod is cut from a well-grassed area, laid in continuous strips across the channel, and thoroughly tamped in place. It is cut generally in strips 1 foot wide and 1 to 3 inches thick. Although thicker sod is more difficult to handle, it provides more topsoil and offers greater initial resistance to washing out. To protect newly laid sod from heavy runoff before the roots have anchored the strips, it frequently is necessary to stake or staple the sod at regular intervals. Anchorage by lightweight woven wire and stakes has even been necessary in some instances. The use of commercial or homemade sod cutters will materially reduce sodding costs. Transportation on 6- to 10-foot boards, rather than attempting to roll it, also has facilitated operations.

Mechanically Protected Outlets. Where mechanical protection is necessary in terrace outlets, low permanent structures of the weir-notch type generally are used. Temporary structures usually have proved ineffective (Fig. 192).

Notch dams may be built of brick, stone, reinforced concrete, precast concrete blocks, or a combination of these materials. The local availability of materials frequently will determine the type of structure used. Where suitable rock is available without expensive quarrying and hauling, rubble masonry construction is usually the first choice. In order to provide a more stable structure and eliminate danger of cracking, reinforced concrete is preferable for the apron slab and cutoff walls, regardless of the kind of material used in other parts of the structure.

FIG. 192.—The use of temporary structures in runoff channels generally gives insufficient protection from erosion. (*Photograph by Soil Conservation Service.*)

The apron slab should have a minimum thickness of 4 inches for dams up to 2 feet high. Above that height, this minimum thickness should be increased about 1 inch for each 2 feet, or fraction of a foot, of additional height.

General specifications for the construction of small masonry check dams are given in Graph 42. Where structures higher than 4 or 5 feet are required, it is advisable to have separate plans and specifications prepared by a competent engineer.

In order to reduce the hazard of washouts, care should be taken to provide (1) ample notch capacity to take care of the runoff from the contributing drainage area, (2) impervious foundations, and (3) checks to seepage around the structures. The headwall should extend well into

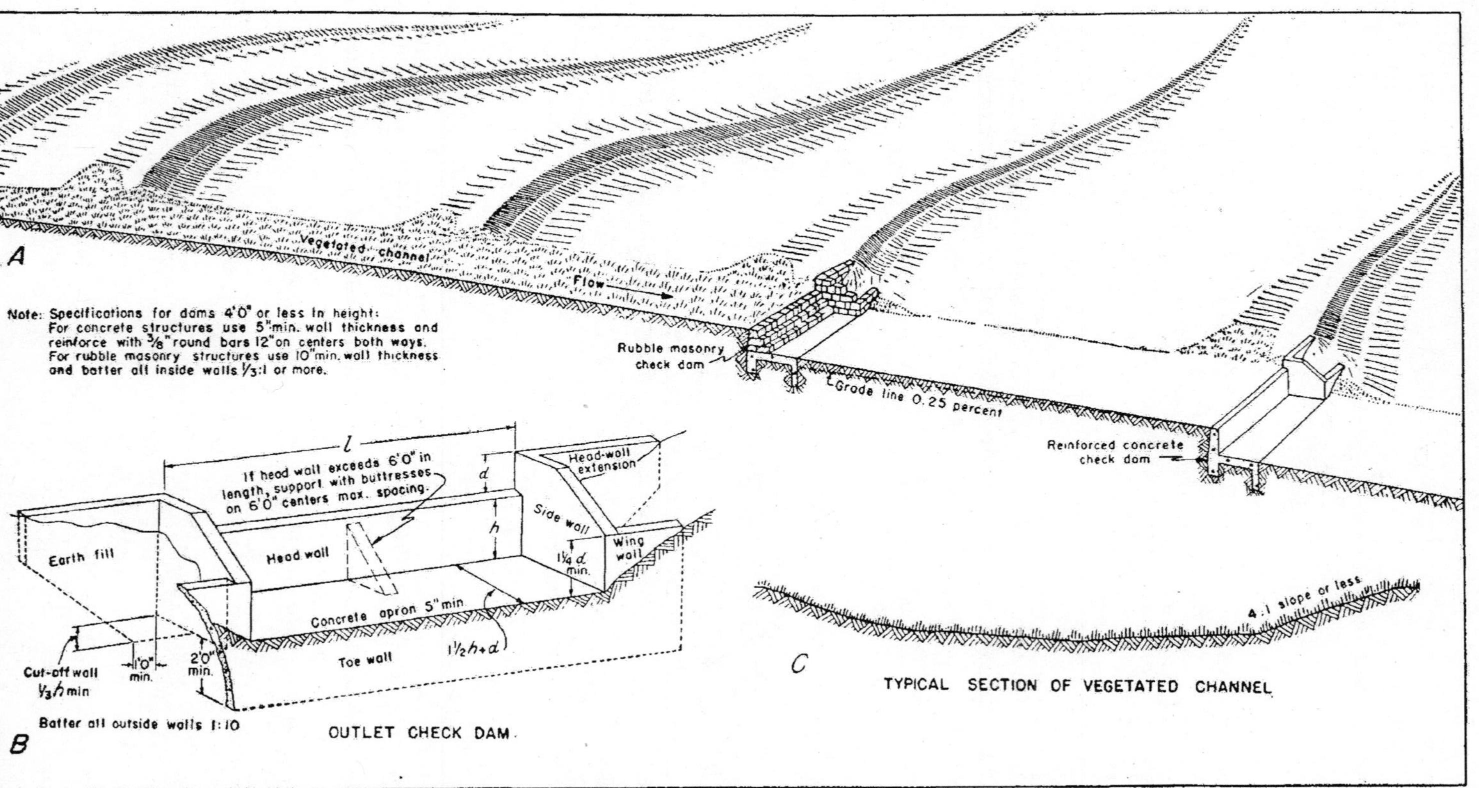


GRAPH 42.—A terrace-outlet channel protected from erosion by a combination of vegetative and mechanical measures. The sectional view in the upper part shows vegetative protection in the upper reaches of the channel and check dams in the lower part. The construction details of the check dams and the cross section of the vegetated channel are shown in the lower part of the illustration. (*Soil Conservation Service.*)

the bottom and sides of the channel excavation, and an impervious type of clay should be tamped firmly around the structure to prevent seepage. Particular care should be taken that the extensions penetrate far enough into the soil to provide adequate protection even after the sides of the ditches slough off to a natural angle of repose.

Outlets to be protected by mechanical structures should be excavated to accommodate the structures and the amount of runoff to be handled. Although full excavation of outlet channels is not always necessary where structural protection is to be provided, it usually is desirable wherever convenient, so that the sides can be sloped and the berms leveled off and vegetated. In some instances, only enough excavation is done to open up the channel, scouring by subsequent runoff being depended on to remove the remaining earth. Structure locations must be excavated at least to the outside dimensions of the installation; and where the concrete cannot be poured directly against the earth, additional excavation usually is required to provide adequate working space around the structure.

The weir notch of terrace-outlet check dams seldom should be less than 18 inches deep, and the elevation of the weir crest should be at least 1 or 2 inches below the channel elevation at the terrace outlet. This drop provides some allowance for rough construction and avoids the possibility of restricted terrace drainage as a result of the crests being finished higher than the bottom of the terrace channel. Where exceptionally large volumes of runoff are expected, it may be desirable to deepen the notch to avoid excessive structure widths, but extreme care must be taken to protect the drop from the terrace channel to the weir crest elevation in order to prevent overfalls from cutting back along the terrace channel. This critical point should be sloped and sodded. A grade of one-fourth of 1 per cent from the weir crest of the lower dam to the bottom of the apron slab of the dam immediately above usually will provide stable channels between structures.

It is generally advisable to use only one outlet structure for each terrace interval. On the steeper slopes where two structures per terrace interval may be necessary, the second structure is usually located midway between the terraces. Where bends occur in structurally protected outlets, the structure should be located so that it will discharge runoff parallel to the direction of the channel below the structure.[1]

[1] For further discussion of runoff-disposal channels and outlets, see *Farmers' Bull.* 1814, U. S. Dept. Agr., by C. L. Hamilton.

# Chapter XXII. Subsoiling and Other Subsurface Tillage Operations

Subsoiling of agricultural land is the process of mechanically loosening or fracturing the subsurface material in order to increase infiltration of rainfall, penetrability to plant roots, and aeration. Several types of machines are employed in the accomplishment of this form of subtillage. One commonly used is the *chisel cultivator* (Fig. 193), a machine carrying one or more points capable of penetrating and loosening the subsoil to depths ranging usually from about 12 to 18 inches. Another more powerful

FIG. 193.—A single-point subsoiler of chisel type, capable of fracturing soil with brittle clay subsoil to a depth of 30 inches, when dry. (*Photograph by Soil Conservation Service.*)

machine of similar type, sometimes referred to as the *pan breaker*, is capable of penetrating to depths of as much as 30 inches. The *gyrotiller* (Fig. 194), with a set of strong revolving points, is used to pulverize subsurface material to depths below that of ordinary plowing.

As used in soil and water conservation work, the primary objective of subsoiling is (1) to conserve water by improving the physical condition of the soil for absorption of rainfall and (2) reduce erosion by decreasing runoff. Until recently, the method was looked upon simply as a mechanical measure for increasing crop yields. These distinctions are of no great importance from the standpoint of the mechanical principle employed, since the three objectives are approached by breaking up intractable subsurface material, which usually is resistant to water penetration as well as to plant-root development. From the standpoint of necessary depth of

FIG. 194.—The gyrotiller, a machine used in England to pulverize subsurface material below depths of ordinary plowing. (*Photograph by Soil Conservation Service.*)

FIG. 195.—Chisel machine used for subsoiling terrace channels. (*Photograph by Soil Conservation Service.*)

mechanical penetration and type of implement required, however, the distinction may take on considerable importance. For example, if rainfall storage is the primary purpose, it may be desirable in localities of high rainfall to fracture the sublayers to maximum practicable depths, whereas shallower depths of subsoil fractionation or pulverization may be preferable if the objective is merely to increase the yield.

The chisel type of subsoiler is employed sometimes to break up tough subsurface materials in order to facilitate the construction of terrace fills and earth dams and to loosen the sublayers of terrace channels (Fig. 195) and contour furrows so as to provide deeper penetration of water and better seedbed preparation.

Subsoiling has been practiced rather sparingly in the United States. In various other countries, it has been used for a good many years. Probably the practice has been followed most extensively in England, Cuba, and Puerto Rico. In the United States, it probably has been used more widely in California, on lands having tough clay or hardpan subsoil. In Louisiana, the gyrotiller has been used to some extent in sugar cane tillage for several years. Subsoiling has been experimented with in many localities, with widely differing results reported. The cost, together with frequent reports of ineffective results, as measured by crop yields, appears to have militated against development of the practice.

Very little information had been accumulated in relation to the value of subsoiling in soil and water conservation until recently; and even now, little in the way of quantitative data is available. Results obtained in various localities on soil and water conservation demonstrations indicate the immediate need for additional information, especially with respect to the technique of subsoiling as well as to soil, slope, and climatic adaptability.

## Adaptability and Results Obtained

Subsoiling, or *knifing*, has been practiced extensively on some of the large *centrals* of Cuba in preparing land for sugar cane. Studies made by the author in collaboration with R. V. Allison revealed that although some soils are benefited markedly by the practice, no striking beneficial effects are discernible on others.[1] It was found that soils like the Truffin clay, which assumes an almost stonelike condition in the dry season, with the loss of almost all available moisture to depths of 4 feet or more, shatter or fracture in a highly favorable way when deeply knifed in the dry condition. The tough subsoil is opened to penetration by plant roots, rainfall, and air to a degree that essentially amounts to changing the soil to a different type. Knifing of such land sometimes results in doubled, trebled, or even quadrupled yields of sugar cane, and the effects last 4 or 5 years or longer.

[1] Bennett, H. H., and Allison, R. V., "The Soils of Cuba." 1928.

On the other hand, those soils having highly plastic, dense clay subsoil, such as the Yaguajay or Bayamo clay, are not benefited noticeably by deep subsoiling. The first soaking rains cause the clay to flow back into its original condition of density and imperviousness, without leaving any observable beneficial effect.

Considerable experimental work has been carried on in England in connection with rotary cultivation and subsoiling. Some farmers are now using this method of soil preparation as a standard practice. Culpin, experimenting with both light and heavy soil in west Cambridgeshire, reports favorable results from deep tillage.[1]

In southeastern United States, results of soil and water conservation work indicate that soils having brittle clay subsoil, such as those of the Cecil group, are benefited by subsoiling on the contour. On the other hand, those with very plastic clay subsoils, such as Iredell, and still others with open and friable subsoil, like the Durham, apparently are not markedly benefited. Experiments in Utah, Illinois, Pennsylvania, Mississippi, and Texas cast considerable doubt on the value of subsoiling on the types of land tested in those states.

Subsoiling has become a rather general practice in California in those irrigated and nonirrigated localities where the subsoils are hard or compact. Here, the land is subsoiled on the contour, at intervals of about 5 feet, in advance of the fall rains. This operation results in shattering the upper 18 inches of the soil profile into large clods which gradually soften and crumble with succeeding rains. The work definitely retards runoff, increases infiltration, and aids in the maintenance of yields.

Investigations of the effects of subsoiling at 12 Great Plains field stations of the Office of Dry Land Agriculture, Bureau of Plant Industry, have not indicated, on the areas treated, any very marked trend toward increased moisture storage.[2]

Investigations conducted in 1936 by Sidney Burton and H. L. Lobenstein of the Forest Service, near Mangum, Okla., indicate that subsoiling previous to the planting of trees (Fig. 196) greatly increases first-year survival and growth.[3]

Measurements of water penetration on the Goshen Hole soil and water demonstration project, near Torrington, Wyo., show that three days after a 0.64-inch, 20-minute July rain, moisture had penetrated to depths ranging from 25 to 50 inches on land contour-chiseled to a depth of 18 inches, whereas no moisture was found below depths of 12 to 14

[1] Culpin, Claude. The Effects of the Fowler Gyrotiller on the Soil, *Jour. Agr. Sci.*, Vol. 26, Part I, Cambridge University Press, London, 1936.

[2] Chilcott, E. C., and Cole, J. S. Subsoiling, Deep Tilling, and Soil Dynamiting in the Great Plains, *Jour. Agr. Research*, Vol. 14, September, 1918.

[3] *Forestry News Digest*, June, 1937.

inches (Graph 43) on adjacent untreated land. Results of a single rain are not at all conclusive, of course, but they may be indicative of possible benefits to be derived from contour subsoiling under conditions like those of this area.

The fracturing effects of subsoiling different kinds of land in the North Carolina Piedmont are illustrated by Graph 44. Profiles *A* and *B* of this graph indicate the probable futility of subsoiling land having sublayers of friable sandy clay and plastic clay, respectively; and profiles *C* and *D*

FIG. 196.—A subsoiling machine used in tree-planting operations. (*Photograph by Soil Conservation Service.*)

indicate the possible value of subsoiling those types having brittle clay subsoils.

The general trend of experimental work in the United States has tended to discourage the practice of subsoiling, especially from the standpoint of its effect on crop yields. Yet definite success obtained under certain conditions indicates that insufficient information on the subject is available for the country as a whole. From the standpoint of soil and water conservation, it appears that although soils having either loose, friable subsoil or highly plastic clay subsoil are not likely to be benefited, generally, by subsoiling, those having brittle clay subsoils that become very hard or compact on drying frequently can be benefited by the practice, if applied when dry.

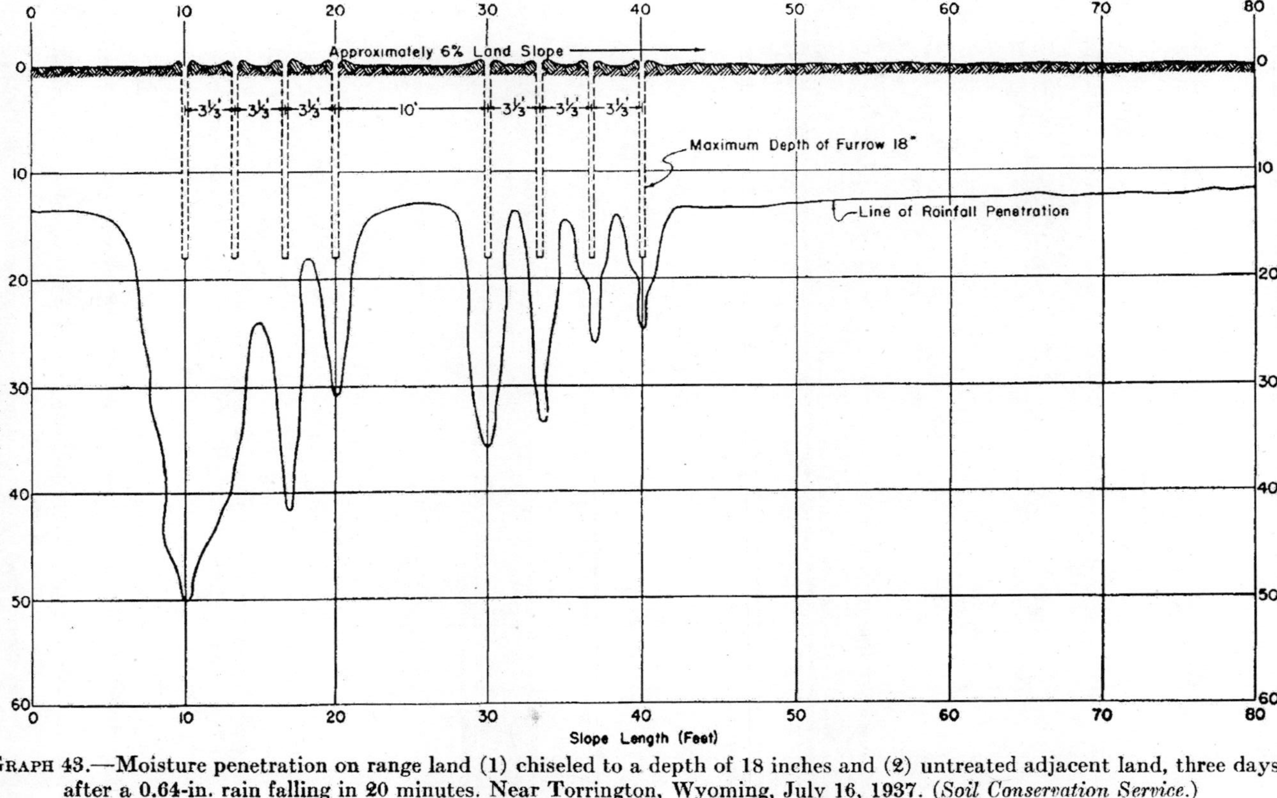


GRAPH 43.—Moisture penetration on range land (1) chiseled to a depth of 18 inches and (2) untreated adjacent land, three days after a 0.64-in. rain falling in 20 minutes. Near Torrington, Wyoming, July 16, 1937. (*Soil Conservation Service.*)

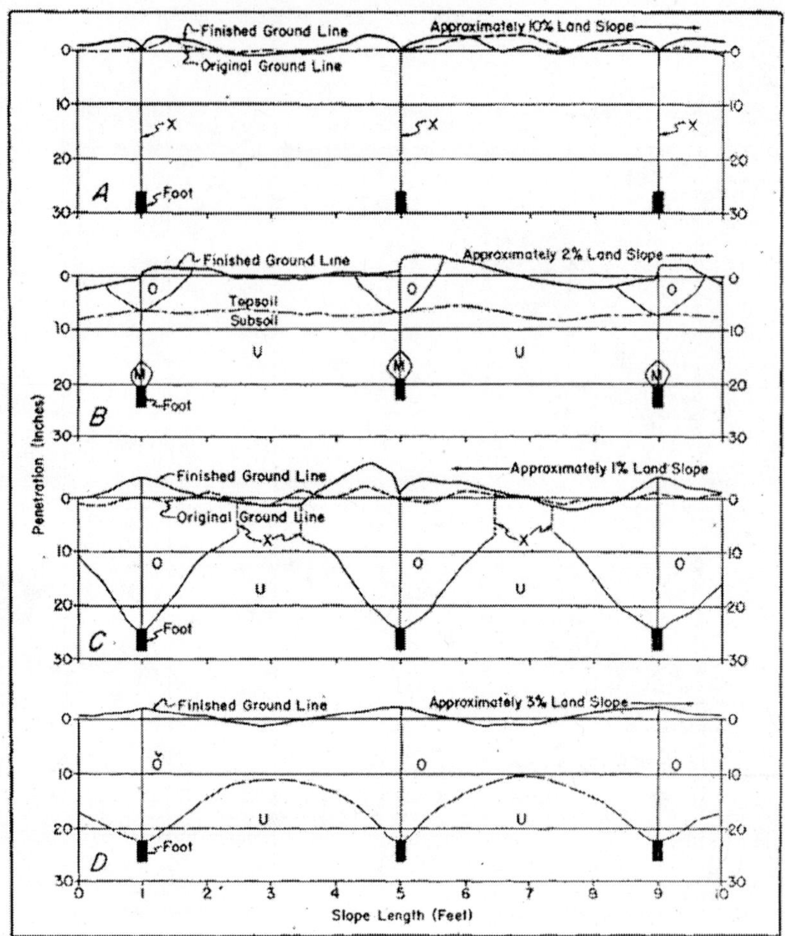


GRAPH 44.—Profiles of soil disturbance by subsoiling, near Greensboro, N. C.:

*A*. Durham sandy loam, with friable subsoil, chiseled to depth of 30 inches without apparent breakage aside from faint cracks near line of penetration by blade. Slight amount of upheaval indicates soil structure was not changed appreciably.

*B*. Iredell sandy loam, with plastic clay subsoil below 7 inches, chiseled 18 to 24 inches. *M* represents cavity left in subsoil by the foot of the blade. *U* represents undisturbed area.

*C*. Appling sandy loam, with moderately brittle clay subsoil, chiseled to 28 inches. *O* represents area of complete breakage, Distinct fracturing from foot to point of transition between sandy loam soil and clay subsoil. Breakage assumed to have extended from this point vertically to the surface, as indicated by dotted lines. *U* represents undisturbed section.

*D*. Cecil sandy loam, with brittle clay subsoil under 5 inches of sandy loam topsoil, chiseled to 26 inches. *O* represents area of complete breakage. Large fragments, with very distinct cracks, except near foot where the fragments were small (much soft rotten rock at this depth). *U* represents undisturbed section.

(*Soil Conservation Service.*)

## Use of Explosives to Break Hardpan

In some localities, particularly in Pacific Coast areas of hardpan soil, explosives have been used successfully to break rocklike sublayers in preparing land for planting of fruit trees. Where soil depth over such impenetrable material is too shallow for cultivation, this seems to be the

Fig. 197.—Upper, rough cloddy surface formed by plowing up clay subsoil, Texas Panhandle country; lower, interception of surface-drifting sand by rough cloddy land surrounding dunes. (*Photographs by Soil Conservation Service.*)

only feasible method for bringing land of this kind into use other than for grazing.

Scarification of shallow rocky land has been practiced with some degree of success for planting subtropical fruits in the southern Florida Peninsula. The limestone areas of this section are so pitted with holes that most of the soil is concealed in the cavities; and in order to prepare any semblance of a seedbed, either dynamiting or scarification is neces-

FIG. 198.—Enormous turning plows, drawn by a tandem of tractors, are used to bring up productive soil that has been covered to depths of 4 feet or more by relatively unproductive debris of erosion. This plow is capable of turning a furrow 6 feet in depth. It is used in parts of California for turning up productive soil covered by overwash or erosion debris. (*Courtesy Post Brothers, Santa Ana, Calif.*)

sary. Scarification is so expensive, however, that it can be practiced only where high-value crops, such as citrus, are grown.

## Upturning of Subsurface Material

The plowing up of relatively heavy subsoil material in order to mix it with loose sandy topsoil, or to safeguard such topsoil from blowing by scattering clods of clay over the surface, has been practiced successfully in parts of the Southern Plains (Fig. 197) and to some extent in areas farther west.

Reclamation of land by turning up soil that has been covered with unfavorable sand, shale, and gravel washed from adjacent uplands or deposited by floods is an important practice in parts of California where high-priced orchard lands are subject to such damage. Enormous turning plows drawn by a tandem of tractors are sometimes used to bring up productive soil that has been covered to depths of 4 feet or more (Fig. 198). This is accomplished with an adjustable blade in front of the big plow. The blade scoops the sand down into the gigantic furrow made by the previous passage of the plow. Then the great turning plow comes along and turns up the old productive soil on top of the unproductive overwash.

# Chapter XXIII. Gully Prevention and Control

Gullying became a serious problem in America soon after the beginning of agriculture. Various methods of control were used in early days, but revegetation generally proved most effective. Washington informed his overseer in 1795 that immediate profit was not so much an objective as the bringing of worn-out and gullied fields into condition to produce grass.

Straw, weeds, and other vegetative refuse were commonly used to fill gullies or to prevent fields from further gullying. In these haphazard attempts to check the growth of ravines, the principal difficulty was to hold in place the materials used. Dams of stone, brush, and logs were tried, but they generally failed, primarily because of faulty construction and failure to provide adequate protection against undercutting at the point of overfall. Most farmers used loose brush alone; others wisely permitted the establishment of volunteer vegetation—weeds, grass, vines, shrubs, and trees. Some aided this process of stabilization or reclamation by planting honeysuckle and other soil-binding vegetation. In a few instances, farmers undertook to hasten stabilization by plowing in the sides of gullies.

From time to time, various organizations, individual farmers, and other interested people attempted to advance and spread knowledge of gully control by education or demonstrational work; but it was not until 1933 that truly widespread action on a nation-wide scale began, under the work program of the CCC camps and the Soil Conservation Service (then Soil Erosion Service). Since then much progress has been made. New techniques have been developed; public understanding and approval have been gained; and several million gullies have been controlled. No longer do these gullies gouge farther into productive farm and grazing land; pour unproductive subsoil over rich farm lands downstream; or fill reservoirs, streams, drainage ditches, and irrigation canals with silt. No longer do they hasten the flow of rainfall into flood streams. Many of them, stabilized with grass and other vegetation, have been converted into useful channels for the orderly disposal of runoff. Some controlled gullies are now utilized, incidentally, for grazing and for the production of hay; others are serving as sanctuaries for wildlife.

Nevertheless, no more than a good beginning has been made. Surveys indicate that more than 200 million gullies have sliced into the landscape of the nation, and new ones are being opened with every heavy rain. Not one of the 200 million gullies was here when white man came to America.

## Causes and Types of Gullying

The principal cause of gullying is the removal of the protective covering of vegetation from the land, followed by cultivation and grazing without provision for retarding the resultant accelerated runoff. Other alterations of natural conditions have contributed much to the spread of gullying, such as cultivation up and down slopes, construction of faulty water channels (ditches, terraces, diversions), roadways, livestock trails, mining operations, and the obliteration of vegetation by smelter fumes.

The types of gully erosion can be classified into three broad groups: *channel*, *waterfall*, and *mass movement*.

CHANNEL GULLYING. Vastly more gullying has had its beginning in artificial channelways, whether starting as a furrow or stock trail (Fig. 199) or in a natural unchanneled waterway stripped of its cover, than in any other way. If not checked, runoff concentrates in such lineal depressions and either slowly or rapidly causes incisions of various shapes and depths, depending principally on the soil (and subsoil) conditions, gradient, rainfall, and size and conditions of the contributing watershed. The depth and type (shape) of channel gullying depend primarily on the character of the soil. If the underlying materials are soft and easily incised to great depths, deep, straight-walled, U-shaped gullies are formed. If the subsoil consists of plastic, resistant clay, relatively shallow, V-shaped gullies are the usual result. Where hard rock occurs near the surface, the incisions will be shallow and usually wide. Other variations of channel gullies are met with, but these major forms illustrate the influence of the material that is gullied.

The influence of certain physical factors on rates of gully development must be weighed carefully in applying control measures. Generally, the principal factors to be considered in this connection include soil resistance to abrasion, speed of runoff as determined chiefly by gradient and vegetative cover, intensity and volume of rainfall, and size and condition of the watershed.

Gullies branch more frequently and become longer, deeper, and wider as unchecked scouring continues—that is, within the limitations of the enclosing watershed or that part of it susceptible to gullying. Often, gullies grow to a mile or more in length and 20 to 40 feet wide and deep. Locally, they expand to much greater dimensions. Extension in length is usually much faster than in width, because the greater volume of water usually enters at the head. After the ravine reaches the watershed divide,

Fig. 199.—Above, a Missouri gully eating its way back into an overgrazed pasture; below, same gully controlled and converted into a stock-water pond. (*Photographs by Soil Conservation Service.*)

extension of length stops, but depth and width may increase still farther. The rate of growth generally is faster along the steeper upper slopes; but when the slopes are uniform, the opposite may be true. Silting, rather than cutting, usually takes place where the channel enters an area of milder gradient. Increased volume of water, however, may overcome the influence of moderate changes in slope along the lower reaches.

WATERFALL EROSION. Waterfall erosion (Fig. 200) is responsible for the formation of many of the deepest and widest gullies. It is also responsible for the branching of many gullies—the cutting out of tributaries, which have their beginnings at vulnerable points along the sides. The points where livestock habitually enter and leave a ravine, for example, are usually vulnerable.

FIG. 200.—Typical example of results of waterfall erosion in an Oklahoma gully. Even a small amount of water falling from such a height develops enough energy to cause rapid erosion at the foot of the overfall. Undermining results, with consequent sloughing of the overhanging lip. The cross section of such gullies is U-shaped. (*Photograph by Soil Conservation Service.*)

A small vertical overfall beginning at the lower end of an eroding depression wears away the underlying material until the top caves in. Continuing by the same process, the waterfall gradually carries the gully upslope. As the point of overfall advances upslope, the height of the vertical escarpment usually increases, since the water cuts deeper and deeper in its attempt to maintain a relatively flat grade at the bottom.

In this way, gullies starting at the banks of natural watercourses often extend far up the slopes, especially where the subsurface material is soft and easily worn away. Such ravines frequently attain depths of 50 to 60 feet or more. As they cut back, they frequently cross lateral or oblique depressions or small waterways, thus favoring the development of tributary gullies. The process of branching may continue until a network of gullies dissects the entire watershed.

Water entering gullies from the sides causes waterfall or undercutting erosion, but the concentration is normally less, and the rate of gully extension correspondingly slower.

Gullies formed by waterfall erosion may extend quite rapidly, even through nearly level land. Their rate of growth often depends more on subsoil characteristics, depth of overfall, and the size of the contributing drainage area than on the slope of the land. Since the characteristic cross-sectional form of waterfall gullies is U-shaped, they are commonly referred to by technicians as *U-gullies*. Once the soft substratum material is incised, growth of the gully by undercutting waterfalls and channel flow (differential erosion), followed by caving of the upper banks, is characteristic of big gullies on land of this kind. Such ravines frequently grow at the rate of 30 to 50 feet a year, and in extreme cases they have been known to advance several hundred feet during a single heavy rain.

Some of the larger gulches formed in this way are known as *barrancas* in the Pacific Southwest; as *coulees* in the Northwest; and as *arroyos*, or *washes*, in the Colorado Basin area.

Often a series of parallel gullies crossing fields of uniform slope mark the sites of furrows, ditches, wagon-wheel ruts, or livestock trails. More often, however, topographic variations; subsoil conditions; and the location, size, and shape of upslope clearings cause marked irregularities in the direction, shape, and size of the ravines.

GULLYING INDUCED BY MASS MOVEMENTS. In localities where mass movement in the form of earthflows, slides, and slumps occur frequently (see Chap. XII, Part 1), a considerable number of gullies begin in these disturbed areas. This is particularly true where the displacement exposes highly erodible subsurface material and where fissure lines and vertical escarpments are formed by such movements.

Sometimes ravines approximating gullies in form are the direct result of the more violent ground disturbances; but more frequently the mass movements of soil merely develop conditions that favor gullying. Many of the slides and earthflows that occur on steep slopes during rains, especially after periods of alternate freezing and thawing, form broad and narrow land scars that are the virtual equivalent of gullies.

Alternate freezing and thawing often loosen the soil of gully sides, permitting the material to slough off, crumble, and be carried away by subsequent rains. This form of gully expansion is particularly active in the southern part of the United States. It may continue for years as the only important form of erosion in gullies that already extend to, or nearly to, the crest of watersheds.

## Prevention and Control

It is much easier to prevent the development of gullies than to control them after they have formed. Gullying usually is preceded by sheet erosion and therefore may be prevented to a large extent by taking the steps necessary to check ordinary surface washing. However, constant precaution is necessary on sloping land to prevent the formation of gullies through the enlargement of minor channels formed by farm roads, wheeled machinery, livestock trails (Fig. 201), and the excavations of rodents. Usually, rather simple measures, such as the relocation of roadways as nearly as possible on the contour, the filling of ruts and rodent tunnels with straw, and the relocating of fences in order to shift stock trails, are sufficient to check incipient gullies, if used in time.

FIG. 201.—Gullies that started in sheep trails. Wisconsin. (*Photograph by Soil Conservation Service.*)

When the cost of gully control exceeds the value of the land protected, the work may not be justified unless it serves to protect adjacent or downstream land, reservoirs, waterways, buildings, highways, bridges, or other property. In appraising the value of such work, or in determining the need for it, it is extremely important to recognize the fact that even a relatively small gullied area may constitute a serious menace to downslope or downstream land as well as to channelways, reservoirs, and harbors. Moreover, it must be remembered that, in relation to flood control and prevention of silting, numerous scattered areas in which gullying is active may, in the aggregate, constitute the sore spot of a large watershed. Also, a single gullied field, or even one large branching gully, may seriously affect the value of an entire farm.

After a gully has eaten its way to the head of the watershed, it usually ceases to be active. If it is merely protected from livestock, volunteer vegetation usually will establish a suitable protective cover. But when a gully has advanced only a short distance into a watershed, there is immediate necessity for employing intensive control measures. The potential damage of such a gully to adjacent fields, if it is allowed to continue uncontrolled, generally will justify the expenditures necessary to assure complete control within the shortest possible time.

## Methods of Gully Control

Usually, the best way to control a gully is to plant the entire ravine to stabilizing vegetation. However, even where it is impractical to plant the entire gully, it is generally essential to stabilize the head. The type of vegetation and the procedure necessary for its satisfactory establishment will vary considerably, according to local soil, climate, and topographic conditions.

The unproductive subsoil commonly exposed in gullies often makes the establishment of vegetation difficult; but with adaptable types of vegetation, aided by fertilization where necessary, effective growths generally can be secured. Runoff on a gullied field, however, must be handled carefully, particularly during the period when stabilizing vegetation is being started in or around the gullies. This runoff hazard increases with the size and condition of the watershed. A heavy concentration of water in a gully under treatment would damage or wash out the plantings. Temporary dams, porous brush or wire *barrages* (sometimes called *porous dams*), and diversion channels will help in many places in the establishment of satisfactory growths. If the volume of runoff is too great for effective vegetative control, permanent structures must be utilized.

From the standpoint of economy and practicability, the principal methods of handling runoff for gully control, in order of importance, are:

Retention of rainfall on the watershed.

Diversion of runoff away from the gully.

Safe conveyance of runoff through the gully.

Frequently, the most practical solution will be a combination of these methods. Whatever method is used should be planned and executed carefully. Haphazard work will lead to failure or even to an aggravation of the original gully conditions. Until vegetation is well established, it is generally very important to exclude livestock from treated areas, whether vegetation or structures are used.

The treatment of gullies varies according to size. Although a gully that would be considered large in the Northeast probably would be characterized as no more than medium-sized in the Southeast, the following classification seems adequate for general use:

Small gully—less than 3 feet deep.

Medium gully—3 to 15 feet deep.

Large gully—more than 15 feet deep.

A similar general classification with respect to size of the contributing drainage area follows:

Small drainage area—5 acres or less.

Medium-sized drainage area—5 to 50 acres.

Large drainage area—more than 50 acres.

Retention of Rainfall. Generally, runoff of rainfall can be reduced materially through the use of ordinary soil and water conservation measures, previously explained. To obtain satisfactory control of runoff under some of the more complicated conditions encountered, however, special measures, such as closely spaced contour furrows, level terraces with closed ends, *syrup-pan dikes* for spreading water, and subsoiling must be used. Where the soils are absorptive, the slopes gentle, and the rainfall light, measures of this kind frequently will eliminate the need for additional mechanical protection in gullies of small and medium size.

Small and medium gullies, with small to medium uncultivated watersheds, frequently can be stabilized by placing a series of earth fills, spaced according to the steepness of the land, across the gully channel.

Diversion of Runoff. In the control of gullies, runoff usually should be diverted from the head of the ravine before control measures are installed within the gully channel itself. This principle generally applies to gullies of all sizes, except those with such small watersheds that runoff is not a serious problem.

By terracing the drainage area above the gully head, much or all of the runoff can be diverted to some other point for safe disposal. Frequently, small gullies can be crossed directly by terraces and so eliminated; but the necessary filling to carry structures across gullies deeper than 4 or 5 feet may entail high costs. Nevertheless, where numerous parallel gullies cross an area, especially where soil conditions make it difficult to establish vegetation, the terracing method or the use of diversion ditches may be the only practicable method of control.

Diversion channels are especially useful for gully control where the watershed is timbered or grass covered. Such structures generally are not effective immediately below unprotected cultivated fields, unless a permanent *filter strip* of vegetation is established above the ditch. If such a strip is used, it should have sufficient width and vegetation density to clear up the runoff and so prevent silting of the diversion channel below.

It is essential that diversion ditches be made large enough to carry all the runoff from the contributing drainage area during periods of maximum rainfall. It is also essential that the gradient and dimensions of the channel be such that the runoff will move through the ditch at nonerosive velocities.

For earth channels without a permanent vegetative cover, the maximum gradient should not exceed 12 inches per hundred feet of length; but with a permanent lining of resistant sod, steeper gradients may be used. It is likely that these steeper gradients will permit serious scouring

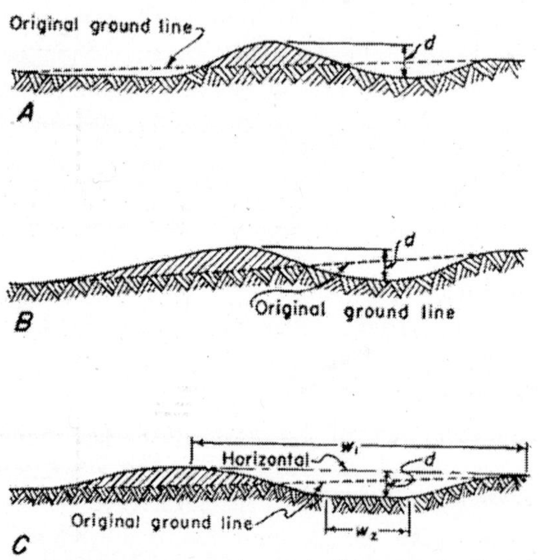


GRAPH 45.—Diversion-ditch cross sections: *A*. Terrace-type diversion ditch for gentle slopes. Construction from both sides. Minimum value of *d* about 18 inches. *B*. Terrace-type diversion ditch for steeper slopes. Construction generally from the upper side only. Minimum value of *d* about 18 inches. *C*. This type of diversion ditch is suggested for watersheds exceeding 10 to 12 acres, especially for the steeper slopes. Minimum value of *d* should be 22 inches. Side slopes should be at least 3 to 1 where land slopes permit. (See *Farmers' Bull.* 1813, U. S. Dept. Agr.) (*Soil Conservation Service.*)

if the vegetative protection of the channel is ever destroyed, and they should be used only where the cover can be depended on during periods of most severe runoff.

The terrace type of diversion ditch is being used extensively where the area of the watershed does not exceed 5 or 6 acres (see cross sections *A* and *B*, Graph 45). On comparatively gentle slopes, this type of diversion ditch is constructed from both the upper and the lower sides; but on steeper slopes (steeper than 4 per cent), construction is generally from the upper side only. There are, of course, many variations from the indicated cross sections, because of local variations of soil, slope, and rainfall intensity.

It is highly important that the ditch be constructed with an ample capacity to handle the maximum expected volume of water. The settled depth of the water channel should seldom be less than 18 inches. A minimum water cross-sectional area of 7.5 square feet is suggested for watersheds of 1 to 5 or 6 acres. Watersheds of 6 to 10 acres will require a channel depth of at least 24 inches and a minimum cross section of 12 square feet. For larger watersheds, cross section *C*, shown in Graph 45, is suggested, especially on steeper slopes.

The diversion ditch should be set back from the gully head a minimum distance of three times the height of the gully overfall. Low points in the ridge and high points in the channel should be corrected before the ditch is put into use. If well-protected natural locations for the disposal of runoff can be found, no special outlet revegetation or construction will be necessary. Even where such natural outlets are not available, it is sometimes possible to discharge the flow from diversion ditches into an outlet or waterway already established for a terrace system. Under such an arrangement, however, it is important to prepare the outlet or waterway for the additional water to be carried.

CONVEYING RUNOFF THROUGH GULLIES. The establishment of protective vegetation in a gully is usually much more difficult if runoff must be handled through the channel at the same time that the plants are being started. Ordinarily, it is necessary to protect the more erodible portions of a gully carrying runoff by transplanting sod, by applying specially anchored mulches, or by installing mechanical structures. Such protective measures need be only of a temporary nature if the vegetation, when established, will provide the necessary protection.

Where mechanical measures are required for satisfactory control, permanent structures, such as masonry check dams, flumes, or earth dams supplemented by vegetation, should be provided to convey the runoff over critical portions of the ravine.

Gullies generally cannot be controlled adequately, however, until a vegetative cover is established over much or all of the exposed parts of the ravine. From the standpoint of control, there frequently is little choice among trees, shrubs, vines, or grasses, because all these types of vegetation provide good protection when well established. Often, however, it is easier to establish satisfactory covers of trees, vines, and shrubs on gullied areas, because of the slowness with which most grasses take hold on such poor soil. Nevertheless, when grass is established, it will withstand greater water velocities and consequently be able to carry more runoff. Grasses are also more satisfactory than trees, vines, and shrubs as a supplement to structures.

NATURAL REVEGETATION. In many instances, natural revegetation will provide a satisfactory cover, if the gullied area is protected from livestock. Quite often, fencing livestock out of the ravine will be all that is needed for the establishment of a satisfactory growth of volunteer vegetation. This opportunity to provide a protective cover at low cost is overlooked frequently, and unnecessary expenditures are made for structures or for planting. Numerous gullies have been controlled completely by volunteer vegetation in many parts of the country, following the exclusion of livestock (Fig. 202).

Artificial Revegetation. Among the plants that have been used successfully in the reclamation of gullies are: trees and shrubs [black locust (Fig. 203), willows, pines, honey locust, poplars, eucalyptus, tamarisk,

Fig. 202.—A Georgia gully originally 25 feet deep stabilized by natural revegetation. (*Photograph by Soil Conservation Service.*)

wild plum, and buckbrush]; legumes and vines [clovers, alfalfa, sweetclover, lespedezas, honeysuckle, kudzu (Fig. 204), Virginia creeper, dewberry, blackberry, and raspberry]; grasses [Bermuda, Kentucky, and Canada bluegrass (Fig. 205), Johnson grass, centipede grass, redtop, orchard grass, brome grass, Sudan grass, Kikuyu, and wheat grasses.] Other plants, such as ice plant (Fig. 206), are used effectively in California.

In gully plantings, it is frequently necessary to slope the banks before most efficient results can be obtained. Figure 207 shows how gully banks can be sloped rapidly with a bulldozer. The use of heavy machinery of this kind is generally beyond the means of individual farmers, but often it can be obtained through some cooperative arrangement.

FIG. 203.—An Iowa gully checked by a 10-year growth of black locust. (*Photograph by Soil Conservation Service.*)

FIG. 204.—A Georgia gully, originally 35 feet deep, being clogged by kudzu. (*Photograph by Soil Conservation Service.*)

Sloping also can be accomplished effectively by hand labor or by a combination of handwork and blasting. Some deep ravines need sloping primarily because of the hazard that they present to livestock. Usually, small gullies in fields can be filled readily with ordinary plows or hand tools, but it is generally advisable to reduce bank-sloping operations to a minimum, because with proper use of vegetation the banks gradually will assume a stable slope.

Occasionally, where suitable sod is available, it may be desirable to make use of the sodding method to stabilize some gullies or parts of them. Generally, however, this method is too costly for extensive use. Its

FIG. 205.—A West Virginia gully stabilized by a combination growth of walnut trees and bluegrass. (*Photograph by Soil Conservation Service.*)

application commonly is restricted to sodding critical sections about gully heads or points along the bank or bottom where protection against waterfall erosion is necessary.

Sodding also can be used to advantage as a supplement to structures. It is the only way to establish a grass cover in those places where it is impractical to seed. Where the amount of runoff is not too large, and good sod is available, it can be used as a substitute for the more costly masonry and concrete materials. *Sod flumes, sod check dams,* and *sod spillways* have all functioned satisfactorily when properly constructed and located.

Sod flumes may be used effectively to control overfalls in gullies where the drop is less than 10 feet and the contributing drainage area is

FIG. 206.—California gully stabilized with plantings of ice plant. (*Photograph by Soil Conservation Service.*)

FIG. 207.—A typical Oklahoma U-shaped gully being sloped with a bulldozer. (*Photograph by Soil Conservation Service.*)

not greater than 25 acres. Where overfalls are not so high, larger watersheds may be considered. Bermuda and bluegrass sod have proved highly effective in adaptable localities.

Where overfall points are to be sodded, the bank must be cut back to a slope gentle enough to maintain a permanent cover. Generally, a 4 to 1 slope should be the steepest used, but the final choice will depend on the soil type, size of watershed, height of overfall, and quality and type of sod used. To avoid destructive water velocities, the flume should be wide in proportion to its depth in order to carry the maximum expected runoff at nonerosive velocities. A width of at least 1 foot per acre of water-

Fig. 208.—Gully head controlled by complete sodding (vegetated flume). (*Photograph by Soil Conservation Service.*)

shed is desirable, and the maximum depth of flow expected over the flume should seldom exceed 1 foot.

Where any considerable amount of runoff is involved, solid sodding is usually necessary for sod flumes (Fig. 208). It is important that the gully channel below the flume be on a stable gradient; otherwise the sod may fail because of the undermining action of overfalls that is likely to develop below the flume. Where erosion-resistant soil and small watersheds are involved, an unprotected gully channel may be reasonably safe for gradients up to about $1\frac{1}{2}$ per cent. Vegetated channels generally will permit the use of considerably higher gradients. Where larger watersheds and more erodible soils are involved, unprotected channel gradients of more than 1 per cent should seldom be used. If it is impractical to provide the necessary channel protection by vegetative measures, unstable gradients may be reduced mechanically.

*Sod checks* are used frequently to assist in the establishment of gully vegetation and to stabilize the channels until the areas between the checks are vegetated. Two of the more important types of sod check are the *sod strip* and the *sodded earth fill.*

The sod-strip check is suitable for small gullies with mild channel gradients (Graph 46) and watersheds of small to medium size. The strips should be placed across the channel so that each lies flush with, or slightly below, the bed of the gully. The strips should have a minimum width of 12 inches and should extend up the gully sides at least 6 inches

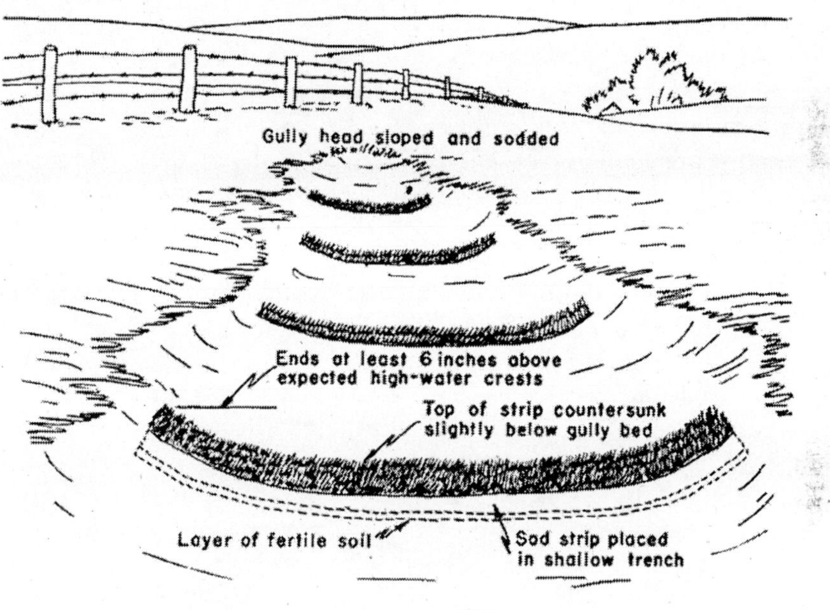


GRAPH 46.—Series of sod-strip checks in small gully. Such checks not applicable to gullies with steep grades. (*Soil Conservation Service.*)

above the level of expected high water. The spacing will depend on the drainage area, the gully gradient, and the spreading characteristics of the sod being used. Intervals of 5 to 7 feet commonly are used.

Low, sodded earth fills often can be used advantageously as a substitute for ordinary brush or wire check dams in stabilizing gully channels. Where suitable sod is available, fills of this type can be constructed at smaller expense. They have been used successfully in small or medium-size gullies with drainage areas of less than 25 acres. The earth fills should be located at strategic points or at regular intervals in the gully, as indicated by individual conditions. Frequently, the fills are spaced in such a way that the top of each will be as high as the base of the next one above. Slopes on the fills steeper than 3 to 1 on the upstream side and 4 to 1 on the downstream face should be used only rarely. The fill should be well tamped and seldom higher than 18 to 24 inches, because of the excessive overfall that may be created. The top of the fill should be low in the center and gradually curved upward to meet the gully sides, with ample provision for weir capacity. The fill should be solid-sodded on top and on the downstream face, and the sodding should extend up the gully sides to a height of 6 to 12 inches above the expected water crest.

STABILIZATION WITH STRUCTURES. Structures are used in gully-control work either to facilitate the establishment of vegetation or to provide protection for those critical sections that cannot be safeguarded adequately by other measures. Where temporary structures are used, they are intended only to function until the vegetation becomes well enough established to provide necessary protection. *Check dams* of this type usually are made of brush, wire, poles, or loose rock. Where it is necessary to make use of mechanical structures as permanent reinforcements, they should be made of such durable materials as reinforced concrete, masonry, metal, or earth.

Temporary check dams thrown up across the bed of a gully have two uses: (1) to collect enough soil and water to insure the eventual growth of protective vegetation and (2) to check head or channel erosion until sufficient stabilizing vegetation can be established at that critical point. Where runoff from the contributing drainage area is diverted from or retained on the watershed, or where the gully has advanced to the drainage divide, any check dams in a gully would be used primarily for the first purpose. Frequently, piles of closely compacted rock and brush across the bottom of a ravine will collect soil and water in a satisfactory manner.

Where temporary structures have been used in gully-control work, it has been found that several low check dams are more desirable than one large structure. Low dams are much less subject to failure than high ones; and after they silt up and rot away, they can be better protected from overfalls with vegetative cover. A temporary dam rarely should have an overfall height of more than 15 inches, and an average effective height of about 10 or 12 inches usually will be more satisfactory.

Temporary dams are especially adapted to gullies with small watersheds. They should be extended far enough into the bottom and sides of the gully to prevent washouts underneath or around the ends and should have sufficient spillway capacity to convey the maximum expected runoff. Generally, an apron will be needed to protect the structure from the undermining action of water discharged from the spillway.

The life of a temporary structure, of course, will depend on the quality of the materials and efficiency of construction; but under ordinary conditions, they should last from 3 to 8 years. Spillway capacities usually are designed for the maximum rainfall intensities that may be expected over a 5- to 10-year period. The requisite capacity, or *spillway notch*, can be determined in the field by estimating the probable rate of runoff on the basis of the size and nature of the watershed. Table 39 gives the spillway dimensions required for various discharge rates, but, as a factor of safety, the notch generally should be made several inches deeper than the indicated depth necessary to avoid overtopping. The notches should be as wide as practical in proportion to depth, in order to avoid unnecessary concentration of runoff as it passes over the structure.

TABLE 39.—APPROXIMATE DISCHARGE CAPACITY IN CUBIC FEET PER SECOND OF RECTANGULAR NOTCHES IN SMALL CHECK DAMS

| Depth of notch $D$, feet | | Spillway length, $L$, in feet required for various discharges indicated | | | | | | | | | | | |
|---|---|---|---|---|---|---|---|---|---|---|---|---|---|
| | | 2 | 4 | 6 | 8 | 10 | 12 | 14 | 16 | 18 | 20 | 22 | 24 |
| 0.5 | Cubic feet discharge per second | 2.3 | 4.5 | 6.8 | 9.1 | 11.3 | 13.6 | 15.8 | 18.1 | 20.4 | 22.6 | 24.9 | 27.2 |
| 1.0 | | 6.4 | 12.8 | 19.2 | 25.6 | 32.0 | 38.4 | 44.8 | 51.2 | 57.6 | 64.0 | 70.4 | 76.8 |
| 1.5 | | 11.8 | 23.5 | 35.2 | 47.0 | 58.8 | 70.5 | 82.3 | 94.1 | 105.8 | 117.6 | 129.3 | 141.1 |
| 2.0 | | 18.1 | 36.2 | 54.3 | 72.4 | 90.5 | 108.6 | 126.7 | 144.8 | 162.9 | 181.0 | 199.1 | 217.2 |
| 2.5 | | 25.3 | 50.6 | 75.9 | 101.2 | 126.5 | 151.8 | 177.1 | 202.4 | 227.7 | 253.0 | 278.3 | 303.6 |
| 3.0 | | 33.3 | 66.5 | 99.8 | 133.0 | 166.3 | 199.5 | 232.8 | 266.0 | 299.3 | 332.5 | 365.8 | 399.1 |
| 3.5 | | 41.9 | 83.8 | 125.7 | 167.6 | 209.5 | 251.4 | 293.4 | 335.3 | 377.2 | 419.1 | 461.0 | 502.9 |
| 4.0 | | 51.2 | 102.4 | 153.6 | 204.8 | 256.0 | 307.2 | 358.4 | 409.6 | 460.8 | 512.0 | 563.2 | 614.4 |
| 4.5 | | 61.1 | 122.2 | 183.3 | 244.4 | 305.5 | 366.6 | 427.7 | 488.8 | 549.8 | 610.9 | 672.0 | 733.1 |
| 5.0 | | 71.6 | 143.1 | 214.7 | 286.2 | 357.8 | 429.3 | 500.9 | 572.4 | 644.0 | 715.5 | 787.1 | 858.6 |

The temporary check dams most commonly used in gully-control work are made of woven wire, brush, loose rock, planks, or slabs. Satisfactory dams may be constructed, however, from other materials or from a combination of these materials. The type selected for use will depend generally on the materials available. *Woven wire, brush, loose rock*, and *slab dams* are illustrated in Graphs 47, 48, and 49 and Fig. 209. Another type of dam frequently used in gullies with small watersheds and moderate slopes, where site conditions are favorable, is the *willow-post dam*. The posts or poles are cut from green willow trees and set in rows across the gully at intervals of about 1 foot. They take root and form a living check across the gully.

Permanent structures are used in gully-control work where temporary measures are either inadequate or impractical. They are frequently necessary in gullies with large contributing watersheds and in gullies that must be retained as permanent waterways. Any gully beginning to cut back into a large watershed usually will justifiy control by the use of permanent structures.

*Weir-notch dams* usually are employed in locations requiring low structures. Where the runoff must be brought down by high overfalls, *multiple-drop dams*, flumes, or *drop-inlet dams* are generally preferable,

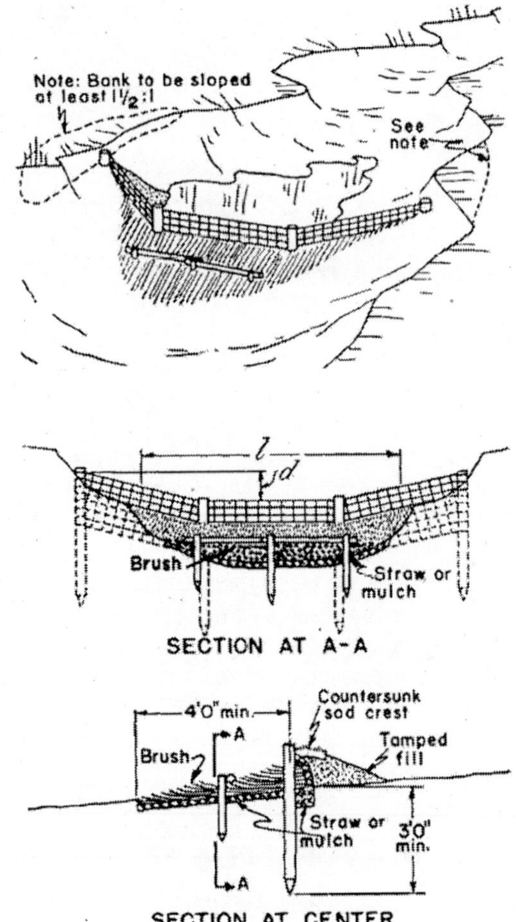


GRAPH 47.—Woven-wire dam. (*Soil Conservation Service.*)

because the relatively high heads that would be required for weir-notch dams would necessitate heavy reinforced concrete construction. Ordinarily, neither masonry nor concrete notch dams should be used to control overfalls in excess of 6 to 8 feet.

Since discharge quantities and velocities are relatively high, channel grades below or between permanent structures should rarely exceed 1 per cent, unless the channel bed contains sufficient rock to prevent harmful scouring. The stability of a permanent gully-control structure depends on stabilized grades in the channel below. An excessive grade leads to the development of an overfall that gradually will undermine the structure. Such a grade can be reduced by changing the location of the structure or by constructing the spillway apron at a lower elevation.

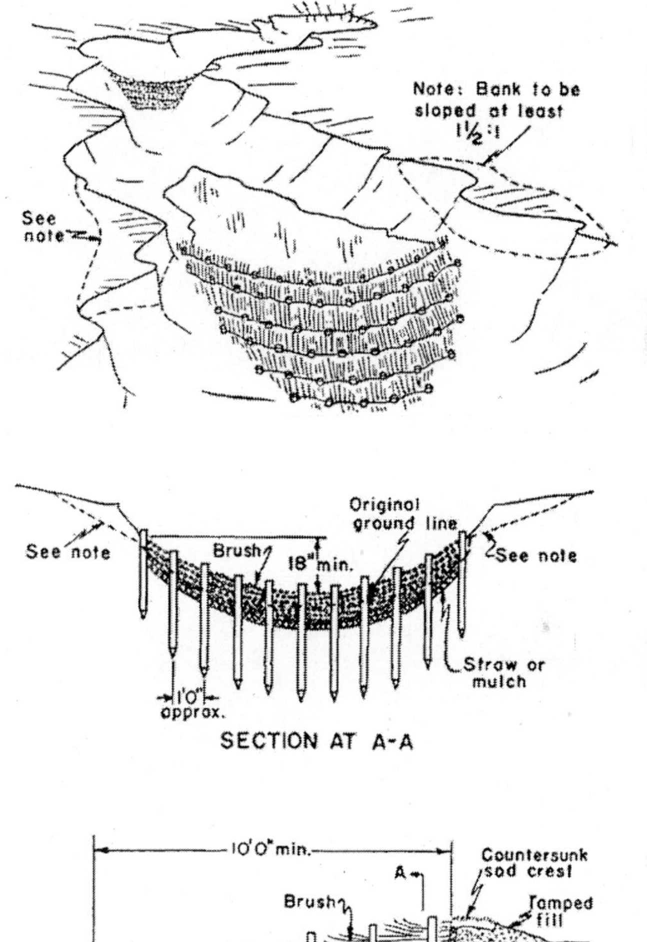


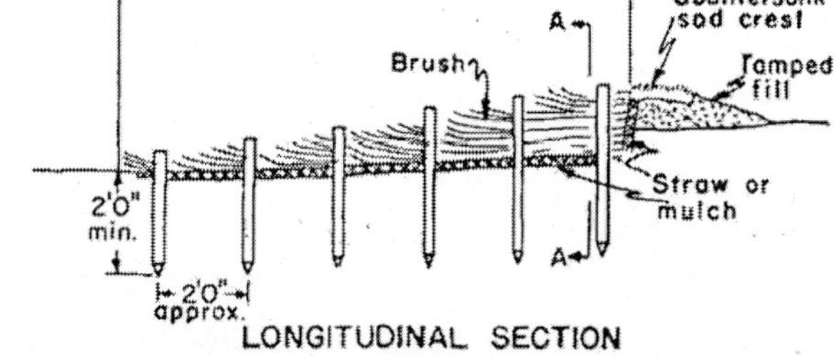


GRAPH 48.—A type of brush dam in common use. (*Soil Conservation Service.*)

Cutoffs to check seepage and washouts around or underneath structures are important. These should extend sufficiently far into the bottom and sides of the gully to provide adequate sealing. Generally, wet foundations should be avoided, especially for earth dams. Surface soil and organic matter should be removed from the dam site so that a good bond can be established between the structure and the impervious foundation material. Below the dam, an apron of sufficient length, width, and depth to prevent undermining is also necessary.

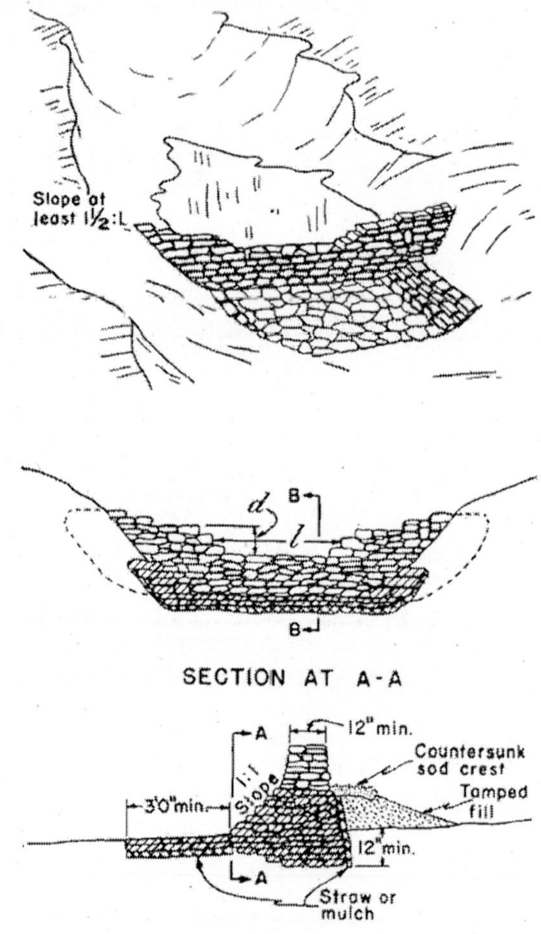


GRAPH 49.—Loose-rock dam. Where the rocks are small, it frequently is necessary to anchor them with woven wire stretched tightly over the structure. (*Soil Conservation Service.*)

The smaller permanent dams should be designed to withstand the maximum runoff from individual storms that may be expected over a 10- to 15-year period; but because of the greater expenditures involved, the larger ones should be built to handle the runoff from the heaviest expected rain over a 25- to 50-year period. Table 39 may be used to determine spillway sizes for small permanent dams with rectangular-notch spillways.

Specifications for Construction of Masonry Dams. Graph 50 gives the standard design for a small *rubble masonry dam*. In general, the minimum thickness of side walls, cutoff walls, and apron should be 12 inches. The main wall from the spillway crest to the top of the dam should also be at least 12 inches thick. Below the spillway crest, a slope of $\frac{1}{2}$ to 1 is recommended on the downstream face. The upstream face should have a batter (angle from the vertical) of about 10 feet fall in a horizontal distance of 1 foot, in order to counteract the pressure of ice and to insure proper settling. An apron length not less than one and one-

Fig. 209.—A slab dam in a small gully. (*Photograph by Soil Conservation Service.*)

half times the total height from the apron floor to the crest of the spillway is recommended. In areas subject to severe cold, the apron, cutoff wall, and toe wall often are constructed of reinforced concrete as added protection against heaving by frost.

Specifications for Construction of Concrete Dams. Concrete dams are used where adequate material is not available for masonry structures. The same general specifications given for masonry dams will apply to concrete dams. Detailed specifications are indicated in Graph 51. It is important to use a good grade of concrete, reinforced with steel bars. These bars should be $\frac{3}{8}$ inch or more in diameter and 12 inches apart from center to center. The specifications indicated in the drawing are for dams under 5 feet in effective height. For structures exceeding this height, special design is necessary. In case the spillway length $L$ exceeds 6 feet, buttresses should be used to brace the headwall.

**Specifications for Construction of Earth Dams.** Earth dams are most practicable in those localities where suitable material for the fill is readily available. These dams often are used both for gully control and as roadways across deep gullies. In some instances, they may even provide a reservoir for stock water. Two types of earth dams commonly

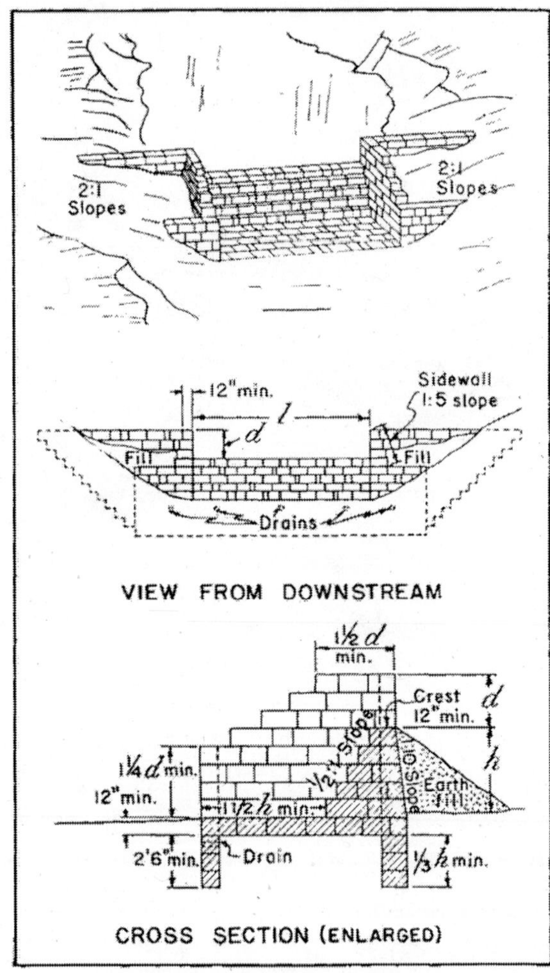


Graph 50.—Design for small rubble masonry dam where the height $h$ does not exceed 5 to 6 feet. (*Soil Conservation Service.*)

are used for this purpose: one where excess runoff is carried around the dam by a side spillway, and the other where it is carried through the dam by a pipe or culvert. Where the latter type has a vertical intake and a horizontal outlet section, the structure commonly is called the *drop-inlet soil-saving dam* (or Adams dam). In many instances, it is desirable to use a combination of side spillway and conduit through the earth fill to by-pass excess runoff. Where this combination spillway-conduit is used, the conduit through the dam is provided with only enough capacity to carry the more common rates of overflow. The side spillway, which is really a safety outlet, usually is constructed with its inlet at a slightly

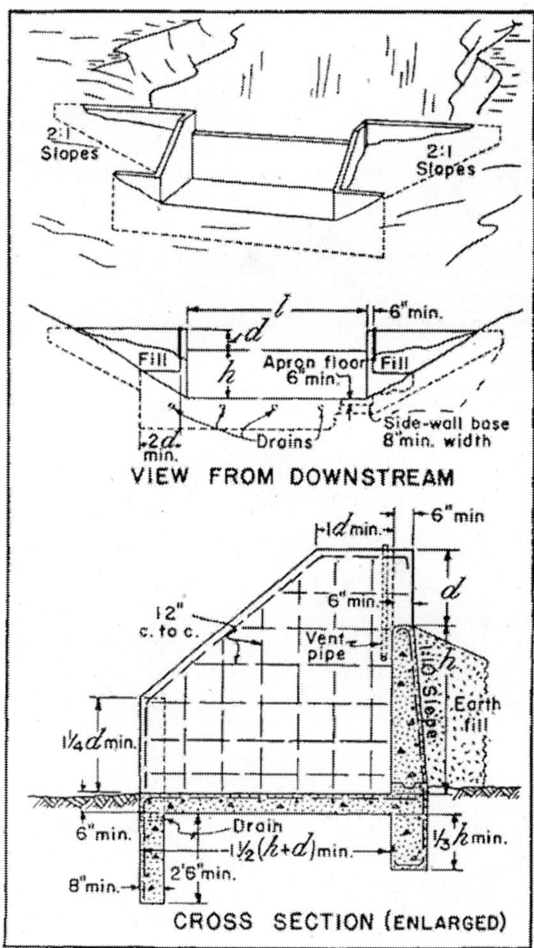


GRAPH 51.—Design for small concrete dam where the height $h$ does not exceed 5 feet. (*Soil Conservation Service.*)

higher elevation than that of the conduit and with sufficient capacity to carry the excess overflow that cannot be carried through the fill during periods of extraordinary runoff.

Graph 52 shows standard design for a soil-saving drop-inlet dam with concrete conduit (*barrel*) and riser (or *well*). It is important that the conduit be incased in stable material and located as near the center of the gully as possible.

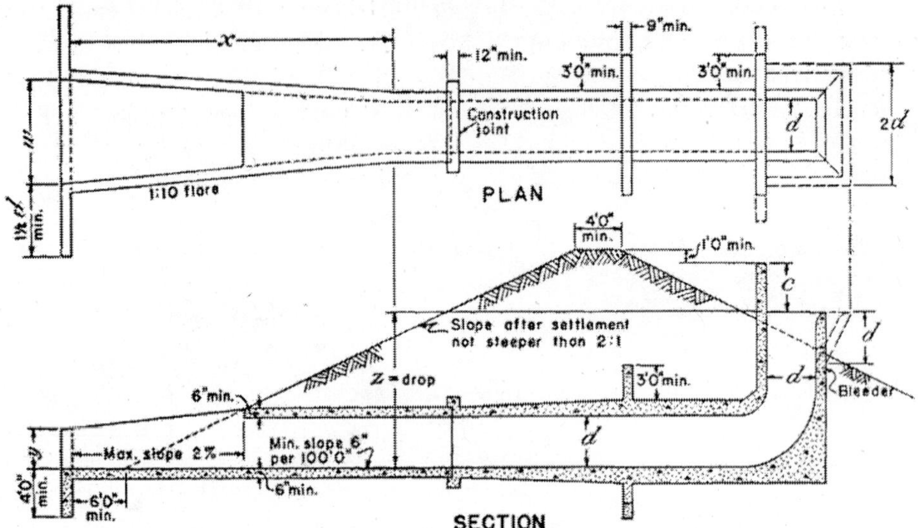


GRAPH 52.—Drop-inlet soil-saving dam with square conduit. The dimensions *c*, *w*, *x*, and *y* vary according to size of conduit used. (See *Tech. Bull.* 122, Univ. Wis.) (*Soil Conservation Service.*)

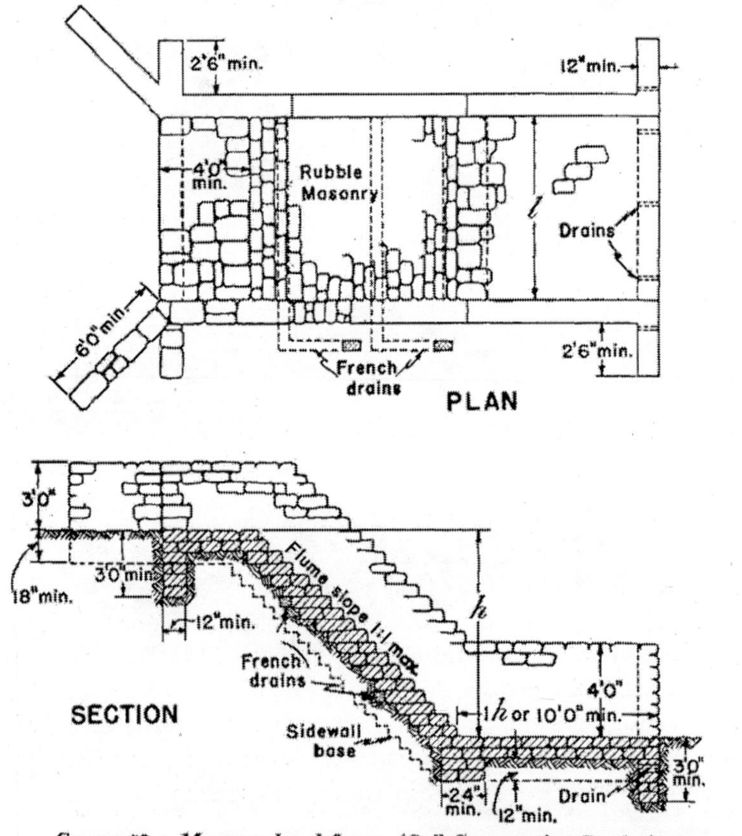


GRAPH 53.—Masonry head flume. (*Soil Conservation Service.*)

For small watersheds, tile or corrugated metal pipes frequently are used for passing the overflow through the fill. Rounded elbow connections between the vertical and horizontal sections of the tube should be used. For drop-inlet dams requiring a discharge pipe greater than 30 inches in diameter, it is usually better to use concrete for the barrel and riser. The concrete should be amply reinforced with steel.

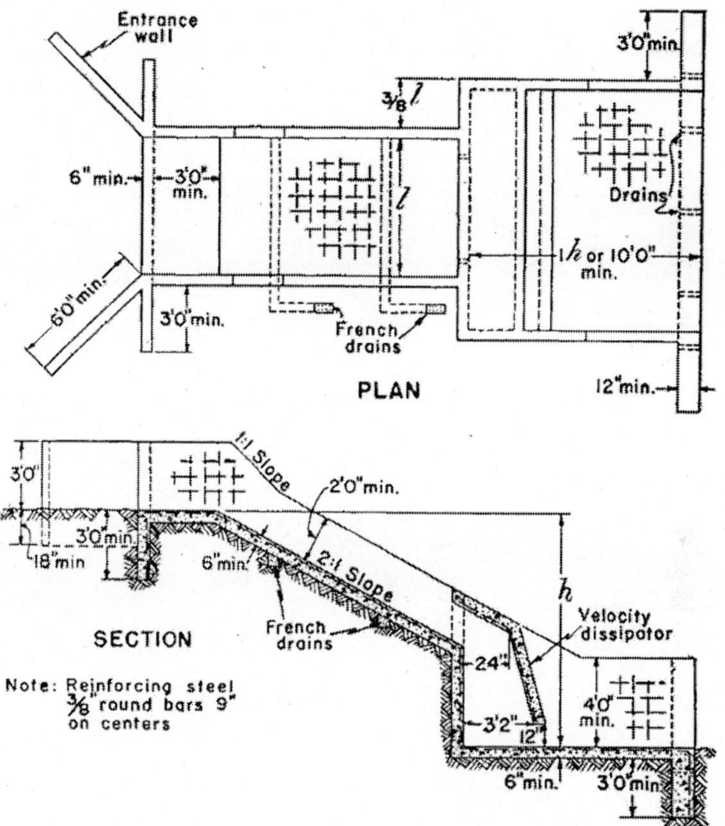


Graph 54.—Concrete head flume. (*Soil Conservation Service.*)

Other Structures. Flumes, multiple-drop structures, head spillways, culvert drop inlets, and other miscellaneous types of structures are used in gully control. The most common structures are flumes and culvert drop inlets.

*Flumes* (sometimes called chutes) are used to convey runoff down steep banks. The steep slopes on which flumes usually are constructed produce high discharge velocities, so that care must be taken to provide necessary apron and grade protection below the structure. Permanent-type flumes generally are constructed of concrete, masonry, or metal. Sod flumes are commonly used in small watersheds.

If concrete is used, the slope of the flume should not be steeper than 2 to 1, in order to facilitate pouring. A masonry flume may have a slope as steep as 1 to 1. Concrete and masonry ordinarily make more desirable flumes than wood or metal, unless permanency is not a consideration. Where applicable, sod should be used in preference to any other material, because of its economy.

Graphs 53 and 54 illustrate the design for masonry and concrete flumes, respectively. The latter should be adequately reinforced with steel bars. The upstream and downstream cutoff walls should be dug to a minimum depth of 3 feet for both types, and the wing walls should

FIG. 210.—Entrance (riser or well) to drop inlet for passing runoff through a highway culvert. (*Photograph by Soil Conservation Service.*)

extend well into the earth on each side. Side walls should be high enough to confine completely the maximum flows expected. The apron length should be not less than the total height from apron floor to spillway crest. It is important that the gully channel for some distance below the flume be on a stabilized gradient; otherwise the structure may be undermined. Ordinarily, flumes are used to control waterfall erosion, not channel erosion. In order to assure adequate spillway capacity for these, the length of $L$ (as indicated in Graphs 53 and 54) should be not less than 4.5 feet, with a minimum of 1 foot for every 15 cubic feet of discharge expected.

*Drop-inlet* highway *culverts* are being used successfully to combat gully erosion along highways. Frequently, the bottom of highway cul-

TABLE 40.—DROP-INLET CULVERT DIMENSIONS REQUIRED FOR THE DISCHARGES INDICATED[1]

| Total drop, feet | Type of culvert | | Watershed discharges, cubic feet per second | | | | | | | | | | | |
|---|---|---|---|---|---|---|---|---|---|---|---|---|---|---|
| | | | 25 | 50 | 100 | 150 | 200 | 250 | 300 | 350 | 400 | 450 | 500 | 550 |
| 5 | Round[2] | Dimensions for drop-inlet culverts | 18 | 30 | 36 | 48 | | | | | | | | |
| | Square[3] | | .. | 2 by 2 | 3 by 3 | $3\frac{1}{2}$ by $3\frac{1}{2}$ | 4 by 4 | $4\frac{1}{2}$ by $4\frac{1}{2}$ | 5 by 5 | 5 by 5 | Use twin culverts | | | |
| 10 | Round | | 18 | 24 | 36 | 42 | 48 | 48 | | | | | | |
| | Square | | .. | ...... | $2\frac{1}{2}$ by $2\frac{1}{2}$ | 3 by 3 | $3\frac{1}{2}$ by $3\frac{1}{2}$ | 4 by 4 | 4 by 4 | $4\frac{1}{2}$ by $4\frac{1}{2}$ | $4\frac{1}{2}$ by $4\frac{1}{2}$ | 5 by 5 | Use twin culverts | |
| 15 | Round | | 15 | 24 | 30 | 36 | 42 | 48 | 48 | | | | | |
| | Square | | .. | ...... | $2\frac{1}{2}$ by $2\frac{1}{2}$ | 3 by 3 | 3 by 3 | $3\frac{1}{2}$ by $3\frac{1}{2}$ | 4 by 4 | 4 by 4 | $4\frac{1}{2}$ by $4\frac{1}{2}$ | $4\frac{1}{2}$ by $4\frac{1}{2}$ | 5 by 5 | 5 by 5 |
| 20 | Round | | 15 | 24 | 30 | 36 | 42 | 42 | 48 | | | | | |
| | Square | | .. | ...... | 2 by 2 | $2\frac{1}{2}$ by $2\frac{1}{2}$ | 3 by 3 | $3\frac{1}{2}$ by $3\frac{1}{2}$ | $3\frac{1}{2}$ by $3\frac{1}{2}$ | 4 by 4 | 4 by 4 | $4\frac{1}{2}$ by $4\frac{1}{2}$ | $4\frac{1}{2}$ by $4\frac{1}{2}$ | 5 by 5 |
| 25 | Round | | 15 | 24 | 30 | 36 | 36 | 42 | 48 | 48 | | | | |
| | Square | | .. | ...... | 2 by 2 | $2\frac{1}{2}$ by $2\frac{1}{2}$ | 3 by 3 | 3 by 3 | $3\frac{1}{2}$ by $3\frac{1}{2}$ | $3\frac{1}{2}$ by $3\frac{1}{2}$ | 4 by 4 | 4 by 4 | $4\frac{1}{2}$ by $4\frac{1}{2}$ | 5 by 5 |

[1] Sizes are for single culverts, inside dimensions. When culverts larger than 5 by 5 feet are required, twin culverts are recommended.

[2] Round-culvert dimensions given as diameter in inches.

[3] Square-culvert dimensions given in feet.

verts is below the bottom of the roadside ditches, so that water carried by the ditches must drop in order to pass through the culvert. Unless protected, severe gullying often has its beginning at the critical point of the drop. The desired protection can be established by installation of a drop inlet (Fig. 210) in the form of a drop box at the entrance of the culvert.

The drop box is essentially the equivalent of a broad-crested weir. The required weir length can be determined by selecting the length $L$ in Table 39 for expected runoff and flow depth and then increasing this by one-fourth. Where the structure is designed in accordance with the specifications for a drop-inlet dam, the size required for different discharges may be taken from Table 40. It is often necessary also to protect the lower end of the culvert in order to prevent gullying. This can be done either by using a suitable drop outlet or by constructing small, permanent dams to establish stable gradients.

# Chapter XXIV. Control of Erosion on Highways

Erosion has long been a serious problem along both primary and secondary highways. The tendency to overlook erosion potentialities through the stages of location and construction appears to have been rather general; and until recently, comparatively little was done to control soil washing along highways after they were built, until conditions forced attention to the problem. In numerous localities, roadside ditches were permitted to enlarge until roadbeds were undermined or threatened with undermining; fills were left bare very often until sloughing became active as the result of gullying and sheet washing. To a very considerable degree, the remedies employed to correct these conditions were usually of a temporary nature, such as resurfacing or filling washouts with earth or gravel, installing flumes or overpasses, and building small check dams. Recently, however, control of erosion has become an important activity in most highway-maintenance programs.

Considerable research and demonstrational work is now being carried on in many states by the Bureau of Public Roads and the Soil Conservation Service and by state highway departments. Upkeep of the far-flung road system of the United States, covering some 3,200,000 miles, constitutes a heavy public expenditure. It has been estimated that, under state systems, maintenance costs approximate $300 per mile annually. Surveys indicate that nearly one-third of this cost may be charged to the direct or indirect effects of erosion. Estimates of damage to county roads, of which there are more than 2,650,000 miles, are not available, but the damage probably is more serious than that along the state and Federal systems.

Too often, the damage has not been restricted to the roads alone; concentrated runoff from highways discharged over sloping farm land, as well as other kinds of land, has caused a great deal of damage. On the other hand, roadside ditches have been washed deeper and wider as the result of runoff from adjacent farm lands.

## Methods of Control

In many parts of the country, satisfactory control of erosion along highways involves a variety of difficult problems. The raw subsoil exposed in grading roads and the steep slopes that usually prevail along cuts and fills often make it difficult to establish a protective growth of plants.

FIG. 211.—Properly graded and grassed roadside in Georgia Piedmont. Previous condition shown in upper photo. (*Photographs by Soil Conservation Service.*)

The use of protective structural measures generally is limited to roadside ditches, and even there they are a hazard to the driving public.

When roads are constructed, they necessarily cut the terrain at all angles and intercept and divert runoff from numerous natural channels. Unless culverts and other structures are located with particular care, they usually cause gullying or other erosional damage to adjacent lands.

In the use of vegetative controls, those plants should be selected which will not develop traffic hazards when full grown. When trees are used, they should be located so that they will not interfere with visibility or invite snowdrifts.

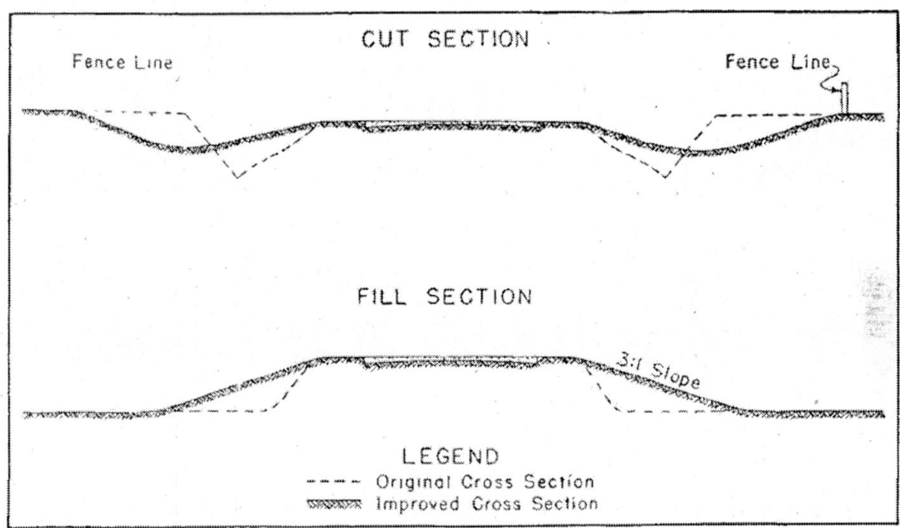


GRAPH 55.—Designs for improved highway cross sections. (*Soil Conservation Service.*)

FIG. 212.—Maintenance of grassed roadside by mowing. (*Photograph by Soil Conservation Service.*)

An important essential in the control of erosion on road cuts, ditches, and fills is to reduce roadside slopes and flatten out ditches, so that a cross section will be developed, which, although conforming with the requirements of sound construction, will favor economical control of erosion

(Fig. 211). Every effort should be made to eliminate steep slopes and angular transitions, not only to facilitate the establishment of an effective

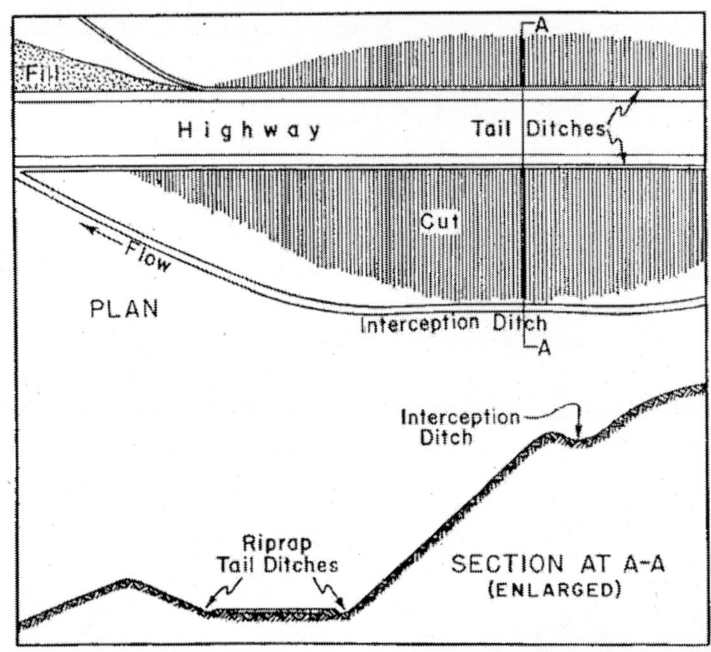


Graph 56.—Construction details for interception ditch. (*Soil Conservation Service.*)

Fig. 213.—Broadcast sodding of road cut with Bermuda grass sprigs. Louisiana. (*Photograph by Soil Conservation Service.*)

cover of protective vegetation but to minimize traffic hazards. Such a cross section is illustrated by Graph 55. Once vegetation is properly established, the maintenance of roadsides is largely reduced to mowing (Fig. 212).

Control on Cut Slopes. Interception ditches above cut slopes are frequently necessary to direct runoff away from the steep, erodible slopes. These must be constructed with ample capacity, gentle grades not susceptible to erosion, and outlets that safely dispose of the runoff (Graph 56).

The usual method for establishing a protective cover of grass on cut slopes is by seeding. Bermuda grass, however, is now commonly planted

Fig. 214.—Slope and fill (left) graded and seeded to grass by drilling (right), Texas. (*Photograph by Soil Conservation Service.*)

by broadcast sodding (Fig. 213). If the work is done at the beginning of a rainy season or during the dormant period of Bermuda, it has been found advisable to seed over the area with a nurse crop. In humid regions, it is generally advisable to make a liberal application of fertilizer.

Good stands of Bermuda have been established in the South by this broadcast method for less than 2 cents per square yard, as compared with 15 to 25 cents per square yard where other methods of sodding are practiced.

Where the slope is not too steep, a good cover frequently can be established by seeding various adaptable grass mixtures (Fig. 214). Lespedeza, clovers, orchard grass, Kentucky bluegrass, redtop, timothy, and rye grass are commonly used, according to local adaptability, in a variety of mixtures for roadside seedings. Fertilizers are generally needed for best results. Where new seedings are exposed to runoff hazards, it is frequently advisable to cover the entire area with a suitable mulch, such as straw, old hay, or forest litter. Generally, seeded areas require some maintenance, and reseeding is frequently necessary to insure a perfect stand.

Vines, such as honeysuckle, kudzu, and Virginia creeper, have provided satisfactory protection in some parts of the East. Planting by the old-fashioned spot method, however, generally has proved ineffective. On steep slopes, most successful results with vines have been obtained by the recently developed trench method of planting (Fig. 215). This con-

Fig. 215.—Trench planting of honeysuckle on highway cuts, Southwestern Virginia. (*Photograph by Soil Conservation Service.*)

sists of excavating trenches in the cut or fill bank approximately 16 inches deep, 12 inches wide, and about 4 feet apart. These are cut on grades ranging from 3 to 1 (drop of 1 foot in a horizontal distance of 3 feet) to 12 to 1, depending on soil conditions. The trenches are filled with vine roots and topsoil. Some fertilizer may be used to stimulate growth, although it is not needed if good topsoil is available. The trenches absorb much of the water that falls on the face of the cut; the excess is slowly drained to the bottom of the slope without causing much erosion.

Control of Highway Ditches. Roadside ditches require protection from the concentrated runoff that tends to convert them into destructive and dangerous gullies. Generally, vegetation will be found the most practical means for their stabilization. The most effective plants for this purpose are those which produce a good cover during the first growing season. Cross sections for roadside ditches should be designed to regulate velocities in such a way as to prevent scouring during intense rains and to hold silting to a minimum during moderate rains.

Supplementary mechanical measures should be used only when it is impracticable to establish a satisfactory cover of vegetation. Unnecessary structures in roadside ditches do not conform with modern trends in highway improvement.

Safe discharge of water at the ditch outlet requires careful design and proper location (see Chap. XXI, Part 2).

Control on Fill Slopes. Protection of slopes on highway fills is not so difficult as on cut banks, partly because the soil material is usually

Fig. 216.—Properly constructed roadside channel and head spillway for passing water beneath road. Arkansas. (*Photograph by Soil Conservation Service.*)

well pulverized and frequently mixed with topsoil derived from excavations. Seeding has proved generally successful, although transplanting is sometimes necessary on raw soil material of low productivity or high erodibility. The use of curbs on the shoulder edge to divert runoff to flumes has been successful on some of the more troublesome slopes, but special care is necessary to control discharge at the end of the flume.

In low-rainfall areas, mechanical measures such as contour furrows, diversion channels, and level terraces have been of much assistance in providing protection for highways. Best results are likely to follow work of this nature where it is done in cooperation with land operators, so that slopes draining into roadways can be treated fully for control of runoff and erosion.

Treatment of Cross Drains. The use of drop inlets and *head spillways* on the upper end of culvert structures has proved particularly effective in controlling gully erosion in highway ditches and on adjacent agricultural land (Fig. 216). *Drop-outlet* structures on the lower end of culverts, to lower the discharge to a stable gradient, are often necessary

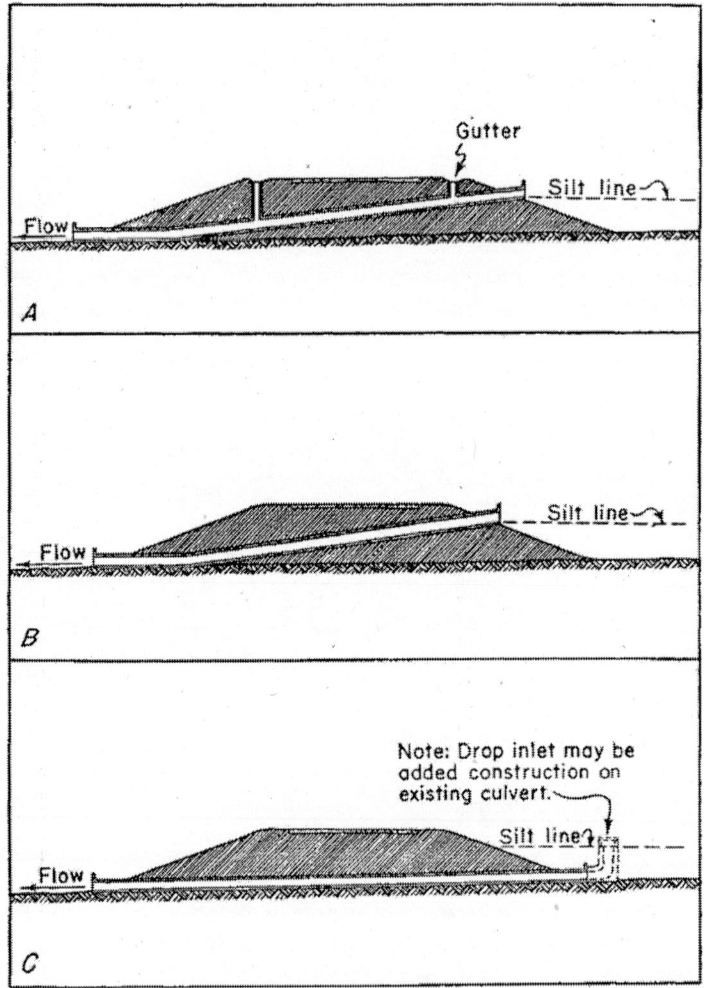


Graph 57.—*A*, boulevard fill section; *B*, broken back culvert; *C*, drop inlet. (*Soil Conservation Service.*)

to prevent gullying. *Broken back culverts* similar to the design shown in Graph 57 (*B*) are sometimes needed to secure proper inlet and outlet elevations for cross drains. They serve the same purpose as drop inlets or outlets installed for existing culverts.

Placement of culverts above the customary location (Graph 58) has proved a desirable practice where the volume of water to be controlled is small and where a good cover of protective vegetation lines the ditch below the outlet (tail ditch). Such culverts help check the development of overfalls and gullying above the structure.

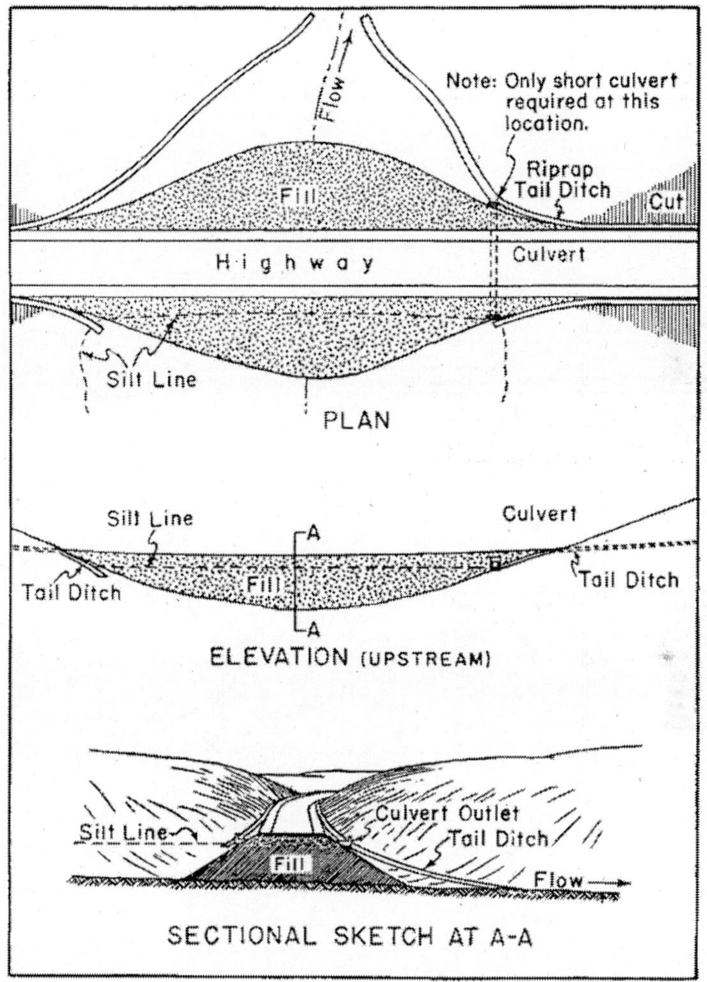


GRAPH 58.—Locating culvert above low point, combining drainage structure, drop inlet, and tail ditch construction. (*Soil Conservation Service.*)

A culvert design and highway cross section that provides road drainage similar to that often employed for city streets is shown in *A* (Graph 57). The runoff from the roadway is collected in gutters and discharged through vertical risers to the culvert. The use of this design is generally limited to roadway sections where the fills are low.

# Chapter XXV. Small Dams for Water Storage

During the past few years, the construction of small dams for water storage has met with considerable favor, especially in the Western and Midwestern States. Although these dams are used primarily to provide a water supply for livestock (Fig. 217), they also aid in the irrigation of small areas and are utilized for recreation (Fig. 218) and wildlife. The

FIG. 217.—An Oklahoma stock pond, with contoured watershed. Frequently, these ponds are the only source of water for livestock. (*Photograph by Soil Conservation Service.*)

value of this type of dam was amply demonstrated during recent periods of extreme drought. With the help of a good stock pond, many farmers were able to furnish water for their livestock and to grow some vegetables and supplemental feed during the hot dry summers, where otherwise they would have been forced to sell their stock or let it die for lack of water.

## Small Dam Essentials

Small dams for farm and range use are usually not expensive, yet they add considerably to the value of the average dry-land farm or ranch.

Besides supplying water and some recreation facilities, they play a part in controlling floods. The cost of such dams is determined largely by their size, the most economical type generally being a low structure with only a small yardage of fill for each acre-foot of storage.

Although the basic requirements with respect to foundation, fill material, and spillway must be met in the construction of small dams, they need not be built according to the precise specifications ordinarily set up for larger structures. Frequently, specifications have been so exacting that construction of small dams has been entirely impracticable. For

Fig. 218.—One of the several types of water-conserving dams constructed by the Soil Conservation Service. Barnes County, N. D. (*Photograph by Soil Conservation Service.*)

example, there are many potential pond locations suitable in every respect other than the availability of material for rip-rap. In such instances, it might be necessary to haul rip-rap material for many miles at a cost entirely prohibitive in terms of the expected benefits. Under such conditions, it is reasonable to use some temporary material for rip-rap or to eliminate its use entirely.

Small dam specifications that demand masonry or concrete for spillways are often unreasonably restrictive. In all probability, many of these dams will serve their intended purpose if a good vegetative spillway is provided. Inasmuch as the spillway is not used ordinarily more than once or twice a year, an adequate native sod should afford ample channel protection. A good point to remember in planning the construction of a small dam is that it must provide an economical service.

Location. The main factors to consider in selecting the site for a stock water storage area are:

Utility.
Foundation.
Storage capacity.
Quality of water.
Type and size of drainage area.
Spillway conditions.
Availability of construction materials.

*Utility.* The purpose for which a pond is built will have considerable influence on its location. If it is to be used for irrigation, the site must be above the area to be irrigated, unless pumping is contemplated. If it is for stock use, drainage from the barnyard or corral should not be allowed to enter the reservoir, and it should be so located that the stock will not have to travel more than about a mile and a half for water.

*Foundation.* The subsoil structure of every potential reservoir site should be tested by borings, as a safeguard against an installation that will lose its water by seepage. Some soil types that are unsuitable for holding water may be made moderately watertight by adding clay and then permitting livestock to trample the surface. The principle of the sheep's-foot roller, a machine used extensively for compacting large earth fills, originated from the practice of using a herd of sheep to trample the bottom of a pond to prevent water loss by seepage. The area behind dams already built may be treated in this manner, but new ponds should not be developed where inspection indicates the probability of excessive seepage.

*Materials.* If a small dam is to be built economically, suitable materials should be readily available near the structure. Many dams fail because a sufficient quantity of compactible soil, of relatively high density, is not found near the dam, and, as a result, less suitable materials are substituted. It is also desirable to have good rock for riprapping available near the dam site; but, as already indicated, the lack of rock for this purpose is not always important enough to prevent the building of a dam.

*Storage Capacity.* The purpose for which a reservoir or pond is to be constructed will determine the amount of storage capacity needed. The need for maintaining an adequate and unfailing supply of water under all conditions will also influence the capacity determination. In making this determination, a sufficient factor of safety should be provided to allow for usage, evaporation, seepage, silting, freezing, and probable future needs. The most desirable type of reservoir is one with good depth and a minimum surface area.

The rate of evaporation is an important consideration. In the Great Plains, for example, an acre-foot of water will be enough to water for a year 100 head of stock requiring about 10 gallons per head per day. That is, it would be enough if there were no evaporation loss. However, records kept by the Bureau of Plant Industry, over the fifteen-year period 1917–1931 show that evaporation takes from about 4 to more than 5 feet from a free water surface in a year.[1] Stock ponds, then, must be deep enough to allow for evaporation losses; and, generally, a depth of double the expected annual evaporation loss is a safe provision. During extended periods of drought the evaporation rate may be increased by something like 50 per cent.

*Drainage Area.* The most desirable drainage area is one covered with vegetation. Many stock ponds, however, have been built where a large part of the drainage is from cultivated or heavily grazed land. Such ponds usually fill with silt in a short time, often losing their usefulness in a few

FIG. 219.—Silt filtered from runoff by a fenced area of grass above a New Mexico stock pond looking downstream, toward pond. This vegetative silt basin is serving admirably to keep the products of erosion out of the reservoir. (*Photograph by Soil Conservation Service.*)

years, unless the drainage area is used in such a way as to prevent erosion. Watersheds should be sufficiently large to provide enough runoff, on the basis of average annual rainfall, to meet the storage demands. Large drainage areas should not be used unless an adequate spillway can be provided economically.

It is frequently possible to control or minimize the silting problem by establishing a grass filter immediately above the pond. This has been accomplished successfully in parts of the Southwest by excluding stock from a suitable segment of land across the drainage depression a short distance above the pond (Fig. 219).

[1] Lowdermilk, W. C., and Barnes, F. F. Stock Ponds in the Great Plains Drought Area, *Soil Cons.*, September, 1936.

*Spillway.* Generally, the spillway should be carefully located and should have sufficient capacity to pass the largest runoff that is likely to occur within a 25- to 50-year period. A grassed spillway is most desirable from the standpoint of cost; but under heavy flows on steep grades, adequate structural protection should be provided. Since it is often difficult to establish good sod under arid or semiarid conditions, the use of natural spillways is desirable wherever possible in such regions. The natural sod in such spillways should remain undisturbed.

Inasmuch as many states have regulations governing the construction of dams for water storage, it is always advisable to obtain advance approval of the plans from state authorities.

*Quality of Water.* Under certain conditions, the quality of water is a limiting factor in site selection. Highly saline water, for example, is unfit for stock or irrigation.[1]

[1] For further details with respect to construction of dams for small reservoirs, see "Low Dams," National Resources Committee, Washington, 1938.

# Chapter XXVI
# Erosion of Stream Banks

Stream-bank erosion is occurring on a large scale in every part of the country, along small trout streams and farm brooks as well as along such great waterways as the Mississippi and Missouri.

FIG. 220.—Channel of Winooski River, Vt., choked with ice during spring flood, March, 1936. The ice caused severe bank erosion by a process of gouging. (*Photograph by Soil Conservation Service.*)

The Winooski River of Vermont furnishes a good illustration of this type of erosion. This medium-sized stream, draining 1,000 square miles, is typical of many waterways in northeastern United States that are subject to severe ice conditions (Fig. 220). Within the boundaries of the soil conservation demonstration area on the lower Winooski, 24 of the 61 miles of river bank are actively eroding. Under average conditions, the quantity of earth that caves into the river each year is approximately 340,000 cubic yards. During a flood in 1936, 50 acres of land were destroyed by bank erosion within this restricted area, and thousands of cubic yards of earth were dumped into the river channel. The land destroyed was the best type of agricultural soil in the state.

In the semiarid West, bank erosion often occurs at a more rapid rate and on a larger scale than in the East, because of the comparative scarcity

Fig. 221.—Santa Clara River at Santa Paula, Calif., at flood stage showing typical bank erosion. Walnut trees being undermined and falling into the stream. (*Photograph by Soil Conservation Service.*)

of vegetation. Stream beds in the West are generally less stable during floods, and this frequently has contributed to severe bank erosion (Fig. 221). It is noteworthy, however, that the floods accompanying the New England hurricane of 1938 caused much stream-bank erosion, even in some well-forested areas. Large quantities of boulders were dumped into some of the streams channels, causing them to resemble the debris-loaded canyons of the West.

## Economic Aspects

Stream-bank erosion is a problem of public concern because it destroys economic values. It ruins extensive areas for further productive or recreational use and contributes to the clogging of stream channels, thus adding to the hazard of floods. A considerable part of the silt deposited in some reservoirs comes from stream banks. Bridges and other structures across or near stream channels are damaged by this form of erosion, and shifting channels increase the cost of their maintenance.

Agricultural areas destroyed by bank erosion usually consist of high-value land. While the bank is eroding on one side of a stream, however, alluvial land sometimes builds up on the opposite side. The comparative age of trees, measured from the shore inland, has shown in a number of instances that land destroyed on one side of a stream was replaced by cultivable soil on the opposite side, within a few decades. Thus it is indicated that approximately the same number of acres in some flood plains have been available for agricultural use many years; and unless sedimentation causes serious economic damage, the public concern is not so great.

Erosion of this type, however, is a matter of immediate importance to the individual farmer, for it means a decrease in the potential farm income. When rich bottomland is destroyed, a vital part of the farm assets is lost.

Frost, water, and ice are the erosive forces that cut away stream banks. Scouring, gouging, sloughing, and mud flows are the principal processes of bank deterioration. These are influenced by such conditions as soil character, presence or absence of a stable cover of trees or shrubs, size and character of floods, the velocity of the current, the land use on adjacent areas, stability of the river-bed material during floods, and climatic conditions.

The stability of the materials underlying the topsoil often varies greatly along the course of a stream. Beds of uncemented gravel or sand are very unstable. Ordinarily, heavy clays do not erode easily, although large chunks often are split off by frost action. Some deep alluvial silts become very unstable when saturated by prolonged inundation and cave readily even though the erosive action of the current may be negligible.

The velocity of stream flow determines to a large extent the rate of erosion; other conditions being equal, swift currents scour stream banks much more rapidly than slow-moving currents. In some instances, water at bank-full stage causes more erosion than floods that deeply inundate the alluvial plain, since the flow at high levels sometimes is impeded by obstacles in the path of the overflow.

Stream-bank erosion has become more severe and more widespread in recent years because of the more intensive use of alluvial plains and their enclosing uplands. In some instances, the process has been accelerated by neglect. Along the Winooski, for example, it formerly was a common farm practice to do some bank-protection work every year. This practice was discontinued about 25 years ago. Subsequently, trees were cut to provide more pasture; cattle were allowed to graze down to the water's edge; and plowing was carried to the immediate banks. At the more vulnerable points, the banks steadily lost stability, and erosion was accelerated. During the largest flood of record, in the fall of 1927, the banks were laid bare in many places; but at that season, vegetation was thick and helped prevent more severe erosion. A winter flood in 1936, however, was unhampered by vegetation, and damage by erosion was enormous. Since then, even minor floods have caused serious washing.

## Stream-bank Protection

Two principal types of stream-bank protection have long been used in various parts of the world. One involves the *blanket* method of protection, and the other is known as the *jetty* method.

The cheapest type of blanket protection is an effective cover of vegetation. Along small streams, a good sod of grass often gives adequate control, except under conditions of severe gouging by ice. Probably, the most effective type of vegetative covering under any condition is a stand of willow trees about 4 to 6 inches in diameter. Trees must be fenced from grazing as thin stands with bare ground between, such as often result from grazing, permit erosion to start. Once started, the trees are likely to lose their foothold and so induce severe bank erosion.

FIG. 222.—Where lower banks are kept soft by fluctuating water level, control frequently requires stone rip-rap below and vegetation above. Winooski River, Vermont. (*Photograph by Soil Conservation Service.*)

Belts of willow trees, planted on the flood plain parallel to the banks, are effective in retarding current velocities, thereby reducing the erosive force of water and floating ice. Where the lower part of a stream bank is under water every year long enough to prevent vegetative growth, it is often necessary to cover this more vulnerable section with stone to necessary depths below the water line (Fig. 222) or to provide adequate

Fig. 223.—Bank treatment with brush matting and timber pile jetties. (*Photograph by Soil Conservation Service.*)

Fig. 224.—Bank caving controlled with trees anchored against bank after sloping. Lower Winooski River, Vt. (*Photograph by Soil Conservation Service.*)

jetty protection (Fig. 223). In deep water, the *timber-pile jetty* is usually more economical and effective. Where current velocities are not too swift (less than 3 miles per hour), the toe of the slope sometimes can be pro-

FIG. 225.—Above, destruction of productive bottomland by bank erosion; lower, same area controlled with brush matting. (*Photographs by Soil Conservation Service.*)

tected by anchoring large trees in a continuous line parallel to the bank to break the velocity of the current and to cause silt deposition (Fig. 224).

Where the banks are so bare that some temporary protection must be provided pending the establishment of a protective cover of trees, compact brush matting, preferably of willows (Fig. 225), generally has proved effective and economical, even under severe ice conditions. Ordinarily, the brush is anchored to stakes with wire; where it is laid parallel to the current, weighting with stone or concrete blocks is generally advisable. Where ice action is not a problem, the brush may be laid perpendicular to the current, as an inducement to deposition of silt.

At extra-vulnerable points, as sharp bends subjected to swift currents, it is sometimes impossible to obtain control through revegetation. Under such conditions, it is necessary to riprap the banks with stone. Where erosion of highly valuable land must be stopped immediately, a blanket

FIG. 226.—Control of bank erosion by riprapping, after sloping. (*Photograph by Soil Conservation Service.*)

covering of stone or concrete is usually the only practicable procedure to follow.

Blanket protection of any type requires some preliminary sloping of the bank (Fig. 226). Under some conditions, the jetty type of control must be used. This can be used alone or in combination with blanket protection. The types of jetties to be used, and the necessary spacing and distance for extension into the stream, will vary with the size of the stream, channel gradient, maximum expected velocity, character of flood flow, ice conditions, and availability of construction materials. Effective jetties have been constructed of barriers of trees, timber piles, stone, pipe and concrete tetrahedrons, automobile frames, and other materials. Timber piles have proved particularly successful, both as jetties and as a means of assisting in the control of especially vulnerable banks (Fig. 227). In some parts of the West, where ice is not a serious problem and where large quantities of silt are transported, loose jetties of brush, piling, automobile frames, and crisscrossed pipe have given adequate protection. *Permeable-type jetties* are generally the most successful, since they not only slow down the velocity sufficiently to cause silt deposition but help

FIG. 227.—Caving bank (above) controlled (below) by sloping and use of piling and wired willow matting. (*Photographs by Soil Conservation Service.*)

stop bank erosion. Impermeable jetties, functioning as dams, are often more expensive and more likely to induce scouring.

Very small streams generally can be protected with such simple and cheap measures as bank sloping and planting of willows. In the low-rainfall areas of the West, particularly in the Southwestern range country, numerous waterways that formerly were well stabilized and of diminutive size have been converted into deep gullies or broad washes, as the result of overgrazing of watersheds.[1] Increased volumes of runoff from such abused areas start erosion along the channels; soon a U-shaped gully is opened. This spreads by a process of bank undercutting until, in many instances, the entire alluvial plain is washed out. The only cure for this type of bank erosion is to rehabilitate the range and protect the wash or gully with plantings of trees, shrubs, or other adaptable vegetation.

The choice of bank-protection measures is an individual problem for each stream and must be based on careful study of existing and expected conditions. Ingenuity is often the keynote of the design and construction of stream-bank protection works.

[1] Bryan, Kirk. Rate of Channel Trenching (Arroyo cutting) in the Arid Southwest, *Science*, Vol. 62, pp. 338–344, 1925. Bailey, Reed W. A New Epicycle of Erosion, *Jour. of Forestry*, Vol. 35, pp. 997–1005, 1937.

# Chapter XXVII. Water Spreading

As a means of providing additional water needed for establishing or restoring a protective cover of vegetation in areas where rainfall deficiency ordinarily restricts the development of an adequate cover for erosion control, *water spreading* can be used effectively on many lowland areas. The practice is proving especially useful in western United States. Not only is erosion controlled, but silting is reduced, more water is held back from flood streams, the underground water supply frequently is replenished, range forage is increased, and opportunity is developed for small-scale production of feed and subsistence crops.

Water spreading has to do with the direct utilization of rainfall runoff for surface irrigation. It is simpler and less expensive than indirect irrigation through the use of runoff stored in reservoirs. Under the direct method, runoff is intercepted, diverted, and redistributed over the downslope surface by various combinations of dams, dikes, ditches, furrows, and other structures.

The utilization of flood water in the production of crops dates back to ancient time. It probably originated in arid regions where the periodic overflow of rivers, such as the Nile, was used directly to water crops on adjacent desert land. Early writings pertaining to Mexico and western South America casually mention the use of irrigation canals as a feature of the agriculture of those regions. In southwestern United States, particularly in Arizona and New Mexico, remains of irrigation dikes and ditches have been found that antedate the advent of the Spanish explorers of the sixteenth century. Flood-water farming was an important practice in the economy of the prehistoric sedentary Indians of the Southwest. Situations favoring the use of this practice largely determined the location of many of the pueblos. Early Mormon settlements also usually centered around small flood-irrigation projects.

The earlier attempts at irrigation were primitive in character and consisted principally in assisting nature to spread water over bottomlands and low flats during flood periods (as under the system of *temporals* used in Mexico and southwestern United States). The same principle is involved in modern water-spreading efforts, but the methods and structures used are changing continually as improvements are developed. The extent to which water spreading can be utilized economically on range and farm lands has not been fully determined. Although distinct benefits have been obtained in many favorable situations, results in other places have been less encouraging. The practice frequently is a precarious one from the standpoint of economy, where ultimate results are dependent solely on the additional moisture provided by water spreading. Adaptable areas are generally isolated, and success often is determined by the success of the entire agricultural enterprise of which water spreading is a part. Involved with the success or failure of the practice, the uncertainty of the weather usually is an important factor.

Water spreading has been limited largely to the arid and semiarid sections of the West. It has been used in connection with the rehabilitation of the range and for crop production. The soil on which water is to be spread should be fertile, of gentle slope, and capable of absorbing and making productive use of the extra water. This method of supplementing rainfall on selected areas should be used only under conditions of proper grazing control and other conservation practices, if needed.

## Adaptability of Land for Flood Irrigation

Where the soils are generally rich and of favorable permeability, and the slopes gentle and smooth, the annual amount of rainfall may provide a general guide in determining the adaptability of land to water spreading. Areas having an annual rainfall of less than 8 inches, or a growing-season rainfall of less than 4 to 5 inches, may not produce sufficient runoff to justify the installation of a water-spreading system. Where the annual rainfall ranges from 8 to 14 inches, the most productive possibilities of water spreading may be for the improvement of range forage or the production of supplemental feed. Under especially favorable conditions of soil and topography, such dry-land crops as the southwestern varieties of Indian corn can be grown successfully. With a rainfall of 14 to 25 inches, the opportunity for successful production of supplemental feed and cultivated crops is generally much more encouraging.

## Water-spreading Principles

Measures designed for water spreading vary with the purpose for which they are to be used. Certain basic principles govern the design of any well-planned water-distribution system. They involve consideration of the minimum, normal, and maximum flows, character of the silt load, diversion sites, size of spreading area, texture and fertility of soil, topography, disposal of silt and waste water, and construction and maintenance of diversion and spreading structures.

When water first comes into contact with most soils, initial absorption takes place quickly. Thus, by the time the water has reached the lower end of a flooded area, the upper end will have absorbed both the initial intake and the continued intake during the full period of flooding. For this reason, the amount of water diverted over a given area should be regulated, as nearly as practicable, at the head of the area, so as not to exceed the amount necessary to cover the tract. If it is impracticable to avoid the use of extra water, provision should be made to handle the waste at the lower end of the tract to be flooded. It is advisable to spread the water as evenly as possible over the area and to control the velocity so that maximum absorption may be obtained with a minimum of soil movement. Distribution is accomplished primarily by proper spacing of contour furrows; by subsoiling; by proper placement of brush, rock, or woven-wire *percolators* (porous structures for retarding and spreading water); and by masonry, concrete, or other type of spreader. Where large quantities of silt are involved, it is also desirable to give some consideration to its even distribution. This may require some adjustment in the gradient or location of spreader structures so as not to concentrate too much coarse sediment near the point of diversion.

As yet, standard specifications have not been developed for water-spreading structures. Methods of construction and location and type of structure are largely controlled by local conditions. The spreading system must, of course, be fitted to the expected amount of runoff, the size and surface configuration of the area, and the soil, for most efficient use of the water. Much depends on the experience and skill of the man who selects the site and installs the system.

## Water Spreading on Range Land

Methods of retarding and spreading runoff have their most extensive application to those smoother grazing lands of the West on which range management alone is either too slow or too uncertain in restoring desirable vegetation. Such measures are particularly adapted to the broad alluvial and valley-fill areas (Fig. 228). Before they were scoured by sheet washing and cut with gullies, these valleys generally were considered the best range lands of the West. Flood waters formerly spread naturally over them, providing sufficient moisture to sustain an adequate cover of vegetation for prevention of rapid runoff and erosion. Now, water is concentrated in the gullies, thus depriving the adjacent flats of needed moisture. The objective of flood irrigation under such conditions is to take the water out of the gullies and spread it artificially where it formerly flowed naturally.

Where runoff, concentrated in a broad, shallow gully or wash, is to be spread over adjacent areas, a series of small graded furrows, ditches, or ridges (small embankments or dikes) provide one of the simplest means for diversion and spreading. Frequently, a combination of graded and level furrows or ridges will give the most effective results. Direct diversion is accomplished by systematically locating the furrows or small dikes so that each turns out a small quantity of flood water, usually not more than 1 or 2 second-feet, from the wash and conveys it over the spreading area. In some instances, small check dams or diversions serve to divert the flows into the graded furrows or ditches or along the embankments; but frequently no such structure is necessary, if the diversions can be extended

Fig. 228.—Recovery of Saccaton and other grasses on a valley-fill area formerly denuded by overgrazing, as the result of flood irrigation with water diverted from a rapidly eroding gully. Near Mexican Springs, N. M. (*Photograph by Soil Conservation Service.*)

across or well into the depression from which the water is to be diverted. Additional furrows or percolators are frequently constructed on the contour below the discharge end of the diversions, in order to spread out the discharged water. With enough diversions of this type on both sides of the channel, large flood flows sometimes may be disposed of entirely, so that water that formerly was running to waste and cutting away rich soil is turned to economic use, and the erosion is stopped.

This type of treatment has been practiced for a number of years by some ranchers in southern Colorado. Here, the primary purpose has been to increase production of range forage, but an incidental effect has been the prevention of gullying.

On uniform slopes, contour furrows and ridges generally can be used advantageously with all types of spreading systems to minimize concentration of water in natural depressions and to increase absorption. Close spacing of furrows and ridges is advisable immediately above gullies, critical slopes, and irrigation canals. Furrows for retaining, rather than conveying, water (Fig. 229) usually are turned upslope as gullies or natural depressions are approached, in order to close the ends. Where soil

**Fig. 229.**—Runoff intercepted and spread by contour furrows and ridges. Dallas County, Texas. (*Photograph by Soil Conservation Service.*)

**Fig. 230.**—A contour water-spreading system made of rock. Texas. (*Photograph by Soil Conservation Service.*)

conditions are unfavorable to the use of furrows, subsoiling is practiced, or brush, rock (Fig. 230), low woven-wire percolators, or concrete *worm spreaders* are used, according to local conditions and availability of materials. Subsoiling on the contour usually results in increased absorption. Contour percolators, constructed of loose rock, brush, or wire, serve to slow up and spread the runoff. Silt usually is deposited immediately above such structures, and this, together with the additional moisture, favors the establishment of vegetation, either naturally or artificially. Thinning of dense stands of sagebrush, combined with the use of the brush to build contour percolators, is giving excellent results in eastern Utah and western Colorado.

Where flattish slopes and heavy flows are involved, concrete worms have been used with good results on the Gila River soil and water con-

Fig. 231.—Earth dike used for diverting runoff from an active gully head, and conveying it to a spreading area for flood-irrigation. New Mexico. (*Photograph by Soil Conservation Service.*)

servation project in Arizona, in spreading water evenly over relatively long distances. The worms, constructed in the form of low, askew sections not exceeding 4 inches in height, impound part of the water and spread the remainder.

In many of the broad valley-fill areas used for grazing, the procedures employed for gully control and water spreading are much the same. A combination earth-fill dam and broad-base earth dike is generally used wherever practicable to turn all the water out of active gullies (Fig. 231). In some instances, the water is diverted with earth dikes before it reaches the gully. By means of carefully constructed openings in the dikes (weirs, or *weeps*), part of the flow is released at strategic points along the structure, and the remainder is discharged at the end. Where necessary, the water is spread still farther by furrows, ridges, or percolators strategically located below the discharge points. Where the gullies have long been established, waterways are restricted to locations where water rights are not involved and where the spreading areas are of sufficient size and permeability to absorb all the diverted water. Locations for diversions

FIG. 232.—Hay produced on Mexican Springs Project, New Mexico, following flood-irrigation of the area with water diverted from an expanding arroyo. Previously, this water ran to waste, carrying large quantities of silt into the San Juan River, thence into Lake Mead. (*Photograph by Soil Conservation Service.*)

FIG. 233.—Navajo corn produced by flood-irrigation near Mexican Springs, N. M., 1935. Yield ran to above 40 bushels per acre on some parts of this field. (*Photograph by Soil Conservation Service.*)

should be selected where deposition of silt is likely to be least troublesome.

Outstandingly successful results have been obtained with complete diversion of gully flood flows on the Navajo and Rio Grande soil and water conservation projects, in New Mexico and Arizona. Here, native vegetation is stabilizing the banks and beds of the gullies below the point of diversion, and the rejuvenation of vegetation over adjacent flooded land has been very marked. In some instances, such flood irrigation has so increased the growth of grass as to make possible the cutting of hay (Fig. 232) and the production of splendid yields of Navajo corn (Fig. 233).

Fig. 234.—A combination diversion and spreading system for rehabilitation of range on rich valley land, New Mexico. The structure in the water channel diverts normal runoff and by-passes surplus runoff during peak flows. A system of loose rock percolators located on the contour spreads the diverted water over the adjacent land, where it is absorbed. *(Photograph by Soil Conservation Service.)*

Where local conditions prevent complete diversion, partial diversions have been made to good advantage. Partial diversion from large gullies may necessitate the construction of a combination type of overflow diversion dam for by-passing minimum or normal flows and for controlled diversion of flood flows (Fig. 234). Since masonry or rock structures may be required for this purpose, the cost sometimes prevents their use. In a number of instances, however, results have justified the expense.

## Water Spreading for Crop Production

In the low-rainfall sections of the West where dry-land farming is practiced, and where the slopes are moderate and the soils deep, fertile, and permeable, many opportunities are afforded for effective utilization not only of the direct rainfall but of runoff from adjacent land. At the Spur branch of the Agricultural Experiment Station, in west Texas, diversion of the runoff from 1,200 acres of pasture land over a near-by 120-acre terraced field of gentle slope resulted in such marked increase in yields that the practice has spread widely within the last few years. Even the diversion of water from highways over adjacent fields and pastures is becoming a common practice in a number of western localities.

Excellent results have been obtained with two methods of water spreading on the soil and water conservation demonstration project near San Angelo, within the drainage basin of the upper Concho River, in west Texas: one by a *syrup-pan* system of terracing or diking, and the other by

Fig. 235.—Water flowing round the end of one of the terraces in a syrup-pan system used for spreading runoff over a 160-acre Texas field. (*Photograph by Soil Conservation Service.*)

*rationing* the diverted water into a system of closed-end level terraces. Frequently, a combination of the two methods is used on the same field.

In this locality, large areas of rich crop and grazing land, in the valleys, are so gently sloping (much of it having a slope of less than 2 per cent) and so permeable that these systems are easily installed and maintained.

The syrup-pan system is so designated because the water is made to flow back and forth across the field in the same manner as cane juice is processed in open syrup pans (Fig. 235). An upper terrace, collecting runoff from a drainageway or adjacent swale, carries it across the field to the opposite end of the terrace, where the unabsorbed water is discharged so as to be intercepted and turned back across the field by the next terrace below. The water is thus directed back and forth across the field until all of it is absorbed or any excess is discharged into a stabilized outlet at the lower side of the field.

Under the rationing system, each terrace receives a controlled amount of flood runoff from either a natural or an artificial drainageway. The purpose is to distribute the water between the terraces in amounts that will be absorbed by each of the interterrace areas. Various methods are used to regulate the amount of water delivered to each terrace interval. Merely regulating the inlet elevations is sufficient in some instances, whereas others require weirs with automatic or hand-operated gates.

The rate at which water may be added safely to an area depends on the infiltration capacity of the soil and the carrying capacity of the terraces. The soil should be capable of absorbing all the water applied within at least 48 hours; otherwise, crops are likely to be injured. On the San Angelo project referred to, the impounding systems providing surface irrigation water for relatively large interterrace areas have a capacity of approximately 2.3 second-feet per acre.

As the acreage rate of water applied increases, the construction cost of the diversion system also increases, because of large-capacity requirements. In order to determine the most economical installation, some balancing as between construction costs and expected increase of yield must be accomplished. Observations of water spreading by these methods indicate that under conditions similar to those on the San Angelo project, the ratio of the contributing watershed area to the area to be surface irrigated should be about 8 or 10 to 1. More effective and uniform flood irrigation may be expected where the system is designed to utilize efficiently the average annual runoff and to by-pass a portion of the peak flows rather than to plan for the utilization of only the peak runoff, since with an insufficient area in the contributing watershed the water supply would be inadequate during normal rains.

The net returns from water spreading are sometimes so modest that improper construction or misapplication of either type or size of system used may easily render the whole enterprise unprofitable.

# Chapter XXVIII. Wildlife and Soil Conservation

Soil conservation and wildlife conservation both depend fundamentally on the reestablishment and maintenance of vegetation. Each, therefore, may not only make important contributions to the other but is actually essential to the other's highest expression. Without soil conservation, climax vegetation with its associated animal life must largely disappear; without wildlife conservation, the vegetation is deprived of important protection, and ultimately the soil itself must lose the benefit of a powerful factor in its upbuilding—the direct influence of animal life.[1]

To insure a clear understanding of the relation between wildlife and soil conservation, it is necessary first to consider the basic requirements of wildlife and some of the factors that have contributed to former wildlife abundance and its subsequent decline.

The term *wildlife*, as used here, is restricted to undomesticated vertebrate animals. The basic essentials of existence for such animals are cover, food, and water—usually of relative importance to land animals in the order stated.

With few exceptions, all wildlife must have cover of some sort—cover in which to conceal its home, cover into which it may dart when attacked by an enemy, cover in which to rest or sleep or in which it may find protection from the elements. Some animals require several kinds of cover, either at the same season or at different seasons of the year; others are less exacting. But all are alike in that, when they do need cover, it must be in such proximity to their food supplies that they will not be unduly exposed when foraging.

Without exception, all wildlife must have adequate food at all seasons in order to survive. Summer and fall are usually seasons of plenty; but unless the supply continues to be available during the critical periods of winter and early spring, the resident wildlife population is doomed. To

[1] Holt, Ernest G. "Influence of Animal Life on Soil and Water Conservation." In "Headwaters Control and Use," pp. 197–198, Washington, D. C., April, 1937. Van Dersal, William R. The Dependence of Soils on Animal Life, 2d North American Wildlife Conference *Trans.*, pp. 458–467, 1937.

be available, food must not only be physically accessible (not covered by deep snow, for example) but must lie within safe foraging distance from habitable cover. Some animal species have different food requirements at different seasons, and the various items of diet must be available when needed. For example, many birds, which when adult live almost wholly on seeds, feed their nestlings on insects and soft fruits. Such birds, though not ordinarily thought of as insectivorous species, serve an important function in helping to hold insect populations in balance. The bobwhite is one of these.

Although free water is not essential to some quail, rabbits, and rodents, it is sometimes the limiting factor in the habitability of areas otherwise favorable for wildlife. Like food, water must be in such relation to cover that animals seeking a drink or bath will not fall too easy prey to their enemies.

The basic requirements of any given species, then, must stand in proper spatial relation one to another. Every element essential to the existence of any animal must be found within the cruising radius of that animal. The cruising radius of the bobwhite, for example, is generally about half a mile. Every need of the bobwhite, therefore, must be found within half a mile of the covey headquarters; but that is not all. The bobwhite will not cross open spaces of any great width, so unless strips of cover (travel lanes) connect the essential elements in a given environment, it is uninhabitable for the bird. Shocks of corn standing in a snow-covered Wisconsin field a hundred yards from a dense plum thicket may not be available to a covey of quail headquartering in the thicket. The value of vegetated fences, or strips of cane or corn across wide fields, in rendering all parts of a farm accessible to quail and other forms of animal life is at once obvious. Save for the fences and corn shocks, only those portions of the fields adjacent to the woods would be available as quail range, and vast quantities of food in the fields would be inaccessible.

Another wildlife fundamental is that animals generally are found in greatest numbers, both species and individuals, along the line of junction between diverse types of vegetation—for example, the margin between fields and woods. This fact has given rise to the expression that wildlife is a phenomenon of edge. It explains why woodland in even-aged pure stand is so notoriously devoid of wildlife. In the encouragement of wildlife, therefore, efforts would be directed to the improvement of *edge* effect and to the maintenance of woodlands of mixed species of all age classes. Such a woodland is shown in Fig. 236, which represents, incidentally, the highest known type of soil protection. Contrast this with the woodlot shown in Fig. 237, almost of an even age and rendered devoid of undergrowth and ground cover by the grazing of cattle. The one will support a normal wildlife population, including ruffed grouse;

the other, nothing save a few small arboreal birds and perhaps a family of squirrels.

Fig. 236.—Unburned, ungrazed woodland of mixed species and age classes, with abundant undergrowth, provides maximum protection for the soil and excellent habitats for wildlife. Illinois. (*Photograph by Soil Conservation Service.*)

Fig. 237.—A grazed woodlot of even-aged trees offers but little soil protection, steadily declines in timber production, and harbors a minimum of wildlife. Pennsylvania. (*Photograph by Soil Conservation Service.*)

Although these fundamentals pertain especially to land forms of animal life, they apply with but slight modification to fish as well. Fish, too, must have cover and food; and of course water is all-important. Protection from enemies (cover) is as essential to fish as to land animals.

Fig. 238.—All that is left of a once lovely Georgia stream utterly spoiled by erosion debris from unwisely cultivated hillsides. Fish have long since disappeared, but the flood hazard grows steadily worse. (*Photograph by Soil Conservation Service.*)

Fig. 239.—Proper treatment of watersheds and protection of stream banks are imperative to safeguard aquatic habitats. Contrast this luxuriantly vegetated Wyoming stream of clear, cool water—alive with fish—with that in Fig. 238. (*Photograph by Soil Conservation Service.*)

Moreover, the quality of the water itself is highly important to fish. The best game fish can live only in clear, cool water practically free of silt or other pollutants. For fish, silt is the most deadly of erosion products; it buries their eggs, clogs the gills of their young, and, by occluding light, destroys aquatic vegetation and the plankton on which they ultimately depend for food. Fish inevitably disappear from a stream well before the deposition of debris and the obliteration of the channel cause its actual physical destruction, as shown in Fig. 238. In Fig. 239 may be seen all the elements of a satisfactory aquatic habitat—clear water, plenty of both marsh and aquatic vegetation to supply ample cover and some food, and banks well shaded and protected against cutting by a heavy growth of shrubbery.

## Abundance and Scarcity of Wildlife

Current ideas concerning the present scarcity of wildlife as compared with the assumed universal abundance of presettlement times are often marked by the acceptance of erroneous premises.

In the first place, scientific evidence does not lend support to the idea that all primeval forest abounded in animal life. Wildlife was doubtless plentiful about the clearings made by Indians, fires, and high winds; yet periods of game scarcity were not unknown to the aborigines, as attested by their legends. Nevertheless, relative to human needs, fish and game were abundant when the first white men came to America and for a considerable period thereafter.

CHANGES FOLLOWING OCCUPATION OF THE LAND. The gradual settlement of the country, however, brought about slow but profound changes. Primeval wilderness gave way before the inexorable march of colonization. Corn and tobacco sprang up where virgin forest had shaded the soil for centuries; wheat appeared where only tall prairie grasses had bowed to the breeze. The effect on the animal population was equally profound. The bison, elk, moose, antelope, panther, wolf, and marten—wilderness creatures all—were slowly but surely, like their habitats, reduced to scattered remnants and will likely so remain because they cannot cope with the conditions imposed by civilization. The prairie chicken and sage grouse are in much the same case, though they might be restored if overgrazing were controlled. Although still existing in considerable numbers, bear, deer, and wild turkey over most of their ranges have been driven to the larger blocks of remaining forest held by Federal and state agencies.

For some wild species, however, agriculture for a considerable time created nearly ideal conditions. The interspersion of swamps and timberland with farms producing small grains and other food crops while tolerating much brushy cover resulted in a great increase in bobwhite, cottontail rabbit, and numerous nongame birds. The bobwhite and cottontail not only increased in numbers but extended their ranges into new territory. At the same time, the vast flocks of passenger pigeons, which Audubon and others say darkened the sky in their flights, were reduced to actual extinction, not, however, through destruction of habitat but by wanton killing.

FIG. 240.—Upper, Lake Como in southeastern Minnesota, once afforded fishing, swimming, and boating to the community of Hokah. Photograph made in 1926. Lower, ten years later the lake shown in upper picture had become a mud flat, with all values destroyed as the result of erosion over the watershed. (*Photographs by Soil Conservation Service.*)

RESULTS OF "CLEAN" FARMING. The heyday for farm wildlife was followed, however, by a period of more intensive use of land, marked by great improvements in farm machinery and the development of clean farming. This cleaning out of fence rows and stream banks and the turning of woodlots into pastures started a decline among the very species that had adapted themselves to an agricultural environment, while the remaining virgin areas continued to shrink with the increased demand for farm land, lumber, and wood pulp and with the expansion of grazing on the Western ranges.

So today, farms that only a few years ago were alive with birds, rabbits, and squirrels can boast little but starlings and English sparrows; once splendid mountain forests of the West, where deer and wild turkey were plentiful, have been reduced in many localities to barren sheep range; where numerous prairie chickens formerly lived in rich grama grass, cattle now compete for the vanishing forage. With this accelerated denudation of the land, and the disappearance of the animals dependent on the destroyed vegetation, have come about a corresponding acceleration of erosion, and the disappearance of much of the topsoil that was protected by the destroyed vegetation. This soil has been washed down into lakes, ponds, and streams; and where once were good swimming, boating, and fishing, there is now only a murky trickle across a mass of mud (Fig. 240). Many natural lakes, springs, and marshes, and even streams, have dried up as the result of (1) too rapid runoff of rainfall from devegetated and soil-stripped watersheds and (2) the expanding of outlets and lowering of the water table by the gullying process.

Such facts have given rise—and not without some reason—to the idea that "wildlife deserts" are left in the wake of destructive erosion. Although it is not definitely known that any species of animal or plant has been exterminated through the agency of erosion alone, it is a matter of common observation that the removal of topsoil by erosion has greatly altered and even rendered uninhabitable certain territory although usually of limited extent. Beaver, for example, have been forced to move to new locations by the sedimentation of their ponds; and profound changes have been brought about in the faunas of some streams by suspended silt. In parts of the Great Plains, wind erosion following cultivation has changed areas of grassland to actual desert where only sand-loving animals and plants can now survive. In the Southwest, many valley-fill and alluvial areas, once richly grassed, have been gutted by erosion channeling to a condition of desert, as the result of accelerated erosion produced by overgrazing. Lakes and marsh pools that formerly were visited by waterfowl have dried up in various parts of the West because of rapid runoff following depletion of vegetation and resultant erosion on the range.

## Importance of the Wildlife Resource

As wildlife has decreased, human demand for it has increased, particularly with respect to game species and fur bearers. Relative to human needs, wildlife is now scarce indeed. Much more is needed than now exists, and the desired increase can be brought about only through positive human effort expressed in fish and game management. Because of the intimate relations between soil, vegetation, and animal life, the needed habitat reconstruction can be accomplished largely by appropriately planned operations for the conservation of soil and moisture.

The importance of wildlife is as far-reaching as it is difficult of evaluation by such a purely artificial standard as money. To the early pioneer, game was vital as food, fur as clothing. But the importance of animal life to human life is far more fundamental than the role that it has played in permitting the white man to occupy new territory; it reaches right back to the soil, on which all life ultimately depends.[1]

Insects are regarded as the greatest of man's enemies save man himself. Unchecked, they probably would render the planet incapable of supporting human life. The role of birds as a constant repressive influence on insects is well known by specialists but not at all adequately understood by the public. The check exerted by other vertebrates is even less widely known although very important.

The biologic importance of wildlife, its position as a vital link in a highly complex chain of interrelations that impinge at innumerable points upon human life, may safely be accepted without further delving into ecology.

The social importance of animal life is second only to its biologic importance. The clear perception of these facts led President Theodore Roosevelt to champion the cause of wildlife conservation; it moved President Franklin D. Roosevelt to call together in Washington the conservationists of a continent for a North American Wildlife Conference; it inspired the Secretary of Agriculture to address the second convocation of that conference in these words:[2]

"As a people we have been intensively engaged in developing and using machinery to subdue nature; we have turned the power generated from other natural resources of coal, petroleum and electricity to the soil and the things which grow from the soil. Wildlife is one of these. It is necessary to readjust our perspective and devote a much larger proportion of our interest to the subject of life itself. Every form of life has value and interest of some sort; even the most insignificant creatures may be found to exercise the most profound influences upon mankind.

"It is particularly important that the restoration and conservation of our wildlife resources be given recognition as a national obligation of great significance

[1] Van Dersal, William R. The Dependence of Soils on Animal Life, 2d North American Wildlife Conference *Trans.*, pp. 458–467, 1937.

[2] Wallace, Henry A. Address, 2d North American Wildlife Conference *Trans.*, pp. 15–19, 1937.

at this time when Americans are earnestly endeavoring to readjust social and economic balances so that the benefits and profits of progress may be fairly distributed among all classes of people."

The economic importance of wildlife is deliberately relegated to third place in this appraisal. Biologic and social considerations outweigh economic reasons for conserving it. Nevertheless, the monetary value of the nation's fish and game resources is enormous. Some states, like Minnesota, now recognize the fact that wildlife is the chief attraction for a tourist trade running into millions of dollars annually. License fees alone from the country's estimated 13,000,000[1] sportsmen are sufficient to put hunting and fishing in the ranks of big business. The additional millions that these same sportsmen spend for guns, ammunition, tackle, camping equipment, travel, and so on is estimated by the Biological Survey to bring to about a billion dollars the amount annually poured into the channels of trade by those who hunt and fish. Remove the incentive, and this flow of money ceases.

These figures refer only to money spent in harvesting game and fish for sport. Although the fur catch of the United States has declined steadily for many years past, its total annual value still runs between 60 and 65 million dollars. A large proportion of the catch is taken by the farmer or his boy in agricultural districts, and the animal topping the entire catch, both in number of pelts and total value of fur, is the muskrat. This is a species that occurs naturally in nearly every state; it is easily managed and may be cropped on any farm fortunate enough to have a permanently wet marsh.

## Relation of Wildlife Management to Soil and Water Conservation

Wildlife management seeks the maintenance and control of wildlife populations. Because of the past general decline in wildlife, control is usually in the direction of higher population levels. The most obvious reason for the decline is the widespread impairment and destruction of habitat. As animal existence is impossible in the absence of adequate habitat, the most essential steps in management are habitat reconstruction and rehabilitation. For the present, therefore, management is concerned more with the manipulation of the animal's physical environment than with direct control of the animal itself. Hence, the most important management tool is vegetation.

Soil conservation seeks the maintenance of soil *in situ*. This implies protection from the destructive forces of wind and water, which, in turn, involves the maintenance of adequate cover. The most effective soil

[1] 71st Cong.. 3d Sess. *Senate Rept.* 1329, p. 4, Jan. 21, 1931.

cover is vegetation. For that reason, vegetation is the most important tool of soil conservation.

As wildlife management and soil conservation employ the same tools, it follows that, with appropriate direction, the objectives of both may be attained through the same operation.

The close relation of wildlife management to soil conservation is further attested by the fact that both must surmount the same obstacle in the employment of their vegetative tools. The land has been laid bare through the necessity of producing crops to support an increasingly heavy human population. The reestablishment of vegetation on these bare soils, therefore, is not a simple matter of planting; it involves the whole question of proper land use.

It follows that soil conservation involves the revision of the farm layout and its cropping system so as to achieve adequate protection for all soil needing it while permitting maximum crop use of all safely tillable land in order to insure profitable operation of the farm.

Animal conservation involves the same problem of attaining its ends without upsetting the farm business. But, just as it is possible to farm the soil and keep it too, so also it is feasible to farm the soil and conserve wildlife. Wildlife is a crop that may be produced over and above other farm crops occupying the same land. It can and should be a valuable by-product of every well-managed farm plant. Although wildlife needs adequate territory—it cannot be crowded—by proper distribution of the essential components of its habitat, the entire farm can be made to serve its purpose. By dedicating specific areas here and there to wildlife, the whole farm may be made available to this resource without sacrificing any appreciable amount of land.

The selection and treatment of these special areas bring about a particularly close tie-up between wildlife management and soil conservation. On every farm, there are numerous areas that may not be safely plowed or that have so far deteriorated that crop production on them is no longer profitable. They present a high erosion hazard or an active erosion problem and cannot be ignored, although advanced erosion usually renders control difficult. Such are galled spots, sinkholes, gullies, eroding field borders, and cutting stream banks. Ordinarily, these places will not support economic types of vegetation; but when planted to pioneering types beneficial to wildlife, they become just the areas needed to effect desired interspersion of habitat elements by providing a protective cover of vegetation to soils on which neither field crops nor climax trees will thrive. Thus, wildlife management steps into an important breach in the soil conservation program.

Wildlife management also makes a definite contribution to water conservation. Water does not long remain in good condition in unprotected reservoirs, farm ponds, and stock tanks. Fence them against trampling and contaminating livestock; protect the shores with marsh plants against eroding waves; insure the purity of the water by the introduction of aquatic plants; shade and cool the water, and reduce evaporation with tree and shrub borders—do such of these things as may be needed; and not only will the quality of the water be improved, but the reservoir will last longer and will become a veritable haven for animal life. This has been demonstrated on the Great Plains from Pierre to the Panhandle.

In the relation between wildlife management and soil and water conservation, the contributions are not at all in one direction; the relation is reciprocal. Besides the elementary fact that without soil there would be no wildlife, nearly all soil conservation operations contribute something to the conservation of animal life. Strip cropping multiplies the amount of edge manyfold. Every measure that holds any soil in place keeps just that much from washing down to pollute streams and clog reservoirs. When the combined measures make it possible for streams again to run clear, fish may be restored to them. Before that time, there can be no fish management.

Wildlife Practices in Soil Conservation. In the vegetative protection of gullies, galled spots, sinkholes, abandoned roadways, eroding field borders, and rough places about the farm as well as cutting stream banks and pond and lake shores, wildlife plantings offer the best solution of the erosion problem. This is true (1) because the plant species adapted to wildlife needs generally are those most likely to succeed under the ecological conditions imposed by such sites and (2) because the areas usually are too small or too inconveniently located for the profitable production of economic plants, even if they could thrive in such places. Incidentally, relocating field fences on the contour and protecting them with plantings of woody vegetation is good erosion-control practice and one of the best possible means of making all parts of the farm available to wild animals.

The type of treatment given these areas is determined by the conditions. Sometimes nothing more is required than protection from fire, livestock, and man; occasionally the engineer's assistance is needed in diverting water or checking its velocity; usually more or less planting is necessary in order to hasten the healing process. Species for planting should be chosen on the basis of their (1) adaptation to the sites, (2) ability to control erosion, (3) value as wildlife food or cover, and (4) usefulness in supplementing existing food and cover. They may be trees, shrubs, vines, forbs, or grasses. Whether entirely new habitats are created by these plantings or old ones rehabilitated, the ideal sought is adequate food and cover at all seasons.

Examples of some of the sites mentioned, their treatment, and the results are illustrated in the following photographs:

| | |
|---|---|
| Gullies | Fig. 241 |
| Galled spots | Fig. 242 |
| Field borders | Fig. 243 |
| Stream banks | Fig. 244 |
| Reservoir shores | Fig. 245 |
| Shrub baffles | Figs. 246 and 247 |

Fig. 241.—Upper, a gullied area in an Ohio pasture, worthless to cattle and obviously barren of wildlife, fenced and newly planted to black locust. Lower, same area as shown in upper picture after two years. Provides cover for many birds and rabbits where none previously could live, and reduces runoff entering flood streams. In a few more years a supply of fence posts will be ready for harvest. (*Photographs by Soil Conservation Service.*)

Occasionally, larger areas may, for various reasons, be unsuitable for profitable farming or grazing or for timber production. These may be managed primarily for wildlife, with excellent results. Examples are steep, severely eroded slopes, rough country on the Western range, Indian

FIG. 242.—Upper, an actively eroding galled spot in the Piedmont of North Carolina, although practically barren of vegetation, becomes a wildlife habitat when properly treated. Lower, a galled spot similar to that shown in upper picture mulched with pine boughs and seeded to lespedezas. By this treatment such places may be revegetated in a single season. (*Photographs by Soil Conservation Service.*)

taboo areas,[1] natural vegetated draws in the Great Plains and elsewhere, and barren or rocky arid land. Developments on such lands must be at once cheaply accomplished and permanent. For the most part, protection from grazing will permit natural plant succession to revegetate such areas.

[1] Areas within Indian reservations avoided because of superstition.

A little judicious planting may be resorted to where erosion conditions are too critical to permit delay in revegetation.

Farm ponds, range water developments, and water conservation reservoirs offer the biologist an especially fruitful field of endeavor. Wildlife returns are rapid and sometimes amazing. During migration periods, ducks stop to rest on even the smaller ponds as soon as the latter impound water, although they are useless to nesting birds until food and

FIG. 243.—An eroding field border adjoining a South Carolina woodland area, protected by a strip of millet and lespedeza to provide soil cover, wildlife food, and to prevent encroachment of the woods on the cultivated land. (*Photograph by Soil Conservation Service.*)

cover are developed. The general practice is to fence such areas (with provision, of course, for watering stock), plant the banks to woody vegetation and the shorelines to marsh species, and introduce aquatics to maintain purity of the water (Figs. 245, 246, and 247). As soon as cover appears within the protected areas, they are populated by nesting ducks, pheasants, and other kinds of birds and often by muskrats and other mammals.

In the establishment of new vegetation or the maintenance of old, protection from fire and livestock is of prime importance. Education and fencing, therefore, rank with planting as wildlife conservation measures.

MODIFICATION OF OTHER SOIL CONSERVATION PRACTICES TO WILDLIFE BENEFIT. Average farms include four types of land: cropland, pasture, woodland, and idle land. The first two, and often the third, cannot be dedicated to wildlife, and there is seldom much idle land on good farms. Hence, the opportunity of improving conditions for animal life would be limited indeed if confined to management measures on idle or waste areas. Fortunately, there is no such limitation, for, by appropriate modification of agronomic, engineering, and woodland practices,

FIG. 244.—Upper, a cutting stream bank in Wisconsin, practically devoid of cover, that has been fenced out of pasture and planted to willows protected by a wattle of willow poles. Lower, same stream bank shown in upper picture 14 months later. The willows already afford protection against cutting, and provide excellent wildlife cover as well as considerable cooling shade for the water. (*Photographs by Soil Conservation Service.*)

a great deal may be accomplished without the least interference with the primary functions of these practices. The margins where fields and woods, fields and pastures, or woods and pastures adjoin are insignificant in area but, because of their great linear dimension, are of enormous importance to wildlife. Modifications of other technical practices, therefore, usually are designed to enhance the value of these borders.

The principle is well illustrated in Maps 5 and 6, showing the layout of a Piedmont farm before and after conservation plans were adopted.

Fig. 245.—Upper, water conservation reservoirs, such as this one recently constructed in South Dakota by the Soil Conservation Service, are potentially the finest wildlife havens, but are practically useless in this condition. Lower, when fenced and planted with protective vegetation to prevent erosion, wave cutting, and undue evaporation, the same reservoirs are eagerly sought by breeding ducks. (*Photographs by Soil Conservation Service.*)

*Agronomic practices* may be turned to wildlife benefit in numerous ways. For example:

1. Strips of certain legumes, grasses, and other plants may be used to protect field borders, where erosion is often severe, especially adjacent to woodland. The encroachment of woodland on cultivated fields is an

important farming problem. Insect depredations begin not at the center of fields or woods but at their edges. This measure, by attracting birds to such margins, contributes materially to the solution of the combined

FIG. 246.—Marsh plants established about the margins of farm ponds prevent wave erosion and afford habitats for water-loving birds and mammals. (*Photograph by Soil Conservation Service.*)

FIG. 247.—Aquatics planted in reservoirs help to desilt and purify the water, improve habitats for fish, and provide food for waterfowl. (*Photograph by Soil Conservation Service.*)

problem of land use, erosion control, and mutual protection of soil, vegetation, and animal life.

2. When appropriate plant species are used in protecting terrace outlet channels, vegetated terraces, and so on, these strips of vegetation

provide food, cover, and breeding areas for certain birds and travel lanes for many forms of wildlife. In the Southern States, use of lespedezas on the banks of terrace-outlet channels serves those purposes admirably.

3. The maintenance of a limited number of trees and shrubs in pastures is of benefit to both wildlife and livestock.

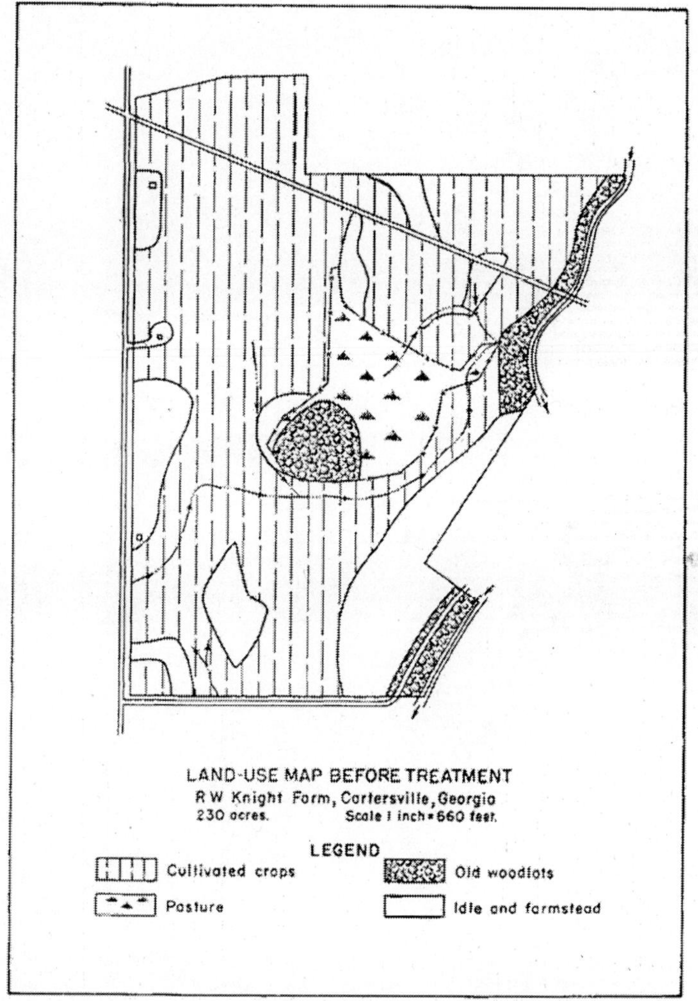


MAP 5.—Land use conditions on an eroding Georgia farm before treatment for conservation of soil, water, and wildlife. (*Soil Conservation Service.*)

4. Where deferred and rotation grazing is practiced, slight modification of dates for turning stock on an area may often permit successful nesting of certain birds.

5. In computing carrying capacities of the Western range, appropriate reservations may be made for wildlife.

6. In regions where large amounts of snow are removed by winter winds, shrub buffer strips or strips of sunflowers or grain sorghums may be used in cultivated fields to hold snow in place. These strips provide cover and food for wildlife, as well as lanes of travel. They also protect fields from wind erosion.

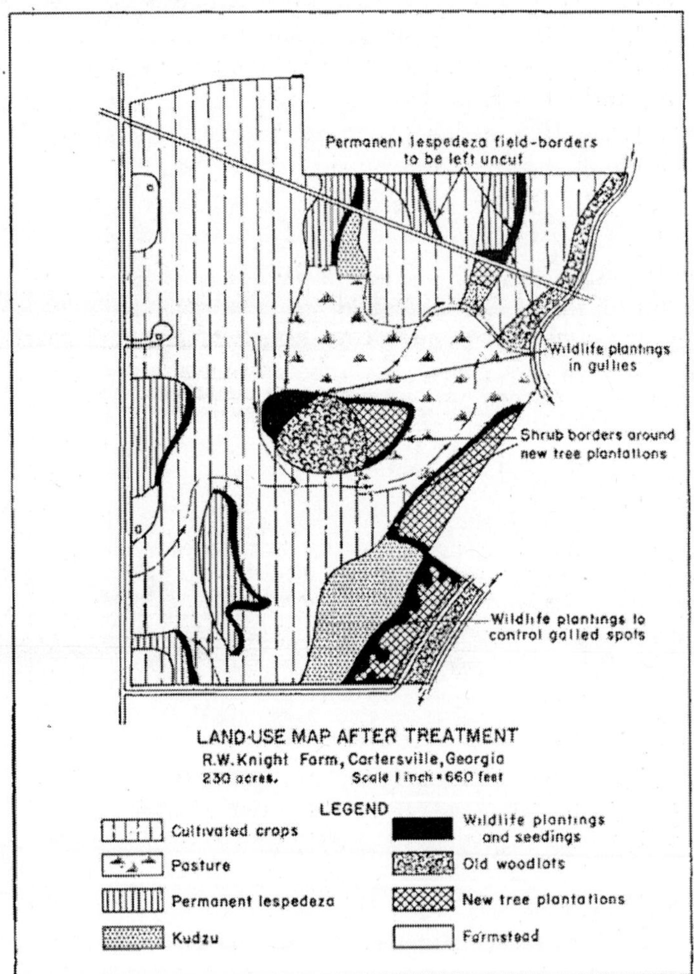


MAP 6.—Same farm as shown on Map 5, after treatment for conservation of soil, water, and wild life. (*Soil Conservation Service.*)

7. In fields of corn, legumes, and various grains, the outermost rows may be left standing over winter for wildlife food, or, if all the crop must be harvested, the farmer may be urged to leave until last those portions adjacent to good wildlife cover.

*Engineering practices* similarly may yield some benefits:

1. Structures such as terraces, terrace outlets, dams, spillways, water spreaders, and others often need the protection or supplemental action

of vegetation for their maximum efficiency. Wherever possible, species furnishing food and cover for wildlife should be used.

2. Almost without exception, stock ponds should be protected by fencing. The shallower portions provide ideal habitats for many animals, especially waterfowl and shore birds, when properly planted. Aquatic and marsh vegetation may be used at inlets to desilt inflowing water and along shores to prevent wave action. Steep banks may be planted to herbs, shrubs, and trees to protect the banks from erosion. In more arid regions, woody plants of value to animal wildlife may be used entirely around the body of water to prevent undue evaporation.

3. Vegetative protection of stream banks, although often done independently, may be carried on in combination with dams, jetties, dikes, riprapping, and other engineering devices.

*Woodland practices* may be particularly beneficial if planned with the welfare of wild animals in mind. Protection from fire and grazing is of tremendous value to the animal populations of woodlands. Other practices may be shaped to wildlife ends; for example:

1. In all operations designed to improve timber stands, the habitats should not be impaired; usually they may be actually improved.

This involves leaving sufficient den trees, not cutting out too many "weed" trees, proper handling of excessive animal populations, maintenance of borders, maintenance of fire lanes, and uneven-aged stands.

2. In new woodland plantations, borders of small trees, shrubs, and vines of food-producing species should be incorporated in the planting. Mixtures of species should be used instead of solid stands. Both practices increase not only the value of the habitat but the value of the forest for the production of wood products.

3. In the planting of windbreaks for control of wind erosion, low-growing shrubs that contribute cover and food should form the outer rows. Where necessary to protect the young trees from soil blowing, pending their firm establishment, rows of sorghum, Sudan grass, sunflowers, or corn may be planted between tree rows and on either side of the windbreak. While protecting the young trees, these plants furnish food for wildlife.

4. In cutting operations, wherever adequate winter cover is not supplied by normal undergrowth, the slash may be piled to afford protection for grouse, quail, rabbits, and other animals.

Special Practices. Planting of beaver has been carried on in several Western areas as a direct aid in the reduction of flood crests and silting (Fig. 248).

A recent survey of results in the Mission Creek watershed of Washington, where 12 beavers were added to a small colony in the east fork, disclosed 22 dams in a 2,040-foot section of the stream, which impound

1.44 acre-feet of water and 5,844 cubic yards of silt. On Ahtanum Creek, in the same state, 12 beavers transplanted in 1936 have constructed several large dams, one of which, in June, 1937, was 90 feet long and 7 feet high. This dam now impounds 5 acre-feet of water in a stream with 65 second-feet of flow. Engineers estimate that it would cost $2,500 to build a dam of similar effectiveness.

Establishment of feeding stations is encouraged in regions of heavy snowfall as a means of tiding wildlife over periods of critical scarcity of

FIG. 248.—Beaver planted by the SCS built this reservoir in a sparsely inhabited Washington watershed. It is equally efficient and less costly than a man-made one would be. (*Photograph by Soil Conservation Service.*)

food until wildlife plantings have time to become effective. Except the few that are thrown together in woodlot management operations carried on as part of the soil and water conservation demonstrations covering the coordinated land-use adjustments of farms made under the program of the Soil Conservation Service, these stations are built and maintained by the cooperating farmers. Grain is supplied by the farmer, the state, or local sportsmen's organizations. Both birds and mammals (rabbits and squirrels) flock to these stations in severe weather (Fig. 249).

Food patches are somewhat similar to feeding stations in their emergency aspects, except that they are regularly maintained by farmer cooperators with a special interest in game-bird production. They are often recommended to complete the wildlife management program.

Game cooperatives seem to offer the best implement by which the farmer may dispose of his game crop for cash and are recommended to farmers interested in a financial return from their efforts to improve wildlife conditions. Often single farms are too small to offer the gunner sufficient acreage for a day's sport, or they may not contain within their boundaries the essentials of more than a limited number of game territories. By banding a number of farms into a cooperative, both handicaps may be overcome, and a better system of game management put into practice on the entire area.

Fig. 249.—Bobwhite utilizing a feeding station built by CCC boys under the direction of SCS in Minnesota. (*Photograph by Soil Conservation Service.*)

Marsh Management. Because of the special importance of marshes in both water conservation and wildlife economy, efforts are made to insure their protection by interesting the owners in a system of management calculated to yield some revenue from these so-called *waste areas.* Muskrats are so widely distributed and so easily cropped that no farm marsh need be without them, if properly managed. Some farmers possessing no old marshes to manage have deliberately created new ones.

Pack-rat Work. After all the obloquy heaped upon the lowly rodent, it is refreshing that at least one humble member of that tribe is now gaining recognition as a soil conservationist. In Arizona, the pack rat's habit of making its nest in almost any pile of debris and then adding to it until the mass reaches huge proportions may be put to work in the building of check dams in the numberless small gullies now riddling so much of the cactus country. Workmen equipped with forks knock branches from the cholla cactus and pitch them into small piles in the gullies. Later, the pack rats move in, adding to and consolidating the piles into effective check dams (Fig. 250).

*Wildlife refuges* may serve no direct function in erosion control, but they are valuable instruments of management in affording decimated

FIG. 250.—This pack rat's check dam, built as a result of SCS encouragement, has arrested nearly 4 feet of silt in an Arizona wash. (*Photograph by Soil Conservation Service.*)

game populations a chance to recuperate. Where eroding land is enclosed by such reservations, protection from cultivation, grazing, and fire aids in conserving both soil and water.

## Plant Species Used in Wildlife Conservation

The procurement of plants useful to wildlife in quantity sufficient to meet the demands of soil conservation operations has heretofore presented difficulties. Most of the more valuable plants had not been in demand in the ordinary channels of trade, and the large-scale propagation of many species had never been attempted. Among the plants so far used in large numbers in present programs of soil conservation work, the following have definite wildlife value:

GRASSES

| | |
|---|---|
| *Dactylis glomerata* | Orchard grass |
| *Panicum adspersum* | Browntop millet |
| *Paspalum boscianum* | Bosc paspalum |
| *Sorghum vulgare sudanense* | Sudan grass |

### Herbaceous Legumes

| | |
|---|---|
| *Chamaecrista fasciculata* | Partridge pea |
| *C. nictitans* | Small partridge pea |
| *Desmodium tortuosum* | Florida beggarweed |
| *Lespedeza sericea* | Sericea lespedeza |
| *L. stipulacea* | Korean lespedeza |
| *L. striata* | Common lespedeza |
| *Medicago sativa* | Alfalfa |
| *M. hispida* | California bur clover |
| *M. lupulina* | Black medic |
| *Sesbania macrocarpa* | Sesbania |
| *Trifolium spp.* | Clovers |
| *Vicia spp.* | Vetches |
| *Vigna sinensis* | Cowpeas (Bradham) |

### Woody Plants

| | |
|---|---|
| *Berberis thunbergii* | Thunberg barberry |
| *Callicarpa americana* | French mulberry |
| *Ceanothus spp.* | Ceanothus |
| *Celastrus scandens* | Climbing bittersweet |
| *Celtis mississippiensis* | Southern hackberry |
| *C. occidentalis* | Hackberry |
| *Cornus amomum* | Silky cornel |
| *C. asperifolia* | Rough-leaf dogwood |
| *C. florida* | Flowering dogwood |
| *C. paniculata* | Gray-stem dogwood |
| *C. stolonifera* | Red osier |
| *Corylus americana* | Hazelnut |
| *Crataegus spp.* | Hawthorns |
| *Elaeagnus angustifolia* | Russian olive |
| *Fallugia paradoxa* | Apache plume |
| *Forestiera neomexicana* | Palo blanco |
| *Ilex spp.* | Wild hollies |
| *Juniperus mexicana* | Mountain cedar |
| *J. monosperma* | Cherrystone juniper |
| *J. scopulorum* | Rocky Mountain red cedar |
| *J. virginiana* | Red cedar |
| *Ligustrum spp.* | Privets |
| *Morus spp.* | Mulberries |
| *Myrica spp.* | Bayberries |
| *Parthenocissus quinquefolia* | Virginia creeper |
| *Pinus spp.* | Pines |
| *Prunus americana* | Wild plum |
| *P. angustifolia* | Chickasaw plum |
| *P. demissa* | Western chokecherry |
| *P. serotina* | Wild black cherry |
| *P. virginiana* | Eastern chokecherry |

| | |
|---|---|
| *Quercus spp.* | Oaks |
| *Rhus glabra* | Smooth sumac |
| *R. trilobata* | Lemonade sumac |
| *Robinia pseudoacacia* | Black locust |
| *R. neomexicana* | New Mexican locust |
| *Rosa spp.* | Wild roses |
| *Rubus spp.* | Blackberries |
| *Salix spp.* | Willows |
| *Sambucus canadensis* | Blue elder |
| *Shepherdia argentea* | Buffalo berry |
| *Symphoricarpos occidentalis* | Snowberry |
| *S. orbiculatus* | Coral berry |
| *Viburnum lentago* | Nanny berry |
| *V. prunifolium* | Black haw |
| *V. trilobum* | High-bush cranberry |
| *Vitis spp.* | Wild grapes |

In addition to the plants listed in the foregoing tables, it seems likely that the following species, when methods of quantity production are developed or more is learned about their site adaptations, will prove even more desirable for the dual purpose of erosion control and wildlife conservation than those now available:

| | |
|---|---|
| *Amelanchier alnifolia* | Western serviceberry |
| *A. canadensis* | Serviceberry |
| *A. stolonifera* | Creeping shadbush |
| *Arctostaphylos spp.* | Manzanitas |
| *Aronia arbutifolia* | Red chokeberry |
| *Atriplex spp.* | Saltbushes |
| *Berberis spp.* | Barberries, hollygrapes |
| *Bumelia spp.* | Buckthorns |
| *Celtis douglasii* | Douglas hackberry |
| *C. georgiana* | Georgia hackberry |
| *C. pallida* | Desert hackberry |
| *Chrysobalanus oblongifolius* | Deer plum |
| *Cowania stansburiana* | Quinine bush |
| *Crataegus spp.* | Hawthorns |
| *Croton spp.* | Shrubby crotons |
| *Elaeagnus argentea* | Silverberry |
| *Ephedra spp.* | Ephedras |
| *Gaylussacia spp.* | Huckleberries |
| *Juniperus spp.* | Junipers |
| *Lycium spp.* | Desert thorns |
| *Malus spp.* | Wild crab apples |
| *Photinia arbutifolia* | Christmas berry |
| *Prunus spp.* | Wild plums, cherries |
| *P. besseyi* | Western sand cherry |
| *Rhamnus californica* | California buckthorn |

| | |
|---|---|
| *Rhus spp.* | Sumacs |
| *R. canadensis* | Fragrant sumac |
| *Sambucus spp.* | Elderberries |
| *Sassafras albidum* | Sassafras |
| *Smilax spp.* | Thornless greenbriers |
| *Sorbus spp.* | Mountain ashes |
| *Symphoricarpos spp.* | Snowberries |
| *Vaccinium spp.* | Blueberries |
| *Viburnum spp.* | Viburnums |
| *Xanthoxylum spp.* | Prickly ashes |

## Wildlife Species Encouraged

No species in a community, unless it happens to live a very isolated life or be very rare, is without effect on the other members of its community. This is ecological law. A corollary is that measures designed to foster the increase of one species automatically affect numerous other species in the same community.

Therefore, habitat improvement to encourage the increase of game birds, game mammals, or fur bearers must automatically have a beneficial effect on song and insectivorous birds and other nongame animals. Wildlife management on the farm should benefit wildlife as a whole. There may seem to be compromises when emphasis is placed on the production of those forms which offer a tangible return to the farmer. For psychological reasons, the production of game animals and fur bearers may be stressed, but the creation or reconstruction of habitats for wildlife in general is the actual goal.

Among the animals normally benefited by wildlife-encouragement practices in connection with soil conservation are the following game and fur-bearing species:

| Game Birds | Game Mammals | Fur-bearers |
|---|---|---|
| Bobwhite | Cottontail rabbit | Beaver |
| Scaled quail | Marsh rabbit | Mink |
| Gambel's quail | Squirrels | Muskrat |
| California quail | Foxes | Opossum |
| Ruffed grouse | Deer | Raccoon |
| Sharp-tailed grouse | Antelope | Skunks |
| Sage grouse | Elk | |
| Prairie chicken | | |
| Mourning dove | | |
| Wild turkey | | |
| Ring-necked pheasant | | |
| Hungarian partridge | | |
| Waterfowl | | |
| Shorebirds | | |

Various species of pond fishes may be established in farm ponds and reservoirs built for water conservation, and trout and many other game fish benefit through the prevention of stream sedimentation.

It has been indicated in the foregoing that wildlife management is an indirect science which seeks increased production of animals through habitat improvement. In attempting to measure the success of management operations, two methods of evaluation may be considered: (1) the

Fig. 251.—A flock of ducks on a reservoir built and fenced by the Soil Conservation Service in Texas. Hundreds of small reservoirs and stock water ponds have been constructed by the Service, particularly in the low-rainfall areas of the Great Plains and Colorado Basin area; and on many of them wildlife is developing. (*Photograph by Soil Conservation Service.*)

counting of wildlife populations before and after operations and (2) the summing up of the different kinds of work that have gone into the improvement of habitats.

The evaluation of accomplishments in terms of absolute increases in wildlife populations involves so many variables that it is practically impossible without employing painstaking research methods and appropriate controls.

Highly cyclic species like the ruffed grouse and snowshoe rabbit periodically exhibit wide fluctuations in numbers that still defy explanation. Catastrophes, such as recent droughts in the Great Plains and the extremely severe winter of 1935–1936 in the Central and Lake States, may reduce pheasants and quail almost to the vanishing point before habitat improvements have a chance to become effective. Where woody cover is the limiting deficiency, several years may be required for plantations to grow large enough to supply the want.

Moreover, wild birds and mammals are such elusive creatures that, except for a few unusually sedentary or conspicuous species, it is extremely difficult to obtain accurate counts of their numbers. Most censuses, therefore, are avowedly estimates and are confined to a few key species like quail, pheasants, and rabbits.

Nevertheless, it is possible to obtain general information on relative increases in wildlife populations, and many data are available to show that quail, for example, have quickly responded to improved conditions by taking up headquarters on treated areas where there had been no quail prior to soil conservation operations.

At Fairbault, Minn., Hungarian partridge have increased under the soil conservation farming programs of the Soil Conservation Service from 6 birds in the winter of 1935–1936 to 50 in 1936–1937. The 1938 population is estimated at approximately 200 birds.

The most striking examples of wildlife response to erosion-control operations, however, are supplied by pond developments on the Great Plains, for here habitats have been created where none previously existed. Just a year after a reservoir was constructed near Huron, S. D., a biologist counted on the water, or about the fenced area surrounding it, 50 ducks, 7 black-crowned night herons, 12 killdeer, 2 upland plovers, about 250 red-winged blackbirds, about 100 swallows, 12 short-eared owls, 4 skunks, and a number of songbirds. In the Panhandle of Texas, ducks have nested on stock tanks the first spring after they were built (Fig. 251); and anywhere in the Great Plains, migrating waterfowl eagerly use reservoirs for resting without waiting for improvements. Ponds formed back of dams constructed to control gullies in the south Palouse River, Washington, project of the Soil Conservation Service, became breeding places for wild duck within a few months after installation.

# Chapter XXIX. Soil Conservation and Flood Control

Conservation farming practices designed primarily to control soil erosion often help to prevent floods.

Floods are caused by runoff of rainfall and snow water at rates so rapid as to overtax the carrying capacity of the streams into which the surface waters are discharged. Soil conservation measures that retard surface runoff and increase infiltration of rainfall into the soil tend to reduce the velocity and volume of water reaching drainage channels. Thus, when all, or a large part, of the land within a watershed is adequately protected from soil erosion, the flood hazard should be reduced.

For generations, much thought and discussion have been devoted to the relation of land use to floods. The literature on the subject covers a period of more than a century. Several thousand separate publications, in all modern languages, deal with the subject. The majority of these discussions are based on empirical investigations, general observations, and historical comparisons without accurate basic information. Comparatively few scientific studies have been made, because of the inherent difficulty of isolating variables from the complexity of interacting factors in watersheds.[1]

George Marsh in his book "The Earth as Modified by Human Action" (1874) summarizes his studies and observations as follows:

> "The felling of the woods has been attended with momentous consequences to the drainage of the soil, to the external configuration of its surface and probably also to local climate, and the importance of human life as a transforming power is perhaps more clearly demonstrable in the influence man has thus exerted upon superficial geography than in any other result of his material effort."

Marsh drew upon a bibliography of 322 works, as well as his own observations over many years. He ascribes the decay of the Roman Empire, for example, to the imprudent removal of forests and consequent

[1] Lowdermilk, W. C. Summary Review of Literature up to 1930 on Forest and Agricultural Influences in Streamflow and Erosion Control (mimeographed), Soil Conservation Service.

waste of the soil. The classic studies of Surrell[1] on the torrents of the High Alps and of Becqueral[2] on the comparative influence of forested and nonforested lands on climatic factors supplied Marsh with supporting data for his deductions.

Gustave Wex,[3] a hydraulic engineer and director of Works of the Improvement of the Danube, in 1873 attributed the recorded diminution of water in wells and streams to the clearing of forests in the basin of the Danube. Belgrand[4] and Valles, in their studies on the floods of the river Seine, arrived at conclusions in substantial disagreement with those of Wex. Although Valles' work discounted the influence of vegetation on flood stages on the lower Seine, the studies of Demontzey[5] indicated a decisive influence on the part of vegetation in the control of torrential floods. He emphasized that it was the fixation of the soil that determined the control of torrents and pointed out that under some conditions grass might be as influential in this respect as forest. This perhaps was the first study showing the relation of erosion control to flood control.

In 1898, Imbeaux,[6] a French scientist, citing the experience of the United States with deforestation, said: "By an inconsiderate exploitation of great areas of primeval forests, the United States has seen their streams, which were formerly quiet and regular, transformed into violent torrents, almost dry all the year, but having sudden and formidable stages during rains."

In spite of these controversies during the past century, European nations have accepted as a matter of governmental policy good land management, including forestation where needed and the building of upstream structures, as an essential part of their water-control programs.

In the United States, however, national as well as local flood-control operations have been confined largely to the construction of levees, floodwalls, floodways, equalizing reservoirs and similar devices, in the larger streams. In other words, the principal effort to control floods has been made downstream, where the water concentrated in trunk channels is most difficult to control. Engineering structures of the type referred to are necessary to provide emergency protection from the extraordinary

[1] Surrell, A. "Étude sur les torrents des hautes-alpes." Paris. 1841.

[2] Becqueral, A. C. Forests and Their Climatic Influence, Smithsonian Inst. *Ann. Rept.*, 1869, Washington.

[3] Wex, Gustave. Diminution of the Water of Rivers and Streams, Smithsonian Inst. *Ann. Rept.*, 1875, Washington.

[4] Belgrand, E. De l'influence des forêts sur l'écoulement des eaux pluviales, *Ann.* soc. météorologique de France, Vol. 1, pp. 176–193, 1853; Vol. 2, pp. 81–87, 1854, Paris.

[5] Demontzey, Louis-Gabriel Prosper. "Étude sur les travaux de reboisement et de gazonnement des montagnes." Paris. 1878.

[6] Imbeaux, Édouard. Essai programme d'hydrologie. *Zeitsch. Gewasserkunde*, Vol. 1, No. 1, Dresden, 1898.

floods that occur from time to time. The application, upstream, of farm, range, and forest measures that retard runoff and increase absorption of rainfall can be made to supplement the protective effect of major engineering operations downstream.

Flood-control programs limited to downstream installations do not provide the benefits possible in a more comprehensive attack on the problem, such as undertakes to conserve rainfall in fields, pastures, and woodlands over the entire watershed. The purely engineering, downstream plan overlooks the possibilities of conserving potential flood waters for useful purposes, on the land; it fails to consider conservation of the land itself and the necessity of preventing damage to downstream flood-control works by the deposition of erosional debris.

## Historical Aspects

Garciliaso de la Vega, a member of De Soto's expedition, recorded in his writings a flood on the lower Mississippi in 1543. From that time until

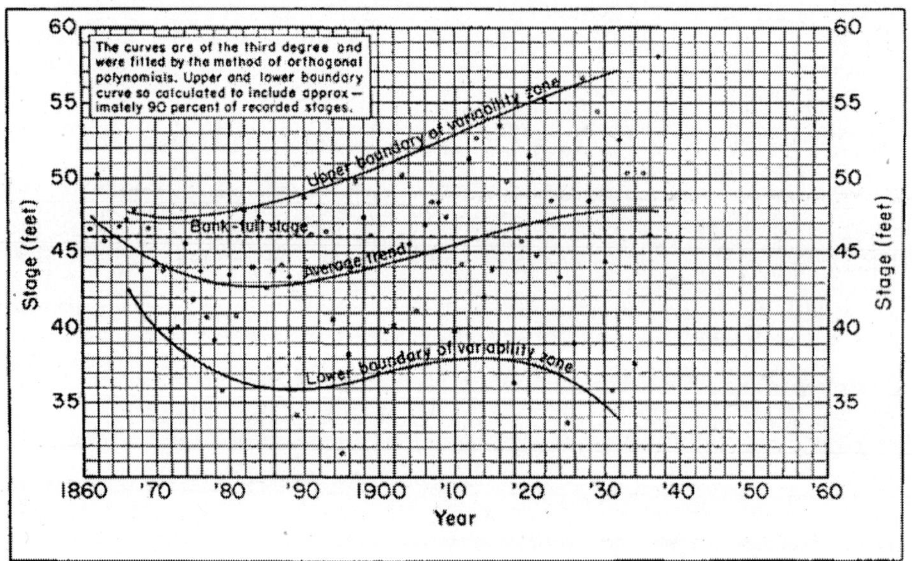


GRAPH 59.—Maximum annual stages, Mississippi River, Natchez, Mississippi, 1861–1937. Curves show average trend and variability of stages.

1799, flood records of the Mississippi are fragmentary. Subsequent to 1799, the records are sufficiently continuous to indicate the occurrence or nonoccurrence of floods but little more. It was not until after the organization of the Mississippi River Commission in 1879 that systematic river stage and discharge records became available for a series of representative points along the river. Unfortunately, by the time these measurements began, agricultural occupation of the land had extended to large areas within the major flood-producing tributaries of the Mississippi. Accordingly, quantitative data necessary for comparing flood flows of the pre-agricultural stage with those following large-scale farming operations are not available for the Mississippi and its tributaries.

Stage and discharge records of many streams indicate that the magnitude of floods is increasing. Taking, as an example, the record of floods at Natchez, Miss. (where levee construction has had a minimum effect on flood stages), it is observed (see Graph 59, showing maximum annual stages since 1861[1]) that the average of maximum yearly stages increased during the latter half of this period. Also, the variation between the maximum stages has widened from year to year. In other words, the

FIG. 252.—A major flood on the Canadian River, Texas, 1935. (*Photograph by Soil Conservation Service.*)

maximum stages have been extending gradually in the direction of both lower and higher levels. This indicates increased irregularity of streamflow. The high stages, of course, are the ones that cause most concern from the standpoint of flood damage.

In considering the flood problem, it is important to distinguish between *great* floods (Fig. 252) and *local* floods (Fig. 253). The former result from storms of wide areal extent, long duration, and generally low to moderate intensity rates. Such storms sometimes cover as much as 400,000 square miles and may encompass all or large parts of major drainage basins. The presence of snow on a watershed may intensify the hazards, for these storms often result in tremendous concentrations of water, with spectacular trunk-stream floods which cause great damage and frequent loss of life.

[1] Data from reports of the U. S. Army Corps of Engineers.

The local type of flood is caused by storms having relatively short duration yet centers of rainfall intensity that frequently reach a rate of 5 inches per hour. This type of storm seldom covers an area of more than 1,000 square miles.[1] It does not produce floods on major waterways but frequently causes serious damage locally. As a matter of fact, the sum total of flood damage resulting from such high-intensity local storms is probably as great over a period of years as the damage caused by major floods on the larger waterways. So far, however, the mitigation

Fig. 253.—Local flood, Stillwater Creek, Oklahoma, 1935. (*Photograph by Soil Conservation Service.*)

of this widespread local damage has not entered into public flood-control programs to any great extent.

Floods of this local type have become more frequent and violent since the country was settled. Residents of many communities subject to frequent local floods verify the fact. The geological and pedological characteristics of stream deposits provide further verification of this fact.

Overlying the old alluvial soils of numerous streams throughout the United States—the material laid down by timeless floods—is a different kind of alluvium. It consists of sediments spread out by flood waters since the beginning of our agriculture. Such deposits reveal unmistakable

[1] Thornthwaite, C. W. The Research Program of the Section of Climatic Research, *Soil Cons.*, April, 1937.

proof that, generally, they were spread over the flood plains by waters much more violent than those which laid down the older material beneath.

That the deposits of the preagricultural state were developed under moderate overflows is shown by their prevailing finer texture and more uniform composition. The line of separation between the two types of alluvium is so sharp that it is usually possible to photograph it without any difficulty. In many instances, the depth of the new material is greater than the entire depth of the older underlying deposit; and generally, it is not only coarser in texture but far more diverse with respect to textural composition and color characteristics through the profile. The new deposits have accumulated in many places within 25 to 75 years, whereas the old deposits required long periods—in some instances probably thousands of years—for deposition.

Soil scientists today are mapping a number of new alluvial soils entirely different in character from those of pioneer days, now entirely buried. We have the history of these soils. We know definitely that they have formed since the agricultural occupation of the country, and there is ample proof that these later deposits were laid down by more violently flowing waters than those of former times.[1]

## Factors Effecting Runoff

It is commonly asserted that during periods of prolonged rainfall, the soil becomes saturated and is unable to absorb more water, so that subsequent rainfall runs off directly into streams. The extent to which this occurs, however, depends on the depth and character of the soil, the character and condition of the plant cover, and the nature of the substrata. Prolonged rains normally charge the upper soil layers with a relatively high content of water, but this does not necessarily mean, as previously pointed out, that the soil is incapable of taking up additional water. As a matter of fact, soils of coarse texture or granular structure, especially those with porous subsoils, continue to absorb rainfall for undetermined periods (see Chap. VIII, Infiltration).

Saturation implies that water has filled all available pore space of the soil. But even when a soil is so wet at the surface as to be miry, some water continues to flow into the substrata by way of the natural pores and through the holes made by roots and insects. Such infiltrating water usually finds its way to ground-water level and rarely reappears in time to add volume to floods resulting from that particular rain.

[1] Bennett, H. H. Management and Use of Agricultural Lands including Farm Woods and Pastures; and Headwaters Control and Use, *Upstream Eng. Conference Proc.*, Washington, 1936.

More than any other single factor, vegetation facilitates the intake of surface water. Lowdermilk[1] has shown that the most important function of forest litter in relation to absorption of rainfall is the protection that it affords against the clogging of the natural water channels leading into the soil.

In the absence of a cover of growing vegetation or a ground cover of vegetal litter, rain striking bare ground picks up fine particles of soil to produce a muddy suspension. As this muddied water sinks into the ground, the fine particles filter out at or near the surface to form a film that tends to clog the smaller surface openings, or soil pores. This reduces runoff, even when the absorptive capacity of the subsurface material has scarcely been affected. This explains why observers sometimes find that during rainstorms most of the water is running off nonvegetated land, whereas the soil 2 or 3 inches below the surface is dry.

As a rule, the clogging of the surface openings of bare land takes place most rapidly on desiccated soil. Many soils on drying, especially where cultivated, assume a powdery condition which facilitates rapid development of a muddy suspension during rains.

The work of S. W. Phillips and I. T. Goddard at the Guthrie, Okla., soil and water conservation experiment station[2] gives some conception of the true significance of ground cover. In the spring of 1930, the runoff and erosion were measured from two adjacent areas occupied by hardwoods, on one of which the forest litter was burned and on the other left intact. During May, the runoff from the unburned plot was at the rate of 250 gallons per acre, whereas the runoff from the burned area of exactly the same size, soil, and slope was at the rate of 27,600 gallons per acre. The excess of runoff from the burned area over that from the unburned was approximately 90 tons per acre. The cover of leaf litter and forest mold was found to have a water-holding capacity of 16.7 tons per acre. Thus, the difference between 16.7 and 90, or 73.3 tons of water per acre represents the additional amount of rainfall that was absorbed by the soil with the ground cover of litter intact. In other words, the important effect of the ground cover, obviously, is to protect and maintain the infiltration rate of the soil, and the *blotter* effect of the cover is of less importance.

An example of the absorptive power of soil after it has reached a state of apparent saturation is provided in the runoff records of a bluegrass plot at the Zanesville, Ohio, soil erosion experiment station. Here, the total precipitation was 10.29 inches during the rainy period of Jan. 2 to 31, 1937, that produced the record flood on the Ohio River. Of

[1] Lowdermilk, W. C. The Scientific Aspects of Flood Control, Supplement to *Science*, Vol. 84, October, 1936.

[2] Bennett, H. H. Relation of Erosion to Vegetative Changes, *Sci. Monthly*, Vol. 35, November, 1932.

this total, only 25 per cent ran off the grassed area; but from the same kind of land in a near-by cornfield, with most of its area bare and subject to unrestrained erosion, 94.7 per cent of the precipitation was lost as immediate runoff (Graph 60). Stated differently, the grass-covered land absorbed 7.72 inches of the total 10.29 inches of rainfall, but the bare corn field absorbed only 0.54 inch.[1]

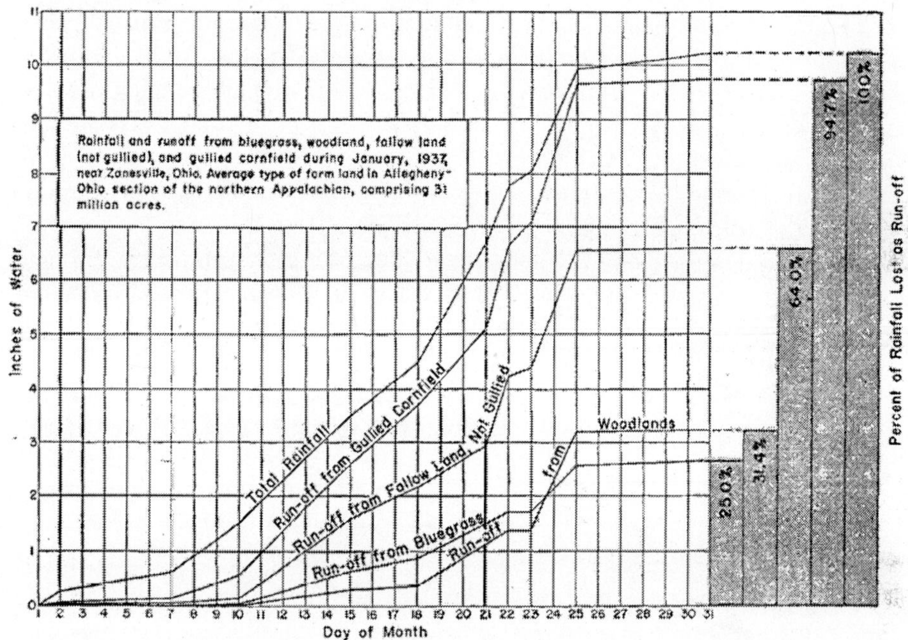


GRAPH 60.—Runoff measurements from average type of farm land, under various conditions of cover and erosion. Near Zanesville, Ohio. (*Soil Conservation Service.*)

Many thousands of similar measurements of runoff produced by individual rains falling on areas carefully selected as representative of many important types of agricultural land throughout the country show that the soil is capable of absorbing large amounts of rainfall when properly safeguarded from accelerated runoff. The same data also show that a

[1] Although these measurements cannot, of course, be construed as representing absolutely comparable rates of rainfall retention, as between bare fields and grassed or timbered areas, for the region that produced the 1937 Ohio River flood, they probably approximate closely the relative rates of rainfall retention over a large part of the affected area. The areas on which the measurements were made were selected, after a careful survey of the general region by engineers, soil specialists, and agronomists, as representing the closest approach to average upland conditions on the predominant agricultural soil of the region. The size of the areas from which the runoff was measured, as well as their location with respect to the Ohio River, enters into the equation; but it is significant that the areal proportion of that part of the Ohio watershed affected by the storm that is steeper (and subject to more rapid runoff) than the areas on which the measurements were made is fully as large as the proportion having milder gradients (and lower runoff rates).

good vegetative cover invariably increases absorption. Good cropping practices; contouring, terracing, water diversion, and other practical farm measures also insure increased infiltration.

These effects have been evident under all conditions of precipitation, temperature, slope, and soil type involved with the measurements, which have been carried on for a period of years under the best known techniques of research.

The tremendous water-storage capacity of the surface and subsurface layers of the land is not generally recognized. If this could be fully utilized, the reservoir capacity of the soil would be exceeded only by that of the oceans. To illustrate: It would take more than 20 inches of water to fill the pore space normally existing in the upper 3 feet of Marshall silt loam, an extensive (covering millions of acres) and important type of farm soil occurring in the valley of the Missouri River.

The storage capacity of most upland, including the underground soil layers, is in excess of the quantity of water falling in the heaviest rain recorded. Only under such situations as flat land with ground water near the surface or shallow soil underlain by impervious rock or clay would this generalization be incorrect. As a supplemental flood-control measure, it is of the utmost importance, therefore, that this great land reservoir for water be utilized by employing, wherever applicable, every practical and economical land-management and water-conservation method that will contribute to the retardation of runoff.

In addition to increased rates of runoff resulting from the stripping of vegetation from agricultural lands, further increases are caused by alteration of surface conditions by erosion: runoff accelerated by (1) washing off the absorptive topsoil down to relatively impervious subsoil, (2) development of millions of gullies that quickly concentrate and discharge rainfall, and (3) increase in the gradient of the land resulting from the gullying process.

Surveys indicate that not less than 200 million gullies have been formed in the United States since its agricultural occupation (Fig. 254). Each of these has its individual watershed; and in most instances, the gradient of each little watershed has been increased by the steep banks of the gullies and the deep sheet washing that so frequently takes place near them. The number of artificial gutters between crop rows that run downslope (Fig. 255) instead of across slopes on the contour (Fig. 256) reaches into the hundreds of millions. Drainage ditches and highways, paved streets and gutters, cattle trails and fire scars, all hasten rainfall to the nearest drainage. Then man-made and man-induced changes in the surface of the land contribute to the hurrying of runoff into streams, to overload them and to accentuate the flood hazard. Figure 257, showing the approximate boundaries of new little watersheds developed within

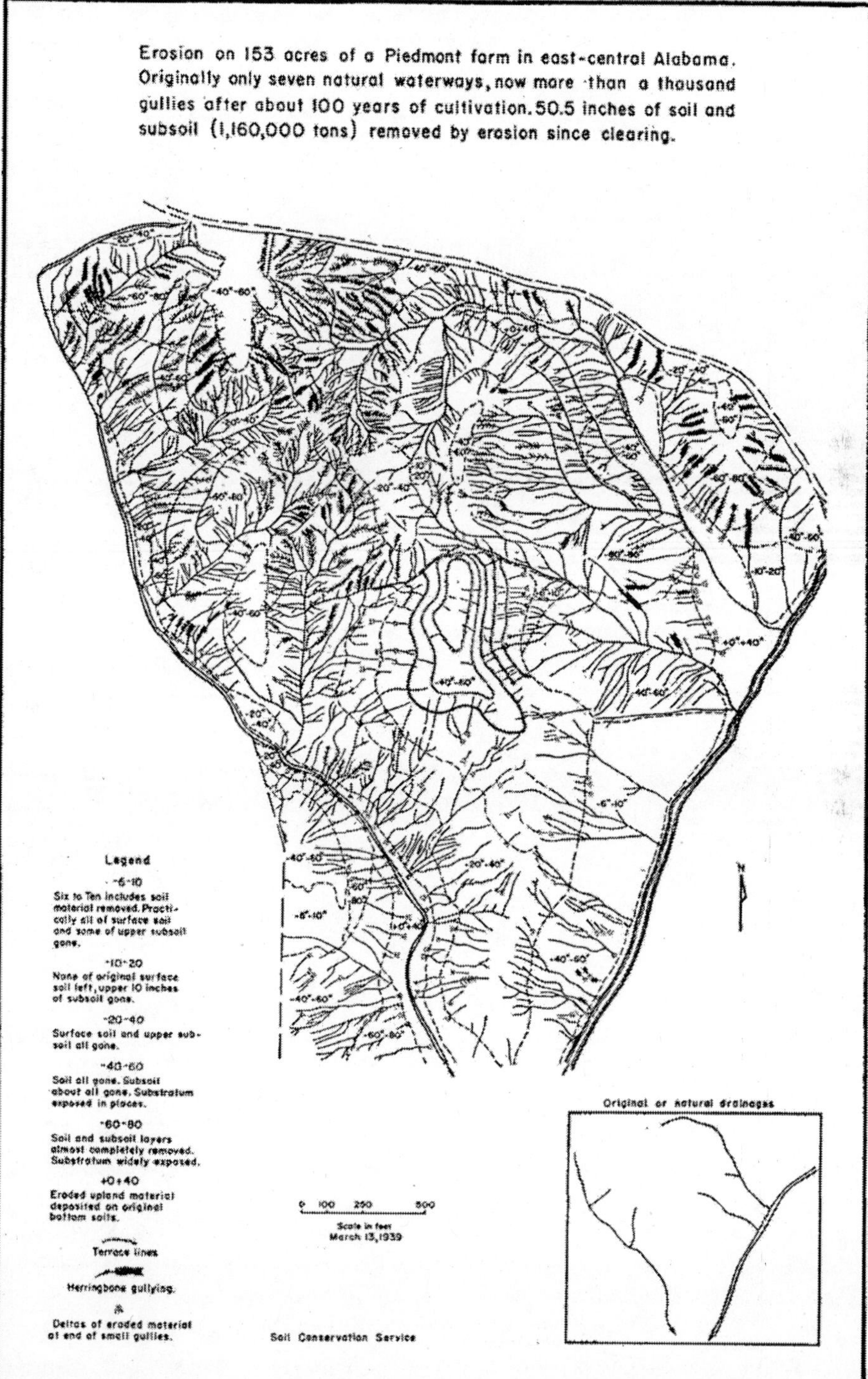


FIG. 254.—Advanced gullying on a Southern Piedmont farm. Near Dadeville, Ala. (*Soil Conservation Service.*)

FIG. 255.—Countless furrows running downhill instead of across slopes, on the contour, hasten flood water into the nearest stream. (*Photograph by Soil Conservation Service.*)

FIG. 256.—Rainfall stored in the reservoir of the soil by contour cultivation. Oklahoma. (*Photograph by Soil Conservation Service.*)

the original watershed by gullying, is illustrative of surface alterations that contribute to this acceleration of floods.

Relation of Soil Freezing to Runoff. Emphasis has been placed on the belief that frozen soil sheds all the rainfall. Under this conception, it has been assumed, apparently, that soils freeze sufficiently deep to prevent infiltration, regardless of cover, wherever the air temperature

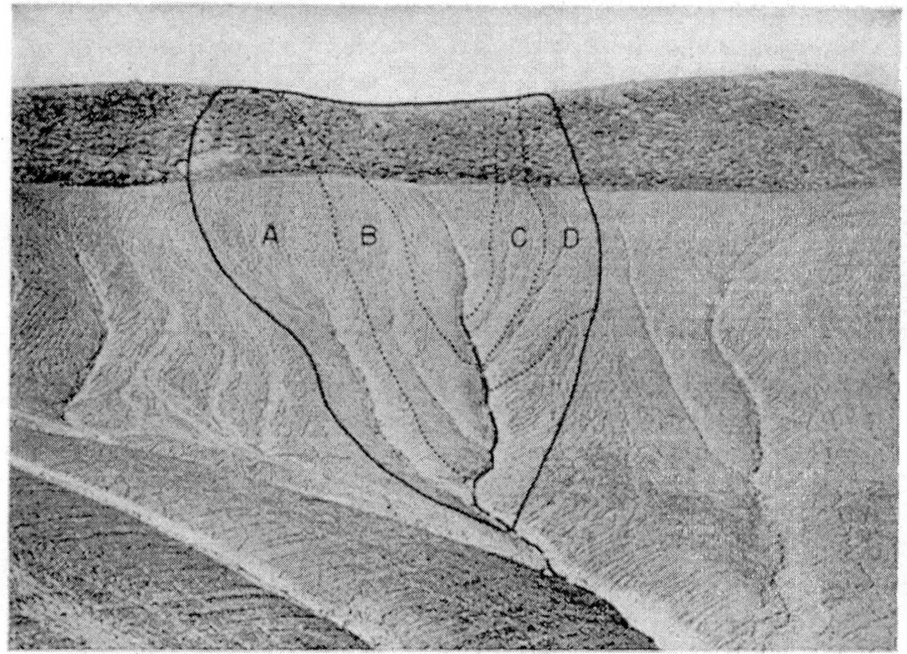


~ Original watershed.
~ New subordinate watersheds.
~ Original waterway.

Fig. 257.—This California slope suffered no visible erosion under the original cover of bush growth, and the small amount of runoff concentrated in the broad, shallow depressions indicated as the original waterway. Following cultivation, the enormously accelerated erosion cut out four major tributary gullies and also converted the original depressional drainageway into a gully. Each of the tributary gullies now has its own watershed as approximately shown by the areas *A*, *B*, *C*, and *D*. Not only has the volume of runoff greatly increased as the result of this accelerated erosion, but the rate of runoff has also greatly increased. (*Photograph by Soil Conservation Service.*)

remains below the freezing point for a considerable period. Although the subject has not been studied adequately, available information indicates that soil under a deep cover of snow or a dense growth of vegetation is not nearly so likely to freeze as exposed land. Absorption is considerably reduced, undoubtedly, when the soil is frozen, especially if it is wet at the time of freezing. Nevertheless, the extent to which soil actually is frozen must be determined carefully before sound conclusions can be reached as to the effect of frozen soil on acceleration of surface runoff in any particular case. Some soils, probably most of them, continue to take in surface water even when frozen. If the water content of the soil is low at time of freezing, the absorptive capacity of the soil may not be modified greatly by the frozen condition.

Snow cover alone has been found to reduce materially the depth to which a soil may freeze; and the depth and distribution of snow, in turn, are influenced by the character of the plant cover.

Bouyoucos[1] found that the soil temperature 3 inches below the surface did not fall much below freezing under a good cover of snow alone and that soil (not previously frozen) did not freeze at all under a cover of snow and vegetation, whereas subfreezing temperatures were almost continuous in comparable bare soil. After four years of observation, he concluded that in exceptionally cold weather soil protected by a cover of vegetation and a layer of snow would have at a depth of 3 inches below the soil surface a temperature lower than bare soil by 25°F.—in the Michigan locality studied. He found also that the soil temperature fluctuated less where the ground is covered with snow than where the soil is bare.

The interrelation of vegetal cover, soil freezing, and runoff is illustrated by results obtained at the soil and water conservation experiment station near Ithaca, N. Y., during the storm period of Mar. 1 through Mar. 19, 1936.[2] This was the storm that gave rise to the serious flood of that year on the Susquehanna River. An area of frozen potato ground, with a slope of 14 per cent, lost 88 per cent of the total precipitation as immediate runoff. A near-by forested area (27 per cent slope) where the soil was not frozen lost only 0.5 per cent of the total precipitation. An unfrozen, heavily grassed area, occupying a 20 per cent slope close by, lost as runoff only 0.2 per cent of the precipitation, whereas another near-by slope with insufficient grass to prevent freezing lost 88 per cent of the precipitation as runoff. It is significant that both of the frozen plots were still able to absorb almost an inch of water.

## Flood Costs

Floods exact a tremendous toll in human life, suffering, disruption of business, and destruction of property. It is extremely difficult to ascertain all the consequences of flood disasters and impossible to put an accurate monetary value on the losses. Property damages caused by floodwaters can be appraised; but loss of life, human suffering, and the effect of floods

[1] Bouyoucos, G. J. Soil Temperatures, Michigan Agr. Exper. Sta. *Bull.* 26, 1916.

[2] Bennett, H. H. Conservation Farming Practices and Flood Control, U. S. Dept. Agr. *Misc. Pub.* 253, 1936.

TABLE 41.—LOSSES[1] BY MAJOR FLOODS BETWEEN MAY, 1903 AND MARCH, 1938[2]

| Time | Location | Number of lives | Property damage |
|---|---|---|---|
| May–June, 1903 | Kansas, Lower Missouri, and Upper Mississippi | 100 | $ 40,000,000 |
| July, '03 | Missouri | ..... | 9,339,000 |
| | Upper Mississippi | ..... | 5,600,000 |
| | Red | ..... | 16,200,000 |
| July, '09 | Missouri, east of Kansas City | ..... | 5,500,000 |
| March, '12 | Lower Mississippi | ..... | 70,000,000 |
| March, '13 | Mississippi and tributaries | 467 | 162,227,293 |
| December, '13 | Texas rivers | 177 | 9,000,000 |
| June, '15 | Kansas | ..... | 5,950,000 |
| August, '16 | Rivers of the Carolinas | ..... | 21,700,000 |
| June, '21 | Arkansas in Colorado | 120 | 25,000,000 |
| September, '21 | Texas rivers | 215 | 19,000,000 |
| April to May, '22 | Upper Mississippi | ..... | 4,700,000 |
| | Ohio | ..... | 4,000,000 |
| | Lower Mississippi | ..... | 7,500,000 |
| October, '23 | Lower Arkansas, including Oklahoma | ..... | 15,000,000 |
| March, '24 | Potomac | ..... | 6,000,000 |
| May, '24 | Potomac | ..... | 1,000,000 |
| Spring, '27 | Mississippi Valley | 313 | 284,117,631 |
| August, '27 | Arkansas and tributaries | ..... | 3,440,000 |
| November, '27 | New England rivers | ..... | 45,577,700 |
| June, '28 | Ohio | ..... | 7,535,500 |
| August, '28 | South Atlantic rivers | ..... | 4,400,000 |
| September, '28 | South Atlantic rivers | ..... | 4,000,000 |
| March to June, '29 | East Gulf rivers | ..... | 8,000,000 |
| | Ohio | ..... | 16,500,000 |
| | Missouri | ..... | 2,000,000 |
| | Upper Mississippi | ..... | 3,600,000 |
| | Lower Mississippi | ..... | 10,000,000 |
| | Arkansas-White | ..... | 2,700,000 |
| | Rivers in Texas | ..... | 8,000,000 |
| July, '29 | Rivers in central Kansas | ..... | 4,000,000 |
| October, '29 | Rivers in Southeastern States | ..... | 9,000,000 |
| January, '30 | White-Wabash | ..... | 7,000,000 |
| May, '30 | Red and tributaries | ..... | 3,000,000 |
| September to October, '32 | Lower Rio Grande | ..... | 2,500,000 |
| March, '33 | Ohio | ..... | 2,000,000 |
| December, '33 | Columbia and tributaries | ..... | 10,000,000 |
| May, '35 | Rivers in eastern Colorado | ..... | 6,000,000 |
| May to June, '35 | Republican and Kansas | ..... | 18,000,000 |
| | Lower Missouri | ..... | 10,000,000 |
| July, '35 | Upper Susquehanna and tributaries | 52 | 26,000,000 |
| December, '35 | Houston, Texas, area | ..... | 2,500,000 |
| March to April, '36 | Rivers in eastern United States | 107 | 270,000,000[3] |
| July, '36 | Rivers in central Texas | ..... | 2,000,000 |
| September, '36 | Rivers in central and northern Texas | ..... | 5,000,000 |
| January to February, '37 | Mississippi and tributaries | 90 | 407,000,000[4] |
| March, '38 | Los Angeles area | 60 | 83,000,000 |
| | Total | 1,701 | $1,697,587,124 |

[1] Not including land damage.

[2] Data on all but the flood of March, 1938, from *Monthly Weather Rev.* The estimate for March, 1938, flood in Los Angeles area from news reports of damage. Loss of life given only where considerable.

[3] Incomplete.

[4] Tentative estimate.

on the morale of the people affected cannot be evaluated in terms of dollars and cents.

Flood losses, moreover, occur not only where the accumulated floodwaters flow but on the lands that shed the water that makes up flood flows. Much of the annual erosion that goes on throughout the country occurs during rainstorms of flood-producing magnitude. This loss of irreplaceable soil from the lands of a watershed is never listed among flood damages. As ordinarily defined, erosion of land shedding the water that goes to make up a flood is not a flood damage; but in truth, the loss of soil from uplands is as much the result of a flood as the damage to a city street or cellar by deposition of the same eroded material from an overwhelming flood. In many instance, the direct erosion losses on uplands actually exceed those inflicted on property downstream.

A tabulation of the flood losses caused by individual great floods along the major streams of the United States, for the 35-year period 1903–1938, is presented in Table 41. This list by no means includes the losses from all the floods that occurred over this period; it gives only the property damage caused by the more spectacular floods. These tabulated damages take no account of the enormous cost of erosion damage to contributing watersheds or the effects of the deposition of erosion debris over agricultural lands downstream or the silting of reservoirs, harbors and stream channels, drainage ditches, and highway and railway culverts.

An illustration of the public tendency to overlook land damage in appraising the cost of floods is presented in the instance of the severe floods resulting from the heavy rainfall that preceded and accompanied the disastrous hurricane that swept across New Jersey and New England in September, 1938. Property damage by wind and floods was reported in such terms as buildings destroyed or damaged, bridges washed out, trees blown down, and damages to highways and communications has been variously estimated at a quarter to a half billion dollars.

The additional cost to farm land by soil erosion was overlooked entirely. On the basis of its permanent effect, this erosion loss may equal the net loss from other types of damage.[1]

## Flood Control in the United States

First efforts toward flood control in the United States were made by local communities for the protection of local interests. Concern on the part of the Federal Government over the flood problem was first actively manifested in the enactment of the Swamp Acts of 1849 and 1850. These acts granted to the several states all unsold swamp and overflowed lands within their limits. The states were authorized to sell these lands and use the receipts for the prosecution of drainage, reclamation, and flood-

[1] Bennett, H. H. A Permanent Loss to New England, *Geog. Rev.*, April, 1939.

control projects. A large number of levees were constructed along the lower Mississippi as a result of these acts; but on the whole, they were largely ineffective for flood protection, probably owing to lack of coordination of plans and work among the states and flood-control districts.[1]

In his report on the control of floods of the lower Mississippi and Ohio Rivers, Charles Ellet, Jr., said 85 years ago:[2]

"Among the causes contributing to the increase of floods in the rivers of the United States in recent years, it is necessary to include an increased discharge of water due to the destruction of the timber and the cultivation of territory which was formerly untilled. It is reasonable to suppose that the removal of the forest growth, and the rank vegetation of the virgin soil will cause the slopes to shed the rain more rapidly into the valleys, and thus produce more sudden and more violent floods than were observed of old."

With reference to the consequences of cultivation and artificial drainage, Ellet stated:

"The immediate consequence of all this, that the water, which, in the original condition of the country, remained upon the surface of the prairies until a portion was evaporated, and a portion absorbed by the earth, to be subsequently given out slowly by the springs, is now hurried along hundreds of thousands of artificial drains into the great rivers which supply the Mississippi . . . for every fifth part of an inch by which the total drainage is increased in the period of sixty days of usual high water, there must be experienced an average increase of about 3 feet of the height of the floods, (in the lower Mississippi). . . . This result may assist the mind in forming some estimate of the consequences which are to spring from the extension of society over the yet un-peopled West, and the cultivation of the vast territory which is drained by the Missouri and its tributaries."

In advocating a system of reservoirs in the headwater streams, Ellet stated:

"It is proposed, in short, to construct new reservoirs to receive the increased drainage produced by the plow, and to compensate for those reservoirs which have been, and are about to be destroyed by the spade . . . "

In spite of the warnings of Ellet and others that levees should not be depended on as the sole means of flood control, operations by the Federal Government were limited largely to this type of control, until recently.

The late Senator Francis G. Newlands contended that flood control was a national problem demanding treatment of the entire river as a unit, from the very springs to the mouth, and that floods could be

[1] The Improvement of the Lower Mississippi River for Flood Control and Navigation. U. S. Waterways Exper. Sta., Vol. 1., Vicksburg, Miss., 1932.

[2] The Mississippi and Ohio Rivers. Report by Charles Ellet, Jr. Philadelphia. 1853.

diverted from destructive to beneficial use.[1] He stressed his conviction not only that the construction of levees and storage reservoirs was necessary in order to regulate and retard the flow of floodwaters but also that other effective measures were available, such as "scientific soil cultivation . . . to absorb instead of shedding moisture" and extension of areas devoted to forests to serve as natural reservoirs. In 1917, a bill incorporating the ideas of Senator Newlands and authorizing a Waterways Commission was passed.[2] It was a compromise and authorized no executive functions.

It did provide, however, for:

"Coordination and cooperation of the engineering, scientific, and constructive services . . . of the several governmental departments . . . that relate to study, development, or control of waterways and water resources and subjects related thereto, or to the development and regulation of interstate and foreign commerce, with a view to uniting such services in investigating, with respect to all watersheds of the United States, questions relating to the development, improvement, regulation, and control of navigation as a part of interstate and foreign commerce, including therein the related questions of irrigation, drainage, forestry, arid and swamp land reclamation, clarification of streams, regulation of flow, control of floods, utilization of water power, prevention of soil erosion and waste, storage, and conservation of water. . . . "

The law authorized the preparation of reports and plans for submission to Congress, but the intervention of the World War caused the Commission to be inoperative. In 1920, the Federal Power Commission was established, and the Waterways Commission was abolished.

Storage reservoirs did not become a factor in national flood-control planning until the passage of the Flood Control Act of May 5, 1928. This legislation directed the War Department to prepare plans for flood control on all tributary streams of the Mississippi River system subject to floods and to determine the effect of the establishment of a reservoir system in the drainage basins of the tributaries.

The Act provided for determination of:

"the benefits that would accrue to navigation and agriculture from the prevention of erosion and siltage entering the streams; a determination of the capacity of the soils of the district to receive and hold waters from such reservoirs . . . and as to their stabilizing effects on stream flow as a means of preventing erosion, siltage and improving navigation."[3]

[1] Darling, Arthur B. "Public papers of Francis G. Newlands." New York, 1932.

[2] Public Law No. 37, 65th Cong., 1st Sess.

[3] Public Law No. 391, 70th Cong., 1st Sess.

The Act provided further:

"That the President shall proceed to ascertain through the Secretary of Agriculture and such other agencies as he may deem proper, the extent to and manner in which the floods in the Mississippi Valley may be controlled by proper forestry practice."

As a result of the 1928 Act, the War Department, in a series of reports, projected a flood-control system involving some 2,000 projects, emphasizing the value of floodwater storage by reservoirs.

In 1933, a far-reaching land and water conservation program was launched. Conservation activities began to receive substantial public support, and the need for integrating conservation work into a comprehensive national policy covering flood control became obvious. The great floods of 1936 on the Connecticut, Susquehanna, Allegheny, Monongahela, and many other lesser streams in the Northeastern States, brought the urgent need for a new national flood-control policy to a climax. On June 22, 1936, President Roosevelt signed the Omnibus Flood Control Act, a remarkable measure, which definitely specified that in planning for flood control consideration be given to those lands from which flood waters originate, as well as to the accumulated downstream flood waters. Section 1 of this Act reads as follows:

"It is hereby recognized that destructive floods upon the rivers of the United States, upsetting orderly processes and causing loss of life and property, including the erosion of lands, and impairing and obstructing navigation, highways, railroads, and other channels of commerce between the States, constitute a menace to national welfare; that it is the sense of Congress that flood control on navigable waters or their tributaries is a proper activity of the Federal Government in cooperation with States, their political subdivisions, and localities thereof; that investigations and improvements on rivers and other water ways, including watersheds thereof, for flood-control purposes are in the interest of the general welfare; that the Federal Government should improve or participate in the improvement of navigable waters or their tributaries, including watersheds thereof, for flood-control purposes if the benefits to whomsoever they may accrue are in excess of the estimated costs, and if the lives and social security of people are otherwise adversely affected."

In Section 2 the following provision is made:

"That, hereafter, Federal investigations and improvements of rivers and other waterways for flood control and allied purposes shall be under the jurisdiction of and shall be prosecuted by the War Department under the direction of the Secretary of War and supervision of the Chief of Engineers, and Federal investigations of watersheds and measures for run-off and waterflow retardation and soil erosion prevention on watersheds shall be under the jurisdiction of and shall be prosecuted by the Department of Agriculture under the direction of the Secretary of Agriculture. . . . "

The original Act specifically names 222 watersheds where studies were to be carried out. In 1937, the Act was amended, adding 119 watersheds to the previous list and authorizing the Secretary of Agriculture to make investigations of all watersheds that previously had been authorized for study by the War Department.

The law provides for two phases of watershed investigations: preliminary examinations, and surveys. The purpose of the preliminary examination is to determine for individual watersheds whether or not significant flood problems exist and, if so, whether or not there appears to be some likelihood of minimizing the flood hazards by treatment of the land with adaptable watershed management practices. The preliminary examination provides a basis for determining the need for and priorities of subsequent or final surveys.

The final surveys of the Department of Agriculture cover in detail the contributing causes of floods, giving consideration to such pertinent factors as topography, geology, soil characteristics, frequency and character of flood-producing storms, normal precipitation, flood and normal flows of streams, land occupancy and economy, and flood and erosion damages. After a careful appraisal of these factors, a detailed plan for watershed protection is developed. This involves a plan that provides for the integration of the various action programs of the Department of Agriculture and the employment of additional upstream control measures that may be necessary to reduce or eliminate flood damages.

In formulating recommendations to Congress for the application of watershed-protection measures, existing or proposed flood-control work by the War Department or other agencies is carefully considered, in order that the proposed land-management practices may effectively supplement the work of other agencies.

In the development of a coordinated land-management program that will serve best the public interests of a watershed, the problem of apportioning costs immediately arises. In this connection, it must be recognized that any such program serves a great number of purposes and that the entire cost, with rare exceptions, cannot be charged against any one objective. Any well-planned watershed program will involve practices that conserve soil, conserve moisture, enhance recreational facilities, promote wildlife development, stimulate increased production of timber crops (where forest lands or farm woodlots are involved), decrease silting of reservoirs and shoaling of stream channels, and lessen flood flows and resultant damage.[1]

[1] A few observations taken from an article on the effects of soil and water conservation work in the Green Creek Watershed, Erath County, Texas, by T. C. Richardson, "Farm and Ranch," Oct. 15, 1938, are pertinent to the subject (see *Education*, Chap. XV, Part 2):

"Passing by increased crop yields, greater carrying capacity of pastures, the checking

In some instances, sufficient benefit will accrue, by the reduction of downstream damage, to justify the entire cost of watershed protection in the name of flood control. In other watersheds, flood damage may be insufficient to justify the expenditure of any flood-control funds for watershed protection. Even in such watersheds, however, there may

---

or complete stoppage of soil losses, and all the other well-known and oft-proven benefits derived from effective conservation practices . . . I went to the Green Creek project to check up on the results in flood control. In the last fourteen days of September 1936 nearly 11 inches of rain fell, which is more than three times the average for that month, and one-third of the average annual rainfall. Most of the conservation work planned had been completed at that time, and all of it on the cooperating farms has since been completed.

"T. J. Mitchell lives at the lower end of the district, about twelve miles from the head branches of the creek. . . . Here is Mr. Mitchell's written statement:

"'I know that previous to 1936, before soil and water conservation work was done on the upper watershed, rains such as we have had would have caused the creek to overflow and do serious damage to my fields. Since the work was done the creek has not overflowed a single time at my place. . . . At times during the summer months of little rainfall, before 1936, the creek had gone dry, but in 1936 and 1937 it flowed the year round.' This despite the fact that five out of twelve months had less than an inch of rainfall and only four months had more than 2 inches. . . .

"Up the creek, seven miles from the head, W. F. Hallmark has farmed for forty years. Here is his testimony: 'Before the installation of soil and water conservation farming on the Green Creek watershed above me, the creek often overflowed on my place and did considerable damage, but since the fall of 1935 overflow water has not damaged my farm in spite of heavy rains. The creek stays up longer, but never gets as high as it did before.'

"Still upstream, three miles from the headwaters, Grover Kiker adds this observation: 'Before the watershed area was set up and the conservation work done, the water ran only a few days after each large rain. Now the creek runs continuously, and a few hours after a rain, the water becomes clear.'

"Now let's go back down the creek to Zeb Tidwell, whose farm lies entirely outside the defined district where work has been done. Mr. Tidwell represents thousands of others . . . whose lands are so situated that they cannot protect themselves from waters that fall miles upstream, outside their own boundaries, may be outside their own counties. Here is what he says: 'I have lived on my place for the past forty years. I have seen Green Creek during high flood stage many times. Previous to the practice of terracing, pasture treatment, and other measures of erosion control on Green Creek watershed flood water was our most serious problem. Even though there have been excessively heavy rains during the past two years, flood waters have not so much as threatened our lands, where heretofore they had been badly damaged. The work done by our neighbors and the Soil Conservation Service on the watershed is worth thousands of dollars to us down here on the lower end.'

"The records show that Green Creek ran more days in 1936 than in 1935, with about 12 inches less rainfall for the year, and that poorly distributed, several months being without enough rain to run, and others abnormally heavy. Two different months since the work was well under way have had about 11 inches of rainfall each. What became of it, if it didn't go down the creek in floods? A lot of it, slowed up by terraces, contours, strip crops and grassed pastures went into the grounds, nobody knows how much. . . . I shall contend that the soil itself is the greatest potential reservoir for water storage, and more water can be stored in the soil body for a dollar expended in labor or cash than on the surface of the land."

be land-use and conservation problems that should be taken care of through governmental action. Each watershed presents a different problem, and careful study is necessary to determine the extent to which public participation in watershed protection is justified.[1]

[1] For further discussion of this problem, see Horton, Robert E., Surface Runoff Control, Headwaters Control and Use, *Upstream Eng. Conference Proc.*, pp. 16–41, Washington, 1936.

# Chapter XXX. Atlantic and Gulf Coastal Plain Region

The low, comparatively smooth country fronting on the middle and south Atlantic and on the Gulf constitutes the *Atlantic-Gulf Coastal Plain Problem Area* (Map 7). Made up mainly of water-deposited material, it extends from the vicinity of Laredo, Tex., to the tip of Long Island, New York, and comprises approximately 284 million acres, or about 15 per cent of the land area of continental United States. It covers, approximately, the southern and eastern third of Texas, all of Louisiana, Mississippi, and Florida, the southern and eastern half of Arkansas, narrow strips in southeastern Oklahoma and western Tennessee, the southern and western two-thirds of Alabama and Georgia, the eastern half of the Carolinas, the eastern third of Virginia, the eastern and southern two-thirds of Maryland, nearly all of Delaware, and the southern two-thirds of New Jersey.

From the point where the Rio Grande flows out of the Edwards Plateau, in the southwestern extension of the region, the interior boundary of the area follows the southerly or easterly boundaries of the following problem areas: (1) Edwards Plateau-Ft. Worth Prairie-Cross Timbers area, (2) Ozark Highlands, (3) Southern Appalachian region, and (4) Northern Appalachian and New England area. The width from Galveston, Tex., northward to the interior border in southern Oklahoma is about 350 miles; from Cape Sable, Fla., to the border in central Georgia is about 565 miles.

Characteristically, the region is plainlike in its prevalent topographic features, ranging from predominantly flat near the coast to rolling or low-hilly near the contact with the Ozark Highlands and the southwestern part of the Southern Appalachian region. In elevation above sea level, the interior boundary ranges from about 200 to 500 (occasionally 600) feet, with culminating altitudes in northwestern Alabama. Prevailing surface features are much smoother than those characterizing adjacent interior regions, yet many small and large isolated areas are decidedly rolling, and some even hilly. Most of the low coastal sections—the Flatwoods east of the Mississippi and the Gulf Coastal Prairie west of that stream—

embracing an area of about 58 million acres, is so nearly level that numerous areas are characterized by slow surface drainage. Portions of the interior, such as the minor flatwoods area of central Mississippi and numerous strips of first- and second-bottom alluvium along the widespread drainage system (alluvial land comprises about 10 to 12 per cent of the area of the entire region), are also of very mild slope (averaging

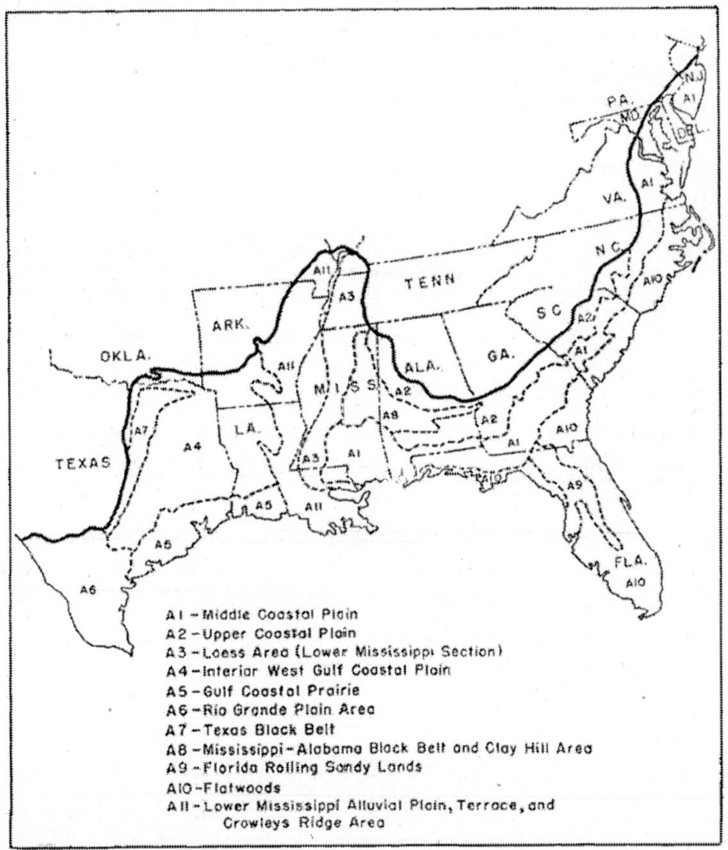


MAP 7.—Problem area of the Atlantic and Gulf Coastal Plain Region. (*Soil Conservation Service.*)

less than 1 per cent). The subdivision, designated the Lower Mississippi Alluvial Plain, Terrace, and Crowleys Ridge area and comprising about 34 million acres, is very largely flatland, with slopes averaging probably not much more than one-half of 1 per cent, except in the instances of the included rolling area of Crowleys Ridge in northeastern Arkansas and southeastern Missouri and occasional strips of silty and sandy soils sloping back from natural levees along the banks of both active and abandoned stream channels.

The soils of the region are highly diversified. East of the loessial area bordering the Mississippi, the predominant types, outside the Mississippi-Alabama Black Belt and Clay Hill area, are sandy loams, with friable sandy clay subsoils. Westward from the alluvial plain of the Mississippi, the more extensive soils are also of sandy loam character, as far as the Texas Black Belt; but here the proportion of land having heavy clay subsoil is considerably larger, especially in the westerly part. In many localities, particularly in the areas of red lands (such as the Orangeburg and Greenville soils) of South Carolina, Georgia, Alabama, and Mississippi, the deep substratum frequently consists of loose sandy geological material which, when incised by gullies, washes easily to develop, by caving and undercutting, deep straight-walled ravines which are difficult to control.

The rainfall is generally heavy, ranging from about 45 to 55 inches (in a few localities over 60 inches) east of the Texas-Arkansas line to 20 inches near the mouth of the Rio Grande. West of Arkansas and Louisiana, average precipitation declines steadily, with an increasing tendency toward longer periods between rains. From south to north, the climate ranges from subtropical to cold temperate.[1]

General farming is the dominant type of agriculture. South of Virginia, cotton and corn are the principal crops. Bright tobacco is extensively grown in southeastern Virginia and eastern North Carolina, in various parts of South Carolina and Georgia, and to some extent in northern Florida. Other types of tobacco, as cigar filler and wrappers, are produced in Maryland and western Florida. Oats are grown through most of the region, usually for hay. Sorghum is rather widely grown, usually in small fields, for syrup, and in many localities in mixture with cowpeas for feed. Peanuts are produced for market in Virginia and North Carolina and to a lesser extent in other localities; the crop is important in Georgia and Alabama as field forage for hogs. Sweet potatoes, truck crops, peaches, berries, and pecans are important specialized crops, produced in many localities throughout the region.

In the section north of the James River in Virginia, wheat, corn, clover, grass, potatoes, vegetables, berries, and fruit are the principal crops. Trucking is the principal farm industry on Long Island and over much of southern New Jersey and other parts of the northerly extension of the coastal plain. The eastern shore of Virginia and parts of New Jersey produce large quantities of potatoes. In Florida and in parts of the Rio Grande plain, citrus and other subtropical fruits, celery, and a large variety of winter vegetables are grown. Winter and early spring

[1] "Atlas of American Agriculture" (Physical Basis). Section on "Climate." U. S. Department of Agriculture.

vegetables and strawberries are also important in various localities near the Gulf and South Atlantic Coast.

Generally, the important matter of adaptability has been given too little consideration in the use of the highly diverse types of land. Over a long period, many kinds of soil, with numerous areas occupying steep, erodible slopes, have been cleared or broken out of native grass and clean-tilled continuously with little or no effort to protect them from erosion. In parts of the older Southeast, however, notably in the Carolinas, Georgia, Alabama, and Mississippi, many farmers have long attempted (more than 75 years in some localities) to protect their sloping lands from erosion with hillside ditching, terracing, and contour cultivation; and in frequent instances, these protective measures have given fair to excellent results. On the steeper, more erodible soils, however, they frequently have proved inadequate. Failures have resulted from such mistakes as cultivation of land too steep for the growing of the clean-tilled crops whether protected or not, improper construction or inadequate maintenance of terraces, and failure to hold cultivation strictly to the contour. But many of the failures have resulted from such causes as continuous cropping without rotation, depletion of the organic supply of the soil, misuse of crop residues or stubble in preparing land for the next crop, and disregard of seasonal cover crops.

From the purely physical aspect, about three-fourths of the coastal plain area is suitable for cultivation, most of it adaptable to the use of improved machinery. Originally, even more of the land belonged in this inadequately defined category. Gullying and severe sheet erosion have reduced the original area of easily cultivable land by about 15 per cent. Of the present area physically suitable to cultivation, it is estimated that upward of 20 per cent consists of inherently poor land and land now entirely unsuitable for cultivation because of swampiness (as tidal marsh and freshwater bog) or susceptibility to overflow (as along the seaward reaches of stream bottoms, like those of the Savannah, Altamaha, and Atchafalaya).

Some 4 million acres were comprised in national forest reservations, as of June 30, 1938. These forests are the Croatan in North Carolina; Francis Marion in South Carolina; Apalachicola, Choctawhatchee, Ocala, and Osceola in Florida; Talladega and Conecuh in Alabama; Bienville, De Soto, Holly Springs, and Homochitto in Mississippi; Kisatchie in Louisiana; and Angelina, Davy Crockett, Sam Houston, and Sabine in Texas. Additional tracts are included in other public land areas, such as state parks and state forests.

More than half, probably 60 per cent, of the area of the Atlantic-Gulf Coastal Plain region now suitable for cultivation without extraordinarily heavy expenditures for drainage, protection from overflow, and fertilization (around 170 million acres) is subject to erosion where used for clean-cultivated crops; and of this, fully 15 per cent is so susceptible to washing under cultivation, regardless of protective measures, that it should be retired at once to the permanent protection of grass or trees. Commercial fertilizers are in general use for nearly all crops throughout that part of the region lying east of the Texas Black Belt.

Thus, erosion is the most critical physical problem relating to land use over some 60 per cent of the Coastal Plain area suitable for cultivation under existing conditions. This estimate does not cover the entire area that would be subject to erosion in some degree if all the tillable land should be put into cultivation. Soil washing and some wind erosion

FIG. 258.—Sand blow on bare field, Coastal Plain of New Jersey. Feb. 28, 1938. (*Photograph by Soil Conservation Service.*)

take place on numerous cultivated areas even in the coastal flatwoods and along some of the gently sloping areas in the alluvial plains of the Mississippi and other streams, particularly on the more sloping soils of the natural levees near stream channels and abandoned channelways.

Although damage by wind erosion has been of small importance in comparison with the effects of water erosion, the problem is of considerable importance on loose sandy soils. Especially are difficulties experienced in trucking and strawberry districts. Young cotton, corn, small grain, and other clean-tilled crops are sometimes severely abraded by ground drift of sand. Truck growers in the vicinity of Charleston, S. C., sometimes strip crop their vegetable fields with rye and, occasionally, with permanent hedges. In parts of Florida, fencing has been used about vegetable fields for protection from soil drifting.

An example of wind erosion was the severe soil-blowing experienced in the vicinity of Moorestown, N. J., in February, 1938. Between Chews Landing and Blackwood, N. J., measurements of drift from an 8-acre field (Fig. 258) indicated that the soil loss resulting from this blow averaged 60 tons an acre (Fig. 259).

The results of soil conservation demonstrations on representative types of erodible land in various parts of the region have shown that it is both practical and economical to curb soil washing on slopes that are not excessively erodible (the critical areas that should be retired to the

Fig. 259.—Effect of soil blowing in Coastal Plain of New Jersey. Sixty tons of sand were blown from an 8-acre field nearby. Feb. 28, 1938. (*Photograph by Soil Conservation Service.*)

protection of permanent cover), through the application of control and corrective measures so coordinated on the land as to meet the needs and adaptabilities of the different parcels of land that make up farms and watersheds. Some of the practices employed and results obtained are summarized below under general discussions of the more important subdivisions of the region. Those subdivisions shown on the problem area map which at present have no erosion or flood problems of widespread seriousness are not discussed in detail either in this or in subsequent discussions dealing with the other problem areas.

## Middle Coastal Plain

Immediately above the South Atlantic and East Gulf Flatwoods is a broad area of predominantly undulating to gently rolling country of some 53 million acres, principally of sandy character, well drained, and to a large extent adapted to both general and specialized farming. The elevation of this Middle Coastal Plain sector rises gradually toward the

FIG. 260.—*Above, sheet erosion and leveling of cotton ridges on unprotected* field of porous sandy loam soil (Norfolk) of only 1 to 2 per cent slope by 4.2-inch rain, June, 1937. Five miles east of Wadesboro, N. C. Below, deposit of clean coarse sand, 1 to 15 inches deep, washed out of same field by same rain. (*Photographs by Soil Conservation Service.*)

interior from about 50 to 150 feet along the border of the Flatwoods to about 250 to 350 feet along the line of contact with the Upper Coastal Plain. Considerable land of the lower border zone is flat and of low erodibility; and many small areas occupying depressions require artificial drainage to fit them for cultivation.

For the most part, the soils are of sandy loam texture, with friable sandy clay subsoil. They are well drained and generally easy to till. Among the most important soil groups are the Norfolk, Orangeburg, Ruston, Red Bay, Tifton, Marlboro, Akron, and Greenville series. By reason of their open sandy character, these soils have high infiltration capacity, a characteristic that gives them considerable resistance, but not entire immunity (Fig. 260), to erosion. In various localities, particularly in the southwestern part, soils with heavier subsoils are of frequent occurrence, such as those of the Susquehanna and Cuthbert groups. These are much less absorptive of rainfall and more erodible on comparable slopes.

| *Type of Land Use* | | | *Acres* | | *Per Cent of Land in Farms* |
|---|---|---|---|---|---|
| Land in farms | | | 23,911,753 | | 100 |
| Cropland | | | 9,723,371 | | 40.7 |
| Clean-tilled | | 7,336,303 | | 30.7 | |
| Erosion-resistant | | 1,579,746 | | 6.6 | |
| Semi-erosion-resistant | | 807,322 | | 3.4 | |
| Permanent pasture | | | 4,494,395 | | 18.8 |
| Woodland | | | 6,636,862 | | 27.7 |
| Livestock: | | | | | |
| Horses | 151,329 | | | | |
| Mules | 352,925 | | | | |
| Cattle | 795,349 | | | | |
| Sheep | 301,974 | | | | |
| Total | 1,601,577 | | | | |

Although erosion is general through most of the area, it is moderate to severe on improperly managed land amounting to about 45 to 55 per cent of all cultivated areas.

The northerly extension of the region, the Long Island-New Jersey-Delaware-Chesapeake Bay Region, is gently to moderately rolling and has much well-drained sandy loam and loam soil of good agricultural value (as the Sassafras and Collington soils), along with considerable imperfectly drained land (as Elkton and Portsmouth soils). The steeper areas are subject to severe erosion under continuous clean tillage. Unfortunately, many areas of gently sloping land, as well as some of deep sandy soil, are suffering from moderate to severe sheet washing under continuous clean tillage, even in orchards.

The agriculture of the area is largely general farming in the south and mixed general farming, trucking, and fruit growing in the north.

Truck crops are also grown in a considerable number of localities in the southern sector, and peaches and pecans are important locally.

Clean-tillage farming predominates very largely over the production of grass, hay, small grains, and other erosion-resistant crops, except in localities such as Delaware and the eastern shore of Maryland, where the growing of wheat and grass and the use of crop rotations have long been important. Recently, the acreage of soil-improving legumes, notably lespedeza, has increased considerably in the more southerly part of the region, where cotton and corn have long been the principal field crops. Considerable lumber is produced, especially from second-growth pine, throughout much of the region south of New Jersey.

Acreages of the more important groups of crops, derived from the Farm Census of 1930, are as shown in the table on page 624.

### MUCKALEE CREEK SOIL CONSERVATION PROJECT

An example of the major land treatment and farm reorganization procedure involved in a representative soil and water conservation project—the Muckalee Creek project, near Americus, Ga.—is illustrative of the type of action necessary to control erosion in the southern part of this problem area.

The Muckalee Creek project is located along the junction of the Middle and Upper Coastal Plain of Georgia. It comprises 62,000 acres in Sumter, Schley, and Marion Counties and covers the entire watershed of Little Muckalee Creek and that of upper Big Muckalee.

Approximately 55 per cent of the land consists of sandy loam, 43 per cent sand, and 2 per cent clay loam, most of it well drained. For the most part, the area varies from undulating to moderately rolling. Although originally forested, all of the land at one time or other has been opened for farming. An erosion survey in 1936 disclosed the following conditions:

Of the land in farms, about 74 per cent had suffered moderate erosion, 21 per cent severe erosion, and 4 per cent had been ruined for cultivation. Many farmers were attempting to control erosion by terracing and contour tillage. Approximately half the cultivated area had been terraced. In many instances, the terraces had been improperly constructed or had unprotected outlets.

Cooperating with the farmers, the Soil Conservation Service has established a complete erosion-control program involving:

*A.* Detailed surveys of 205 farms, comprising 62,122 acres, and actual cooperative work on 149 farms, as of Mar. 31, 1938.

*B.* Preparation of control plans for each cooperating farm, covering the following principal methods and procedures:

1. Retirement of severely eroded cultivated land to permanent protection of grass, kudzu, sericea lespedeza, or trees.

2. Strip cropping.
3. Terracing of adaptable slopes.
4. Contour cultivation.
5. Stabilization of terrace outlets with grass or other suitable measures.
6. Contour furrowing.
7. Rotation of crops.
8. Development of vegetated channelways and meadow strips for safe disposal of water from terraces.
9. Use of seasonal cover crops.
10. Gully control.
11. Grazing control.
12. Soil improvement by increasing area devoted to legumes and other soil-building, dense vegetation.
13. Pasture improvement by contouring and reseeding.

The successful use of these conservation practices within the project area has had a marked influence on the spread of such practices to farms outside the area. A survey of the surrounding area indicates, for example, an increase in the use of winter legumes amounting to 8 per cent, of terracing 12 per cent, and of small grain 6 per cent.

Moreover, the *Middle Western Ocmulgee River Soil Conservation District,* comprising 1,195,114 acres in Wilcox, Dooley, Houston, Pulaski, Peach, Marion, and Crawford Counties, has been established as the result of the effectiveness of the Muckalee project

Following is a list of the major conservation accomplishments on the Muckalee Creek project, as of Mar. 31, 1938:

| *Area* | *Acres* |
|---|---|
| In cooperating farms | 36,867 |
| In cultivation at beginning of work | 19,297 |
| Retired from cultivation under the cooperative program | 4,629 |
| Approved rotations adopted on | 4,507 |
| Contour tillage adopted on | 5,177 |
| Adaptable terraces constructed on | 4,629 |
| Strip cropped | 195 |
| Pasture improvement work on | 1,537 |
| Trees planted on | 511 |
| | |
| Linear feet of water-diversion structures installed | 2,756 |
| Number of gully-control structures installed | 3,930 |
| Number of terrace outlets stabilized | 4,013 |

## Upper Coastal Plain

The Upper Coastal Plain, with an area of about 23 million acres, swings in a crescentlike belt around the Southern Appalachian region from central North Carolina to the locality of contact between the Alabama-Mississippi and Tennessee state lines. Rising inland from a general elevation of about 200 feet above sea level along the interior boundary of the Middle Coastal Plain to elevations ranging from about 350 to as high as about 600 feet in a few places in the northwestern extension, the area is of dominantly gently rolling, rolling, and hilly topography, good drainage, and sandy loam soils generally similar to those of the adjacent lower division. Agriculture is essentially the same as that of the southern part of the Middle Coastal Plain. In the northwestern section, however, the proportion of land in cultivation is not so great because of the more rolling surface and the larger extent of erosion-depleted land.

Contact with the Piedmont subdivision of the Southern Appalachian region is marked by a nearly continuous strip of deep sand (principally Norfolk sand), about 5 to 30 miles wide and extending from the vicinity of Sanford, N. C., nearly to Tuskegee, Ala. Originally, this Sandhills area was densely forested with mixed pine and hardwoods. Most of the original growth has been removed, but numerous tracts have reforested naturally with second-growth pine, together with considerable blackjack and other scrub oaks. Here, also, the proportion of land in cultivation is less than in the lower adjacent country, mainly because of the deep sandy nature of the soil.

Erosion is a problem over most of this upper sector of the Coastal Plain and a very difficult one on the steeper lands of the northwestern extension, as well as in belts bordering most of the streams. Isolated hilly areas, such as Pontotoc Ridge in Mississippi, have suffered seriously from washing. Fully three-fourths of the uplands is subject to impoverishing erosion under continuous clean tillage. The low organic matter content of the soils has considerable to do with the generally high susceptibility to sheet washing. Gullying is a serious problem locally, but sheet erosion is by far the more serious menace to soil stability.

Soil conservation demonstrations have shown that it is entirely practicable to control erosion adequately on most of the soils having slopes not greater than about 8 to 10 per cent—that is, where erosion already has not disposed of the topsoil. Surveys and experience indicate that approximately a fifth of the uplands should be retired from cultivation permanently, the percentage ranging from around 10 per cent in the eastern half to about 30 per cent in the western part.

## Mississippi-Alabama Black Belt and Clay Hill Area

An area of about 12 million acres extending in a crescent-shaped belt from the vicinity of Columbus, Ga., nearly to the Tennessee line in northeastern Mississippi is roughly designated the Mississippi-Alabama Black

Belt and Clay Hill area. This subdivision covers a wide diversity of soil and surface features, principally (1) the Alabama-Mississippi Black Belt of undulating to gently rolling topography and highly erodible heavy soils (Houston, Sumter, Oktibbeha groups); (2) clay hills of predominantly rolling to hilly topography and highly erodible soils with stiff clay subsoils (Susquehanna and Oktibbeha); (3) the interior flatwoods of Mississippi, with mainly grayish heavy soils overlying stiff clay (Susquehanna, Montrose, and Lufkin series); and (4) associated areas of rolling to hilly sandy soils (Ruston, Akron, Orangeburg, and Cuthbert).[1]

Except in the comparatively small interior flatwoods area, erosion is decidedly the most serious problem relating to land use. Large areas have been abandoned as the result of deep sheet washing and gullying. In the instance of the Houston clay of the Black Belt, most of the original 8 to 10 or 12 inches of dark-colored, rich topsoil has been removed by sheet erosion during years of continuous use for cotton and corn. As a result, the yellowish-brown clay subsoil has been exposed, and a new and less productive soil—Sumter clay—has appeared in place of the original Houston clay. In turn, over numerous areas, continuing erosion has washed off this exposed subsoil, to lay bare whitish, chalky limestone (Selma Chalk formation). Thus, an even poorer soil than the Sumter clay has made its appearance as the result of soil removal.[2] In a single county—Lowndes County, Alabama—56 thousand acres were mapped as Sumter clay in 1916,[3] all of which probably was Houston clay originally.

According to the soil survey of 1905, Montgomery County, Alabama,[4] had 86,000 acres of Houston clay; a resurvey of the county in 1926[5] showed only 10,000. This indicates rapid erosion over the intervening period of 21 years, as the result of which the dark topsoil was stripped from a large part of the Houston to expose brownish clay subsoil, such as is classed Sumter clay. The later survey shows some 60,000 acres of Sumter soil in Montgomery County.

[1] For details, see soil map in Bennett, H. H. "Soils and Agriculture of the Southern States." Also soil survey reports covering the region.

[2] In 1935, and again in 1937, the author crossed different parts of the Black Belt of Alabama, investigating erosion conditions. He had studied the region while making soil surveys some 30 years previously, when the area for the most part was a fine farming country, and inability to find an area of even approximately virgin soil was a revelation. In other words, within a third of a century, the surface of an entire geological formation had been violently altered by soil erosion—an alteration that very largely could have been prevented.

[3] Soil Survey, Lowndes County, Alabama. Field Operations, Bur. Soils, U. S. Dept. Agr., 1916.

[4] Soil Survey, Montgomery County, Alabama. Field Operations, Bur. Soils, U. S. Dept. Agr., 1905.

[5] Soil Survey, Montgomery County, Alabama, (Series 1926) No. 12, Bur. Chemistry and Soils, U. S. Dept. Agr., 1926.

Remnants of old furrows and farm buildings in the rolling to hilly section north of the Black Belt in Perry County, Alabama, show that all or the larger part of several hundred thousand acres of land, now largely covered with second-growth pine, was once cultivated. Much of it formerly consisted of Akron soil, principally a reddish sandy loam originally of excellent productivity. Cultivation of sloping areas, without adequate protection from erosion, brought about the depletion and abandonment of this large area.

Summarizing conditions in this problem area, it can be said that at least 80 per cent of the entire area (nearly all of it except part of the flatwoods and the bottoms and second bottoms of streams, such as the Alabama and Tombigbee rivers) is subject to rapid decline by erosion wherever cultivated without the application of a complete, properly coordinated and adapted erosion-control program. Although much land has been essentially ruined for further practical use for cultivated crops, a considerable aggregate area remains that can be protected and continued in use for farming if properly protected. Conservation demonstration work has shown this to be a fact. In the Black Belt, much of the eroded land is now used for hay and grazing. A good cover of grass will hold the soil and slowly improve it.

## Loess Area

The 18 million acres of predominantly undulating to rolling upland extending in a narrow to broad belt from southeastern Louisiana along the eastern edge of the Mississippi bottoms, across western Mississippi and Tennessee into southwestern Kentucky consists largely of brown silt loam of the Memphis, Lexington, and Grenada series.

The material from which the predominant soils of this region have been derived is of wind-blown origin, having accumulated over the area in past geological time. This loess deposit is deepest in the bluff areas nearest the Mississippi River, some exposures showing depths of more than 50 feet. Toward the east, the wind-blown material thins out gradually, the area finally merging with soils derived from water-lain deposits along a rather indefinite line of separation. Topographically, much of the area fronting the Mississippi bottoms has been severely affected by erosion and is locally so rough that cultivation is impracticable. Eastward, the relief becomes milder. More than 50 per cent of the eastern half of the loess area has an undulating to gently rolling surface, the more rolling lands occurring as narrow to wide strips bordering drainage lines.

The entire area is underlain by sedimentary deposits consisting mainly of interbedded layers of sandy material and clay. Erosion is active on most of the cultivated land, the exceptions being flat bottoms, second bottoms, and occasional rather small tracts of upland. Fully 75 per cent of the entire area is subject to erosion where cultivated without protection. In the deeper deposits of loess, the underlying unconsolidated silty strata (often containing more than 80 per cent silt) are so soft and unstable that where gullies cut into the material, they advance at a destructive rate and are very difficult to control. Many areas having a sublayer of tough sedimentary clay gully in a peculiar manner, particularly where the loess is shallow or only moderately deep. Here, the ravines frequently expand laterally as fast as and sometimes faster than the cutting proceeds in the direction of the upslope heads. In northwestern Mississippi are found many large and small areas of this type of highly destructive erosion, which can be controlled only with trees and grass or expensive engineering works.

It is estimated that about one-fifth of the cultivated upland has been stripped of the normal topsoil of 7 or 8 inches of brown silt loam, down to the reddish-yellow silty clay loam subsoil.

General farming constitutes the principal type of agriculture. Recently, the importance of livestock appears to be increasing steadily, along with an increase in the area devoted to grass and lespedeza, both of which grow splendidly on these silty lands. In a number of localities the production of strawberries and truck crops is important.

Soil conservation efforts in the area have definitely shown that erosion on cultivated land can be controlled in a very practical way, except on the steeper slopes, such as should be retired to permanent grass, lespedeza, kudzu, or trees. The results have been accompanied with marked reduction of flood flows in some of the streams where the entire watershed has been treated, as in the instance of the Bakers Creek project, at Port Gibson, Miss.

## Interior West-Gulf Coastal Plain

The Interior West-Gulf Coastal Plain comprises approximately 48 million acres, with large areas in (1) southern Arkansas, (2) that part of Louisiana west of the Mississippi alluvial plain and north of the coastal prairie, and (3) that part of east Texas north of the coastal prairie, extending southwesterly to the Guadalupe River. A comparatively narrow strip extends from the Arkansas line to the vicinity of Ardmore, in southeastern Oklahoma.

For the most part, the elevation above sea level ranges from about 100 feet along the boundary of the coastal prairie to around 400 feet along the boundary of the Texas Black Belt. An occasional hill or ridge has an elevation of 500 feet or a little more. In the main, the topography is moderately rolling to rolling. Toward the Gulf, the relief becomes milder, finally grading into prevailingly undulating country, with an occasional almost level area of flatwoods. Many of the streams in the more northerly part are bordered by strips of rolling to broken land, most of which is unsuitable for cultivation because of susceptibility to rapid erosion. In so far as surface configuration is concerned, by far the greater part can be cultivated, where slopes have not been riddled with gullies. With respect to adaptability, however, about one-fourth to one-third of the land is not suitable for cultivation at present, because of the effects of erosion in the past, susceptibility to erosion, poor soil, and unfavorable drainage conditions. In detail, surveys indicate that the slope gradient of about 30 per cent of the area is less than 2 per cent; approximately 60 per cent has slopes ranging from 2 to 8 per cent, and 10 per cent has slopes exceeding 8 per cent. Most of the land steeper than 8 per cent cannot be used safely for cultivated crops because of erodibility and should be retired to the protection of a permanent cover of vegetation. Also, a considerable area having impervious clay subsoil (as the shallow types of the Susquehanna soils) erodes so rapidly under use for the clean-tilled crops that it should go into permanent grass or trees, even on slopes of less than 8 per cent gradient.

The soils are prevailingly of sandy loam texture. Important areas, however, particularly along the southern and western border, are of heavier texture, ranging from very fine sandy loam to clay. The subsoils, with the exception of occasional hillocks or belts of sand (mostly Norfolk sand), are of clay texture, ranging from absorptive, friable sandy clay to stiff heavy clay of low infiltration capacity. The more sandy types usually are light in color, low in content of organic matter, and generally thoroughly leached of lime carbonate. Near the contact with the Texas Black Belt, considerable land is of darker color and higher content of organic matter and lime. The principal soil series are the Kirvin, Ruston, Cuthbert, Susquehanna, Nacogdoches, Bowie, Lufkin, Norfolk, and Caddo. Some borderline areas consist of dark limy clay of the Houston series and brown clay loam and heavy sandy loam of the Crockett series.

Precipitation ranges from around 45 inches along the Arkansas-Louisiana line to 30 inches in the southwestern extension.

About 35 per cent of the area is in cultivation; 7 per cent is idle; 13 per cent is in pasture; and 45 per cent in timber. Of the timbered area, a considerable proportion, particularly in Arkansas, Louisiana, and northeast Texas, was cultivated at one time, the present stand of trees consisting of second-growth pine. A reconnaissance erosion survey indicates that about 45 per cent, or some 7 million acres, of the area now in cultivation has suffered seriously from erosion, and about 6 per cent, or a million acres, has suffered severely. Of the second-growth pine areas, much more of the land has suffered severely from erosion. The low figure for severely eroded land is partly accounted for by the comparatively large area in the southerly border zone that has never been plowed,

**Physical condition of land, and land use before and after application of erosion-control program**

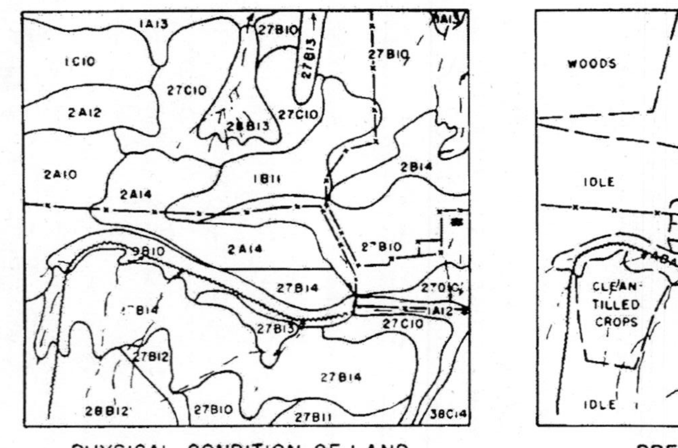


## PHYSICAL CONDITION OF LAND

**EROSION**

1-No apparent erosion
2-Slight sheet erosion
27-Slight sheet erosion, occasional shallow gullies
28-Slight sheet erosion, severe gullying
38-Moderate sheet erosion, severe gullying
9-Ruined by gullying

**SOILS**

10-Kirvin fine sandy loam
11-Kirvin stony loam
12-Nacogdoches gravelly fine sandy loam
13-Bowie fine sandy loam
14-Ruston fine sandy loam
15-Hannahatchie fine sandy loam

**SLOPE**

A-0 to 3%
B-3 to 8%
C-8 to 10%
D-Over 10%

Shallow gullies
Deep gullies
Fences
Field boundaries

EXAMPLE

2A14 = Slight sheet erosion-0 to 3% slope-Ruston fine sandy loam

## PRECEDING LAND USE

**LAND USE BEFORE EROSION-CONTROL TREATMENT**

Cultivated land 62.6 Acres.
Sheet erosion serious on 79% of farm, 1/4 of all top-soil removed.
Gullies serious, 12% of land ruined for cultivation.
No definite crop rotation.
Extremely poor pasture.
High percentage tilled crops.
Eighteen irregularly shaped fields.

Check dams
Terrace outlet
Farmstead
Terrace outlet onto woods
Terraces
Strips close-growing annuals
Hay and legume strips
Abandoned land revegetated

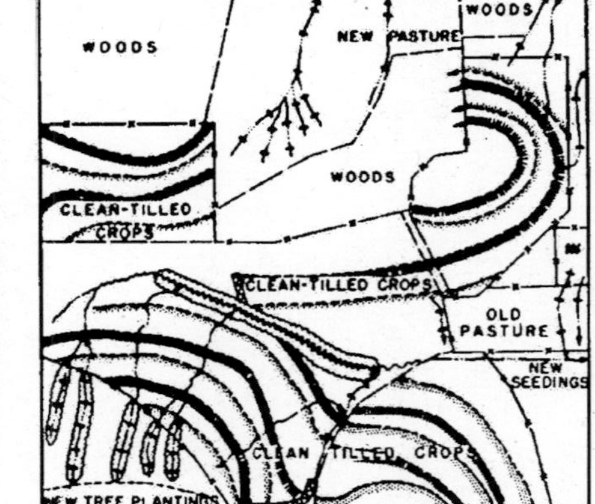


## REORGANIZED LAND USE

**LAND USE AFTER EROSION-CONTROL TREATMENT**

Retired from cultivation to permanent vegetative cover, 10 Acres.
Sheet erosion controlled on entire farm by strip cropping, seeding, terracing, rotations, and contour farming.
Gullies controlled on 64 Acres with soil-saving dams and vegetation.
Rotations established on contour basis.
All pastures reseeded and contour furrowed.
Reduction of acreage of erosion-producing crops, (cotton reduced 40 percent).
Nine fields rearranged for efficient handling and cropping (fences relocated).

FIG. 261.—Steps in the application of a completely coordinated erosion-control and land-use readjustment program on an eroding farm. Representative farm of 100 acres, Smith County, Texas (Lindale Project). (*Soil Conservation Service.*)

together with a rather large area of flat land in this section. Over a considerable proportion of the more rolling part of this problem area, the percentage of severely eroded land, as well as that subject to moderate erosion and to slight erosion, is much higher. The most erodible of the extensive farm soils is the Kirvin sandy loam (or fine sandy loam). On this soil, which predominates in northeastern Texas and adjacent portions of Louisiana and Arkansas, it is generally unsafe to cultivate slopes steeper than about 8 per cent. According to measurements made on Kirvin fine sandy loam at the soil and water conservation experiment station near Tyler, Tex., the entire depth of topsoil would be washed from an 8.75 per cent slope under continuous production of cotton in about 46 years. Loss of rainfall as immediate runoff accompanying such soil wastage amounts to nearly a fourth of the entire precipitation. Dense cover and crop rotations, however, greatly reduce the rates of erosion and runoff (see Graph 12, Chap. V, Part 1). The Nacogdoches fine sandy loam is very much more resistant to erosion than the Kirvin. The time required to remove the topsoil from Nacogdoches, having a slope of 10 per cent, would be about 160 years, according to measurements made at the same station.

Cotton is the most important crop, but considerable corn and some oats are grown. Livestock and dairying are becoming more important, accompanied by an increase in the area devoted to grass.

Soil and water conservation demonstrations have been highly successful in various parts of the area, covering the principal types of land and farming. Results on the project at Lindale, in Smith County, Texas, have thus far proved about 85 to 90 per cent effective with respect to control of erosion. Here, the land is as erodible as any of the important farming types in the entire problem area. Figure 261 shows the conditions of erosion and other physical characteristics, and previous land use, on a representative farm in the Lindale demonstration project, together with the principal measures employed in defending the land against further erosion.

## Texas Black Belt

The wedge-shaped belt of black prairie land extending from near the Red River in northeast Texas southwesterly to the vicinity of San Antonio comprises approximately 8 million acres. Outlying areas to the east and north of the main belt amount to perhaps 2 million acres. The elevation ranges from about 500 to 700 feet, and the surface is prevailingly undulating to gently rolling. Some isolated areas are nearly level, and there is considerable flat bottom- and second-bottom land along the stream courses. As indicated by conservation surveys, about two-thirds of the area is characterized by slopes not exceeding 2 or 3 per cent; more than one-fourth (30 per cent) ranges from 2 to 6 per cent in gradient; and about 5 per cent ranges up to 8 per cent or steeper. Something over 75 per cent of the area was under cultivation in 1935, mostly to cotton. Considerable corn, oats, and, in the northern part, wheat are grown. About 15 per cent of the Black Belt is in pasture, and the remainder consists of idle or abandoned land. Twenty-five per cent of the area has lost half of its topsoil (the range of loss being from one-fourth to all of the topsoil); 15 per cent either has lost all of the surface soil or has been ruined for further cultivation. Much of the remainder is suffering in milder degrees from sheet washing. About 5 per cent of the area is too steep for cultivation under any kind of treatment; and a considerable area, probably something over 5 per cent, has been so severely damaged that cultivation is practically out of the question.

At the erosion experiment station near Temple, Tex., the principal soil of the area, Houston black clay, has been losing topsoil by sheet washing at the rate of nearly 24 tons an acre annually from slopes of 4 per cent under continuous clean cultivation, along with a loss of 14 per cent of the total precipitation. This means that 7 inches of topsoil, which closely approximates the average depth on 4 per cent slopes, would be completely stripped off under continuous clean tillage in approximately 34 years. On slopes of 2 per cent, it would take about 75 years to lose 7 inches of soil under comparable conditions of soil, rainfall, and clean tillage. Grass and crop rotations have been highly effective in controlling both soil and water losses (see Graph 19, Chap. V, Part 1).

The prevalent soil of the region—Houston clay and black clay—is rich in lime and organic matter and, where not severely eroded, is highly productive. The rather extensive associated Crockett clay loam is also a productive soil, but it is not so rich in lime and is more rolling and more erodible.

Demonstrations carried on by the Soil Conservation Service in various parts of the region have shown that the greater part of the Houston is amenable to a high degree of effectiveness with respect to the use of practical erosion-control measures. Fairly simple measures, such as strip cropping, terracing, and contour cultivation, have proved very effective. Strip cropping alone has given especially good results on gentler slopes.

An illustration of the types of measures used, together with some of the accomplishments, is summarized below for a typical project—the one at Garland, Tex.:

### DUCK CREEK SOIL CONSERVATION PROJECT

The 24,806-acre watershed, comprising the Duck Creek soil and water conservation project at Garland, Tex., is representative of a large portion of the northern section of the Texas Black Belt.

The predominant soils are Houston black clay and Houston clay (black and ashy gray, respectively), both of which in the virgin condition are deep, rich in organic matter and lime, and highly productive. They are underlain by lighter colored calcareous clay, which, in turn, normally overlies cream-colored marl or chalk (parent material) at depths of about 3 to 4 or 5 feet. The deeper soil occurs on the gentler slopes. On drying, the soil shrinks, cracks deeply, and assumes a favorable granular-fragmental structure; with rain, it swells and slicks over, causing high runoff and intensive sheet washing, even on moderate slopes. The northern part of the watershed is generally undulating, with long, uniform slopes; the southern part is moderately rolling with short, steep slopes. Erosion progresses more rapidly on the steeper land, of course; but most of the upland, even that sloping no more than 1 or 2 feet in a hundred, is subject to soil decline by sheet washing.

The following erosion conditions were found by a survey of the project area in 1935:

| | *Acres* |
|---|---|
| Little or no erosion on flat stream bottoms, native pasture, and meadows | 2,000 |
| Slight erosion: up to 25 per cent of topsoil removed | 12,400 |
| Moderate erosion: 25 to 75 per cent topsoil removed | 6,900 |
| Severe erosion: more than 75 per cent topsoil removed, together with some subsoil | 2,000 |
| Land essentially ruined by erosion | 1,500 |

Originally grass-covered, most of the area was used for raising cattle. With the advent of railroads and barbed wire, cotton soon became the principal product of the farm. Continuous clean cultivation over a long period has induced excessive erosion. Frequent patches of the brownish subsoil and even of the light-colored parent material are exposed on many of the steeper slopes and locally on areas of mild gradient.

In the demonstration of erosion-control practices carried on here, 132 farmers operating 13,276 acres are cooperating with the Service. Complete treatment for each farm, planned on the basis of individual surveys of all the farms, involve such conservation measures as crop rotations; strip cropping; contour cultivation; terracing; contour furrowing and reseeding of pastures; use of winter cover crops; development of stable waterways for disposal of runoff from terraces; gully control; protection of terrace outlets; fencing to encourage better use of the land; construction of stock-water ponds; and retirement of severely eroded land to the permanent protection of grass, trees, and shrubs.

In addition to the 132 cooperating farmers, 11 noncooperating farmers owning 1,793 acres in the Duck Creek watershed, are carrying on a complete program with only technical assistance from the Soil Conservation Service. The general attitude of the community toward the soil and water conservation program has been good. Local civic organizations, the Garland High School, the Parent Teachers Association, and other organizations and institutions are cooperating with the Extension Service, Vocational Agricultural teachers, and the Duck Creek Soil and Water Conservation Association in sponsoring field days and encouraging visitors to inspect the work on the land.

Some of the major accomplishments of the demonstration project, as of June 30, 1938, are:

| | *Acres* |
|---|---|
| Retired from cultivation to permanent cover | 1,143 |
| Approved rotations introduced on | 7,664 |
| Contour tillage introduced on | 7,070 |
| Terraced | 2,283 |
| Strip cropped | 6,943 |
| Pasture improvement | 3,314 |
| Planted to trees | 17 |

Strip cropping has proved one of the most economical methods for controlling erosion on gently sloping land. Alternate strips of small grain, sorghum, Sudan, alfalfa, Hubam clover, and grasses have fitted in satisfactorily with the clean-tilled crops. Applied on a rotation basis, and in conjunction with contouring, erosion has been controlled to a degree of more than 90 per cent effectiveness in many fields, along with marked reduction of runoff. It is believed that as the humus supply of the soil increases under proper rotations, the general effectiveness of these three complementary measures will provide even greater protection and further diminish runoff.

Terracing, supplemented by these measures, has given good protection in fields of steeper slopes.

On Jan. 20 to 24, 1938, 5.84 inches of rain, part of it falling at an intensity of 1.55 inches in 30 minutes, occasioned only slight erosion on the steeper protected lands and no visible erosion in the mildly sloping fields.

Quantitative comparison with the old system of disregarding the contour is found in the results obtained on Houston soil at the soil and water conservation experiment station at Temple, Tex. Here, on land having the same slope as many of the fields in the Duck Creek project area, the soil loss from a 3-inch rain on Apr. 28, 1936, was at the rate of 17 tons an acre where the rows ran down the slope, as against only 0.63 ton where a comparable field was strip cropped on the contour.

Bermuda and buffalo grass have proved very effective in controlling washing at the outlets of terraces and in waterways established to carry runoff from upper to lower slopes. Little bluestem has given good results in pasture improvement work and in strip cropping (Fig. 262). Hubam

FIG. 262.—Little bluestem, a native wild grass, worked into strip-cropping program of central Texas. This grass also being used extensively in Black Belt of Texas for rehabilitation of pasture. (*Photograph by Soil Conservation Service.*)

FIG. 263.—Hubam clover, by reason of its rapid growth and heavy production of seed, which mature before or about the time the effects of cotton root rot begin to develop, is fitting well into rotations for the Black Belt of Texas. (*Photograph by Soil Conservation Service.*)

clover, by reason of its rapid growth and heavy production of seed, which mature before or about the time the effects of cotton root rot (which affects legumes) begin to develop, is fitting well into Black Belt rotations (Fig. 263).

## Rio Grande Plain Area

The Rio Grande Plain area, including the flat alluvial lands bordering the Rio Grande and other streams, comprises about 22 million acres of prevailingly undulating to gently rolling country lying to the south of the Edwards Plateau and to the west of the Guadalupe River. This subhumid area, with an average rainfall of 20 to 25 inches, has long been an important cattle country. Recently, however, there has been a gradual increase in farm operations, with intensive truck and fruit growing on the smooth lands of the Rio Grande Valley proper and to some extent in other localities. Locally, the surface is rolling; along some of the streams, it is decidedly broken. The alluvial lands are usually rich in lime and highly productive. Uplands are also productive, except for some of the shallower soils and loose sands.

The principal soils outside the areas of "red lands" are the gray, limy types of the Brennan group; the dark-colored, limy soils of the Maverick; and the gray, loose sands of the Nueces. In the red lands sections, the more reddish soils are those of the Duvall series; the darker lands with reddish subsoils belong in the Goliad group. Originally, most of the area was covered with a dense bush growth (chaparral), with a considerable admixture of prickly pear and, on the alluvial lands, small trees.

Erosion is becoming a problem of increasing seriousness on the steeper cultivated areas. Much gently sloping land, also, is beginning to wash. Control measures should be established as rapidly as possible. Contouring, strip cropping, terracing, rotations, and revegetation are the principal adaptable measures.

## Areas of Relatively Slight Erosion

Approximately 100 million acres of the 284 million in the Atlantic and Gulf Coastal Plain are comprised in four of the areal subdivisions where erosion is much less serious than in the subdivisions discussed above, chiefly because of mild slope and relatively small extent of cultivated land.

### Flatwoods

The 45 million acres of low plains country bordering the Atlantic and Gulf from southeastern Virginia to southwestern Alabama is prevailingly so flat that it is generally known as the *Flatwoods*. Millions of acres consist of tidal marsh and swampland, reclamation of which is now regarded as impracticable; other millions are so flat that artificial drainage would be required for use other than grazing or forestry. It is estimated that approximately a third of this subdivision could be brought into cultivation without expensive drainage operations. A considerable part of this, however, consists of such inherently poor soil that it is not likely soon to be used to any large extent except for forestry purposes.

Of the area reasonably suited to farming without heavy drainage costs—probably about 20 million acres—two-thirds is too nearly level to have any serious erosion problem, except for occasional blowing on the more sandy lands. The remainder is susceptible to sheet washing, such as can be controlled largely by strip cropping or terracing. Occasional belts along streamways, with slopes of 4 or 5 per cent or more, are too erodible for safe cultivation and should be used for grass or trees.

This area includes some of the best forest land in the United States. Those tracts which have no hardpan or claypan near the surface and are not too swampy or boggy are splendidly adapted to fast-growing pines, particularly slash pine. A large total area is now in forest, with an important output of turpentine, resin, lumber, and pulpwood. The opportunity for successful large-scale forestry operations is particularly promising on soils like the Norfolk, Ruston, Caddo, Kalmia, and Ruston fine sandy loams.

A considerable area of swamp has been drained for agricultural use, such as portions of the Florida Everglades and the swamps of eastern North Carolina. Much of this drained land consists of peat, a humus soil which burns readily when dry. Fires are very destructive during protracted dry seasons, being difficult to overcome, especially where the peat is deep. Numerous tracts of sandy soil, together with some heavy land, have been drained by individual farmers for general farming, trucking, mixed farming, and the production of specialty crops, such as strawberries, potatoes, celery, and citrus. Numerous knolls, ridges, and other relatively high areas are also used for the same purposes, without necessity for drainage. Cattle raising is important in parts of Florida.

## Gulf Coastal Prairie

The flat country bordering the west Gulf Coast, from south-central Louisiana to the Guadalupe River in southern Texas, has about the same topographic characteristics as the Flatwoods, averaging a little more nearly level. The eastern half is largely imperfectly drained. Here, many tracts of pine are interspersed with the grasslands. Strips of hardwood or mixed hardwoods and pine border most of the major drainages in the sector east of the Brazos River; westward, the proportion of natural prairie increases. The soils average much heavier than those of the area's eastern counterpart—the Flatwoods. Gray, brown, and black clays, silt loams, silty clay loams, and very fine sandy loams predominate.

West of the Brazos River, drainage conditions are much better, and a large area is devoted to general farming, with cotton as the principal crop. Cattle raising has long been important throughout the coastal prairies. Rice is locally important in the Louisiana and southeast Texas sections. Some sugar cane is also grown in the eastern part of the Louisiana section.

The Gulf Coastal Prairie includes about 12 million acres. Little serious erosion is found except on the more rolling areas near streams and the higher interior part. Strip cropping and contouring generally will give adequate control.

## Lower Mississippi Alluvial Plain, Terrace, and Crowleys Ridge Area

This large area of some 34 million acres includes the alluvial lands bordering the lower Mississippi and its tributaries, the flat terraces bordering these first-bottom lands, and the rolling ridge in northeastern Arkansas and adjacent Missouri known as *Crowleys Ridge.* Although much of the land is low and swampy, such as the great swamp along the Atchafalaya River in Louisiana, a large total area is used for cotton, corn, sugar cane, and other crops. Sugar-cane production is limited largely to the highly productive lands of the first bottoms—the lower Mississippi "Delta" area. Rice is an important crop on the heavy second-bottom or terrace lands near Lonoke, Ark., and on similar lands in southern Louisiana. Cotton and corn are grown on all grades of soil, ranging from heavy "buckshot" clay (Sharkey clay) in the Delta country to the Memphis and Grenada silt loam of Crowleys Ridge.

Extensive drainage operations have brought much former swampland into cultivation. The great levee system along the Mississippi and its tributaries has made farming far less hazardous than it could possibly be without protection from overflow.

In the flat Delta and Terrace sections, erosion is of no great importance except on some of the more sloping lands bordering stream channels. This is mostly sheet washing and can be controlled by strip cropping, contour cultivation, and crop rotations. During the great inundations that sometimes sweep over large areas of the lower alluvial plain, considerable local damage is done both by erosion and by deposition of sand. Grooves and potholes are gouged out in places, and sand is laid down as "sand blows" and extensive sheets near breaks in the levee and about trees and other obstacles to the current.

The rolling lands of Crowleys Ridge have suffered severely from washing, both sheet erosion and gullying. Conditions here are much the same as in the loess area immediately east of the Mississippi. Terracing, contouring, strip cropping, crop rotations, and revegetation of severely eroded areas and steep lands are the principal remedies.

### Florida Rolling Sandy Lands

This roughly defined area occupies some 7 or 8 million acres in the north-central part of the Florida peninsula. For the most part, the soil consists of deep sand (Norfolk sand) of undulating to rolling topography. Numerous areas of flatwoods are included, and lakes, "cypress ponds," "bays," peat bogs, and strips of stream swamp are of frequent occurrence In the vicinity of Gainesville and Brooksville, heavy soils (sandy and clay soils with heavy clay subsoils), forested with hardwoods, occur as isolated tracts, known as *hammock* land. Such areas are conspicuous because of their sharp contrast to the prevailing deep sandy lands with their characteristic growth of pine. The deep sand is used in many localities for citrus and other specialty crops, as well as for general farming. Erosion removes considerable topsoil from the more sloping fields on these sandy lands during long heavy rains. Sloping fields of the heavy hammock land are affected more seriously than the deep sand. Generally, erosion is not recognized as a problem, and little or nothing is done to check it. It is not, however, a very serious problem, partly because of the highly absorptive nature of the dominant sandy soil and partly because of the relatively small area in cultivation. Contour cultivation, strip cropping, and terracing will give a large measure of control.

# Chapter XXXI. Southern Appalachian Region

The Southern Appalachian Region, comprising approximately 115 million acres, extends from Maryland southwesterly to central Alabama and from the Coastal Plain region in central Virginia, North Carolina,

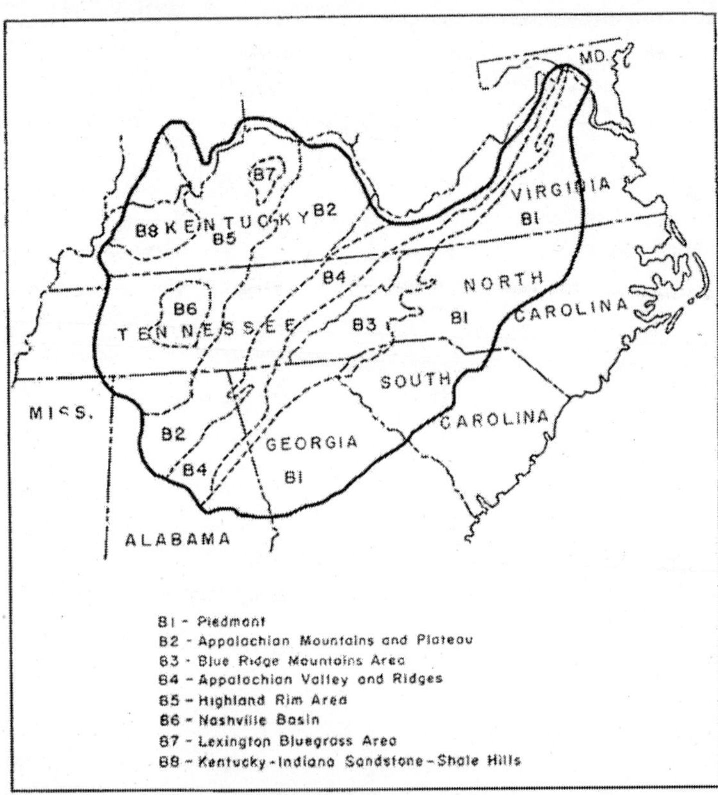


MAP 8.—Problem area of the Southern Appalachian Region. (*Soil Conservation Service.*)

South Carolina, and Georgia northwesterly to the Central Prairie region (the timbered border portion) in southern Indiana and Ohio (Map 8). It covers most of Kentucky and Tennessee, the western two-thirds (approximately) of Virginia, about half of North Carolina, more than a third of

South Carolina, about half of Georgia, and a large area in northeastern Alabama. A comparatively narrow marginal strip extends into southern Indiana and Ohio.

The principal subdivisions are (1) the Piedmont Plateau, (2) Blue Ridge Mountains, (3) Appalachian Valley and Ridges, (4) Appalachian Mountains and Plateau, (5) Highland Rim, and (6) Kentucky Sandstone-Shale Hills.

The region is largely hilly to mountainous. Stream bottoms and second bottoms (alluvial terraces) are generally flat, and here and there flat lands occur in the limestone valleys and Highland Rim subdivisions. Little more than half of the area is sufficiently smooth to permit plowing; the remainder is too steep, rough, stony, or severely gullied for cultivation. A considerable part of the land that is now plowed or that could be plowed is unsuitable for cultivation because of its susceptibility to rapid decline by erosion. The elevation ranges from about 200 to 500 feet above sea level along the outer boundary of the Piedmont, to a maximum of 6,684 feet at the summit of Mt. Mitchell in western North Carolina. The elevation of the northwestern boundary, in Ohio and Indiana, is approximately 500 feet.[1]

The upland soils are derived from the underlying rocks, consisting for the most part of the granitics, rocks of the basic igneous group, micaceous schists, sandstone, shale, slate, and limestone. Texturally, the more important soil types vary from sandy loam to silt loam or clay loam at the surface, with subsoil usually ranging from friable sandy clay to stiff clay of low sand content. Drainage generally is well established throughout the uplands, although some seepy strips occur along the foot of slopes in the valleys, and occasional depressions and level tableland areas in the Highland Rim country are imperfectly drained.

Rainfall is heavy throughout the region, ranging from around 50 inches along the outer boundary to 80 inches in a small mountain area near the point of contact between Georgia and the Carolinas. Generally, the heavy rains of spring and summer cause considerably more erosion than the storms of fall and winter, but bared land is susceptible to washing throughout the year, except on the higher elevations subject to long periods of frozen soil.

The principal type of agriculture is general farming, but fruit growing is important in various parts of the region, especially apples in the Blue Ridge and limestone valleys, and apples and peaches in parts of the Piedmont. Cotton and corn are the principal crops in the Piedmont from about central North Carolina southward as well as in the valleys and smoother highlands of northern Alabama and Georgia. Tobacco is a crop

[1] See map of physical features of the United States, "Atlas for American Agriculture," U. S. Dept. Agr., 1936.

of major importance in the Piedmont sector from about central North Carolina to middle Virginia, as it is also in Kentucky and adjacent parts of Tennessee. Corn, small grain, and grass are the principal general farm crops from middle and eastern Tennessee and western North Carolina northward. Dairying is of some importance near the larger cities, and the raising of cattle is of considerable importance in southwestern Virginia and western North Carolina.

About a fourth of the region is in cultivation. Considerable land, previously cultivated, is now idle or abandoned, principally because of erosion. About 85 to 90 per cent of the cultivable upland land is subject to erosion under cultivation, and the stream bottoms (less than 10 per cent of total area) are subject to injury by the deposition of sand and finer material washed out of the uplands. Reservoirs have suffered seriously from silting; some have filled completely over a period of less than thirty years; others over periods of less than fifty years. Erosion is the most serious problem relating to the use of farm land over probably 85 per cent of the area. Practical farm measures, demonstrated by the Soil Conservation Service, have greatly reduced, or controlled, erosion in a number of representative areas in various parts of the region. Adequate control of erosion in all instances has required the retirement of many steep and highly erodible areas to grass or timber. Reconnaissance surveys indicate that about 20 to 25 per cent of the land now in cultivation should be retired to permanent cover.

Work on farms of the Southern Appalachian region under the provisions of the Agricultural Conservation Program of the Agricultural Adjustment Administration provides for a number of the practices that have been found effective in soil and water conservation.

## Piedmont Region

The southern Piedmont region embraces the moderately elevated, highly dissected plateau bounded, roughly, on the west by the foot of the Blue Ridge Mountains and the southern extension of the Appalachian Valleys and Ridges and on the east by the Atlantic Coastal Plain. Southward from central Virginia, the average width is about 115 miles, the greatest width being approximately 200 miles along an east-west line through Raleigh, N. C. This southern extension of the Piedmont comprises about 41 million acres. The country is rolling to hilly, rising gradually from an elevation of about 200 to 500 feet above sea level, along the contact with the Coastal Plain, to an elevation ranging from about 700 to 1,000 feet or a little more at the foot of the Blue Ridge. Although the region really comprises the foothills of the Blue Ridge Mountains, the line of contact is rather sharply defined, topographically, in most places. Occasionally, the two regions merge into one another in such a way that it is impossible to fix a precise line of separation. Here and there, hills of more resistant rock, such as those near Kings Mountain along the North Carolina-South Carolina line, rise conspicuously above the general upland level.

So intricately has the region been ramified by its drainage system that no large areas are without drainage outlets or slopes that quickly lead to drainageways. Along the many deeply eroded valleys, the topography is rolling to hilly, with numerous slopes too steep for cultivation. By far the greater part of the area, however, is, or originally was, cultivable. Erosion is active almost everywhere—numerous areas, large and small, have been so gullied that they have no further practical use for crop production. Approximately 10 million acres have been essentially ruined by erosion as the result of plowing slopes too steep and erodible for cultivation. In general, farming has been forced out of these areas by erosion, and pine trees are taking them over. Most of the land that has lain idle some 10 years or longer has overgrown with "old-field" pine, which, in spite of frequent fires, usually maintains a ground cover of litter sufficient to arrest severe soil washing. For the greater part, the older trees are valuable for lumber, pulp, or firewood. On some of the most severely washed land, however, they have made very poor growth. For this reason, some gullies are still eroding.

Over the broader stream divides, the land is smoother of relief, with undulating to gently rolling surfaces predominating. This is where the better farms are.

The soils of the Piedmont are formed of products derived from decay of the underlying rocks. The mantle of soil and decomposed rock varies from less than 3 feet over some of the quartzite, schist, and slate areas to more than 50 feet in some of the southerly areas of granitic rocks. Here and there, gullies have cut into this granitic material to depths exceeding 50 feet. The most abundant soils are the red lands of granitic rock origin, such as the Cecil, Helena, and Madison soils. Other important types are the gray soils, over yellowish subsoils, derived from granites, such as the Appling and Durham; the red lands derived from basic igneous rock, such as the Davidson and Mecklenburg groups; brownish-yellow, stiff lands derived from basic igneous rock, as the Iredell and Wilkes; red and yellow slate lands of extensive occurrence in the Carolinas and south-central Virginia, as the Georgeville and Alamance; and grayish soils with red subsoil, overlying schist, as the Louisa. A number of important areas of reddish and yellowish soils, such as the Wadesboro, White Store, and Granville, have been formed from the shale and sandstone beds of Triassic age in Virginia and North Carolina.

In the southern portion of the Piedmont, cotton has long dominated the agriculture, although corn is an important crop for farm consumption. In the central part, tobacco is an important crop; to the north, wheat, corn, and grass are the principal farm products. Locally, apples and peaches are grown commercially, the former mainly north of North Carolina, and the latter principally in South Carolina and Georgia. Not nearly enough livestock or dairy products are produced for home use. Many farms have no cows; a large number have neither chickens nor hogs; and far too many are without gardens. Fertilizers are commonly used for practically all crops.

Approximately 77 per cent of the cultivated area is used for the clean-tilled crops that give no protection from erosion. This is entirely too large a proportion of row crops for a region with so much erodible land. It accounts to a very large degree for the extensive area of severely eroded land.

Acreages of the more important groups of crops, derived from the census of 1930, are as follows:

| | | *Acres* | *Per Cent of Land in Farms* | |
|---|---|---|---|---|
| Land in farms | | 29,281,000 | | |
| Cropland | | 9,056,000 | | 31 |
| Clean-tilled | 7,005,000 | | 24 | |
| Erosion-resistant | 1,756,000 | | 6 | |
| Semierosion-resistant | 295,000 | | 1 | |
| Permanent pasture | | 6,614,000 | | 23 |
| Woodland | | 8,531,000 | | 29 |

There were in the region, according to the census of 1930, approximately 150,400 horses, 416,100 mules, 948,100 cattle, and 142,200 sheep.

Protection of the soil from erosion is the most difficult and important land problem confronting the farmers of the Piedmont. Already the process has violently upset the economic and social life of many communities, literally crowding people off the land. Pressure for good land has been felt for some time in a number of localities. Much good soil is left, but most of it is subject to erosion, and an additional large area will be lost if soil washing is not controlled. As it is, so much land has become unfit for cultivation that every remaining acre of productive arable land must be conserved if the rural economy of the region, on which urban welfare largely depends, is to be preserved.

At the erosion experiment station near Statesville, N. C., the principal soil type of the area, Cecil sandy clay loam, has been losing soil from slopes of 10 per cent at the rate of 25 tons an acre annually, under continuous clean cultivation to cotton, along with a loss of over 10 per cent of the total precipitation as runoff. This means that 7 inches of topsoil would be completely stripped off in 46 years under continued production of a clean-tilled crop. According to the measurements that have been made, it would take around 80 thousand years to strip off the same amount of soil under a dense cover of grass (see Graph 16).

The Soil Conservation Service has carried on many demonstration projects of erosion control throughout the Piedmont. Excellent results have been obtained on lands not too steep for cultivation. Examples of the major land treatment and farm reorganization procedures involved in two representative soil and water conservation demonstration projects—the Little Beaver Dam Creek project of Anderson County, South Carolina, and the Brown Creek project in Anson County, North Carolina—are illustrative of the type of program necessary to control erosion in the Piedmont problem area.

### LITTLE BEAVER DAM CREEK PROJECT

The Little Beaver Dam Creek project, comprising 35,500 acres, is located northeast of the city of Anderson, in Anderson County, South Carolina. The principal soils are the clay loams and sandy loams of the Cecil, Appling, Durham, Madison, and Davidson series. The topography varies from undulating and gently rolling to strongly rolling near the streams. With respect to soils, topography, type of agriculture, and erosion, the watershed of this stream is representative of a large and important Piedmont area.

On the basis of surveys made prior to the establishment of the project, about a third of the entire area is too steep for cultivation. Less than 10 per cent of the cultivated land occupies slopes under 3 per cent. Such gently sloping land, under normal conditions of tillage, is subject to a minimum of erosion. Approximately 50 per cent of the cultivated land occurs on slopes ranging from 3 to 10 per cent. For the most part, land this steep is suitable for the clean-tilled crops only where effective erosion-control measures are used. Some areas of coarse texture, such as the highly absorptive Durham sandy loam, can be safely cultivated, under good conservation practices, on slopes with gradients ranging up to about 15 or 16 per cent. Nearly 30 per cent of the cultivated land has slopes ranging from 10 to 14 per cent. Most of this is fine-textured, highly erodible, and unsuitable for clean-tilled crops, although some of it may be used for forage and pasture crops where the land is covered throughout the year. Something over 10 per cent of the cultivated land is suitable only for forests, shrubs, or other permanent vegetation.

About 60 per cent of the cleared land has lost from half to three-fourths of the topsoil; 20 per cent, from a fourth to half; 10 per cent, less than a fourth; and 10 per cent, more than three-fourths.

Approximately 75 per cent of the land is farmed by tenants. Erosion is a serious problem on most of the land. It has been the custom to plant even the steepest slopes to clean-tilled crops year after year without attention to soil-conserving rotations. When an area has washed so severely that cultivation does not pay, or when it has become so dissected with gullies as to preclude farming operations, it is turned to pasture or abandoned to voluntary restocking with second-growth pine. For many years, some farmers have terraced the more erodible fields as well as they could, but frequently the slopes were too steep, or the terraces were improperly constructed or maintained, so that severe erosion continued between the structures in spite of their efforts. Many terraces have broken, with the result that adjacent lower fields frequently have suffered from gullying. Also, failure to provide properly protected outlets has caused much gullying at the ends of terraces.

The program of erosion control carried on here by the Soil Conservation Service, cooperating with the farm operators, involves 172 farms, comprising 19,116 acres. Each farm has been covered by a detailed survey of physical and economic conditions. Plans developed on the basis of these surveys have included the following principal measures:

1. Conversion of severely eroded land to permanent protection of grass, trees, kudzu, or other dense vegetation.
2. Strip cropping.
3. Contour cultivation.
4. Terracing.
5. Protection of terrace outlets.
6. Rotation of crops.
7. Development of grassed waterways to remove excess rainfall.
8. Use of green manure crops.
9. Protection of fields with seasonal cover crops.
10. Pasture improvement.
11. Gully prevention and control.
12. Woodland improvement.
13. Development of wildlife resources.

Some of the major accomplishments, as of June 30, 1938, are as follows:

| | *Acres* |
|---|---|
| Retired from cultivation to the protection of permanent cover | 1,302 |
| Approved rotations adopted | 3,229 |
| Tilled on contour | 9,052 |
| Protected by terraces | 5,791 |
| Strip cropped | 2,752 |
| Pasture treated | 1,578 |
| Planted to trees | 1,209 |

### BROWN CREEK SOIL CONSERVATION PROJECT

The Brown Creek project covers parts of Anson and Union Counties in North Carolina, together with a small portion of Chesterfield County,

South Carolina. The area comprises approximately 65,000 acres in the upper watershed of Brown Creek. It is located entirely within the Cotton Belt and is representative of conditions over a large area within the Carolina slate and southern Triassic sandstone-shale sections.

The principal upland soils are (1) those derived from Carolina slates, as the silt loams, clay loams, and slate loams of the Georgeville and Alamance groups; (2) those derived from sandstone, shale, and claystone, as the sandy loams and clay loams of the Wadesboro series and the clay and clay loam of the White Store series; and (3) those derived from diorite, as the Iredell clay loam. The topography varies from gently to moderately rolling, interrupted by an occasional conspicuous hill, such as Gordon Mountain.

Surveys have disclosed the following erosion conditions, based on studies of 461 farms: 4 per cent not affected by erosion, 15 per cent only slightly eroded, 38 per cent affected with moderate erosion, 32 per cent with moderately severe erosion, and 11 per cent very severe erosion. Approximately 30 per cent of the area was damaged to various degrees by gullying.

Fourteen per cent of the project area has slopes of less than 3 per cent. On these, erosion ordinarily is not serious. Moderately sloping land (3 to 7 per cent), such as can be farmed safely where erosion-control measures are used, comprises 68 per cent of the area. Nine per cent is too steep (7 to 12 per cent gradient) for intertilled crops, such as cotton or corn, but can be used to some extent for small grains, legumes, or pasture. The remainder consists of slopes of 12 per cent or more and is too steep for cultivation. It should be utilized as woodland or pasture.

Under simple working agreements, the Service and the farmers cooperatively have applied an integrated erosion-control program to 461 farms, comprising 53,687 acres. Each farm was surveyed in detail; and on the basis of the physical conditions of the land, a coordinated plan of treatment was worked out for all the farms. These plans were adjusted to the economic situation. The principal measures employed were as follows:

1. Conversion of severely eroded land to permanent protection of grass, trees, kudzu, perennial lespedeza, or other dense vegetation.
2. Strip cropping.
3. Contour cultivation.
4. Terracing.
5. Protection of terrace outlets.
6. Rotation of crops.
7. Development of grassed waterways to remove excess of rainfall.
8. Development of meadow strips to control runoff.
9. Use of green manure crops.

10. Protection of fields with seasonal cover crops.
11. Pasture improvement.
12. Gully prevention and control.
13. Woodland improvement.
14. Development of wildlife resources.

Traditional methods of farming in this area have not taken into consideration adequate plans for controlling erosion. Easily eroded soil, the continuous cultivation of large acreages of cotton and corn, and the prevailing system of tenancy have been the most important obstacles to the establishment of effective control measures. Sixty-seven per cent of the land is farmed by tenants. On the sandstone-shale soils of the Triassic belt, the percentage of tenancy is higher, the cotton acreage larger, and susceptibility to erosion greater than on the slate soils. General dependence on cotton is indicated by the large quantities of farm products bought from outside sources. It was estimated in 1915 that the following items[1] were shipped into Anson County annually: 100,000 bushels of corn; 100,000 bushels of oats; 600 tons of hay; 30,000 barrels of flour; 250,000 pounds of lard; and 1,000,000 pounds of pork. Recently, the situation probably has changed considerably.

As compared with the Wadesboro and White Store soils of the Triassic belt, control of erosion is simple on the slate lands of the western part of the area. Erosion has long been severe on the sandstone and shale soils. Deep gullies had cut into the slopes long before the War between the States. In 1865, as Federal cavalry passed through the White Store locality, much livestock is said to have been saved from capture by being driven into the severely gullied areas where it was difficult for mounted troops to operate. A considerable area in this section has been essentially ruined for cultivation, and many farms include large or small tracts of severely eroded land.

The common practice was to clear off virgin forest and use the land without regard to conservation of the soil until it washed to a condition of excessively low productivity or was severely dissected by gullying. Then it was abandoned to broom sedge and old field pine, and fresh clearings made. Soon little virgin land was left. As the result, much abandoned land subsequently was brought back into cultivation by reclearing. In some instances, this process has been repeated a number of times.

In the western, or slate-land, section, farming has been more diversified, and a larger proportion of the farmers have remained on the land and operated on a "live-at-home" basis, producing more small grain, hay, and hogs along with their cotton and corn.

Evidence of the influence of the demonstration work is the increase of interest in soil conservation shown by landowners throughout the sur-

[1] Soil Survey, Anson County, North Carolina. Field Operations, Bur. Soils. 1915.

rounding territory. When CCC erosion camps were established in Anson and Union counties, in which the Brown Creek demonstration project is located, the immediate demand for assistance by farmers throughout these counties greatly exceeded the technical and labor facilities of the SCS and CCC. Calls for technical assistance in the various phases of the program, particularly for terracing, timber-stand improvement, advice on timber sales, and wildlife conservation ran far ahead of immediate capacity to handle the work.

Recently, the Brown Creek Soil Conservation District, covering the 115,000 acres within the Brown Creek watershed, was organized and established, principally as the result of general acceptance of the work carried out under the direction of the demonstration project technicians.

Following is a list of the major accomplishments of the Brown Creek demonstration project, as of June 30, 1938:

| | *Acres* |
|---|---|
| Retired from cultivation | 2,793 |
| In approved rotations | 12,453 |
| Tilled on the contour | 13,463 |
| Protected by terraces | 12,305 |
| Strip cropped | 1,622 |
| Pasture treated | 2,743 |
| Trees planted | 1,363 |

## Conditions in the Old Tobacco Belt

The watershed of Reedy Fork Creek, comprising 47,483 acres near Greensboro, N. C., is fairly representative of soil, topographic, erosion, agricultural, and economic conditions in the Piedmont section of the Old Tobacco Belt.

Agriculture began in this area before the American Revolution. In 1935, when operations were started in the Reedy Fork soil and water conservation demonstration project, the population was between 5,000 and 6,000, of which approximately 3,000 were rural. Of the 834 farms, averaging 56.3 acres in size, about 25 per cent were operated by tenants, a proportion somewhat smaller than the average for the southern Piedmont.

In this section, flue-cured tobacco ("bright" tobacco) has long been the principal cash crop. Corn, wheat, oats, rye, clover, and hay are produced on most farms, principally for home consumption or as cover or rotation crops. Fertilizers are used on most crops, heavy applications of 800 to 1,000 pounds per acre being made for tobacco. In 1935, approximately 37 per cent of the land was used for crops, and 8 per cent for pasture; 50 per cent consisted of woodland, and 5 per cent was idle.

Erosion, including areas affected by deposition (8.3 per cent), had damaged all but 4.2 per cent of the area. More than 94 per cent of the uplands, which comprise 92.2 per cent of the entire area, had suffered from erosion in some degree, 81 per cent seriously. Of the 81 per cent affected seriously, about half had been so severely damaged that it had little further use for really efficient cultivation, and about one-fifth had been essentially ruined for cultivation. Less than 1 per cent of the flat alluvial lands had been affected by erosion, but most of them had been damaged by deposition of materials washed out of the uplands. Many formerly cultivated areas had been buried by 3 feet or more of erosional debris.

Most of the uneroded uplands were areas that had never been cultivated. All but 6.3 per cent of the total area in crops, alluvial land and upland, had been affected by erosion, 70 per cent moderately and 19 per cent severely.

The principal soils are Cecil sandy loam and sandy clay loam (28 per cent) and Appling sandy loam (34.5 per cent). Other important soils are Cecil clay loam and loam, Wilkes and Helena sandy loams, Iredell loam, Davidson clay loam, and alluvial bottomlands. Of the Iredell, Helena, and Wilkes soils, all but 2 per cent had been affected by erosion, 87 per cent seriously; of the Cecil, Appling, and Mecklenburg, all but 5.5 per cent had been affected, 81 per cent seriously; of the Durham-Davidson group, 66 per cent had been affected seriously.

Approximately 22 per cent of the area is too steep for cultivation (*D* slopes: generally 12 per cent or steeper); 30 per cent is too steep for clean-tilled row crops (*C* slopes: generally of 7 to 10 per cent gradient); nearly 41 per cent is of moderate slope on which cultivation is safe with the use of conservation practices (*B* slopes: generally of 3 to 7 per cent gradient); and 7 per cent ranges from level to gradients of less than 3 per cent (*A* slopes) and is safe for cultivation without conservation practices, where properly drained and protected from overwash.[1]

Approximately 16.7 per cent of the pasture and woodland in the area had not been affected by erosion (part of this never cultivated) at the time work began, but about 73 per cent of the former and 64 per cent of the latter had suffered seriously.[2] Most of the pasture and woodland had been cultivated at some time and abandoned because of erosion.

Operations in the Reedy Fork soil conservation demonstration project had extended to 365 farms by June 30, 1938. The principal measures employed were contour cultivation; strip cropping; terracing; crop rotation; seasonal cover crops; gully control; wildlife planting; retirement of steep, erodible land to the protection of trees or grass; and establishment

[1] The gradients thus defined vary somewhat with degree of erosion and the soil type.

[2] Stevens, W. W., Bragg, H. V., Sease, E. C., and Lewis, O. C. Erosion and Related Land Use Conditions on the Reedy Fork Demonstration Area, North Carolina, U. S. Dept. Agr. *unnumbered Pub.*, 1938.

of stable waterways and meadow strips. Some of the more important accomplishments were the contouring of 9,283 acres; strip cropping of 4,266 acres; terracing of 6,015 acres; protection of 3,516 acres with adaptable rotations and timely seedings of cover crops; and retirement of 4,881 acres to permanent pasture, trees, and wildlife plantings.

## Blue Ridge Mountain Area

The southern Blue Ridge Mountains area, covering about 13 million acres, extends southwesterly from Maryland across the western parts of Virginia and North Carolina, the eastern margin of Tennessee, extreme western South Carolina, and northern Georgia to the vicinity of Talladega in central Alabama. The northern and southern extremities are narrow; the maximum width, about 125 miles, is attained in southwestern North Carolina. Southwesterly from the point in north Georgia where the mountain belt narrows sharply, elevations are considerably lower, and the range, merging with the associated highlands, almost loses its identity as a separate problem area. It is in the Blue Ridge, in western North Carolina, that the highest point east of the Black Hills is attained.

The greater part of the area is forested, although much of the timber has been cut over, and a considerable part consists of second growth.

The principal soils are those of the reddish Porters and yellowish Ashe series, derived from granitic rocks, and the comparatively shallow reddish schist-derived soils of the Talladega group. Rock outcrop and stony land are surprisingly scarce, and a large proportion of the soil is remarkably deep (to bedrock) for such mountainous terrain.

General farming is the dominant type of agriculture. Apples are important, especially in the northern part; and the raising of livestock, principally cattle, has attained a place of considerable local importance, as in various parts of western North Carolina. Dairying has been successfully developed in the vicinity of some of the larger towns.

Temperature has considerable effect on crop adaptability. In general, seasons are too short or cool for cotton, even in the southern extension, but grass does better than on comparable areas within the adjacent warmer Piedmont. Timothy and bluegrass and some buckwheat are successfully grown as far south as North Carolina.

Typically mountainous country throughout, except in the southwestern extension, the prevailing slopes are too steep for easy cultivation. But there are numerous areas of good farm land, occurring principally in the basins and broader valleys, on shouldering positions and lower slopes, and as narrow strips of alluvium. It is estimated that about 75 per cent of the area is too steep for safe cultivation.

Some $1\frac{3}{4}$ million acres are included in the Chattahoochie, Cherokee, Nantahala, and Pisgah national forests, together with a large area lying within the George Washington and Jefferson national forests, portions of which are in the Appalachian Valley and Ridges area. The Shenandoah and Great Smoky Mountains national parks comprise some 570,000 acres. Still other areas are included in various tracts of public lands.

Of the land that can be cultivated safely, fully three-fourths will require protection from erosion, if productivity is to be maintained. With respect to severity of erosion, degree of slope is the principal determinant, although soil character locally affects erosion rates to a marked degree. Some of the more gravelly land and most of the darker reddish soils, such as the Porters clay loam, are strongly resistant to washing, relatively, because of high infiltration capacity.

Although gullying is severe in many places, sheet washing is by far the most extensive and troublesome type of erosion. Because of this, superficial observation has led to frequent misjudgment with respect to the resistance of the steeper lands to erosion. Moreover, the casual observer may fail to discern the fact that the frequent stands of second-growth timber generally occupy land abandoned for cultivation because of impoverishment by erosion. One sees many such reforested fields, for example, along the highway from Asheville to Murphy, N. C.

Erosion generally can be controlled on slopes of not over 10 to 12 per cent gradient with contour cultivation, strip cropping, terracing, good crop rotations, seasonal cover crops, and preservation of crop stubble. Some of the more gravelly land and the deeper more loamy phase of the Porters and Ashe soils can be farmed safely, if defended with a properly applied system of erosion control, on slopes up to 16 or 18 per cent. The steeper lands should either be kept in grass or returned to timber. Pastures should be carefully protected from overgrazing and unseasonable use. Fertilizers are usually needed for all crops.

## Appalachian Valley and Ridges

The southern extension of the Great Appalachian Valley, together with its associated ridges, extends in a troughlike strip of irregular width, southwesterly from Maryland, through western Virginia, eastern West Virginia, eastern Tennessee, and northwestern Georgia, to the vicinity of Birmingham, Ala. It is roughly enclosed by the Blue Ridge Mountains on the east and the Cumberland-Allegheny Mountains on the west. The area is approximately 11 million acres. In different localities, it is known by different names, such as the Shenandoah Valley or The Valley in Virginia and West Virginia, the Valley of East Tennessee to the south, and the Coosa Valley in Georgia and Alabama. Some detached areas are practically identical in general characteristics within a main valley, as the Sequatchie Valley of Tennessee and portions of the irregularly shaped valley along the Tennessee River in northern Alabama.

The more typical limestone valley areas are characteristically undulating to gently rolling over their main floors and are admirably suited to cultivation. About 65 or 70 per cent of the area is readily tillable. Generally, the floor of the valley ranges from about 500 to 3,000 feet below the bordering mountains. The difference is about 2,700 feet, on the average, in the center of the area along the Virginia-Tennessee line. Southwestward, the difference in elevation falls to around 500 feet or less in north Alabama, and northward to around 700 to 1,000 feet near the Maryland line. Locally, streams have cut valleys within the valley to depths of more than 200 feet.

Hills and ridges rise abruptly from the valley floor to elevations of around 750 to 1,000 feet or more. Comparatively narrow ridges, underlain by cherty limestone, shale, and sandstone paralleling the general northeast-southwest trend of the main valley are of frequent occurrence in east Tennessee and southward. The detailed valley characteristics of this Appalachian trough are due primarily to the fact that the dominant rock, limestone, has decayed more rapidly than the rocks of the bordering ranges.

The most important soils of the valley area are the brown loam and silt loam of the Hagerstown; the red loam, silt loam, and clay loam of the Decatur series; and the grayish-brown gravelly silt loam of the Frederick and Clarkesville series. The subsoil of the Hagerstown consists of light-red clay; of the Decatur, red clay; and of the Frederick, yellowish-red silty clay. At the surface the Clarkesville soils (mainly gravelly silt loam) are like the Frederick, but the subsoil is of yellow color and more cherty. These are much more productive and, collectively, more extensive than the associated grayish to light-brown shale and sandstone soils of the Hartsell and Berks series (with yellow clay subsoils), which normally are of rolling to hilly topography, often shallow and frequently stony or shaly. The cherty (gravelly) limestone soils, such as the Frederick and Clarkesville, are prevailingly more rolling than those of the Hagerstown and Decatur groups. The last two represent the outstandingly productive lands of the region, although under good treatment the lighter colored Frederick soils are also of good productivity.

Sheet erosion has taken a much heavier toll of nearly all varieties of soil throughout the area than has been generally realized. Gully erosion has been serious locally in the northern part and generally serious in the southern part, except on the more gravelly areas. South of the Virginia line, the inroads of erosion have been and continue to be very heavy—considerably more so than to the north. Many areas, including some of the most productive land of eastern United States, have been forced out of cultivation solely because of gullying, deep sheet washing, or both. At present, soil wastage in the southern extension is proceeding so rapidly that land depletion continues to outstrip all efforts at stabilization and improvement. To a lesser extent, this is true also on the more rolling valley lands of the northern extension. Here and there, land is still being cleared for cultivation on slopes so steep that fields cannot possibly be protected adequately under any type of utilization except for permanent grass or trees. In 1934, the author saw slopes in cultivation on Clynch Mountain, in Tennessee, having gradients up to 80 per cent. The operators frankly stated that such fields would normally be expected to last about 5 to 8 years from the time of clearing if used for corn or tobacco or only about 3 years if exceptionally heavy rains should fall.

Probably not less than 75 or 80 per cent of the cultivated area is subject to impoverishing erosion under clean tillage. About half of the entire area is too steep for cultivation under any kind of protective system but prohibitively expensive bench terracing and should be kept in such cover as grass, lespedeza, kudzu, or woods. With respect to the better limestone land, about 20 to 25 per cent of that now in cultivation should be retired to the protection of permanent cover. Some very gravelly areas of limited extent, such as the Frankstown gravelly silt loam in the vicinity of Winchester, Va., have such high absorptive capacity that cultivation is feasible on gradients up to 20 or 25 per cent. On the shale and sandstone lands and the better limestone soils, however, cultivation generally should be limited to slopes of not over 8 to 10 per cent. Gradients of about 12 to 15 per cent should be the limit of cultivation on the lighter colored gravelly limestone land (Frederick and Clarkesville). Even with these slopes, a completely coordinated treatment of erosion control must be employed if the soil is to be conserved. This will include, among other measures, contour cultivation; crop rotation; strip cropping; terracing; use of seasonal cover crops; and, frequently, application of lime and fertilizers.

General farming is the dominant type of agriculture, with corn, clover, grass, and small grain predominating north of the center of the Valley of East Tennessee, and cotton, corn, lespedeza, and oats predominating to the south. Apples are highly important in Virginia and West Virginia. Some tobacco is grown in east Tennessee.

About half the area is occupied by forest or brushy growth and pasture. Large portions of the Chattahoochee, George Washington, and Jefferson national forests lie in the Appalachian Valley and Ridges area.

## Appalachian Mountains and Plateau

The southern Appalachian Mountains and Plateau region parallels the Appalachian Valley and Ridges area on the west. It comprises about 17 million acres, including the Kentucky-Indiana sandstone-shale hills area. From the strictly mountainous country—Cumberland Mountains—immediately bordering the valley, the slope falls gradually westward to the Plateau sector, and the surface becomes generally smoother.

In the mountainous eastern sector, elevations of 2,500 to 3,000 feet are not uncommon; in the Big Black Mountains of Kentucky and Virginia, some points rise to approximately 4,000 feet. In the Plateau proper—the relatively smooth country west of the rough lands bordering the Valley—the dominant skyline is fairly even although interrupted by local eminences. Deep valleys have cut into or across the Plateau. All of these are bordered by rolling to rough lands largely unsuitable for cultivation. Undulating to almost level areas, well suited to cultivation, are encountered frequently on the interstream divides of the smoother westward portion of the area.

Although the upland soils are derived from the underlying sedimentary rocks—principally sandstone and shale—these have been so resistant to weathering that depth to bedrock averages very much less than in the Blue Ridge division of the Southern Appalachians, where many of the rocks are tough, but more readily weathered crystallines. Over much of the country, bedrock is reached at depths of about 3 feet, and numerous slopes are occupied by rock outcrop or very thin, unstable layers of material more of the character of rock debris than of true soil. Lower slopes in the mountainous section have been covered in places by accumulations of rock talus to such depth as to preclude any kind of use.

The principal soils, outside the areas of *rough stony land* or *steep mountain land*, are the gray sandstone or shale lands with yellow subsoils, such as the Hartsell (formerly Dekalb) and Berks. Some areas derived from similar rocks have a red clay subsoil (Hanceville soils).

General farming, frequently of the subsistence type, is carried on throughout the area. Considerably more than half the area is in some kind of forest. Much of the country has been cut over. Stands range from small second growth to occasional remnants of heavy timber. Farms generally have relatively small areas in cultivation. The Black Warrior and Cumberland national forests, comprising some half million acres, lie largely within this problem area. Corn, small grain, and grass are the principal products. Tobacco is of some importance locally, as are apples and peaches.

Erosion is widespread, and generally little has been done to check its invasion of the average field. Much land has been abandoned following removal of the topsoil, and other areas have been forced out of cultivation by gullying. Fortunately, some form of vegetation usually takes hold of such "thrown-out" land—weeds, native grasses, bushes, trees—and slowly provides a degree of stabilization. Although soil recovery is exceedingly slow or negligible on all severely washed areas, such land is sometimes recleared after "resting" for about 5 to 10 years. It is planted for a year or two, then "turned out" again. Even after this, some areas are cleared a third time, following a "rest" period of 12 to 20 years or longer, and put into cultivation. Erosion continues with each successive clearing almost as if nothing had been done to check it; and after one crop, usually a pitifully small one, the land finally goes back to whatever protection nature can provide. Such rotation of "bush fallow" and cultivation is one of the economic evils resulting from uncontrolled erosion. Where it is followed, subsistence farming is the usual result. It is on the order of the patch agriculture found in jungle clearings in outlying areas of Central America. This futile type of farming is fairly common in the rougher parts of the Southern Appalachian region. However, it is somewhat less futile on eroded limestone soil of the Hagerstown and Decatur groups, because the subsoil of these lands is more productive than the highly acid subsoil of the shale and sandstone lands.

Most of the cultivated land in this mountain-plateau country is in need of protection from erosion. Demonstrations have shown that much of it can be protected with rather easily applied, practical farm measures, where they are individually or collectively used to meet not only the requirements of each field and parcel of land but the needs of entire farms. Cultivation should be discontinued generally on slopes steeper than about 10 or 12 per cent and at about 8 per cent on the shallower soils overlying clay. Applicable control measures will be much the same as those recommended, in the next chapter, for similar lands of the southwestern portions of the Northern Appalachian area—the Muskingum soils of the Allegheny-Ohio section.

The Sandstone-Shale Hills region of northwestern Kentucky and adjacent southern Indiana are much the same as portions of the southern Appalachian Plateau. Soils and topography are very similar, and erosion conditions and needs are of the same general order.

## Highland Rim Area

The large area of predominantly limestone land lying west of the southern Appalachian Plateau, in Kentucky, central Tennessee, and north-central Alabama, is designated the Highland Rim area. Although the surface has a prevailingly smoother skyline, and some areas of plateau-like characteristics are undulating to nearly level, much of the land is of rolling to hilly or ridgy topography. Some localities, as in southern Kentucky, have a distinctly pitted surface (karst topography) due to sinkholes leading to underground openings into the basal limestone. Streams generally have entrenched deep valleys, often enclosed by vertical limestone cliffs or steep, rough escarpments.

The general upland level averages in the neighborhood of 1,000 feet above sea level, with a gradual slope westward. Approximately 23 million acres are comprised within the Highland Rim, including the Nashville Basin and Lexington bluegrass area. In general, the soils are much more productive than those of the sandstone-shale plateaus to the east. General farming prevails, with tobacco, small grain, corn, and grass the principal crops. Dark export and Burley tobacco are extensively grown in Tennessee and Kentucky. Special crops, such as strawberries, are locally important, and stock raising has a prominent place in the agriculture of some farms. Probably at least 50 per cent of the area is in cultivation, the remainder consisting chiefly of woodland, permanent pasture, and abandoned eroded land.

The most important soils are the limestone lands of the Hagerstown, Frederick, and Decatur series. The shale lands are much like those of the Appalachian Plateau.

Erosion is a problem over about 80 per cent of the cultivated area and generally a serious one. About 15 to 20 per cent of the land in cultivation should be retired to a permanent cover of grass or trees, and most of the remainder will need treatment with control measures fitted to the needs of the diverse topographic and soil conditions. The ease with which grass, lespedeza, and clover are grown makes the problem of control easier than on the thinner soils of the neighboring sandstone and shale portions of the area. Fertilizers are in general use for most crops. Grass and legumes give good results in soil-saving rotations, especially with applications of lime and phosphatic fertilizers.

The Nashville Basin and Lexington Bluegrass area include principally good limestone soils, such as the Hagerstown and Maury loams, silt loams, and clay loams. Here the agriculture is about like that on the better limestone lands of the Rim Country. Erosion, however, because of milder slopes and better land use, is not so serious generally in these basin areas as in the surrounding country, but it is active on much land, especially in the Nashville Basin. Control measures of the same order as those needed in the Highland Rim Country would apply to these predominantly high-grade farm lands.

# Chapter XXXII. Northern Appalachian and New England Area

The Northern Appalachian-New England area consists of a highly complicated arrangement of mountain masses, plateaus, and deeply infolded and stream-cut valleys. It comprises approximately 127 million acres, covering all of New England and West Virginia, nearly all of New York and Pennsylvania, the northern third of New Jersey, the western half of Maryland, a large area in southeastern Ohio, and a small part of western Virginia (Map 9). Elevations range from sea level along the New England coast to more than a mile above tide level in the White Mountains of New Hampshire.

In detail, the region embraces a large number of subdivisions; in general, it comprises (1) the Piedmont Plateau, lying along the eastern base of the Appalachian Mountains; (2) the northern extension of the Appalachian Mountains and Valleys, extending northeasterly from the Potomac River through eastern New York and western New England; and (3) the northern Appalachian Plateau, which declines westward from the Appalachian Mountains into the Ohio Valley and northward nearly to the Great Lakes.

## Climate

Annual precipitation in the form of both rain and snow varies from a maximum of 50 inches in the mountainous parts of the Northern Appalachian-New England region to a minimum of 30 inches along Lake Ontario and in the Finger Lakes region of central New York. At all elevations above 1,000 feet in the more northerly localities, and above 1,500 feet in the south-central portion, the winter snow cover is usually adequate to provide protection for winter grasses, legumes, and grains. Snow is less continuous in the southerly and lower lying sections, more of the precipitation taking the form of rain. Persistency of snow prevents frequent freezing and thawing of surface soils and so reduces winter erosion. Concurrently, those areas which experience adequate winter snow cover are the ones that still retain a large proportionate area in forest, and within which a grass cover in the form of pastures and meadows is most generally present. Over a large part of the region, winter snow persists from the middle of December until the end of February. Snow melts most rapidly and over the widest areas during March and April. This is a period of flood hazard and severe erosion, especially when rains add to the volume of runoff.

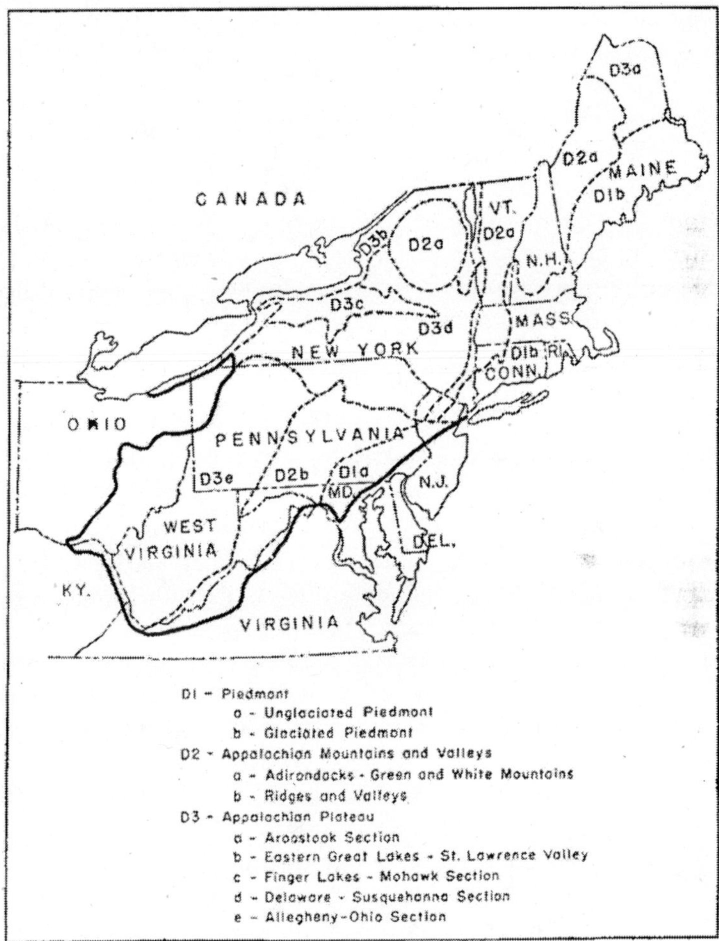


MAP 9.—Northern Appalachian and New England Problem Area. (*Soil Conservation Service.*)

A second period of frequent severe runoff comes during June and July. At this time, many fields are bare of vegetation, and spring-sown grains afford but a minimum of vegetative protection. Throughout the region, local thunderstorms are likely to cause much washing and even gullying.

Length of the growing season exerts a strong influence on erosion. At the higher altitudes and in the more northerly section, the average of this period is less than 130 days. This restricts agricultural opportunity, even on the more favorable lands. The eastern Great Lakes section, the St. Lawrence and Hudson valleys, and many of the interior valleys of Pennsylvania have a longer and more favorable growing season. In southern New England, through most of New Jersey, Maryland, Delaware, and West Virginia and all of Pennsylvania southeast of the Appalachian Ridges, an average frostfree period of more than 150 days provides greater agricultural opportunity.

The period of land preparation and planting is the time of maximum exposure to erosion. It extends from March well into May in the more southerly localities and at lower elevations and from April into June farther north and at higher elevations. In the fall, the harvesting of potatoes, late truck crops, and buckwheat and the preparation of land for seeding to winter grain and cover crops again expose considerable areas to severe erosion. Clean cultivation of orchards, vineyards, and all intertilled crops continues through the summer. These conditions, partly attributable to the climate, have a marked influence on erosion.

## Erosion

The Northern Appalachian-New England area, with its cool climate and normally abundant precipitation, is generally adapted to easy establishment of grass and other forms of protective vegetation. In its natural state, it was almost entirely covered with forest, usually mixed coniferous and deciduous growth. With the advance of occupation for agricultural purposes and for lumbering, the smoother, less stony tracts were cleared and cultivated. In the early days, upland grasses were not available for the seeding of grasslands, either for pasturage or for hay. Consequently, grain crops were grown extensively, and the surface was exposed to erosion through those periods when there was little or no vegetative cover. Accelerated erosion began almost at once with the clearing of the land; and even in some forested areas, it was promoted by the burning, accidentally or purposely, of cut-over lands. It was not until well along in the nineteenth century that crop rotations began to be adopted by seeding land to grasses between periods of occupancy by corn, wheat, or oats.

In that part of the area used for farming, it is estimated that from a fourth to half of the original surface soil has been washed out of fields. Approximately a fifth of the cultivated area has been affected in some degree by gullying (Fig. 264).

The most widespread form of soil erosion in the Northeastern States is sheet washing. This form of erosion has been found on surfaces having slopes of not over 1 per cent, but usually serious effects are restricted to gradients of 3 per cent or more. It becomes increasingly serious on slopes up to about 25 per cent, where the use of the land for production of intertilled crops ceases on the better-managed farms. Slopes greater than 25 per cent are used most frequently for permanent pasture or as farm woodlots. On the more erodible lands, cultivation should stop generally at gradients of about 10 to 15 per cent. All fields should be given adequate protection with sound conservation measures.

Soil erosion in this general region has long been unnoticed; its influence on decreasing farm efficiency has not been fully appreciated, partly because the most seriously affected fields usually are turned out to grow over with brush, grass, or weeds, which quickly hide the effects. A primary need of the area, consequently, is a more general recognition of the necessity for taking steps leading to the conservation of soil resources and

Fig. 264.—Pasture showing results of gullying. Newer gullies in grain field below. Steuben County, New York. (*Photograph by Soil Conservation Service.*)

reduction of the flood hazard through control of runoff from the land. It is also essential that sheet erosion be recognized as an important cause of declining agricultural efficiency, since this form of land depletion is one of the principal reasons why millions of acres of land, especially in the Appalachian Plateau, have become submarginal or have ceased to be occupied as farm lands.

In practically all parts of the general farming and dairying sections, many slopes in excess of 10 per cent are used for cultivated crops. On most farms, all of the processes of preparing and cultivating cropland are performed in rectangular fields without regard to the direction or degree of surface slope (Fig. 265). This practice has been one of the chief causes of soil loss by erosion. To permit early spring seeding of small grain, fall plowing of corn stubble and even of sod land is a common practice at the higher altitudes of the dairy farming section. This leaves the surface exposed to fall and spring rains and the runoff of melting snow before the grain crops have attained sufficient height and density to provide adequate protective cover. Also, lands from which potatoes and vegetables have been harvested are left without winter cover, with attendant soil losses in fall and spring. Vineyards, peach orchards, and some apple orchards usually are clean cultivated and left without fall or winter cover crops, giving rise to serious soil washing. In sections used intensively for vegetables and canning crops, the land is cultivated the greater part of the year. Grain and grass cover are provided only for limited areas. Cover crops are not commonly employed. Under these conditions, erosion is serious even on gentle slopes.

Fig. 265.—Up- and downhill farming promotes erosion. Centre County, Pennsylvania. (*Photograph by Soil Conservation Service.*)

Recently, many upland pastures have become thin of cover on a large eroded area; and lime and fertilizer, along with water conservation, adjustment to carrying capacity, seasonal grazing, and rotation grazing, are necessary to reestablish a good sod. Sheet washing has removed so much soil that many pastures are now covered with a pavement of stone. Much land of this kind has been allowed to revert to brush.

More than 16 million acres of farm land are devoted to pasture in addition to 8 million acres of pastured woodland. A large part of this occupies slopes in excess of 20 per cent. Few pastures have been fertilized, limed, or reseeded, so that the greater part of this land does not support sufficient grass to protect it from washing. Retirement of the steeper and more impoverished tracts to forest and the general improvement of the better lands is an important and widespread need.

Contour planting of new orchards and terracing of existing orchards are essential steps in controlling soil erosion in the orchard belt of south-central Pennsylvania, western Maryland, and eastern West Virginia.

Many of these orchards occupy rather steep slopes of erodible soil. Water from higher elevations is discharged across many of them. Clean cultivation is a common practice. In some instances, diversion terraces are required, and increased use of cover crops is desirable.

The orchard and vineyard areas of the Hudson Valley, the Finger Lakes region, and western New York generally need terrace planting, terracing of existing plantings, and more extended use of cover crops.

The Problem on Croplands. Twenty-six million acres in the Northern Appalachian-New England area are classed as cropland. This includes all lands used for hay, grains, corn, potatoes, tobacco, orchards, vineyards, small fruits, vegetables, and canning crops. Retirement of the steeper slopes used for these crops to permanent pastures or the more permanent forms of hay production is urgently needed. Equally necessary is a general readjustment of field boundaries to permit contour cultivation and strip cropping. Diversion terraces are needed to break up the accumulated flow from long slopes. These slopes sometimes exceed a half mile in length, and destructive gullying as well as sheet washing results from unimpeded runoff across them. On more gentle slopes, especially in areas of intensive cultivation, field terraces help reduce erosion. Greater use of fall and winter cover crops and conservation of crop residues and stubbles should be encouraged. In both dairying and general farming sections, adjustment of the general procedure in crop succession is needed. At present, some rotations favor too prolonged use of meadows, and several of the customary short rotations permit too frequent exposure of the soil to rainwash. Much land should have its organic supply replenished, and this can be accomplished through increased use of legumes and grass in the rotations.

The Problem of Farm Woodlands. The existence of more than 20 million acres of woodland on the farms of the Northern Appalachian-New England area provides an excellent opportunity for the spread of proper methods of farm-woodlot management, which would contribute to better control of erosion. When properly cared for, these woodlands have a marked influence in retarding and distributing the runoff of melted snow, thus giving added protection to lower lying, more valuable farm lands and probably causing some reduction of floods. Eight million acres of farm woodland in the region is classed as *pastured woodland.* In practically all instances, the water- and soil-conserving possibilities of such woodland would be improved by fencing out livestock. This also would permit cheap reproduction to perpetuate the wooded area. The replanting of steep slopes, especially in pastured areas, leads both to the protection of the soil and to the retirement of hazardous land from grazing use. Probably more than a third of all of the land in farms is capable of improvement from the standpoint of soil and water conservation.

The Problem on Pasture Lands. More than 16 million acres, in addition to the pastured woodland, are classed as *land in pasture.* A large proportion of this occupies slopes steeper than 20 per cent, some of it having a declivity of more than 45 per cent. Generally, farmers have not maintained their pastures by such needed treatment as fertilization, liming, and reseeding, so that the greater part is inadequately protected from erosion—principally sheet washing. Retirement of the steeper and rougher areas to forest and general improvement of the better parts are a widespread need.

## Four Years of Work

The first erosion-control work by the Soil Conservation Service in the Northeastern States was carried out on demonstration projects in various problem areas.

Forty-five of these projects have been established in nine states since the beginning of operations in 1934. By Jan. 1, 1938, work was under way or completed on 4,260 cooperating farms, covering 558,462 acres. A work plan was prepared for each farm, and the different kinds of land were treated according to their individual needs and adaptabilities, with careful consideration given at the same time to the economic needs of the farmer.

On these farms, the soil and water conservation program had brought an increase of more than 43,000 acres to the area of woodland. In addition, woodland plantings were completed on 21,121 acres, and forest stand improvement was demonstrated on 22,609 acres. Eleven hundred thirteen miles of fence were built, chiefly for the purpose of excluding grazing from farm woodlots. Plantings for the benefit of wildlife covered 3,372 acres.

Nearly 13,000 acres of pasture had been retired to other uses. The area in permanent hay had been increased from 16,000 to more than 45,000 acres. Approximately 37,000 acres of pasture and meadow had been treated with lime and fertilizer and reseeded to grass and legumes where necessary. An area of three thousand acres was contour-furrowed.

The introduction and demonstration of contour farming in its different forms is one of the chief features of the program in the Northeastern region. This practice was practically unknown until 1934, when the demonstration work of the Soil Conservation Service (then known as the Soil Erosion Service) began. Now, nearly 75,000 acres are cultivated on the contour. Strip cropping is practiced on virtually the same acreage. Contour planting of orchards was practically unknown, but some 200 acres of new orchards have been planted in this way since the system was first demonstrated in the Blue Ridge fruit belt of southern Pennsylvania and western Maryland in 1935 (Fig. 266).

Terracing was not practiced in the region before 1934, but nearly 2,000 acres had been protected by 157 miles of these structures by 1938 (Fig. 267). The use of diversion ditches to intercept and carry away

FIG. 266.—Young peach orchard, 1938, planted on the contour in 1936 under the demonstration of the SCS, Adams County, Pennsylvania. Cover crop of rye disked in. (*Photograph by Soil Conservation Service.*)

FIG. 267.—Diversion terrace, Harford County, Maryland, strip cropped with barley above and below terrace. (*Photograph by Soil Conservation Service.*)

excessive accumulation of water on long, erodible slopes was being demonstrated by 250 miles of such structures. Forty miles of outlet channels for disposing of water from terraces and diversion ditches had been constructed. The practical efficiency of gully-control measures was being demonstrated by 2,000 permanent and 13,000 temporary gully-control structures and by the sloping and seeding of 765,000 square yards of gully banks. Stream-channel and stream-bank improvements had been made along 272,000 linear-feet of streams, where floods seriously endangered or partly destroyed valuable bottomlands in the past.

As a result of the interest aroused by these demonstrations, three states in the region already have passed soil conservation districts acts, and a number of districts have been established. Thus, within 3 years from the inception of work in the Northeastern region, the soil conservation program began to pass from the government demonstration stage to large-scale district programs, legally constituted and governed through community action.

## Work of Other Agencies

Other agencies than the Soil Conservation Service have had a part in furthering soil and water conservation in this region. The Forest Service is one. National forests comprise over 3 million acres in Maine, New Hampshire, Vermont, Pennsylvania, Virginia, and West Virginia. Within these areas, the standard practices and regulations of the Forest Service with regard to forest protection and soil and water conservation are being applied. In addition, the 12 states in the region have set aside state forest tracts aggregating nearly 5 million acres. Still other lands have been included in state and municipal parks. Their protection represents a valuable contribution to soil and water conservation.

Some of the states have provided for the gradual acquisition and conversion of marginal and submarginal farm lands for purposes of reforestation and recreation. Part of the land so acquired is replanted to trees; part is already wooded; and part is allowed to revert naturally to a wild state. A natural or voluntary cover of trees, brush, and other vegetation aids in the stabilization of soil and the conservation of water in headwater areas.

It is estimated that some 37 million acres of wooded land in the region is privately owned, in addition to some 20 million acres classed as *woodlands in farms*. Some of this privately owned woodland is being carefully and systematically used to perpetuate the stand. Much of it, however, is being cut over for various lumbering purposes, without adequate provision for fire protection and other good forestry practices such as should aid in the control of erosion and floods.

Provisions of the Agricultural Conservation Program of the AAA have encouraged the use of many of the practices that have been found effective in soil and water conservation. In every state in the region, AAA payments are made for the improvement of meadows and pastures by liming, fertilization, and seeding. The use of cover crops is encouraged. Woodland management for the improvement of stand, the exclusion of grazing from farm woodlots, and planting of forest trees are included in the program. In several states, provision is made for replanting steep slopes to trees. Strip cropping and contour cultivation have been set up as standard practices in New Hampshire and New York. Contour furrowing in pastures is an approved practice in Pennsylvania. Terracing is specially included as a part of the program in Delaware and Pennsylvania.

All these practices assist materially in soil and water conservation, although when used separately they fall short of a complete, effective treatment of the individual farm.

## Floods

The principal rivers have their headwaters in the Appalachian region. Those forming in the more mountainous sections originate chiefly in forested territory, although some of the tributaries and parts of the trunk channels flow through cultivated valleys. The Merrimac, Connecticut, and Delaware Rivers are examples. The north and west branches of the Susquehanna and the Juniata River rise in localities partly forested and partly cultivated. The Allegheny, Monongahela, Cheat, Great Kanawha, Little Kanawha, and Big Sandy Rivers of the upper Ohio drainage originate in country that is largely forested but flow through areas encompassing much farm land.

Flood damage in the Northeast has been most spectacular along the trunk streams. The total damage caused by these big river floods, however, may be less than that caused by the more frequent overflows every year along the numerous tributaries. These tributaries, frequently overflowing their alluvial plains, cut new channels and deposit worthless sand, gravel, and inferior subsoil material in countless places, thus damaging, annually, a large aggregate body of the most productive agricultural land in the region. General adoption of erosion-control and water-conservation practices on upstream lands needing treatment would give some degree of protection to downstream areas.

## Economic and Social Implications

The area of land in farms in the Northeastern region increased until about 1880. Since that time, a gradual and rather general decrease has ensued. A considerable part of this reduction may be charged to changing economic conditions and to the shifting of the agricultural population to Western lands. Part of it is due to the concentration of agriculture on lands capable of intensive cultivation to specialty crops and in localities near metropolitan markets. But an important part of the decrease must be attributed to decreased productivity of sloping lands (Fig. 268). This impaired capacity to produce is due chiefly to long-continued and unrecognized sheet erosion, frequently accompanied in its later stages by gullying. The extent and rate of such soil losses have been widely observed and accurately measured only within recent years. Yet the accumulation of soil materials on the uphill sides of stone fences and hedgerows, the removal of soil from orchards and vineyards, and the increasing number and extent of areas where raw subsoil clay is exposed in sloping fields all bear evidence to the cumulative effect of soil erosion. Because of lowered soil productivity, the area of land that can be farmed successfully has been reduced. Hundreds of thousands of acres of farm

FIG. 268.—Farm completely abandoned as result of erosion. Steuben County, New York. (*Photograph by Soil Conservation Service.*)

land now classed as marginal or submarginal are testimony to the combined effects of soil erosion and adverse economic conditions.

The preservation of the better farm lands from further decline or ruin by erosion constitutes a major problem of the region, affecting not only the men who till the land but the people of the cities who depend to an important degree on the farms of the Northeast for dairy products, fruits, and vegetables.

Each interruption in the steady flow of food products from farm to city market, whether caused by deep winter snows or by floods, emphasizes the necessity of preserving the soil resources of the region. Protection of the intensively farmed areas is imperative, not only because of their importance to those living on the land but because so many local communities depend to such a large degree on the welfare of the agricultural population. Soil conservation is required on a large area of farm land to prevent its slipping across the border from agricultural occupation to a condition of marginality or abandonment. Over the greater part of the region, proper methods of soil and water conservation are still capable of maintaining and improving the productive capacity of the agricultural land.

## Piedmont Plateau

The northern Piedmont Plateau, with an area of some 22 million acres, extends from the Potomac River northeasterly to the St. Croix River, which forms the boundary between southeastern Maine and New Brunswick. From the Potomac to Raritan Bay, in New Jersey, its eastward margin is overlapped by sedimentary deposits of the coastal plain. This outer border varies in elevation from a few to about 200 feet above sea level. Practically the entire New England sector borders Long Island Sound or the Atlantic Ocean. The inner boundary contacts the Appalachian ridges at elevations rising locally to about 1,000 feet, but the line of demarcation is indefinite in many places. A number of ridges and many isolated hills rise above the general level of the Plateau. The surface is rolling to hilly and is crossed throughout its extent by the major stream drainages of the Northeastern States. Numerous tributary streams have cut the surface into an intricate pattern of rolling hills and comparatively narrow valleys.

In general, the soils of the southerly sector of the northern Piedmont are derived directly from the underlying crystalline rocks, including granites, gneisses, schists, slates, and basic intrusives, and from the sandstones and shales of the Triassic basins of Maryland, Pennsylvania, central New Jersey, and the lower Connecticut Valley. They range from sandy loams to clay loams in texture. All of these soils are utilized almost to the limit of arability by a general type of agriculture. Forests are confined chiefly to the rougher and more stony parts of the Plateau. Corn, tobacco, potatoes, wheat, and hay constitute the most important crops, with small areas of pasture on nearly every farm.

The principal soil types are loam and clay loam of the light brownish Chester group, derived from granitic rocks; loam and silt loam of the purplish-red Penn group, derived from Triassic sandstone and shale; and red clay loam of the Montalto group, from basic igneous rocks. All these are moderately productive. They are gently rolling to rolling or hilly and usually of rounded configuration. On slopes greater than 2 or 3 per cent, they are subject to erosion in unprotected fields, especially where crop rows follow the direction of slope rather than the contour. Many of the steeper slopes have been stripped of surface soil by sheet washing, and a considerable area has been affected by gullying.

The northern extension of the area is more rolling and hilly, and the soils have been formed from glacial material derived largely from the underlying rocks, with some contribution of material from outside areas.

The soils consist principally of sandy loams, gravelly loams, and loams, with numerous stony areas, some very rough. Gloucester and Troy are the principal soil groups. Because of their relatively high porosity, due to rather high content of rock fragments, erosion of these soils is not so severe as on the soils of the southerly sector of the northern Piedmont. Locally, however, many areas are susceptible to serious washing. Even the gently sloping loams and fine sandy loams of the Connecticut Valley wash seriously during heavy rains (Fig. 269).

Fig. 269.—Erosion of bare field by rain preceding and accompanying hurricane of September 21, 1938, near Rockville, Connecticut. (*Photograph by Soil Conservation Service.*)

Only the smoother, less stony lands are cultivated. A large part of the area remains in forest. The smooth stream terraces or *intervales* are the best and most intensively farmed areas.

## Appalachian Mountains and Valleys

Immediately to the west of the Piedmont are the Appalachian ranges and their valleys. The great mountain masses of the Adirondacks in New York and the White-Green Mountains of New Hampshire, Vermont, and western Maine, together with the higher ridges of northern New Jersey and those along the New York-New England line, rise to elevations ranging from about 2,000 to a maximum of 6,293 feet (Mt. Washington, New Hampshire) above sea level. Because of elevation, latitude, rugged topography, and rocky surface conditions, agriculture is not extensive. Large tracts are occupied by national and state forests and state parks. Farming is confined to the gentler, lower slopes and consists chiefly of dairying, with some localized potato production. This area includes approximately 39 million acres.

From northern New Jersey southwestward through central Pennsylvania, west-central Maryland, and eastern West Virginia, the Appalachian region consists of high, nearly parallel or interlocking mountain ranges, with deep valleys.

The ridges rise to extreme elevations of from 2,500 to 3,000 feet and overtop the enclosed valleys by elevations of 500 to 1,200 feet. These ridges are steeply sloping and prevailingly rocky and rough. Consequently, only small tracts of the dominant loam and silt loam soils have been used for agricultural purposes. The remainder is covered chiefly by brush or timber.

The floors of the valleys, whose soils are derived from limestone, shale, and sandstone, are generally rolling and rather deeply cut by the main streams crossing them as well as by some of the local tributaries. These valleys have long been used for a type of general farming that includes the production of corn, wheat, oats, and hay. Many farms support a mixed dairy- and beef-cattle industry. The limestone soils of the main Appalachian Valley, as well as a large number of smaller valleys, are among the most productive in the Northeastern States. Some of them have sustained a high type of general farming for more than 200 years.

The principal limestone soils—Hagerstown silt loam and loam—are naturally highly productive. Unfortunately, they are subject to impoverishing erosion on all unprotected slopes. Already, many fields and parts of fields have been changed from silt loam or loam to less productive clay loam. Erosion is also active on the cultivated sandstone and shale soils of the mountains and ridges—principally the Dekalb and Berks shale loams. These are shallow soils, which too frequently are cultivated on slopes entirely too steep to avoid destructive washing. In all instances, careful defense measures must be used if these mountain and sloping valley lands are to be preserved.

## Appalachian Plateau

The northern extension of the Appalachian Plateau stretches from the southwestern boundaries of West Virginia and southern Ohio in a northeasterly direction through Pennsylvania to the Adirondack Uplift in northern New York. Its eastern boundary, known throughout a considerable part of its extent as the *Allegheny Front*, rises to elevations of 2,500 to 4,000 feet above sea level, overlooking the mountain ridges and valleys on the East. Throughout West Virginia and western Pennsylvania, the Plateau slopes toward the valleys of the Ohio and Allegheny Rivers, its low elevation of about 600 feet occurring near the mouth of the Big Sandy River, near Huntington, W. Va. The section in southeastern Ohio also slopes gradually, both east and west, from a maximum elevation of about 1,300 feet. The northern part of the Plateau, which extends from the Catskill Mountains westward to Lake Erie, slopes from elevations of about 2,500 to less than 200 feet along the St. Lawrence River. Several distinct subdivisions of this northerly sector of the Piedmont are described below. The northern Plateau in its entirety comprises approximately 65 million acres.

### Allegheny-Ohio Section

The more southerly part of the Allegheny-Ohio section is rolling to hilly along the Ohio River and its smaller tributaries. The higher country presents more of a true plateau aspect, although it is deeply cut by such major streams as the Allegheny, Monongahela, and Kanawha. At the higher elevations along the eastern and northern boundaries of this section of the northern Appalachian Plateau, much of the country is forested or grown up to second-growth timber and brush. Locally, in the vicinity of coal mines and industrial centers, dairying, potato growing, and subsistence farming have given rise to considerable agricultural development.

In a broad belt along the Ohio River and its principal tributaries in West Virginia, western Pennsylvania, and southeastern Ohio, most of the land is rolling to hilly. Much of it is used for corn, oats, wheat, hay, and pasture, with local areas devoted to vegetables for the industrial centers. The soils consist chiefly of silt loams, loams, and clay loams derived from the underlying sandstone and shale, with an intermixture of limestone in the Ohio Valley. The principal groups are the Muskingum, Westmoreland, Meigs, and Upshur.

Thousands of people in this area combine some farming with work in the mines or industrial plants. The continuing ability of this land to help support the population depends to a considerable degree on effectively checking soil erosion. Farming, to be conducted at all, must in many instances be done on slopes in excess of 10 per cent. Severe sheet erosion and a considerable amount of gullying already have occurred. To insure permanency of agriculture in this dairying and general farming area, soil and water conservation measures are greatly needed. Chief among these is contour strip cropping of all cultivated sloping land. This should be supplemented by diversion terraces as required. Land steeper than about 15 per cent should not be cultivated, and slopes in excess of 25 per cent should be retired to woodland. Intermediate slopes should be used for pasture, and the pastures improved to provide better grazing as well as better protection against sheet washing. Contour furrowing in steep pastures is needed to save soil and to retain moisture in the soil as a reserve against midsummer periods of low rainfall. Gullies should be controlled by temporary and permanent check dams.

The Moundsville, W. Va., demonstration project, located in the panhandle of West Virginia, illustrates how soil and water conservation practices can be used on the highly dissected terrain of the Allegheny-Ohio section of the Appalachian Plateau.

Here, the principal soils are Westmoreland silt loam and gravelly silt loam, derived from interbedded sandstone, shale, and limestone of the Coal Measures. They are well drained, except for occasional seepy spots. Although the soil mantle is normally shallow, natural productivity is comparatively high. Topographic features are very irregular. Most of the land is steep, slopes ranging from 25 to 40 per cent or more over about two-thirds of the project area and from 15 to 25 per cent over about 30 per cent of the area. The valleys are V-shaped, with narrow flood plains along the streams. The lower slopes, usually representing the steeper areas, are mainly wooded. Concentration of farming on ridge crests gives the landscape a striking feature.

Dairying is practiced extensively, and the Westmoreland soils are well suited to this type of farming, which is so largely dependent on pasture and the production of hay. Favorable local markets for dairy products are provided by the cities along the Ohio River, near the project.

Forty per cent of the area was under crops at the time work began; 30 per cent was occupied by pastures; 24 per cent was wooded; and 6 per cent was idle or in urban or miscellaneous occupancy.

Sheet erosion had severely damaged about 10 per cent of the area, and moderate to severe sheet erosion was found on an additional 83 per cent. Thus, approximately 93 per cent of the land had been seriously affected by soil losses. In addition, gullying was found on 28 per cent of the area, and slides on 16 per cent.

Sheet washing had removed so much of the topsoil from many farms that they could not be cultivated profitably. A considerable acreage of formerly tilled land has been turned back to pasture or brush within the past 15 years. To replace this land, additional areas of pasture and woodland have been brought into cultivation. This process of shifting land use exposes additional land, most of which is less resistant to washing than the "farmed-out" tracts were originally.

Soil slips or landslides are a characteristic feature on the Westmoreland and associated soils of West Virginia, southwestern Pennsylvania, and southeastern Ohio. These displacements occur on moderate to steep slopes, irrespective of cover. During protracted rains, great masses of waterlogged soil slide and flow downhill, exposing raw subsoil and forming mounds, ridges, and benches where the soil piles up along the lower part of the slip. These conditions frequently result in accelerated runoff and gullying and usually make cultivation impracticable.

A variety of defense measures have been found necessary for effective erosion control in this part of the Appalachian Plateau. Woodlands must be protected from fire and overgrazing. Steeper lands of high erodibility must be replanted to trees or retired to permanent pasture. Pastures, thinned of grass and eroded because of overuse and neglect, require improvement by liming, fertilizing, reseeding, and better grazing management as well as moisture conservation by contour furrowing (Fig. 270).

Fig. 270.—Soil and water conserved by contour furrows, in a West Virginia pasture. (*Photograph by Soil Conservation Service.*)

Contour strip cropping is proving one of the most effective and popular methods of soil defense, because it permits the cultivation of steep land with a minimum of soil and water losses. Supplementary measures that have proved their worth are (1) long crop rotations which permit a considerable part of the land to remain in grass cover, (2) diversion ditches, and (3) permanent strips of sod. Alfalfa, on suitable soils, is helping considerably.

These and other conservation measures, with only minor modifications, are applicable to a broad belt, at least 200 miles wide, along both sides of the Ohio River from Pittsburgh to the mouth of the Scioto River.

### Delaware-Susquehanna Section

That part of the Appalachian Plateau north of the glacial line in Pennsylvania, together with most of the New York portion of the

Plateau, constitutes the Delaware-Susquehanna section. This area is characterized by rolling to nearly level plateau country, deeply cut by the headwaters of the Delaware River, the north branch of the Susquehanna, the Genesee, and some of the tributaries of the Allegheny. Long, moderate to steep slopes separate the general upland level from the stream bottoms or benches (Fig. 271). The soils have been formed from shallow glacial till derived from sandstone and shale. They are prevalently loams and silt loams of the Lordstown, Canfield, Volusia, and associated series, usually containing a high percentage of a platy sandstone or shale chips. At the higher elevations, quite extensive tracts are

Fig. 271.—Pattern of farming in hill country of west-central New York. East side Cohocton Valley, Steuben County. (*Photograph by Soil Conservation Service*)

marked by the occurrence of a distinct claypan at depths ranging from about 3 to 18 inches below the surface. They are also distinctly acid. On the steeper slopes, the soils consist of stony loams of depth frequently less than 2 feet to shale or sandstone. Erosion is serious on much of the cultivated land. On the more gentle slopes, various soils, chiefly stony and gravelly loams, are derived principally from glacial till having a depth ranging from about 3 to 150 feet in extreme cases. The outwash soils of the valleys are usually gravelly loams or loams at the higher levels and sandy loams, loams, or silt loams within present range of annual or occasional overflow. Swampy areas, caused by inadequate outlets or overflow, occur both on the top of the Plateau and in the larger valleys.

Abundant precipitation and a short growing season have favored the development of dairying as the principal type of agriculture. On the majority of farms, the rougher land remains in woodlots; steeply sloping areas are held in permanent pasture; and the arable land is used for corn, potatoes, oats, and hay. At the higher elevations and on poorly drained lands, buckwheat is the principal cash crop. It is about the only one that can be matured in the short growing season.

The Cohocton River demonstration project, within the watershed of Cohocton River, near Bath, N. Y., comprises about 152,000 acres. The northern part has a large area of well-drained land, much of which is intensively used for potatoes. The southern part, with a larger proportion of imperfectly drained land, is devoted almost entirely to dairying, and much more grass is grown here. Thus, the project embraces types of agriculture that are representative, in a general way, of the principal farming methods in effect over much of the northern part of the Appalachian Plateau. Erosion-control measures employed here are applicable to much of this northern sector.

About half of the project area is devoted to crops, and a fifth to woods; the remainder is about equally divided between idle land and pasture.

The soils are largely gravelly silt loams derived from glacial material. About a third of the area is covered with soils having a compact, impervious subsoil which seriously impedes underdrainage. About a sixth of the land has bedrock within 3 feet of the surface. Surface soils are generally shallow, and soil profiles are imperfectly developed. Except in a few instances, weathering has extended downward not more than about 30 inches.

The topography is that of a dissected plateau, with level valley floors and long slopes of varying steepness extending up to gently rolling ridge tops. About 10 per cent of the land has slopes steeper than 35 per cent. At the beginning of work on the Cohocton project, approximately 30 per cent of the cropland was on slopes steeper than 15 per cent.

Agricultural settlement began about 150 years ago, and much of the land has been farmed for at least 100 years. The rectangular system of land lines, with fences running according to the compass, rather than slope contour, has resulted in considerable off-contour cultivation as well as destructive concentration and rapid discharge of water from the lower sides of fields. Shallow spring thawing of bare topsoil over frozen subsoil favors severe sheet washing at this time of the year. Frequent and intense thunderstorms during the summer cause considerable gullying, with attendant sheet erosion in most of the cultivated fields.

According to a survey made in 1935 and 1936, very little land had been completely ruined by erosion at that time. Forty-five per cent of the area had suffered moderate sheet erosion, and 12 per cent had been severely washed. About 10 per cent had suffered from gullying, part of it to such a degree that cultivation was restricted on some farms by gullies less than 100 feet apart. Some lower slopes and stream bottoms

FIG. 272.—Above, steep hillside of formerly productive land, reduced to a condition of general crop failure by erosion, west-central New York, 1935. Below, same farm stabilized and improved by rotation strip cropping, 1938. (*Photographs by Soil Conservation Service.*)

had suffered by deposition of gravel and poor subsoil material washed out of gullies.

A considerable part of the area is in grass and trees that afford considerable protection against erosion. In spite of the ease with which vegetative methods of controlling erosion can be established, only three or four farmers were making a conscious effort to utilize such methods prior to the beginning of the project in 1935. They were using strip cropping with good success, except in one instance where improper selection and rotation of crops resulted in failure (Fig. 272). Few instances were observed where eroding waterways across fields were protected with permanent sod. Many of the long slopes are fairly regular and admirably suited to contour tillage, but the practice was not being followed. A few of the steeper slopes were being cultivated approximately on the contour.

Generally, the defense measures adopted for this project have emphasized the use of permanent close-growing cover for the steepest, most erodible land; the employment of proper agronomic and engineering practices on cultivated slopes; and the closing of gullies. Retirement of highly erodible land to permanent cover has called for the planting of trees and shrubs suitable for wildlife feed and cover; the establishment of grass; and improvement of old pastures by reseeding, fertilizing, and liming. Woodlots have been improved; pastures containing droughty soils have been contour-furrowed for conservation of rainfall; and long-time meadows including alfalfa have been established.

The control measures adopted for sloping cultivated lands represent a judicious combination of agronomic and engineering practices. The former include the use of a good crop rotation, contour tillage, strip cropping on the contour, sodded waterways, and winter cover crops. Engineering structures include diversion ditches to break the off-flowage on long slopes and to remove excess water from soils with impervious claypan subsoil; and some field terraces. Gullies are controlled by planting erosion-resistant vegetation on the banks, by using dams or other structures where needed, and by diverting water where practicable.

In the three years prior to Apr. 1, 1938, the 335 cooperating farmers, operating 43,000 acres, adopted a variety of erosion-control practices that have resulted in a reduction of 1,300 acres in the area devoted to clean-tilled crops, the improvement of 1,400 acres of pasture, and the introduction of contour strip cropping on 11,000 acres. Approximately 2,600 acres of steeply sloping and rocky lands have been planted to trees and shrubs, and 11,000 acres of farm woodlots have been placed under improved management. Many miles of diversion ditches have been constructed to stabilize gully heads and protect long cultivated slopes.

## Finger Lakes-Mohawk Section

From the vicinity of Utica westward to Buffalo lies the rolling to hilly Finger Lake-Mohawk country. Its southern boundary follows an extremely irregular line south of the Finger Lakes of central and western New York, at elevations between 500 and 1,000 feet. The northern boundary, at 300 to 500 feet above sea level, parallels the southern shore of Lake Ontario from the vicinity of Oswego, N. Y., almost to Lake Erie. The surface is extremely varied, ranging from hilly and rolling to nearly level, with many abrupt descents to the numerous lakes and drainageways (Fig. 273).

Fig. 273.—Characteristic steep lake border topography, Keuka Lake, New York. (*Photograph by Soil Conservation Service.*)

The soils prevalently consist of loams and clay loams containing varying amounts of stone and gravel. They are principally of glacial origin, although small areas of primary soil derived from local limestones and calcareous shales are found within the area. The principal series are the Honeoye and Ontario.

The excellent productivity of much of the land, together with the good transportation facilities and proximity to industrial centers, has prompted the development of a high type of general farming over the greater part of the area. Corn, beans, wheat, oats, alfalfa, and mixed hay constitute the principal crops. A vineyard industry of considerable magnitude has been established in the Finger Lakes region. Apple orcharding is also important over many parts of the area.

Erosion is serious, locally, on sloping cultivated lands where contour tillage is not practiced.

## Eastern Great Lakes-St. Lawrence Section

The Eastern Great Lakes area comprises a narrow strip through northeastern New York along the southern shore of Lake Erie. Eastward, this is succeeded by a broader plain, which skirts the south shores of Lake Ontario and the St. Lawrence River to the Champlain Valley, where it extends southward to and along the upper Hudson River as far as the mouth of the Mohawk. Its elevation is never higher than 600 feet, the greater part being only about 250 to 350 feet above sea level. The surface is dominantly gently rolling to nearly level. It is crossed by many streams directly tributary to the Great Lakes and St. Lawrence drainage, some of which have cut deep channels, although many of them flow through nearly level, somewhat swampy bottomlands.

The soils of the western part of the area, from Oswego, N. Y., to Ashtabula, Ohio, consist chiefly of sands, sandy loams, loams, and stony loams, largely derived from glacial lake material. Extensive bodies of calcareous till have given rise to loam and silt loam soils.

Along the St. Lawrence River and in the Champlain Valley, considerable level and undulating land is imperfectly drained. The principal soils are derived from water-laid deposits and consist of dark silty loam overlying stiff, gray clay loam or clay.

The western sector supports a diversified agriculture, with orchard and vineyard crops dominating. Apples, peaches, pears, and grapes are the most prominent fruit crops. Vegetables for direct marketing and for canning are produced extensively. Corn, wheat, and hay are important crops. In the St. Lawrence-Champlain area, oats and hay are the principal crops, with corn and potatoes grown to some extent.

In this area erosion is of local importance only.

## Aroostook Section

The Aroostook Area occupies the northeastern part of Maine. It is characterized by a gently rolling to hilly surface with moderate slopes. The principal soil is a mellow silty loam derived from material formed by glaciation of the underlying calcareous shales and sandstones (Caribou loam). Favorable soil and climatic conditions account for the extensive production of potatoes. As large an area as possible is planted to the crop every year. Heavy fertilization and intensive cultivation are depended on to maintain yields.

Intensive cultivation of the largest possible acreage to potatoes makes the Aroostook section of Maine a special problem area. The loamy soil, occupying slopes up to 15 per cent, is subject to severe erosion on the steeper land. Broad-based terraces are being used effectively to reduce soil loss without reducing the area to be planted. Strips of sod on the contour and contour cultivation materially supplement the efficiency of terraces.

The Presque Isle, Maine, project, covering approximately 36,300 acres, is located in the potato district of Aroostook County. The nature of the soil and the cropping practices in this potato-growing area present a serious problem of soil conservation.

The soils have been formed by weathering of glacial till composed largely of material derived from shaly limestone. For the most part, they are well-drained, friable, and open and have an acid reaction near the surface. Loam and shale loam predominate. They are characteristically shallow, overlying unweathered till at depths of 18 to 24 inches.

The land is undulating to strongly rolling, with generally long, fairly uniform slopes. Here and there, abrupt slopes and limestone outcrops make cultivation difficult, but most of the land is easily cultivable. Most slopes having a gradient of 20 per cent or more are in woods, small pastures, or idle. The gentler slopes, ranging from about 5 to 10 per cent, are more extensive, especially back from the Aroostook River and through the central and southern portions of the project area.

Estimates of land use, in percentage of the area, are as follows: potatoes, 30; hay, 25; oats, clover, and barley, 20; pasture, 8; forest and waste land, 17.

The growing of potatoes is highly specialized, and the returns from this crop represent the chief cash income of farmers. Scientific measures of fertilization, seed selection, and disease control are in general use, and modern implements are employed. Because of the short growing season, potatoes are planted as soon as the soil is in condition, and harvesting is so late in the fall that it is difficult to establish a winter cover crop afterward. Fertilizers have aided in the maintenance of yields where erosion has not been too active. It is generally recognized that the soil supply of organic matter must be kept up if good crops of good quality are to be grown. The better farmers are using more organic matter; but in too many instances, potatoes are planted year after year in the same fields with no addition of organic matter.

Such crops as hay and the small grains are not generally considered by the farmers to have great importance in the farm economy.

Most of the cultivated land has been damaged in some degree by erosion or by deposition of eroded material. Most of the soil loss is from potato fields. Where rows run directly up and down the slope, the depression between the ridges deepens quickly and readily during heavy rains. Serious sheet washing takes place after the potato harvest on sloping land permitted to remain uncovered. Snow protects the land in winter, but much washing takes place in spring before small grain or crimson clover becomes well enough established to afford proper protection. Fields used consecutively for potatoes have no protection, of course, during these critical periods.

Deep gullies sometimes form during heavy summer rainstorms; but since they interfere with cultivation and the use of spraying equipment, they usually are filled in promptly. As a result, gully erosion is not readily apparent except on neglected farms. Gullies not filled in or controlled with vegetation or other measures soon cut down to bedrock beneath the till. Some of these are 6 to 10 feet deep; a few 20 feet deep have been observed.

FIG. 274.—Contour cultivation and a sodded waterway to control erosion in Maine potato district. (*Photograph by Soil Conservation Service.*)

Since the establishment of the project in April, 1936, cooperative work has been carried on with 66 farmers, operating nearly 10,000 acres. Various measures, such as contour cultivation, terracing, strip cropping, pasture improvement, and water diversion (Fig. 274), are being employed effectively. The steeper lands are being retired to protective plantings; rotations are being improved; gullies controlled; and water outlets installed.

# Chapter XXXIII. Central Prairie and Eastern Timbered Border Region

The Central Prairie and Eastern Timbered Border region includes all of Michigan, Minnesota, Wisconsin, and Iowa; nearly all of Illinois and Indiana; about two-thirds of Ohio; approximately the northern half of Missouri; the eastern half of Kansas; the eastern third of Nebraska; part of northeastern Oklahoma; and a relatively narrow strip across the eastern edges of North and South Dakota (Map 10). In its broadest dimensions, this complicated region extends southward from the Canadian boundary about 950 miles to the Ozark Highlands in northeastern Oklahoma and from the Great Plains in east-central Nebraska some 950 miles eastward to the Northern Appalachians in northwestern Pennsylvania. It comprises an area of approximately 288 million acres and supports a dense population. A vast network of railways and highways crosses it to connect all sections with industrial and commercial centers, and an enormous commerce has been established on the Great Lakes.

Broadly, the region embraces the lower part of the floor of the upper Mississippi Basin east of the Plains. The central and western parts are prevailingly prairie country, but a large area along the northerly and easterly borders originally was timbered. Predominantly, the region is a nearly level to undulating or rolling plain. Over the smoother terrain, the slopes seldom exceed 4 or 5 per cent; but over the more rolling or hillocky morainic portion, gradients often reach 8 to 10 per cent, and locally 15 to 20 per cent. The prevailing surface of the old lacustrine deposits and the alluvial terraces and flood plains is nearly level, or rarely steeper than one-half of 1 per cent. Over much of the upper lake region and in the territory adjacent to the Appalachian Highlands on the east and southeast, the topography ranges from rolling to hilly. The larger streams usually are bordered with strips of rolling to broken or hilly country, with many slopes too steep for cultivation. The average elevations above sea level range from about 500 feet in the south to 1,000 feet or more in the north. Drainage of the upland soils is generally well established, except in some of the nearly level bodies, such as the claypan soils (Putnam) and the heavier types of prairie (as the Webster soils of Iowa). The drainage of much of these level to slightly depressed areas has been improved by the establishment of artificial outlets. About 60 per cent of

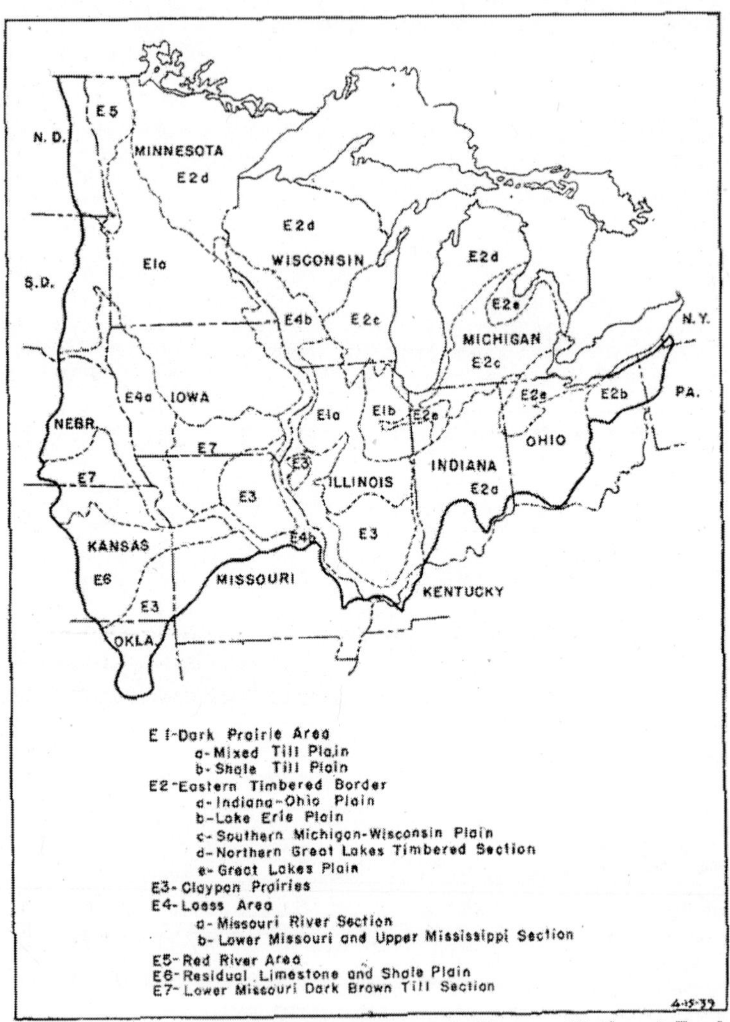


MAP 10.—Problem area of the Central Prairie and Eastern Timbered Border region. (*Soil Conservation Service.*)

the land is under cultivation. In the prairie areas, about 80 per cent is cultivated.

The climate covers wide seasonal extremes of temperature. The annual precipitation varies from about 20 inches over portions of the northwestern section to 45 inches in the more southerly part. Distribution is fairly uniform throughout the growing season. Spring rains are often heavy when the ground cover is sparse, resulting in large runoff and much erosion on the steeper lands. The average growing season ranges from about 180 days in the southern part to 110 days in the northern part.

The soils of the region include a variety of loams, silt loams, silty clay loams, and sandy loams, with considerable deep sand in the Great Lakes section. They are derived from glacial till; loess; lake and river deposits; and disintegrated shale, sandstone, and limestone. The region is divisible into numerous land areas, with more or less common characteristics with respect to soils, topography, and agricultural adaptations.

As a rule, the upland soils are absorptive and retentive of moisture, rich in organic matter and plant nutrients, well drained, and easy to cultivate. Extensive bodies of the more friable, dark-colored loams, silt loams, and silty clay loams of both the prairies and the timbered areas are among the most productive upland soils of the world. Such important soil groups as the Webster, Marshall, Carrington, Clarion, and Tama are representative of these highly productive lands, all of which are splendidly adapted to corn, soybeans, grass, and other crops. Probably 90 per cent of these superior types of farm land is in cultivation, devoted mostly to the growing of corn for feeding hogs and cattle. Beef cattle are important in the southern and central parts of the region, whereas dairying dominates the agriculture of the Lake States.

Wheat is produced extensively on the better lands of eastern and central Kansas and adjacent Oklahoma, such as the Summit and associated soils. Other good soils, such as the Grundy, Shelby, Clinton, and Fayette groups, are used extensively for general farming, dairying, and the raising of hogs and cattle.

In the claypan sections, the flat areas are best suited to winter wheat and grass, and the more rolling lands to various small grains and grass.

Of the more rolling lands bordering the streams, about 25 to 30 per cent should be retired from clean-tilled crops and used for grass, alfalfa, or other protective thick-growing crops or for trees. About 10 to 25 per cent of the rolling lands immediately back from the drainages should be retired from cultivation.

Stream bottoms are cultivated extensively throughout the region, principally to corn and alfalfa. The better drained loams, silt loams, silty clay loams, and some of the clays, of such series as the Wabash, Sarpy, Osage, and Verdigris, are highly productive although subject to damage by floods and by outwash from the hills.

The agriculture of the region is based on growing corn, the small grains, and hay, including legumes, for sale and for feeding hogs, beef cattle, and dairy cows. Localities of limited extent specialize in fruits, vegetables, sugar beets, potatoes, and tobacco.

Much land is farmed under systems of intensive crop rotation and thorough tillage. In some of the corn-hog sections, clean tillage prevails in too many sloping fields. Contour tillage is practiced too little over most of the region. Per-acre yields have declined in many localities, largely because of soil losses and impaired soil structure by sheet erosion.

Farms, for the most part, vary in size from about 80 to several hundred acres, and about 50 per cent are operated by owners. Efficient labor is generally available for all needs, and equipment is adequate on most farms.

Nearly all the prairie area, immediately back from comparatively narrow strips of rolling land along the drainages, varies from nearly level to gently rolling and is admirably suited to the use of any kind of farm machinery. Most of the timbered portion is also topographically well suited to cultivation, although the erosion hazard is too great for clean tillage on the steeper lands. Probably more than 80 per cent of the land is easily cultivable. However, in spite of the prevailingly smooth surface, about 60 per cent of the entire area slopes sufficiently to make erosion a definite problem under conditions of clean-tillage farming. Nearly half of this erodible area already has suffered from moderate to severe washing, and numerous areas have lost nearly all of the topsoil; many individual tracts have become so badly gullied that further cultivation is impracticable. Sheet washing is far more prevalent than gullying, although some of the more rolling sections, particularly where comparatively shallow soil overlies stiff clay subsoil, are gullied very seriously. V-shaped gullies have become almost a characteristic feature of some of the more rolling localities, as in many of the areas of Shelby and Lindley soils in north-central and northeastern Missouri and neighboring parts of Iowa.

Some of the more important areas suffering from erosion, sheet or gully washing, or both, are the rolling glaciated lands of northern Missouri, southern Iowa, southeastern Nebraska, and northeastern Kansas; the unglaciated rolling to hilly country of largely loessial soil bordering the lower Missouri and the upper Mississippi Rivers; and the rolling sections of claypan soil in southern Illinois, northern Missouri, southeastern Kansas, and northeastern Oklahoma. Sheet washing gradually is changing more and more of the dark-colored soils of the more rolling corn-growing sections of Illinois, Indiana, Iowa, Missouri, eastern Kansas, and eastern Nebraska to a light-brown or yellowish-brown color and is decreasing crop yields on these extraordinarily productive lands.

The Soil Conservation Service has installed erosion experiment stations in a number of agricultural districts in the region to collect information with respect to rates of erosion and runoff for special groups of erodible soils and to work out practical control measures. A large number of erosion-control and water-conservation demonstration projects and

CCC camps have been strategically placed over the region, where practical measures and systems for soil conservation could be applied for the benefit of the people of the respective sections. The improved cropping systems and land-treatment practices recommended and demonstrated are based on experimental research and actual field experience. Legally constituted soil conservation districts are now being established in various localities, in which the farmers, cooperating with state and Federal agencies, will work out their own land-use problems.

The state agricultural experiment stations have contributed much to the development of effective conservation measures and practices, such as adaptable systems of crop rotation, cover cropping, and soil building. In Missouri, the State experiment station has developed an efficient method for maintaining a nearly permanent cover for grazing by seeding grain on lespedeza without plowing out all of the lespedeza stubble.[1]

## Dark Prairie Area

The Dark Prairie area comprises about 54 million acres in the southern and southwestern half of Minnesota, the north-central half of Iowa, and the central and northern half of Illinois, together with an important strip across eastern South Dakota. It is interrupted by a belt of loessial deposits bordering the upper Mississippi River. The area includes two major divisions: (1) The extensive Mixed Till Plain and (2) the relatively small Shale Till Plain, in northeastern Illinois and northwestern Indiana.

### Mixed Till Plain

Primarily, the area is a smoothly glaciated plain, with portions of the surface covered by a thin mantle of loess. Topography ranges from nearly level to undulating or gently rolling, and prevailing slopes seldom exceed 5 per cent, except over the rolling morainic terrain, where 8 to 10 per cent slopes are common. Broken to hilly topography is characteristic of many narrow strips bordering the major streams, with some slopes ranging to a declivity of as much as 25 or 30 per cent. In the broad aspects of the area, erosion has not greatly modified surface features; most of the streams cross the undulating country in relatively shallow valleys.

The soils are derived, in the main, from glacial till and loess. They are relatively immature and, in general, deep, permeable, and highly productive. Among the principal series derived from glacial material are the Carrington, Clarion, Webster, and Dickinson; and of the loessial or partly loessial groups, the Tama, Muscatine, and Grundy. The Muscatine and

[1] Etheridge, W. C., and Helm, C. A. Korean Lespedeza in Rotations of Crops and Pastures, Missouri Agr. Exper. Sta. *Bull.* 360; Wheat in Missouri, Missouri Agr. Exper. Sta. *Bull.* 398. Etheridge, W. C., Helm, C. A., and Brown, E. Marion. Winter Barley, a New Factor in Missouri Agriculture, Missouri Agr. Exper. Sta. *Bull.* 353.

Webster soils occur on broad, flat plains or in depressions or swales of imperfect natural drainage. These are dark-colored soils, often almost black; and because of their immaturity, the horizons are not well defined. Topsoils are deep and rich in organic matter. The other groups, Carrington, Clarion, Dickinson, and Tama, are chiefly deep, friable, dark grayish-brown loams and silt loams. They are well drained, easy to cultivate, and very productive. These soils, together with some of the associated groups of lesser extent, are among the most productive upland soils of the world.

The average rainfall varies from about 22 inches in the northern part of the area to about 40 inches in the southern part, and usually it is well distributed over the growing season. Prolonged spring rains and the flashy downpours of summer and fall produce serious erosion on the sloping areas. Floods of minor importance occur almost yearly over the lowlands of some of the larger streams, and serious overflows at rarer intervals. Summer drought sometimes does serious damage to crops.

The predominant type of agriculture is general farming and hog raising, with corn, oats, and legumes as the principal cash and feed crops. Soybeans are important in the central and southern sections. About 90 per cent of the land is under cultivation, with a very large proportion devoted to intertilled crops. The average-size farm is about 160 acres, and about 60 per cent of all farms are operated by the owners. Frequently, the tenants are good farmers.

Slight to moderate sheet erosion is general throughout the area, except on the level areas and depressions. On much of the more rolling land, sheet washing is severe, and gullying locally serious. Clean cultivation without regard to the contour is the cause of most of the erosion damage. The steeper areas should be retired to close-growing grasses and legumes. The general erosion problem calls for an improved soil management system, including such measures as crop rotations, production of legumes, liming, contour farming, contour strip cropping, gully control, managed grazing, and pasture renovation. The indications are that general application of these needed soil- and water-saving practices not only will control erosion adequately but will materially reduce the hazards of floods and silting of bottomlands, stream channels, stock ponds, and large reservoirs. No important economic or social obstacles would be encountered by the general adoption of farm programs of this character.

Four demonstration projects for control of erosion and conservation of rainfall have been established within the area, in addition to a number of CCC projects. Excellent results have been achieved on all of these projects, and widespread interest aroused on the part of both cooperating and noncooperating farmers. The work has been carried on in cooperation with interested state agencies.

### SANGAMON RIVER DEMONSTRATION PROJECT

Some of the principal measures employed in the control of erosion, conservation of rainfall, and reorganization of land use in the Dark Prairie area are given for a representative demonstration area in central Illinois.

The Sangamon River project, located in the vicinity of LeRoy in McLean County, Illinois, comprises approximately 150,000 acres and is representative of the dark-colored glacial prairie soils of north-central Illinois in particular and of a large portion of the Dark Prairie area in a general way. The soils are among the most productive of the United States, and their defense against deterioration by erosion is of large importance to the locality, the state, and the nation. In the prevailing system of land use, losses by soil erosion have not been given due consideration as a production menace. The result has been a marked reduction in average crop yields.

An inventory of physical land conditions within the project area reveals the following erosion conditions: 70 per cent of the land has lost topsoil by erosion in amounts ranging from small quantities to 25 per cent of the entire surface layer; 20 per cent has lost from 25 to 75 per cent of the topsoil; and 3 per cent has lost more than 75 per cent of the topsoil and, in places, some of the subsoil.

Formerly, McLean was one of the outstandingly productive Corn Belt counties. The people generally believed that the soils represented a definitely permanent and productive natural resource. It was not until 1933 that erosion received serious recognition as a land-depleting problem, even though per-acre yields were known to have decreased. The type of farming is best described as *cash grain*. Corn is the principal crop, although soybeans and oats are important. About three-fourths of the corn produced is shipped out of the area. Farms are operated in relatively large units through the use of labor-saving machinery.

The topography is gently undulating to slightly rolling, and drainage is generally well established. The predominant soil, Clarion silt loam, is a deep, dark-colored prairie type, high in organic matter and very productive.

Many farms are continuing the use of soil-depleting practices, but the cooperative soil conservation demonstration has impressed many operators with respect to benefits to be derived from the use of conservation practices. Natural and economic conditions, coupled with traditional farming customs, have operated to retard somewhat the general adoption of better land-maintenance methods. Farm lands changed hands in McLean County during the boom years at very high prices, frequently under heavy encumbrances. About 65 per cent of the farms are tenant operated, mostly on a one-year lease. These tenants have been forced to devote large acreages to cash crops of corn and soybeans in order to meet the rents and family needs. Many fields have been planted to clean-tilled crops so long that their present productivity is very low. On some of this depleted land, it is now difficult to produce good crops even of needed soil-building legumes.

The erosion-control program has been extended to 576 farms, aggregating 98,594 acres. Physical surveys were made of the entire area, and all programs for individual farms were drawn up by specialists, in collaboration with the farmers and on the basis of the facts ascertained through the surveys and the work of the State College of Agriculture. The following soil-building and conservation measures and practices were incorporated in whole or in part in every farm plan:

1. Rotation of crops to include at least one year of legumes.
2. Contour tillage and strip cropping.
3. Location of fences on the contour.
4. Retirement of severely eroded cropland and steep erodible slopes to grasses and legumes.
5. Pasture improvement by reseeding to grasses and legumes, control of grazing, and contour furrowing.
6. Liming of acid lands.
7. Terracing of fields having slopes less than 10 per cent.
8. Installation of terrace outlets and field waterways, stabilized with vegetation.
9. Gully control with check dams, vegetation, and diversion of water from the heads.
10. Stream-bank protection.
11. Woodland management.
12. Rehabilitation of wildlife by planting critically erodible areas to stabilizing vegetation capable of producing needed food and protective cover.

The following statistics from the farm accounts of 164 operators in the project area are indicative of changes in the practices of the cooperating farms and of increased financial returns (the list includes 139 cooperators and 25 noncooperators):

| Kind of Crop | Per Cent of Farm Area: Cooperators | Per Cent of Farm Area: Noncooperators |
|---|---|---|
| Corn | 46.8 | 54.4 |
| Small grain | 21.7 | 26.0 |
| Soybeans | 7.5 | 4.2 |
| Legume hay and pasture | 10.3 | 3.9 |
| Nonlegume hay and pasture | 9.7 | 10.5 |
| In fallow | 4.0 | 1.0 |
| Livestock investment per acre | $ 6.21 | $ 4.34 |
| Net return per acre | 17.77 | 15.79 |

Not only has erosion been largely reduced on the cooperating farms but, with the additional moisture stored in the soil, crops have suffered much less from the effects of drought.

## Shale Till Plain

This subdivision of the Dark Prairie area lies immediately back of the lacustrine deposits of lower Lake Michigan, in northeastern Illinois and adjacent northwestern Indiana. It comprises some 7 million acres.

Surface relief is undulating to gently rolling. Along some of the streams and on the moraines, the topography is locally rolling. Gravelly and sandy hills and ridges occur in some localities. The annual precipitation of about 36 inches is fairly well distributed.

The soils are derived largely from glacial till and assorted glacial outwash material. The till consists of grayish, compact, plastic material, almost free of pebbles. It is derived largely from the Maquoketa shale formation, which lies immediately to the northeast of the area and over which the ice passed before depositing its till load. The derivative dark-colored soils consist mostly of silt loams and silty clay loams, identified in Illinois as the Clarence and Elliott series. An outstanding characteristic of these groups is the plastic clay subsoil. Lighter colored soils of similar characteristics occur along the streams and on the more rolling areas. Small bodies of glacial outwash, lacustrine and alluvial deposits, and peat and muck are associated with the glacial drift soils.

The principal crops are corn, wheat, oats, barley, rye, buckwheat, and hay. Potatoes and dairying are important in some localities. Yields are generally reported as being considerably below original production levels, and the trend still seems to be downward.

A significant characteristic of the sloping lands of this subdivision is their exceptional erodibility under the prevailing type of land use (see "Rates of Erosion and Runoff," Chap. V, Part 1). Even some of the nearly level land has suffered seriously from sheet erosion during the past 10 years. Under the original condition of prairie, with a constant supply of decayed vegetation, dark to black, deep loamy topsoils were developed, which favored ready absorption of rainfall. Little surface runoff occurred, and there was practically no erosion. But when the land was plowed out of the prairie sod for cropping purposes, a vastly different condition was established. The constant source of organic matter was destroyed; and as the result of continuous cultivation, the original supply of organic matter was gradually dissipated. This has resulted in a profound structural change in the surface soil. Apparently, the material has increased in density, causing a reduction in the rate of absorption and in the water-holding capacity, which, coupled with the impervious nature of the subsoil, has increased the runoff and accelerated both sheet washing and gullying.

Most of the measures for control of erosion, soil building, and better land use recommended for the Dark Prairie area are generally applicable to the erodible lands of the Shale Till Plain. It is likely, however, that the use of vegetation in protecting the soil from further erosion will become much more important than mechanical treatment.

## Eastern Timbered Border

The Eastern Timbered Border subdivision, comprising some 117 million acres, extends from the Northern Appalachian area on the east to the Dark Prairie area on the west. It narrows to a wedge-shaped point in southwestern Indiana but is very broad in the region of the Great Lakes, extending from the southwestern corner of New York nearly to the northwestern corner of Minnesota. The area covers about two-thirds of Ohio, nearly all of Indiana, all of Michigan, all of Wisconsin except a narrow belt along the Mississippi River, and a large part of northern Minnesota.

Originally, the greater part of this plain was heavily timbered. Most of it ranges from nearly level to gently rolling. Strips of relatively steep land, however, border most of the important streams. The more rolling portions, occurring principally in the northern peninsula of Michigan and the extreme northern parts of Wisconsin and northeastern Minnesota, are characterized by low ridges and hillocks. The Mississippi, Wisconsin, Wabash, Miami, and Scioto Rivers and their main tributaries have cut deep trenches along their courses. Surface relief has been otherwise modified by one or more periods of glaciation. Along the contact with the Appalachian Plateau, some areas are quite hilly. The more southerly areas are generally of smooth surface, whereas considerable portions of the northerly sections are ridged and hummocky, with many scattered sandy plains and bogs. Along the lake borders, particularly in the Saginaw Bay and Maumee Basin sections, are extensive flat areas of old lake deposits. These lacustrine lands, generally, have been reclaimed from their naturally wet condition for farming purposes. Other smaller bodies of marshy land throughout the region have been similarly treated.

On the basis of differences in topography, soils, and types of agriculture, the region includes the following principal subdivisions: the Lake Erie Plain, the Indiana-Ohio Plain, the Southern Michigan-Wisconsin Plain, the Northern Great Lakes Timbered Area, and the Great Lakes Plain.

### Lake Erie Plain

The Lake Erie Plain subdivision merges with the Appalachian Plateau, along the Pennsylvania and Ohio fronts, at elevations of about 1,500 feet.

The country is rolling to hilly, with broad-topped divides and relatively steep slopes along the drainageways. It falls to an elevation of about 600 feet near the southern shore of Lake Erie, from which it is separated by a narrow belt of lake plain. Upland slopes commonly range from 5 to 15 per cent (Fig. 275) but locally run up to 30 per cent or more along the deeper stream entrenchments. The steeper slopes are commonly wooded, but those of intermediate gradient are largely used for pasture. Most of the land having slopes less than 15 per cent is cultivated.

Fig. 275.—Irregular slopes, typical of more rolling section of Lake Erie Plain, northeastern Ohio. (*Photograph by Soil Conservation Service.*)

The soils have been formed chiefly from glacial till derived from the sandstones and shales underlying the area. The brown Wooster soils, derived largely from sandstone material, are well drained and moderately productive. The loam and silt loam are the principal types. The less productive grayish Mahoning soils are generally of heavier texture, grayish in color, and not so well drained. Claypan soils of still poorer drainage occur to some extent.

Corn, oats, hay, and pasture occupy the major portion of the farm land and form the basis for the prevalent dairy type of farming. The common rotation is corn, oats, wheat, and 1 year or more of hay. This maintains a fairly good vegetative cover, except when corn land is fall-plowed for spring oats.

## Indiana-Ohio Plain

The Indiana-Ohio Plain embraces most of Ohio west of the Scioto River and all of Indiana except the extreme northerly and southerly portions. Over the greater extent, the surface ranges from flat to gently sloping (Fig. 276). In some localities, particularly in southwestern Ohio and southeastern Indiana, the topography is more uneven. Local morainic deposits are characterized by rounded ridges. Over the more nearly level portions of the Plain, slopes usually range from about 3 to 5 per cent; in the morainic sections, some slopes are as steep as 15 per cent. It is only near the southern margin and along the larger streams that the declivity exceeds these gradients. The general level of the country rises from about 500 feet in southern Indiana to about 1,000 feet in west-central Ohio. More than two-thirds of the area lies between these extremes.

The soils are derived from glacial material, much of which comes from the regional limestone, shale, and sandstone. The ridges are commonly more gravelly and better drained than the more prevalent smoother

Fig. 276.—Erosion on long gentle slope characteristic of much land in the Indiana-Ohio Plain, northwestern Indiana. (*Photograph by Soil Conservation Service.*)

lands. Much of the level land has been drained artificially. Slight differences in elevation and slope have produced decided differences in natural drainage conditions and soil color (Fig. 277). The soils are distributed in intricate patterns of the well-drained grayish uplands represented by the Miami series; the intermediate gray soils of the Crosby group, with their mottled subsoils; and the dark soils, with mottled or gray subsoils, of the Brookston series, occupying depressions and level areas. The Brookston and Crosby soils are locally known collectively as *black and white land*. Much of this upland has a calcareous subsoil. Over the flat uplands of southwestern Ohio and southeastern Indiana, the principal soils are the grayish, imperfectly drained silty lands of the Clermont and Vigo groups. These lands are too acid to produce good yields of clover or alfalfa without treatment. At present, there is a tendency to reduce their use for corn and wheat in favor of hay and pasturage.

The Indiana-Ohio Plain country constitutes an eastward extension of the great Corn Belt of the Central States. Corn and winter wheat are the most important crops, although oats are important in the more northerly sections. Hay and grass for pasture are the other important crops. Corn is sold as a cash crop or used for feeding beef cattle and hogs. Dairying is important where concentrations of population provide satisfactory markets. The crop rotation usually consists of one or several successive years in corn, followed by winter wheat and, quite commonly, by one or more years of mixed hay. Some localities alternate corn and wheat. In others, the rotation is extended frequently to a 5-year period of corn, wheat, and 3 years of grass. Some of the grassland is utilized as rotation pasture.

FIG. 277.—Section of aerial photograph of a typical area in the Indiana-Ohio Plain, showing intricate association of Crosby (intermediate gray), Miami (light), and Brookston (dark) soils. Clumpy black areas represent woods. Darke County, Ohio. 1 inch equals 1,667 feet.

Sheet washing is common over a large part of the area, even on quite gentle slopes. Many of the steeper slopes have been affected by gullying. Only the nearly level and depressed areas have negligible erosion.

Most severe erosion occurs in summer and during periods of melting of snow. Soil washing in winter and spring is generally most severe in fields of fall-seeded wheat and on bare land. Where corn follows corn, erosion is especially severe; it is frequently serious where corn is alternated with wheat. A standard three- or four-year rotation of corn, wheat, and grass or a legume is a much better cropping plan from the standpoint of erosion control as well as maintenance of soil productivity.

Large gullies are rare, but thousands of small gullies, or rills, are formed annually between corn rows where cultivation is up and down the slope.

## Southern Michigan-Wisconsin Plain

This area, covering, approximately, the southern half of the lower peninsula of Michigan and the southeastern portion of Wisconsin from Green Bay to northeastern Illinois, comprises nearly 22 million acres. Predominantly, the country is of gently rolling to rolling topography.

There are fewer flat and poorly drained areas and more rolling, morainic country than in the Indiana-Ohio Plain. The greater part of the area lies between 600 feet elevation near the lake borders and 1,200 feet in parts of southern Michigan and in the elevated section between Lake Michigan and the Green Bay-Fond du Lac depression of southeastern Wisconsin.

In spite of the prevailing mild relief, many ridges and rolling areas rise in complicated pattern above the lower levels of the plain. These usually consist of morainic materials, with occasional limestone outcrops as cliffs in southeastern Wisconsin. The glacial ridges are interspersed generally with flat areas of glacial till and sandy outwash. Within the till plains areas, many bodies of naturally wet or swampy land have been drained and brought into a condition of high productivity.

The soils have been derived directly or by water assortment (outwash sandy flats) from glacial material of limestone, sandstone, and shale origin. They are prevailingly acid in the surface horizons, but the glacial till types overlie calcareous materials.

The Miami loam and Hillsdale fine sandy loam are extensive soils, occurring on the low ridges and gently rolling till plains of southern Michigan and southeastern Wisconsin. They are well drained and have light-brown, friable surface soils and yellowish-brown permeable subsoils of heavier texture. They are rich in lime at a depth of about 2 to 4 feet, but the surface is generally acid. The Kewaunee soils of Wisconsin and the Isabella of Michigan are associated with the Miami but have somewhat heavier subsoils and are more productive.

The Kent soils of southern Michigan and the Superior of Wisconsin are less permeable and not so well drained. On the cultivated sloping lands, sheet erosion is generally active, especially where the gradient exceeds about 5 per cent. Gullies are formed on some of the steeper areas.

The sandy and gravelly materials of the Southern Michigan-Wisconsin Plain give rise to soils of the Plainfield series on the gravelly ridges and to the Coloma on the more nearly level outwash areas. Both groups are highly permeable and less productive naturally than the Miami soils. Erosion by water is of very little importance, generally. Blowing is sometimes severe. High sand dunes have formed here and there along the shore line. Associated with these soils are those of the dark-colored Brookston group, occupying the lower flats and depressions. These are rich in organic matter and, with the establishment of adequate drainage, are highly productive. They are not subject to erosion.

Dairying is the dominant agricultural industry of the Southern Michigan-Wisconsin Plain. Corn, oats, barley, and hay occupy the largest crop areas and are grown chiefly for consumption on the farm. Alfalfa does well on the better drained uplands, especially where the subsoil is calcareous within 12 to 18 inches of the surface. Soybeans are grown to some extent. Fruit is locally important near the lake shores, especially in southwestern Michigan and in the Door peninsula between Lake Michigan and Green Bay. The feeding of cattle and hogs is important in some localities. Beans and sugar beets are important cash crops. Beans are grown chiefly on the Miami soils, and sugar beets on the moist lands of the Brookston and related types.

The crop rotation commonly used on the dairy- and beef-cattle farms consists of 1 year or more in corn, followed by oats or mixed oats and barley, succeeded by 2 years or more of hay. Considerable areas in the rolling sections are kept in permanent pasture. Many farms have woodlots on the steeper lands.

## Northern Great Lakes Timbered Section

This subdivision comprises the upper peninsula of Michigan and all of central and northern Wisconsin and Minnesota lying east of the Mississippi River. It includes nearly 55 million acres. The topography on the whole is the most varied of any part of the Eastern Timbered Border region. Nearly all of it lies at elevations in excess of 1,000 feet above sea level, with extremes of 2,000 feet or more in some parts of the northern peninsula of Michigan.

The entire area has been glaciated, giving rise to a wide variety of surface features within short distances. High morainic ridges rise above the till plains and sandy outwash areas. Level tracts of ancient lake deposits occur here and there. Interrupted ancient mountain ranges occur in the northern peninsula of Michigan and to the northwest of Lake Superior. In general, the surface of the morainic ridges varies from about 10 to 15 per cent in slope, whereas the till plains range from level to less than 10 per cent. The outwash plains and old lake beds are level or nearly so.

The soils of the region are extremely varied. They range from rather poorly drained, heavy lake deposits to well-drained gravelly and stony loams on the ridges. Extensive areas of well-drained sandy soil occupy the flat outwash plains. Loam soils of intermediate-to-good drainage predominate over the till plains. Large areas of muck and peat occur in the low-lying, poorly drained situations.

A large part of the area lies in the cut-over Timber Belt of the upper Great Lakes. Originally, it supported a varied stand of coniferous and hardwood trees; and for 50 years or more, lumbering was the chief industry. Some farming to supply lumber camps with food and forage began almost with the logging operations. As lumbering proceeded, agricultural activity extended over much of the better land and some of the poorer. At present, state withdrawals of lands less suitable for agricultural pursuits are being made in Wisconsin. Throughout a considerable part of the region, less than 10 per cent of the land is farmed; in some counties, not more than about 3 per cent of the area is cropped. Some important tracts are still timbered. Much of the cut-over area is now occupied by volunteer brush, aspen, and other second growth.

Subsistence farming is the dominant type of agriculture, although commercial production of potatoes and fruits is locally important. Hay, corn for silage, oats, rye, and field peas are grown in connection with dairying.

Because of mild gradient, the presence of a protective cover of timber or brush on much land, and the general tendency to cultivate only the smoother areas, erosion has not become a serious problem on much of the area. This is particularly true of the sandy and gravelly ridges and the more porous sandy flats. Wind erosion has affected some fields where sandy land is used for intertilled crops. Sheet washing is limited to the steeper slopes.

### Great Lakes Plain

Some 10 million acres of flat lake deposits occur along the southern shore of Lake Erie; in the Maumee basin of northwestern Ohio; in the Saginaw Bay area of Michigan; around the head of Green Bay in Wisconsin; along the shores of the upper peninsula of Michigan; and at the western extremity of Lake Superior, near Duluth, Minn. The material was deposited when the lakes stood at higher levels. Originally, it was very wet and had to be drained artificially for agricultural occupation.

The soils are dark colored and rich in organic matter in the surface and varicolored and calcareous in the subsoil. They are highly productive where properly drained.

Erosion is generally of little importance, although some of the more sloping areas are beginning to wash rather seriously. This condition is found in a number of places even on dark-colored, heavy soil, where the land has been in cultivation for a considerable time. Erosion seems to be associated with an increased density of the soil, such as probably has developed because of partial depletion of the organic matter.

### Soil Conservation

Farming began in the southern part of the Eastern Timbered Border region more than a century ago and has spread gradually northward until most of the smoother, better lands have been occupied. Some degree of erosion has followed rather generally on the steeper lands, and blowing has affected many tracts of sandy soil. Sheet washing has taken heavy toll where corn is grown successively for several years or where the rotations are too short to include adequate periods of grass cover.

Conditions of erosion shown by detailed surveys near Hamilton, Butler County, Ohio, and in Cass County, Indiana, are fairly representative of large farming areas in this region.

Approximately 29 per cent of the 30,000 acres surveyed near Hamilton, Ohio, has a surface slope of 3 per cent or less; over 47 per cent of the area has slopes ranging from 3 to 8 per cent; 10 per cent has slopes ranging from 8 to 15 per cent; and 14 per cent is characterized by slopes in excess of 15 per cent. About 25 per cent of the land is too steep for cultivation to the clean-tilled crops and should be retired to the permanent protection of grass or trees. This area is typical of much rolling land in southwestern Ohio and adjacent Indiana.

Of the 23,000 acres surveyed in Deer Creek Township, Cass County, Indiana, which is representative of much of the gently rolling to undulating portion of the Indiana-Ohio Plain, nearly 84 per cent has a slope of 2 per cent or less, nearly 11 per cent ranges from 2 to 5 per cent, and only 5.5 per cent has slopes in excess of 5 per cent. Here only about 6 per cent of the land is too steep for the production of clean-tilled crops.

The contrast in erosion conditions as between these two areas, although influenced by soil factors and cropping practices, is largely accounted for by differences in surface gradient. In the Hamilton area, only 5.5 per cent of the cultivated land has been unaffected by erosion, although an additional 17 per cent has suffered but slightly. About 42.5 per cent has lost from 25 to 50 per cent of the topsoil; 32.5 per cent has lost considerably more than half of the topsoil; and 2.5 per cent has been practically ruined by sheet erosion and gullying. In other words, erosion is a serious problem on more than three-fourths of the farm land.

In contrast, 55.5 per cent of the Deer Creek Township area has suffered little or no erosion, and 29 per cent more has been but slightly affected. The area affected by moderate erosion amounts to nearly 12 per cent of the cultivated land, with only 3.5 per cent ruinously washed. Here, erosion is a serious problem on only 15.5 per cent of the land.

Farther north, where the country is more rolling and a great dairy industry has been established, the cropping system is based largely on the production of corn for silage, oats for feeding on the farm, and hay. Where the land is covered with grass or alfalfa, it is safeguarded from erosion; but on the areas used frequently for corn, sheet washing is generally in evidence, particularly on the heavy sloping lands. As a rule, the washing is much more severe on the steeper lands. Already, a large aggregate area has been stripped of the more productive topsoil, and small gullies are developing in too many fields.

In Wisconsin, a considerable area of submarginal land has been closed to farming during the past 8 years through the operation of county ordinances under the state zoning law. It is estimated that upward of five million acres eventually will be returned to forest, park, and other uses. Included in this are large areas of the poorer sandy soils now under cultivation and susceptible to serious wind erosion.

In the Lake Erie Plain, where dairying is the dominant type of farming, it is necessary, in the interest of proper soil defense, to devote the steeper slopes to farm woodlots, making whatever plantings that may be necessary to establish an adequate protective cover. The more gentle slopes, still too steep for safe cultivation, should be used for pasture grasses. Adequate stands should be developed by liming, fertilizing, reseeding to suitable grass mixtures, and properly managed grazing. Where the more erodible slopes must be cultivated, they should be held in hay as long as feasible, under a rotation plan of corn, small grain, and hay. Contour strip cropping and contour row crops are advisable. Grass stands generally can be improved by liming.

In the Indiana-Ohio Plain, contour strip cropping, contour cultivation, and terracing of adaptable slopes have been combined to control soil erosion on the gently rolling uplands. The lengthening of crop rotations to increase the proportion of grassland or legumes is generally helpful. Liming and fertilizing usually aid in the establishment of better stands of legumes for hay and soil building. Contour farming is advisable, and gully control is necessary on some farms. Grassed waterways are usually needed for the safe disposal of water from terrace outlets. Diversion ditches are needed on some of the longer slopes to protect lower lying cultivated land from excessive runoff.

In the dairying sections of the Southern Michigan-Wisconsin Plain, crop rotations generally should include longer periods for grass. Steeper slopes of the morainic ridges should be retired to trees or to permanent pasture, and grazing should be excluded from farm woodlots. Pasture improvement is generally needed. On the more rolling, heavier soils, contour strip cropping and contour tillage should be practiced.

## Claypan Prairies

The Claypan Prairies area of southern Illinois, northeastern and west-central Missouri, southeastern Kansas, and northeastern Oklahoma includes about 27 million acres. The topography is predominantly nearly level to undulating or gently rolling. The natural drainage, especially of the more nearly level areas, is imperfect.

Among the predominant claypan soils of the smoother areas are those of the Putnam, Oswego, Parsons, Cory, and Cherokee groups. The true claypan soils have gray to very dark-gray silt loam surface soils, overlying light-gray silt loam. The subsoil begins abruptly at depths ranging from about 10 to 16 inches below the surface and consists of yellow to dark-brown, stiff, impervious clay. This tough material is locally known as "claypan," "pan," or "hardpan."

The agriculture of the area consists primarily of general farming, with livestock and grain as the main sources of income. The principal crops are wheat, oats, sorghum, and grass.

Because of high-rainfall intensity and the lower absorptive capacity of the dense clay subsoil, both sheet and gully erosion (Fig. 278) have become serious on the more rolling areas. As the depth of topsoil is thinned by sheet washing, the volume and rate of runoff have increased, still further accelerating erosion.

FIG. 278.—Destructive erosion in claypan section of southern Illinois. (*Photograph by U. S. Forest Service.*)

The need of protection for these erodible lands is urgent. Productivity is not naturally very high. Improvement is feasible through the application of practical soil conservation measures and by additions of lime, organic matter, and phosphates. Conservation of water is especially important in the western part of the area, where serious droughts frequently damage the crops. Plowing under cover crops and barnyard manures is helpful, but all land having any considerable slope is in need of sound conservation treatment, such as contour farming, terracing, and strip cropping on a rotation basis. Much pasture land needs contour furrowing and controlled grazing.

The application of conservation measures has demonstrated effectively the benefits to be derived from such work. The accomplishments on the McCredie demonstration project, near Fulton, Mo., are indicative of the possibilities of practical soil and water conservation in this problem area.

### MCCREDIE DEMONSTRATION

The McCredie project, comprising approximately 25 thousand acres, is located in Callaway County, Missouri. The area is representative of the Putnam-Lindley group of claypan soils and, in a general way, of all the claypan soils. About 70 per cent of the project area is covered by the dark-grayish Putnam silt loam and about 20 per cent by the yellowish Lindley soils of more uneven surface. The surface of the area varies from level to gently undulating, with rather steep breaks to the drainageways. The Lindley soils occupy the steeper slopes.

As late as 1871, McCredie was described as being situated "in the midst of extensive pasture lands." This implies that tillage had not become general at that time. Corn was the first important crop grown in the region; and even with crude tillage practices and inferior varieties, yields of 50 to 60 bushels per acre are said to have been common. At the present time, after less than 70 years of cultivation, these lands are reported as producing an average of no more than 10 or 12 bushels to the acre. Corn acreage decreased 45 per cent from 1909 to 1929 and then increased by 9 per cent from 1929 to 1934. It seems significant that the late increase in acreage was attended by a decrease in per-acre yield. Apparently, the farmers were expanding acreage to retrieve a living out of constantly falling productiveness.

Statistics reveal a steady decrease in population and in the number of farms and an increase in the size of the farm units. The average farm mortgage debt increased from $1,720 in 1910 to $3,685 in 1930. The average mortgage debt amounted to 60.7 per cent of the value of the land in 1930.

The serious advance of erosion was disclosed in 1934, when it was shown by an erosion survey of the project area that three-fourths of the land had lost from small amounts to 75 per cent of the original productive topsoil. One-fourth of the land had lost from 75 per cent to all of the topsoil. Gullies were not general although occasionally in evidence. The erosion condition explains the low yields.

Cooperative arrangements for conservation treatment of 100 farms, representing 70 per cent of the project area, were in effect by June, 1937. Under these arrangements, a complete treatment for soil and water conservation was planned for each farm, according to the needs and adaptability of the various types of land on the farm. Among the practices found necessary to meet these various needs were: contour tillage; contour strip cropping on a rotation basis; use of buffer strips; terracing of slopes up to 9 per cent; contour furrows in pastures; gully control; diversion of water from gullies and other critical areas; fencing; construction of vegetated channels for disposal of runoff; application of limestone, barnyard manure, and other soil amendments; cover crops; green manuring; and retirement of badly eroded lands and steep slopes to trees, grass, or wildlife plantings. The good results already obtained are recognized by farmers within the project boundaries as well as by many of those outside. Business and professional men have expressed their recognition of the conservation program as a vital need for the long-time welfare of the region.

Among the major accomplishments on this project are the retirement of 4,326 acres of erodible land to the protection of permanent cover, terracing, strip cropping, and contouring of 3,121 acres; installation of 800 temporary and permanent gully structures; the building of 25 stock-water dams; and extensive pasture improvement.

## Loess Area

The Loess area is composed of two main sectors; the Missouri River section of 23 million acres, extending in broad outline down the river from southeastern South Dakota into northwestern Missouri; and the Lower Missouri and Upper Mississippi section of 22 million acres, extending along the Mississippi from west-central Wisconsin and adjacent Minnesota to southern Illinois and along the Missouri from northwestern Missouri to the Mississippi River.

The topographic features vary widely. The northern part of the Missouri River sector is undulating to smoothly rolling, with considerable rough land immediately skirting the alluvial plain; the remainder, including all of the Upper Mississippi River sector, is predominantly gently rolling to broken. Slopes vary from around 4 or 5 to more than 40 per cent. Drainage is well established.

Over most of the area, the loess was laid down over glacial drift material. In a large section of southwestern Wisconsin and adjacent Minnesota, where there was no glaciation, the loess overlies limestone, sandstone, and shale. The mantle of loess shows wide variations in thickness. In parts of the Missouri River sector, the depth may exceed 100 feet; in other localities, it thins down to less than 3 feet. The heaviest deposits are those of the more westerly areas.

The predominating soils are the Marshall silt loam, developed under conditions of grass-covered prairie; and the Clinton, Fayette, Dubuque, Memphis, and Knox silt loams, developed under forest. The Marshall silt loam is a highly productive, dark soil, rich in organic matter and readily permeable to rainfall. The timbered soils average more rolling; locally, they are broken to hilly. They have, generally, light-brown or yellowish-brown, floury topsoils and yellowish silty clay loam or silty clay subsoils, usually more compact than the similarly textured subsoil of the Marshall. In the rougher sections along the upper Mississippi, stony slopes and rock cliffs are common.

FIG. 279.—Upper, soil removed from typical cornfield, Missouri Valley loess area, by a 4.5-inch rain (in 2 hours) and deposited over a formerly gullied area that was controlled by diking. In places the eroded material deposited during this rain was 2 feet deep. Lower, cornfield on same type of land as cornfield above, protected by wheat seeded between corn; very little erosion by same 4.5-inch rain. Both pictures taken April 22, 1937, near Tarkio, Missouri. (*Photographs by Soil Conservation Service.*)

The agriculture is based principally on the production of grain and on livestock industries. Corn and oats are the principal crops. Hogs, beef cattle, and dairying are important. Sweetclover and alfalfa are grown for hay and for soil improvement. The Marshall silt loam is utilized very largely for the production of corn, as a cash crop. Much of the rolling and more broken land is in pasture, although the acreage of good grassland is relatively small. The more broken sections of country along the rivers, particularly the Mississippi, are largely timbered.

Both sheet (Fig. 279) and gully erosion are general on unprotected slopes—in many places, severe (see "Rates of Erosion and Runoff," Chap. V, Part 1). Much of the more rolling farm land has lost most of the original topsoil, with many areas down to the subsoil, and some badly gullied (Fig. 279). In general, the soils are very erodible. On some of the valley bench lands having sandy subsoil, many large and destructive

Fig. 280.—A formerly highly productive valley farm, in drainage basin of Buffalo River, Wisconsin, almost ruined by gullying. This recently formed gully is 50 feet deep in places. Sand washed from this ravine, and many others like it in the unglaciated region of southwestern Wisconsin and adjacent Minnesota, repeatedly damages highways and downstream farm lands and much of it goes into the channelway of the Mississippi River. (*Photograph by Soil Conservation Service.*)

gullies have developed within the last few decades (Fig. 280), as in the drainage basin of Buffalo River.

Erosion control is a farm problem of major importance, generally. All land-use practices should be in line with this objective. Because of inherent productivity, even the more rolling land is generally capable of producing fair to good yields under conditions of a stabilized surface. On the other hand, continuous cultivation without adequate protection will lead to eventual gullying and abandonment. Many of the steeper slopes and bluff areas should be kept in trees.

Practical work in the conservation of soil and water is proving highly effective. Demonstrations in Coon Valley, Wis.; at Winona, Minn.; and McGregor and Shenandoah, Iowa, are outstandingly successful. The desire for large returns on the productive lands of this problem area is quite natural. However, the practical value of reorganized land use and the installation of conservation practices, including such beneficial measures as crop rotation, contour farming, strip cropping (Fig. 281), terracing, gully control, cover crops (Fig. 282), and pasture improvement, have been so impressively demonstrated on nearly 200,000 acres in the various important subdivisions that the trend appears to be

FIG. 281.—Highly effective results have been obtained with strip cropping in the Upper Mississippi section of the loess area, southwestern Wisconsin. (*Photograph by Soil Conservation Service.*)

definitely in the direction of general adoption of a soil-conserving type of agriculture.

## GILMORE CREEK DEMONSTRATION

This project, in Winona County, Minnesota, comprises approximately 6,000 acres and is fairly representative of the steeper river lands of the Upper Mississippi River sector.

The soils are predominantly light-colored silt loams with friable silty clay subsoils. Broadly, the topography is rolling, with considerable steep to broken land. Some of the steeper land consists of shallow soil over rock, and other areas are very stony. Both sheet and gully erosion are active on all land from which the protective cover of trees and grasses has been removed. Continuous clean-tillage operations everywhere causes severe sheet washing and in many places rapid gullying. Overgrazing has resulted in damage to much of the pasture land, including woodland pasture.

A survey has shown that half of the topsoil has been removed from all cultivated farm land and that most slopes having a declivity exceeding 6 per cent have gullied to some extent. Accelerated floods along the streams have damaged much valuable bottomland by erosion and by deposition of sand, gravel, and rock. Crop yields have decreased as ero-

FIG. 282.—Upper, formerly productive field ruined for crop use by gullying; lower, reclaimed in 2 years for pasture (foreground) and production of black locusts for fence posts (center). Lower Missouri Section of loess area. (*Photographs by Soil Conservation Service.*)

sion has increased. Between 1880 and 1930, the population of the county decreased by about one-third.

A total of 5,600 acres, or 94 per cent of the total project area, representing 51 cooperative farm agreements, has been treated for the control of erosion and conservation of rainfall. Nearly 30 different measures of land treatment have been used to meet the menace of erosion on the varied conditions of soil, slope, and land use. These measures involve such major types of treatment as contour tillage; strip cropping; terracing; crop rotation; cover crops; utilization of crop residues; replenishment of the soil supply of organic matter; liming; improvement of pastures and woodlands; retirement of critically erodible slopes to the permanent protection of grass, trees, and wildlife plantings; gully control; and streambank protection. The results have been a very large reduction in the loss of soil and a marked improvement of flood conditions.

## Red River Area

The Red River area comprises some 12 million acres in the Red River Valley of northwestern Minnesota and eastern North Dakota. It lies within the basin of ancient Lake Agassiz.

The valley covers three divisions: an outer morainic belt of rolling character, sloping lake deltas, and the flat lake bed proper. The streams have cut narrow grooves, seldom exceeding 10 to 15 feet, below the general level of the land. Old beach lines of sand and gravel ridges extend longitudinally along the valley; they vary in height from 3 to 15 feet and in width from about $\frac{1}{4}$ to $\frac{1}{2}$ mile. Drainage is well established on all sloping areas but is deficient on some of the level lowlands.

The predominating soils are the Fargo and Bearden series; both are derived from lake material and have developed under prairie conditions. They are dark colored and rich in organic matter. About one-half of the land consists of heavy clay, which, though absorbing rainfall slowly, has high moisture-holding capacity. The other half consists of permeable sandy and loamy types. Some of the more sandy land is very droughty.

The average annual precipitation ranges from about 19 inches in the northern to 25 inches in the southern part. Rainfall is usually well distributed throughout the growing season, although summer droughts are not uncommon.

The agriculture of the valley is primarily cash-grain farming, with corn, wheat, oats, and barley the principal crops. Potatoes are locally important. Roughly, about 75 to 80 per cent of the area is farmed, and part of the remainder is tillable. A very intensive type of cultivation is practiced; much land is left relatively bare of cover much of the year and so is vulnerable to wind action. Wind erosion has affected much of the area, particularly the sandy lands. In general, adverse weather conditions and faulty systems of cropping are responsible for the severity of soil blowing. Ordinarily, the damage thus far has been confined to single fields. Among the largest single areas that have been severely affected are those found in parts of Kittson and Marshall Counties, Minnesota.

Better planned rotations to provide increased production of legumes as sources of organic matter, to counteract wind action, and to maintain productivity are plainly needed. Supporting practices should include strip cropping, cover crops, and cultural practices that leave the surface cloddy and at least partially covered with crop residues. The lighter sandy soils are largely unsuited to general farming; the greater part of them should be used for hay, pasture, or close-growing crops. The heavier lands are well suited to intensive cultivation and are not much affected by wind erosion.

Not enough demonstration work in erosion control has been carried out in the area. The Minnesota Agricultural Experiment Station is starting operations having to do with the control of wind erosion on a number of farms in Swift and Lac Qui Parle Counties. This work is likely to point the way to practical measures of control of both moderate and serious wind action in those localities.

## Residual Limestone and Shale Plain

The Residual Limestone and Shale Plain area is largely restricted to east-central Kansas and west-central Missouri. It comprises some 13 million acres, bordered by the Great Plains on the west, the Brown Till area on the north, and the Claypan Prairies on the southeast and south.

Surface relief is generally undulating to rolling, and drainage is well established. Relatively deep valleys have been cut by the principal streams: the Osage, Neosho, and Verdigris Rivers and their main tributaries. Strips of rolling to broken country, largely timbered, border the larger waterways. Some of the hills and ridges are strewn with rock fragments. The natural vegetation of the prairie areas consists largely of bluestem and other grasses. The steeper slopes are covered in many places with a generally small growth of such plants as dogwood, sumac, wild cherry, oak, hackberry, and persimmon. The timbered alluvial areas support a variety of trees, such as hackberry, elm, ash, oak, walnut, locust, maple, cottonwood, sycamore, hawthorn, and willow.

Average precipitation is about 30 to 35 inches, with extremes as low as 25 and as high as 55 inches. A large proportion of the rainfall occurs during the growing season from April to October. Crop failures are rare, although short midsummer droughts, accompanied by hot winds, occasionally damage the corn crop.

The underlying rocks consist of limestones, sandstones, and shale. A number of distinctive groups of soils have been formed through the breaking down of these rock formations. The Summit, Labette, and Bates are the most important series. The soils of the first two groups are predominantly heavy in textures, usually ranging from silt loam to silty clay loam. They are derived largely from limestone and shale. The principal types of the Bates group are fine sandy loam and loam. These are derived chiefly from sandstone and shale. On the slopes bordering drainageways, much of the land is thin and stony.

Approximately 60 per cent of this problem area is under cultivation. Originally, most of the region back from the main drainages was grassed prairie country. It was used first as range land for cattle. The agricultural development of the region has centered around some type of livestock industry. Corn production was expanded for feeding. Gradually, wheat, oats, rye, barley, flax, alfalfa, and kafir were introduced into the general farming scheme. Today, kafir, corn, wheat, and hay are the principal crops. They are grown on many farms in combination with the raising of cattle and hogs. Dairying is important in some sections. Some localities emphasize the production of grains; others specialize in livestock and hay. The alluvial soils ordinarily are used for corn, alfalfa, and clover.

The average size of farms is about 160 acres; about 40 per cent of them are operated by tenants.

Erosion is active on nearly all sloping lands used for the cultivated crops, especially where plowing is not on the contour. Many fields, even on gentle slopes, show sheet erosion losses, as evidenced by the lighter color of the soil and decreased yields. Gullying has done considerable damage on some areas. Some overgrazed land has also suffered from erosion.

Only recently has the maintenance of land against erosion been given serious consideration. A number of demonstrations have shown the value of soil and water conservation practices. There is evidence that increased use is being made of such practical farm measures, although the spread is still too slow for the safety of the land and those who live by the land.

## Lower Missouri Dark Brown Till Section

The Lower Missouri Dark Brown Till area, covering some 18 million acres, occurs in two principal bodies, one in south-central Iowa and adjacent Missouri, and the other in southeastern Nebraska and adjacent Kansas.

For the most part, the topography is gently rolling to rolling, but many flat areas occur over the broader interstream divides. The uplands have been deeply dissected by V-shaped valleys, forming an intricate pattern of stream activity. Slopes vary widely in length, direction, and declivity. Most of the large streams are bordered by strips of strongly rolling or rough country, where the slopes are short and steep, with gradients commonly ranging from 8 to 20 per cent. Back from the main drainages, the land is characteristically smoother, with generally long slopes, ranging from about 5 to 15 per cent. The higher interstream plateau areas are still smoother, with slopes seldom exceeding 5 per cent.

Annual precipitation ranges from about 30 to 40 inches. The heaviest rainfall usually occurs in early summer. At that time, fields are scantily protected with plant residues, and much erosion results.

The area is underlain by glacial till, consisting of clayey materials containing varying amounts of sand and gravel. Much of it originally was covered with a thin mantle of loess. Grundy and Carrington silt loam and Shelby loam are the dominant upland soils. The undisturbed topsoil consists of about 6 to 14 inches of very dark-brown loam to almost black silt loam or silty clay loam. The subsoils are yellowish brown and vary from moderately stiff to very stiff clay. The subsoil of the Shelby is a stiff yellowish clay, composed of till material containing some sand and gravel. Lime concretions and frequently layers or pockets of softer calcareous

FIG. 283.—Severe rill erosion on glacial soil (Shelby loam) northern Missouri. (*Photograph by Soil Conservation Service.*)

material are usually present at depths ranging from 2 to 5 feet below the surface of these soils.

Scattered through the region, particularly the Missouri-Iowa sector, are tracts of Grundy silt loam, capping the smoother divides. This nearly level to gently rolling land appears to have been derived from loessial material. The surface is dark gray to dark grayish-brown silt loam, underlain at about 10 inches by lighter colored silty clay loam, which frequently grades below into silty clay. The Grundy soils and most of the Shelby were developed under prairie conditions. Some of the steeper areas of Shelby were timbered at the time of settlement; occasional small tracts are still forested.

The associated yellowish Lindley soils are locally extensive. They occupy timbered slopes which frequently extend from the higher upland levels down to the streamways. The surface is usually steep and frequently rough. Most of the land is unsuitable for crop production. Erosion has long been active on these steeply rolling lands.

Between 50 and 60 per cent of the farm lands of the Dark Brown Till area is used for crops, and most of the remainder for pasture. Farm woodlots are customarily restricted to the rougher areas. Corn, the principal crop, is frequently grown for several years in succession in the same field. Wheat and oats are grown in some localities, particularly in the Kansas-Nebraska sector. About 30 to 40 per cent of each farm is regularly in pasture grasses or legumes.

The Shelby soils are very erodible (Fig. 283) (see Chap. V, "Losses from Representative Erodible Land, Lower Missouri Dark Brown Till Section"). Both sheet and gully washing have seriously affected a large aggregate area. Much of the less rolling Carrington and Grundy area is affected also but not so seriously. The practice of growing corn two or more years in succession on the same land has markedly increased erosion. Because of this soil wastage, more and more of the cultivated acreage of Shelby loam is being shifted from corn to pasture or woodlots. Practical measures for the control of erosion, such as those referred to under other subdivisions of the Central Prairie and Eastern Timbered Border region, have proved highly successful on these lands. These measures must be applied intensively on practically all rolling farm land if erosion losses in the Dark Brown Till area are to be held to a minimum. Already many fields and parts of fields have been ruined; some farms have been abandoned; and a very large total area has suffered varying degrees of impoverishment, solely because of erosion. The effectiveness of the application of a completely coordinated plan of farm treatment for soil and water conservation has been so outstanding that it is believed that most farmers are likely to put erosion under control as rapidly as they can do the necessary work.

# Chapter XXXIV. Ozark Highlands

The Ozark Highlands includes an extensive region of elevated country in southern Missouri, northwestern Arkansas, and eastern Oklahoma, aggregating approximately 40 million acres (Map 11). Although the region varies widely as to details of topography, soils, erosion condition,

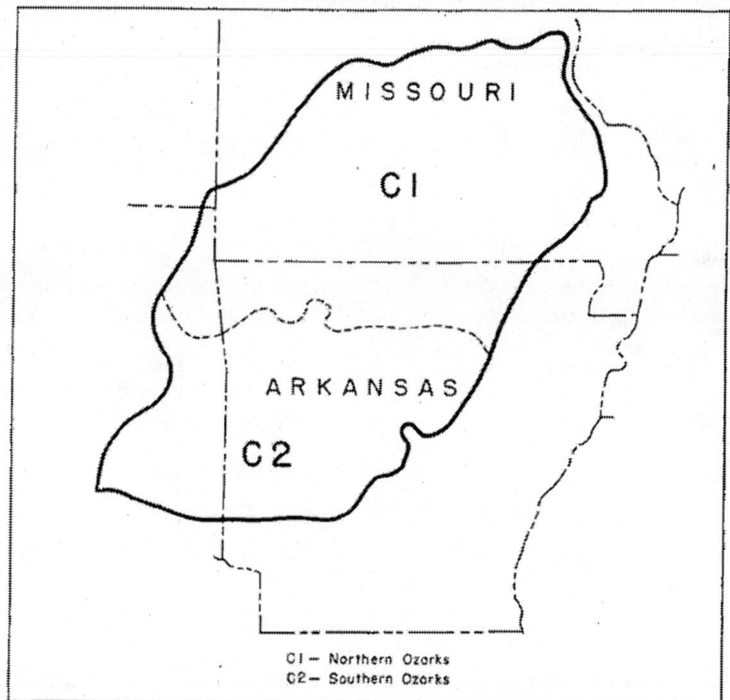


Map 11.—Ozark Highlands Problem Area. (*Soil Conservation Service.*)

and agriculture, it is delineable into two main subdivisions: (1) the Northern Ozarks, comprising about 25 million acres in Missouri and northern Arkansas and representing that portion of the Highlands underlain principally by limestone formations; and (2) the Southern Ozarks, comprising 15 million acres in the Boston Mountain sector in northern Arkansas and the Ouachita Mountains of central-western Arkansas and eastern Oklahoma. This southerly area is underlain by sandstone and shale.

The region as a whole is conspicuously elevated above the smoothly undulating to rolling country that immediately surrounds it. The topography may be characterized as generally hilly to mountainous. Many valleys, however, include large areas of relatively smooth land, well-suited to cultivation. Some basinlike areas also are characterized by favorable topography. Much of the border country lacks the rugged

Fig. 284.—This moderately steep slope of Baxter gravelly silt loam, in the Northern Ozarks sector, was successfully used for grass twenty years, and then plowed for tomatoes. Severe sheet washing resulted. Reseeded 1938 to orchard grass, bluestem, and timothy, within a short time the washing was largely checked. (*Photograph by Soil Conservation Service.*)

surface conditions of the higher central sectors. Elevations above sea level range from about 300 to 800 feet in the lowlands of the Ouachita Mountains to 2,500 feet or more on the high ridges (Magazine Mountain is 2,800 feet). In the Boston Mountain sector, the higher elevations range to about 2,300 feet and in the northerly limestone area, to about 1,100 feet.

The mean annual precipitation of 40 to 50 inches is well distributed normally. Maximum precipitation generally occurs during the period of April to July. Records indicate that the maximum expected rainfall intensity for a 10-minute period on the basis of a 10-year frequency is 1 inch; and for a 60-minute period, 2.25 inches.

In the virgin condition, most of the Ozark Highlands was forested. The growth varied considerably with soil conditions. On the limestone lands of the Northern Ozarks, the principal upland trees are black oak, white oak, and hickory, with an occasional intermixture of pine. In the valleys, sycamore, hackberry, elm, black walnut, butternut, gum, ash, and other deciduous trees were abundant before the spread of agriculture. On the more sandy lands of the Southern Ozarks, pine is more abundant although frequently mixed with hardwoods. Some isolated prairie areas in the northern sector supported good stands of big and little bluestem and other grasses.

Unfavorable topography, together with shallow, stony soil, has prevented cultivation of much land. A large part of such land, however, has, at one time or another, produced considerable income from forest products even though trees have been cut indiscriminately from many areas, with little or no provision made for the reproduction of desirable species. Careless lumbering practices, improper location of trails, and the burning of woodlands, intentionally or otherwise, have not favored the maintenance of the productive value of woodlands.

The agricultural occupation of the Ozarks has resulted in the clearing and cultivation of a large area that was too steep and erodible for farming. Erosion began immediately with the clearing of such sloping land as well as on much land of moderate slope (Fig. 284). It is not uncommon, especially in the Northern Ozarks, to cultivate land with slopes of 25 to 40 per cent. As erosion progressed, new land was cleared, and impoverished, soil-stripped fields were abandoned.

Surveys of representative areas indicate that in the northern sector approximately 27 per cent of the land of the better farming localities is in clean-tilled crops, 10 per cent is idle land, 30 per cent is in pasture, and 33 per cent still supports a stand of timber and that in the southern part approximately 26 per cent is cultivated, 14 per cent is idle, 14 per cent is in pasture, and 46 per cent remains in timber.

The extent to which erosion has progressed on farm lands within the Ozark Highlands is illustrated by the results of surveys of farm lands in various representative localities covering some 200,000 acres (Table 42).

The surveys show that for the areas covered, about 58 per cent of the cultivated land in the Northern Ozarks has been seriously affected by erosion and about 63 per cent in the Southern Ozarks. The corresponding percentages of serious erosion on pasture land are about 75 and 54, respectively. This is partly explained by the fact that much of the pasture land consists of formerly cultivated areas which were abandoned because of erosion.

Throughout the Ozark Highlands, there is general need for better protection of the land from erosion. Retirement of the steeper, severely eroded areas to pasture or trees, improvement of pastures, and careful management of woodlands would aid materially in stabilizing agriculture. Much cropland lying immediately below steep, thinly covered pasture or timber receives heavy runoff from these areas. Protection of such land by diversion of runoff is a special measure of much importance to the region.

In addition to damage to uplands by erosion, much formerly productive stream bottomland has been impoverished by overwash of sand, gravel, and stone. Deposits of sand 2 to 6 feet deep are not uncommon on the alluvial soils.

## Northern Ozarks

The Northern Ozarks section includes that part of the Ozark Highlands generally referred to as the Ozark Mountains, Salem Plateau, and Springfield Plateau. It covers a large part of Missouri lying south of the Missouri River, the northern tier of counties in Arkansas, and the northeastern corner of Oklahoma. Surface relief is extremely variable, ranging from undulating on the border prairies to extremely hilly over much of the interior country. Drainage outlets reach every square mile. Except in the smooth border and interior basin areas, slopes are short and steep. Many of the ridges are narrow and often continuous, separating streams that meander through valleys often several hundred feet below their crests.

TABLE 42.—RELATION OF LAND USE TO EROSION IN THE NORTHERN AND SOUTHERN SECTIONS OF THE OZARK HIGHLANDS

| | Land use | Extent of erosion, per cent | | | |
|---|---|---|---|---|---|
| | | None | Slight | Moderate | Very severe |
| Northern Ozarks | Cultivated | 23.4 | 18.3 | 56.5 | 1.8 |
| | Idle | 11.8 | 9.2 | 65.4 | 13.6 |
| | Pasture | 7.8 | 17.4 | 68.5 | 6.3 |
| | Farm woodland | 13.6 | 54.2 | 29.1 | 3.1 |
| | All land | 14.3 | 28.8 | 52.0 | 4.9 |
| Southern Ozarks | Cultivated | 21.5 | 15.6 | 58.1 | 4.8 |
| | Idle | 14.0 | 14.5 | 56.1 | 15.4 |
| | Pasture | 19.1 | 27.0 | 45.3 | 8.6 |
| | Farm woodland | 29.8 | 62.5 | 6.7 | 1.0 |
| | All land | 12.8 | 38.5 | 32.5 | 5.2 |

The Baxter and Clarkesville cherty silt loams and stony silt loams, derived from limestone, are the most important of the upland soils. The former have red clay, and the latter yellow silty clay subsoils.

Corn, small grain, orchards, vineyards, small fruits, truck crops, hay, and various forage crops occupy most of the cultivated area. Erosion is serious on most of the upland. Table 43 shows the degree of erosion on the various slope classes for the major types of land use in a number of erosion-control project areas.

FIG. 285.—Gently sloping (3 per cent) 51-acre field of valley land, protected by contour strip cropping of oats and red clover, on rotation basis, near Harrison, Arkansas. (*Photograph by Soil Conservation Service.*)

FIG. 286.—Bermuda grass seeded in corn, spring of 1937; 33 beef steers were turned in for fattening, after harvesting a good crop of corn in October, same year. This field, which was subject to impoverishing sheet washing, will be kept in grass. Near Harrison, Arkansas. (*Photograph by Soil Conservation Service.*)

The effectiveness of soil conservation measures has been demonstrated in representative areas and includes woodland and pasture improvement; stabilization of tilled land by contouring, terracing, and strip cropping (Fig. 285); seeding highly erodible land to permanent grass for

FIG. 287.—Plowing under rye and vetch on eroded field of Baxter soil, 7 per cent slope, to replenish humus supply. Northern Ozarks. (*Photograph by Soil Conservation Service.*)

TABLE 43.—EXTENT OF EROSION BY SLOPE CLASSES FOR THE MAJOR TYPES OF LAND USE IN THE NORTHERN OZARKS

| | Slope range, per cent | Per cent of total | Extent of erosion, per cent | | | |
|---|---|---|---|---|---|---|
| | | | None | Slight | Moderate | Very severe |
| Cropland | 0 to 3 | 37.4 | 13.1 | 33.7 | 52.2 | 1.0 |
| | 3 to 12 | 55.1 | 1.4 | 14.6 | 76.7 | 7.3 |
| | 12 to 16 | 6.0 | 0.8 | 10.3 | 78.9 | 10.0 |
| | Over 16 | 1.5 | 0.5 | 3.8 | 73.8 | 21.9 |
| Pasture | 0 to 3 | 14.7 | 24.0 | 27.6 | 47.1 | 1.3 |
| | 3 to 12 | 59.2 | 1.2 | 14.8 | 75.8 | 8.2 |
| | 12 to 16 | 19.3 | 1.2 | 17.8 | 74.6 | 6.4 |
| | Over 16 | 6.8 | 0.9 | 33.9 | 63.3 | 1.9 |
| Woodland | 0 to 3 | 9.1 | 39.0 | 44.7 | 15.5 | 0.8 |
| | 3 to 12 | 34.4 | 10.0 | 59.3 | 28.1 | 2.6 |
| | 12 to 16 | 28.2 | 6.5 | 52.4 | 37.3 | 3.8 |
| | Over 16 | 28.3 | 6.7 | 60.4 | 29.1 | 3.8 |

pasture (Fig. 286); replenishing the supply of organic matter (Fig. 287); and rotating with soil-building legumes.

## Southern Ozarks

The Southern Ozarks area comprises that part of Arkansas lying between the northern or limestone portion of the region and the Coastal Plain, together with an extension into eastern Oklahoma. The Boston Mountains, Arkansas Valley, and Ouachita Mountains are the three most important subdivisions. The country is characteristically hilly to mountainous over the greater part of the interior of the Boston and Ouachita Mountains and rolling to hilly over most of the marginal area, interior basins, and broader valleys. Numerous scattered areas, in both the interior and the border sections, have favorably undulating or gently rolling topography, and some of the mountain masses are capped with relatively smooth land. The Arkansas Valley, a broad structural depression, contains the important alluvial plain of the Arkansas River, together with considerable smooth land in the bordering strips.

In general, the Southern Ozarks consists of a series of roughly parallel divides of an east-west trend, separated by narrow, smooth-floored valleys. It is thoroughly stream dissected. The rugged topography, in combination with numerous areas of stony, shallow soil, has restricted cultivation, generally, to the lower slopes; the stream bottoms; and the smoother portions of the valleys, basins, and marginal areas. Most of the rougher country is forested. In many localities, stave and lumber mills serve as an outlet for merchantable timber and add materially to the farm income.

The upland soils are derived from the weathering of sandstone and shale and for the most part have grayish sandy loam to silt loam surface soils and red clay subsoils (Hanceville group). Some of the smoother valley-floor areas, derived from fine-grained material, have yellowish subsoils (Conway soils). Depth to bedrock is frequently less than 3 feet, sometimes only a few inches over the steeper slopes. The smoother lands are much deeper. Generally, these soils are of moderate virgin productivity. Under the conditions of erosion prevailing over most of the cultivated slopes, much land has been lowered in productivity, in degree ranging from moderate to severe. Not only is land stripped of its topsoil much less productive, but crops suffer seriously on it from the effects of even short periods of dry weather. The alluvial soils are generally highly productive and are largely in cultivation. Some tracts of imperfect drainage are best suited to grass.

Farming is based chiefly on the production of cotton, with enough feed and food crops grown on many farms to supply home needs.[1]

[1] For further discussion of physical, economic, and social conditions in the Northern Ozark area, see Carl O. Sauer, The Geography of the Ozark Highland of Missouri, Geog. Soc. of Chicago, *Bull.* 7, 1920.

On the basis of data obtained from farm surveys in the soil and water conservation projects, Table 44 shows the condition of erosion on the various slope groups for the major types of land use. The survey results indicate that about 56 per cent of the cropland (including idle land) has

Fig. 288.—Above, new terrace in eroding field of sandy loam soil, Southern Ozarks; below, same terrace stabilized with rye and vetch. (*Photographs by Soil Conservation Service.*)

a slope of less than 2 per cent and that 59 per cent has lost half of its surface soil. The indicated decrease of erosion on the steeper areas is due to the fact that most of the cultivated steep land was recently cleared and has not been subjected to erosion long enough to produce severe soil washing.

## Soil Conservation

Interest in erosion control has greatly increased as the result of the accomplishments in the soil and water conservation demonstrations recently established in various parts of the Ozark Highlands. In addition

Fig. 289.—Above, rapidly enlarging gully on Hanceville fine sandy loam soil, near Waldron, Arkansas, 1936; below, same gully, 1937, filled in and stabilized with grass, serves (1) as a channel for safe disposal of water from terraces on adjacent land and (2) as a source of hay. (*Photographs by Soil Conservation Service.*)

to this demonstrational effort, in Arkansas the farmers have organized a number of soil conservation districts.

In the fight against erosion, farmers not only are readjusting their fields and practices, but many are even changing their type of agriculture. Such measures as retirement of steeply sloping, highly erodible land to the protection of grass or trees; improvement of pastures and woodlands; rotating, contouring, strip cropping, and terracing cultivated land (Fig.

288); and control of gullies are being installed over entire farms. Numerous gullies have been effectively controlled, some of them converted into

FIG. 290.—Bermuda grass hay cut from waterway, which was sodded to take care of runoff from ends of terraces on adjacent slopes. Southern Ozarks. (*Photograph by Soil Conservation Service.*)

TABLE 44.—EXTENT OF EROSION BY SLOPE CLASSES FOR THE MAJOR TYPES OF LAND USE IN THE SOUTHERN OZARKS

| Land use | Slope range, per cent | Per cent of total | Extent of erosion, per cent | | | |
|---|---|---|---|---|---|---|
| | | | None | Slight | Moderate | Very severe |
| Cropland | 0 to 2 | 55.9 | 10.6 | 26.2 | 59.1 | 4.1 |
| | 2 to 8 | 38.1 | 1.9 | 6.4 | 76.8 | 14.9 |
| | 8 to 12 | 3.7 | 0.9 | 6.0 | 67.5 | 25.6 |
| | Over 12 | 2.5 | 4.5 | 60.2 | 23.9 | 11.4 |
| Pasture | 0 to 2 | 53.4 | 14.0 | 44.6 | 38.7 | 2.7 |
| | 2 to 8 | 31.7 | 0.8 | 15.9 | 67.1 | 16.2 |
| | 8 to 12 | 7.3 | 0.0 | 15.7 | 69.0 | 15.3 |
| | Over 12 | 7.6 | 0.0 | 38.1 | 46.6 | 15.3 |
| Woodland | 0 to 2 | 23.2 | 37.9 | 56.9 | 4.7 | 0.5 |
| | 2 to 8 | 19.3 | 28.1 | 56.8 | 12.7 | 2.6 |
| | 8 to 12 | 12.0 | 31.8 | 56.4 | 9.5 | 2.3 |
| | Over 12 | 45.5 | 14.5 | 79.8 | 5.3 | 0.4 |

waterways for conducting runoff water from terraces safely downhill (Fig. 289). Also, many unproductive depressions have been sodded with grass for water-disposal outlets (Fig. 290).

By September, 1938, work of this kind had been completed or was under way on 652 cooperating farms, comprising almost 82,000 acres in the Northern Ozarks area, and on 1,200 farms, comprising 155,000 acres in the Southern Ozarks. Crop rotation and contour cultivation have been established on more than 50,000 acres; 50,000 acres have been strip cropped; many thousands of acres have been terraced; 14,000 acres have been turned from clean-tilled crops to permanent hay; 13,000 acres have been planted to trees; and more than a hundred stock-water ponds have been constructed. Many miles of fence have been constructed to protect new plantings of trees, grass, and wildlife shrubs.

In addition to this work on farm land, state and Federal forestry agencies have done much to protect forests from fire. Approximately a million and a half acres of forest land in Arkansas are owned by the United States, and two million by the state. These lands are being managed on a basis of furnishing a permanent output of wood products.

The program of the AAA has resulted in a direct contribution to soil and water conservation on much farm land throughout the Ozark Highlands. Benefit payments to individual landowners for planting grasses and legumes and for reducing acreages of erosion-permitting crops have done much good. The work of the FSA in assisting tenant farmers has started many poorer farm families in the direction of better farming and a better opportunity in life.

# Chapter XXXV. The Great Plains

## General Features

The Great Plains region embraces the vast area of plains country lying between the Rocky Mountains on the west and, roughly, the 98th meridian on the east (Map 12). Extending from Canada to the Edwards

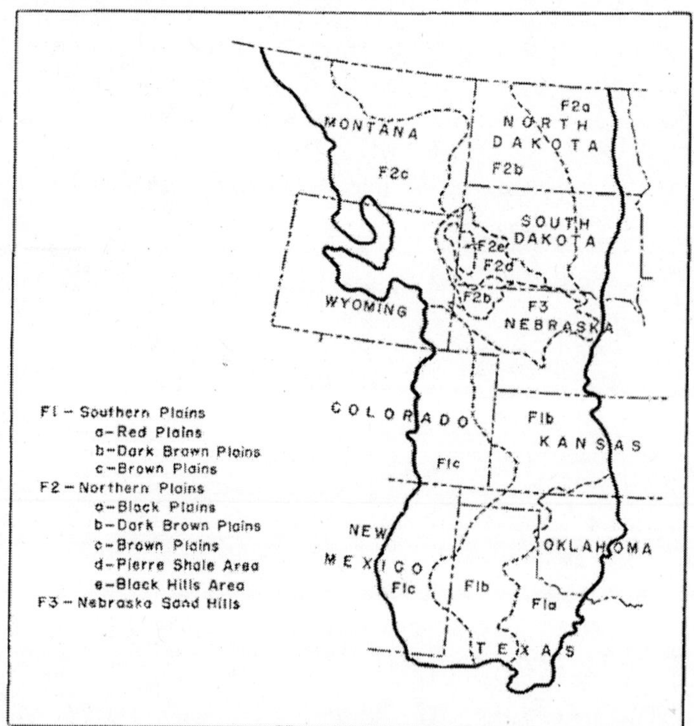


MAP 12.—Problem area of the Great Plains Region. (*Soil Conservation Service.*)

Plateau, near San Angelo, in central Texas, it is approximately 1,200 miles long and in width varies from about 380 miles along the northern New Mexico-Oklahoma line to something over 700 miles along the Canadian boundary from northeastern North Dakota to the Rocky

Mountains. The eastern foot of the Rockies forms a generally distinct western boundary, but the eastern border is not so sharp, having been variously defined in terms of climate, soil, and natural vegetation.[1] The region, including the Black Hills, comprises approximately 352 million acres, or about one-sixth the area of the United States.

From west to east, the color of the soils usually becomes darker, the rainfall heavier, and the subsurface layer of lime carbonate accumulation thinner. Finally, a line is reached east of which the leaching effect of the heavier rainfall has prevented the formation of a definite caliche layer. This line marks the boundary between the pedalfers, or humid soils, and the pedocals, or subhumid soils. Although not everywhere sharply defined, it also marks the boundary between the plains and prairies and delimits the western edge of the eastern humid region.

Over the less dissected, or major, portion of the Great Plains, the surface is prevailingly level to undulating or gently rolling. Along many of the major drainages, such as the Canadian, Cimarron, and Arkansas Rivers, the topography is characteristically rolling or hilly to rough and broken. Such areas, together with the badlands of the Dakotas, Montana, and Wyoming, are too rough generally for cultivation. They are best suited to grazing. In the more southerly areas, as along the drainage lines of the Panhandle section of the Southern Plains, the rough lands, or "breaks," constitute some of the best range lands of the entire region. Sand-dune areas of moderately rolling to hillocky topography are found in a number of localities, especially in the southern sector. Because of their susceptibility to blowing, these areas also are best suited to grazing.[2]

In general, the surface rises gradually toward the west from an elevation of about 1,000 to 2,000 feet above sea level along the eastern border to about 3,500 to 6,000 feet along the base of the Rockies. The southern and central portions of the region are considerably higher than the Northern Plains. For example, along the northern New Mexico-Oklahoma line, the elevation rises from about 1,200 feet on the east to approximately 6,000 feet on the west, whereas the corresponding range along the Canadian line is from about 1,000 to 3,500 feet. In the Southern Plains, the proportion of smooth land is considerably higher than in the northern sector, where, in addition to many small and large isolated areas of badlands, every major stream and mountain area is bordered by a belt of hilly or severely broken country, usually of badland character. About half the area of the Great Plains is sufficiently smooth for easy use of any kind of improved farm machinery, having a flat or nearly flat surface,

[1] The Future of the Great Plains: Report of the Great Plains Committee. 75th Cong., 1st Sess. *House Doc.* 144.

[2] Whitfield, C. J. Sand Dunes of Recent Origin in the Southern Great Plains, *Jour. Agr. Research*, Vol. 56, No. 12, 1938.

often featureless with respect to relief as far as one can see. The other part is too steep or rough for any practical use aside from grazing.

Most critical among soil characteristics in their effects on land use are texture, structure, and depth. These greatly influence the water-absorbing and water-holding capacity of the soils. Differences in texture account for important local variations in availability of soil moisture and crop adaptability. Soils with fine-textured surface material, such as clay or clay loam, have high power to retain water once it is absorbed, but infiltration is so slow that much of the precipitation serves no productive purpose, because it is lost in runoff or through evaporation. Hence, yields from "heavy" soils have a tendency to be more seriously affected by prolonged drought than those from the more permeable coarser grained soils. Those soils having shallow depth over beds of gravel are of low productivity because of excessive droughtiness resulting from rapid loss of moisture by percolation. Also, shallow soils over rock or caliche are inclined to be exceptionally droughty.[1]

For the most part, the soils are relatively high in plant nutrients. As regards productivity, water is the principal limiting factor. Occasional areas, usually impermeable shale-derived clay, are too saline for the growth of anything but salt-tolerant plants. The coarser sands and shallower soils are the least desirable of the smoother, uneroded lands because of their lower productivity and greater susceptibility to blowing. The tough clay lands ("gumbo") are not only difficult to handle but susceptible to gullying and are often excessively salty. Wind erosion reduces the productivity of all soils slowly or rapidly, depending on their susceptibility to blowing. If uncontrolled, it not only strips off the more productive topsoil but piles troublesome hummocks and sheets of drift over neighboring areas to damage or smother plant life and to reduce productivity, especially where the finer wind-assorted particles have blown away. Drifting material also blocks highways and accumulates around buildings and over farm machinery.

Generally, the Great Plains in its virgin condition was a region of grasses, devoid in most places of trees. Recent trials have indicated, however, that certain species of trees will survive and mature in the more favorable situations if not destroyed by grazing, fire, or other means. Early pioneers noted a decrease in the height of vegetation as they moved westward toward the mountains. Predominantly, the vegetation consists of drought-enduring species able to enter a moisture-scarcity rest stage, when necessary, and to produce seed in a remarkably short time in seasons of favorable moisture. Grama and buffalo grass ("short grasses") are the most conspicuous species, but the taller grasses characteristic of the

[1] See Leighton, M. M. Geology of Soil Drifting on the Great Plains, *Sci. Monthly*, July, 1938.

region of more abundant moisture to the east are present in a subordinate role. They become prominent, as a rule, only during the occasional humid years, if not overgrazed. In the sandy areas where the soil moisture is more readily available, the tall grasses of the prairies are commonly found, together with such woody plants as sage, wild plum, and shinnery oak.

The taller grasses suffer most during years of deficient rainfall as well as from overgrazing and other abuses. The present dominant species—the short grasses—exist probably because they alone were able to survive misuse. In any event, attempts to delimit the boundary between short grasses and tall grasses have not been very successful because of its indefiniteness and its tendency to shift widely with variations in rainfall. This makes it impracticable to draw a definite "grass line" through the area where the Plains merge with the central prairies of the Mississippi Valley.

Land-use problems of the Plains arise largely from the climate of the region. The most important factor is the light and uncertain rainfall which impedes or precludes the establishment of a sound agricultural economy except through careful adjustment to physical conditions. The irregular but persistent recurrence of severe drought, the susceptibility of bare or thinly vegetated land to wind erosion, and the general failure of settlers to adjust their agricultural practices to these hazardous conditions are chiefly responsible for the peculiar problems and the economic maladjustments that characterize the Great Plains.

The average annual precipitation is less than 20 inches, whereas in most of the other large agricultural areas of the United States the average is 30 inches or more. This general deficiency in moisture, rainfall variability, and susceptibility to soil blowing are the most critical factors in the permanent agricultural occupation of the Plains, since these constitute the basis of most of the agricultural risks of the region. Sections that are humid or subhumid in some years are semiarid or arid in others.[1] Marked variations occur not only from year to year but also in irregular periods, or cycles, of several years' duration.

Were it possible to foresee shifts in climatic conditions and to adjust cropping plans accordingly, the risk involved in the operation of farms and ranches could be reduced notably. This is not yet possible, but it is possible and highly practicable to use measurements of the moisture content of the soil at seeding time as a guide for cropping plans.

In no other part of interior United States are prevailing wind velocities so high as in the Great Plains. Throughout most of the region, velocities average 10 to 12 miles per hour, but the average may run as

[1] Thornthwaite, C. Warren. "The Great Plains." Chap. V, in Goodrich, Carter, *et al.* "Migration and Economic Opportunity," 1936.

high as 16 to 18 miles an hour in some localities. This high rate of air movement increases evaporation, intensifies droughts, and promotes blowing in bare fields and overgrazed areas.

Climatic variability affects not only current yields but the extent of plantings in following years. Crop yields fluctuate widely in the Plains. The yield of corn ranges from almost nothing to 40 bushels or more per acre, and that of wheat varies from around 5 to more than 30 bushels per acre, over entire states. In the drier areas, yields show even greater variations, with a higher percentage of failures or near failures.

The unreliability of the water supply is one of the major hazards in many parts of the area. The drying up of springs, wells, and water holes; the necessity of hauling water for stock; the removal of stock from drought-stricken areas; and the final abandonment of farms for want of water have recently presented evidence of the serious effects arising from this hazard.

Contrary to popular opinion, there is no evidence that breaking out of the range in the Plains country or the drying up of lakes and ponds has reduced the average precipitation. Modes of cultivation dominant in the region have tended, however, to intensify the serious effects of periods of deficient precipitation. Physical alteration of the soil and depletion of organic matter through deterioration and erosion have resulted in increased moisture losses by runoff and evaporation, thus reducing the amount of moisture available for plants. Moisture-conserving methods of farming will serve on nearly every area adaptable to farming to ameliorate the effects of drought.

Recent experiments indicate that periodic measurements of soil moisture and planting only when moisture is adequate will generally result in higher average yields in most sections of deficient rainfall.[1] On some of the heavier soils, alternate planting and fallowing have given favorable crop returns in certain localities. It cannot be emphasized too strongly that the success of an agricultural economy in the Great Plains area pivots primarily on effective utilization of all the rainfall that can be conserved by practical cultural measures.

Among various other problems of the Plains calling for solution are the need for marked reduction in the acreage of cash crops along with increased production of livestock, in such localities as eastern Colorado and western Kansas; increased size of the farm unit to permit economic livestock operations; reduction in the size of some large holdings of

[1] Hallsted, A. L., and Mathews, O. R. Soil Moisture and Winter Wheat with Suggestions on Abandonment, Kansas Agr. Exper. Sta. *Bull.* 278, 1936. Clements, F. E., and Chaney, R. W. Environment and Life in the Great Plains, Carnegie Inst. Washington *Supplementary Pub.* 24, 1936. And Thornthwaite, C. Warren. "The Great Plains." Chap. V, in Goodrich, Carter, *et al.* "Migration and Economic Opportunity," p. 221, 1936.

cultivated land to permit better care of the land; and adjustment of fettering debt and tax conditions.

## Water in the Great Plains

Paucity of water is the most striking characteristic of the Great Plains. Husbandry and intelligent use of available supplies of water are necessary if the Plains area is to sustain an economic development permanent in character, free from violent fluctuations, and conducive to wholesome conditions of life.

SURFACE WATERS. The principal streams of the Great Plains originate in the Rocky Mountains and flow in a southeasterly direction. Such rivers as the Missouri, the Arkansas, and the Canadian are perennial in character although subject to great fluctuation in volume. Many of the Plains streams are of intermittent flow, carrying flashy floods in spring and summer but usually dwindling or drying up in winter.

Those streams of substantial perennial flow are a dependable source of water for irrigation; but in the basins of most such rivers, additional storage would be required for the reclamation of new lands. Generally, the streams originating in the Plains are not dependable sources of water for highly developed irrigation practice, unless large amounts of storage are provided.

Irrigation with surface water has been undertaken chiefly along the perennial streams, and it is the opinion of specialists that development in this direction pretty largely has been carried to a point where the minimum flow is scarcely sufficient to provide the required water, except perhaps on the Missouri. Storage, however, has not been developed to its maximum in most sections. Irrigation with surface water has not been practiced generally in the eastern portion of the Plains, although dependable supplies of water could be obtained in some instances. Surface water impounded in small reservoirs is the only practical source of water for stock over large areas of the Northern Plains.[1]

Results obtained on clay loam soil of 2 per cent slope at the Spur, Tex., Experiment Station in the Red Plains area of the Great Plains show the following average annual losses of soil and water for the 12-year period 1926–1937 (total average annual rainfall for period, 20.73 inches):

| | *Water, Inches* | *Soil, Tons per Acre* |
|---|---|---|
| Buffalo grass | 0.90 | 1.29 |
| Milo | 2.36 | 3.77 |
| Cotton | 3.22 | 7.03 |
| Fallow | 4.28 | 8.67 |

[1] Report of the Great Plains Committee. 75th Cong., 1st Sess. *House Doc.* 144.

These results show that water erosion is destructive on very gentle slopes and that loss of rainfall as runoff runs to a high percentage of the total rainfall. From a 1 per cent slope of the same soil, the losses for the corresponding period at Spur were 3.13 inches of water and 5.27 tons of soil per acre under cotton production. This easily avoidable waste of water from a 1 per cent slope if utilized by cotton plants would be sufficient to produce an average of about 95 pounds of lint cotton per acre annually. Water now running to waste throughout the region generally could be saved or largely saved and turned to practical use in fields and on the range.

The results above indicate that under conditions similar to those obtaining at Spur, fallowing is a comparatively extravagant practice with respect to both erosion and loss of water by runoff. In other localities, summer fallowing appears to be a good practice on soils heavy enough to develop a cloddy surface or where stubble of the preceding crop can be so handled in plowing that a considerable part of it remains above the surface as a wind screen. In all water-conservation plans, especially on sloping land and in the construction of reservoirs or stock-water ponds, provisions must be made for taking care of runoff from the intensive rains ("cloudbursts") so characteristic of the greater part of the area. These intensive rains seem to be especially typical of a belt immediately east of the higher mountains.

GROUND WATERS. Ground-water supplies in the Great Plains are found both near the surface and at considerable depths. The shallow supplies generally are found in alluvial areas, glacial drift, and sedimentary beds other than clayshales such as the *Pierre*. The deep-water deposits are encountered in water-bearing sands confined either by rock or by clay beds. The depth of such water varies considerably through the Southern Plains, being exceptionally near the surface in such areas as the Plainview and Hereford sections of the Texas Panhandle.

Ground water is used extensively for domestic, municipal, and stock-water purposes in some parts of the region and locally for irrigation in the Southern Plains. Future irrigation with water from alluvial sands in river valleys can be expected in those areas where pumping is now taking place. In addition, some irrigation with ground waters as well as surface waters may prove feasible in certain areas along the eastern margin and in the northern part of the Great Plains. Water for livestock can be obtained generally over the southern half of the region from deep wells; but over great areas of the northern half, the depth to water-bearing formations is prohibitive under present conditions, so that dependence must be very largely on small artificial storages of surface water.

The problem of depletion of water supplies is discussed in the *Report* of the Great Plains Committee referred to previously.

## Historical Aspects

When Denver was settled in 1858, a vast sea of grass extended eastward for hundreds of miles, short grass predominating immediately to the east, and beyond it the tall grass of the Plains border and prairies. Herds of buffalo migrated with the seasons from northern to southern pastures. With the coming of white man and the extension of railroads across the plains, these animals were slaughtered in countless numbers for their hides and tallow. Wanton destruction of the buffalo was a prelude to the land exploitation that followed.

First came the cattlemen. After the Civil War, the central and northern Great Plains were stocked with longhorn cattle from the ranges of Texas; thousands of animals were driven over long trails to the markets or shipping points of the North. Profits were good at first; speculation followed. By 1890, much of the Plains was stocked to capacity; and during the 5 subsequent dry years, the range was seriously depleted. Cattle died by the thousands, and investors both in the East and in Europe suffered tremendous losses.

Farmers gradually followed the cattlemen. As the humid lands farther east were settled, agriculture pushed westward, encroaching on the range despite protests from the ranchers, who temporarily convinced many persons that the Plains area was not all suited to agriculture. In some localities, the land was divided into small farms to preclude speculation, but the 160-acre tracts allotted under the Homestead Act generally were too small to support a family under subhumid climatic conditions. During periods of heavier than average rainfall, however, farmers forgot past droughts and ignored the limitations imposed upon them by small farms. They often stated, and frequently believed, that "rainfall followed the plow" and that each new mile of railroad constructed was an added guarantee of moisture. The Timber Culture Act of 1873 was passed in the hope that tree planting also would increase rainfall. Farming steadily moved westward; more and more sod was broken; and the organic content of the soil was dissipated by cultivation and oxidation.

When the sod was first broken, wind erosion was negligible, and crops usually were good; but during the first drought, thereafter, drifting and dust storms, "sand storms," frequently plagued the West. For more than 50 years, individual Plains farmers have faced the menace of recurring drought and wind erosion. Recent dust storms are merely the accumulated result of long-continued exploitation, intensified by an unusually severe and protracted drought.

Most of the Plains was settled so slowly that soil blowing developed gradually, but Oklahoma presents a sharp contrast to this step-by-step progress. Its agricultural development waited until the various Indian reservations were opened to settlement. The land then was taken up rapidly, was cultivated, and presented a serious problem of wind erosion within a period of less than 10 years. During this short time, the farmers of what was then the Indian Territory developed through necessity a variety of means for checking soil drift. Most of them served good purposes in some places, and some proved helpful rather generally, yet the problem increased to proportions beyond control by individuals. This rapid development emphasized with special clarity the relationship between land utilization and the erosion hazard.

Soil drift was aggravated by economic necessity. Most of the original homesteaders lacked capital and had to grow cash crops, which usually were conducive to erosion. In periods of drought, crop failures were frequent. When the homesteaders failed, they were followed by short-period tenants to whom immediately profitable harvests also were necessary. Neither owners nor tenants could survive without cheaply produced cash crops.

The hazard introduced by cultivation and destruction of the native sod was increased by climatic conditions. In Oklahoma, as in other sections of the Great Plains, droughts and periods of more than average rainfall recur at irregular intervals. Much of the year's precipitation frequently falls in a single month of spring or early summer; and in some localities, 15 per cent or more of the year's total commonly falls in an hour. Within a single season, farmers encounter problems of both water and wind erosion. In general, however, wind erosion is a major menace only in the western part of the state, whereas water erosion is state-wide although most destructive in the eastern and central sections.

Wind erosion attracted attention first because dust clouds were conspicuous. In 1893, when the Cherokee Outlet was opened, incoming settlers were greeted by dust storms that began about Sept. 10 and lasted 2 weeks. The dust evidently was blown from older cultivated areas beyond Oklahoma, but the storm itself should have served as a warning. Local drifting frequently began within 2 or 3 years after the land was plowed, especially where soils were sandy. From the first, warnings were sounded by a few experienced men, but they generally went unheeded. During the period 1887–1901, the rainfall was heavier than average. Dust storms occurred, but they were less frequent, less severe, and briefer than they had been.

During the dry period of 1901–1904, dust storms increased in numbers and extent, affecting much of the plowed land in western Oklahoma. Local soil movement increased to such an extent that the numerous swirls merged into dust storms of large proportions. Farmers who had never before experienced soil drifting began to complain. No longer could the problem be regarded as one to be solved by individuals.

From 1905 to 1908, the rainfall again was above normal, and again the reports of drifting decreased, only to be revived once more during the dry period of 1909–1914. Drifting soil frequently covered entire fields of growing crops. In addition, the sand blew with such force that it often cut down young plants. A farmer in Major County reported that in 3 different years his alfalfa crop had been ruined in this manner.

The evils of wind erosion were not entirely ignored. As early as 1900, Professor A. M. Ten Eyck described the processes of deterioration that led to widespread soil drift:

"When the wild prairie is first broken, the soil is mellow, moist and rich, producing abundant crops. After a few years of continuous cropping and cultivation, the physical condition of the soil changes; the soil grains become finer; the soil becomes more compact and heavier to handle; it dries out quicker than it used to; bakes worse. . . . After a soil has been cultivated and cropped a long time, it tends to run together and is very sticky when wet, but when dry the adhesive character disappears almost entirely. The grass roots which formerly held it together are decayed and gone, and now when loosened by the plow it is easily drifted and blown away."[1]

In general, cover crops were regarded as the best means for controlling wind erosion. Various grasses, drought-resistant forage crops, and dry-farming cultural practices, such as listing, were used. These did some good, at least locally. Bermuda grass, which was esteemed chiefly for its ability to hold the soil and prevent washing, was used by an occasional farmer to control both wind and water erosion. Alfalfa was regarded with special favor because it produced several crops of hay during a single year; but it was difficult to establish. Fall and winter winds often blew the seed from sandy fields; when good stands were obtained, the blowing sands of spring sometimes cut the plants to pieces. This was overcome, to some extent, by the use of a nurse crop. Many farmers sowed oats or sorghum in the spring and harvested the feed in August, leaving a stubble 5 or 6 inches high. Alfalfa was then sowed in the stubble. The fall growth usually was killed by early frost but was left standing for protective purposes. Some farmers combined oats with alfalfa in spring planting, but others preferred cowpeas because they both held the soil and improved it. Kafir, sweetclover, and even wheat were tried. In some places, cropped fields were alternated with fields of native meadow or tame grass. In other regions, 10 rows of kafir were alternated with 10 rows of some contrasting crop.

Dead covers were used as well as living ones. The farmers of some sections spread manure on plowed or harvested fields. A smaller number of operators scattered straw and other trash, pressing them into the

[1] Ten Eyck, A. M. Grasses, Kansas State Agr. Exper. Sta., *Bull.* 175, January, 1911.

soil with disks. Stubble proved to be effective and cheap, yet neither it nor the other dead covers achieved general acceptance. The majority of farmers raked together all such litter and burned it, thus both destroying the protective cover and diminishing the humus supply of the soil.

Thousands of forest plantings were made in Oklahoma before 1907, originally in the hope that trees would bring rainfall. By the time that that theory was exploded, the habit of planting trees was somewhat ingrained, and the use of trees as windbreaks was continued by a considerable number of farmers. Among the trees most used were Russian mulberry, black locust, osage orange, and tamarisk. Much care was used in getting the trees started and, later, in cultivating them, but water was not artificially applied. A considerable number weathered the droughts, especially in situations most favorable to retention of rainfall and melting snow.

Cover crops generally were not cash crops, and the latter, as stated above, were necessary to the Oklahoma farmer. The dry-farming boom greatly increased the acreage of cultivated plants. Dry farming, introduced at the end of a period whose rainfall was below average and coincident with a period of heavier rainfall, captured the imagination of the Oklahoma farmers and was hailed as the savior of agriculture on the Plains. The decrease in blowing during the second half of the nineties was, to a considerable extent, credited to dry farming; and it was said that H. W. Campbell, the method's chief advocate, had proved that drifting soils could be made practically stationary.[1]

Campbell's methods of dry farming, whose principal element was soil mulching by cultivation, succeeded in some places. Its big mistake was the assumption that a single method of cultivation was applicable universally, under all climatic conditions and on all types of soil. When dry periods recurred, blowing was increased, not only because the cultivated area was extended but because the soil too often was left in a bare and finely pulverized condition favorable to wind movement. Campbell had recommended that the soil when plowed be left in a "small cloddy" condition, but that was impossible on loose, sandy land. Although the system as a whole fell into disuse, certain elements of it continued in general practice and were modified to permit cultivation with a minimum amount of blowing. Deep fall and winter plowing thus persisted, though it was not recommended for loose, sandy land. Even on other lands, its effectiveness frequently was modified by climatic conditions during the winter. If the season turned out to be relatively wet, fall plowing proved satisfactory. If it did not, danger of blowing increased.

Gradually, the lister replaced the turning plow, especially in spring planting. Crops were planted in the furrows which, where they followed

[1] Campbell, H. W. "Soil Culture Manual." Billings, Mont. Revised 1917.

the contours, served as reservoirs for rainfall. The intervening ridges made miniature windbreaks and were not entirely cultivated down until growing crops protected the soil.

The disk harrow was almost as popular as the lister. Both were originally designed to conserve moisture, and both emerged as tools for controlling soil drift. The disk was used to roughen the soil; as a substitute for a packer; and to press stubble, trash, or manure into the ground. In many places, the use of the disk temporarily checked blowing, and it was considered the best tool available for that purpose.

During and immediately following the World War, mechanized agriculture developed, spreading rapidly across the Plains. Machines enabled the dry-country farmer to grow wheat at less than half the cost of production in the rougher but more humid lands to the east—at least, in years of favorable moisture. Low costs encouraged land speculation, and enormous new areas were planted to wheat. With the development of labor-saving machinery, farming often became a part-time job. A crop could be planted and harvested in 6 weeks, and farmers could spend the remainder of the year in other work. With two sources of income, they could afford to gamble on the weather in areas where full-time farming had not been economically possible.

Between 1920 and 1930, land values fell throughout the country; but in the High Plains of Kansas, Oklahoma, Texas, and other parts of the Plains, prices soared, and population increased. From 1924 to 1929, cropland in the Great Plains increased by nearly 15,000,000 acres. This expansion was the immediate prelude to the major dust storms that accompanied drought in the 1930's.

Although some Oklahoma farmers curbed or attempted to curb soil drift, their example was not followed widely, even in that state. No attempt was made to develop a coordinated plan of erosion control involving treatment of the various types of land, existing under different climatic conditions, according to their individual needs and adaptabilities. Some of the measures that were used extensively were used improperly. Listing, for example, was seldom on the contour, except by accident, so that one of its greatest values was missed—that of conserving rainfall.

In other parts of the Plains, such protective measures as strip cropping were employed to some extent. On the whole, however, attempts to control erosion were pitifully meager. Countless farmers did nothing to hold the soil or conserve the rainfall. On the other hand, a great deal was done in the opposite direction, such as cultivating up and down the slope, burning or overgrazing crop stubbles, and plowing for seeding at times when the meager content of moisture in the soil indicated little opportunity to produce a reasonable yield.

Although surveys of soil and erosion conditions, together with farm experience and soil and water conservation work on many widely distributed typical farms, show that a large aggregate area used for wheat should be retired permanently to grass and that other areas devoted to continuous wheat cropping should be farmed under a rotation system that would further reduce wheat acreage, Census figures indicate that these necessary changes have not been made. Table 45 below, based on census records of four Plains states where wheat is grown widely and where the livestock industry has long been important, shows (1) the trend with respect to acreage in wheat for the years 1924, 1929, 1934, and 1938 and (2) the number of sheep and cattle on the farms of these states.

Drought, low prices, and the program of the AAA had much to do with the reduction in wheat acreage in 1934. Moreover, the increased wheat acreage of 1938 would have been even greater but for the AAA program. Drought conditions were largely responsible for the livestock reduction between 1934 and 1938.

TABLE 45.—ACRES PLANTED TO WHEAT, AND NUMBERS OF CATTLE AND SHEEP

| | | Montana | North Dakota | South Dakota | Nebraska |
|---|---|---|---|---|---|
| 1924 | Wheat | 3,199,000 | 8,674,000 | 2,417,000 | 3,182,000 |
| | Cattle and sheep | 3,510,000 | 1,652,000 | 2,666,000 | 3,930,000 |
| 1929 | Wheat | 4,771,000 | 10,694,000 | 3,758,000 | 4,047,000 |
| | Cattle and sheep | 5,174,000 | 2,063,000 | 2,797,000 | 3,151,000 |
| 1934 | Wheat | 3,737,000 | 9,210,000 | 3,035,000 | 3,373,000 |
| | Cattle and sheep | 5,353,000 | 1,959,000 | 2,952,000 | 3,920,000 |
| 1938 | Wheat | 4,879,000 | 10,025,000 | 3,701,000 | 5,041,000 |
| | Cattle and sheep | 3,783,000 | 2,052,000 | 2,891,000 | 3,708,000 |

## Needs of Region from Standpoint of Soil and Water Conservation

Emergency measures for controlling wind erosion on the Plains usually have been practiced *after* the soil begins to move. They take effect immediately and afford protection for indefinite periods, which generally are brief. They consist principally of tillage operations which roughen the surface by listing or plowing up resistant clods from the compact or fine-textured subsurface.

It is very nearly true in the Plains that as long as the organic content of the soil remains low, all tillage measures for wind-erosion control are emergency measures. It is also generally true that only a permanent cover of vegetation can afford more than temporary relief on deep, loose, sandy land and on shallow soil.

Replenishment of the organic supply in cropped soils is greatly needed. Probably the most practical way to meet this need will be through crop rotation. Thus far, the principal rotations that have been practiced on the demonstration projects of the Southern Plains are (1) sorghum-cotton (or corn) for the lighter soils and (2) wheat-sorghum for the heavier lands. Owing to the highly variable climate, a certain degree of flexibility in these rotations is necessary. Undoubtedly, rotations would be more effective with respect to restoration of organic matter if grass could be introduced into the crop succession. Further experience is needed, however, as to methods of seeding and the best grasses to use.

Under conditions of the Northern Plains, the problem is not so difficult, since it appears to be generally easier to establish grass. Good rotations, including small grain, corn, and grass, have been established recently on many farms in the soil conservation projects of Montana and North Dakota. Crested wheat grass, western wheat grass, and brome grass are giving good results here.

Permanent control of erosion in the Plains involves a combination of precautions that will prevent the soil from getting into a condition that favors erosion; or, if it is eroding, practices and measures that will restore soil stability must be installed. For the most part, such precautions and installations are less immediate in their effect, but far more lasting, than those of emergency tillage. Unfortunately, permanent control measures may not take full effect within a year. Nor is every measure included in a permanent system for controlling erosion individually effective over the years. Some, such as contour listing and range contouring, may need to be repeated or reinstalled; others, such as retirement of erodible land to permanent protective cover and the building of dams, are individually planned for permanency. Specifically, permanent control involves (1) water conservation by detention, diversion and spreading structures, and contour cultivation of fields and contour furrowing of range land to promote continuity of vegetative cover; (2) the use of protective vegetated strips and borders of grass, crops, shrubs, or trees; (3) the adaptation of crops and cultural practices to varying topographic, soil, moisture, and seasonal conditions; (4) the conservative utilization of increased organic residues produced by these measures; (5) the retirement of critically erodible areas to permanent vegetative protection; (6) a shift from extensive cash crop farming to (*a*) balanced types of livestock operation and farming or (*b*) livestock farming plus the growing of supplemental feed crops only; and (7) the proper distribution, rotation, and deferment of grazing on range lands and adjustment of numbers to carrying capacity.

The application of these principles involves the determination of the different needs of individual areas; the selection of a particular device or combination of devices to afford protection within the limitations of a particular tract of land; the selection of a combination of devices to meet jointly the needs of associated areas of land of different characteristics; and, finally, efficient application and maintenance of the selected devices. Also, readjustments in the size of the farm unit will be necessary in many instances to establish an economically sound enterprise, especially where there is a shift from emphasis on crops to emphasis on livestock.

The most widespread opportunities for water conservation are to be found in runoff prevention on the heavier soils and diversion of runoff from wasteland areas to sites of usefulness. Effective use of vegetation involves conservative regulation of grazing range land, crops, and stubble; the prevention of burning of crop residues; the use of tillage methods that allow trash to remain on the ground; strip cropping; and protective use of grass for the retirement of critically erodible areas. The greatest chance to use emergency cover crops probably occurs in early summer or midsummer when rains are most likely to fall. The best opportunity to utilize border and strip designs is found on soils of favorable moisture retentiveness or in sites where waste water can be utilized in the support of windbreak tree plantings and in fields where the erosion-inducive or clean-tilled crops are grown or summer fallow is practiced. Emergency tillage operations are most likely to be needed in bare or thinly vegetated fields and in places where neighboring lands have been permitted to reach a condition that favors excessive blowing. In every case, when these measures are applied, the Plains conservationist must determine what combination of methods is most economical and effective. Very rarely can adequate control be achieved by using a single method.

The ability of an economical combination of methods to effect reasonable control should determine changes in land use. For example, soils too sandy, too shallow, or too infertile to permit economical maintenance of adequate vegetative cover from the residues of annual crops must be retired from cultivation and turned into grassland if adequate erosion control is to be achieved.

With proper land use and normal soil stability, emergency control measures should rarely, if ever, be needed. Farmers facing the problem seriously for the first time have found that they need emergency measures for temporary respite or until opportunity permits something more lasting. In such circumstances, emergency methods are the first step toward permanent control.

Unfortunately, however, the magnitude of the immediate emergency problem in the most seriously affected areas has partly obscured the need for permanent control through a farsighted precautionary program. Also, such repair work as the leveling of soil drifts and hummocks has been looked upon as final. These efforts are worse than useless, however, unless steps are taken to prevent recurrence of troublesome drifts. Indeed, this may be said of all emergency control.

Unless vigorous steps are taken to effect permanent control, emergency efforts cannot be justified long. The fallacy of reliance on temporary methods alone has been demonstrated amply in many parts of the Plains during the recent long drought. The fact that farmers have been unable to control the situation with these temporary measures shows the need for a coordinated permanent program, which, properly directed and extended, will prevent recurrence of conditions now spreading in many localities.

Demonstrations carried out on a large scale by the Soil Conservation Service in a number of representative localities in the Great Plains show definitely that runoff water can be stored by practical farm and ranch methods in the more favorable soils in amounts that will contribute substantially to increased production of grass and farm crops. Observations on both experiment station plots and Soil Conservation Service demonstrational areas have shown as much as 300 per cent increase in grass production after the installation of complete contour-furrowing systems on buffalo-grass land. Experimental data indicate that level terraces and contour tillage may increase the available soil moisture for plant growth as much as 50 per cent. Experiment station observations show as much as 49 per cent increase in wheat production over a period of 6 years and a 40 per cent increase in lint-cotton production over a period of 11 years, as the result of level terracing and contour tillage. One year's observations on 9,143 acres within the Dalhart, Tex., demonstrational area show a 28 per cent increase in production of grain sorghums, as the result of contour tillage, and a 57 per cent increase as the result of level terracing and contour tillage. Every practicable effort should be made to encourage the spread of such proved methods to adaptable lands for stabilizing and conserving water. The conservation of rainfall and the protection of soil from erosion are basic procedures for most farm and range lands of the vast Plains region and indispensable to the permanence of agriculture. For cultivated land, such conservation measures include, among others, contour cultivation and listing, level terracing, contour strip cropping, rough tillage, preservation of crop stubbles and residues, building up of the organic content of soils, water spreading, crop rotations, and summer fallowing.

Water, in a sense, is the beginning and the end of agriculture on the Plains. Plants cannot be grown without it, of course, and vegetation is needed to anchor Plains soil against wind. Cultivated soil, depleted of binding grass roots and spongy vegetable matter, without water is turned into a powdery substance susceptible to easy shifting by wind. The plants of the Plains will not grow in shifting soil; shifting of soil cannot be stopped in the Plains without plant coverage. Although rainfall is scant, usually there is enough to make a crop on the more favorable lands if none is wasted. Good land use, therefore, requires the retention of all possible rainfall on the land as the first step in protecting the soil from erosion or restoring it to use after erosion has started; and for this reason, conservation of rainfall is the first step toward the maintenance of a stable agriculture in the region. ✓

## Some Measured Results of Program

The following accomplishments, as of June 30, 1938, are indicative of the progress of the work of the Soil Conservation Service in the Great Plains:

Since the spring of 1934, the Soil Conservation Service and some 6,200 farmers and ranchers of the Plains have been cooperating in a series of soil-protection and land-use demonstrations embodying the actual application of conservation principles to more than 2½ million acres of typical Plains land, not including a considerable number of outlying single-farm demonstrations conducted in cooperation with State Extension agencies. Seventy-nine separate demonstrations have been developed in scattered areas throughout the region. In effect, they provide a network of conservation "show windows" in which farmers of the Plains country may see conservation practices in operation, learn how to apply them to the land, and judge their effectiveness in terms of actual results on the land itself.

Each demonstration area is a distinct entity, selected because of the very severity and the representative nature of its land problems. In each, a physical inventory of the land has been made by survey experts of the Soil Conservation Service. Erosion conditions, as well as the factors influencing erosion—topography, soils, and land use—have been mapped in detail. Complete plans for a coordinated program of land treatment, involving the application of conservation practices in accordance with the differing needs and adaptabilities of each parcel of land have been made for individual farms. Farmers and ranchers voluntarily are putting them into effect under the guidance of conservation specialists.

The primary purpose of the work in these selected areas is one of demonstration. The introduction of conservation farming practices on 2½ million acres in less than 3 years' time, however, represents no inconsiderable advance in the better utilization of Great Plains land.

Before the initiation of the demonstration program, only a few acres in the 79 project areas of the Plains region were farmed with a view to conservation of water and soil. Today, conservation-farming practices are being applied to more than 2,500,000 acres in the same areas and on the outlying farms. About 200,000 acres are being strip cropped to retain water and protect the soil from wind. Nearly 460,000 acres are being tilled on the contour to conserve rainfall and impede the sweep of wind. Contour furrows have been run on some 260,000 acres of grassland to conserve water. More than 16,000 miles of terraces—enough to stretch more than halfway around the world have been built to hold the moisture on some 330,000 acres; when present agreements are completed, nearly 510,000 acres will be terraced. In the 79 project areas, the acreage devoted to clean-tilled, erosion-inducing crops is being reduced 29.7 per cent; and the acreage of dense, erosion-resisting crops, increased about 30 per cent. Nearly 1,400,000 acres of grassland on which grazing was formerly uncontrolled are now being managed carefully to prevent overgrazing and consequent erosion.

All of these practices are designed to retain rainwater on the land, force it into the reservoir of the soil, build up the underground reserve of moisture for the nourishment of crops, and increase the production of protective vegetative cover for the land. They are not difficult of application; the average farmer on the Plains can easily follow them in the culture of his land.

For surface storage, more than 1,388 permanent dams have been built in the Great Plains region by the Soil Conservation Service alone. Most of these dams were constructed on the lands of cooperating farmers and ranchers as part of the normal demonstration program of the Service; others have been built outside the demonstration areas as part of the emergency drought-relief program. Their total storage capacity is estimated at nearly 21,000,000,000 gallons—a material addition to available water supplies for stock, irrigation, and other purposes in the Plains.

More important than this land-treatment work itself, however, is the effect that has been made in convincing Great Plains farmers that soil and water conservation is both practical and advantageous. Some 25,000 farmers and ranchers outside the boundaries of the demonstration areas are known to have applied one or more of these conservation practices to more than 3 million acres in the Plains. Such rapid and widespread acceptance of sound conservation principles would appear to indicate not only the effectiveness of the program but the readiness of Plains farmers to accept advice and to adopt, voluntarily, proved methods of soil defense. So vast is the area, however, and so great the amount of work to be accomplished, that only a beginning has been made.

## Problem Areas of the Southern Plains

The prinicpal subdivisions of the Southern Plains (see *Problem Area* map) are the Red Plains, Dark Brown Plains, and Brown Plains.

### Red Plains

The Red Plains subdivision of the Southern Plains sector of the Great Plains region comprises about 30 million acres in west-central Texas and adjacent central Oklahoma. Lying at a lower elevation, it is separated from the main body of the Southern Plains to the west, frequently referred to as the *High Plains*, by a narrow to broad zone of rolling to rough country, which generally falls rapidly from the characteristically flat plains (the Dark Brown Plains and Brown Plains subdivisions). This intermediate belt of predominantly rolling to rough country is locally referred to, at least in part, as the "Breaks."[1] The line of separation from the higher plains country on the west usually is clearly defined, either by a sharply descending surface or by a definite escarpment. This is particularly true in Texas and Oklahoma, where the escarpment ranges to several hundred feet in height and is visible many miles from the lower country. In places, the escarpment is a zone of rough land, often stony, 2 or 3 miles wide, abrupt along the upper part near the "cap rock" and less steep along the lower part.

On the east, the Breaks merge with the Red Plains in such a way that any line of separation must be more or less arbitrary. As a matter of fact, numerous scattered areas through the western part of the Red Plains are essentially the same as much of the more easterly portion of the Breaks country.

Predominantly, the Red Plains area is of rolling topography, although numerous areas are characterized by mild relief—nearly flat, undulating, or only gently rolling. In some localities, especially along drainage lines, the surface is severely broken or rough as the result of advanced erosion; and other areas are distinctly hilly. The Wichita Mountains in Comanche and Kiowa Counties, Oklahoma, represent an inclusion of distinctly different topography and soil characteristics (soils derived from crystalline rocks). Isolated areas of wind-blown sand of dune topography occur in small and large bodies, mainly in the vicinity of major drainages in the western part of the Oklahoma sector. In the western half, occasional steep-sided, islandlike buttes or mesas are conspicuous features, visible from long distances. They are remnants of a former higher surface, which has been generally lowered by the geologic process of erosion. The larger streams, especially those of Oklahoma, usually are bordered by broad strips of level bottomland and second bottoms.

The surface of the Red Plains rises gradually from about 1,000 feet along the eastern edge to about 2,000 or 2,500 feet on the western border (eastern border of the Breaks).

The upland soils, derived from underlying beds of shale, claystone, and sandstone, are usually of purplish-red color ("Red Beds") and of fine sandy loam and clay loam texture. The most extensive types are the Vernon fine sandy loam and the Miles fine sandy loam and clay loam.

[1] See soil map in Bennett, H. H. "The Soils and Agriculture of the Southern States."

The more nearly level areas usually are occupied by darker colored heavier soils, such as the Kirkland and Abilene clay loams.

Almost the entire upland area, representing about 90 per cent of the region, is subject to erosion in some degree by water or wind, whenever cultivated to the clean-tilled crops without the use of conservation farming measures. Overgrazing is locally responsible for considerable washing and blowing, especially in the western part. Water erosion is the most serious problem, although blowing is bad enough, especially on the old dune areas which never should have been broken out of their native cover of vegetation.

Not many parts of the United States have suffered more severely from erosion than a large aggregate portion of this area, whose extensive agriculture is largely of very recent development. As in most parts of the United States, farmers with large supplies of productive virgin land at their disposal thought little of conservation. They cultivated downhill, seldom practiced crop rotations, and paid little or no attention to the maintenance of the organic-matter supply of the soil.

Geological erosion, going on for ages, had dissected so much of the Red Plains country by the time white men arrived on the scene that some specialists designated the area as the "Eroded Plains." But vegetation had stabilized surface conditions pretty generally. Then, the growing of cotton, corn, sorghum, and wheat year after year in the same fields, and liberal grazing without much regard for carrying capacity, rotation, or seasonal conditions, started a process of accelerated erosion which has affected seriously a large area. Now the region is presented with a difficult task of land defense. Work carried on in a number of soil and water conservation demonstration projects has shown that the problem can be met in a very adequate way through the use of practical farm measures, aided by retirement of the more erodible areas to the permanent protection of grass or, in some places, trees.

Experiments carried on at the soil and water conservation experiment station near Guthrie, Okla., have revealed the disastrous rate at which erosion proceeds on the predominant type of erodible farm land under continuous cultivation and, in contrast, the comparatively low rates under crop rotation and contour cultivation, as well as the insignificantly slow rate under a cover of grass (see Table 7, Chap. V, Part 1).

### Dark Brown Plains

The southern Dark Brown Plains area (including the "Breaks" along the southeastern border) extends from the Nebraska Sandhills on the north to the Edwards Plateau on the south and lies between the Brown Plains on the west and, roughly, the 98th meridian and the Red Plains on the east. It comprises approximately 60 million acres, or about one-fifth the area of the entire Great Plains region. The elevation ranges from around 2,000 feet above sea level on the east to about 4,000 on the west. The surface is characteristically smooth, although there are rolling to rough, broken areas along streams and in the Breaks section bordering the southeastern part of this problem area. The smoother areas are varied locally by roundish depressions which hold water in wet seasons and dry out at other times (*playa* lakes). Most of the area, excluding the associated Breaks, is ideally suited to the operation of agricultural machinery.

The soils in the northern part are derived to a large extent from loessial material; whereas in the southern part, the parent material is largely of water-transported origin, having been spread out over the plains from mountainous areas to the west in Tertiary time. Generally, the soils are of silt loam, silty clay loam, and clay loam texture, well drained, and topographically well suited to cultivation. Sandy loams are important locally. In the southern part, the Pullman, Richfield, and Potter soils are representative of the most extensive groups; and in the northern section, the Keith, Holdredge, Hays, Colby and related soils are widely distributed.

The average annual precipitation ranges from about 26 inches along the eastern edge to about 18 inches along the western boundary.

Approximately 50 per cent of the entire area is cultivated. The agriculture is based principally on wheat production over the predominantly smooth lands, along with a substantial acreage devoted to the feed sorghums. Livestock is of importance on a considerable number of farms. The rougher lands of the Breaks are used almost exclusively for grazing and are admirably adapted to such use. Although those methods of land use which have proved impractical do not involve such extensive areas as in the adjoining Brown Plains on the west, many small areas of inferior soil are still in cultivation.

Wind erosion is a major problem. The practices most widely adapted to its control have to do with water conservation and crop management. Outstanding among these for the croplands are contour tillage, supported by level terraces and the growing of erosion-resistant crops in a flexible rotation, along with conservative use of all stubble and residues. Seeding of wheat should be adjusted to the supply of moisture in the soil at seeding time. Contour furrowing is an effective remedial practice on thinly vegetated grazing areas. Grazing of crop stubble must be carefully restricted to insure maintenance of an adequate protective cover.

Results obtained and field methods employed on the Hereford, Tex., demonstration project are summarized below, as an illustration of the practical possibilities for controlling erosion and conserving rainfall in the southerly part of the southern Dark Brown Plains, where conditions are more favorable to grain-sorghum farming than in other parts of the Southern Plains.

### RESULTS OF CONSERVATION WORK ON THE HEREFORD, TEX., DEMONSTRATION PROJECT

The Hereford soil and water conservation demonstration project, comprising approximately 32,000 acres in Deaf Smith, Castro, and Parmer counties, is located in the High Plains of the Texas Panhandle, at an elevation of approximately 3,900 feet above sea level. Although the long-time average annual precipitation is 17.5 inches, rainfall was below normal for the years 1933–1936, inclusive (7 inches in 1933, 12.4 inches in 1934, 13.9 inches in 1935, and 14.9 inches in 1936). In 1937, however, the precipitation was about normal (17.67 inches), and rainfall was favorable in 1938.

Wind velocity is characteristic of the southern High Plains, reaching a maximum in March, April, and May. Predominantly, the surface is flat, with an almost uniform ascent of about 12 feet to the mile from southeast to northwest. Small lake basins of about 5 to 40 acres extent dot the land here and there. Their drainage areas range from 200 to 1,500 acres. The principal soils are clay loams and silty clay loams of the Pullman series, together with considerable fine sandy loam, clay loam, and clay of the Richfield, Zita, Potter, and Spur groups. Most of the area is topographically well suited to cultivation.

Wind erosion was serious in September, 1935, when conservation demonstration operations began on the project. Much cropland had lost a large part of the topsoil, and many areas of range land had been almost denuded of native vegetation. Much wind-blown soil had accumulated in and around buildings and along fence rows, and numerous hummocks and small dunes were in evidence in many fields that formerly had been cultivated profitably. In 1935, drought and erosion had resulted in a shortage of both grass and feed crops. The conservation program has resulted in the removal of hazards of this nature, and land that was not producing in 1935 is now being farmed or grazed in a satisfactory manner.

Because of unfavorable economic conditions resulting from drought and erosion, some financial institutions had ceased to lend money for the purchase of farms within the project area. Choice land that was offered for sale for $8 an acre in 1935 was selling in 1938 at $25 to $35 an acre. Many farm operators who had been forced to seek employment as relief laborers in 1935 and 1936, were making a substantial living from their stabilized farms in 1938. Blinding dust storms were common from 1933 to 1936; but since the spring of 1936, little soil has moved on the Hereford demonstration project, even though blowing took place in 1938 on untreated lands adjacent to the project area. However, as the result of the spread of the conservation practices, only 80 sections of land suffered from serious blowing in Deaf Smith county during 1938.

In order to establish an effective, practical program of erosion control under conditions obtaining on the Hereford project, which is representative of a large area in the southern Dark Brown Plains, the following practices are among the more important put into effect:

1. Strip cropping where applicable.
2. Planting erosion-resistant crops, such as the grain sorghums and forage sorghums, Sudan grass, and broom corn; and the sowing of winter wheat on good wheat land, under favorable soil-moisture conditions.
3. Conservation of practically all rainfall by contour farming supported by level terraces with closed ends and by contour pasture furrowing.
4. Tillage of the kind that produces a rougher cloddy surface condition.
5. Protection of crop stubble from excessive grazing.
6. Establishment of crop rotations.
7. Establishment of a protective cover of grass on areas severely damaged by erosion or susceptible to such damage.
8. Maintenance of native sod in grass, and improvement, where needed, by water-conserving contouring.
9. Protection of grazing areas by deferred and rotated grazing, adjustment of numbers to carrying capacity, and establishment of properly distributed watering places.
10. Installation of windbreaks of trees in favorable and advantageous locations.
11. Timely contour tillage for efficient distribution and utilization of rainfall.

Some of the accomplishments on the 50 cooperating farms are listed in the table on page 749.

Pasture furrowing has held most of the rain, distributed it fairly evenly over the pastures, and resulted in marked improvement of the grass. Tests made on Pullman silty clay loam in a contoured pasture of 1 per cent slope, following a 9-inch rain of two weeks' duration, in the spring of 1937, showed that moisture penetrated to a depth of 42 inches in the furrows and to 18 inches 7 feet away from the furrows. The furrows, in pairs, were 4 inches deep, 8 inches wide, and 7 feet apart, with a 14-foot spacing between each pair. Penetration by the same rain in neighboring untreated pastures, including the same kind of soil, nowhere exceeded 18 inches. The furrowed part of the contoured area produced 1,761 pounds of air-dried grass per acre; and the unfurrowed part, only 704 pounds.

Before demonstration operations began, farmers generally felt that terracing and contour cultivation were unnecessary because the land appeared to be so nearly level that loss of water by runoff was not recognized as of any importance. Although much of the land has a slope of less than one-half of 1 per cent, the long slopes, coupled with the slow

FIG. 291.—Rainfall conserved on level-terraced (with ends closed), gently sloping land—water spread from terrace to terrace 15 hours after 2.2-inch rain. Soil and water conservation demonstration, near Hereford, Texas, April, 1938. (*Photograph by Soil Conservation Service.*)

| | *Acres* |
|---|---|
| Treated or planned for treatment (average size of farm units, 490 acres) | 24,500 |
| Protected by terraces | 9,046 |
| Protected by contour tillage alone | 4,774 |
| Pastures contour-furrowed | 3,300 |
| Planted to trees | 47 |
| Wind-blown hummocks leveled | 843 |
| Diverted from continuous wheat to flexible rotations including wheat | 18,116 |
| Strip cropped | 1,750 |
| Gullied areas controlled | 247 |
| Retired from cultivation to permanent grass | 1,850 |
| Improved range management installed | 6,535 |
| Drift material removed from buildings and fences, cubic yards | 318,191 |

absorptive capacity of the heavier soils, frequently cause torrential runoff from heavy rains. Practically all of the rainfall has been held in fields with level terracing (Fig. 291) and with contouring (Fig. 292), resulting in largely increased yields.

Wherever practicable, water is diverted from roadside ditches and distributed over pastures, and in some instances over fields, by the aid of

Fig. 292.—Contour listing saves rainfall. Deaf Smith County, Texas. May, 1937. (*Photograph by Soil Conservation Service.*)

Fig. 293.—All of rainfall conserved in this level-terraced field. This, plus the water diverted from a roadside ditch, was utilized for wheat production. This field, in the Hereford, Texas, soil and water conservation project, produced 23 bushels per acre in 1938 and 36 bushels in 1939. (*Photograph by Soil Conservation Service.*)

small distributary furrows. On one farm, such added water increased the 1938 yield of wheat 6 bushels an acre over the yield from that part of the farm where only the direct rainfall was conserved. The fields were terraced and contour-tilled in both instances; the former produced 23 bushels an acre in 1938 (Fig. 293), and the latter 17 bushels. An adjacent

FIG. 294.—Wheat failure on land like that of Fig. 293, where no conservation practices were followed, near Hereford, Texas, June, 1938. (*Photograph by Soil Conservation Service.*)

FIG. 295.—Wind erosion so impoverished this area in Deaf Smith County, Texas, that no vegetation was produced in 1938. (*Photograph by Soil Conservation Service.*)

untreated field of comparable soil produced only 4½ bushels of wheat an acre in 1938 (Fig. 294). In 1939, with better rainfall the untreated field produced 16 bushels an acre and the treated field across the road 36 bushels an acre.

On some of the farms, soil drifting had become so active by the end of May, 1936, that it was difficult to establish sufficient vegetative cover to hold the land (Fig. 295). Runoff from heavy rains was high, owing both to deficiency of organic matter in the soil and to a lack of cover. The land was listed on the contour and level-terraced. From a number of fields thus treated, wheat produced in 1938 from 17 to 21 bushels an acre (Fig. 296). The average 1938 wheat yield from 2,640 acres on 26 terraced farms in the Hereford project (having an average slope 0.65 per cent)

Fig. 296.—Twenty bushels of wheat per acre was produced as result of water conserved by contouring and level-terracing. Hereford, Texas, soil and water conservation project, June, 1938. (*Photograph by Soil Conservation Service.*)

was 10.7 bushels per acre, as against an average of 7.8 bushels from 2,903 acres on 19 neighboring farms on land of the same kind except that the average slope was considerably milder (0.42 per cent). The average yield for the county (Deaf Smith) was 3.7 bushels on 1,100 farms, comprising 215,000 acres.

## Brown Plains

The southern Brown Plains, comprising about 50 million acres, lie between the Rocky Mountains on the west and the Dark Brown Plains on the east. Elevation above sea level ranges from about 4,000 feet along the eastern boundary to about 6,000 feet at the foot of the Rockies. Average annual rainfall varies from about 18 inches along the extreme eastern and western edges to about 12 inches in parts of the interior. Variability of precipitation is a hazard that must be met by conserving all possible moisture with such proved methods as contouring, terracing, and restoration of grass on the more unstable areas.

Most of the land is of favorable topography, having in the southern part the characteristic flat surface of the Panhandle country and in the northern part the typical undulating to smoothly rolling features of the wind-lain soils of the Plains. A belt of rolling country skirts the Rocky Mountain front along the entire distance from southeastern New Mexico to southeastern Wyoming.

In the southern sector, the soils are derived mainly from water-transported material, and much of the land is light textured, such as the sandy loam and sand types of the Amarillo and Springer groups. Areas of old sand dunes occur in various parts of the southerly extension of the subdivision. Heavier soil, such as the loam, silty clay loam, and clay loam of the Zita, Baca, and Potter soils, is fairly abundant. Toward the north, a larger proportion of the land is of silt loam, silty clay loam, and loam texture, much of it belonging to the Colby and Weld series. Some areas derived from shale consist of tough clay, such as the Pierre soils in the vicinity of Hugo, Colo.

Repeated efforts to develop cash grain farming generally have failed in the southern Brown Plains. Usually, conditions are most favorable for livestock and the production of supplemental feed crops, although at the present time many farm units are too small for economic livestock production.

Wind erosion has been exceedingly severe over a large and widely scattered area. Much land left barren of vegetation by the failure of wheat and other crops has been stripped of its topsoil, in some instances down to the caliche subsoil; and numerous areas have been so deeply covered with sheets and hummocks of wind-blown sand and silt that every vestige of plant life has been smothered. Such critical conditions can be remedied only by the restoration of a protective cover of vegetation, and this can be done, generally, only by conserving every possible drop of rainfall. Certain areas of loose sandy soil and exposed caliche, or thin soil over caliche, should be retired immediately to the permanent protection of vegetation—even to weeds, if grass cannot be established at once.

Violent floods originating in the Brown Plains area and adjacent mountains are frequently destructive of property and the irrigated farm lands of the alluvial plains within and to the east of the area. Control of erosion and runoff over the upper watersheds presents the opportunity for alleviating the flood hazard.

Field methods of soil and water conservation applicable in the southern Brown Plains, together with results obtained, are illustrated in the following summarization of the work of the Smoky Hill River demonstration project in Cheyenne County, Colorado.

## RESULTS OF CONSERVATION WORK IN THE SOUTHERN BROWN PLAINS SMOKY HILL RIVER PROJECT, COLORADO

The Smoky Hill River soil and water conservation project, comprising 160,000 acres in east-central Colorado, is fairly representative of predominant physical, agricultural, and economic conditions involving a large area, probably 15 million acres at least, in the northerly part of the southern Brown Plains (eastern Colorado and adjacent Kansas).

The most important soils in the project area are loams and clay loams of the Colby and Weld groups. Generally, these are undulating to gently rolling. For the most part, they absorb rainfall at slow or moderately slow rates, but their moisture-holding capacity usually is good. Runoff is normally rapid. Deposition of fine material of a colloidal nature from the dust storms of recent years appears to have decreased infiltration capacity and increased runoff.

Formerly a livestock country, farming has been carried on rather extensively the past 20 years. During the period of agricultural development, much of the land was divided into comparatively small tracts and sold to nonresidents. At present, about 80 per cent of the project area is owned by nonresidents.

A survey of the project area shows that an area of 112,400 acres was in sod of widely varying density in 1937; 9,500 acres were cultivated; and 38,100 acres of former cropland were idle or abandoned. Abandonment has increased steadily since 1931. When the conservation program started in 1935, most of the abandoned land supported little or no vegetation; nearly all of it had been severely eroded. In addition, much of the sparsely vegetated grazing land had suffered, both by direct blowing and by deposition of material from neighboring areas. In 1937, only 72 ranch and farm operators were living in the area.

Damage by erosion and the deposition of wind-blown materials, together with a resultant intensification of drought effects, has occasioned an acute need for a more stable type of agriculture—one based primarily on grass, livestock, and feed production. The Soil Conservation Service has directed its demonstration program toward stabilization of all blowing areas through the conservation of rainfall, retirement of around 40 per cent of the cultivated land to grass, and encouragement of livestock farming by combining small holdings by lease or purchase into units large enough for economic operation.

Prior to the systematic fencing of agricultural lands under this program, much of the grazing land in the Smoky Hill area had been utilized as "open range." Numerous tracts of formerly cultivated land were idle or abandoned, and many fences had been removed or destroyed.

When work began in September, 1935, it was impossible to accomplish very much in the direction of improving grazing practices, because of inadequate fences resulting from the spread of farming over the area, and the open-range grazing methods generally followed. Outside stockmen moved in with their herds wherever grass and water were available and so aided the development of an exceedingly inefficient and destructive grazing system. This highly competitive use of the range, coupled with the effects of drought and erosion, resulted in destruction or impoverishment of much of the grass, thus favoring waste of rainfall, further decline of cover, and increased erosion by water and wind. Since precipitation for the 7 years preceding 1937 had not exceeded 9 inches for any one year, crops had largely failed through this period.

Such losses and failures, together with a normal rainfall of only about 17 inches, revealed the need for conserving all possible moisture in the program of soil defense. A survey of the physical and economic conditions at the time work began on the project pointed to the need for a completely coordinated and adaptable program of stabilization and conservation, involving the following major procedures:

1. Establishment of an effective vegetative cover on all abandoned lands.

2. Retirement of all severely eroded areas and the most erodible cultivated land to permanent cover.

3. Application of practical moisture-conservation measures on all cultivated land—principally contour cultivation and terracing.

4. Installation of effective agronomic practices on all cultivated lands for control of erosion (both wind and water)—principally strip cropping, flexible crop rotations, and the growing of erosion-resistant crops.

5. Establishment of fenced grazing units by consolidating land under different ownership into an economically effective operating unit.

6. Use on range land of such practical mechanical measures for conserving water as contour furrows, contour ridges, and water diversion and spreading structures.

7. Construction of stock-water reservoirs at locations favorable to good grazing practice.

8. Erection of fences needed for proper range management.

9. Adoption of a sound range-management program.

10. Planting of trees and shrubs in adaptable situations for gully control and shelter purposes.

With less than 10 inches of rain, and in spite of a severe infestation of grasshoppers, excellent recovery in the conditions of both range and cropland was made during 1937. For example, contour-furrowed plots of range land showed as much as 112 per cent increase in grass density over similar untreated areas, and considerable feed was produced in contoured or terraced fields similar to untreated fields that produced nothing. Drainageways originating within the area have carried only small flows since installation of the protective measures, in spite of a number of summer rains of high intensity. Previous to treatment of their watersheds, these dry channels always had flooded after rains of equal intensity, the water running to waste as unutilized flood flows.

On Apr. 15, 1938, 162 active cooperative agreements, covering about 78,000 acres, were in effect on the Smoky Hill River project. Below are listed some of the more important accomplishments of the work within the project area.

| | *Acres* |
|---|---|
| Erosion controlled by retirement to permanent vegetation | 8,008 |
| Terraced and contoured for continuing cultivation | 4,543 |
| Contoured for cultivation (much of this was abandoned in 1937) | 15,383 |
| Strip cropped | 7,333 |
| Crop rotation (flexible) installed on | 20,230 |
| Planted to erosion-resistant crops | 7,307 |
| semierosion-resistant crops | 8,156 |
| Former overgrazed open range consolidated into 14 units to be operated as combined livestock enterprises, with small cultivated areas for supplemental feed | 38,740 |
| Range land contour furrowed or contour ridged | 41,953 |
| Improved range program devised for (including part of acreages listed above) | 56,260 |
| Trees and shrubs planted for shelter belts, wood lots, and gully protection | 311 |

| | |
|---|---|
| Water-spreading and diversion structures (earthen) built on range land | 33 miles |
| Stock ponds completed (combined storage capacity, 447 acre-feet) | 110 |
| Fences built for range rehabilitation and management | 65 miles |

### DALHART, TEXAS, PROJECT

The Dalhart soil and water demonstration project, comprising approximately 48,000 acres in Dallam County in the northwestern corner of the Texas Panhandle, is located along the boundary between the Dark Brown and Brown Plains sectors of the Southern Plains area. The most important soils are the sandy loams of the Amarillo and Springer groups, together with the loams, silty clay loams, and clay loams of the Pullman, Richfield, and Potter groups. The surface is flat to undulating, with slopes seldom exceeding gradients of 2 or 3 per cent. In the absence of streams, the area drains into shallow "wet-weather," or intermittent, lakes locally referred to as *playas*. The average annual rainfall is 17.6 inches; but for the period 1933–1937, inclusive, it was only 11.5 inches. Characteristically, the winds reach their maximum velocity during March, April, and May—the severe "blow season."

When conservation demonstration operations began on the Dalhart project in August, 1934, much of the land was in a critically eroded and eroding condition. More than 50 per cent of the area had been damaged. Some 11,000 acres of cultivated land had been badly hummocked by accumulation of sand; 9,000 acres had lost from 2 to 4 inches of topsoil; and 2,500 acres had lost more than 4 inches of topsoil, with some areas stripped to plow depth. Troublesome drifts had accumulated along fences and about farm buildings. Many pastures adjoining cultivated fields had been severely damaged through the abrasion of grass by shifting soil and by smothering accumulations of wind-blown debris.

Economic and social consequences of erosion were probably commensurate with damage to the land itself. Some financial institutions had ceased to make loans for the purchase of farm land. Many tracts within the project area were not salable at any price; homes were abandoned, and health authorities reported an increase in respiratory diseases during the blow months of winter and spring. Men found it difficult to work during blinding, choking dust storms. In Dallam County, the number of farm operators decreased from 700 to 443 between 1935 and 1936, and the farm population declined from 2,500 to 1,500.

Agricultural decline during this period resulted from a combination of improper land use, inefficient water conservation and utilization, and poor crop management.

In order to carry out an effective and practical demonstration in the control of erosion under conditions obtaining at Dalhart, and other similar areas throughout the Brown Plains sector of the Southern Plains, the following practices were installed:

1. Planting erosion-resistant crops, such as grain sorghums, forage sorghums, Sudan grass, and broom corn, as an aid to control of soil blowing.

2. Conservation of the entire limited rainfall by contour farming supported by level terraces with closed ends and by contour pasture furrowing.

3. Tillage practices that leave the surface in a roughened condition.

4. Protection of stubble on all cropland from close grazing.

5. Retention of present grass cover and retirement of land unsuitable for cultivation to the protection of permanent vegetation.

6. Protection of pastures by deferred and rotation grazing, adjustment of grazing to carrying capacity, and development of properly distributed stock water and salting places.

7. Establishment of windbreak tree plantings in favorable locations.

8. Timely contour-tillage operations for proper distribution and maximum utilization of rainfall.

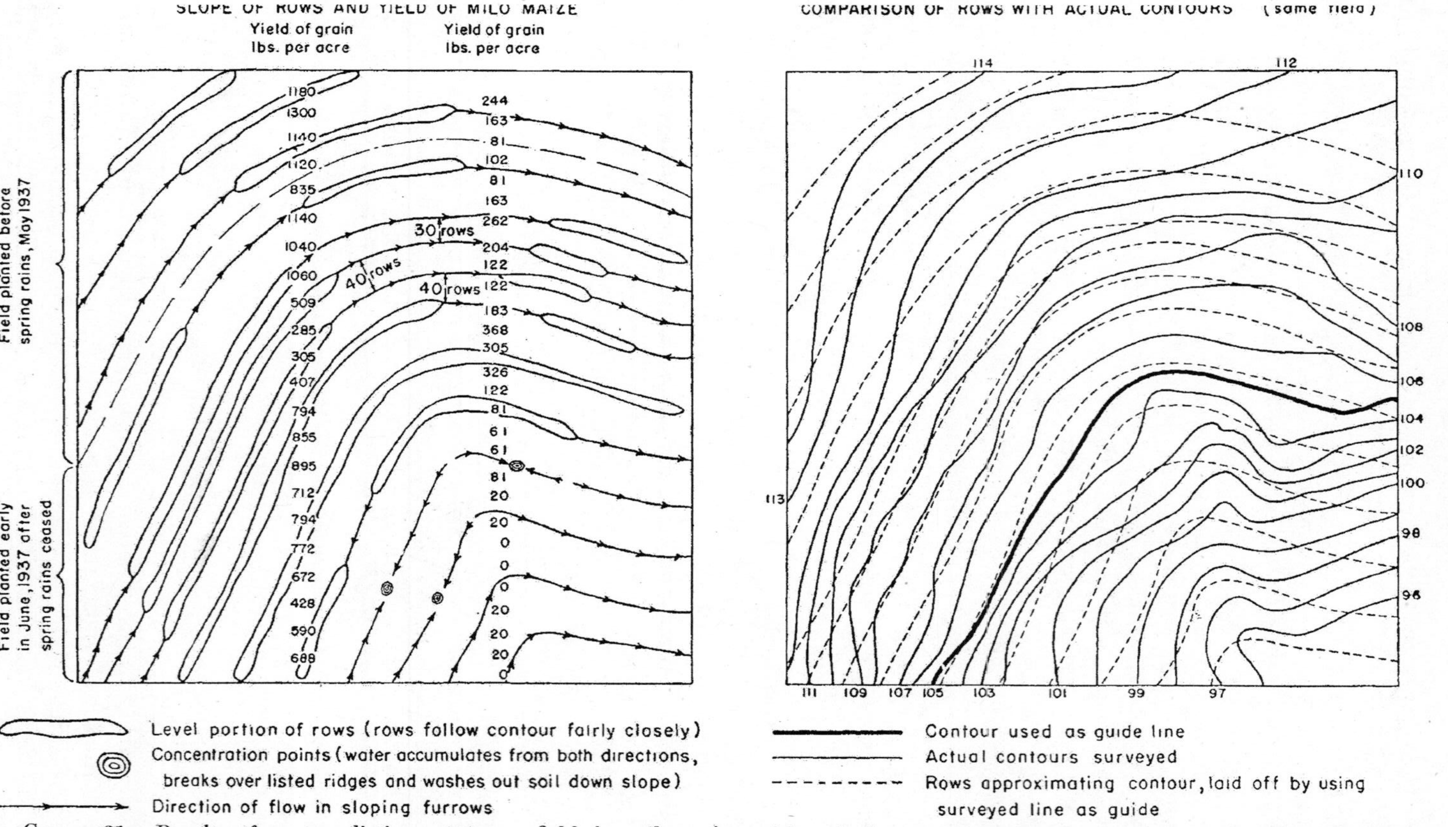


GRAPH 61.—Results of contour-listing a 160-acre field, for milo maize, with a single surveyed guide line, 6 miles south of Conlen, Texas. Pullman silty clay loam, average slope 0.5 per cent. (Interval between lines equals 60 rows, except as indicated.) Scale 1 inch = 165 feet. Average yield, 160 acres—399 lbs. per acre; average yield on contour, 80 acres—662 lbs. per acre; average yield off contour, 80 acres—206 lbs. per acre.

For maximum efficiency, then, contouring must be properly done. (*Soil Conservation Service.*

9. Restriction of wheat to the most favorable soils, and seeding only when soil moisture is sufficient for reasonable assurance of a crop.

10. Leveling of field hummocks and fence-line drifts to enable establishment of stabilizing vegetation.

11. Stabilization of drifting dunes by leveling, contour listing, and planting of sorghums on both the dunes and the surrounding lands.

12. Introduction of flexible crop rotations to restore needed organic matter.

These practices are fast becoming generally recognized and accepted by farmers as a safe and practical way for reestablishing vegetation and controlling wind and water erosion. And this is assisting materially in stabilizing agriculture in this part of the Southern Plains. Farmers are learning that erosion-resistant crops can be produced with the scant precipitation, provided all rainfall is effectively conserved through use of the practical farm and ranch methods demonstrated on the project. They are learning also that such work as contouring and terracing must be done correctly if best results are to be had (Graph 61).

Some accomplishments from 44 cooperating farms in the Dalhart project are:

| | *Acres* |
|---|---|
| Under agreement (average size of units, 653 acres) | 28,765 |
| Protected by terracing | 11,025 |
| Protected by contour tillage alone | 3,940 |
| Pasture contour furrowed | 463 |
| Diverted from continuous wheat to flexible rotation, including wheat | 14,800 |
| Reduction of wheat acreage between 1934 and 1937 | 14,100 |
| Diversion from continuous corn to erosion-resistant crops and strip-protected corn, 1934 to 1937 | 4,000 |
| Increased acreage of sorghums, 1934–1937 | 12,640 |
| Retired from cultivation to permanent vegetation | 2,746 |
| Planted to windbreak trees, miles | 23.6 |
| Hummocks leveled | 8,352 |
| Drift material removed from buildings and fences, cubic yards | 210,000 |

## Northern Plains

The Northern Plains, embracing most of the Dakotas and large portions of Montana, Wyoming, and Nebraska, are occupied by extensive bodies of glacial soils, wind-blown sand, residual soils, and badland areas where the basal rocks have been exposed by geological erosion. Although not so elevated as the Southern Plains, a much smaller proportion of the area is flat. The surface of a large part of the Northern Plains ranges from undulating to rolling or hillocky. There is much more rough country—badlands and broken foothills—than in the Southern Plains. Such rough lands are especially extensive in the western part. Aside from the isolated bodies of badlands, such as those of the western Dakotas, every major stream is bordered by long stretches of severely dissected terrain, and similar rough country skirts the base of most of the mountains.

As in the Southern Plains, too much of the northern division of the Great Plains has been broken out of native grass, with little regard for soil type, slope, or rainfall intensity; and, as in the south, wind erosion and, in lesser degree, water erosion have plagued many areas, large and small. Depletion of cover by overgrazing has bared the surface of much land, especially in the western part, to both water and wind erosion. Even irrigation farming has encountered erosion difficulties on sloping land, as in parts of the Wind River Basin of central Wyoming.

Betterment of the erosion situation in the Northern Plains calls definitely for conservation of rainfall on all dry-farmed land steep enough to permit any considerable runoff; a return of a large total area of erodible soil to a protective cover of grass; a reduction of the wheat acreage on thousands of farms; and the introduction of rotations that provide protection from wind and replenish the supply of soil organic matter. Areas returned to grass can be utilized advantageously for building toward greater security for the livestock element of farming and for rebuilding wildlife resources. Still further security on all farms can be gained by storing supplies of feed in good years, such as 1938, for years of lean feed production, such as 1934 and 1936. Many fields can be given a considerable measure of protection from wind by appropriate plantings of hardy trees and shrubs, and homes can be made more livable with shelter belts established about the farm buildings. Strip cropping is an important measure for reduction of wind erosion. Wind stripping (stripping across the prevailing direction of wind during the blow season) is adaptable to areas of such gentle slope as to have little or no water erosion. Where water erosion is a problem, strip cropping should be on the contour. Considerable soil washing was observed in 1938 on many of the wind-stripped (alternating fields of grain and summer fallow) areas in the rolling sections of Montana.

The principal subdivisions of the Northern Plains are discussed briefly below, together with a summarization of ways and means for better conservation of water and soil.

## Black Plains

The Black Plains comprise 31 million acres of predominantly productive black soil in a broad belt crossing the eastern Dakotas. The average annual precipitation is around 24 inches along the eastern boundary, decreasing gradually to an average of about 16 to 18 inches along the western edge. The elevation ranges from around 1,000 feet above sea level on the east to about 2,000 feet on the west. Occasional morainic hillocks, numerous basins and depressions, and a generally billowy or undulating surface give the area the characteristic topography of glacial terrain. Most of the land is well suited to cultivation, in so far as surface features are involved.

The principal soil groups are the Barnes, Fargo, Bearden, and Moody. All have black surface soils and, generally, dark-brown to black subsoils. Texturally, they consist for the most part of clay loam, silty clay loam, and clay. Sandy soils, even sand dunes, occur here and there. In the main, these soils absorb rainfall readily, and runoff is light except during unusually heavy rains. Although some areas are affected by water erosion, the chief physical problem requiring immediate attention in the Black Plains subdivision is wind erosion. Where lacking a protective cover of vegetation, these lands, especially the sandy types, are susceptible to severe wind damage.

A large proportion of the area has been broken out of the native grass cover and turned to wheat production. Some corn is grown, chiefly for forage. Farm units generally are small.

The general need for soil and water conservation can be met by such major practices and measures as strip cropping, contour listing and cultivation, water spreading, preservation of crop residues, retirement of inferior and highly erodible land to permanent cover, stock-water development, pasture contouring, improvement of grazing practices, and adjustment of size of farm units. These and other measures are being used successfully in the program of the Soil Conservation Service in this area. In the southeastern section, gully control, pasture improvement, soil improvement, and a decrease in the area devoted to clean-tilled crops are some of the more important practices.

Livestock should have an important place on most farms. Abundant feed, such as grass, sweetclover, corn, forage, and alfalfa, is easily produced in years of favorable rain. In good years, sufficient reserves of feed can be stored for use during years of below-normal rains. Where farming is of the balanced livestock-crop type, or where any important number of animals are kept, it would be advisable generally to store enough feed for one or two dry years. Adaptable feed crops can be worked into helpful rotations and strip-cropping systems for preventing soil blowing.

As an illustration of the field methods employed in the soil and water conservation program in the Black Plains area, results of the work on the Bottineau, N. D., demonstration project are summarized as follows:

## Results of Conservation Work in the Northern Black Plains Bottineau Demonstration Project

The Bottineau demonstration project, in north-central North Dakota, is fairly representative of the Black Plains area. The most extensive soil types within this area of approximately 48,000 acres are the Bearden very fine sandy loam (occupies 57 per cent of project area) and Fargo clay and silty clay loam (20 per cent of area). The surface varies from nearly level to undulating, slopes seldom exceeding a gradient of 3 per cent. More than 90 per cent of the area has slopes of 0 to 3 per cent, whereas 6 per cent ranges from 3 to 10 per cent.

Fig. 297.—Above, large area of Barnes fine sandy loam terrifically damaged by wind erosion. At the beginning of work on the Bottineau Project, North Dakota, this land was abandoned. Lower, stabilization of land like that above with sweetclover. 1936. (*Photographs by Soil Conservation Service.*)

The average annual precipitation at Bottineau is 15.24 inches, about 2 inches of which is derived from snowfall. Wind velocities are relatively high.

The project includes 180 farms, averaging 265 acres in size. A large part of the land was broken out of virgin grass in 1890, with a considerable increase during the World War period. Wheat has been the principal crop from the beginning. Following the World War, however, increased attention was devoted to livestock. The most successful farmers are those who have included livestock as a part of the farm enterprise.

At the beginning of project operations in 1935, 60 per cent of the area had suffered slightly from erosion, 28 per cent had been affected moderately, and 3 per cent had been affected severely (Fig. 297). Troublesome wind-drift material had accumulated along fences; many fields were covered with sheets and hummocks of such material, and numerous "blowouts" had developed holes 4 to 6 feet deep over formerly cultivated land. Major factors contributing to the erosion problem have been: con-

Fig. 298.—Protection from wind erosion by plowing wheat stubble so that part of the straw remains above the ground as effective mulch. North Dakota, 1938. (*Photograph by Soil Conservation Service.*)

tinuous wheat farming, with summer fallowing; and fall plowing for spring seeding.

In order to carry out a completely coordinated and practical demonstration of wind-erosion control and soil and water conservation under conditions prevailing in this area, the following practices were initiated on the project:

1. Strip cropping, with strip widths varying from 100 feet on the lighter soils to about 200 feet on the heavier types.

2. Introduction of rotations, such as fallow, corn or other feed crop, wheat, wheat and sweetclover, sweetclover, wheat.

3. Retirement of severely eroded and inferior land to permanent cover, using such adaptable grasses as western wheat, slender wheat, brome, and crested wheat.

4. Deferred and controlled grazing.

5. Development of properly distributed stock-watering places.

6. Discouragement of fall plowing to prevent baring of erodible soil to winter and spring blowing.

7. Delaying fallow operations until after the blow season of spring.

8. Substitution of duckfoot type of cultivation, where practicable, to reduce excessive pulverization of soil affected by moldboard or other types of plows.

9. Maintenance of "trashy" surface conditions to impede blowing, by special disk cultivation, such as leaves a considerable part of wheat or other stubble projecting above surface (Fig. 298).

10. Use of the damming or regular lister on contour to conserve water.

11. Encouragement of type of farming based as nearly as economically practicable on proper balance between crops and livestock, for adequate conservation of soil and rainfall.

On Feb. 1, 1938, conservation-farming arrangements were in effect on 61 cooperating farms, covering 14,780 acres, on the Bottineau demonstration project. Some of the accomplishments resulting from this cooperation are as follows:

| | *Conservation Work, Per Cent* | |
|---|---|---|
| | *Before* | *After* |
| Cultivated land in project area | 79.7 | 54.1 |
| Cultivated land retired to grass and hay crops | .... | 19.0 |
| Cultivated land retired to trees and shrubs as wind shelters and for wildlife | .... | 3.9 |
| Clean-tilled crops | 18.0 | 11.7 |
| Semierosion-resistant crops | 58.4 | 32.4 |
| Erosion-resistant crops | 3.3 | 10.0 |
| Idle cropland | 3.0 | 0. |
| Pasture or range | 11.4 | 21.6 |
| | *Acres* | *Acres* |
| Legumes | 591 | 1,778 |
| Rotated pasture or range, temporary | 53 | 576 |
| Rotated pasture or range, permanent | 0 | 1,010 |
| Cultivated land retired to pasture | ... | 1,525 |

## Dark Brown Plains

The northern Dark Brown Plains, with an area of 59 million acres, borders the Black Plains on the east. It includes, approximately, the western two-thirds of South Dakota, the western two-fifths of North Dakota, a large part of northeastern Montana, and a small part of northeastern Wyoming. It extends from the Canadian line to the southern boundary of the Nebraska Sandhills, which separates the area from the Southern Plains.

Predominantly, the surface of the northern Dark Brown Plains varies from gently rolling to rolling, although numerous areas along drainageways are steeply rolling to extremely broken and rough (badlands). The average annual precipitation ranges from about 16 to 18 inches along the eastern side to about 12 to 15 inches at the western edge. The soils, for the most part, are dark brown in the surface and lighter colored, calcareous, and compact in the subsoil. The principal series are the Williams, Scobey, and Arnegard. In the main, they range from clays and clay loams to sandy loams. The predominant heavy types absorb water slowly; and with heavy rains, loss by runoff is high wherever there is much slope. Wind erosion severely affects most open land having inadequate cover of vegetation, and water erosion is a problem on steeply sloping areas similarly thinly vegetated. Very little true soil occurs in the severely dissected badlands, such soil being restricted to flat-topped remnants of the former surface and intervening depressions. These broken lands consist very largely of "raw" geological beds, mostly steep and thinly vegetated. Much of the exposed shale and clay is of saline character and contributes locally to unfavorable alkali conditions. Floods are not a serious problem on the streams rising within this subdivision of the Great Plains. Long periods of low water seem to be a characteristic of such streams as the Souris, James, and Cheyenne Rivers in North Dakota.

The principal native grasses are western wheat, blue grama, "niggerwool," needle and thread, and buffalo.

Types of agriculture vary from wheat farming in the eastern part to both livestock and wheat farming in the western part. The average size of farms varies from about 320 to 640 acres from east to west.

The following practices are among the most important employed on the soil and water conservation projects:

1. Retirement of the more severely damaged and more erodible areas permanently to grass.

2. Improved management of pastures and ranges, adjustment of grazing to carrying capacity, and development of stock-water ponds for pastures and range land. (On 8 of the 13 stock ponds established on the Culbertson, Mont., project, 12 broods of wild ducks, totaling 96, were raised in 1938).

3. Decrease of cash-crop acreage and increase of livestock farming.

4. Contour farming and contour furrowing on pasture and range land to reduce runoff.

5. Changes in cropping practices in order to produce more feed for livestock, to increase soil productivity, to make use of strip cropping

and crop rotations, to provide for utilization of crop residues, to avoid fall plowing, and to develop cloddy surface conditions in plowing.

6. Development of wind-shelter and wildlife plantings of adaptable shrubs and trees.

7. Adjustment in the size of farms.

Generally farmers have approved these changes resulting from the introduction of conservation-farming practices in the Culbertson soil and water conservation project, in northeastern Montana.

## Brown Plains

The northern Brown Plains extend southeasterly through central and eastern Montana from Canada to the North Platte River in east-central Wyoming and from the base of the Rocky Mountains to the Dark Brown Plains on the east. The area covers about 61 million acres; and over the greater part, the surface varies from gently rolling to rolling or hilly. Along the major drainages and in the badlands sections, the country is characteristically extremely rough, as the result of geological erosion.

The topography, on the whole, is considerably more rolling than that of the Dark Brown Plains, and there is proportionately much more broken country. Hilly and severely dissected broken country, much of it typical badlands, is extensive along the base of the Rockies and Big Horns and along the principal streams, particularly the Big Horn, Yellowstone, Powder, Tongue, Missouri, Little Missouri, Cheyenne, Belle Fourche, Cannonball, Big Muddy, and Wind River. The rougher country, although providing fair grazing in some parts, is scantily vegetated, generally, because of low rainfall, rapid runoff, and scarcity of productive soil. Better grazing is found in the areas of more rounded topography, even where the country is hilly, because of the usual presence of a covering of true soil.

Nearly a fourth of the area is so rolling, hilly, or broken that it has no other value than for grazing, and part of the remainder is too steep for safe cultivation because of susceptibility to water erosion, especially near the mountains, where rainfall is frequently of the "cloudburst" type. The smoother lands of both sandy and heavy soil are subject to severe wind erosion under bared or thinly vegetated surface conditions, whether resulting from plowing or from overgrazing.

The soils of the smoother areas vary from sandy types to clay loams and clays, the heavier types predominating. Although mainly of light-brown color, the dry surface generally has an ashy-gray cast. Some of the more important upland soil groups are the Morton, Arnegard, and Jordan series. The Flasher soils are mainly sandy and extremely susceptible to wind erosion. Many of the low flats, especially those in and near areas of rough, broken country, are occupied by tough clay, frequently with high content of water-soluble salts. Many such areas, probably thinly grassed under virgin conditions, have suffered from gullying as the result of overgrazing. Salt sage and greasewood are the principal plants on the more salty areas.

Most of the smoother lands, outside the limits of the saline flats and depressions, and occasional bodies of tough shale-derived clay, originally were grassed, as shown by the present good cover on areas that have not been plowed or overgrazed. Among the principal grasses are blue grama, western wheat, and niggerwool. Sagebrush is widely distributed.

Rainfall varies from around 12 to 15 inches in the eastern part to less than 10 inches in the Big Horn and Wind River Basins of Wyoming. Floods are much more serious than in the Dark Brown Plains, especially along streams rising in the mountains. Water erosion is also more serious, mainly because of higher intensities of rainfall and steeper slopes. Although much wheat continues to be seeded, it appears that increasing interest is being taken in livestock. As compared with the eastern portion of the Northern Plains, land holdings are larger, and the livestock industry is of greater importance.

Need for water conservation on range land is general, and soil erosion by wind, and to some extent by water, is a local problem of outstanding importance, even on the irrigated lands. A great deal of silt is contributed to a number of streams in the "return" water from sloping irrigated lands. For example, the Big Horn River carries much silt into the Yellowstone throughout the irrigating season, gathering its load from both the Big Horn and the Wind River Basins. As an illustration, samples collected during the irrigating season of 1938 indicate that Teapot Creek, a typical small draw, in Wind River Basin, Wyoming, delivers to Wind River about 20 second-feet of return water, containing around 6 per cent silt, over approximately a five-month period.

These problems are being dealt with in the program of soil and water conservation through such procedures as shown at the bottom of p. 770.

As an illustration of improved range management, the main features of range practices instituted on a typical area are shown in Fig. 299, covering 44,801 acres (the Riverton Tract) on the Shoshone Indian Reservation, Wind River Basin, Wyoming.

### Pierre Shale Area

The Pierre Shale Area comprises some 13 million acres surrounding the Black Hills, in southeastern South Dakota, northwestern Nebraska, and northeastern Wyoming, with a relatively small segment in southeastern Montana. Similar shale areas occur as smaller isolated tracts in other parts of these states and in eastern Colorado. The prevalent surface configuration varies from nearly level to undulating and gently rolling. Small and large tracts of badlands are scattered through the area. Characteristically these are severely dissected and too rough for any use

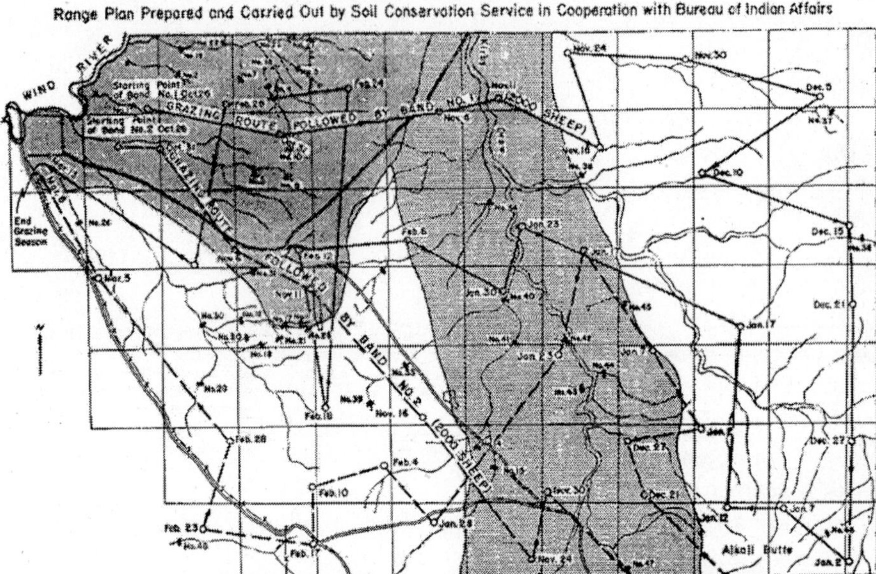


Fig. 299.—Demonstration of Range Improvement, Riverton Area, Shoshone Indian Reservation, Wyoming.

**Before improvement**

Use — 3335 sheep from Sept. 15 to May 31, in addition to 598 trespass sheep. Use of much of area wholly dependent on water from snow or small temporary ponds, resulting in waste forage over parts of range, and severe overgrazing on other parts.

[dark shading] With Wind River furnishing the only permanent water on the entire range, the sheep were held on this area approximately 3 months, resulting in serious erosion and numerous over-grazed bedgrounds.

[medium shading] A few small water holes, cleaned out whenever used, allowed intermittent use of area during lambing season.

[white] Without permanent water, nearly 60 per cent of the range could be used only intermittently, when snow was available.

Proper management impossible. Lack of watering facilities and control of range enforced improper season of use and concentration of stock, contributing to unsatisfactory condition of range.

**After improvement**

Survey of the soils, conditions of erosion, and amount, value and distribution of the forage furnished the necessary basic information for the improved plan of management, and indicated, for each portion of the range, the proper period and intensity of use.

Use — Entire area grazed by two bands of 2000 sheep each from early Oct. to April. Under improved management the carrying capacity has increased 25 per cent in 3 seasons to 29,736 sheep months. The present rate of stocking of 26,000 sheep months should insure continued improvement on the range.

— Dams for stock water and erosion control located throughout the area.

——— Route of Band No. 1
— — Route of Band No. 2
With available water at grazing intervals, based on carrying capacity of the land and its susceptibility to erosion, the entire range is now uniformly and properly grazed.

○ Camps and bedgrounds showing date of use. Salting place at bedgrounds.

-x-x- Fences constructed to keep out trespass stock.

but grazing. The most outstanding soil type is brownish stiff clay or clay loam derived from the extensive yellowish and dark-gray shales. Pierre, Boyd, Winnifred, and Lismos are among the most important soil groups.

Elevation from east to west ranges from about 3,000 to 4,500 feet above sea level; and rainfall, from about 13 to 19 inches.

Primarily, the Pierre Shale country is used as range for livestock. Farming is hazardous because of low rainfall, unfavorable soil, and high susceptibility to erosion. Some wheat is grown under dry-farm methods, and such crops as sugar beets, small grain, and alfalfa are produced in the irrigated section in Butte County, South Dakota.

The present average carrying capacity of the range has been estimated at about 60 acres per animal unit. Range improvement methods indicated as helpful for the associated Dark Brown and Brown Plains areas are applicable to this area; but because of unfavorable conditions, it may not be possible to increase the grazing capacity very much. Holdings vary from 320 to several thousand acres.

The characteristic heavy soil has such low infiltration capacity that the ground, especially under thin cover, may not be wetted to more than about 1 inch by a heavy rain. The dry soil quickly runs together during rains to form a sticky, slippery, impermeable sheet. Runoff is consequently high; and wherever the gradient exceeds 1 or 2 per cent, erosion by sheet washing or gullying takes place. New areas of badlands are being formed continuously by a combination of overgrazing and trailing on many of the steeper slopes.

Wind erosion is a problem also. By a process of fragmentation, the clay soil tends to break down into small pellets which are rolled readily by wind, where vegetation is scant or absent.

Aside from these difficulties, there is the problem, in parts of the area, of excessive content of selenium.[1]

## Black Hills

The Black Hills comprise 2 million acres of hills and mountains in western South Dakota and adjacent northeastern Wyoming. The elevation rises to something over 7,000 feet, and most of the area is too steep or stony for important use aside from forestry. The soils are of entirely different origin from those of the surrounding plains, being derived principally from granite and other igneous rocks. Some of the depressions and shouldering positions are used for small grain, corn, and feed crops. Soil washing is rather severe in some fields, especially on the red loam and sandy loam soils derived from sedimentary rocks. Gullying, sheet washing, and crescent slips or "catsteps" are common. Terracing, contour tillage, strip cropping, basin listing, crop rotations, and retirement of the steeper slopes to grass or trees are the remedies.

[1] Byers, H. G. Selenium Occurrence in Certain Soils in the United States with a Discussion of Related Topics, U. S. Dept. Agr. *Tech. Bull.* 482, 1935; *Tech. Bull.* 530, 2d Rept. 1936; and *Tech. Bull.* 601, 3d Rept. 1938.

## Nebraska Sand Hills

The Nebraska Sand Hills occupy an area of approximately 11 million acres, situated mainly in central and north-central Nebraska. Predominantly, the area is composed of old sand dunes which have long been naturally stabilized, very largely, by a cover of grass—bluestem and other species. Through experience, farmers have learned that it is unsafe to plow this deep, loose sand which begins to blow with removal of the protective cover. Accordingly, most of the area is used only for grazing. Cattle trails or excessive grazing have started blowing here and there; but for the most part, such blowouts are of small extent and appear not to be expanding under the prevailing system of land use.

Some small grain, corn, and feed crops are grown on the moist floors of the frequent interdune depressions. Occasionally, farmers have extended tillage operations too far up the slopes, with the result that some fields have suffered from both wind and water erosion.

---

### Farm Land

1. Adoption of farm practices, including strip cropping, contour listing, contour cultivation, conservation and utilization of crop stubble, reduction of area devoted to cash crops, and increase of area devoted to feed crops.

2. Reestablishment of a cover of grass on bare or thinly vegetated areas not adapted to cultivation.

3. Adoption of rotations, including grass and, where practicable, legumes.

4. Improvement in irrigation practice on sloping land in order to reduce erosion and to make more economical use of water, through such procedures as distribution of water as nearly along the contour as practicable, better location of distributing ditches on the farms, and repeated use of any excess of applied water.

### Range Land

1. Adjustment of stocking to carrying capacity, and improvement of other range practices.

2. Conservation of water by contouring and other measures for holding and spreading.

3. Development of stock ponds and springs.

4. Adjustment of size of operating units to an economic basis.

5. Encouragement of adjustment of farm debts.

6. Encouragement of adjustment of taxation to productivity and stability of land.

7. Encouragement of increase in owner operators or adoption of long-time leasing practices.

8. Accumulation of reserve feed supply.

9. Education of ranchers to an appreciation of benefits to be derived from sound conservation practices on range.

# Chapter XXXVI. Edwards Plateau-Fort Worth Prairie-Cross Timbers Area

The Edwards Plateau-Fort Worth Prairie-Cross Timbers area comprises approximately 52 million acres of prairie, timbered, brush-covered, and cultivated country of generally rolling to rough topography, extending from southwestern Texas in an easterly-northerly direction across central Texas and Oklahoma (Map 13). Elevations range from about 800 feet above sea level in the Fort Worth Prairie to approximately 4,000 feet in parts of the Edwards Plateau.

This problem area comprises three major subdivisions: (1) the West Cross Timbers, bordering the eastern edge of the Red Plains subdivision of the Great Plains; (2) the Fort Worth Prairie and East Cross Timbers, lying between the Texas Black Belt and the West Cross Timbers; and (3) the Edwards Plateau, covering the southerly-westerly extension and bordering on the Rio Grande Plain area and Mexico.[1]

Normal annual rainfall ranges from 15 inches or less in the extreme western part of the Edwards Plateau to about 35 inches in the eastern part of the Fort Worth Prairie and East Cross Timbers sector. The expected rainfall intensity for a 10-minute period over a 10-year period (10-year frequency) in the western part of the area is from 1.12 to 1.25 inches, whereas the corresponding expected intensity for the eastern part of the area is from 2 to 2.5 inches. The expected rainfall intensity for a 60-minute period, 10-year frequency, ranges from about 2 inches in the western part to about 2.5 inches in the eastern part.

Much of the rain over the eastern and northern parts of the area falls during autumn and winter, when the land ordinarily has the least protection against erosion. At this time, many fields are thinly covered with small grain or are in a condition of bare or nearly bare fallow. Clean cultivation of erosion-permitting crops—chiefly cotton, corn, and sorghum—together with the lack of systematic rotations, is conducive to serious soil and water losses.

In the Edwards Plateau sector, relatively less land is cultivated, a large proportion being used for range. Although erosion is active over part

[1] Carter, W. T. The Soil of Texas, Texas Agr. Exper. Sta. *Bull.* 431, July, 1931.

of the range, much of it has a fair to good protective cover. The same is true over a considerable part of the timbered and brush-covered rougher lands of this and the other subdivisions of the region.

For the most part, the East and West Cross Timbers area originally was forested to deciduous trees, mainly blackjack oak and post oak of

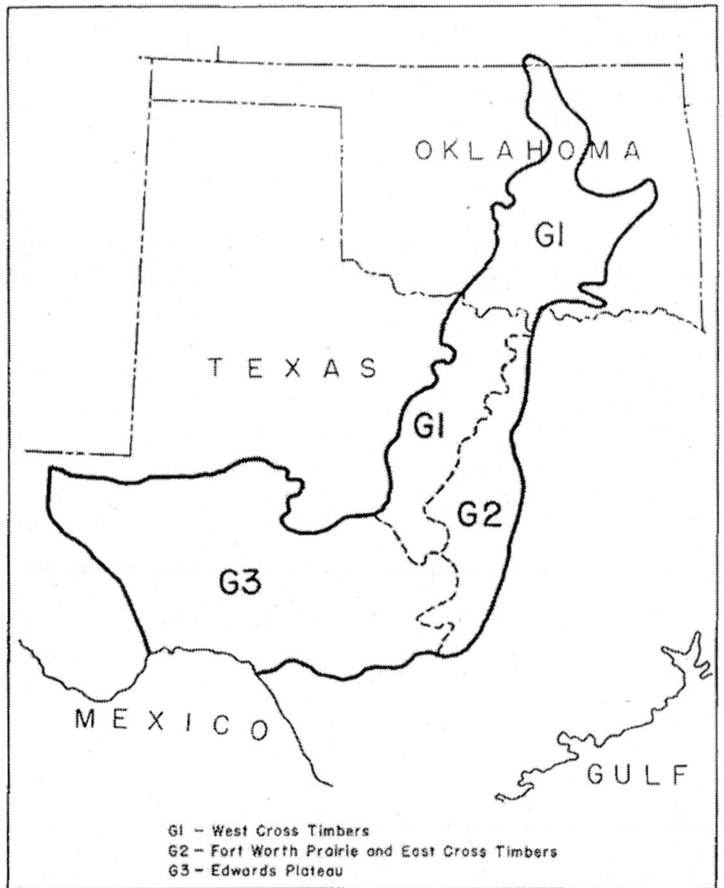


MAP 13.—Problem area of Edwards Plateau—Ft. Worth Prairie—Cross Timbers Area. (*Soil Conservation Service.*)

small size. Many small and some rather large bodies of prairie were scattered through the timbered country, particularly the West Cross Timbers.

Accelerated erosion has followed cultivation quickly and generally throughout most of the region. It has been a common practice to cultivate land until the yields dropped to unsatisfactory levels and then abandon it or permit it to lie idle and grow up to weeds and brush. After the idle land was "rested" for three or four years, it was again put into cultivation with the result that erosion was started again. This practice has continued until numerous fields have been forced into final abandonment by severe erosion. Following abandonment, additional tracts, usually steeper land, have been cleared or broken out of the prairie sod for crop production. Generally, erosion has attacked these newly plowed lands almost immediately.

The result has been the development of a very general and serious erosion problem on most of the tilled land of the entire region. Water erosion, accompanied by loss of needed water, as runoff, has been most destructive; but wind erosion has been active on much of the sandy land.

FIG. 300.—Gully converted into water channel and stabilized with buffalo grass. Takes care of runoff from 50 acres of terraced and strip-cropped land. No damage occasioned by an 8-inch rain, April, 1938, falling in 2 hours. Near Denton, Texas. (*Photograph by Soil Conservation Service.*)

Probably the most outstanding need of the regional farm land, therefore, is control of erosion and conservation of rainfall. Effective control has been brought about through the use of such protective measures as crop rotation, strip cropping, contour cultivation, terracing, safe disposal of runoff (Fig. 300), pasture contouring and rehabilitation, improvement and protection of woodlands, and retirement of steep or badly eroded areas to permanent vegetation. Application of these and other practical farm measures to each acre of land needing protection, in accordance with its adaptability and the economic feasibility of the practices, has not only brought about a large degree of stabilization to the land and the farmer on some of the most representative types of crop and grazing land but has resulted in conservation of needed rainfall which otherwise would have run to waste, in reduction of floods on small streams, and in diminution of silting.

Overgrazing has resulted in excessive erosion on many parts of the rougher lands so extensively used for livestock, especially where the slopes are steep, and goats and sheep are raised in large numbers. As erosion has advanced, the cover has become thinner, the land poorer, and yields lower. Controlled grazing and contour-furrowing of the smoother land have proved effective in remedying the situation. Stock-water ponds, strategically located, also have aided.

The extent to which accelerated erosion has progressed on farm lands in the Fort Worth Prairie and West Cross Timbers subdivisions is illustrated by data compiled from detailed surveys of representative farm lands involving approximately 120,000 acres. These data showing the relation of land use to erosion, are given in Table 46.

Of 93,000 acres of farm land surveyed in the West Cross Timbers sector, approximately 47 per cent is cultivated, 6 per cent is idle, 37.5 per cent is used for pasture, and 9.5 per cent is woodland.

The data shown in Table 46 indicate that erosion has affected most of the land, whether cultivated, idle, pastured, or wooded. Even woodland has suffered almost as much as the cultivated area, the proportion of uneroded land being approximately 22 per cent of the former and 18 per cent of the latter type. This is due largely to overgrazing of the comparatively thin soil prevailing over the rougher lands of the timbered type.

TABLE 46.—RELATION OF LAND USE TO EROSION IN THE WEST CROSS TIMBERS AND FORT WORTH PRAIRIE

| | Land use | Extent of erosion, per cent | | | |
|---|---|---|---|---|---|
| | | None | Slight | Moderate | Very severe |
| West Cross Timbers | Cultivated | 17.9 | 48.0 | 31.7 | 2.4 |
| | Idle | 17.3 | 29.6 | 34.3 | 18.8 |
| | Pasture | 23.2 | 29.4 | 33.3 | 14.1 |
| | Farm woodland | 21.6 | 24.8 | 50.1 | 3.6 |
| | All land | 20.2 | 37.7 | 34.2 | 7.9 |
| Fort Worth Prairie | Cultivated | 3.3 | 43.1 | 50.2 | 3.4 |
| | Idle | 11.5 | 34.6 | 49.1 | 4.8 |
| | Pasture | 30.1 | 37.7 | 27.8 | 4.4 |
| | Farm woodland | 14.0 | 32.2 | 52.0 | 1.8 |
| | All land | 17.1 | 39.7 | 39.3 | 3.9 |

The data indicate that erosion on cultivated land in the Fort Worth Prairie area is not greatly different from that on cultivated land in the West Cross Timbers. In the former subdivision, detailed surveys of

25,000 acres of representative farm land indicate that approximately 44 per cent is cultivated, 2 per cent is idle, 49 per cent is in pasture, and 5 per cent is in woodland.

## Fort Worth Prairie and East Cross Timbers

The Fort Worth Prairie and East Cross Timbers, with a total land area of about 8 million acres, extends from the Red River about 300

Fig. 301.—Characteristic topography of the better farm land of the Fort Worth Prairie (Denton clay). This entire farm of 81 acres of pasture and 140 acres of cultivated land protected from erosion by contouring, strip cropping, terracing and other adaptable measures. Runoff disposal channel in center. (*Photograph by Soil Conservation Service.*)

miles southward to the vicinity of the Colorado River, where it merges with the Edwards Plateau. It is a smoothly rolling to hilly limestone plain, deeply cut by numerous streams. The thick beds of limestone have been cut entirely through in places, leaving rough stony ridges, steep escarpments, and high remnants of land on the order of mesas and buttes. Much of this steeper land is stony and essentially barren. Numerous areas consist of rock outcrop, and many others have so little soil that vegetation is very scant.

On the average, that part of the area lying north of the Brazos River is considerably smoother, and a larger acreage is used for crops (Fig. 301). The southern half, known as the *Lampasas Cut Plain*, is predominantly rough or rolling. Much of it is covered with juniper and scrub oak. The raising of cattle, sheep, and Angora goats is an important agricultural enterprise. Cotton, wheat, oats, corn, and feed crops are the principal farm products.

The predominantly dark, productive soils of the northern half are derived from interbedded limestone and marly material. A larger proportion of the lighter colored, shallower soils has been left in the native grass. The upland soils are largely clays and clay loams of the San Saba, Denton, Crawford, and Brackett series. Bell, Milam, and Lewisville are the important stream terrace soils, and Frio and Catalpa are the principal first-bottom soils. In the Lampasas Cut Plain Section, much soil consists of light-colored, shallow, calcareous material overlying limestone (Brackett soil).

Fig. 302.—Rejuvenation of overgrazed pasture by contour furrowing, East Cross Timbers section, Texas. Buffalo grass rapidly spreading as result of water conserved by the furrows. Durant soil of 2 per cent slope. (*Photograph by Soil Conservation Service.*)

The native prairie vegetation consists of palatable grasses, such as buffalo, bluestems, side-oats grama, and curly mesquite. Mesquite trees are scatteringly present throughout the Fort Worth Prairie and East Cross Timbers Sections. Juniper and oak are common, often in dense stands, on the steep, thin lands of the southern part. The grass cover frequently is thin among these trees.

As the result of improper grazing, weeds and annual grasses are crowding out the better range plants. Erosion has rather severely affected much of the cultivated and grazed lands, as indicated in Table 46.

In the East Cross Timbers section, erosion is very severe on the steeper sloping lands, such as the red post-oak sandy land. Even the smoother areas, including pasture (Fig. 302) and cropland, generally need some form of protection.

## West Cross Timbers

The West Cross Timbers area, comprising about 17 million acres, is predominantly gently rolling to hilly. Although large areas consist of hilly to rough stony land, many of the broader divides are occupied by relatively smooth country. The principal streams crossing the area are the Red, Brazos, and Colorado Rivers. These and some of their tributaries have cut deep, narrow valleys, bordered in many places with strips of rough country.

Post oak and blackjack oak are the principal trees. The soils are derived largely from sandstone and are predominantly sandy. The most important upland type is the light-colored Windthorst fine sandy loam. This has a red, moderately stiff sandy clay subsoil. The Nimrod soils are lighter colored and more sandy.

Cotton is the principal cash crop on most farms. Corn and feed crops are also grown by most farmers, chiefly for consumption on the farm. Peaches, melons, vegetables, and peanuts supply a part of the farm income in many localities. Small grain is grown on some of the heavier land of the included prairie areas. The raising of livestock is important in some localities, particularly on the included prairies.

Erosion has severely affected much of the cultivated land, as indicated in Table 46. Control measures, such as those referred to above, are proving highly effective not only in checking erosion but in conservation of rainfall and reducing floods along streams whose watersheds are largely covered by the work (see footnote, Chapter XXIX).

## Edwards Plateau

The Edwards Plateau is a high, rather intricately dissected limestone plateau, with a total land area of some 27 million acres. It is bounded on the north by the Great Plains, with a fairly sharp line of separation. On the east, it merges with the Lampasas Cut Plain sector, generally without any very distinct boundary. On the south, however, the Balcones Escarpment sharply defines the boundary with the Rio Grande Plain. West of this Plain, the Plateau crosses the Rio Grande into Mexico.

Many of the streams have carved steep-walled, narrow valleys, frequently bordered with rough country. Rock outcrop in the form of cliffs and steep, rough slopes are of common occurrence near the principal drainages. Remnants of the higher plateau level are flat to undulating or gently rolling on top. Runoff usually is rapid, and soil losses are often heavy on exposed land, even where the slopes are mild.

The soils are underlain at comparatively shallow depths by limestone or interbedded limestone and chalky material. The principal types in the eastern part are the brown Denton clay and the black San Saba clay. With the decreasing rainfall toward the west, first the Valera soils are encountered, and then the Reagan and Ector groups, all of which are highly calcareous and of an ashy color.

Included with the Edwards Plateau is an area known as the *Central Basin*, or *Llano Basin*, in which erosion has exposed a considerable area of crystalline rocks, such as granite, gneiss, and schist. Characteristically, this is a rolling to hilly country, with rather broad basins. The

FIG. 303.—Above, pasture contouring, with ends of contour ridges turned upslope to hold all rainfall, 1936; below, grass recovery on same area, 1937. Thus, rainfall that would have run to waste is stored in the soil for the production of grass. Near San Angelo, Texas. (*Photographs by Soil Conservation Service.*)

rougher lands are usually stony and best suited to grazing, but there is considerable good farm land on the smoother areas.

The native vegetation consists largely of tall and short grasses, along with many small and some fairly large stands of mesquite trees, oak, juniper, and, in the extreme western part, desert shrubs. Although post oak is plentiful in the eastern part, live oak is conspicuous over much of the central area, along with considerable mesquite and some shin oak.

In the western sector, the larger growth consists principally of such shrubs as mesquite, catclaw, sumac, buckthorn, and agrito, with sotol, lechuguilla, yucca, cacti, and other desert plants in the extreme western section. Buffalo grass, curly mesquite, little and big bluestem, side-oats grama, and other grasses grow luxuriantly, where not excessively grazed, on the deeper soils of the eastern and central sections and on the rich

FIG. 304.—Above, contour-listed range, near San Angelo, Texas, 1936; below, same area 1937, with buffalo grass predominating. (*Photographs by Soil Conservation Service.*)

valley lowlands. Grasses are scarcer on the shallower soil and in the drier western part of the area. Overgrazing appears to have caused the spread of useless weeds. Conservation of moisture by contouring (Figs. 303 and 304) and adjustment of grazing to carrying capacity are giving splendid results.

The principal agricultural enterprise of the Edwards Plateau is the raising of livestock—cattle, sheep, and goats. Locally, on the deeper soils, cotton, corn, small grain, and the sorghums are grown. Erosion is an important problem on much of the cultivated land and a considerable part of the range.

## Soil and Water Conservation Work

Much effective work in soil and water conservation has been carried out recently in the Edwards Plateau-Fort Worth Prairie-Cross Timbers area. A number of demonstration projects and CCC camps in various parts of the problem area have contributed to this. More than 600 operators have entered into cooperative arrangements with the projects and camps to develop on their farm and ranch enterprises a complete program of soil and water conservation. As an illustration of the accomplishments in the Fort Worth Prairie and West Cross Timbers portions of the region, 48,000 acres have been put under contour tillage, with strip cropping on 42,000 and terracing on 23,000 acres; 66,000 acres of pasture have been improved by contour furrowing and other treatment; and a large area of highly erodible cropland has been shifted from cultivation to permanent grass and trees.

# Chapter XXXVII. Colorado River Basin Region

The Colorado River Basin Region (Map 14) is a roughly defined soil and water conservation problem area of such vast extent that only a few of its more outstanding characteristics can be summarized here.[1] Under five major subdivisions, some of the more important physical aspects and problems are briefly treated below. Actually, each subdivision individually includes a variety of distinct soil, topographic, climatic, vegetative, erosion, and agricultural conditions. These numerous areas of distinctive characteristics and special problems cannot be treated separately in this volume, but the problems of major or average importance are pointed out in sufficient detail to give a fairly broad conception of the regional seriousness of range conditions, erosion, floods, and siltation, and the practical possibilities of remedying these conditions.

Some 241 million acres are comprised within the region, which covers practically all of the Colorado River drainage, together with the basins of the Gila and Rio Grande and portions of other headwater drainages flowing out of the Colorado Basin region. Broad plateaus of both smooth and severely dissected character, mountains, basins without drainage outlets, deep canyons, cliffs, breaks, mesas, broad and narrow valleys, strips of alluvium, extensive areas of valley-fill land, and bare salt flats are some of the more prominent features of the terrain. Climatically, the range is from absolute desert to timbered elevations of high rainfall and short growing seasons. Land utility varies from valueless desert and inaccessible rough lands to the most productive type of irrigation farming known. With respect to the area utilized, grazing outranks all other agricultural interests manyfold.

Accelerated erosion is a major problem over at least three-fourths of the range country. Erosion by water is the principal form of soil stripping and the major cause of land depletion and siltation of reservoirs. Range depletion is both the cause and the result of erosion over a vast grazing area. Erosion and range depletion have a clear causative relation

[1] For details see Bowman, Isaiah, "Forest Physiography," pp. 246–297. New York, 1911.

to accelerated floods, and all three are intimately associated with the impoverishment of occupants of the land. Wind erosion is also a serious problem locally, especially on sandy land. In parts of the Southwest, as in the watersheds of the Rio Grande and San Juan, "sands on the march"

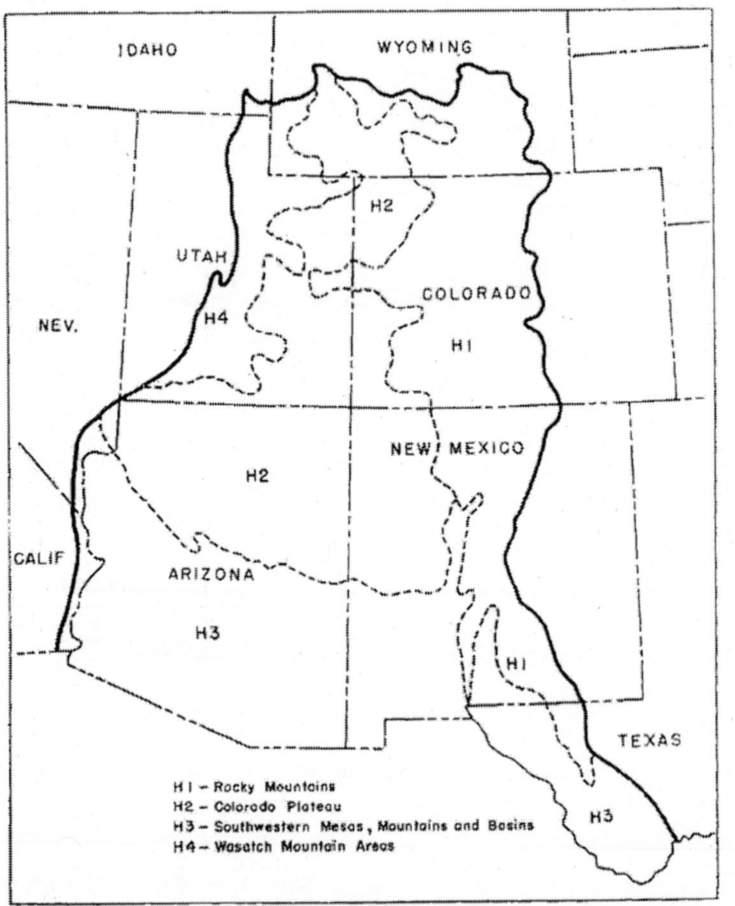


MAP 14.—Problem area of the Colorado River Basin Region. (*Soil Conservation Service.*)

—dunes recently developed as the result of overuse of the land—are covering good lands to the lea and even ascending the sides of mountains.

The problem now is to find practical ways to cope with the evil of accelerated erosion before additional areas of desert have been formed and the economic situation becomes worse. The Soil Conservation Service, the Forest Service, the Division of Grazing, the Bureau of Indian Affairs, the Agricultural Adjustment Administration, and other Federal agencies as well as state agencies, community organizations, and individuals are striving cooperatively to conserve the principal resources of the region—the land, water, and vegetation. Slowly, headway is being made in numerous localities, but the major task is still out ahead.

In setting up three great watershed projects on the Rio Grande, Gila, and San Juan-Little Colorado drainages, comprising over 59,000,000 acres of land, the Soil Conservation Service took a step beyond a purely demonstration program. In recognition of both a high degree of public responsibility, particularly federal responsibility, and the closely knit relationships between range use and erosion on upper watersheds and damage to reservoirs, rivers, and irrigated farm lands below, the Service undertook the application of an extensive erosion-control program, which at the same time has served as a demonstration program. In this program, not only the Soil Conservation Service but other public agencies are expected to take their share of responsibility for erosion-control work on a large scale on public lands.

Watershed surveys, covering all phases of land use and condition and human need, provide an over-all watershed picture for these areas. This broad appraisal of conditions and needs is the basis for finding and selecting for immediate work the most critical erosion and silt-contributing areas within the watershed—areas where erosion control is badly needed and which, in the light of the entire watershed picture, must come first in erosion-control operations. In this way, the lands throughout these large districts are treated as part of a unified plan for the whole instead of as separate units without relation to one another. Thus, farm and ranch, range land and irrigated land fall together into a living mosaic which forms the basis for the operations of the Service in the Colorado Basin area.

Before taking up the individual problems of the major subdivisions of the region, some of the principal measures employed generally over the area in controlling erosion through the use of various land treatments and water-retardation devices, so integrated as to meet the varied needs of the highly diversified land, and climatic, vegetative, and economic conditions, will be discussed. The principal types of soil and water conservation work engaged in are at least broadly applicable to large areas in most or all of the subdivisions.

## Some of the More Important Conservation Measures Applicable to the Colorado River Basin Region

### Range Management

The key to control of erosion on the extensive range lands of the Colorado River Basin Region is proper range management. The principal measures for successful results are (1) proper stocking of the range as to numbers, (2) proper stocking as to class of livestock, (3) correct seasonal use, (4) proper distribution of stock over the range, and (5) conservation of rainfall.

Proper Stocking. Of these measures, proper stocking as to numbers is the most important, since without it any effort toward forage protection and erosion control is futile. Until proper stocking is achieved, it is scarcely worth while to apply additional erosion-control measures. To do so is to ignore the experience of 50 years.

Fig. 305.—Range management, as on the right of the fence, results in better vegetative cover and more productive stock. Dona Ana County, New Mexico. (*Photograph by Soil Conservation Service.*)

Quantitative proof that conservative stocking pays in the short run as well as over the long pull has long been needed. Results obtained on the demonstration projects of the Soil Conservation Service indicate that conservative stocking pays almost immediately (Fig. 305). Gains in weight of animals, heavier wool clips, larger calf and lamb crops, and smaller death losses are the results that have been achieved by conservative stocking within the demonstration areas, as compared with results on overgrazed, eroding ranges outside.

In these demonstration areas, surveys are made to determine the grazing capacity of each distinctive parcel of land. The density of the plant cover is determined, and estimates are made as to the proportion of perennial grasses, browse plants, and weeds present, taking into account the palatability of the various species; and on the basis of these and other factors, the number of stock that the range can properly support is determined. These findings are submitted to the user of the range;

and if he agrees to adjust his stocking to the estimated grazing capacity, a range-management plan for that particular ranch is developed.

The grazing-capacity figure is a guide to proper stocking only; corrections upward or downward are made to hold stocking in line with the changing condition of the range during usage, due to improvement of the forage by favorable rains or to decline as the result of drought. If forage growth and density of cover improve rapidly, the grazing-capacity estimate may be increased; if improvement is slow because of drought, stocking is decreased further.

Correct Seasonal Use. Wide local and general variations occur in the amount of forage produced yearly as the result of seasonal factors, principally variations in rainfall. Such changes necessitate a system of grazing that provides for alternating periods of use and protection such as favor increased vigor or density of the forage. On range suited to year-long use, the deferred-rotation system of grazing provides protection during the seed-producing period on a different portion of the total area each year.

The rotation plan is so arranged that, on those areas where the amount of forage produced is the least, additional protection from grazing may be obtained. On range land that is distinctly favorable to seasonal use, control of the opening grazing dates is important. Vegetative readiness is an important factor in the proper use of a seasonal range. Total exclusion for any one or all of the seasonal ranges is advocated as a part of the rotation system only when the range is in such condition that complete rest is desirable to promote vegetative recovery.

Aids to Distribution of Livestock. A more nearly uniform distribution of livestock is obtained mainly by building fences, developing water, herding, and proper use and distribution of salt.

These aids favor uniform grazing, which, in turn, serves to protect plant cover by preventing overuse of one part of the range through concentration of stock. Drift and pasture fences enable the rancher to adjust grazing to the proper season, to avoid grazing before the range is ready, and to practice deferred and rotation grazing. This allows the grass to gain vigor and to produce more forage as well as seed. Fencing off bull and buck pastures permits a rancher to regulate the breeding season and thus avoid high death losses from lambing and calving in unseasonable weather. Fences also hold down losses from straying and stop trespass.

Watering places are developed not more than 3 or 4 miles apart, where practicable, so that stock will not need to travel more than 2 miles, at most, for water. An adequate number of watering places holds trailing to a minimum.

Salting in little-used areas between established watering places helps to spread cattle over the range. Cattle crave salt, and they tend to graze from water to salt and from salt to water. Salting near water tends to hold cattle at the watering places, where they remain concentrated until often the cover is completely destroyed, and serious erosion sets in. Salting places are changed from time to time to prevent denuding an area by attracting the animals to other parts of the range.

### Structural Aids in Erosion Control

Where erosion is controlled in its early stages, when the vegetation first starts to give way, when the first small rills are forming in the upper reaches of a watershed, the use of structures generally is unnecessary. But over many millions of western acres, erosion has gone far beyond the early stage. Over numerous areas, vegetation has been depleted or largely changed in composition and value, and the scant supply of water drains rapidly from the surface, cutting away soil and gouging rills and gullies.

The balance in many places has turned too far downward, so that structures are needed to tip it upward. Water is the key to a better cover of grass and other vegetation. By holding and spreading water, by checking erosion until vegetation can get a new foothold, structures speed recovery and erosion control. By increasing the water supply and vegetation in depleted areas, structural aids bring a better support to dependent stock and people.

Water Spreaders. Water spreaders have been extensively used in the region to divert water from gullies out over near-by flats to stop blowing or washing and to increase the vegetative cover. In this way, water that would run to waste is put to use. The spreader may be a small or large structure made of earth, rock, brush, or a combination of these materials.

Hundreds of such spreaders are in use on the erosion-control areas. Most of them are of the smaller type and divert water from the heads of small gullies. Some are larger installations, or spreader systems, which distribute water over large areas of meadow or cropland where it once naturally spread from the original shallow channel. The smaller structures, which cost little to build, have proved their worth when built of suitable material and placed in properly selected locations. The larger ones, in view of their cost and great loss in case of failure, need longer trial before they can be recommended for general use over the ranges of the Colorado Basin region. Time alone can answer some of the questions that have been raised about water spreading. Will deposition of silt from the flood waters spread over the treated area eventually kill the grass or build up the flood area so that spreading becomes impracticable? Meanwhile, grass waves luxuriantly over formerly denuded areas that have been treated, and many millions of tons of soil are being held out of costly reservoirs.

Some of the larger spreaders installed in this region divert water accumulated from watersheds of several thousand acres. Thus far, after four years, most of them are working satisfactorily. From 62 acres of land below one dike and spreader system at Mexican Springs, on the Navajo Reservation, 40 tons of hay were cut in 1936 from land that two years before had been practically bare of grass. This yield is about half that which may be expected from irrigated land. Above another spreader distributing water accumulated from a 2,552-acre watershed, 4,600 cubic yards of silt were caught and retained in the gully channel. Behind a dike across Deer Springs Wash near by, 28,760 cubic yards of silt collected from 3,600 acres. These diked gullies, instead of growing deeper and wider with each storm, are healing and revegetating.

Water spreading by means of large structures appears to be an especially promising measure for ranges of low grazing capacity, where most of the feed is grown in small, favored areas. In the lower lands on the Upper Gila watershed, much of the forage formerly grew on the broad *saccaton flats* in the major valleys and in the shallow *tobosa swales* between ridges sparsely covered with creosote shrubs. Although saccaton and tobosa grasses are not highly palatable all year, they furnish during spring and summer a luxuriant growth relished by most livestock. And saccaton, even when mature, will be eaten in times of drought, when the grama grasses on the hillsides are dry and tough.

In the lower, hard-used portions of the valleys, these grasses, robbed of flood water from surrounding slopes by quick drainage through erosion-gouged gullies, have died out. Here, by use of dikes and water spreaders, effort is being made to restore grass to the flats. The method is illustrated on the Freeman Flat experimental area at Safford, Ariz., comprising 800 acres in the flat and 3,000 acres over the watershed above. An arroyo winds through the flat for several miles; at its lower end, on the outskirts of Safford, it forks, and two washes, 25 feet deep and 40 to 100 feet wide, guide flood waters to valuable irrigated lands along the banks of the Gila River.

By agreement with cooperating ranchers, control of stocking was obtained over the whole of this little watershed, and stocking was adjusted to grazing capacity. Some small earth diversion dikes were thrown across the smaller washes. At the upper end of the flat, a major structure, a dike, was built across the head of the arroyo. This diverted the water out from the deep channel of the arroyo and carried it to the flats on both sides, where it was spread in a thin sheet over some 450 acres by a series of rock and brush spreaders. Perennial grasses—saccaton and Rothrox grama, principally—were seeded on the flat in moist places, and trees and shrubs—willows, tamarisk, and *batamoté*—were planted in the arroyo.

Three years ago, before treatment, the flat was practically bare of grass. A sparse growth of burroweed, salt bush, creosote bush, and mesquite covered some areas. After water was spread, some annual grasses and weeds appeared the first year, then much salt bush, followed by perennial grasses that had been broadcast behind the spreaders. It is estimated that two years after seeding, the grass will yield $1\frac{1}{2}$ tons to the acre and pasture for 30 cows for the four spring months.

The use of the larger types of water-spreading systems, such as that of the Freeman Flat experiment, is restricted to areas where there is a clear legal right to the water. The vested rights of irrigation districts below make this necessary. The policy of the Soil Conservation Service is to confine use of large spreaders to areas that have been severely depleted by erosion and that have been natural spreading areas within the memory of living men. The practice also is limited by soil type and rainfall. If runoff is light and infrequent, spreaders are not installed. If the soil is impervious and incapable of holding sufficient moisture to start and maintain growth, such structures cannot be used successfully.

Contour Furrows. Contour furrowing is considered one of the most effective soil and moisture conservation practices now being employed in the Southwest. Such furrows have proved especially beneficial in areas having a rather high vegetative potential. On soil types of low moisture-holding capacity, on the other hand, contour furrowing appears to have doubtful value. On stiff clays of high salt content, the practice has given poor results; the surface tends to seal over and prevent infiltration.

Most of the contour furrowing in this region has been functioning not longer than two or three years, and it is perhaps too early to determine definitely its place in an erosion-control program for all types of land. In the buffalo-grass country to the east, excellent improvement in contour-furrowed pastures has occurred during this period. In the grama-grass and bunch-grass areas of the Southwest, the response of vegetation has not everywhere been so pronounced, although the furrows have effectively controlled runoff and erosion. In most places, the response in the first year is a good growth of weeds. In the second and third years, perennial grasses begin to appear, and it remains to be seen how soon the perennials will displace the weeds.

As a means of controlling runoff in the Colorado Basin region, contour furrowing needs no defense. In all instances, the furrows have retarded runoff and arrested erosion effectively. During the first week of March, 1938, at St. George, Utah, for example, a furrowed area was subjected to severe test by a rain of approximately $6\frac{1}{2}$ inches which fell on a denuded area furrowed six months previously. Little runoff occurred, whereas surrounding untreated range lands contributed largely to a severe flood in the Virgin River. Tests following the storm revealed the fact that moisture had penetrated to depths of 27 to 30 inches immediately above and below the furrows and only 15 to 18 inches on untreated areas. Runoff from the untreated areas carried away much of the surface soil and contributed to a flood which ruined many acres of productive river-bottom land, drowned livestock, destroyed roads and bridges, and added to the silt load carried toward Boulder Dam. On the furrowed area, there was little soil movement; most of the water was absorbed by the soil.

PERCOLATORS. Percolators, or pervious dams, are used widely to retard runoff and erosion. Their use is restricted generally to land sloping not more than 8 or 10 feet in 100 and to areas of thinly vegetated, dense soils which have not eroded to the stage of gullying. On these lands, vegetation has not recovered quickly under improved range management, but percolators have proved effective in reducing erosion and increasing growth of vegetation.

Percolators are constructed of rock or brush, or a combination of these materials, in such a way as to impede but not completely halt runoff. Spacing is adjusted to slope, the structures being placed 30 feet apart on the steeper slopes and 100 feet or more on the gentle slopes.

CRESCENTS. On steeper slopes and rougher lands, crescents are more applicable than contour furrows or percolators. These small, crescent-shaped structures are 15 to 18 feet wide from tip to tip and 12 to 18 inches high on the lower side. The points of the crescent are turned uphill. They are used primarily to protect lower lying lands from runoff and overwash. Usually, from 20 to 25 are installed on each acre.

GULLY DETENTIONS. In the early formative stages of the erosion-control program in the Colorado Basin region, gully detentions (frequently called *gully plugs* or *check dams*) were used extensively. Generally, these have arrested gully cutting effectively where used on slopes of not more than 15 per cent; but they cannot be recommended for general use. The cost is too great for the good that they do, except under unusual circumstances. Gully detentions appear to have a worth-while place, however, when used, for example, to protect irrigated land or irrigation works immediately below or to stabilize the bottoms and banks of gullies. Only on the flat, gentle slopes do they catch enough silt to justify their cost.

STRUCTURES TO CONTROL HEAD CUTTING. Once a gully has started, it usually continues to cut headward through a valley, even where the valley is fairly well covered with grass. Head cutting of this kind has destroyed much of the most valuable range land. Control of such ravines yields large returns in proportion to expenditures.

In many instances it is possible to stop head cutting of large gullies by constructing a diversion dike about 50 feet above the head so as to take the water out of the gully and spread it to either side. Often, however, the spreading area is not sufficiently extensive to take care of the diverted water, or an occasional flow may be greater than the spreader system can handle. In such instances, drop structures made of masonry or concrete, or chutes made of these materials, have proved very satisfactory in head cuts of more than 6 or 7 feet deep. If the drop is less, plugs or brush or rock may suffice.

On cattle ranges, use of the range often is concentrated along streams, ephemeral arroyos, and washes. Concentration and heavy grazing along channels tends to prevent reproduction of small willows, cottonwoods, and other trees and vegetation which normally stabilize the banks. Such situations are often the key to the control of erosion and silt in large watersheds.

Fencing of strips along the waterways a mile or two long and a hundred yards to a half mile wide is proving very successful in controlling erosion. It gives natural vegetation a chance to take new hold on the banks and provides protection to replantings and seedings in these areas. A relatively inexpensive process, it often takes the place of structures or provides added protection for structures that may be placed in the channels.

Stock is excluded from these protective strips for 2 years after fencing. Then, limited use of the wider strips as reserve or holding pastures may be allowed in some cases.

## Colorado Plateau

The Colorado Plateau, comprising some 81 million acres, occupies roughly the northern third of Arizona, a large area in northwestern New Mexico, a large part of southeastern Utah, and a strip across Colorado west of the Rocky Mountains. A small portion crosses the southeastern corner of Nevada.

Although the region has the general appearance of a plain, numerous areas are severely dissected. Here and there, a mountain, such as San Francisco Peak near Flagstaff, Ariz., rises conspicuously above the general upland level.

Although much of the land is of relatively smooth plateau character, numerous areas of rough, broken country occur throughout the Plateau region. Probably about half of it is rough terrain, covering such broadly defined types as *rough broken land*, or badlands; canyons and breaks; painted desert; mesas; cliffs; and intricately dissected flatlands. Although detailed surveys had not been extended far enough to provide an accurate statement of the proportion of land unsuitable for important practical use, it is estimated that about half of this rougher country is too broken, too thinly or scatteringly vegetated, too susceptible to erosion, or too far from water for economical grazing. If these estimates are approximately accurate, about a fourth of this problem area is very nearly worthless in its present condition. Some parts of these rougher areas, however, can be improved by good range management and water development.

The elevation of the greater part of the plateau ranges from about 4,000 to 8,000 feet above sea level.[1]

The Colorado and its principal tributary Green River have their headwaters in the enclosing mountains. Emerging from the steeper highland, their valleys broaden in the foothills only to be restricted again in

FIG. 306.—The channels of major streams of the Colorado Plateau are deeply entrenched, and much of the adjacent country is exceedingly rough. Arizona. (*Photograph by Soil Conservation Service.*)

their course through the Colorado Plateau, where the channels become deeply entrenched (Fig. 306). The Grand Canyon, with a depth exceeding 6,000 feet, illustrates the erosive power of flowing water through the softly consolidated formations of the Plateau. The breaks leading to these great canyons, as well as many of the tributary canyons, are excessively steep and include numerous cliffs and inaccessible areas.

The climate is semiarid, with an average annual precipitation of about 10 inches over the greater part of the area. Rainfall (including snow) ranges as high as 20 inches over some of the higher elevations, usually of small areal extent, and as low as 7 inches in some other local-

[1] See map *Physical Features of the United States*. "Atlas of American Agriculture." U. S. Department of Agriculture 1936. Gregory, H. E., U. S. Geol. Survey, *Water Supply Paper* 380, 1936. Fenneman, N. M. "Physiography of Western United States." New York. 1931.

ities. Precipitation is about equally divided between winter and summer. Winter rains, however, usually are gentle and of broad extent, seldom resulting in marked runoff or severe erosion. In contrast, summer rains characteristically are of much more local character; they are more intense and produce a great deal more erosion.

The growing season through most of this high country averages about 4 to 5 months. Within this area occur many bodies of irrigated land. Very little dry farming is attempted, but irrigation farming is of large importance locally. Patches of flatland at the base of highlands and along some of the streams are farmed here and there through the use of diverted flood waters. About 95 per cent of the area has no other present use except for grazing of sheep and cattle. To some extent, even the rougher lands are grazed, but forage is scant and much more scattered, so that the industry is of much less importance in proportion to area than on the smoother lands.

Because of the low and uncertain character of precipitation, the moisture requirements of useful vegetation usually range close to the danger line in this arid to semiarid region. This marginality of the moisture supply is further intensified by unfavorable soil conditions over a large aggregate area. Whenever the moisture supply is favorable, however, the warm summers are conducive to maximum production of grass and crops.

Winds of high intensity are common during the early spring and fall months throughout the Colorado Plateau.

Some of the most serious erosion of the entire Colorado River Basin region is encountered in the Plateau section. Sheet erosion is most serious, but gullying is destructive locally, particularly on the more productive alluvial and valley-fill lands (Fig. 307). Over many large areas, gullying has become a characteristic feature of the valleys and basins. Most of them are of the straight-walled, undercutting type which branch and expand rapidly wherever the watersheds are overgrazed. Gullying, however, is not restricted to the valley lands; many overgrazed slopes have been scarred with short ravines, especially on the heavier shale-derived soils of high salinity. But sheet washing is the principal menace to the more sloping areas. It is taking place over the greater part of the uplands, most seriously where the surface has been bared or nearly bared of vegetation. Because of the relatively larger area of sandy land, wind erosion is most serious in the southern sector.

The soils of the northern part of the Plateau are derived largely from the underlying shales and fine-textured sandstones. Many of the shales, such as those of the Mancos formation, contain relatively large quantities of salt and give rise to soil much of which is excessively impregnated with water-soluble salts (alkali). In general, the soils are calcareous and of light-brown or grayish-brown color. The subsoils generally are only slightly heavier than the material of the surface. They are distinguished chiefly from the topsoil by their accumulation of lime carbonate. Saline shales have given rise to rather broad belts of heavy grayish soils along the San Juan River throughout most of its course in New Mexico, Arizona, and Utah.

South of the San Juan, a larger proportion of the soils is derived from yellowish to reddish sandstone, and is of sandy texture. Clay areas

Fig. 307.—Gullying is destructive in the alluvial or valley-fill areas. Oraibi Wash, Arizona. About four miles above Tolani Lake. (*Photograph by Soil Conservation Service.*)

here and there are derived from shale, and some of these contain considerable salt. For the most part, the soils of the southern sector are brown, light brown, or grayish brown in the surface, moderately calcareous, and of good absorptive capacity. The subsoils are of darker brown or more reddish color, usually higher in content of clay, and carry more lime carbonate. In general, the soils of this southern area are more favorable to plant growth with the same rainfall because of better permeability and lower salinity.

The three most important groups of soils in the Colorado Plateau are those derived, respectively, from sandstone, shale, and valley-fill materials. Sandstone-derived soils are of general occurrence throughout the Uncompahgre Plateau in west-central Colorado and over much of the Little Colorado watershed west of the Arizona-New Mexico line.

In general, these sandstone lands exhibit a number of common features. The surface soils are yellowish to light reddish brown, friable, and usually relatively low in content of lime carbonate, whereas the sharply defined subsoil is darker colored, hard, moderately to highly calcareous, and of prismatic structure. Generally, the soils have good water-holding capacity, except where the parent sandstone occurs at shallow depths. Where the cover of grass, sage, and other low growth has not been disturbed by grazing, erosion is of no importance; but under conditions of depleted vegetation, as the result of overgrazing, sheet erosion, gullying, and wind erosion are active over large areas.

The principal soils are the Shavano, Defiance, Pinedale, and Lobe. The Shavano soils occupy the lower elevations and are the least resistant to erosion. The others are more or less similar, differing from the Shavano chiefly in rather minor profile aspects. Locally, the soil has washed off to bedrock as the result of overgrazing, such as the areas found here and there on formerly timbered slopes of the Shuska Mountains, near Mexican Springs, in northwestern New Mexico.

Shale-derived soils are about as extensive as those developed over sandstone. They are of broad occurrence in eastern Utah and occupy most of the lower elevations of western Colorado and northern New Mexico. Mainly, the surface soils consist of light-brown to dark grayish-brown (nearly black when wet) clay, overlying similarly colored, hard subsoils which show a cloddy structure on disturbance. At depths seldom greater than about 20 inches, partly decomposed shale fragments are usually present. The soils are almost universally calcareous, frequently alkaline, and contain gypsum crystals in many localities. Vegetation is generally thinner than on the sandstone soils of the region. Erosion is severe nearly everywhere. Gullies are numerous but generally shallow because of the compact, underlying shales. These soils are adapted only to grazing. The principal series is the Chipeta.

The valley-fill or water-transported soils are generally quite similar. Most of them are light brownish in color, showing little change through the profile, except that at depths usually exceeding 3 feet sandy or gravelly strata are frequently encountered. In some localities where the uplands are occupied extensively by reddish sandstone soils, the color of the valley-fill areas is light reddish to dark reddish brown. Originally, the larger proportion of the area occupied by these soils was well vegetated; but with the introduction of grazing, erosion has become generally active, especially channel cutting or gullying. The principal series are the Winslow, Shiprock, Weepo, and Dinnehotso. Some extensive valley-fill areas derived largely from shale outwash, such as those in northwestern New Mexico, western Colorado, and eastern Utah, consist of grayish-brown to light-brown clays and clay loams. When dry, these usually are loose or granular to depths not exceeding about 6 inches, below which relatively hard clay of about the same color is encountered. These soils are highly calcareous; they shrink and crack deeply on drying. On wetting, the clay swells to a tough, dense mass of impervious nature. Erosion has produced numerous deep, wide-bottom gullies. Flood flows rapidly undermine the soil, cutting away the unstable lower strata and causing great masses of material to cave in. Usually, sheet erosion is also serious in all overgrazed areas. The principal soil group is the Billings.

Local areas of clay loam and clay of brown to reddish-brown color have been derived from basalt, as in parts of northwestern Arizona. These are frequently stony but produce good grass as well as timber where the rainfall is adequate.

The northern extension of the area included under the Colorado Plateau is known as the *Wyoming Basin*. This is prevailingly undulating to rolling country in south-central Wyoming. A small part extends into northern Colorado. Most of the Basin drains to the Colorado River via Green River. The general elevation is about 6,000 to 7,000 feet, although some included mountainous areas are much higher. Precipitation normally is less than 13 inches. Alkali and near-alkali conditions are widespread. The soils predominantly are light-brown to red fine sandy loams and loams of fairly good depth. Most of the area is adapted only to grazing, although a number of alluvial areas with available water are cultivated under irrigation. Dry farming is carried on in some localities, producing principally wheat and feed for livestock.

Ordinarily, erosion by wind is more serious than by water, although summer rains of higher intensity sometimes remove large quantities of soil from sparsely vegetated areas. In the southern, or Red Desert, section, it is not uncommon to find clumps of sage and tufts of grass standing on conspicuous pedestals as the result of both wind and water erosion.

Sagebush (*Artemisia tridentata*) is so abundant that much of the country is designated locally as "sagebrush." Saltbush and greasewood occupy many of the saline flats, where the heavier soils usually occur. Blue grama, western wheat grass, spiked wheat grass, and niggerwool are fairly abundant in localities not depleted by overgrazing. Galleta is one of the dominant grasses in the southern sector. Much of the area has been overused for winter range. Portions of the Red Desert section require from 100 to about 1,200 acres to support an animal unit one year. The western and southern borders extend into mountainous country of higher rainfall and better range.

From the standpoint of erosion control and water conservation, commercial production of small grain by dry farming should give way more to the growing of livestock feed. With respect to the range, such large acreages of cheap land are utilized that it is commonly felt that comparatively little expense is justified for conservation measures. The most economical procedure includes range improvement by proper stocking, better distribution of the animals, and adjustment of grazing to seasonal conditions, together with adequate stock-water development, water spreading, and conservation of rainfall by contouring in the more favorable areas.

In cooperation with those using the range, the Division of Grazing, Department of the Interior, is undertaking here, as in many other parts of the Western range, to improve conditions through administrative grazing districts.

A considerable part of the area—most of the rougher lands—consists of exposed geological material on which erosion has long been too active for the development of true soil.

A very large part of the land of the Colorado Plateau is under public ownership or control, representing principally public domain, national forest, Indian reservations, and state lands. Privately owned lands consist chiefly of the large irrigated areas, "railroad" lands, and relatively small tracts of range. The Federal Government has made large investments in this region, as in the Indian reservations and Boulder Dam.

For these and other reasons, responsibility for erosion control, conservation of water, and better land use is largely that of the Federal Government.

The San Juan, draining directly into Lake Mead, probably carries the heaviest silt load of any stream of equal size in the United States. It drains most of the 16 million acres included in the Navajo Indian Reservation, where approximately 50 thousand Indians are faced with the menacing problem of decreasing possibilities for a livelihood because of the decline of their range under the impact of overuse and increasing erosion.

Accordingly, the Navajo soil and water conservation project was established in 1934. Although the project was extended in 1937 to include some 12 million acres of non-Indian land on the San Juan and Little Colorado drainages, the major erosion-control program of the Colorado Plateau has been carried forward on the Navajo Reservation.

### NAVAJO RESERVATION EROSION-CONTROL PROJECT

The Soil Conservation Service has joined forces with the Bureau of Indian Affairs in an effort to develop an over-all program of land management and erosion control for the entire Navajo Reservation. The plan of land management is based on every phase of resource use on the reservation. As a first step, 12 demonstration areas of from 4,000 to 38,000 acres each were selected in various representative parts of the reservation to serve as proving grounds. These have been fenced, a variety of erosion-control and water-spreading structures installed, new watering places developed to permit uniform grazing, and stocking adjusted to grazing capacity.

For example, the 8,000-acre enclosure on the Ganado demonstration was stocked to estimated grazing capacity in the fall of 1935. Twelve Navajo sheep raisers furnished 400 ewes. The lamb crop from 400 ewes for the 1935–1936 grazing season was 93 per cent. The lambs averaged 67 pounds and brought 7 cents a pound, a premium price. At marketing, the oldest lambs were only a few weeks over five months of age. All were born within a thirty-five day period. Traders in the vicinity paid a premium because they had contracted to supply buyers with lambs that averaged 55 pounds. The heavier animals from the demonstration area were in demand to bring up the average weights.

Outside the area, the corresponding lamb crop, produced under prevailing Indian practices, was only 61 per cent—a third smaller. The average weight of the lambs was 45 pounds—a third less. They brought only 5 to 6 cents a pound. Since the bucks had run with the ewes yearlong, the lambs were of various ages, from six to eight months old, and lacked uniformity.

Inside the demonstration area, the average wool clip from the ewes was 8.17 pounds per sheep unit; outside, 5 pounds, a difference of 3.17 pounds. At 25 cents a pound for wool and 6 cents a pound for lambs, the income per sheep unit inside the area was $5.78; outside the area, $3.08—a difference of $2.70 in favor of the demonstration sheep. Actually, the prices received for fleeces and lambs from inside the area were higher than prices received for fleeces and lambs produced outside the area.

For the grazing season of 1936–1937, the record was even better. The lamb crop was 98 per cent, 5 per cent higher than the year before, and lambs averaged 73 pounds weaned for market (Fig. 308). Traders said lambs outside the area were averaging 55 pounds that season, or 18 pounds less than demonstration lambs.

By demonstrations such as these, it is hoped to convince the Navajos that improvement of the range by conservation of rainfall, together with conservative stocking and better range management, will pay. There is evidence that their interest is becoming aroused. When the first demonstrations were set up, it was difficult to induce the Indians to part with their sheep. They did not trust the Government or the white man's technique and feared that they might lose their stock. This attitude is changing since news of the results at demonstration centers has passed around the reservation. At present, more sheep are offered by the Indians for demonstration than can be accepted.

Although the demonstration areas were being established, surveys were started over the reservation to determine the condition of the range, the uses that were being made of the land, and the specific locality needs with respect to range-management and farming practices. The information collected was used to devise a reservation-wide erosion-control and range-management program. The reservation subsequently was divided into 18 districts, or range-management units, varying in size from 350,000 to 2,000,000 acres. Livestock-management plans and other erosion-control measures were drawn up district by district. Now, in each of these districts, the Navajos are gradually bringing about an adjustment of livestock numbers to the grazing capacity of the reservation lands.

FIG. 308.—On the Ganado demonstration project of the Navajo Reservation the lamb crop for 1936–1937 was 98 per cent, and the lambs averaged 73 pounds, as compared with 55 pounds for animals raised outside under prevailing practices. Arizona. (*Photograph by Soil Conservation Service.*)

Reduction in numbers of livestock does not come easily to the Navajos, for the Indian stockman, even more than the white stockman, measures wealth in terms of livestock, not dollars. So firmly fixed in the mind of the Navajo is the idea that wealth is measured by the number of livestock owned that any sale is regarded as a reduction, even where sales result only in culling the herds or disposal of natural increases by normal stock sales. His point of view is that only sheep and goats provide against the rainy day.

Close culling of livestock would in itself provide some measure of relief to the range without curtailing the Navajo's income from his flock. The attachment of the Indian to his stock is such that he holds on to many more horses than he needs, aged ewes, barren cows, wethers, and undesirable bulls and bucks long after the stockman with a keener profit-taking sense would have disposed of them. Unfortunately, the present grazing capacity of the range is so low, as the result of overgrazing induced by a rapidly increasing population within a restricted area, that deeper cuts than mere culling are required to bring livestock numbers down to the point at which vegetation will improve sufficiently to stabilize the land and the economy of the Indians.

Navajo sheep trail the reservation much more than is necessary. The Navajo builds his corral at his hogan (house) and drives his sheep home each night and sometimes at noon. He drives them to water and to certain browse plants for salt. His band is constantly on the move, often at a lope. By this excessive trailing, he harms both his range and his sheep. Navajo sheep, competing for the little grass ahead, seem to run even as they graze. This habit holds over, according to herdsmen, after the sheep are run on a demonstration area where the grass has partly recovered. They will run for about 2 weeks in such new environment; then they seem to learn that it is not necessary. After that they calm down and graze quietly.

A certain amount of trailing has been eliminated by providing several hundred new watering places and by rehabilitating some of the old. More trailing would be eliminated if the Navajos could be induced to bed their sheep out at night, as the Service is urging.

Breeding, particularly in the small bands, is usually uncontrolled. Bucks run with the ewes at all times. Most of the lambs are born around the first of April which is at least a month too early. Off-season lambs and lambs of all ages are common in most bands. Bulls run with the cows the year round, and calves may be born at any time. Mortality of calves and lambs is therefore usually high. In the Navajo conservation program, the attempt is being made to solve this problem by inducing the Indians to set aside pastures for bucks and bulls and withhold them from the breeding herd except at certain seasons of the year.

The Navajo livelihood by no means depends altogether on stock raising. In fact, about a fourth of the families own no stock of any kind. Farming, rugmaking, silversmithing, and the collection of piñon nuts make important contributions to the tribal welfare. In the last 3 or 4 years, it has been necessary to augment their incomes by programs that provide emergency relief work; and this has helped to carry them through troublesome times. But wage income from soil conservation work and building of roads, fences, stock tanks, and other improvements cannot be relied on as a permanent source of income. Yet, as long as the range continues on the downgrade, the Navajos may be expected to earn less and less from stock. Even though they should adjust their stocking to the grazing capacity of their range, they would probably earn less from livestock than they now do, at least until the range recovered sufficiently to permit heavier stocking. This is the Navajo dilemma.

The Indians spend a large part of their income for food. By farming more land—farming it properly—they could produce more of their food at home. Approximately one-fourth of 1 per cent of the reservation is now farmed. An agricultural survey in 1935 located 42,300 acres of cultivated crops. Divided among the present population, this is nine-tenths of an acre per person. Land suitable for farming, however, is not uniformly distributed. Certain areas, such as the Chin Lee Valley, are highly developed agriculturally, but the typical situation for the reservation as a whole is one of small fields widely scattered. The 1935 survey located some 11,000 acres of potential farm land in scattered tracts, and more intensive surveys may be expected to increase this figure considerably.

Along most small drainages, a little delta, or alluvial fan, occurs at or near the point where the water issues from a steep area to one of mild gradient. These areas of alluvial soil are potential fields. They are small—usually not more than 2 or 3 acres—and slope gently. The method of developing them is illustrated on the Navajo experimental area at Mexican Springs. Several delta areas were leveled by the Navajos working with slip scrapes and teams. Low dikes were built so that fields could be flooded to a depth of about 4 inches. Water from recent rains was diverted from the drainages to the fields by spreaders, and rock spillways were provided to carry away any excess amount. Nine fields containing 6 acres were developed in this locality. More beans and corn were produced from these fields than a Navajo family could use. Corn produced up to 60 bushels, with an average of 45 bushels an acre. Yields of pinto beans amounted to 10 bushels an acre.

The possibilities of increasing the production of grass and food crops for the Indians with water utilized in this way appear to be large enough to bring considerable relief to the severely grazed uplands.

Ill-equipped to work the heavier soils, the Indians select the lighter lands for their farm sites. Sandy soils blow easily, and wind erosion is severe in many fields. Such lands should be protected by planting trees for windbreaks and by strip planting. In the strip-planting plan, crops are alternated so that only a narrow band of soil is exposed to wind action, whereas the intervening band is growing a dense protective crop. This method of planting checks blowing sufficiently to hold the soil until the windbreak trees are grown. In 1936, approximately 1,300 acres were strip planted, and the practice proved effective.

Unless it is adequately protected from erosion, much more of the Navajo Reservation will be lost. Already a large total area has been ruined or severely damaged. In places, acres of land have been stripped of soil down to bedrock. There is no possible way of conserving these lands except by widespread adoption of the practical measures for soil and water conservation adapted to the peculiar physical requirements of the region.

## Southwestern Mesas, Mountains, and Basins

The large area covering the southern parts of New Mexico and Arizona, and roughly designated the Southwestern Mesas, Mountains, and Basins, comprises about 81 million acres of highly diversified country. Broadly, the area is characterized by a series of parallel, discontinuous mountain chains having a northwest-southeast trend. These mountain masses attain elevations of 5,000 to 10,000 feet. The intervening valleys and basins (Fig. 309) have elevations varying from about

FIG. 309.—The higher elevations are generally timbered. Mora County, New Mexico. (*Photograph by Soil Conservation Service.*)

1,000 to 4,000 feet. The higher elevations are generally timbered; much of the lower areas is sparsely vegetated with hardy xerophytic plants.

A striking feature of the area is the frequency of alluvial or talus fans. Between the more widely separated ranges, these fans extend for many miles with progressively decreasing gradient to axial valleys or playas. In many places, the fans begin almost at the top of the mountains. Such extensive fan development is probably due largely to the coarse character of the rock debris that develops under the wide diurnal temperature changes characteristic of the region. Coarse material is swept down from the rugged, steep-sided mountains by the intense summer rainstorms, aided by gravitational creep. Because of rapid absorption of runoff by the coarse material, the transporting capacity of the off flowage diminishes rapidly. The abundant coarse material, being deposited first, results in the formation of very steep slopes at the base of the mountains. The relatively small amount of fine material is carried farther and farther out over the fan, according to degree of fineness, resulting in the development of a progressively flatter gradient. The encroachment of alluvial fans on valley plains has blocked natural drainage in places, thus forming playas, or ephemeral lakes.

Summer temperatures are high, and winters moderately cool. Snow falls only at the higher elevations and seldom accumulates. The average annual rainfall varies from less than 3 to about 20 inches on the higher ranges, the average being about 10 inches. Something over 55 per cent falls during summer as local thunderstorms of intense character. Intensities of 3 to 5 inches an hour occur during this period. Winter rains are

FIG. 310.—This southern New Mexico area formerly was well grassed. Overgrazing has largely replaced the grass with less nutritious weeds. Bernalillo County. (*Photograph by Soil Conservation Service.*)

of more general distribution and gentle character. Because of the spotted nature of summer rains, an area may receive its entire average annual precipitation within an hour. Other localities may go without rain for a year or more.

Under the arid conditions prevailing over large areas, vegetation is scant and of desert character. Where irrigation water is available, a variety of crops is successfully produced, such as cotton, citrus, melons, and vegetables. Grazing is the only agricultural activity outside the irrigated areas.

Owing to high intensities of rainfall and sparse vegetation, the normal rates of erosion have been high over a large aggregate area as the result of overgrazing. Grasses that were common in many localities have been replaced almost entirely by less nutritious plants or unpalatable weeds having little value for retarding erosion (Fig. 310). Some areas having

5 inches or less of annual precipitation have long been stabilized more by rock or gravel pavement than by vegetative cover. Other associated areas have not yet developed sufficient pavement to control normal erosion completely. Here accelerated erosion generally has been serious only on the alluvial valley lands, where water formerly accumulated to support good stands of grasses. Deep, fingering gullies are rapidly invading such bottomlands. A considerable part of the silt contributed to the main streams originates in these areas.

The soils of this great Southwest area are derived principally from basic and acid igneous rocks, such as basalt and the granitics. Surface

Fig. 311.—A sparse growth of creosote bush presents little resistance to erosion. New Mexico. (*Photograph by Soil Conservation Service.*)

soils, as a rule, are mildly calcareous; the subsoils contain an abundance of accumulated lime carbonate. Soil color, for the most part, is inherited from the parent material. Only under the heavier rainfall of higher elevations has any considerable humus formed from the available organic matter. Most of it is dissipated by oxidation under the intense heat. Moisture absorbed by the soils derived from rocks of the basic igneous group and from quartzite usually is readily available to plants. The favorable structure of such types provides exceptional resistance to erosion. The coarser textured soils, as those derived from granitic rocks, generally have heavier and less permeable subsoils and are highly erodible. The extensive, unleached, highly calcareous soils occurring throughout the area support only a sparse growth of shrub, mainly creosote bush, such as presents very little resistance to erosion (Fig. 311). Dense clays occupy the playas almost completely. They usually are highly impervious, contain large amounts of alkali salts, and are devoid of vegetation.

The upper and middle portion of the Gila Basin is one of the more important problem areas of the southern New Mexico-Arizona region. This drainage, particularly the San Simon, contributes large quantities of silt to Coolidge Reservoir, on which more than 80,000 acres of irrigated lands depends for water. Salt River and the Rio Grande also embrace large irrigation developments and constitute important problem areas. These drainages enclose the most highly developed commercial irrigated agriculture in the Colorado River Basin. Irrigated lands within the basins of these three rivers aggregate nearly half a million acres. They produce large quantities of supplemental feed for range livestock, cotton, citrus fruit, melons, vegetables, and miscellaneous grains. The products of these highly productive lands are largely responsible for the cities of Phoenix, El Paso, and Albuquerque as well as many smaller towns of 1,000 to 10,000 population.

To a large degree, the population of this southerly area is dependent on the regional agriculture, and crop production is, for the most part, dependent directly on the life of the great federally owned storage reservoirs: Roosevelt, San Carlos, and Elephant Butte. Protection of these reservoirs from silting can be achieved only through adequate protection of the watersheds that feed them. Since the greater part of the area consists of federally owned or controlled land, the responsibility of the government for an effective program of erosion control is obvious.

As a result, two large erosion-control projects are now in operation on the watersheds of the Rio Grande and Gila.

Rio Grande and Gila Erosion-control Projects. The Middle Rio Grande Valley in central New Mexico furnishes a striking example of the problems arising from accelerated erosion in this region.[1] Here about 55,000 rural people depend chiefly on subsistence agriculture for a livelihood. The pressure of population on the available area of productive land is very great. Indian farmers have irrigated here for 1,000 years, Spanish-Americans for 300 years, and Anglo-Americans for nearly a century. The irrigable lands were developed long ago. Accordingly, farms are small, many of them little larger than good-sized garden patches; and they are handed down from generation to generation. Seldom is a farm offered for sale. Until recently, few were mortgaged. Agriculture is not of the commercial farming order. Farmers sell or barter a little chili and a few head of cattle, but most of the products coaxed from their precious soil are eaten at home.

During the past half century, the land situation has grown distressing. Many of the irrigated fields have become waterlogged, impregnated with

[1] Cooperrider, C. K., and Hendricks, B. A. Soil Erosion and Stream Flow on Range and Forest Lands of the Upper Rio Grande Watershed in Relation to Land Resources and Human Welfare, U. S. Dept. Agr. *Tech. Bull.* 567, 1937.

alkali, and useless, or they have been covered by sand and gravel, as the result of accentuated flood conditions. In 1800, about 100,000 acres were irrigated. By 1880, about 125,000 acres had been developed; but by 1934, the cultivated area had shrunk to 40,000 acres. Two-thirds of the irrigated acreage, roughly, has been abandoned in the last 50 years. To eke out a living, many of the families migrated seasonally to the beet and potato fields of Colorado or to the smelters of the region and labored by

FIG. 312.—High water at San Marcial, flood of 1937. Rio Grande, Socorro County, New Mexico. (*Photograph by Soil Conservation Service.*)

the day; but even this opportunity was largely closed to them by the recent depression. The relief load in the valley has been heavy since 1934.

Immediately above the head of Elephant Butte Reservoir, shoaling of the Rio Grande with sand has been responsible for such severe flood damages and swamping of valley lands that the town of San Marcial (Fig. 312), which had a population of over 1,000 prior to 1929, has been almost abandoned. Since 1914, when Elephant Butte Reservoir was built, the bed of the Rio Grande a short distance above the reservoir has risen an average of about 13 feet, according to surveys made in 1936. Two miles farther upstream, and just above the San Marcial townsite, a rise of about 7 feet occurred during the same period.

Largely because of this rise in the stream channel, the dikes built to protect the towns, and irrigated lands adjacent to the Rio Grande have been repeatedly overtopped and broken by floodwaters, with disastrous results. Over considerable areas, sand was deposited by the flood of 1929 to depths of about 1 to 6 feet; and clay, to depths exceeding 2 feet.[1] Many of the irrigated fields west of the river at San Marcial became swamps after the 1929 flood. In the spring of 1937, the Rio Grande broke out of the floodway to which it had been confined by dikes and followed a new course directly across the remaining irrigated lands lying east of the river.

The Rio Puerco joins the Rio Grande some 50 miles above San Marcial, delivering sediment to the Rio Grande much faster than the river can move it. Consequently, as shown by comparative surveys in 1927 and 1936, the Rio Grande channel has aggraded at rates as much as 9 inches a year. As a result, floodwaters in the spring of 1937 overtopped and washed out the levee and a paved highway and converted hundreds of acres of pasture land into a sandy waste. Large parts of the formerly irrigated lands about the village of Contreras have become swamps because water can no longer be drained into the river, which is now flowing at a level above that of the bordering fields.

Some of these damaged lands may be reclaimed. At present, the largest single effort is the Rio Grande Conservancy District, which was set up in 1927 to protect cities, towns, and utilities against flood damage and to reclaim some 70,000 acres by construction of drainage canals, storage works, and floodways.

At best, the prospects for success of this program are dubious unless the whole of the watershed, from the forests down to the banks of the river and its tributaries, is hereafter conservatively managed. Testifying at a flood-control hearing at Santa Fe in April, 1938, Stanley Phillippi, assistant chief engineer of the conservancy district, said: "The maintenance of the channel with present conditions of silting is gradually becoming an impossibility." The whole project, he pointed out, is endangered by the great loads of sediment that are washing into the river from overgrazed range lands, filling the river channel at an alarming rate.

The Rio Grande is a perennial stream flowing through an arid and semiarid region. It derives its water largely from rain, melting snow, and springs in the forested mountains of Colorado and northern New Mexico. When it leaves the upper reaches, its waters are clear. It receives little further contribution as it flows south through New Mexico, except from the Chama, Galisteo, Jemez, and Puerco, all of which course through overused range lands. Discharge from these tributaries is small except during storms, when they pour huge quantities of silt into the Rio Grande. Although the Puerco supplies only 2 to 4 per cent of the water carried by the Rio Grande, it contributes 40 per cent of the silt load. When the

[1] Bennett, H. H. Relation of Erosion to Vegetative Changes, *Sci. Monthly*, Vol. 25, 1932.

Rio Grande is in flood, it carries, below the Rio Puerco, ten times as much silt as an equal volume of flood water flowing down the Mississippi.

The grazing capacity of that part of the Rio Grande watershed between Elephant Butte Reservoir and the Colorado line is estimated at 140,000 cattle units yearlong. (One cattle unit equals five sheep units.) The estimate of the number of sheep and cattle now using this range is equivalent to 250,000 cattle units yearlong, which is about 80 per cent in excess of present grazing capacity. This figure is based on all lands,

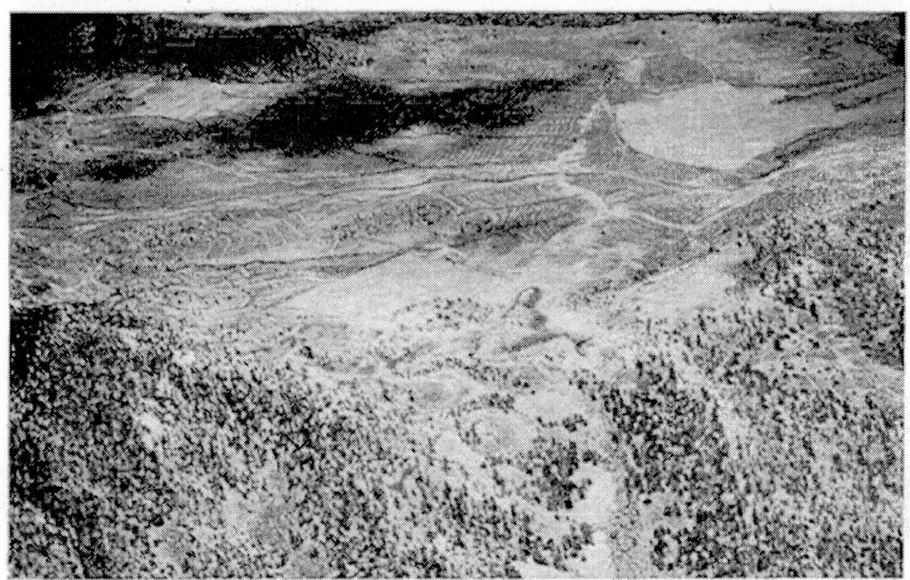

FIG. 313.—By 1938 nearly two million acres of range and farm land were being treated by the Soil Conservation Service for control of erosion, conservation of water, and rehabilitation of vegetation, in the watershed of the Rio Grande. Rio Arriba County, New Mexico. (*Photograph by Soil Conservation Service.*)

13 million acres—the public domain, Indian lands, privately owned land, state lands, national forests, and railroad lands. Since some parts of the range are managed conservatively, large areas therefore are stocked two or three times their grazing capacity. If the ancient agriculture along the Rio Grande is to be saved, some drastic adjustments in range use and management must be made over the greater part of the watershed.

The program of the Soil Conservation Service on the Rio Grande is directed toward the amelioration of these conditions as well as the immediate rehabilitation of the forage and control of erosion on the millions of acres of range land comprised within the watershed.

The creation of an *Interdepartmental Board* in 1938, composed of representatives from a number of interested agencies, including the Soil Conservation Service, Forest Service, and Bureau of Indian Affairs, indicated public recognition of the complexity of the problem and need for its solution.

By 1938, approximately 2 million acres of range and farm land had been put under better management in the Rio Grande project (Fig. 313). This acreage included: Indian reservations and grants, "resettlement" land, public domain, grazing districts, private land, and state land. Results pertaining to erosion control and vegetative recovery are reflected in the increased stock yields on the treated areas.

Work on the Laguna Indian Reservation and the near-by Montano grant provide excellent examples of this. Some 2,200 people live in the seven villages scattered over 237,000 acres of the tribal land of the Lagunas. In 1935 these Indians agreed to reduce their sheep by 75 per cent. About the same time the Soil Conservation Service started an erosion-control program on the Montano area, an old Spanish grant of 44,000 acres, which had been purchased recently by the Federal Government. The Service found that the grant would carry 3,000 sheep for seven months out of the year if properly managed.

To facilitate the Laguna stock adjustment, the Indian Service and Soil Conservation Service permitted 3,000 Laguna sheep to graze the Montano grant area for seven months of 1935 and 1936. On the properly managed area, forage and reserve feed increased, the vitality of the range improved, and soil washing decreased markedly.

The lamb crop from the Laguna sheep grazing on the grant was 95 per cent, as compared with only 65 per cent on near-by overused ranges. The wool clip averaged a pound more per sheep on the properly grazed range.

In November, 1927, a representative of the Kansas Agricultural Experiment Station selected 518 lambs from the Laguna and Acoma flocks to be placed on experimental feed lots. These were selected out of several thousand lambs looked over on many New Mexico ranges. At the end of the feeding period, the Laguna lambs averaged 90 pounds. The packer reported that 75 per cent graded "Premium" compared with 40 per cent of Premium grade for other western-fed lambs for that season.

Throughout the Rio Grande District and the entire southern Arizona-New Mexico area, range management, water diversion, fencing, contour furrowing, and other measures are bringing about erosion control, water conservation, and better stock production through revegetation of the ranges. At the same time, transportation of silt is being cut down.

Dry farming in this area involves a relatively small acreage. However, this small area occupies an important position in the upper watersheds. Human dependence on such areas is high, as they are largely restricted to crowded localities of meager resources.

The chief crops grown are corn and beans. Steep slopes and clean tillage have led to severe erosion on much of this dry farming land. Terracing and contour cultivation have resulted in marked reduction of erosion and increase of yields.

## Wasatch Mountain Area

The Wasatch Mountain problem area occupies 18 million acres in central Utah. Including the Bear River Mountains, the range extends into southeastern Idaho. An included projection, the Uinta Mountains, extends along the Wyoming line into northwestern Colorado. South of this, another easterly projection practically joins with the lower portion of the Rocky Mountains in Colorado. The Wasatch Mountains proper rise to elevations in excess of 11,000 feet. Their crest is 5,000 feet or more

FIG. 314.—Range management and erosion control on the steep slopes of the Wasatch. Box Elder County, Utah. (*Photograph by Soil Conservation Service.*)

above the lacustrine plain of ancient Lake Bonneville, on the west. The slopes are very steep, especially on the west, with cliffs of 1,000 feet or more in places. Narrow valleys and gorges incise the mountain mass in many places. Considerable portions of the crest, as well as parts of the eastern slope, are relatively smooth.

Precipitation varies from 20 to 35 inches or more, including snow, over the higher elevations. Summer storms are extremely intense. In watersheds denuded by excessive grazing or burning, destructive floods and severe erosion are the result, not only in the highlands but on the adjacent lowlands as well.

Where dry farming is carried on over some of the smoother parts of the eastern slope, under a system of alternate years of fallow, erosion has become an extremely serious problem. In many places, both surface soil and subsoil have been washed off within the last 20 years. Stock trails, over which thousands of sheep are driven to and from the high summer ranges every year, are especially subject to severe washing.

Recently, farmers dependent on irrigation water from this mountain area have begun to realize the importance of watershed protection (Fig. 314). It is highly important that erosion be arrested on the western slope as speedily as possible, since a very important share of the agriculture of Utah is developed on the alluvial fans at the base of the mountains. Almost annually, some of these lands, and the improvements on them, are damaged to the extent of thousands of dollars by flood flows and overwash of rock debris. The heavier floods sweep boulders that sometimes weigh many tons over the valuable lands at the base of the mountains. Moreover, the highlands are the source of irrigation water, and for that reason every effort must be made to maintain an adequate cover of water-conserving vegetation.[1]

Along the crest and the western slope of the Wasatch, the principal soils are derived from granite, rhyolite, and other igneous or intrusive rocks; on the lower elevations of the eastern slope, they are derived from sedimentary rocks. The more extensive types are brown to dark brown in the surface, friable, and fairly high in organic matter. Over a considerable area, the subsoil is of heavier texture and dark brown or dark reddish brown in color. On the steeper slopes, the depth to bedrock averages comparatively shallow, often less than 2 or 3 feet. All types are subject to erosion under exposure, but the heavier types appear to wash faster.

Most of the watersheds draining the western slope of the Wasatch represent problem areas, varying only in degree. A considerable part of the range is included in national forests. Within the reservations, where grazing is now under management, vegetation appears to be making satisfactory recovery from previous conditions of erosion. A good cover has been reestablished over some areas after only a few years of protection. Some of the more seriously affected areas, however, require planting, the construction of check dams, and various types of contouring. The Forest Service is carrying on erosion-control operations, including contour trenching, planting, and other measures, on a number of critical, high watershed areas in the Wasatch Mountains.[2]

Of the Soil Conservation Service activities in this area, the Weber River project is the most important. Here, extensive areas of range

[1] Forsling, C. L. A Study of the Influence of Herbaceous Plant Cover on Surface Runoff and Soil Erosion in Relation to Grazing on the Wasatch Plateau in Utah, U. S. Dept. Agr. *Tech. Bull.* 220, 1931; see also Floods and Accelerated Erosion in Northern Utah, U. S. Dept. Agr. *Misc. Pub.* 196, 1934.

[2] Bailey, R. W. Epicycles of Erosion in the Valleys of the Colorado Plateau Province, *Jour. Geology*, Vol. 43, May, June, 1935; and Contour-trenches Control Floods and Erosion on Rangelands, Emergency Conservation Work, Forestry *Pub.* 4, Washington, 1937.

border the strips of productive valley land along Weber River and its tributaries. The population is dependent on the products of both farm and range. Irrigation agriculture prevails on the alluvial lands, and dry farming on adjacent lower slopes. Although livestock depend largely on the range in summer, they must be fed in winter.

Much of the range has suffered marked depletion from improper use. Here, as elsewhere on the Western range, rejuvenation depends primarily on adjustment to carrying capacity and good management. Where individual holdings are composed of both range and farm land, adjustments

Fig. 315.—Contour irrigation of a young orchard in the Wasatch Mountains, Utah. (*Photograph by Soil Conservation Service.*)

are easier to make, as a rule, than where holdings include only range land. In some instances, continued use by one kind of livestock, as by cattle, has changed vegetative conditions from an original cover of grass, browse, and weeds to one of browse and weeds. Under such conditions, those operators capable of handling more than one kind of livestock find that a change from cattle to sheep has often resulted in increased carrying capacity.

Proper use and management of irrigated and dry-farming lands also constitute an important phase of the soil conservation program. All proposed changes in the use of irrigated lands are based on the needs of soil conservation, with due regard for the economic limitations of the operator. Retirement of eroded lands to permanent cover, proper use and disposal of irrigation water, cropping systems, soil fertilization, and improved methods of cultivation are among the measures employed on irrigated lands.

Solution of the erosion problem on lands used for dry farming calls for gully control and water conservation, with such measures as contouring, dam listing, strip cropping of both the permanent and rotation types, and conversion of the more erodible areas to permanent vegetation. Cooperating with the Forest Service, the Soil Conservation Service is also attempting, by controlled grazing and other practical means, to overcome the severe mud flows originating in some of the headwaters of privately owned range lands.

Another critical problem of erosion in this region has to do with severe washing caused by irrigating steep slopes, as in the orchards in the vicinity of Willard, Utah. Progress is being made with this difficult phase of erosion control, especially in the direction of contour cultivation and contour application of water by means of underground pipe lines with risers (Fig. 315). Another promising measure is the distribution of water through open, concrete-lined ditches with provisions for diverting at intervals into contour furrows.

## Rocky Mountain Area

That part of the Rocky Mountains included as a problem area in the region of the Colorado River Basin comprises 58 million acres. Only part of this large sector of the continental divide lies within the Colorado River drainage; the waters of its eastern slope drain to the Gulf by way of the Mississippi. Elevations vary from 8,000 to 14,000 feet. The eastern slope drops abruptly to the Great Plains; west of the divide, the mountains, in the northern part, fall away more gradually to the low country east of the Wasatch.[1] Near the crest and over the generally smoother westward slope occur many mountain meadows, or "parks," such as North Park, in the vicinity of the Colorado-Wyoming line. A number of plateau areas, remnants of what formerly was a much more extensive plateau, are scattered through the country below the zone of mountain meadows. These are separated by steep-sided valleys formed by glacial or stream action. One of the more noteworthy plateaus is the Grand Mesa, in west-central Colorado, reputed to be one of the largest mesa-form eminences of the world.

During the period of active glaciation, ice floes developed the common U-shaped valleys of the upper elevations. Below this area of glaciation, the valleys are narrow and generally V-shaped. Farther down, many of these become broader and are bordered by a belt of coalescing alluvial fans

[1] See physical features map in "Atlas of American Agriculture," U. S. Department of Agriculture, 1936.

developed by tributary drainages issuing from the major interstream areas.

Timber line is at about 11,600 feet. In the higher forests, spruce, aspen, fir, and lodgepole pine predominate; below, western yellow pine comes in; then, lower down, scrub oak, juniper, or piñon. Above timber is an abundance of grass-sedge vegetation and numerous exposures of rock and accumulations of rock talus. Toward the south, there is not so much difference between the eastern and western slopes of the Rockies.

The main backbone of the mountain, as well as the eastern slope, is underlain principally by granites, granodiorites, and related rocks. Uplifted sedimentaries of various composition and age form the general base of the western slope, with minor areas, as the Grand Mesa, capped by lava flow. Soils are generally shallow and loose, except on the smoother areas, as the mountain meadows. They absorb water readily, but to a considerable degree they are not very retentive of moisture. Because of shallowness, steepness, and fragile structure, they wash readily under conditions of bared or thinly covered surface. At the higher elevation, much of the soil is of coarse texture, often consisting of *rough stony land.*

Precipitation ranges from 20 to 50 inches, more than half of which falls as snow. In contrast to conditions prevailing over most of the general region, summer rains at higher elevations usually are prolonged and of rather gentle character. Cool temperatures restrict agricultural use of the area largely to grazing. At the lower elevations, quick-growing forage crops can be matured. Snow begins to accumulate over the high mountains in early fall and continues to accumulate until early May. It melts quickly with advent of spring, producing rapid runoff, frequently of destructive intensity. Only a good cover of vegetation will prevent destructive washing and wastage of rainfall, as both soils and slopes are conducive to severe erosion and rapid runoff. This great mountain mass is the principal source of water for the Southwest. Accordingly, the economic welfare of the majority of those living in the arid parts of the Colorado River Basin problem area is dependent on the maintenance of an adequate protective cover.

Fortunately, the greater part of the mountain area is in national forests, where land-use regulations have been in effect for 30 years, at least over much of the included area. As the result, erosion is generally not severe. Outside the forest reserves on the public domain, state lands, railroad lands, and private holdings, overgrazing, unregulated timber cutting, and burning have brought about highly unfavorable conditions of erosion and accelerated runoff over a large aggregate area. Conditions are particularly critical in the valley and foothill area through which stock are driven to and from summer range in the higher mountains. Such depleted areas are outstandingly productive of floods and high silt loads.

# Chapter XXXVIII. Pacific Northwest Region

The Pacific Northwest Region, embracing about 198 million acres, covers all of Washington, most of Oregon and Idaho, the western third (approximately) of Montana, and a considerable area in northwestern

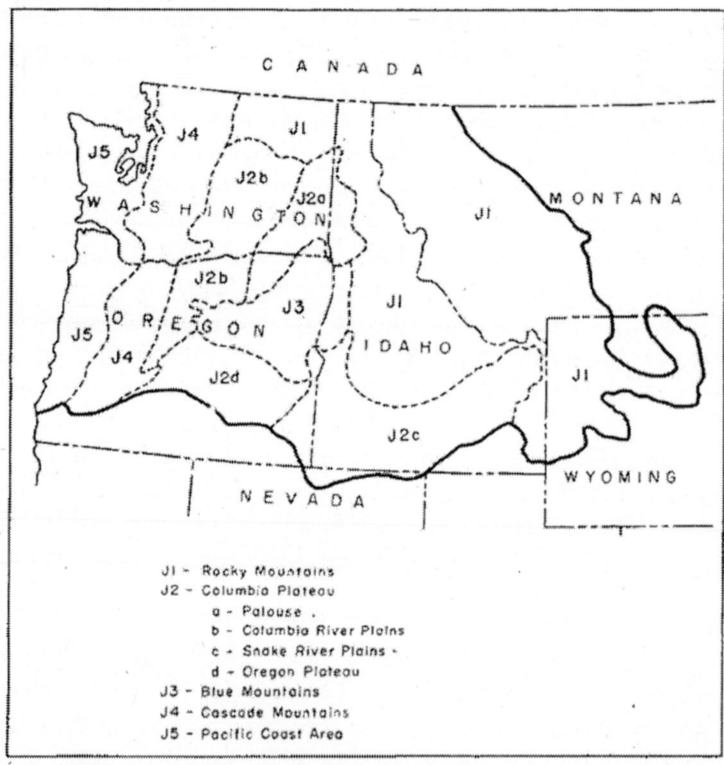


MAP 15.—Problem area of the Pacific Northwest Region. (*Soil Conservation Service.*)

Wyoming (Map 15). Most of the area drains into the Columbia River or the Pacific Ocean. It is crossed from north to south by three mountain chains: the Olympics and Coast Range, the Cascades, and the Rocky Mountains. A considerable area near the center is occupied by the Blue

Mountains. Roughly, these mountainous areas comprise nearly 130 million acres, or more than 65 per cent of the region.

A large plateau or basin area of relatively smooth topography is almost enclosed by the Cascades, Rockies, and Blue Mountains. By a narrow connection, this lowland area swings around the southwestern extension of the Blue Mountains and extends in a broad belt south of the Blue, Wallowa, and Salmon Mountains easterly across the Oregon line, thence along the Snake River, nearly to the Wyoming line. To the south, this southerly sector crosses into northeastern Nevada for a short distance to merge imperceptibly with the Great Basin area.

This roughly defined intermountain area constitutes the Columbia Plateau. It is composed of a large number of distinct subareas, or divisions, the more important of which are: the Palouse, Columbia River Plains, Snake River Plains, and Oregon Plateau (part of Harney Basin). In detail, the surface varies from almost flat, as in the western part of the Big Bend of the middle Columbia River, to undulating and rolling, as in the plains of the southerly sectors. The Palouse country, to the north of Blue Mountains, is characterized by a peculiar dunelike hill topography not duplicated on a large scale elsewhere in the United States.

The Plateau has been deeply incised by canyons of the Snake and Columbia Rivers and some of their tributaries. The Snake River Canyon between Seven Devils Mountains in Idaho and the Wallowa Mountains of Oregon averages 5,500 feet deep for a distance of 40 miles. It is the deepest canyon in the United States.[1]

Another relatively low area is approximately shut in between the Cascades and the chain of hills and mountains along the Pacific Coast. This trough extends from Puget Sound across Washington, through the flat-floored Willamette Valley, into the highlands of southwestern Oregon. Through the heavier rainfall portions of these intermountain areas, as well as some of those of comparatively low precipitation, various important types of agriculture have been developed, principally wheat and fruit farming, trucking, irrigation farming, and dairying. The drier localities are used for range purposes.

Most of the intermountain lowlands west of the Cascades is less than 1,000 feet above sea level. The northern part, the Puget Sound section, borders the sea. The northern sector of the Columbia Plateau ranges in elevation from less than 1,000 feet near the Columbia River to about 3,000 feet in the Blue Mountain foothills. In the Oregon Plateau and Snake River Plains, the elevation ranges from about 2,000 to 6,000 feet.

The Rockies and Cascades are generally rugged mountains, although a large part is sufficiently accessible for summer grazing. High elevations

[1] See Freeman, D. W. The Snake River Canyon, *Geog. Rev.*, Vol. 28, No. 4, October, 1938.

are attained, especially in the Cascades. Rainier, the loftiest peak, rises to 14,408 feet above sea level. The Olympic Mountains, with a maximum elevation of 8,150 feet, and the Coast Range, with a maximum of about 4,000 feet, generally are of a more rounded configuration. Over much of the Coast Range of Oregon, both climate and soil are especially favorable to the growth of either grass or trees.

Nearly all the major streams and many minor ones have their headwaters high in the mountains. With the exception of some cut-over areas, the upper watersheds are forested, principally with coniferous trees. West of the Cascades divide and in the white-pine type of forest in the northern Rockies, much of the timbered country is of such dense stand that it is poorly suited to grazing. Elsewhere, however, the forests are more open; and nearly all land not in farms, whether forest, brush, or grass, affords some grazing.[1]

The climate of the Pacific Northwest is characterized by humid winters, dry summers, and a wide range in precipitation. Rainfall is heaviest near the ocean and over the higher elevations. It decreases progressively inland and downward from mountain crests, except for local modifications due to topography. Precipitation in the "shadow" of the mountains is consistently lower than on the exposed slopes. West of the Cascades, the average amount received annually varies from 15 to more than 100 inches on the lowlands and from 60 to more than 120 inches at observation stations on the mountain sides. East of the Cascades, the range is from a few inches over some of the lowlands to about 60 inches on mountain crests. Nearly three-fourths of the precipitation occurs during winter, most of it as snow at the higher elevations. Annual snowfall west of the Cascades varies from a few inches in the valleys to nearly 50 feet at Paradise Inn, in Mount Rainier National Park. East of the Cascades, the moderating influence of the ocean is less; winters are more severe; and the proportion of precipitation occurring in the form of snow is greater than to the west. In general, only a few inches of snow falls over the lower valleys of this lower rainfall part of the region; in the mountains, falls ranging from 5 to 20 feet have been recorded frequently at observation stations; and much deeper falls occur on the higher peaks.

Rains usually are gentle. Intensities greater than ¼ inch in 5 minutes or ⅘ inch per hour have been recorded in winter only a few times west of the Cascades. During about fifty years of records, rainstorms exceeding these rates have occurred only five times at Roseburg, Ore.; nine times at Portland, Ore.; and five times at Seattle, Wash. East of the Cascades, such storms occasionally occur in summer in the Blue Mountains and near the Rockies, but they are not at all characteristic of the climate.

[1] For further detail with respect to the physical characteristics of the region, see N. M. Fenneman, Physiography of Western United States, pp. 183–273.

Within a period of about fifty years, intense rains of this nature have been recorded seven times at Baker, Ore., and three times at Walla Walla and twice at Spokane, Wash. Eastward through the Rockies, the climate becomes increasingly continental in type, with greater frequency of intense summer showers.

Precipitation rates exceeding 2.5 inches in 24 hours are seldom experienced east of the Cascades; west of the range, however, such rains have fallen somewhat more frequently. Locally, they have occurred at rather frequent intervals, as at Tatoosh Island, on the northwestern coast. Here, 65 such storms have been recorded in 40 years.

Maximum seasonal runoff occurs at times of rapid melting of snow. Heaviest runoff generally comes in early spring at lower elevations and in early summer on the higher elevations. Occasional heavy winter rains, accompanied by snow-melting temperatures, produce excessive runoff, accelerated erosion, and sometimes severe floods.

Annual runoff per square mile from different watersheds varies widely. For example, the per-square-mile water delivery from the Malheur River in southeastern Oregon is 94 acre-feet as against 8,160 acre-feet from the Wynoochee in western Washington, or, in precipitation equivalents, from less than 2 inches in the dry Oregon Plateau to nearly 13 feet in the humid coastal mountains of Washington.

Of the approximately 198 million acres in the region, nearly half is in forest (largely national forests), almost a third is used for open grazing (range), and about a fourth is in farms. Only about one-third of the land in farms is used for crops. There are about 10 million acres additional of publicly controlled lands, such as national parks, state forests, Indian reservations, military reservations, and similar areas.

Erosion, by both water and wind, is a serious problem on a large area of farm and grazing land in the Pacific Northwest region. Because of differences in soil, topography, climate, and use, the problem is discussed under five major areas: the Rocky Mountains, Columbia Plateau, Blue Mountains, Cascades, and Pacific Coast.

## Rocky Mountains

The Rocky Mountain area, comprising about 86 million acres, includes the mountains and valleys of western Montana, northern and central Idaho, and large bodies in northeastern Washington and northwestern Wyoming. Maximum elevations generally range from about 10,000 to 12,000 feet. Fremont Peak in the Wind River Range of northwestern Wyoming has an elevation of 13,730 feet; and Cloud Peak in the Big Horns, 13,165 feet.[1] The valley floors, which occupy a relatively

[1] "Atlas of American Agriculture" (Physical Basis), U. S. Department of Agriculture, 1936.

small proportion of the total area, typically are narrow and enclosed by rugged mountains usually of precipitous slope. They range from about 1,000 to 4,500 feet above sea level. Some of them broaden out locally or are bordered by hilly to low mountain country, frequently well grassed (mountain meadows). The mountains, for the most part, are timbered, but many areas are rocky and barren or nearly barren; others, cut over or burned, now support very little timber. A large part of this mountainous country is forested. Nearly all accessible areas are used for summer grazing, mainly cattle and sheep from outside.

The total precipitation varies from about 11 to 20 inches in the valleys to over 60 inches at established mountain-slope observation stations and larger amounts on the mountain tops. The annual snowfall varies from about 30 inches in the valleys to over 200 inches on the upper mountain slopes.

The greater part of this mountain mass is too rough or precipitous for cultivation. Above the valley floors, much of the land consists of coarse gravelly, sandy, and stony land, with scattered tracts of loam and stony loam, interspersed with rock outcrop. On the floors and rolling fringes of many of the valleys, the soils are derived from glacial till or outwash and alluvial or alluvial-fan material. For the most part, they consist of gravelly and sandy loams, sands, and stony loams. On the basis of available information, about a third of the valley lands have open, friable subsoils; another third has compact subsoils of relatively fine material; and the remainder, heterogeneous mixtures of sand, gravel, stone, and clay.

The greater part of this northwestern sector of the Rocky Mountains is in national forests, national parks, Indian reservations, and other reservations. The principal national forests included entirely or largely within the area are: the Absaroka, Beaverhead, Bighorn, Bitterroot, Boise, Cabinet, Caribou, Challis, Clearwater, Coeur d'Alene, Colville, Deerlodge, Flathead, Gallatin, Helena, Idaho, Kaniksu, Kootenai, Lemhi, Lewis and Clark, Lolo, Nez Perce, Payette, St. Joe, Salmon, Sawtooth, Shoshone, Targhee, Teton, Washakie, Weiser, and Wyoming. In the aggregate, these forest reservations comprise some 42⅓ million acres. Other large reservations are the Yellowstone, Glacier, and Grand Teton National Parks, with a combined area of more than 3¼ million acres. Only about one-fifth of the land is in farms, and of this about 20 per cent is cropland. Not more than a fourth of the cropland is used for tilled crops, the remainder being devoted to grass and pasture. The large area of noncropland on farms and the nearness of range lands, including both forest and nonforest areas, favors a livestock type of farming. Slight to severe erosion is widespread on both cropland and range areas. Forage on the ranges has declined in overstocked areas, with the result that erosion has damaged numerous small and some large areas locally. Often uncontrolled sheet washing is a problem on the distinctly sloping areas of both irrigated and nonirrigated land.

## Columbia Plateau

The Columbia Plateau, with an area of about 55 million acres, embraces four principal subregions: the Palouse, Columbia River Plains, Oregon Plateau, and Snake River Plains. Soil, topographic and climatic conditions, and types of use vary widely. The features of greatest areal similarity are those of topography and elevations, described on page 815.

### Palouse

The Palouse, together with a marginal strip of foothills adjacent to the associated mountains, comprises something over 8 million acres. Most of the wheat belt of eastern Washington and adjacent parts of Idaho and Oregon is included in this subdivision. The country is predominantly hilly, with characteristic long southerly slopes and short, steep slopes toward the northeast. Most of the land has been broken out of the original stand of bunch grass and put into cultivation.

In detail, the western fringe of the Palouse is undulating to rolling, with slopes generally less than 15 per cent. Eastward, this lower smoother country grades into typical dunelike, hilly topography of the Palouse, with windward (southwesterly) slopes as steep as 30 per cent and leeward (northeasterly) slopes as steep as 65 per cent. Many of the hills range from 100 to 200 feet above the bottoms of the valleys. In the foothills along the northwestern part of the Blue Mountains occur many long, flat-topped ridges with uniform slopes, interrupted by occasional deep canyons.

In the north-central part of this section is an area, called the *channeled scablands*, where the former soil covering has been eroded off to expose the basalt base.

The climate varies from semiarid at the western edge of the Palouse to subhumid near the bordering mountains on the east. The entire area, however, has humid winters and dry summers. From west to east, annual precipitation varies from about 12 to 35 inches. Snowfall varies from about 2 feet in the west to nearly 10 feet in the southeastern foothill section. At lower elevations, snow frequently melts during winter, sometimes permitting the soil to freeze before subsequent falls or to melt on frozen soil. Such conditions contribute to accelerated runoff and sheet washing. The snow cover is more at higher elevations, affording better protection from intermittent freezing and thawing.

The most extensive soils are considered of loessial origin, are generally deep, and are friable or moderately friable in the substrata. They range from brown grassland types in the west to dark brown in the east. From west to east, the principal groups are the Ritzville, Walla Walla, and Palouse.

The brown Ritzville soils, chiefly loam, are underlain by basalt at depths ranging from a few inches on steep slopes of the major drainageways to 40 feet on some of the hilltops. Below about 30 inches, lime carbonate is usually present in fairly abundant quantity. The area over which this soil prevails is a gently rolling plain.

The dark-brown Walla Walla soils differ from the Ritzville principally in their greater depth to bedrock, higher content of humus, darker color, and greater depth to the zone of lime. The Palouse soils are dark brown to black, high in content of silt and humus, and of granular structure. They usually range from 10 to 100 feet deep over bedrock. Flecks and concretions of lime occur locally in the deep subsoil. Practically all of the virgin soil consists of silt loam, but erosion has exposed locally the yellowish silty clay or silty clay loam subsoil. The Palouse silt loam has a dunelike surface, with individual hills more than 100 feet above the hollows.

Other less extensive and shallower soils are the dark-brown to black grasslands of the associated foothill and mountain plateau type of the Southwick, Nez Perce, and Waha series. Timbered foothill soils are the light-colored loessial Helmer series and the residual granitic Moscow series. Bottomlands usually consist of narrow strips of dark-colored alluvium, frequently gravelly or stony.

Most of the Palouse originally was covered with bunch grass, with a little sagebrush along the western edge and scattered patches of timber, mostly ponderosa pine, along the eastern border. Farming has been going on from 40 to 60 years, with wheat as the principal crop. The common practice has long been to alternate wheat with summer fallow. In recent years, field peas, both green peas for canning and mature peas for market, have been substituted rather generally for fallow over an important extent of the more humid section.

About four-fifths of the land is in farms. Of this, nearly three-fifths is devoted to crops. About half the area consists of scabland, pasture, and range. In some localities, nearly all the land is under the plow. The average size of farms varies from about 1,200 acres in the lower rainfall sections to about 300 acres in the more humid localities.

As the result of erosion, nearly a fifth of the cultivated area has lost all or most of the topsoil. Also, large losses in nitrogen and carbon have occurred.[1] The eroded soil is partly deposited on lower slopes, where it is not needed and over highways (Fig. 316) and sometimes on the streets of towns.

[1] Sievers, F. J., and Holtz, H. F. The Significance of Nitrogen in Soil Organic Matter Relationships, *Bull.* 206, Washington State Coll. Agr., 1926.

Under the summer-fallow system of producing grain, the land is plowed in the spring, kept free of weeds through summer, worked down

FIG. 316.—Soil eroded by melting snow deposited over an important highway. Southeastern Washington. (*Photograph by Soil Conservation Service.*)

FIG. 317.—Erosion on bare, summer-fallowed land during snow-melting season. (*Photograph by Soil Conservation Service.*)

to a seedbed of fine tilth in the fall, and planted. Sometimes, depending on fall rains, planting is delayed until spring. In either case, the land is

TABLE 47.—EFFECT OF COVER ON RUNOFF AND SOIL LOSSES FROM PALOUSE SILT LOAM ON 30 PER CENT SOUTH SLOPE[1]

| Condition of cover during winter | Average runoff, per cent | | Average soil loss per acre, tons | |
|---|---|---|---|---|
| | 1933–1934 | 1931–1935 | 1933–1934 | 1931–1935 |
| Bare, untilled | 26.17 | 24.90 | 38.29 | 28.55 |
| Winter-wheat seedlings on summer-fallowed ground | 19.48 | 12.30 | 29.58 | 16.79 |
| Spring-wheat stubble from continuous seedings | 4.53 | 3.21 | 3.96 | 1.66 |
| Winter-wheat stubble, summer fallowed previous season | 0.31 | 5.91 | 0.08 | 0.27 |
| Grass (continuous) | 0.01 | 3.69 | 0.001 | 1.47[2] |

[1] Pullman, Wash., soil and water conservation experiment station. (Year of measurement July 1 to June 30.)

[2] Most of this loss occurred during the first year of measurements when grass was not well established. Eliminating that year, the average annual loss from grass is 0.18 ton per acre.

FIG. 318.—Control of erosion on excessively steep upper slope, formerly cultivated, in the Palouse section of Washington, by planting to mixture of alfalfa and grasses. (*Photograph by Soil Conservation Service.*)

exposed to the winter precipitation season without an adequate cover of protective vegetation. These conditions give rise to a very serious hazard of erosion. With approximately half the cultivated land going through winter nearly bare of cover, smooth of surface, and the subsoil partially filled with water, runoff and erosion are greatly accelerated (Fig. 317).

Soil and water losses from bare, thinly covered, and fairly thickly covered (stubble) areas of typical Palouse silt loam, as measured at the Pacific Northwest soil and water conservation experiment station at Pullman, Wash., are shown in Table 47.

The extent of erosion damage for the entire Palouse, estimated on the basis of a reconnaissance survey, is summarized below:

| Type of erosion | Per cent of land affected | |
|---|---|---|
| | Cultivated | Pasture |
| Mainly sheet erosion: | | |
| Slightly damaged—less than one-fourth topsoil lost | 10 | 50 |
| Moderately damaged—about one-fourth topsoil lost | 10 | 25 |
| Seriously damaged—about half topsoil lost | 35 | 5 |
| Severely damaged—more than half topsoil lost | 25 | 3 |
| Affected by gullying: | | |
| Moderately to severely gullied | 30 | 40 |
| Affected by wind erosion: | | |
| Moderately to severely affected by removal or deposition | 40 | 35 |

The erosion problem on the cultivated lands of the Palouse is intimately related to the fallow system of farming, with its deficiency of cover and smooth surface condition of fields during the winter rainy and spring blowing seasons. In the more humid localities, the problem is also involved with a relatively high carry-over of soil moisture, which tends to minimize absorption of winter rainfall following summer fallow. Solution of the problem involves, among other conservation practices, minimum use of summer fallowing on the more erodible lands of humid areas, increased use of perennial grasses and legumes (Fig. 318), extension of crop-rotation systems of farming, utilization of crop residues, and cultivation aimed at the establishment of a rough (cloddy) surface condition.

Below are summarized some of the practices employed for conservation of soil and water:

1. Contour tillage under all conditions.

Fig. 319.—Upper, deep snow drift on brink of very steep slope of Palouse section of southeastern Washington; center, destructive slide and erosion caused by rapid melting of accumulated snow; below, snow evenly distributed over critical slope as result of tree and shrub plantings. (*Photographs by Soil Conservation Service.*)

2. Strip cropping where applicable.

3. Use of straw scatterers rather than windrowers on combine harvesters.

4. Utilization, wherever practicable, of stubble, straw, or other crop residues, which formerly were burned, to provide protection from washing and blowing.

5. Prevention of overgrazing of field stubble.

6. Extension of use of trashy fallow wherever adaptable (as turning under part of grain stubble, with part left aboveground to impede erosion).

7. Restriction of intensity of tillage to avoid excessive pulverization of soil, such as favors erosion by both wind and water.

8. Increased seeding of permanent grasses and legumes on steep erodible slopes and areas of shallow soil.

9. Increased use of rotations including grasses and legumes.

10. Extension of use of sweetclover on cultivated land as green manure.

11. Protection of hayfields from erosion and from drifting of snow by leaving part of crop.

12. Restriction of grazing in pastures in order to leave about one-third the season's growth for prevention of erosion and drifting of snow.

13. Planting of windbreaks to prevent wind erosion and accumulation of snow along the brink of steep slopes, where subsequent melting develops special hazards of erosion and sliding (Fig. 319).

14. Control of gullies with dams, sloping and seeding of banks, or filling and seeding.

15. Use of dikes and small dams to control silt and impound water for flood irrigation of adaptable lowland.

16. Use of interception or diversion terraces to reduce flow of water from long slopes and to prevent gullying.

17. Provision of adequate protection for outlets from diversion dikes and terraces.

18. Contour furrowing of pasture land for conservation of water and soil.

19. Improvement of pastures and ranges by reseeding and better management of grazing.

## Columbia River Plains

The Columbia River Plains include about 15 million acres in central Washington and north-central Oregon. Part of the northern extension is known as the *Big Bend area* because of its location within the great westerly bend of the Columbia River, in central Washington.

This subdivision of the Columbia Plateau embraces the relatively smooth plains country east of the Cascades associated with the Columbia

River and its tributaries. The area has been incised locally by these drainages. Deep and moderately deep valleys, gorges, and canyons are conspicuous features in various parts of the region. Grand Coulee is a rock gorge of outstanding prominence. Locally, the general plains aspect is interrupted by such prominent elevations as Yakima Ridge, Rattlesnake Hills, Horseheaven Hills, and Saddle Mountains. Much of the smoother part of the area is of undulating to gently rolling topography.

FIG. 320.—Severe erosion caused by downhill irrigation and excessive use of irrigation water on steep land in Kittitas County, Washington. In some fields from 2 to 3 inches of the rich topsoil, which averages only about 10 to 12 inches deep in many places, is washed off annually as the result of this practice; and the material fills up irrigation ditches and covers highways. (*Photograph by Soil Conservation Service.*)

Elevations vary from about 100 feet immediately along the Columbia River in the southwestern part to about 2,000 feet on the higher part of the plains proper. Some of the included elevations range up to 5,000 feet.

The climate varies from semiarid to arid, precipitation ranging from 3 or 4 inches at the lower elevations along Columbia River to about 12 inches at the higher elevations. Although nearly all the precipitation occurs during the winter, snowfall is light. Snow cover is frequently intermittent, sometimes melting with warm winter rains and sometimes evaporating without melting. Because of light rainfall and absorptive soils, runoff seldom occurs except from spring melting of snow when the ground is frozen. Water erosion, therefore, is less severe than in the Palouse. Soil blowing is a serious problem, however, especially in the spring.

The most extensive and important soils are the light-brown, friable loam and very fine sandy loam of the Ritzville and the Portneuf series.

Three types of agriculture dominate the area: the growing of wheat under dry-farming methods, irrigation farming, and grazing. About half of the 15 million acres comprised in the Columbia Plains is included in farms. Approximately a third of this is used for crops, some 500,000 acres being irrigated, and 2,250,000 acres dry farmed. Hay and fruits are the principal irrigated crops, and wheat practically the only crop grown without irrigation. Scablands (areas with frequent outcrop of basalt), hills, and mountains are used for grazing, principally sheep.

Erosion results chiefly from the summer-fallow system used in the production of wheat, excessive downhill use of water on irrigated slopes (Fig. 320), overgrazing, and the breaking up of remaining native stands of stabilizing bunch grass.

A reconnaissance soil conservation survey of the area shows the following approximate conditions of erosion:

| Type of erosion | Per cent of land affected | |
|---|---|---|
| | Cultivated | Range |
| Water erosion: | | |
| Slightly damaged—less than one-fourth topsoil lost | 13 | 40 |
| Moderately damaged—about one-fourth topsoil lost | 25 | 20 |
| Seriously damaged—about half topsoil lost | 25 | 20 |
| Severely damaged—more than half topsoil lost | 25 | 10 |
| Wind erosion: | | |
| Area affected | 61 | 42 |

Methods for controlling erosion are about the same as those employed in the Palouse. Legumes cannot be grown so successfully, however, so that trashy fallow and cloddy tillage must be depended on to a much greater extent. Maintenance of the organic supply of the soil can best be effected by growing perennial grasses and legumes in long rotations. Severe wind erosion has resulted from recent plowing out of remaining stands of native bunch grass for grain production, as in the Horseheaven Hills.

### Snake River Plains

The Snake River Plains comprise some 22 million acres in the drainage of Snake River in southern Idaho and adjacent parts of Oregon and Nevada. A large part of the area is of relatively smooth plains topography, dissected by a number of deep canyons. Along the outer margin, belts of foothills are included. Elevations range from about 2,000 feet near the confluence of the Owyhee and Snake Rivers to about 6,000 feet in the northeastern extension, near Yellowstone National Park. In the southeastern extension, portions of the Minidoka National Forest and Fort Hall Indian Reservation are included.

The climate is semiarid, precipitation occurring largely during late fall, winter, and early spring. Average precipitation varies from less than 10 inches over some of the included valley floors to nearly 20 inches in marginal belts near the mountains. Generally, thundershowers occur at rare intervals in summer. Along the southeastern border, under the influence of the near-by mountains, rainfall is sometimes heavy enough to produce local floods and considerable erosion as well as damage to valley lands by overwash from neighboring highlands. Snowfall is generally light over the entire area.

Light-brown to ashy-gray soils of the Portneuf series are the most extensive. Smaller bodies of Ritzville soil occur in the moister situations near the enclosing highland. Dry farming is confined principally to the Ritzville types and the more favorable tracts of the looser Portneuf. Soils of the latter group usually are so dry that crop failures are too frequent to encourage their extensive use for dry farming. But in the central part of the area, they are successfully utilized under irrigation.

Three principal types of agriculture are found: irrigation farming, cash-grain farming under dry-land methods of culture, and grazing. Of the approximately 23 million acres in the Snake River Plains area, some 5½ million acres are in farms. Nearly half the farm land is used for crop production. About two-thirds of this, or nearly 2 million acres, is irrigated. Along the lower Snake, near points of confluence with the Boise, Payette, and Weiser Rivers, irrigation farms are devoted largely to tree fruits, whereas upstream they produce chiefly specialty crops, such as potatoes, sugar beets, beans, and alfalfa. Stock ranches are located on the fringes of the main plain, where some sheltered mountain valley with water supply and adjacent to public domain or national forest forms the base of operations. A very large proportion of the Snake River Plains is used for livestock. Grazing districts organized under the Taylor Grazing Act cover most of the area. Operations on the more successful ranches have been based on favorably sheltered valleys, water supply, and proximity to summer range in national forests.

Erosion problems and adaptable measures of control are much the same as in the Columbia River Plains. Adjustments in practice are necessary to meet conditions of a shorter growing season imposed by the higher elevations. Greater emphasis on forest protection and management of grazing on the open range will be necessary, from the standpoint of both better forage and protection of reservoirs from silting.

### Oregon Plateau

The Oregon Plateau is a roughly defined problem area, comprising about 10 million acres of high, dry plateau country in central Oregon. Scattered mountains and ridges interrupt the prevailing comparatively smooth surface features. A number of areas are without drainage outlets, such as the country about Malheur Lake. The prevailing upland level is about 4,000 feet above sea level, but occasional prominent elevations rise from 2,000 to 3,000 feet above this general level.

Over most of the Plateau, precipitation ranges from 6 to 12 inches. This falls very largely as gentle rains and light snows during late fall, winter, and early spring. Local summer showers sometimes cause considerable runoff and some erosion on the higher lands. Some years, frost occurs in some localities nearly every month.

Extensive plateau and mountainous areas are too porous of soil, too shallow, or too rocky for any use except grazing. Around many of the lakes or basins, soil distribution is about as follows, from center outward: peat, successive flats of clay loam and fine sandy loam, frequently of high salt content; then benches or terraces of sandy loam and coarse sandy loam. The rougher lands are predominantly stony residual soils. In the western part, pumice sands are of extensive occurrence.

Agriculture is centered about the raising of livestock. The entire area is included in grazing districts organized under the Taylor Grazing Law. Most of the irrigated and dry-farm crops are part of a stock-feeding program, except on the pumice-sands area, where some potatoes and clover seed are grown. The area used for grazing is about eighty times that used for crops.

The principal erosion problem is associated with overgrazed range land. About half the range has lost more than a fourth of the topsoil as the result of erosion by water and wind. Grazing management, including adjustment to carrying capacity, rotation, and water development, offers the most practical solution.

## Blue Mountains

The Blue Mountains, together with Wallowa and other neighboring mountains and associated valleys, embrace something over 12 million acres in northeastern Oregon and adjacent parts of Washington and Idaho. Nearly one-half of the area is included in the Malheur, Ochoco, Umatilla, Wallowa, and Whitman national forests (5½ million acres). The higher mountainous country attains elevations of 6,000 to 9,000 feet. Valley-floor elevations average about 3,000 feet above sea level. Slopes are generally precipitous. Most of the mountain land is forested.

Yellow pine predominates on the lower southerly exposures, and Douglas fir on the more humid northerly slopes.

The mountains have a higher precipitation than the valleys—some 12 to 30 inches, mostly in the form of snow.

The principal valley soils include those derived from residual basaltic material, terraces and alluvial fans, wind-modified material, and recent alluvial and lake-laid deposits. The residual soils are dark- to light-brown silt loams and loams. The terraces and alluvial fans are occupied principally by dark-brown or brown clay loams and loams, generally gravelly in both surface and subsoil. The wind-modified soils are dark-brown sandy loams, loams, and silt loams. The valley-floor soils are brown to light-brown loams and clay loams, productive where well drained but often poorly drained and alkaline.

Farming is restricted largely to the production of supplemental feed crops on the irrigated lands of the valleys. During summer, livestock are grazed on the mountain ranges. Alfalfa and native grass hay are fed in winter.

Range land outside the forests has suffered considerably from erosion produced by overstocking. Also, some timbered areas have been similarly damaged. Placer-gold operations have damaged portions of the valley lands.

## Cascade Mountains

The Cascades include some 23 million acres of mountains and foothills extending across Washington and Oregon from British Columbia to California. The foothills on the western slope descend to elevations of only a few hundred feet above sea level. Except for a few lofty peaks, the main mountain mass is something under 10,000 feet in elevation. It is crossed, approximately in the center, by the Columbia River, at a level of about 100 feet above the sea.

The Cascade Mountains present an effective barrier to the prevailing moisture-laden winds from the Pacific. On the western side, annual precipitation varies from about 30 to 50 inches over the foothills to more than 10 feet at the crest; on the east, the corresponding range is about 12 inches in the foothills to about 60 inches at the crest. Snowfall ranges up to about 50 feet on the mountain tops. A larger proportion of the precipitation occurs as snow on the eastern slope, temperatures being milder on the western side. Accumulated snowfall is of considerably more importance to the country east of the mountains, because of its need for the more extensive irrigation operations on that side. Snow surveys on the eastern slopes are proving helpful to the agricultural interests.

Dense forests of Douglas fir and associated species occupy much of the western slope; on the eastern side, the forests are more open and for this reason much more extensively grazed. About 12 million acres consist of national forests, including the Chelan, Columbia, Deschutes, Mt. Baker, Mt. Hood, Snoqualmie, Umpqua, Wenatchee, and Willamette. Mount Rainier National Park and the Yakima and Warm Springs Indian Reservations enclose approximately $1\frac{1}{2}$ million acres.

The principal erosion problem is on the eastern slope, where both grazing and clean cutting in lumbering operations have contributed to local soil washing. Grazing management and improved methods of cutting are the answer to the main part of the problem. Complete cutting of entire watersheds undoubtedly accelerates runoff. Stream-bank erosion and deposition of erosional debris over alluvial farm lands during spring floods are said to have increased in recent years.

## Pacific Coast Area

The Pacific Coast area embraces some 21 million acres west of the Cascade Mountains, in Washington and Oregon. The Olympic Mountains, Puget Sound basin, and basins of the Chehalis and other tributaries of the lower Columbia River are included in the Washington sector; the comparatively low Coast Range, the Willamette Valley, and various other valleys, such as those of the Umpqua and Rogue Rivers, are included in the Oregon sector. Some 2 million acres are included in the Olympic and Siuslaw national forests and the Olympic National Park.

Precipitation in the Olympic Mountains section ranges from 80 inches at sea level on the western coast to over 140 inches on the higher mountain slopes. In the Puget Sound basin, the range is from 15 to 50 inches at sea level to about 60 inches in the surrounding foothills. On the westward slopes of the Oregon Coast Range, rainfall varies from about 70 inches at the coast line to more than 100 inches at the higher elevations. The range in the Willamette Valley is from about 30 inches at the interior to 70 inches in the foothills. More than three-fourths of the annual precipitation of the coastal area occurs during the six months of winter; midsummer is usually rainless. Snowfall is light in the valleys, and the soils seldom freeze.

On the rolling uplands of the Puget Sound area, the soils are predominantly sandy or gravelly loams, underlain by sand, gravel, or compact silty clay material. Drainage is frequently excessive because of the porous character of the soils, and erosion generally is not serious.

In the southern part of the Sound section, the soils occupy glacial outwash plains and contain much sand and gravel. Even the finer textured types require some irrigation for successful farming.

Stream-bank cutting and deposition of sand in the areas of productive alluvial land are the principal erosion problems of this section.

The principal soils of the Willamette Valley of Oregon, which covers some 5 million acres, are the alluvial lands along the streams (about 13 per cent of the area); old valley-filling lands (about 22 per cent); residual foothill lands (about 32 per cent); and rough, stony, and mountainous lands (about 33 per cent).

The most important of the alluvial lands are the silt loam or silty clay loam of the Newberg series. These occur immediately along the streams and are subject to stream-bank and overflow erosion. On the second bottoms or stream terraces, the Chehalis soils are the most extensive and productive. Generally, erosion is not a serious problem on these alluvial lands.

FIG. 321.—Orchard erosion, in the belt of rich, rolling country surrounding the Willamette Valley, Oregon. (*Photograph by Soil Conservation Service, March* 11, 1937.)

Of the old valley-filling soils, those of the brown, well-drained Willamette series are the most extensive and important. Two other groups, the Amity and the Dayton, include gray and dark-gray poorly drained soils overlying mottled, impervious claypan subsoils. The Dayton soils are known as *white land;* and the Amity, as *half white land.* The principal problem connected with erosion has to do with overwash from adjacent higher lands.

Two groups of residual hill soils are important, one derived from sedimentary rocks—Melbourne, Carlton, and Sites series—and the other from basaltic rocks—Olympic, Aiken, Cascade, Polk, Viola, and Waldo series. Some 500,000 acres of the first group occur in the Willamette Valley and additional areas in southwestern Washington. They are among the most erodible soils west of the Cascades (Fig. 321). The Melbourne soils, the most extensive series, are light reddish brown in the surface, with yellowish-brown subsoils overlying sandstone or shale. On the basis of sample surveys, it is estimated that about 80 per cent of the Melbourne soils has lost three-fourths of the topsoil by erosion. Tests indicate that the infiltration rate has been reduced about 50 per cent where excessive erosion has occurred.

Soils derived from basaltic rocks are of extensive occurrence, the Olympic, Aiken, and Cascade groups comprising more than a million acres in the Willamette Valley. These related soils are red or reddish-brown loams and clay loams with red clay subsoils overlying basalt. They are rolling to hilly, have good surface drainage, and for a period following clearing are markedly resistant to erosion. Those areas of the Aiken soils which have been cultivated several decades have lost more than a fourth of the topsoil, and the Olympic about a third.

Of the nearly 5 million acres of western Oregon in farms, about 1,750,000 acres are used for crops. Nearly a million acres of this are used for cultivated crops, including both field and orchard products. Much of this is rolling hill land of the Olympic, Aiken, and Melbourne soils, and erosion is becoming a serious problem on much of it. Willamette Valley is relatively densely populated. The average size of farms is 97 acres, with 32 acres devoted to crops. According to the Census of 1930, 22 per cent of the farm land was used for general farming; 13 per cent for fruit; 13 per cent for dairying; 20 per cent for part-time, or self-sufficing, farming; and 32 per cent for specialty crops, poultry, etc. Farm families have continued to enter the valley, recently at a rate of more than a thousand a year; and because new land has not been cleared rapidly enough to provide adequate land, the tendency has been to subdivide further existing holdings of cleared land. This appears to be increasing the use of steep erodible land.

Contouring, strip cropping, terracing, rotation of crops, and the planting of steep lands to a protective cover of vegetation are among the measures that promise relief for the increasing erosion.

Drifting sand dunes along the coast of Oregon and Washington present a special problem of erosion control, the solution of which, as demonstrated by work in northwestern Oregon, is to stabilize the dunes by the establishment of a soil-holding cover. Holland grass has given excellent results in the first steps of control. When this is established, sand movement is slowed down so that other grasses and plants come in to assist with complete control.

# Chapter XXXIX. Pacific Southwest Region

Wide variations in erosion, soil, relief, climate, and agriculture characterize the Pacific Southwest Region, embracing California, Nevada, and contiguous portions of Oregon, Idaho, and Utah. Its 205 million acres cover high rugged mountains, rocky, timbered, and brush-covered; timbered foothills and footslopes; smooth fertile valleys; sparsely vegetated desert plains; coastal terraces; alluvial fans; and wet lowlands.

In order properly to describe the region, its varied characteristics and erosion conditions will be treated under six major problem areas: the Great Basin, Sierran Area, California Valley, Coast Range and Lowland area, Sonoran area, and Southern California Coastal Plain area (Map 16).

In each of these broad areas, changing patterns of contrast occur within short distances. One soil type may be damaged severely by sheet and gully erosion on less than a 4 per cent slope, whereas another, used on the same farm for the same purpose, may exhibit only slight erosion, even on slopes up to 40 per cent. Acid soils dominate the western part of one coastal county (San Luis Obispo), whereas alkaline soils predominate in the eastern part. Snow-clad Mt. Whitney, with an elevation of over 14,500 feet, is 75 miles west of hot, dry Death Valley, some 300 feet below sea level. Less than 3 per cent of one county (San Bernardino) is in farms, the remainder consisting of sparsely settled mountain slopes and desert plains. Yet on this 3 per cent, nearly $35,000,000 of agricultural commodities are produced annually.

Climatic types range from alpine to sonoran. Localities with damaging frost in midsummer contrast with those having a year-round growing season; sections that have recorded 70 inches of rain in a single month contrast with those recording no "trace" of rain in more than a year. Xerophytic, mesophytic, and hydrophytic vegetation is represented in large and small occurrences; sparse desert shrubs and cacti, sagebrush, chaparral, grassland, marsh, deciduous and coniferous forests, all are present.

More than 270 series of soils have been established and mapped.[1] Most of these are pedocalic (alkaline) in character, although many are pedal-

[1] Mainly in California. Less than 1 per cent of Nevada has been covered with soil surveys.

feric (acidic); some even possess lateritic, podsolic, and renzinic attributes. Although some parts of the region are suffering serious damage through wind erosion, soil conservation, in general, primarily concerns the problem of man-accelerated water erosion. In varying degree, the effects of all classes of erosion are in evidence, ranging from surface stripping by sheet washing to spectacular "barrancas" 50 feet deep;

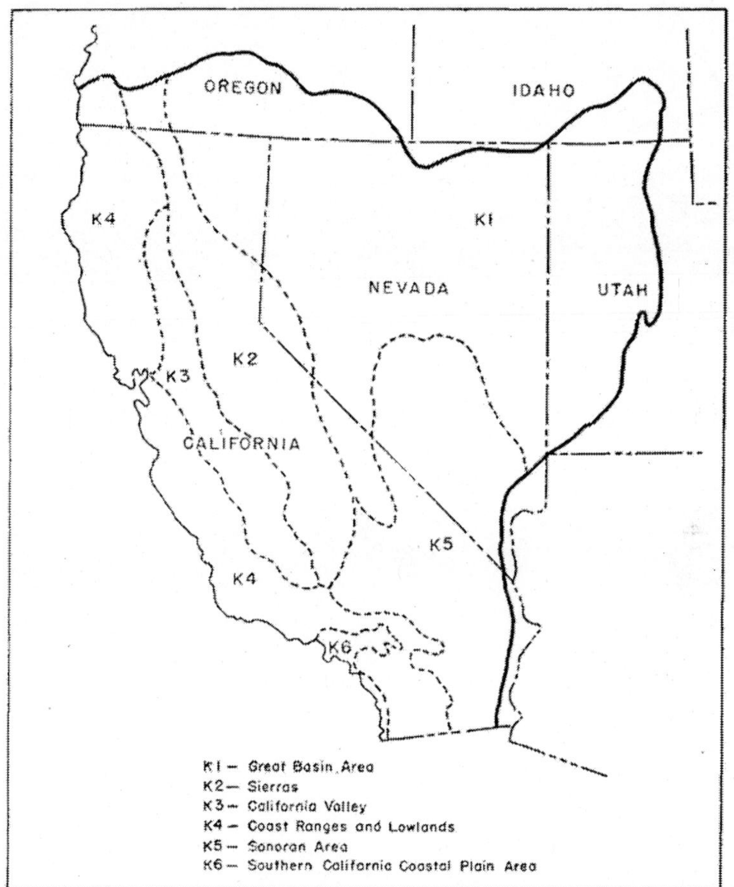


MAP 16.—Problem area of the Pacific Southwest Region. (*Soil Conservation Service.*)

from the little recognized wastage of soil and water through improper irrigation practices to confusing processes of erosion approaching geologic denudation.

A diversity of agricultural activities has developed on the basis of these varied physical conditions. Great expanses of open range land, both public and private, are used for the grazing of cattle and sheep. Extensive acreages are utilized for the production of grain, forage, and other field crops by dry-farming methods. An additional large acreage supports a relatively dense farm population, actively engaged in growing more than 200 commercially important crops by irrigation or other intensive practices. Small subsistence farms dot the sparsely settled districts; orchards and vineyards invade the bordering, steeper slopes; and in certain thermally favored localities, subtropical fruits are produced successfully as far north as the latitude of Baltimore or Denver.

Complete data as to current land use are not available, either for the region or for the individual problem areas. Based on state and county records as reported in the 1935 Census of Agriculture, and allocated in accordance with general knowledge of conditions and areal extent, a reasonably close approximation of the agricultural use of land is possible. This indicates that some 38 million acres, or 20 per cent of the land area in the region, are in farms. Of this, about 30 per cent, or 11 million acres, is classified as cropland, exclusive of plowable pasture. Approximately 66 per cent of the 175,000 farms are owner-operated, and 20 per cent are tenant-operated. The total value of all farm land and buildings approaches $2,500,000,000. This is a valuation of about $14,000 for an average farm of 220 acres, or about $65 an acre.

A large part of the farm products is shipped to consuming centers some 2,000 miles away. This handicap has largely been met by highly developed, intensive systems of production and marketing. Distinctive environmental conditions are utilized to the utmost, and many of the better favored localities are characterized by one or more closely related specialty crops.

Out of these circumstances has developed not only a distinctive landscape of environment and land use but also an economic and social pattern evidencing a high appreciation of the value of cooperative endeavor among the farmers themselves. Some 80 farmer-operated organizations are in existence. These have placed on the markets of the world a variety of standardized farm products.

Agricultural production in the Pacific Southwest region has an annual value of more than $750,000,000. Livestock accounts for more than $250,000,000; oranges, for $35,000,000; tame hay, for $30,000,000; and 40 other crops contribute between $1,000,000 and $15,000,000 each.

Arising out of the wide diversity of environmental conditions, land use, and resultant economic and social circumstances, the problems of soil and water conservation are equally diverse and numerous. The characteristic rainfall regime of the coastal belt, with its short season of high-intensity precipitation followed by 6 to 9 months of drought, added to the diversified farming methods generally practiced, is sufficient to illustrate the complexity of the problems and indicate that they may not, in all instances, be solved by simple, standard practices.

## Great Basin

This Great Basin, with an area of approximately 96 million acres, is bordered on the east by the Wasatch Range, on the west by the Sierra Nevada mountains, and on the north and south by a series of low ranges or divides. It is characterized by a succession of northeast-southwest mountain ranges with intermountain alluvial valleys, many of which represent sites of ancient lakes. Below the precipitous crests of these mountains, long, gentle fans of detritus descend to broad, flat-bottomed valleys, usually broader than the ranges, so that the extent of level and gently sloping land exceeds that of steep land.

About 46 per cent of the area has an average precipitation less than 10 inches, and only about 6 per cent has more than 20 inches. The eastern and southern portions of the Basin, and to some extent the western portion, are subject to erratic precipitation from summer convectional storms. These sometimes develop to cloudburst proportion and are locally very destructive. In the northern part, below zero temperatures are common during winter, extremes of 30 and 40 degrees below having been recorded.

Native vegetation ranges from desert species to those found only under alpine conditions, a variation due largely to differences in altitude and resultant climate. The growth at higher elevations is characterized by western yellow pine, scattering fir, spruce, and aspen, interspersed with sagebrush. Bunch grass and sagebrush predominate on some mountain areas. Over the floor of the Basin, sagebrush, greasewood, and other desert plants are the characteristic growth. The piñon-juniper type of vegetation dominates many of the higher mountain areas.

Most of the Great Basin consists of gray desert soils, such as (1) residual shallow soil of the hilly to mountainous blocks; (2) deep, permeable soils derived from old water-transported material, usually with strong lime accumulation, occurring on broad gently sloping alluvial fans extending downward from the base of the mountains; and (3) soils formed of recent water-transported material, on the valley floors. In addition to these broad groups, shallow highly erodible soils occur on dissected elevated benches of lacustrine origin, and heavy soils high in alkali are found on bare desert flats.

Approximately $9\frac{1}{2}$ million acres, or 11 per cent of the total area of the Great Basin, are included in farms; and about $2\frac{1}{2}$ million acres, or 3 per cent, consist of cropland. National forests (Dixie, Fremont, Humboldt, Modoc, Nevada, and Toiyabe) contain about $7\frac{1}{2}$ million acres; an area of 4 million acres is in railroad land grants; and about 6 million are in Indian and other Federal reserves. The balance, about 66 million acres, is in public domain, of which some 40 million acres are in established grazing districts.

Farming is confined almost wholly to the lands close to waterways with a supply of irrigation water. Dry farming is practiced to only a minor extent. Since farming is dependent on water, concentrations of agricultural activity occur in the better watered localities near the foot of the Wasatch and the Sierra Nevada Ranges. Elsewhere irrigated land is widely scattered.

In these two concentrated areas, general farming dominates, with dairy, crop specialty, and truck farming as the major types. Even in these areas the ranging of sheep and cattle is an important enterprise. Throughout the remainder of the area, livestock production predominates.

Range and farm are closely integrated and mutually interdependent in this subdivision of the Pacific Southwest region, the farm being headquarters for all operations. Hay and pasturage are provided on the farms for livestock during winter and other periods when range forage is not available or when grazing would be detrimental to the ranges. Farms without livestock sell hay to neighboring farms that carry livestock and so are indirectly dependent on animal industry. The ranges supply feed during the grazing season, which normally extends from early spring to late fall. Although in some instances range operations can be carried out wholly on the range, the most effective procedure involves combining the use of range and farm.

The population of the area is about 650,000, largely concentrated along the eastern and western borders. Much of the interior is very sparsely inhabited. Settlement, which began about 1847, took place along those streams which provided a readily available supply of irrigation water. Various localities today have inadequate supplies of water. In the absence of water measurements during early settlement, many streams were over appropriated for direct flow. This, together with insufficiently developed supplies, inadequate storage, and inefficient diversion facilities, accounts for present shortages. In addition to the major streams, minor streams, springs, and artesian supplies are scattered widely over the Basin, each furnishing water for one or more livestock farms. The total of these small, scattered water supplies, although constituting only a small part of the visible water, is the key to the economic use of a considerable part of the vast area of range land.

Most of the winter precipitation on the high mountains falls in the form of snow, and it is on these limited natural storage areas that most of the Great Basin depends for summer stream flow and replenishment of the underground water supply. Most of the flow issues from mountain streams as flood water in May and June, whereas the maximum need for irrigation is during July and August. A great need is for adequate reservoir capacity to store unused spring runoff for use during the growing season.

Over a large area in northern Nevada and adjacent Oregon, as well as Utah, the drainage is internal, or of the basin type; that is, it eventually disappears in the ground locally. Great Salt Lake, for example, receives drainage from the surrounding country but has no outlet.

Recreation and wildlife activities are desirable land uses and form important sources of income in a number of localities. Lakes and streams of the eastern slope of the Sierra Nevada and streams falling from the western slope of the Wasatch and other ranges are well stocked with fish. They attract thousands of anglers. The Bear River Wildlife Refuge on Great Salt Lake and the Malheur Refuge in Oregon are two of the larger refuges in the United States for migratory water fowl. Eleven organized Grazing Districts, administered by the Division of Grazing under the Taylor Grazing Act, comprise 40 million acres of public domain devoted to grazing.

Erosion is a problem of varying seriousness over most of the range land as well as on considerable sloping irrigated land. Much of the cropland is on the flood plains of streams, adjacent terraces, and alluvial fans, where the slopes are commonly gentle and erosion usually is not serious. Here the situation can be controlled by better leveling for irrigation, by construction of adequate headgates and turnouts, by the designing and building of proper channels for waste-water collection, and by the installation of proper drainage outlets. Stream-bank erosion is severe in many localities and will require various structural treatments, as drops and checks, jetties, and tetrahedrons, together with some vegetative plantings. To a considerable degree, successful use of such measures depends on control of erosion over adjacent range lands.

Erosion is active in some degree on nearly all the range land of the Great Basin. Over considerable areas, sheet washing has removed a large part of the surface soil from the uplands, and gullies have incised numerous lowland areas as well as portions of the uplands (Fig. 322). Usually, range-land erosion has been preceded by depletion of the native plant cover through overgrazing.

The value of farm land depends to a considerable extent on the method of utilizing adjacent range; returns from the former usually decrease with impoverishment of the latter by overgrazing and resultant erosion and water wastage through accelerated runoff. Farm income thus reduced makes increasingly difficult the maintenance of necessary structures and a balanced program. A decadent rural area results.

The control most needed on both private and public land is a grazing program that conforms with the carrying capacity of the range. Such a program may involve deferred and rotation grazing; correct seasonal use by the type of livestock to which the range is best adapted; or, in some

Fig. 322.—Gullies that were cutting headward into this valuable area of Nevada meadow have been controlled by a series of masonry "drops." When not protected, numerous areas, such as this, are completely ruined by the encroachment of gullies and consequent lowering of the water table and replacement of meadow grass by worthless rabbit brush. (*Photograph by Soil Conservation Service.*)

Fig. 323.—Small check dams of this type control gully erosion caused by "return" irrigation water, and facilitate more uniform distribution of water to lateral ditches. (*Photograph by Soil Conservation Service.*)

instances, exclusion of all livestock for varying periods, depending on the condition of the particular tract. To effectuate the program may necessitate such supplementary practices as fencing, development of stock-water facilities, small dams for gully control, contour furrowing, and water pocketing or spreading (Fig. 323).

The Soil Conservation Service, during its limited period of operation in the Great Basin, has demonstrated measures for control of stream-bank erosion, gully and water-table control on farmed areas, and canal protection by overpasses and diversion channels (Fig. 324). On range lands, the advantages of water spreading by means of grade ditches are

Fig. 324.—Black willow cuttings and rock jetties are used to control bank cutting along the Virgin River, Nevada. (*Photograph by Soil Conservation Service.*)

being demonstrated. The Panaca CCC project affords an excellent demonstration of water conservation and erosion control by the use of small dams and water pockets.

Legally constituted soil conservation districts, now being formed, present increased opportunity for control of erosion on private lands through better community cooperation and more effective relationships with Federal and state agencies.

Despite an inadequate water supply, floods are of common occurrence in the Great Basin. In some of the more critical areas, such as those along the base of the Wasatch Range, erosion and floods have severely damaged farms, highways, railroads, and other public works. To a lesser degree, similar floods occur throughout the Basin. Not only have farm lands and public and private improvements been damaged, but the carrying capacity of the range has been reduced. For many farm areas, loss of range resources means economic ruin, whereas for other areas it means an entire readjustment of farming operations.

## Sierran Area

Rugged, snow-capped mountain masses, ranging from 2,000 to more than 14,000 feet above sea level, characterize the Sierran Area, which extends from the Tehachapi Mountains, in Kern County, California, northward to the somewhat lower and less rugged Cascade Range, near the California-Oregon line. The area lies almost entirely in California. The intermediate and lower slopes are covered with dense forests of

FIG. 325.—Dense growth of coniferous forest typical of the Sierran area, California. (*Photograph by Soil Conservation Service.*)

ponderosa pine and other conifers, with occasional grassed areas (Fig. 325), whereas the higher elevations are sparsely vegetated or permanently snow-covered.

In the central and southern two-thirds of the area, the crest is within 15 miles of the eastern and about 40 miles from the western boundary. The westward drainage concentrates in a dozen or more large streams, such as Feather, Yuba, American, Mokelumne, Stanislaus, Tuolumne, Merced, San Joaquin, Kings, Kaweah, Tule, and Kern Rivers, fed by storm waters and melting snow. The resultant perennial flow of these streams has cut many spectacular, steep-walled canyons to depths of 1,000 feet or more below the adjacent terrain.

Relatively warm, dry summers and cold, wet winters characterize the climate. Temperature ranges from well below freezing in winter to more than 100°F. during the hotter middays of summer. The diurnal range is likewise great. Length of growing season varies with local conditions of topography, elevation, and exposure but seldom exceeds 200 days. Annual precipitation is high, particularly in comparison with that on the adjoining lowlands of the California Valley and Great Basin. Ranging, for example, from about 30 inches at Sonora at an elevation of 1,800 feet in southern California to more than 75 inches at Inskip at 4,800 feet, it tends to increase with elevation up to about 6,000 feet. At corresponding elevations, it is usually greater in the northern sector of the Sierras.

About three-fourths of the annual precipitation falls during the five winter months, November to March, usually in the form of snow. Owing

FIG. 326.—Severe sheet and gully erosion have forced abandonment of this formerly productive field, near Placerville, California. (*Photograph by Soil Conservation Service.*)

to prevailing low temperatures, this tends to accumulate in an extensive deep snow mantle, usually 10 feet or more.

Primary acid soils of semimature profile are typical of the Sierran subdivision. To the north, the parent rocks are largely basic igneous, principally basalt, and the Aiken clay loam is a representative soil type; to the south, granitic rocks are extensive and give rise to the Holland sandy loam and associated soils.

Throughout its full depth, the extensive Aiken clay loam normally is permeable and has a moderately high water-holding capacity. However, its permanent wilting point is unusually high, and the small amount of soil moisture available for plant use constitutes a problem of great importance to farmer and orchardist. Where the Aiken soils have the full protection of native cover, soil loss by erosion, even on moderately steep slopes, is imperceptible. Damage is generally severe, however, wherever the cover has been reduced or destroyed by fire, overgrazing, or clearing.

Most serious damage occurs on cultivated land (Fig. 326). The Holland soils are even more susceptible to damage by erosion.

Eleven national forests (Mono, Inyo, Rogue River, Shasta, Lassen, Plumas, Tahoe, El Dorado, Stanislaus, Sierra, and Sequoia) and four national parks (Lassen, Yosemite, General Grant, and Sequoia) comprise within their boundaries the greater part of the area. Lumbering and summer grazing of livestock, largely under national forest regulation, are two leading activities. Neither, however, exceeds the importance of

Fig. 327.—This burned area in which most of the trees and the ground litter were destroyed is exposed to accelerated erosion and runoff. Northern California. (*Photograph by Soil Conservation Service.*)

the area as a year-round source of water for both domestic use and irrigation.

Except for a few widely separated farms in the marginal belt of lower elevation, agricultural activities are concentrated chiefly in the small mountain valleys of Plumas, Shasta, and Siskiyou Counties. Some valley land is planted to alfalfa and small grain. Most of the remainder is in native grass, which likewise is irrigated and cut for hay to provide necessary supplementary winter feed.

Forest fires, overgrazing, lumbering, and farming have played important parts in disturbing the natural balance that formerly existed in the Sierran area (Fig. 327). In recent years, the Forest Service and other Federal agencies have recognized the problem and have endeavored to correct the situation.

The mountain snow is at once a great benefit and a hazard. If it melts slowly, a continuous supply of water is available for hydroelectric plants, irrigation districts, and metropolitan areas. When it melts quickly in spring or early summer, runoff is rapid, and erosion severe. Raging torrents boil down the mountain canyons to overflow reservoirs and expend their destructive force on intensively farmed lower lands in the Sacramento and San Joaquin Valleys. Adequate justification for soil and water conservation in the Sierra Mountains could be found solely in the protection afforded to these valley lands. However, there is great need for protecting and preserving the watersheds, forests, and recreational values within the mountain area.

Most of the area has a natural forest cover, protection and improvement of which constitute a first basic step in any program of control. Over a large area, a cover of grasses, weeds, and shrubs supplements the trees in reducing erosion and runoff. This cover should be given full protection, even though it may mean changes in season of use as well as in numbers of using animals. Riding, open herding, salting, fencing, and other range practices should promote a more uniform distribution of grazing. Portions of the area that cannot be used advantageously by animals, or where the damage to reproduction is too great, should be closed to grazing as long as these conditions continue.

For the scattered tracts of cultivated land, certain changes in cropping and cultural methods, augmented by simple structures, are the recommended procedure. On the erodible soils, excessive slopes, those over 15 per cent generally, should not be cultivated. In general, mechanical treatment for the protection of mountain slopes will not be economical. Structures, such as revetments, levees, detention and check dams, and other forms of mechanical treatment can be used effectively for control of flood runoff, bank cutting, and other types of stream erosion.

The economic implications of a program of flood and erosion control in the Sierran Area are far-reaching. As indicated, the greatest benefit to be gained is protection of the watershed values, although the recreational values are scarcely less important. It would not be inconsistent with benefits to assign to downstream areas a portion of the cost of certain large upstream structures and works having measurable protective effect on the lower areas.

## California Valley Area

The great Interior Valley of California, comprising 12 million acres, is a troughlike depression, about 450 miles long and more than 50 miles wide in places. It encompasses almost as great a variety of physical conditions as are present in the entire Pacific Southwest region and is one of the most extensively farmed areas in the region.

For adequate presentation, important differences in the physical conditions pertinent to local problems of soil and water conservation necessitate subdivisional treatment of the area.

The lower *Sierran slopes and foothills* extend in a narrow belt from a point north of Redding southward to the vicinity of Bakersfield. Crossed by numerous major and minor streams, a few of which occupy deep canyons, this subdivision is characterized by a succession of transverse interstream ridges, the tops of which are smoothly rolling to hilly. These ridges, sloping toward the valley through moderately steep gradients, merge at the lower levels with the upper alluvial plains. The lower slopes are covered with grass and oaks; the intermediate portions, with grass, brush, digger pine, and oak; and the higher, better watered sections with dense forests of pine and fir. Many of the ridges have better air drainage and higher rainfall[1] than adjacent lowlands and have achieved distinction in the production of early deciduous fruits. Irrigation is necessary during the dry summer season, in many instances water being diverted and sold for this purpose after it has passed through hydroelectric plants. An important part of the lower slopes is dry-farmed to grain. Both orchards and grain fields, however, occupy a relatively small acreage, more than 80 per cent of the foothill area being in native cover and utilized for grazing.

As in the adjacent Sierras, soils of the Aiken series, derived from basic igneous rocks, predominate in the northern portion, whereas soils of the Holland series, derived from granitic rocks, prevail in the southern portion. The most serious damage to these lands has occurred in the irrigated, clean-cultivated orchards that have been square-planted on slopes up to 25 or 30 per cent. Despite the justifications offered for this practice, better cultural methods must be adopted if these lands are to be protected. Irrigation and other operations along slight gradients approximating the contour will go far toward reducing erosion. A complete, permanent cover of grasses or legumes in the orchards will entirely stop it, but this frequently involves a question as to the adequacy of water supply.

The *northern valley floor* extends from the vicinity of Redding southward to the Sacramento-San Joaquin delta. Bordered on the east by the Sierra Nevada foothills and on the west by the Coast Range, this subdivision of the California Valley includes the lowland portions of a dozen counties. The eastern and western sides are bordered with belts made up of a series of coalescing fans. The central axis of the valley rises almost imperceptibly (about 1 foot per mile) as far north as Red Bluff, above which point the gradient steepens. A wider range in temperature char-

[1] Oroville, Auburn, Placerville, and Auberry have average annual precipitation of 37.07, 31.93, 41.47, and 22.29 inches, respectively.

acterizes the northern end of the area, which likewise has a higher rainfall.[1] A great variety of crops is grown. Some localities have achieved recognition as important producers of prunes, peaches, grapes, hops, rice, and other special crops. Irrigation agriculture dominates the floor of the valley; the higher, bordering alluvial fans are dry-farmed to small grains or used for grazing.

The soils are almost exclusively of the secondary group, derived from water-transported (valley-fill) material. A wide variety of recent, or immature, soils occupies the lower lying valley plain; and older soils, many of which have heavy, compact subsoils, dominate the steeper marginal slopes around the valley. Although most of the older secondary soils generally occur on slopes of less than 15 per cent, they are moderately susceptible to erosion because of relatively shallow depth to impermeable subsoil. Both overgrazing and clean cultivation result in loss of soil from the sloping lands, with consequent depositional damage to somewhat more valuable, lower lying areas.

The *Sacramento-San Joaquin delta* comprises a distinctive and agriculturally important area consisting principally of sea-level peat located at the confluence of the Sacramento and San Joaquin Rivers, some 40 miles east of San Francisco Bay. The two rivers and their distributary channels interlace this 250,000-acre tract, dividing it into a hundred or more interfluvial units, or "islands." The central portion of each is at or below sea level, with the water in adjacent channels several feet higher. Surrounding each island, and giving it a saucer-like configuration, is a higher stream-built ridge of mineral alluvium.

By means of strong protective levees along these ridges, and adequate drainage and irrigation systems, almost the entire delta has been brought under cultivation during recent years. Asparagus, sugar beets, potatoes, celery, onions, and other crops, to the value of nearly $30,000,000, are produced annually by intensive methods of both hand labor and machinery.

Fire is the chief hazard to farming activities. Like wind and water in the case of mineral soils, it is a destroyer. Once they are burned, there is no practicable method of restoring these valuable organic soils. Fires in the delta are of two types: accidental conflagrations and "cultural burnings." The latter are a long-established practice, designed to give better tilth and prevent weed and pest infestations. However, each cultural burning lowers the land surface between 1 and 6 inches, and each constitutes an additional step toward ultimate destruction of the inflammable organic soil.

[1] Redding, Red Bluff, Chico, Marysville, and Sacramento have average annual precipitation of 37.32, 23.56, 19.59, and 18.50 inches, respectively.

A novel problem in soil conservation is presented by this area, and appropriate corrective measures need to be developed. The elimination of cultural burnings and the inauguration of better irrigation and fertilization programs are indicated.

The northern and east-central portion of the *San Joaquin Valley floor* lies immediately south of the delta and includes most of the more important agricultural land in seven counties (San Joaquin, Stanislaus, Merced, Madera, Fresno, Kings, and Tulare). At the north, this area extends across the valley, reaching from the Sierra Nevada foothills to the base of the Coast Range. Southward, however, it covers only the eastern half of the valley, the boundary separating it from the even more arid southern part of the San Joaquin Valley swinging diagonally across the central trough. Coalescing alluvial fans form moderately steep, broad plains along both the eastern and the western margins. These grade imperceptibly into the basinlike alluvial bottom of the valley trough. The longitudinal axis of the valley, like that of Sacramento Valley to the north, has little gradient; within a distance of some 200 miles, the elevation increases less than 400 feet, an average grade of about 0.03 per cent.

This unit is of great agricultural diversity and importance. Irrigation agriculture is highly developed wherever water is available, most of the normal stream runoff being impounded in reservoirs to meet water requirements. Vineyards; orchards of deciduous, citrus, and subtropical fruits; and other intensive perennial plantings are numerous and extensive. Dairying and alfalfa production give distinction to the environs of Modesto and Turlock, and citrus to the Porterville district. Cotton is rapidly dominating the plains to the south of Fresno. Owing to the extent of nonirrigable lands, grain production by dry-farming practices has long been important. Also, the grazing of sheep and cattle is an important seasonal activity on these lands, the livestock usually moving into the adjacent foothills and high Sierras during summer and fall.

The climate is characterized by hot, dry summers and mild winters. Precipitation decreases from about 15 inches in the northern part to less than 10 inches in the southern, the greater part falling during winter. The southern end is marked by high temperatures, much sunshine, and aridity. These conditions have favored the maturing and curing of several important crops, particularly raisin grapes.

A wide diversity of secondary soils characterizes this area. The parent materials range from recent deposits to old valley-fill beds. Some of the older soils are characterized by rocklike hardpan in the subsoil. The older soils dominate the upper, more steeply sloping plain; and the younger types characterize the less sloping to nearly flat, lower alluvial plain. The former have many pedalferic attributes, whereas the latter are definitely pedocalic. The San Joaquin series is extensively developed on the upper plain and may be considered representative of the more erodible group of older soils.

Although the San Joaquin and the other old soils generally occur on slopes of 15 per cent or less, they are markedly susceptible to erosion. This is primarily due to their shallow topsoil and dense, relatively impermeable subsoil. Overgrazing and the prevalent practice of clean cultivation have increased the erosion hazard and have led to serious losses of soil. Through deposition, many lower lying areas of more productive land have been severely damaged. A definite program of better farming practice and erosion-control work—one in keeping with land values—is needed.

The southern, upper part of San Joaquin Valley is definitely arid, being situated within the "rain shadow" of the central Coast Range. This area includes the western part of Fresno, Kings, and Kern Counties. It is characterized by wide variations in diurnal and seasonal temperatures and by an annual rainfall of less than 10 inches.[1]

The dominant physiographic feature is the broad, moderately sloping piedmont plain, marked in its upper, steeper part by sparsely vegetated low hills and ridges and, in its lower part, by intermittent lakes. The native vegetation is xerophytic, typified by short-lived annuals and low desert shrubs. Petroleum is a chief source of income, although relatively large acreages of land in the vicinity of Bakersfield have been brought under irrigation and successfully utilized for cotton, alfalfa, and other crops. However, the greater part of the area has no adequate supply of water, and the grazing of livestock is practically the only agricultural activity.

The soils are highly calcareous, mainly secondary in origin, and dominated by the Panoche series. Because of their high absorptive capacity, these soils are not particularly susceptible to erosion. They have, however, a pronounced tendency to puddle and pack. In localities heavily tramped by livestock, the infrequent rains of high intensity usually result in serious erosion.

The *northern coast-range foothills* extend along the eastern margin of the Coast Range, between Trinity and San Benito Counties. Separated from the ocean by high mountain ridges, this narrow belt of foothills is characterized by a relatively low rainfall[2] and by high summer temperatures. For the most part, the foothills are characterized by a cover of grass, with scattered oak and pine. Some of the shallower and drier portions have a cover of chaparral. The greater part of the belt is held in

[1] Coalinga, Dudley, Bakersfield, and Maricopa have average annual precipitations of 6.55, 6.40, 5.62, and 5.54 inches, respectively.

[2] Fruto, East Park, and Guinda have average annual precipitation of 21.33, 17.14, and 21.48 inches, respectively.

large units, utilized mainly for grazing or dry farming to small grains. Some small areas have been planted to vineyards and deciduous orchards.

Soils of the Altamont series are characteristic of this foothill belt. They are best suited to grazing. Moderate erosion is noticeable in many localities, and land slips are of rather frequent occurrence. Heavily overgrazed areas, as well as those plowed for winter grain, show serious losses through erosion.

The *central coast-range foothills* represent a southern projection of the northern foothill subdivision. Extending southward to Los Angeles County, it includes portions of San Benito, Monterey, San Luis Obispo, Santa Barbara, and Ventura Counties. This area occupies an even more pronounced "rain-shadow" belt and is characterized by a relatively low annual precipitation[1] and high summer temperatures. The uplands support a cover of grasses and herbaceous annuals, with scattered areas of chaparral and forests of oak and pine. Large land holdings prevail; and although a few minor plantings of grain and other crops are made on the better situated lands, the area is utilized primarily for the grazing of beef cattle and sheep.

The Altamont soils occur throughout the northern foothills, whereas the Kettleman soils occupy the drier, southern part. The latter, derived from sandstone and shale, have developed under conditions of low rainfall and sparse cover of grass and weeds. Although they absorb and hold water well, their shallowness and thin cover expose them to serious erosion during the occasional heavy rains. These soils are of limited value even for grazing; and if they are so used, care must be taken to preserve enough of the native cover to protect the soil from washing during winter.

Generally, throughout all six subdivisions of the California Valley area, precipitation is of high intensity, with resultant severe sheet washing and gullying on both the moderately and the steeply sloping lands. In several localities, such as the intensively farmed delta lands near the city of Stockton and the sandy areas near Turlock and Arvin, soil blowing is a problem.

On the important mixed annual and perennial grass range and the dry-farmed hay and grain lands on both sides of the valley, the severe sheet erosion and active gullying are largely due both to lack of winter cover and to depletion of organic matter. General and severe overgrazing of the annual vegetation leaves the soil practically bare in many places at the beginning of the rainy season.

Continuous single cropping on dry-farmed lands has so depleted the soils that they are excessively erodible, the late-planted cover of cereal for hay or grain providing an inadequate protection, even on moderate

[1] Antelope Valley and Pattiway have average annual precipitation of only 8.20 and 9.33 inches, respectively.

slopes. Many large grain fields occupy long slopes, a condition especially conducive to gully formation.

The intensively developed flat central plain primarily requires protection from flooding and deposition. This can best be effected by stabilization basins and dams, downstream and upstream; and by an extension of grass, shrub, and tree cover in the upper areas. On the areas being damaged by wind erosion, protection likewise should include provision for a satisfactory ground cover. Water conservation by such methods as contour listing and furrowing is needed to enhance plant growth over practically the entire area, as are terracing and gully stabilization to prevent further dissection of farm and orchard land.

Once excellent grazing land in the vicinity of Fruto and Stony Creek is being destroyed rapidly by erosion. Deferred and rotation grazing, and possibly pasture furrowing, supplemented by gully control and planting, would effect marked improvement of present conditions. In the vicinity of Exeter, Lindsay, and Porterville, erosion damage is taking place over important areas of diversified agriculture, including orange groves and grazing.

If soil and water conservation is to become a fact, and soil productivity is to be maintained, areas of intensively operated farms and orchards must be protected by such measures as cover crops; the use of green manures or mulches; rotation of crops; and contour planting, cultivation, and listing.

Methods of deep tillage and furrowing or listing are most effective when the contour is closely followed. Such methods of moisture conservation, in combination with increased vegetative cover, can be extended effectively to the heavier soils and steeper slopes. Holding rainwater on the land by these and other practices will, in most instances, provide sufficient moisture to produce a better protective cover. Contour planting of orchards and vineyards is necessary on sloping land.

The Soil Conservation Service has two soil and water conservation projects in the California Valley area. These have demonstrated the basic importance of vegetative cover and the frequent necessity for supplementing such cover with other control measures, determined by soil character, density of cover, and degree of erosion as the result of past treatment.

At the Placerville demonstration project, located near the eastern edge of the Sierra Nevada foothills, the value of a permanent sod crop in irrigated orchards on steep slopes has been shown. This cover effectively controls erosion, promotes penetration of water, and is reported by farmers to reduce cost of crop production. Today, there is an important trend toward the greater use of permanent cover crops in irrigated orchards of all kinds throughout the state (Fig. 328).

At the English Hills erosion-control project, many areas have been producing above the level of sustained yield, largely because too much emphasis has been placed on maximum immediate returns from the land.

Fig. 328.—Permanent cover crops of this kind (grass in 9-year old avocados) aid in the control of erosion by winter runoff and irrigation water. California. (*Photograph by Soil Conservation Service.*)

Fig. 329.—A portion of a severely eroded 900-acre fruit farm shifted to pasture and supplementary feed production, for sheep. English Hills section, California. (*Photograph by Soil Conservation Service.*)

Much of the inherent productivity of the soil has been lost, and a lower sustained-yield level is all that can be expected under continuation of present farm practice. Had due consideration been given to proper erosion-control practices, much of the land now severely damaged might well have continued in use on a sustained-production basis.

The concept of soil and water conservation is based on sustained productivity of the land. To accomplish this, changes in land use are frequently necessary. Examples of such changes are fairly numerous. For example, at English Hills, several hundred acres of deciduous orchard trees were removed, and a livestock enterprise set up (Fig. 329). Changes from field crops to grazing have also been successfully made here.

Any effective flood-control program for this area will require large-scale treatment of watersheds to protect and increase vegetative cover, together with the building of retention reservoirs to retard runoff and reduce peak-flood flows. Part of the Central Valley has insufficient water for irrigation of all agricultural land.

## Coast Ranges and Lowlands

### Northern Part

Rising rapidly from the seacoast, across a narrow belt of marine terraces and low hills to the steep, rounded mountains of the interior, the northern part of the Coast Ranges and Lowlands area is characterized by high rainfall, dense forest cover, and highly erodible soils. The typical virgin cover consists of a dense growth of redwood, fir, and broadleaf trees, with a thick undergrowth of shrubs and other perennials. Lumbering already has removed much of the original timber; a large part of the present forest cover is second growth. Gradually, the agricultural population is increasing.

Annual precipitation, three-fourths of which usually occurs during winter, ranges from about 30 inches at Santa Rosa (elevation 167 feet) to 109 inches at Monumental (elevation 2,750 feet), near the Oregon line. Fogs are common in summer. Because of the climatic conditions, cut-over lands normally revegetate rapidly.

Most of the valley lowlands of the coastal belt and part of the gently sloping uplands are used for dairy farming, the dominant agricultural activity. Lumbering, still the most important industry, has moved inland along with a scattering of subsistence farms. The heavy growth of fir and pine on the higher hills and mountains in the central and northern parts of the area is largely included in five national forests (Siskiyou, Shasta, Klamath, Trinity, and Mendocino). In them, lumbering and grazing are conducted under Forest Service regulations.

The principal soils are derived from sandstones and shales; most of them are of acid character. Hugo clay loam is the most extensive type, except immediately along the coast, where Melbourne clay loam predominates. Both types erode readily where bared of cover. But for the rapidity of vegetative recovery, much more land would be severely eroded. This is shown by the damage that has occurred in a number of fields and overgrazed areas. Sheet and gully washing of an advanced stage has

Fig. 330.—Serious erosion in a clean cultivated square-planted vineyard near Ukiah, California. (*Photograph by Soil Conservation Service.*)

Fig. 331.—Annual ditches and cover crops are proving very effective in controlling erosion in deciduous orchards of California. (*Photograph by Soil Conservation Service.*)

affected a considerable number of orchards, vineyards, and diversified farming areas in the four southern counties (Mendocino, Lake, Sonoma, and Napa). Vineyards are especially in need of erosion control. General practice has been clean cultivation, both across and down the slopes. This has resulted in loss of soil and depletion of organic matter, together with markedly reduced yields (Fig. 330). Conservation measures recently initiated in a number of vineyards in Mendocino and near-by counties, with emphasis on contour cultivation and contour planting, are showing promising results. Annual cover crops with annual ditches or basin listing in orchards and, in some instances, in vineyards are necessary for best control. Large acreages of steeply sloping, clean-cultivated hill land, now farmed to various field crops, should be shifted either to improved pasture or to woodland. Where soil conditions are suitable, pasture furrowing retards runoff and improves the grass. Burning of range land should be stopped, and farm woodlands improved by planting and proper management.

Work of the Soil Conservation Service CCC camp at Sebastopol has demonstrated the value of annual cover crops, augmented by annual ditches (Fig. 331), for control of washing in orchards. Here a luxuriant cover has been developed on lands that had been severely impoverished by erosion.

Under the depleted conditions of much range and farm land, rapid runoff follows storms of cloudburst character; and destructive floods occur on the Eel, Russian, and other rivers. Application of conservation measures to the lands within these watersheds should retard a large part of this runoff and aid materially in reducing flood damage.

### Central Part

The central part of the Coast Ranges and Lowlands area, including the densely settled district around San Francisco Bay and extending southward to encompass the important agricultural valleys of Santa Maria and Lompoc, in Santa Barbara County, is marked by a complex physiographic pattern of high mountain ranges, low hills, and intervening lowlands.

The climate is of the Mediterranean type, with a somewhat narrower temperature range and lower rainfall than in the northern part of the problem area. Precipitation varies greatly, depending on location, the average annual rainfall ranging from about 10 to 55 inches. Winter is the season of highest precipitation, January being the wettest month. Heaviest rainfall occurs over the higher mountains north and south of San Francisco and on the coastal slopes of the centrally located Santa Lucia Mountains. These highlands generally support relatively dense stands of redwood and other forest growth and are becoming important recreational areas. The lower rainfall of valley and plain is evidenced by a native cover of grasses and live oak, whereas chaparral is the characteristic vegetation in the even drier localities to the south.

Under the stimulus of important local markets, most of the valley lands and many of the steeper slopes have been brought under cultivation. Owing to the limited precipitation in localities of arable land, irrigation is generally necessary for all but winter-grown crops. Where water is available, a wide variety of deciduous fruits, nuts, and field crops is produced. Lands without water for irrigation are usually dry-farmed to small grains or utilized for grazing, principally for beef cattle. Fairly extensive areas are also used for grapes and deciduous fruits under dry-farming conditions. In the southern part of this central sector, favored by nearly frostfree winters, garden peas are a common winter-grown crop, whereas dried beans are an important summer crop.

Of the more than 9,000,000 acres in this central subarea of the Coast Ranges and Lowlands country, some 5,500,000 acres (58 per cent) are in farms. Of this, nearly 1,700,000 acres (18 per cent of the total) are classed as cropland. The 25,000 farms average 220 acres in size, including about 70 acres of cropland. This is the third most important of the major problem areas of the Pacific Southwest. Total value of all farm land and buildings exceeds $400,000,000; the average farm valuation is nearly $16,500, or about $75 an acre.

Most of the land is privately owned, public land being largely restricted to the relatively small acreages in national forest and in state and national parks. Specialty farms predominate, although considerable diversification and even extensive operations are prevalent in some sections. Napa, Sonoma, and other northern valleys include important vineyards, orchards, and poultry farms; Ignacio and San Ramon Valleys specialize in walnuts and vegetables; Livermore Valley produces principally grain hay and livestock; Santa Clara Valley, prunes and apricots; Pajaro Valley, apples and vegetables; lower Salinas Valley, lettuce, other truck crops, and sugar beets; and central and upper Salinas Valley, livestock, hay, pink beans, and almonds. Santa Maria and Lompoc Valleys are largely devoted to the production of truck crops, sugar beets, dried beans, and flower and vegetable seed. Most of the seed used for planting mustard as a cover crop in California orchards are grown in Lompoc Valley, in rotation with small grain.

The principal soils are those of the Yolo, Botella, Rincon, Los Osos, Cayucos, and Arnold groups. Because of their good water-holding capacity, the Cayucos and Los Osos are only moderately susceptible to erosion. In spite of this, however, considerable damage by erosion has resulted from overgrazing and improper farming practices. In marked contrast, the soils of the less extensive Arnold series are highly susceptible to erosion. Some spectacular examples of sheet and gully washing are found on this shallow, sandy soil (Fig. 332). Most of the severe erosion of the Arroyo Grande vicinity has occurred during the past 10 to 20 years, or since the beginning of large-scale use of this highly erodible land for winter peas.

Most range land is overgrazed, and little is being done to retard the resultant erosion. Many farms are inadequately covered with vegetation during the rainy season, and no provision is made for removing excess rainfall without damage to the land. Ranges are generally in need of grazing management to maintain a proper ground cover, involving adjustment to carrying capacity, contour furrowing where soil type permits, development of water holes, and installation of check dams. The

FIG. 332.—Continuance of improper land use takes its toll. This abandoned field and farm house are all that remain of a once productive farm near Arroyo Grande, California. (*Photograph by Soil Conservation Service.*)

upland areas are devoted principally to livestock production; and to achieve maximum returns over a period of years, it is necessary that land-use programs for crop production be closely integrated with those for the range. The natural forage of highest value is composed predominantly of annual plants. This is a circumstance that necessitates an even more carefully planned grazing program than in the case of a perennial cover. In order to relieve the range during periods when native forage is not available in satisfactory amount and quality, a supplementary supply of forage must be produced. This can be accomplished by devoting those parts of the cultivated area better suited for pasture and forage crops to the production of feed.

The erosion problem cannot be wholly solved by protection of range land alone. Adjoining cultivated land will require the use of such measures as cover crops, strip cropping, contour furrowing, tree planting, and mechanical structures in order to provide defense for the varying types of land used for various crops.

Cover crops in deciduous orchards serve two purposes. They provide protection during heavy rains and add organic matter to the soil. Basin listing and annual furrows add to the effectiveness of cover crops in checking erosion and promote infiltration (Fig. 333). In many instances, this increased moisture supply more than meets the need of the cover crop. In addition to the need for annual and, under some conditions, permanent cover in orchards and vineyards, crop rotations are helpful aids in controlling erosion on these lands. Rotations may include cereals, legumes, and grasses, such as will provide supplemental feed and build up the supply of organic matter in the soil.

Fig. 333.—Basin listing in a deciduous orchard near Watsonville, California. (*Photograph by Soil Conservation Service.*)

The work of the Soil Conservation Service in this section has demonstrated: the basic importance of cover crops; the necessity of combining cultural and vegetative aids for conservation of rainfall; the feasibility and need of strip-crop rotations; the effectiveness of properly utilized crop residues or stubble; the necessity for control of surplus rainfall; the value of pasture furrowing; and the definite correlation of soil, slope, and erosion as determinants of proper land use. As a basis for determining correct procedure in adjusting grazing to carrying capacity, a range survey recently was completed in this section through cooperative action of the Soil Conservation Service, Forest Service, AAA, Farm Security Administration, California Agricultural Experiment Station, and Division of Grazing. (In this survey, effective integration of range survey and soil conservation survey techniques were coordinated.)

## Southern Part

Precipitous mountains and rolling to gently sloping intermountain valleys, with a cover of small trees and shrubs locally termed *chaparral*, characterize the southern part of the Coast Ranges and Lowlands area. The series of parallel transverse ranges, together with narrow intervening valleys, extends southeasterly from Point Conception to culminate in San Bernardino Peak (10,666 feet), near San Bernardino. Here they join the north-south Peninsular Ranges which extend beyond the Mexican boundary. The altitude of San Jacinto Peak, in this group, is 10,805 feet. The elevated peaks and higher crests of San Gabriel, San Bernardino, and San Jacinto Mountains support little or no vegetation and are characterized by bare, rocky slopes. The more arid exposures, chiefly those facing the Sonoran Area, have a partial cover of sage and other desert shrubs.

Drainage of the mountains is along steep-gradient tributaries of such major streams as the Santa Ynez, Santa Clara, Los Angeles, San Gabriel, and Santa Ana Rivers and numerous shorter streams. All these are subject to heavy flood flows of short duration. They have built up relatively steep fans at the points of emergence from their mountain canyons.

The major portion of the area lies within the boundaries of the Los Padres, Angeles, San Bernardino, and Cleveland national forests. Although grazing and recreation are important activities, the maximum benefits to society arise from the value of the mountains for irrigation and domestic water supply.

The Santa Ynez Mountains to the north comprise sedimentary formations that give rise to the Altamont and Diablo series, whereas the San Gabriel, San Bernardino, and other mountains to the south are predominantly characterized by granitic and metamorphic rocks from which the soils of the Vista, Fallbrook, and Holland groups have been derived.

Agriculture is of secondary importance. Of the total of 7,775,000 acres, only 1,251,500 are classified as *land in farms*, and 232,500 acres as *cropland*. Agricultural operations are confined principally to scattered farming in the intermountain valleys, such as Santa Ynez to the north and the broad valley and hill belt crossing central San Diego and Riverside Counties. Most of the cultivated land is dry-farmed to grain. Cattle and sheep are grazed on the stubble of harvested fields and over the extensive areas of open woodland range. Alfalfa and other irrigated crops are grown in a few favorable situations where water is available.

Hot, dry summers and cool, wet winters characterize the southern coast range area. The average annual precipitation is 20 inches; most of it occurs as rain during the five-month period from November to May. Snowfall is confined to the higher mountains.

The destructive flood of March, 1938, was preceded by two storms, the first on Feb. 27 to Mar. 1; the second, and more intense, storm on Mar. 1 to 3. Although rainfall in the valleys was comparatively light, 20 inches or more fell in the mountains during the five-day period—as much as 35 inches in places. Typical intensities measured instrumentally

Fig. 334.—A typical barranca cuts its way through valuable farm lands in Las Posas Valley. California. (*Photograph by Soil Conservation Service.*)

on Mar. 2, at Opid's Camp (elevation 4,380 feet) in the San Gabriel Mountains and at Lake Arrowhead (elevation 5,000 feet) in the San Bernardino Mountains were:

1.25 inches in 30 minutes.
2.00 inches in 1 hour.
10.89 inches in 8 hours.
15.06 inches in 24 hours.

Forest fires have been the greatest hazard to watershed efficiency in this locality. The chaparral cover is highly inflammable, and the long, dry summers greatly increase the danger of ignition. Experience has shown repeatedly that burns in this area are followed by severe erosion, heavy runoff, and disastrous floods. Moreover, it is difficult to bring back an efficient cover once an area has burned over.

Erosion is also a serious problem on cultivated valley land, especially in the southern portion of the area. Highly erodible soils, cultivation of excessively steep slopes, and improper land-use and cultural practices are all responsible causes. In many instances, deep, broad gullies (barrancas) are cutting headward into adjacent slopes and tending to destroy an existing equilibrium (Fig. 334).

In this southern area, protection from floods, maintenance of groundwater supplies, and permanent agriculture are dependent on soil and water conservation. On the basis of direct effects, land values and returns in many instances will not justify the costs necessary for control. However, where the benefits from the application of such measures effect a reduction of flood damage and debris deposition on the more valuable, lower lying areas, the work will be justified, and costs should be apportioned accordingly. It is only through such a coordinated plan and proper distribution of costs that a completely efficient program of soil and water conservation can be put into effect in southern California. And the cost frequently will be high, especially in those localities where large detention reservoirs are required. Moreover, the life of some reservoirs is likely to be comparatively short because of the difficulty of controlling outwash of debris—soil, gravel, and boulders. Following fires that destroy supporting vegetation, the slightest disturbance of the surface will, in some localities, set in motion avalanches of rock and soil which frequently travel to the foot of slopes, where the next flood moves the accumulated material rapidly downstream.

## Sonoran Area

Comprising some 33,000,000 acres in southern Nevada and southeastern California, the Sonoran area is characterized by extreme aridity; high temperatures; sparse vegetation; and a series of parallel rocky ridges, low mountains, and broad intervening plains. In its northern and central portions, the general elevation is about 1,500 feet above sea level, although some of the mountain crests rise to 10,000 feet, and two major depressions (Death Valley and Salton Sink) drop to nearly 300 feet below sea level. The southern part is of lower elevation and, although similarly crossed by ridges and low mountains, presents a more rounded and milder relief. Annual precipitation seldom reaches 10 and in most places is less than 5 inches. The rains are extremely variable; a storm of high intensity may occur at any time, or it may be localized or general. Yucca (*Y. brevifolia*), creosote bush, atriplex, bunch grass, and various short-lived herbs constitute the characteristic growth (Fig. 335).

The mountains are extremely rocky, with very little vegetation. The usual scanty growth offers little protection to the land. Much of the area is covered by "desert pavement"; sand dunes are common, and bare alkali spots occur in many places. Most of the palatable vegetation occurs in occasional waterways.

Among the extensive soils are those of the reddish Mohave series. In the northern part gray soils of the Daggett series predominate.

Farming is limited to a few areas in the northern part, where water is available for irrigation. Small areas watered by springs and artesian

FIG. 335.—Typical vegetation and general relief characterizing much of the Sonoran area, California. (*Photograph by Soil Conservation Service.*)

wells are farmed in a few localities. The greater part of the irrigated land is devoted to the production of forage, chiefly alfalfa, part of which is consumed locally and part shipped to dairies in the vicinity of Los Angeles. Although the ranges are dry and of low carrying capacity, they are utilized for grazing by both cattle and sheep whenever growth conditions permit.

Erosion in the northern section is almost wholly confined to range land. As vegetation is depleted in dry washes, erosion is accelerated by flood flows. In this way, a cycle of erosion is started which extends out from the main washes. Despite the limited rainfall, erosion is a serious problem in the southern part of the area. Desert torrents, resulting from rains of cloudburst type, constitute a definite hazard to structures as well as to soil. High winds from the West and Northwest during spring and summer cause considerable blowing of sandy land. Small hummocks and dunes move across fields regardless of individual efforts to check them.

The southern part of the area, comprising the Imperial and Coachella Valleys, has a growing season that permits production of a wide range of crops. Vegetables, alfalfa, and cotton are important crops. The most extensive plantings of dates in the United States are found in the Coachella Valley.

## Southern California Coastal Plain Area

In a broad curve from Point Conception to the Mexican boundary, the coastal mountains swing inland some 60 miles. Between these high-

Fig. 336.—Marked erosion contrast between clean-tilled orchard land and adjacent chaparral-covered slope: no erosion on latter; destructive washing (by a single rain) on the former. Southern California. (*Photograph by Soil Conservation Service.*)

lands and the ocean is a crescent area of marine terraces, coastal plains, narrow valleys, and low hills, constituting the Southern California Coastal Plain area. Much of the lower land once supported a native cover of live oak and grass, with interspersed desert vegetation on the more sandy soils. Intermediate elevations and lower mountain slopes are covered with chaparral. The mild climate has favored the development of an intensive agriculture. With a rainfall of 10 to 20 inches, irrigation is necessary for the more important crops.

The Altamont and Vista are the principal groups of residual soils; the Las Flores, Huerhuero, and Aliso occupy the older marine terraces;

FIG. 337.—Debris basins near foot of the Sierra Madre Mountains. These basins greatly reduced the flood damage in March, 1938. Southern California. (*Photograph by Soil Conservation Service.*)

FIG. 338.—Orchard on bank of a bench-terraced gully. With this method erosion on slopes up to about 40 per cent can be controlled effectively. (*Photograph by Soil Conservation Service.*)

and the Yolo, Sorrento, and Hanford are the more important series of soils of the recent deposits of the valleys.

The marine-terrace soils are characterized by friable sandy soils underlain at depths ranging from about 8 to 18 inches by intractable, heavy clay, impervious and resistant to penetration by water or plant roots. The surface soil is separated from the subsoil generally by a light-gray porous layer about 2 to 4 inches thick. The upper subsoil has a definite columnar structure, with grayish "biscuit tops" typical of solonetz soils. Erosion is very severe on land of this kind, since heavy rains quickly saturate the surface soil, especially where shallow, causing rapid runoff. Although the Vista soils are only moderately susceptible to erosion on gentle slopes, many steep areas wash severely where cultivated without protection. The major erosion problem is on land left fallow during

Fig. 339.—Basin listing and annual cover crop in a young, dry-farmed deciduous orchard at Cucamonga, California. Increased moisture stored in this manner more than compensated for that used by the cover crop. (*Photograph by Soil Conservation Service.*)

the rainy season. Much of the grain, hay, and bean land has suffered so seriously that it can be farmed for only a short time longer, without protection. In many localities, deposits of eroded material from this land constitute a serious menace to lower, more valuable lands.

An adequate cover of vegetation is the most important general need for soil conservation. Vegetation incorporated with the soil promotes moisture penetration and sufficient root development to help hold the soil in place (Fig. 336). Basin listing, combined with winter cover crops, is an important measure for control of erosion. Water conservation is necessary to establish proper vegetative protection.

The intensively developed agricultural areas on the lower portions of alluvial fans need protection chiefly from floods and deposition of outwash material. Debris basins (Fig. 337), dams, and stabilizing vegetation in the upper parts of the drainages; and provision for adequate downstream channels with freedom from constrictions are the most important measures for flood control.

Annual cover crops have proved very effective in citrus and other irrigated orchards. Permanent cover crops provide excellent protection, although they require more water and fertilizer than an annual cover. These requirements may possibly be reduced considerably by mowing. A permanent cover crop eliminates the need for cultivation, with its cost and attendant disadvantages, such as root pruning, breaking of branches, and baring the soil. Bench terracing, where the soil is suitable, has proved

FIG. 340.—Erosion typical of that occurring on the steep slopes used for lima beans in the Las Posas project, California. (*Photograph by Soil Conservation Service.*)

particularly adapted to citrus and avocado trees grown on steep slopes. These terraces consist of a comparatively flat top, or "tread," with a steep outer side, or "riser" (Fig. 338). The trees are planted at the outer side of the "step," or bench, and the relatively flat area between the row of trees and the toe of the next terrace above is furrowed to apply irrigation water. A cover crop is usually grown on this narrow strip. On the steep riser, a permanent cover of either annual or perennial vegetation is used for protection.

These terraces may be constructed with a grader or terracer before the trees are planted; or the trees may be planted in "grade-contour" rows across the slope, and the bench terraces built up by cultivation in a few years.

In the work on the soil and water conservation demonstrations of the Soil Conservation Service, attention has been devoted largely to orchard problems. The various methods referred to above have been applied effectively in a number of localities, on typical types of land.

On dry-farmed orchards and on cropland, basin listing, as a moisture-conserving measure, has been employed successfully (Fig. 339). In irrigated orchards, basin listing should not penetrate to a depth greater than other farm tools because of danger of pruning shallow roots. Listing may have some disadvantages where it is necessary to spray and where smudging is carried on. At Las Posas, definite results show the necessity of a change in land use in the case of the steeper cultivated fields. These areas (representative of much of the land used for lima beans) have been subjected to an extreme type of clean cultivation, and rapid loss of soil has resulted (Fig. 340). Many of the fields have lost all the topsoil within a period of 20 years. Now these steep, eroded areas are valuable only for grass and trees. With productive topsoil stripped off, yields are too meager for crop use. Aside from this, gullying usually begins on these lands at this stage of erosion.

# Chapter XL. Early Efforts toward Erosion Control

## Work in New England

At the time the Eastern seaboard of America was being settled, Old World farmers, confronted with a scarcity of land, were changing from a system of extensive to one of intensive agriculture. The work of Jethro Tull, in the eighteenth century, bore fruit, and European farmers turned from the old system of noncultivation to the new horse-hoe husbandry. The old fallow system of resting the land began to give way to crop rotations. By the end of the eighteenth century, agriculture in England had made great progress. But in the American colonies were found, as one traveler said, "the worst slovens in Christendom" whose only object was "to wear out one piece of land in order to plow up another."

When the Colonials settled in New England, they found a primitive system of agriculture. The only conservation measure used was planting a fish in a hill of corn to fertilize the soil. Captain John Smith reported that this practice was adopted to some extent by the colonists in New England but not in Virginia.

The colonists grew corn, wheat, and tobacco on the virgin land until it was so depleted it would produce no longer. Then they moved to a new piece of ground. This continued as a common practice until after the Revolution, a period of almost 150 years. (Even today, some operators continue the practice.) By 1763 a strip almost 100 miles wide had been cleared along the tidewater.

Danckaerts[1] reports that in Maryland in 1679 one piece of ground had been cropped for 30 successive years and still retained its fertility; but by 1700 land around Boston and New York was reported to be worn out; and by 1750 areas in Maryland, Virginia, Connecticut, and near Philadelphia had been "cropped to death." By the beginning of the eighteenth century wheat production was largely eliminated from eastern Massachusetts; and in New York it had been pushed gradually farther and farther west as the land was reduced in productivity.

[1] Danckaerts, Jasper, and Sluyter, Peter. Journal of a Voyage to New York, 1679–1680, Long Island Hist. Soc. *Mem.*, Vol. 1, p. 133.

No thought was given to maintaining the humus content of the soil; rotations were unknown; and the only meadows were natural ones. Labor and capital were scarce, and land was plentiful. It was less expensive, as Thomas Jefferson said, to clear a new piece of ground than to fertilize an old field. The wooden plows only scratched the surface; the rows were run up and down the hills, leaving the loose topsoil an easy prey to heavy rains. As time went on, the soil lost its humus, gullies appeared, and seldom was effort made to stop them. Most of the stock ran in the woods on the open range, and their manure was lost. That retained in the barns was not used; frequently farmers moved their barns so that they would not have to contend with the manure.

The natural grasses in the farming areas were soon eaten by the rapidly multiplying stock; the timber was cleared out, deadened, or burned; and in some areas, at a relatively early date, cattle were dying of starvation, and the people of some localities suffering for lack of fuel. English grasses, although introduced in the seventeenth century, were not used to any great extent. The grass seeds that were sowed came mostly from the sweepings of haymows and so were mixed with weed seed or were poor in quality.

The recurring floods probably were increased by the continued clearing of the land and by the shallow plowing which increased runoff. William Byrd reported a violent flood in Virginia in 1685, and John Bartram attributed an increase in both floods and erosion to the clearing of the land. John Bartram said:

> " . . . above 20 years past when y^e^ woods was not pastured & full of high weeds & y^e^ ground light then y^e^ rain sunk much more into y^e^ earth and did not wash & tear up y^e^ surface (as now). y^e^ rivers and brooks in floods would be black with mud but now y^e^ rain runs most of it off on y^e^ surface is colected into y^e^ hollows which it wears to y^e^ sand & clay which it bears away with y^e^ swift current down to brooks & rivers whose banks it over flows. . . . "[1]

The gullies began to assume the aspect of "yawning caverns," which afforded "not little inconvenience."[2]

Agriculture reached its lowest ebb in the first half of the eighteenth century. Farmers emigrated by the thousands, but many were deterred from going because of lack of roads and markets and fear of hostile Indians. By the middle of the century, certain signs of improvement appeared. Here and there, a farmer reported that he had shortened the fallow period or had substituted clover for bare fallow. Clover was grown in Maryland in 1635 and in New England in 1670 and was reported

[1] Bartram, John. Letter to Jared Eliot, in "Essays upon Field Husbandry in New England and Other Papers, 1748–1762," by Jared Eliot, p. 204. New York. 1934.

[2] Deane, Samuel. "New England Farmer." Worcester, Mass. 1790.

around New York and Boston in 1700. Timothy was known in New Hampshire in 1720; and Peter Kalm, the Swedish botanist and traveler, reported that he saw a number of clover fields in 1749.

The German settlers (100,000 were in Pennsylvania by the time of the Revolution), compared to the English colonists, were excellent farmers. At an early period, they grew clover and occasionally used it to improve the land. They turned under buckwheat for green manure, saved the dung from their stock, and transported rich mud from ditches and creeks to spread on the worn hillsides. A system of contour ditching was used by which excess rainfall was led downhill, in some instances for more than a mile.[1]

Substitution of rotations for the fallow system developed slowly. About 1750, the planting of clover became more common as a means of "resting" the land following one of the more soil-exhausting crops. After the Revolution, the practice of sowing clover or some kind of grass in the fields that were to be pastured became more general in the Middle Atlantic and New England States. The fallow system, however, was practiced widely throughout most of the eighteenth century, but the fallow period was shortened somewhat, and in many areas clover spread naturally.

A few men read works on modern agriculture and horse-hoeing husbandry by Tull and others, familiarized themselves with the changing European practices, and conducted experiments with clover and grasses to improve the soil. The most important member of this early group of conservationists was Jared Eliot, a minister and physician of Killingsworth, Conn., who contributed the first book on American agriculture. His first essay on husbandry was published in 1747 and was followed by six others. Many of the practices that he advocated are regarded as sound today.

At the time Eliot was writing, soil depletion had become a menace in Connecticut. He noted the constant emigration to the West because of worn-out land and deprecated the system that caused it. The fallow system, whereby the land was made a "sheep's walk" for 10 or 15 years, was criticized severely. The systematic wearing out of the land, Eliot felt, was equivalent to spending more than one earned. The new horse-hoeing husbandry introduced by Tull was recommended highly, provided plenty of dung was used. This was based on Eliot's own experience with a horse hoe that he had imported from England.

The keynote of Eliot's agricultural teachings was the use of red clover, which gave the farmer a hay crop and at the same time built up the land. One of his favorite rotations was: turnips, barley or oats with

[1] Kalm, Peter. "Travels in North America, 1748–1750." Vol. I, p. 162. New York. 1937.

clover, clover, and wheat. In general, he recommended clover for two successive years.

It was observed that some of the land was so poor that it would produce turnips no larger than buttons. To enrich it, Eliot suggested manure and clover. He also encouraged the building of more fences so that the manure could be saved and the neighbors' stock prevented from trampling and eating the provident farmers' pastures. Not only was dung recommended, but many other kinds of fertilizer, such as ashes, peat, limestone, creek mud, shells, clay, and other materials. He built a long pen in which he mixed soil, dung, and various farm wastes.

Eliot invented a drill that would open a furrow, plant seed, and drop manure in a single operation. In designing it, he obtained the aid of President Clap of Harvard and a village blacksmith. The drill was sent to William Logan of Philadelphia to test. It is possible that Eliot was influenced by Logan's experiments with rotations. Logan's rotation, which had some followers in Virginia and Maryland, covered a period of 9 years and included corn, potatoes, flax, wheat, wheat, winter barley, buckwheat with clover and timothy, and two crops of clover.

Eliot was also an advocate of deep plowing. He relates in one of his essays that an old farmer in East Jersey told him "about 10 years ago" about the benefits of deep plowing. Green manuring also was practiced by Eliot. Although he preferred millet, he also used buckwheat, oats, and rye.

It is doubtful if Eliot had much influence outside a very small circle, because farmers of his time did not read. Significant, however, is the fact that his "Essays on Field Husbandry" was more widely circulated than he had anticipated. Benjamin Franklin, a gentleman of "refinement and learning," ordered 50 copies, and there is some evidence that they were rather widely read. In a letter to Eliot in 1749, Franklin remarked[1]:

> "I perused your two Essays on Field Husbandry, and think the public may be much benefited by them; but, if the farmers in your neighborhood are as unwilling to leave the beaten road of their ancestors as they are near me, it will be difficult to persuade them to attempt any improvement."

Franklin had purchased a 300-acre farm and conducted experiments with clover and other grasses, no doubt using some of Eliot's ideas. He was more impressed, however, with plaster than with any other soil improver. He decorated one of his fields with it so as to form the words "This field is plastered." The following year, it was reported that the outline of the words he had written on his hillside in this manner was rich green, whereas the surrounding vegetation was of a lighter shade.

[1] Eliot, Jared. "Essays upon Field Husbandry in New England and Other Papers, 1748–1762," pp. 223–224. New York. 1934.

Franklin, like Eliot, advocated deep plowing, as is indicated by the following extract from "Poor Richard's Almanac"[1]:

"Plow deep while sluggards sleep
And you shall have corn to sell and to keep."

Eliot also corresponded a great deal with John Bartram, the Pennsylvania naturalist and farmer. The former used clover extensively, as well as creek mud, to enrich his hillsides but allowed his fields to stay in clover for 3 years at a time.

One of the most outstanding conservationists of the following generation was Samuel Deane. Like Eliot, he was a minister, but he also found time to familiarize himself with almost all of the authorities on agriculture, including those of ancient Rome. His little book "The New England Farmer," published in 1790, contained references to the best agricultural practices known to the world at that time.

Deane himself was a practical farmer and was disgusted by slovenly methods of farming. He felt that the best farmers were those with the fewest acres; that the plenitude of land was the curse of American agriculture.

Deane was not so confident of the value of clover as was Eliot; but though he doubted that it was better than other "grasses," it is significant that he recommended it in all his rotations. He suggested that fallow with deep plowing was as good as anything to renovate worn-out lands but admitted that Europeans had found that land could be improved while producing crops. Because of soil and climatic differences, he showed some of the same skepticism toward adopting European rotations in the United States. On the basis of the limited experience of the Colonials, he made certain tentative suggestions concerning rotations.

Fall plowing and deep plowing were highly recommended; even trench plowing was suggested for some soils. This was a practice in which the plow was run in the furrow to turn up additional soil—a type of subsoiling. One of the handicaps to deep plowing was the clumsy plow with its wooden moldboard. Deane suggested several improvements to reduce friction, among them an iron moldboard which would turn the ground more smoothly and with less effort. He advised against plowing up and down the hill, suggesting that it be done in only one direction across slopes, so as not to form a water furrow. An alternative practice which he called *ribbing* consisted of running furrows horizontally across the hill to catch rains and prevent washing and gullying of the hillsides. On tilled land, he said the furrows should be 3 feet apart; and on pasture land, about 9 or 10 feet.

[1] Franklin, B. "Proverbs from the Almanac of One Richard Saunders," p. 6. New York. 1908.

Probably the reason that Deane emphasized fall plowing was that he believed in building up the soil with green manures, such as barley, buckwheat, rye, peas, or oats. Other fertilizers were recommended, including dung, ashes, plaster, creek weed, leaves, stalks, trash, and waste of all kinds. In order that none of the liquid or solid manure should be washed away should the barnyard be located on a declivity, it was often advisable, he said, to mix clay with dung for sandy soils and sand with dung for clayey soils. Dung was the common denominator for the several composts and mixtures that Deane recommended. Rich mud which had been washed from the hills into the ditches and creeks was to be mixed with manure and restored to its former resting place. The banks of the ditches were to be planted in a grass with strong roots to prevent their caving and filling up.

The wasteful practices of the early agriculturists had denuded many of the hills of timber. In a few areas, this caused "sand floods" which piled against the fences and hedges. The drifting of sand near the coast had also long been a source of annoyance to the settlers. The town of Truro, Mass., incorporated in 1709, had suffered so much from sand drifting by 1739 that a law was passed prohibiting the grazing of stock in certain areas because[1]: "The sand was being driven, in storms and high winds, from the beach upon the meadows . . . a great part of the meadowland was already buried and useless for grass . . . the whole was likely to be covered with sand if the drifting were not prevented. . . ."

The same year a similar act was passed to protect the meadows of Plum Island in Ipswich Bay. Other acts were passed at intervals throughout the eighteenth century for the same purpose. In Provincetown, Mass., the drifting became so bad that the town was almost depopulated. Sand storms had been noticed in other colonies as well. Bartram had observed them on the coast of North and South Carolina. They were referred to in a geography of Maryland and Delaware published in 1807 as if they were a familiar occurrence. "It is not uncommon, in great droughts, when high winds arise, to see the sand raised from the fields . . . thirty or forty feet high, and carried a considerable distance. If corn fields be situated near the dwellings of planters, the drifting sand, in dry weather, and in high winds, enters every door and crevice, in like manner as drifting snow."[2]

For these blow areas, or "bald" spots, Deane recommended locust trees, noting that the falling leaves eventually would enrich the ground, forming a mold so that it would be capable of supporting vegetation and

[1] Kinney, J. P. Forest Legislation in America Prior to March 4, 1789, Cornell, N. Y., Agr. Exper. Sta. *Bull.* 370, pp. 358–405, 1916.

[2] Scott, J. "Geographical Description of the States of Maryland and Delaware," pp. 22–23. Philadelphia. 1807.

that the sand storms would cease. Although the small group of scientific and professional men read Deane, he had little influence on the mass of farmers.

## Work in the South before 1860

The system of Colonial American agriculture in the North was bad, but that in the South, particularly Virginia and Maryland, was worse. The agriculture of New England sank to its lowest level before the Revolution, but that of Virginia and Maryland had depreciated even more by the end of the eighteenth century. Tobacco for a long time had been the great staple crop. It was customary to grow it until the land was exhausted and then clear a new field. In this manner, tobacco culture spread through the tidewater section during the 150 years preceding 1789.[1] The Revolution had closed to the Colonials the tobacco market of the West Indies, and James Madison declared that after 1776 the Americans were greater slaves to the English than ever before. To some extent, wheat was substituted for this crop; but wheat was a land impoverisher, also. Considerable corn was produced, particularly on the land not rich enough for tobacco. Tobacco usually was grown in small patches, and the fields were manured or fertilized to some extent. But the corn and wheat fields covering larger areas were, as a rule, not fertilized.

Planters blamed their agricultural ailments on the lack of a market or the tariffs levied by the rising commercial class, which was beginning to dominate the Federal Government. In this they were partially justified; lack of transportation facilities and the various embargoes imposed during the Napoleonic period were important factors in the agricultural decline.

Thomas Moore of Maryland observed that erosion started the second year of cultivation and that it was caused by the system of shallow plowing, which was not over 4 inches deep. Land sterility was attributed not so much to soil exhaustion as the actual removal of the soil itself. His remedy was deep plowing, supplemented by manure and proper rotation of crops.

Another writer paints a vivid picture of the severe effects of erosion in Virginia in 1799[2]: "At that time the whole face of the country presented a scene of desolation that baffles description—farm after farm had been worn out, and washed and gullied, so that scarcely an acre could be found in a place fit for cultivation."

This writer goes on to say that the cause was a most horrible system of land butchery. The rows were run up and down the hill; the land was

[1] Hall, A. R. Early Erosion-control Practices in Virginia, U. S. Dept. Agr. *Misc. Pub.* 256, 1937.

[2] Craven, J. H. System of Farming, *Am. Farmer*, Vol. 15, p. 130, 1833.

scratched instead of plowed; and nothing was thought of rotations or soil improvement of any kind. This situation was general throughout most of the farming sections of Virginia and Maryland.

In some areas, agricultural improvement had begun by this time. This was due, in the first place, to improved market conditions, which made wheat growing profitable. Secondly, a remarkable group of public-spirited men lived in this section. Many of them were prominent nationally and internationally and had made great contributions in other than agricultural fields. Among these were Washington, Jefferson, Thomas Mann Randolph, and Madison.

FIG. 341.—The Randolph hillside plow of the type used on Thomas Jefferson's farm in the Virginia Piedmont, near Charlottesville. Inserts show moldboard (left) and share (right).

Washington, as a farmer, had long studied the problem of erosion. He had experimented as early as 1769 to determine if ridging or level plowing reduced erosion, had filled gullies on his farm with trash and brush, had tested the effectiveness of rotations, and had attempted to improve his land in various ways. Finding tobacco hard on the land, he determined to grow only what he needed. It is doubtful if many of Washington's ideas were put into practice before 1800. Like Jefferson, he failed to stem the tide of agricultural depression and could not, in the end, make his farm a paying proposition.

James Madison also analyzed the reasons for agricultural decay. He felt that the system of plowing employed in Virginia tended to press the soil down and make it more compact. Air was a good manure; therefore, the ground should be stirred often and should be left in ridges so more of it would be exposed to beneficial aeration. Since the destructive practice of plowing up and down the hills caused gullying, Madison advocated deep plowing on the contour as a remedy.

In the tidewater section there was a severe shortage of timber and fuel in Madison's time because the trees had been destroyed. Grass was insufficient to feed the cattle for more than two or three months during the year, and overgrazing of the natural pastures increased erosion. As a remedy, Madison suggested the planting of trees and the seeding of enclosed pastures in clover and timothy. He advocated the feeding of as much grass as possible so that the supply of manure would be increased.

More important and more influential than Madison was Thomas Mann Randolph, a distant relative and son-in-law of Thomas Jefferson.

Fig. 342.—Erosion on land that was once farmed by Thomas Jefferson near Charlottesville, Virginia. (*Photograph by Soil Conservation Service.*)

It is probable that some of the soil-conservation measures used by Randolph were Jefferson's ideas, and vice versa.

Randolph invented a hillside plow (Fig. 341) in 1808 and did much to popularize horizontal plowing. Horizontal plowing was not new in Virginia and had been practiced in some parts of Pennsylvania for many years. The principal feature of Randolph's plow was a reversible moldboard which could be shifted at the end of every row so that the dirt would always be thrown one way.

By 1810, Jefferson had become an ardent believer in horizontal plowing. He contended that it practically eliminated erosion and that little water ran off his son-in-law's fields, whereas those of his neighbors were washed and gullied by heavy rains. In laying off the furrows around the hill, the rafter level was used every 30 or 40 yards. The earth was thrown into beds 6 feet wide, with water furrows in between. These held the water that collected on the fields or forced it to drain off slowly. Since Jefferson's time, much of his land has suffered seriously from erosion (Fig. 342).

John Taylor of Caroline County, like Jefferson, was aroused over the increasing inroads of the commercial class on the rights of the farmers. He became an exponent of states' rights to protect the landholders. Taylor regarded agriculture as the basis of the whole economic system and recognized the undermining effects of tobacco cultivation. The central idea in his book "Arator" was the restoration of eroded soil by what he called *enclosing*. The system included the enclosure of all the arable land and a process of penning, bedding, or feeding, so that the maximum of manure, stalks, litter, and other matter could be turned under. He advocated contour plowing, deep plowing with four good horses, and turning under a coat of dry vegetable matter to create a cover drain which "vastly obstructs the formation of gullies in hilly land." Clover as well as artificial fertilizers, such as gypsum, lime, and marl, were recommended for thin land.

The work of these early soil builders had a salutary effect on Southern agriculture, and from 1790 to 1815 some improvement ensued. Abandoned farms were reclaimed and old eroded areas were planted in clover and treated to gypsum and manure. Fertilizers, though little used in the Carolinas, were widely used in Virginia and Maryland. These included marl, lime, plaster, marsh mud, ashes, fish, and almost every kind of vegetable matter, including oak leaves and pine needles. Green manuring also, was widely practiced in various parts of the country, and more care was taken of manures, especially dung. This was supplemented by the increasing use of crop rotations.

Horizontal plowing became quite common. It was reported that "thousands of farmers practiced it in Virginia, though few had hillside plows." The practice spread south to the Carolinas and west to the area near Natchez, Miss.

In spite of the adoption of soil-saving measures by some farmers, poor farming prevailed throughout the South in 1815. Many made no attempt to control erosion, and others were discouraged by partial failures. The state of agriculture, however, did not sink to its former low plane. The increase in agricultural societies, the relatively widely distributed periodicals, and the premiums offered by voluntary groups for excellence in agricultural essays were all important factors in continued soil improvement.

The old wooden plows were supplanted by iron plows; by cast-iron moldboards; and finally, in 1835, by steel moldboards, which gave the farmer, for the first time, a moldboard that would scour in clay soil. All the new plows were conducive to better and deeper plowing. In 1845,

Gideon Davis of Georgetown, D. C., made a subsoil plow. Subsoiling soon came to be generally recommended.

Farmers were warned that they should run furrows not on the contour but across the hill at a gentle incline in order to carry the water off gradually "and prevent the washing of the ploughed land into gullies."

Trench plowing was practiced in the South as well as in Pennsylvania. Some suggested that this method of deep plowing was less laborious than the older way. Two plows were used, one following the other in the same furrow.

In 1848, prizes for good farming were given by the *American Farmer*, a periodical published at Baltimore. Among the recipients was Colonel Nicholas M. Bosley. He had moved on to a gullied, washed, and worn-out farm in 1811, in the coastal plain section of Maryland, and in 12 years had used 60,000 rails for fences and 21,000 bushels of lime. He built up the farm by deep plowing and by using timothy and clover.

To fill gullies, the Southern farmers used trash, rocks, logs, and dams of various kinds. An occasional operator plowed down the steep banks. In Maryland, a special plow designed for this purpose was used by one farmer, who also ran water furrows 1 to 3 yards apart around the contour of the contributing watershed. In Tennessee, locust trees and herd's grass were planted in the washes. Hedges of brush were used on the eastern shore of Maryland to prevent ditch banks and creeks from caving. They were extended into the shallow water and held down with stakes.

Various methods of strip cropping were employed. This technique was not new, but it became more varied and better designed to prevent water erosion. In 1824, mention is made of permanent rows of grass grown horizontally across hills. On steep hills the rows were 6 feet apart; on gentle slopes a wider space was left. Winter reed grass and meadow oats were used because the plants grew in winter and would not interfere with the tilled crop.

In Maryland, crimson clover was used to prevent washing, since it did not have some of the objectionable qualities of the other clovers. Turnips, an integral part of the old English rotation, were grown and fed to sheep on the land. The many varieties of clover, long grown in the North, found their way into the Southern region, becoming a part of the various rotations adopted by some of the more progressive planters.

Possibly the most important development after 1820 was the rapid increase in the use of fertilizers throughout the entire South. In December, 1824, John Skinner, the editor of the *American Farmer*, introduced Peruvian guano into Maryland. This became very popular; and by 1850, the supply was not equal to the demand.

The person most responsible for the use of lime or marl was Edmund Ruffin, a farmer of Prince George and Hanover Counties, Virginia. After

"Arator" was published in 1813, Ruffin put into practice the principles of Taylor but was later convinced that the system was a failure. About 1818 he began experimenting, using marl containing fossil shells for fertilizer. This marl, found near his farm in large quantities, contained approximately one-third carbonate of lime. A liberal application of this material, along with manure, increased his corn yield 40 per cent. He also grew clover, cowpeas, and other soil builders. Ruffin's farm soon attracted attention and in 1832, when he became editor of the *Farmers' Register*, his essay on "calcareous manure" was published. His later expansion of this treatise was said to be the most thorough piece of work of the nineteenth century on the subject. A great many farmers were persuaded to marl their eroded farms. Ruffin probably was read more widely and had a greater influence on agricultural practice than any other writer of the period. Craven says that this influence can hardly be overestimated in the history of agricultural improvement of the section[1]: "The stories of changes in Prince George County, Va. and other eastern counties in both states often read like fairy stories. Fields once galled and gullied were now growing rank with clover; lands once abandoned, now brought forth abundant yields of wheat and corn."

According to Craven, the lands in eastern Virginia increased in value by 23 million dollars. Although in North Carolina in 1828 little horizontal plowing, rotation, or deep plowing was practiced, and numerous old fields were cut by deep ravines, a great deal of improvement was noted between 1845 and 1855. Ruffin's system was given credit for the change, but other practices probably helped bring about this improvement.

The hillside ditch was used widely throughout the South during the 20 years preceding the Civil War. Hillside ditching was practiced in the southern Roanoke Valley by Humberston Skipworth from 1815 to 1831. His ditches were 3 feet wide and 2 feet deep and were spaced from 20 to 50 or 100 yards apart. Skipworth used ravines at the ends of the ditches to carry off the surplus water. In 1830, this method was improved on by James C. Bruce of Maryland. His hillside furrows were protected by an embankment thrown up with the plow and the hoe. This may have been the beginning of the American type of field terrace. The rafter method was a popular one for spacing and grading ditches and field terraces. Naturally, many variations characterized the design of hillside ditches. This was especially true in construction of the outlets. Redtop was sowed occasionally in the outlets, or the drains were paved with rock. The subject of hillside ditching occupied much space in Southern periodicals for a long time, and at least one book[2] was devoted to it.

[1] Craven, A. O. "Soil Exhaustion as a Factor in the Agricultural History of Virginia and Maryland, 1606–1860," p. 143. 1925.

[2] Sorsby, N. T. "Horizontal Plowing and Hillside Ditching." Mobile, Ala. 1860.

The hillside ditch was unsatisfactory for several reasons. In numerous instances, the gradient was so steep that the ditches were gradually converted into gullies by channel erosion.

Hillside ditching was not popular in Virginia after 1850, although it was used widely farther south. Many of the old terraced hillsides in the South today are monuments to the era of hillside ditches, which continued to be used to a considerable extent until around 1900. The modern terrace has proved to be more efficient in controlling soil erosion. A few farmers, however, still use the hillside ditch, as in parts of the Savannah River section of the Georgia Piedmont.

By 1860, Virginia and Maryland had turned to the diversification of crops to some extent.

Much land in the Deep South was susceptible to severe washing. In the area around Natchez, Miss., for example, the fine-grained loess soil (Memphis silt loam) was so erodible that after a few years of cultivation, much of it was washed away. Here, cotton, instead of tobacco, was the crop whose cultivation resulted in so much soil washing. By 1842 the planters in parts of the loessial area of Mississippi, much disturbed about erosion, tried deep plowing. The conditions were little better in many other sections. In 1832, the farmers of Georgia and South Carolina were advised that only manure could save the sterile field from "our wretched system of agriculture." Sorsby, in his book published in 1860, noted hillside ditches in all the Southern States. This was the only thing he found to commend "the murderous system" of agriculture employed. Immense tracts of land had been thrown out of cultivation because of severe soil washing.

Before the Civil War the system of agriculture showed some signs of improvement. By 1854 contouring was used by many of the planters in Mississippi and other Southern States. Many of them used the eye to lay off the contour lines, but the more careful used a triangular spirit level.

The cowpea, used by Washington and recommended by John Taylor and others, was probably the main soil builder of the South. By 1845, John C. Calhoun had reclaimed a worn-out farm in South Carolina chiefly by the use of cowpeas and "guard" drains on his upland fields to prevent erosion. Near Natchez, Miss., some of the planters troubled with erosion sowed winter oats between the cotton rows. Bermuda grass also was used with success. A city ordinance of Natchez, in 1822, required that Bermuda be planted in the public square to prevent it from being gullied.

Bermuda was used to some extent throughout the South. The warm, moist climate was conducive to its rapid growth. It spread in many of the abandoned fields and was a problem in the cultivated ones. In Hinds County, Mississippi, shortly after the Civil War, it was reported to be "everywhere."

Although Southern agriculture improved, it was still far below that of the North. It was still enslaved by a one-crop system. Some land was reclaimed, and some improved, but agriculture had sunk so low in some localities that improvement was slow.

## Work in the North after 1800

Numerous travelers contrasted the agricultural practices of the North and the South before the Civil War. They invariably found agricultural improvement carried on in a greater degree in the North. In 1819, Maryland was contrasted with York and Lancaster Counties, Pennsylvania. Land sold for five to six times as much in Pennsylvania as in the more southern area. It was declared that it was as difficult to sell land in Maryland as to draw a prize in a lottery, although improved farms in Pennsylvania were in great demand. The Middle States, as a whole, had improved their lands with clover, with manure, and with many of the other practices already described as being used in the South.

At the time Samuel Deane was writing in New England, a number of public-spirited men were experimenting in New York and Pennsylvania. Robert Livingstone wrote an essay on sheep and experimented with lucerne (alfalfa), having, in 1793, 15 acres of it growing in New York. Another was John Beal Bordley who, although he lived in Maryland, was prominent in the Philadelphia Society of Agriculture in the 1790's and had considerable influence on Northern agriculture. He contributed a pamphlet on rotations (1784) and a book on husbandry and rural affairs (second edition 1801) which was, perhaps, the most complete American treatise on rotations up to that time.

Bordley's system was based on the Norfolk rotation of England. One of his favorite rotations consisted of: first year, buckwheat to be turned under; second year, rye and clover or timothy; and third year, clover. He favored green manures (many of his rotations provided for the plowing under of at least one crop) and the use of stock manure. Grass was important because it gave dung, and dung gave fertility. He admired the agriculture of the Pennsylvania farmers with some reservations, but, as other writers, condemned the soil-exhausting system of Maryland and Virginia. In general, it was agreed that the Germans of Pennsylvania were the best farmers. As early as 1753, a premium was offered in Pennsylvania for the best field of clover.

More important than Bordley was Richard Peters, a lawyer by profession but a practical farmer as well, who specialized in improved breeds of sheep and cattle, pioneered in soil improvement, and was the first president of the Philadelphia Agricultural Society. Peters published a pamphlet on the use of plaster in 1797, which had a wide circulation. It was said that this work caused old, worn-out estates to be made productive through the use of gypsum, clover, and grasses. He advised against the "overwhelming reliance" on plaster alone. Peters first observed the effects of gypsum on the Commons of Philadelphia in 1777 and after years of experimentation recommended it highly if used in conjunction with deep plowing, rotations, clover, grass, and green manure.

He was a friend of George Washington and corresponded with him on the subject of agricultural improvement. It was at Washington's suggestion that he sent out a questionnaire to farmers in connection with his research on gypsum, and he dedicated his publication on plaster to Washington.

Peters experimented with other soil-saving techniques as well. He tried John Taylor's system "repeatedly" and found that turned-under crop residues would not rot quickly but that green manure never failed to do so. Therefore, on this subject he differed from his "much respected revolutionary friend." Peters felt that clover grown on the land for a long period would impoverish the land. Some other "grass," or forage crop, preferably millet, should be substituted occasionally.

Liming was to be followed by a summer crop, then fall plowing followed by a cover crop, and then an application of dung as a preparation for wheat. Deep fall plowing was also practiced by Peters, which with regular alternations of tilled crops, pasture, and meadow would conserve the fertility of the soil.

Not only did Peters employ soil-saving methods himself, but he required his tenants to do so. An extract of his lease to tenants published in 1811 in the Memoirs of the Philadelphia Agricultural Society is as follows[1]: "A fixed quantity of *dung*, from the tenants' muck heap, may be taken if the landlord chooses . . . every kind of economy, as to timber duly to be exercised. The tenant is to procure, and sow *clover seed* and *plaister* of Paris. The landlord to pay half the cost of purchase. The tenant must leave the fields of winter grain sown or not with *clover seed*. . . . "

After 1800, Northern agriculture began to improve, partly as a result of the work of these men. More important influences, however, were the economic changes. Before this time, New England farmers were prodigal in their use of land. In 1801, the average Massachusetts farm, according to valuation returns, had 4 to 6 acres in row crops, 8 to 10 acres in upland grass, and about the same area in natural meadow. The remainder of the farm seems to have consisted of old, exhausted fields. The Connecticut tax list of 1796 distinguishes between clear and brush pasture, the latter according to Bidwell and Falconer being old, worn-out fields.[2]

[1] Peters, R. Leases to tenants, extracted from the lease for Belmont Farm, Philadelphia Soc. Promoting Agr. *Mem.*, Vol. 2, pp. 263–265, 1811.

[2] Bidwell, Percy Wells, and Falconer, John I. "History of Agriculture in the Northern United States, 1620–1860." Washington. 1925.

About 1820, the farmers began investing more capital in land improvement. The first four decades of the nineteenth century saw a change in the economy of the Northern farms that was almost revolutionary. Roads were extended gradually into the backwoods, and stream navigation increased. Between 1810 and 1840, a new industrial population came into being as a result of the industrial revolution. This created new markets and tended to break down the old, self-sufficing economy and bring about commercial farming. Trade with the West developed rapidly from 1800 to 1860. The North and Northwest were enabled to supply the Southern market and allowed the South to concentrate on a cash crop. Cotton factories were developed to take care of this trade.

Mechanical inventions on the farm also facilitated development. In 1797, Charles Newbold of Burlington, N. J., obtained a patent for a cast-iron plow. Jethro Wood improved the moldboard in 1819 and developed a plow not cast in one piece. In 1839, Samuel Witherow and David Pierce invented a moldboard that would bend the furrow slice so as to fracture it. Between 1800 and 1830, 124 patents for plows were issued by the United States Patent Office.[1]

The awakening in the field of agriculture also resulted from the rise of agricultural periodicals and societies. The *Columbian Magazine* (Philadelphia) published an article in 1787 that described experiments with clover and wheat and listed a 4-year rotation. After that time, articles on the improvement of worn-out lands were published occasionally. The Society for the Promotion of Agriculture, Arts, and Manufacturing, organized in 1791 in New York, discussed among other topics in its 13 meetings: improving poor land by sowing red clover seed, perennial grasses, experiments and observations on lucerne, and cast-iron plows. In 1788, the Philadelphia Society for Protecting Agriculture, recognizing the importance of soil conservation, offered a premium for the best discussion of various erosion problems, among them "the best methods of restoring old gullied fields to a hearty state."

Beginning with 1819, several agricultural journals were established which, no doubt, were important factors in agricultural improvement. The *American Farmer* edited by John Skinner was first issued Apr. 2, 1819. The same year, Solomon Southwick's *Ploughboy* began publication at Albany. The *New England Farmer* was published at Boston by Thomas Green Fessenden, with Timothy Pickering, David Webster, and John Lowell as contributing editors.

Grazing early became a main industry in some areas. By 1800, it was established in southeastern Pennsylvania and the Connecticut Valley. Connecticut seemed to have started improving earlier than the other New England States. Colonel Humphreys observed in 1816 that lands

[1] Agriculture of the United States in 1860. U. S. Census Off. 8th Census, Vol. 2, p. XVIII. Washington. 1864.

worn out 40 years ago were now productive. Manure, rotations, and artificial grasses had brought about this change. After 1800, dairy herds in many sections of New England increased in size; improvements were undertaken on the land itself; per-acre yields increased; fields were made smaller; and manuring, rotation of crops, and thorough plowing were practiced more widely. By 1855, a large percentage of the improved land was in timothy and clover. The rural population declined steadily. Marginal land and steep slopes (where erosion was worst) were abandoned when production proved unprofitable.

Thomas Green Fessenden served as a sounding board for all the progressive agricultural ideas of his time. In his book "The Complete Farmer and Rural Economist" (Boston, 1839), he gave all the modern practices, frequently quoting arguments from both sides when authorities disagreed.

Fessenden became convinced from his associations and observations that deep plowing was not beneficial in all instances and disapproved also of the custom of dressing lands in the fall with manure, because the rains washed away the manure. In some parts of Vermont and New Hampshire, Fessenden observed, the farmers no longer attempted to grow wheat. He felt that wheat should not be grown except in rotation with clover and that this should be supplemented by fertilizer—dung, ashes, or rich earth.

Jesse Buel, who established the *Cultivator* at Albany in 1834 and wrote the "Farmer's Companion," a book that went through at least six editions, personified the movement for agricultural improvement from 1820 to 1840. The principles, which he called the *new husbandry*, consisted of saving all fertilizers and wastes of the farm, rotating crops, plowing, and draining to retain the fertility and increase the profits. He criticized severely the old methods of successively cropping until the humus of the soil was exhausted and was one of the first to recognize the limits of artificial fertilizer. Exhausted lands must first be built up with animal and vegetable matter, he pointed out.

The weaknesses of Buel's principles were that they ran counter to the system of commercial agriculture by requiring that farms be self-sufficient. He insisted, however, that one of the chief causes of deterioration was the lack of money spent for farm improvements.

Buel stressed deep plowing as well as subsoiling and trench plowing. Everything possible should be turned under that would increase the humus content of the soil, including greensward, buckwheat, clover, and leguminous crops. Underdrainage also was advocated. This could be done by frequent plowing and lapping the furrow slice so as to leave a place for the water to flow.

Few of Buel's ideas were new, but he was successful in getting many of them put into practice where his predecessors had failed. For this reason, he was possibly the most important Northern soil improver before the Civil War.

The raising of livestock became increasingly profitable; fantastic prices were paid for improved breeds. Colonel David Humphrey of Connecticut and Robert Livingstone of New York imported sheep from Europe, and the raising of merinos became a fad. By 1840, the old race of inferior swine had been almost supplanted by superior breeds. Oxen were no longer used in the Middle Atlantic States but were retained by some New England farmers.

Hay became the staple crop in New England, New York, Pennsylvania, and New Jersey. According to E. Emmons, who wrote a natural history of New York, grass and clover, supplemented by plaster, were used to build up the wheat land, especially in western New York. In the North, red clover was the most popular cover crop and continued to be throughout the century, but white clover also was used, as well as other legumes and various grasses.

The use of fertilizer increased, and lime was said to have brought to life many farms in Pennsylvania and New York after 1810. Limestone and gypsum came into general use in Connecticut. Perhaps most representative was the report of Elisha M. Bradley of New York, in 1852, insisting that manure was most important, and deep plowing and subsoiling next, in the renovation of worn-out land.

Green manure also was used widely. Buckwheat straw was turned under on lands too poor for clover. Rye and oats were turned under as well. Emmons, who conducted an agricultural survey of New York in 1846, claimed that turf turned was almost as good for land as clover.

By 1850, legumes and grass husbandry were well established in the North, although lucerne, sanfoin, and the vetches were not popular. The New England system was to produce the maximum amount of hay. Emmons also recommended a rotation for New York wheat growers, which consisted of clover, treated with plaster, and an occasional year of fallow. By 1860, it was reported that rotations and deep plowing had worked wonders on the long-worn farms of New England. A common rotation was potatoes, oats, grass for 3 to 6 years. Manure or compost was applied at the time the grass was sowed.

Deep plowing and subsoiling were considered a part of the "new husbandry." A report to the Wheatland Agricultural Society of New York in 1833 emphasized fall plowing, deep plowing, and subsoiling. In 1839, deep plowing was recommended as the sole remedy to restore old fields in Massachusetts. In Lebanon, N. H., in 1852, it was reported that "many now plow deep and also manure." Deep plowing was practiced also near Rutland, Vt. The practice of plowing diagonally down the hill, inclining to the left hand and "returning light," was reported in Connecticut and

Massachusetts. Unless this practice was followed on sloping lands, it was said to be useless to apply manure, since it would wash away.

Soil-improving practices in New Jersey and Delaware were said to be decidedly inferior to those of Pennsylvania. Statements regarding the improved agriculture of the North usually excepted Delaware and occasionally New York.

James P. Mapes, an educated, practical farmer, was responsible for much of the improvement in New Jersey after 1847. He opened a school for agricultural education, which was responsible for a large increase in the use of soil-saving practices.

John Johnson, a Scotsman of New York, advocated and practiced crop rotation and grazing. The failure of wheat in New York had brought the farmers face to face with the problem of soil depletion. By 1855, due largely to the activities of Johnson, the agricultural decline was checked.

In Pennsylvania, the early German farmers were reputed to be the best in the United States and "a great blessing for this country." In addition to irrigation and hillside ditching, closed drains also were used on the hillsides to prevent gullying, and stone walls were built across gullies to check them. The lands were plowed horizontally, and various modifications of hillside ditches were used. In one instance, furrows were run diagonally across the hill. The 20 years before the Civil War saw subsoiling advocated and practiced widely in Pennsylvania, New York, and other northern areas.

It would be erroneous, however, to assume that Pennsylvania farming was perfect. There were poor farmers here as elsewhere. One Pennsylvania writer regretted that manure was used only on cropland and not on pasture. Here and there in searching the files of the *Farmers' Cabinet*, a periodical published in Philadelphia, one finds reports of poor farming and of gullied and washed lands, but such instances are rare as compared with the number of eroded farms reported in journals in other sections of the country.

By 1860, the agricultural economy of the North to some extent had been stabilized. Much of the land had more or less successfully weathered the normal cycle of frontier soil exploitation. The frontier had moved slowly westward, subjecting new agricultural areas to the destructive system of frontier agriculture. Although the farmers of the North from that time on had to face a tremendous western competition, which could undersell them because of cheap land and large-scale farming, nearness to the markets gave them an advantage in supplying the great cities with vegetables, milk, and other foodstuffs. In New England, land abandonment continued throughout the nineteenth century, although the land retained in cultivation received better care.

## Work in the South after 1860

The Civil War inflicted blighting hardships on the agriculture and entire economic structure of the South. Many owners of large estates, returning from 4 years of conflict, found their farm buildings burned and their land in a run-down condition. The currency of the Confederate States was valueless; and numerous planters sold some of their land in small tracts to white farmers or leased it to renters or sharecroppers. This resulted in abuse of the land by a progressively increasing tenant and sharecropper class who did not know or care about soil conservation.

A second result was the development of the one-crop credit system. Without money, planters had to depend on local merchants to "furnish" their renters or sharecroppers. But the merchant was not a man of large means. He was financed by the wholesaler or jobber, who, in turn, frequently was supplied capital by the Northern or English banker. To protect their interests, these groups required the sharecropper to produce a specified number of bales of cotton. The amount advanced by the capitalist to the jobber, to the merchant, to the landlord, and to the sharecropper or tenant was determined by the number of bales of cotton to be grown. Often, the quota was not fulfilled. Tenants usually got as many supplies as possible and as a result were generally in debt and constantly called on to grow more cotton.

Deterioration of Southern agriculture during the postwar period is revealed by the fact that not until 1878 did the cotton crop equal that of 1860. Erosion, which had been a serious problem before the war, continued on a widespread scale. The rolling uplands of the South, particularly in Tennessee, Georgia, Alabama, Mississippi, and the Carolinas, were generally very susceptible to severe water erosion under the heavy precipitation. In 1870, numerous areas in Georgia were reported to be red and seamed with gullies. By this time, many areas in middle Virginia and the Carolinas were abandoned as wastelands, too eroded to produce a crop. It was estimated that in 1877, millions of acres in the South were in this condition.

Nearly 50 years ago, Shaler had the following to say about erosion in the Southern States[1]:

"Although the evils arising from the washing away of the soil in America have not as yet been very serious, a close reckoning of the loss would probably show that it already amounts to the practical destruction of that coating over an area some thousands of square miles in extent. These depauperated districts lie

[1] Shaler, N. S. Origin and Nature of Soils, U. S. Geol. Survey 12th *Ann. Rept. for 1890–91*, Vol. 1, pp. 331–333. Washington. 1891.

almost altogether in the region to the south of the glacial belt, and mainly in the hilly portions of the so-called Southern States, especially in Virginia, the Carolinas, Kentucky, Tennessee, and Mississippi. There is scarcely a county in these States where it is not possible to find a number of areas aggregating from 300 to 500 acres where the true soil has been allowed to wash away, leaving exposed to the air either bare rock or infertile subsoil. Where subsoil as well as the truly fertile layer has been swept away the field may be regarded as lost to the uses of man, as much so, indeed, as if it had been sunk beneath the sea, for it will in most instances require thousands of years before the surface can be restored to its original state. . . .

"Besides this annihilation of the earth resources in the area where the soils have been allowed to wash completely away, a vastly more important though less visible damage has been done by the partial destruction of the nutritive layer, in the course of which it has been thinned and worn to a point where it will no longer pay the cost of tillage. When brought into this impoverished condition it is, in the common phrase "turned out," or, in other words, committed to the slow process of redemption which the natural agents of soil-making may bring to bear. . . .

"The total area of these abandoned fields which lie in the States of Virginia, Tennessee, and Kentucky alone amount, according to the estimate I have made with some care, to between five and six thousand square miles, or about one-thirteenth of the total tillable surface of these States. . . . "

Before the Civil War, some attempt at diversification had been made.[1] In the upper country of North Carolina and Virginia and in Tennessee, Kentucky, and Missouri, many farmers broke away from the one-crop system. Richmond, Petersburg, and Louisville became small industrial centers, and various industries sprang up that helped agriculture. Farmers in the Shenandoah Valley of Virginia, the Bluegrass region of Kentucky, and the Nashville Basin of Tennessee generally diversified their agriculture. Peanuts were grown in southeastern Virginia and adjacent North Carolina. In small areas, here and there in the Carolinas, Georgia, Mississippi, Arkansas, Louisiana, and Texas, farmers grew vegetables for sale to near-by cities. Under the protection of levees, rice was a major crop near the coast, on the alluvial lands of such streams as the Cape Fear in North Carolina; the Santee and Pee Dee in South Carolina; and the Savannah, Ogeechee, and Altamaha in Georgia.

Here and there, farmers in the Deep South continued to use erosion-control measures developed before the war. A considerable number of farmers practiced hillside ditching. Of these, William Durham Lee, a Civil War veteran, was probably representative. He had seen the results of hillside ditching when marching over the war zones between Richmond and Gettysburg before he bought his farm in Hinds County, west-central

[1] Gray, L. C. "History of Agriculture in the Southern United States to 1860." Vol. II. pp. 909–923. Washington. 1933.

Mississippi, which had been neglected for 7 years. Only the rich valley land had been cultivated during this period; the hillside fields were eroded badly, and old-field pines had covered most of them. Gullies were treated with trash and pine boughs. On the contour of the uplands, he dug hillside ditches about 18 inches deep and 100 feet apart. Guard drains were constructed at the ends of the ditches, and rock dams or baffles were placed at some of these points to break the force of the cascading water.

In some of the gullies, he planted Bermuda sod, hauling it by the wagonload. Cowpeas were grown for green manure or as a soil-building crop. Manure and cottonseed were used as fertilizers. But in his community, Lee appears to have been the only farmer who really took care of the land.

Farmers in other sections of the South also found Bermuda excellent for eroded hillsides. One writer recommended it for the area around New Orleans, and it was used as far north as Maryland to prevent erosion. In the northern part of the South, crimson clover was used to some extent.

The remarkable spread of lespedeza, or "Japanese clover," was noted in 1884. It was growing in many of the woodlands and idle fields of the South by that time, being particularly widespread in Alabama and Mississippi. Some of the northern clovers also had made their way south. Frequent references to their use as soil binders are found in Maryland and Virginia farm journals. Generally, grasses and clovers came into greater use in the northerly part of the South for permanent pastures.

Legumes were increasingly recommended. The Commissioner of Agriculture of Georgia, in 1881, said that the hope of Georgia lay in legumes. The cowpea became the most popular of these soil-building crops. It was sowed between corn or planted alone for hay or food or to be turned under as green manure. The few farmers interested in soil conservation emphasized the value of a ground cover to prevent erosion during winter.

Plows were improved for deeper tillage with a minimum expenditure of effort. Several varieties of hillside and subsoil plows were developed. Plowing was employed occasionally to fill gullies. J. Wilkerson, writing in Maryland in the 1870's, describes a gully 300 feet long and 1 to 10 feet deep that he eliminated by plowing in soil from the sides.

Rotations and diversification were practiced by only a few farmers. In Maryland, a reference in an agricultural journal, published in 1886, probably gives a cross section of the practices of the more progressive farmers. Five farmers in the Deer Creek farmers' club listed their rotation systems as follows:

1. Wheat, corn, grass for several years.
2. Corn, wheat, clover, wheat, wheat, grass 5 years.
3. Seven-year rotation of corn, wheat, oats, and 4 years of grass.

4. Grass, top-dressed and turned under for corn.

5. Eight-field rotation of grass and wheat (several years at a time, fertilized each year).

As crop diversification increased, hillside ditching grew in disfavor. David Nichols of Alabama was among those who found the old-style hillside ditches unsatisfactory, reporting in 1877 an improved method of constructing them. He ran his ditches through a clover field and as an experiment decided to leave the vegetation above the ditch for a width of about 8 feet. For six years, he experimented with "stripping" his ditches with grass or some other vegetation and found it so satisfactory that he felt it would revolutionize agriculture. The results of his experiments were published in a small book concerning his "system."[1]

Of more importance, however, was the work of Priestly Mangum[2] near Wake Forest, N. C. He, too, had found hillside ditches unsatisfactory and had experimented with strips of vegetation (some of them as wide as 50 feet), which entirely replaced the ditches. But the system proved unsatisfactory for controlling erosion on all lands. In 1885, Mangum constructed his first terrace, which was about 1 or 2 feet high, 10 feet wide, and so designed that all the land could be cultivated, with provision for the water to drain off slowly without washing the field. His type of terrace was constructed by turning an equal amount of soil from both the upper and the lower sides. Although he used a steeper grade than approved for modern terraces, some of the structures that he built have been used for 50 years.

One farmer in Newberry County, in the Piedmont section of central South Carolina, started building terraces similar to the Mangum type 50 years ago, using a terracing machine which was patented in 1883. His terraces were constructed on the contour, with a 3-foot vertical interval. Some of these are still in use.

The effort of these postwar soil conservers availed little to stem the tide of soil erosion. Generally, too few farmers adopted the approved practices or employed only a single method. In the southern Piedmont and some of the adjacent areas, particularly in Georgia and South Carolina, both hillside ditching and terracing were used extensively by 1880. Of the entire country, by far the greatest conscious effort to control erosion was made in this locality on up to sometime after the close of the century, when terracing began to spread through the Cotton Belt and, in a small way, to other regions. But the system, although successful in many instances, failed to hold the soil over extensive areas because of excessive

[1] Nichols, David. "Preservation and Protection of Cultivated Land from Surface Washing." Atlanta, Ga. 1883.

[2] Mangum, P. H. My Father Invented It, *Country Gentleman*, pp. 14, 73, November, 1937.

steepness of the land, faulty construction, or failure to maintain the terraces. No adequate experiments were carried out anywhere to determine the proper grades, dimensions, or adaptable slopes and soils until the latter part of the 1920's.

The one-crop system of agriculture tended to preclude any substantial improvement of erosion-control practices in general. With the turn of the century, however, the long agricultural depression ended, and the South shared in the upturn. The efforts of agricultural journals, agricultural colleges, and the new experiment stations resulted in some spread of crop diversification; but in spite of this, the proportion of tenant farmers grew larger. Erosion increased at an appalling rate, devastating large areas. Millions of acres were abandoned, and much land was retained in cultivation only through the use of increasingly large quantities of commercial fertilizer.

In general, it might be said that, although erosion proceeded to take its heavy toll of soil on most rolling farms, those practices which were retained were improved as the older, deficient ones were gradually abandoned. The soil-saving remedies advocated and practiced to a slight extent slowly became more varied.

## Conservation Efforts in the West

Emigration of farmers from settled communities to unsettled areas probably has been the most persistent characteristic of American agriculture. From the earliest Colonial times, the frontier was constantly pushed westward. The influence of this movement can hardly be overemphasized, because frontier agricultural practices have had most profound effects not only on the land itself but on the whole American economy. To these pioneer practices, under conditions of an abundant supply of land, may be laid, perhaps, the main contributing cause of erosion. The dominant motive in gaining a livelihood from the frontier farm was to extract from the soil a maximum of crops with the least effort. The abuse of the land was a cause of amazement not only to English travelers who visited the American colonies but also to the Easterners, who later visited the Western frontier.

In Kentucky in 1838, it was reported that 50 years of poor farming methods were wearing out the land. The "skimming system" was almost universally practiced in Missouri in 1849, "not one farmer out of fifty having ever hauled a load of manure." By 1850, the best farmers of Kentucky grew clover, although little fertilizer was used. In Dane County, Wisconsin, in 1851, the belief that the soil could be cropped indefinitely without becoming exhausted was beginning to be questioned. Farmers of this area frequently dumped their barnyard manure in the creek; the one-crop system was practiced generally; and only a few rotated their crops. Little livestock was raised except by European immigrants. By 1854, soil depletion was causing great concern in eastern Ohio. It was noted by one writer in 1870 that thousands of farmers along the frontier never thought of manure; that much western wheatland had deteriorated; and that the remedy was underdrainage, rotation with clover, and use of manures. It was reported in 1878 that many gullied fields and even entire farms, near Elkton, Ky., had been abandoned and allowed to grow up in broomsedge.

This ruthless waste of the land was followed to some degree by improvement. Parts of Ohio had been settled before the nineteenth century, and by 1840 the practice of bare fallow began to be abandoned, and cowpeas or sod sometimes were plowed under. By 1851, clover was coming into use. As a rule, wheat was of first importance on Western farms; then oats, barley, corn, and finally hay and forage crops. Perhaps the experience of John Muir, the celebrated naturalist who spent his boyhood on a Wisconsin farm, was typical of much Midwestern farming. He relates that, at first, wheat, corn, and potatoes were the principal crops but that in four or five years the soil was so exhausted the yield of wheat declined to 5 or 6 bushels an acre. Then more attention was paid to the corn crop. When its yield became meager, English clover was grown on the exhausted land and turned under before corn was planted again. Later, a complete change of farming methods was brought about; the farmers grew clover, corn, and crops which are fed to the stock in the field.

The inefficient cropping methods of older areas were frequently transplanted to the frontier. Dicken and Brown[1] describe two areas in western Kentucky that are typical of such conditions. The Lincoln Farm area in Larue County was settled by Virginians between 1780 and 1810; corn became a staple crop, and later wheat and tobacco. Row crops predominated, and poor methods of plowing were used. Erosion and land abandonment followed, and, as newer fields became exhausted, some of the abandoned areas were cleared and cultivated again. In some instances, bushes and briars partially conceal the scars of soil erosion in old fields that have not been cultivated for a century. South-facing slopes were particularly susceptible to erosion because of the frequent alternation of freezes and thaws.

At the close of the Civil War, the western half of the Mississippi Valley was largely unsettled, although before this time farmers had pushed into the prairies of Iowa and eastern Kansas. To the settlers from the East, the absence of trees in the prairie states indicated unproductive land. Accordingly, woodlands were the first cultivated, and a shortage of

[1] Dicken, S. N., and Brown, H. B., Jr. Soil Erosion in the Karst Lands of Kentucky, U. S. Dept. Agr. *Circ.* 490, 1939.

timber resulted. Later, after the advent of the steel plow, the fallacy regarding the productivity of the prairies was dispelled.

The land policy of the Federal Government also stimulated settlement of Western lands through the Preemption Act of 1841 and the Homestead Act of 1862, with its subsequent modifications. In 1870, two railroads were extended across the Great Plains. These encouraged further settlement. By 1885, a wedge of settlement spread across the prairies into western Kansas and Nebraska. The frontier had advanced into eastern Colorado by 1890; and between 1889–1906, Oklahoma Territory was opened to farmers. Thus, the last great area of restricted public farm land was thrown open to settlement.

Water erosion in the Midwest caused much damage. For more than a century, erosion-control methods had been publicized but little practiced, until soil exhaustion became critical. A typical example of frontier-land exploitation was found in the Blackjack or Cross Timbers section of central Oklahoma. This area was settled about 1889; and within a few years, most of the upland was ruined for farming purposes or severely impoverished. After the trees were removed from the sandy hillsides the uplands revealed their peculiar susceptibility to water erosion. Little livestock was raised. The principal crops were corn and cotton. Rows were run up and down the hill, and it was customary to burn crop residues in the spring.

From 1905 to 1908, precipitation was above normal throughout Oklahoma; in 1908, the Blackjack section had a rainfall of 60 inches. This resulted in excessive sheet and gully erosion and caused widespread land abandonment. The situation on the land was referred to as a "succession of gullies and bottom patches made useless by the sand washed over them from the hillsides."[1] The abuse of the land had been accentuated by tenants who seldom lived on a farm more than one year and consequently had little or no interest in the welfare of the soil.

Among those who deprecated the poor farming methods that caused erosion in the Blackjacks and other localities was John Fields, editor of the *Oklahoma Farm Journal*.[2] Fields felt that the whole system of farming based on the one-crop-cash-crop system was ruinous. He pleaded with the farmers to turn to livestock and grass. In his experiments with grass, he had found Bermuda well suited to Oklahoma conditions and in 1902, when he became editor of the *Oklahoma Farm Journal*, started on his long crusade to popularize the growing of this Southern perennial.

[1] Anonymous. Profits from Sand, *Oklahoma Farm Jour.*, p. 2, Mar. 1, 1908.

[2] McDonald, A. Erosion by Wind and Water in Oklahoma, *Soil Cons.*, Vol. 2, pp. 233–235, 1937; Erosion and Its Control in Oklahoma Territory, U. S. Dept. Agr. *Misc. Pub.* 301, 1938.

One of Fields' supporters was Harry E. Kelley,[1] who had moved to Fort Smith, Ark., in 1887. Shortly after his arrival in Arkansas, Kelley decided to devote his life to propagating Bermuda grass, in order to save the country from erosion. As a large landholder, he set out hundreds of acres of Bermuda, for many years gave away sod for planting, and continuously encouraged its use. Kelley was instrumental in having it set along the highways of Arkansas and induced Fields to promise never to publish an issue of the *Oklahoma Farm Journal* unless it contained an article on Bermuda grass. F. A. ("Bermuda Grass") Mitchell, who preached and farmed in Oklahoma, was also one of the pioneer advocates of Bermuda for erosion control.

Bermuda grass probably has been used more than any other type of cover to control erosion in some parts of the Midwest. It has failed repeatedly in the more northerly areas, although some has been grown successfully as far north as Nebraska. To a large extent, Bermuda failed in western Oklahoma in 1910 and 1911 because of dry and cold winters.

Sweetclover has been used successfully and extensively to control both wind and water erosion in many parts of the country.

By 1913, John A. Adams of Johnson County, Missouri, had developed a system of gully control, the main principle of which was a dam with a drain-outlet pipe beneath. This pipe was so arranged as to allow the water to drain away after being impounded to a certain height behind the dam. This principle continues to be used extensively in erosion-control operations.

During periods of drought, the settlers in the Plains and other parts of the West were faced with the problem of wind erosion. Wide fluctuations in the rainfall of the Great Plains resulted in variations in crop yields from good to complete failure over large areas. Such conditions have imposed upon the region an unstable type of agriculture and have encouraged speculative farming during periods of heavier than average rainfall, followed by depression in times of drought. In wet years the crop acreage was increased, and in dry years crop failure and soil blowing plagued the farmer. Wind erosion affected many localities when the soil favored blowing shortly after cultivation was introduced. Drifting of the soil was reported in parts of Oklahoma four years after the sod was broken and in North Dakota 6 years after settlement (1888).

In dry years, crops frequently were ruined over wide areas throughout the West. In Greer County, Oklahoma Territory, crops were destroyed by drought repeatedly from 1902 to 1906. Thousands of acres of cotton were ruined, and nearly every field of sandy land that was broken flat and planted level suffered from blowing. In the Oklahoma Panhandle, wheat blew out repeatedly.

[1] McDonald, A. The Bermuda Grass King, *Soil Cons.*, Vol. 3, pp. 97–99, 1937.

The problem of wind erosion became serious in western Kansas; northwestern Texas; western Oklahoma; eastern Colorado; and parts of the Dakotas, Nebraska, and other Western States.[1] Wind erosion became such a menace that in some areas, whole communities were mobilized to fight it. In Thomas County, Kansas, the citizens organized and raised $8,000 through contributions from bankers, merchants, professional men, and farmers. The life of the community was threatened. Strips 250 to 330 feet wide were listed and planted to sorghum or corn; unworked strips were left between. In some instances, it was considered necessary to list only three or four furrows at intervals of 165 to 330 feet. The result was that blowing was checked on 65,000 acres.

The disk was also used as an implement for the prevention of wind erosion under conditions of dry-land agriculture. The disks were set straight to force grain stubble and other crop residues into the loose soil. The disk was also used to some extent for flat breaking when it was desired to develop a cloddy wind-resistant surface. Some farmers plowed alternate rows as an aid to wind-erosion prevention. The use of harrows and rollers was discouraged, because the leveling, smoothing, and pulverizing of the surface was found to be conducive to blowing. Under conditions of clean-culture summer fallow, as extensively practiced in the Wheat Belt of eastern Washington and adjacent Oregon and Idaho, particular effort was made to maintain a cloddy surface. Rod weeders, of the subsurface type, came to be an important implement for this purpose.

More effective than other methods were the cover crops sometimes used for control of blowing. These were stripped or interplanted with corn or cotton. Occasionally, wheat, oats, rye, and (in Oklahoma, Texas, and Arkansas) cowpeas were planted between the rows; and in some localities, Bermuda grass or crab grass was allowed to grow.

Frequent failures of corn, as the result of drought, probably contributed to the increased importance of the grain sorghums. These crops became especially important in the Southern Plains, not only as a source of feed but, on some farms, as a means of checking soil blowing, particularly on sandy land. Sometimes, the stubs were left on the ground as a soil stabilizer; and in other instances, the crop was cut in summer for feed, and the second growth left to hold the soil through winter. Where the soil was not too "blowy," the second crop was turned under by some operators as green manure. Occasionally, a nurse crop was used to prevent blowing until the main crop could establish itself.

Rotations practiced here and there usually consisted of several years of grass followed by some other crop. By 1907, an eight-year rotation was practiced by some Kansas farmers, which included 3 to 4 years of grass or alfalfa. A. M. Ten Eyck and W. M. Jardine of Kansas, and other

[1] Hazen, T. E. Blowing Soils, U. S. Dept. Agr. Bur. Plant Ind. *Bull.* 130, 1908.

authorities on semiarid farming, felt that grass was the only solution to the erosion problem.

Trees were used to some extent to check wind erosion, especially in the Great Plains, in parts of California, and along the Columbia River in Oregon and Washington. Some of the plantings were put out in the hope that they would cause an increase in rainfall.

Other measures for minimizing the effects of wind erosion were used, but the number of farmers employing precautionary measures was far too restricted to have any very pronounced effect.

THE RANGE. The 20 years following the Civil War saw an unprecedented growth of the cattle and sheep industry in the Western States. Ranching spread over parts of western Kansas, Nebraska, eastern Colorado, the Dakotas, Montana, Utah, southern Wyoming, and western Canada. Soon, much of the range land was overstocked.

By 1885, grass was becoming scarce in many localities. The hard winter of 1886 virtually wiped out the cattle on numerous ranches in the Northwest, to which region livestock had spread previously. Somewhat later, the Southwest went through similar experiences. Between 1890 and 1910, large areas throughout the West were overgrazed.

Prior to this large-scale development of the Western cattle industry, complaints of overgrazing were occasionally heard. The secretary of the California Wool Growers' Association wrote, in 1863[1]: "Where the lands have been so persistently overstocked . . . the herbage has necessarily become thinner and thinner. . . . This process of depasturage, though not confined to any one species of herbage, is most strikingly exhibited in the great goat ranges. . . . This system of stocking the grazing lands must ultimately result in their entire depasturage."

In 1880, the census reports that some ranges were overstocked and depleted even then. It was reported, for example, that in 1880 a horse could step over the tiny stream that flowed through the valley of Cañada de los Alamos, in the Santa Barbara National Forest of California. After years of overgrazing of its drainage area, increased runoff resulted in so much erosion that the little drainageway grew into a gully at least 100 feet deep and 200 feet wide. The water table was lowered to such an extent that groves of cottonwoods died of thirst.

In 1862, Mountain Meadows, Utah, was a fertile valley supporting a rich growth of grasses. Today most of the topsoil and nearly all the grass are gone, so that grazing is largely a thing of the past. In 1897, the National Academy of Science, in an investigation authorized by the government, found much overgrazing by sheep throughout the area between the Columbia River and the southern boundary of Washington as well as in other areas. The report also mentioned the occurrence of

[1] United States Congress. The Western Range, 74th Cong., 2d Sess. *Sen. Doc.* 199, 1936.

frequent forest fires started by ranchers, travelers, miners, and coal-burning locomotives.

Overgrazing and fires, together or separately, resulted in increased floods. A notable example is that of Manti Canyon, Utah, which was overgrazed for many years. Much gullying resulted, even on the higher crest of the Wasatch. Floods gradually increased, until in 1900 they were disastrous. The range was closed, overgrazing was stopped, and by 1910 the floods apparently had ceased. In 1912, a fire in the Piru region, in Colorado, resulted in severe erosion and marked increase in the runoff. Not only did floods become a menace, but valuable orchards were covered to depths of 5 feet with soil material washed out of the burned area. In the "Happy Hill" grazing district of northern Washington, repeated fires and overgrazing resulted in severe impoverishment of productive hillside cropland. By 1897, many parts of the ranges in San Saba, Tom Green, and Taylor Counties, Texas, were reported to be almost ruined, where thirty years before they were said to be covered with grass 1 to 3 feet high. A stockman in this locality showed the attitude of some ranchmen by introducing facetiously the following resolution at a meeting at which a visitor had expressed some concern for the grass[1]: "Resolved, that none of us know, or care to know, anything about grasses, native or otherwise, outside of the fact that for the present there are lots of them, the best on record, and we are after getting the most out of them while they last."

But many stockmen finally changed their attitude. In 1905, Theodore Roosevelt sent out a circular to 1,400 stockmen throughout the Western States, requesting information about range conditions. It was the consensus of opinion of those who answered that millions of acres had been ruined by overgrazing and that only the Federal Government could remedy the evil. In 1878, Lieutenant Powell, then chief of the Geological Survey, had advised revision of land laws to fit conditions in the semiarid West, but his advice was ignored. However, a beginning in conservation was made when the forest reserves were set aside in 1891. In 1896, the Northern Pacific Railroad attempted to restore its lands by means of leasing under a system of fencing to insure proper distribution of the cattle.

With the turn of the century, the Federal Government took more interest in range conservation. This was due chiefly to the interest and leadership of Theodore Roosevelt and Gifford Pinchot. At this time, the livestock of the West were grazed for the most part on public lands. During Theodore Roosevelt's administration, an area of 148 million acres was withdrawn from the uncontrolled public domain and included in national forests. Regulated use of the forest reserves had been authorized

[1] Bentley, H. L. Cattle Range of the Southwest, U. S. Dept. Agr. *Farmers' Bull.* 72, 1898.

in 1897. In 1900, investigations of range use were initiated by both Federal and state agencies; in 1910, the state agricultural experiment stations started work along the same line; and in the same year, the office of Grazing Studies was established in the Forest Service. The investigations covered such problems as grazing capacity; seasonal use; class of stock; systems of grazing; and influence of grazing on soil, erosion, runoff, and stream flow.

The Forest Service, by its control of the national forests, began improving the ranges. It enforced better distribution of cattle by means of drift fences and by other methods. In 1907, Senator Newlands introduced his bill to prevent floods by means of dams at headwaters of streams, reforestation, and storage of water to prevent erosion.

The National Conservation Commission (1909) collected much valuable information regarding erosion and conservation in general and stimulated interest in the national problem of wasteful use of natural resources. It found that the 300 million acres of public range, on the whole, were in bad condition and recommended reforestation, regulation of grazing, reseeding, and control by the Federal Government of all public lands.

# Chapter XLI. Erosion Problems in Foreign Countries

Man in his primitive stage was a part of nature; but as he became "civilized," he failed to recognize that the world "was given to him for usufruct alone, not for his consumption, and still less for profligate waste."[1] Erosion is one form of toll that man is paying for his extravagance. The extent of the penalty did not become evident until the last of the agricultural frontiers began to feel the effects of accelerated soil wastage.

Erosion operates locally, but the problem of its control is national, or even international, in scope. Farmers do not willfully destroy their land; they are simply the tools through which society operates. The mills of Manchester and New England must share with the farmers of the South the responsibility for erosion caused by continuous cotton production, and the populations of our great cities and the Allied armies must assume a large share of the responsibility for the exploitation of the Great Plains of America and the wheat lands of Australia.

Nations now are beginning to balance the value of continued cultivation against immediate profits obtained by land exploitation accompanied by the impoverishment or destruction of the soil itself. But no one country has experienced all the social and economic conditions that lead to the development of erosion; and in the New World, agricultural history is too short to show the effects of continued erosion or the value of its control. All countries should, therefore, draw upon the vast store of international experience in building long-range programs for soil conservation.

## Erosion in the Old World

### The Mediterranean Region

Erosion in Italy. Italy is a leader in the movement for land reclamation and conservation in Europe. Under Rome, agriculture had flourished. The decline after the fall of Rome was caused largely by neglect of the land. The Italian slopes, denuded of trees, became pastures for

[1] Marsh, George P. "The Earth as Modified by Human Action," p. 33. London and New York. 1874.

sheep and goats. Floods increased in number and became more destructive; the soil washed from the hills covered the fertile fields below, blocked the river channels, and increased the area of malarial swamps. The Roman Campagna and the Pontine marshes, which had supported flourishing populations, were practically deserted. Until the last decade, the few farmers of the lowlands remained only long enough to sow and harvest their crops, returning thereafter to their highland homes.

Farther north, the valley of the Po experienced equally radical changes. Ravenna, which was once an environmental counterpart of Venice, is today 4 miles from the sea; and Adria, a port at the time of Augustus, is more than 14 miles inland. Until the initiation of reclamation, the new land formed was chiefly swamp which in no way compensated for the destruction caused by erosion.

The removal of forests from the central and southern Apennines began in remote antiquity. The forests of the northern Apennines, the eastern Alps, and the Maritime Alps were spared until the Middle Ages, when they were cut to provide timber for the navies of Venice and Genoa or to remove the ambuscade that might provide shelter for the invading Turks. Much of the desolation evident in the karst region north of Trieste is definitely attributable to the washing away of the soil after the cutting of the native pines.

In 1862, the Karst was described as[1]:

" . . . a picture of the dreariest waste, for only the scantiest vegetation flourishes here. With the exception of a few places where oak and hornbeam grow, not only is forest lacking but one might almost say that the karst is entirely without vegetation. Only here and there stands out a little weed, or a thicket from the limestone piles. The land is scarcely suitable for the cultivation of grain or raising cattle. In the depressions only a few spots can be found which afford a meagre return or here and there a favorable place for a single grape-vine. . . . "

Similar conditions still prevail.

Not until the initiation of the *Bonifica Integrale* in 1929 were all phases of land reclamation in Italy united under a single authority and directed toward the accomplishment of a single goal—the improvement of the national economy by raising the standards of living and agricultural production of the rural population. Unless the various aspects of reclamation were consolidated, there was danger that valuable water resources might be lost in draining the land or fighting malaria or that, in the regulation of water courses, forestry or grazing restrictions might be enforced without regard to their effect on the mountain population. The problem of land reclamation is essentially one of controlling erosion on the

[1] Anonymous. Der unterirdische Lauf der Recca, *Aus der Natur*, Vol. 20, pp. 250–254, 263–266, 1862.

slopes in order to prevent loss of soil in the highlands and to reduce floods and sedimentation in the lowlands.

In 1929, a special Under-secretariat of State was established to supervise the national reclamation program. Most of the actual work of reclamation is carried on by nongovernmental agencies, known as *consorzi*, to which treasury grants are made. Plans initiated by the *consorzi* are examined and approved by the government before grants are allotted, and at intervals the work in progress is inspected. Within reclamation areas the state determines, on the basis of importance to the nation as a whole, the method and extent of state aid to be granted.

It is estimated[1] that about one-third of Italy will be improved at a cost of over 7,000 million lire under the *Bonifica Integrale* during the fourteen-year period ending in 1944. Of this, 4,300 million will be assumed by the state, and the remainder is to be repaid over periods varying from five to fifty years by the farmers benefited. To preclude injustices to certain population groups, many of the changes are being made gradually. On the Lower Piave project, for example, the size of landholdings will be reduced gradually as the tenants become financially able to purchase their farms. In other areas, such as the Roman Campagna, the land was subdivided into small farms as soon as the installation of reclamation works made close settlement possible.

The problems of agricultural improvement are of two general types: reclamation and protection. On the Littoria project near Rome, inferior pastures and malarial swamps that had been prosperous farming land in Roman time have been restored to productivity. Before reclamation could be started, roads, railroads, and villages provided with necessary sanitation had to be constructed. In the new towns of Littoria, Sabaudia, and Pontinia, houses and farm buildings were erected, and the land broken before the farms were assigned.

The Lower Piave project is chiefly protective in nature, although it includes a considerable amount of reclamation. The regulation of rivers, started in the Middle Ages by the Venetian Government, was continued intermittently under the kingdom of Italy. Water regulation in the upper Ongaro area was initiated in 1903 and progressed with occasional interruptions until the World War. The present agricultural development is basically a resumption of prewar work.

In the mountain regions, no sharp line can be drawn between reclamation and protective works. Highland agriculture must be maintained in such a manner that soil and water will be conserved and floods reduced without resorting to complete reforestation, which would displace large farm populations. The Brisighella project near Florence consists chiefly

[1] Longobardi, Cesare. "Land Reclamation in Italy. Rural Revival in the Building of a Nation." Trans. by Olivia Rossetti Agresti. London. 1936.

of hilly and mountain land. Its most representative erosion feature is the *calanche*, which is also characteristic of the Pliocene clays throughout the foothills of the Apennines. As water drains off the slopes, it carries away large amounts of soil, and the gullied clay is left in ridges too steep for cultivation.

Because of the increased population and somewhat reduced foreign trade, Italy's need for agricultural land is sufficient to justify the expense of reclamation on the Brisighella project amounting to $250 or $350 per acre (Fig. 343). On the *calanche* lands, the much dissected slopes are

Fig. 343.—Land that was ruined for cultivation by severe erosion restored to agricultural use by elaborate smoothing, terracing, and planting processes. (*Courtesy, M. H. Cohee; compliments of the Italian Government.*)

dynamited and then partially leveled by hand (Fig. 344). Dams are built in the deeper gullies; some areas are seeded to grass or grain, which may be superseded by vineyards when satisfactory stabilization is accomplished. In this way, the clay material can be built up to a fair state of productivity within a few years.

The recently developed *gradoni* system has considerably reduced the problem of reforestation.[1] The *gradoni* are broad steps generally about 4 feet in width and spaced about 20 feet apart. They follow the contour and are inclined inward at a slope of about 30 per cent to prevent sheet washing or gullying on the outer rim. The system is rather expensive

[1] Ringland, Arthur C. Notes on Soil Erosion and Reforestation in Italy: Suggestions for American Application. U. S. Forest Service. 1934. Processed.

(costing about $70 an acre for construction and planting); it is particularly well adapted to steep slopes. At the suggestion of Mussolini, the reforesta-

FIG. 344.—Upper, formerly forested land (right background) ruined by erosion (calanche); foreground shows same type of land brought into agricultural use following blasting and smoothing. Lower, blasting operations on calanche land same type as above. Northern Italy. (*Courtesy, M. H. Cohee; compliments of the Italian Government.*)

tion of Mount Subasio by means of the *gradoni* system was initiated in 1930 as a memorial to Saint Francis of Assisi.

Farther south in Italy, the *Bonifica Integrale* is more concerned with irrigation. Although the rainfall is inadequate to maintain agriculture, the water flowing off the denuded mountain slopes during the heavy spring rains causes disastrous floods. After the destruction of the hydraulic works of the Greeks and Romans, the lowlands were converted into marshes, and the wooded or cultivated slopes into eroded pastures. By reducing the runoff and storing the water for irrigation, Italy is now recreating the agricultural productivity of classical times.

Erosion in Greece. Although there has been a continuous history of forest removal and soil wastage in Greece from ancient times to the present, the farmers continue to use obsolete tools and to alternate cultivation and fallow, as was done at the time of Pericles, 400 years before Christ. On the eroded slopes, pastures replaced cultivation; and when the deteriorating vegetation could no longer support sheep, goats increased in number. Today, there are more goats than people in Greece. The apathy of many of the people may be partly explained by the prevalence of malaria, which became endemic after the fourth century B.C., because of the increase in swamps. Since the World War, an influx of refugees from Bulgaria, Thrace, and Asia Minor brought new vigor to the areas that they settled. In Macedonia, the grain-producing area increased over 70 per cent. Grazing declined proportionally, and diversification of crops and intensive farming were introduced.

Spain. Most of the early accounts describe Spain as a farmer's paradise, with rivers that "do not govern but serve the land with their waters." The traveler of today is more conscious of the large stretches of desertlike waste and of the havoc wrought by seasonal torrents. Much of the country has always been unsuited to agriculture, but the unproductive areas have been increased as a result of unrestricted cutting of forests, overgrazing, and neglect of agriculture. In northern Spain—Catalonia in particular—where small diversified farms predominate, agriculture is based on human labor. The fields are carefully tilled, and erosion is less general.

After the discovery of the New World, emigration was rapid, and the smaller farms fell into the hands of large landowners, who obtained special privileges from the government. Their profit was derived chiefly from Merino wool; and the sheep that grazed in the mountains in summer, and in the lowlands during winter, were granted the rights of way and of water on their migrations. Fences were prohibited, and the flocks devoured the crops as well as the grass and scrub vegetation. Today, goat grazing in the forests of Spain is prohibited, but attempts at law enforcement frequently have met with open hostility.

**Palestine.** Much land in Palestine has been seriously damaged by erosion.[1] Especially serious from the standpoint of soil conservation and land reclamation are the increasing population and social conflict. Before

[1] Strahorn, A. T. Agriculture and Soils of Palestine, *Geog. Rev.*, Vol. 19, pp. 581–602, 1929.

the World War, the size of the population and herds was held in check by raids and disease. Since the establishment of the British Mandate, the natural increase in population has averaged about 24,000 a year. That reclamation may be practical on lands previously regarded as valueless is illustrated by the Birkar Ramadan area. Along the seacoast, orange groves have replaced the sand dunes; and at an average cost of $80 an acre, the malarial swamps have been converted into irrigated farms. Other areas have been similarly reclaimed through the application of capital and intelligent enterprise; but in consequence, the feeling prevails that there is discrimination in the distribution of land.

When the Hebrews first migrated to the Promised Land, the mountains were wooded. Today, only 76 square miles of forest remain in all Palestine, although another 450 square miles that have been closely cut over or overgrazed still retain some forest characteristics. The government has estimated that approximately 10,400 square miles are suitable for cultivation or afforestation.[1] Over 6,000 square miles are barren rocky hills or coastal dunes which have been classified as "unfit for cultivation." The Forestry Department recommends that 14 per cent of the land suitable for farms and forests be reserved for afforestation, fodder production, and erosion control; but without the cooperation of landowners, the government alone could plant only some 4 square miles of woodlands annually.

EROSION IN NORTH AFRICA. North Africa, which was an integral part of the ancient Mediterranean world, is again today more closely related to Europe than to the rest of the African continent, for Algeria is a province of France, not a colonial possession, and Lybia is increasingly important to the Italian Empire. But during the interval between the fall of Rome and the beginning of territorial expansion by the European powers, much of the cultivated land of North Africa reverted to a desert or semidesert state. The restoration of agriculture in French territory in North Africa, which began about 1870, after the establishment of the Third Republic, has continued with few interruptions to the present. The Italian reclamation in Lybia is a postwar development. In both instances, the goal is the restoration of agriculture to the heights that it attained under Roman dominion.

## NORTHERN AND EASTERN EUROPE

### TORRENT CONTROL

In the mountainous sections of Europe, erosion control developed during the Middle Ages but was secondary to the more immediate problems of torrent control and maintenance of vegetation in the mountain pastures. Although floods are the direct cause of property damage both

[1] Palestine Roy. Comm. *Rept.*, London, July, 1937.

on the mountains and on the adjacent lowlands, erosion and deforestation, which are contributory causes of the floods, threaten the permanence of agriculture and grazing in the highlands.

Grazing in the mountain forests of Europe is an integral part of the traditional rights, or *servitudes*,[1] which were developed to provide for the domestic requirements of local populations. As early as 1453, however, the grazing of sheep and goats was prohibited in the Canton of Freiburg; and during the sixteenth and seventeenth centuries, most of the Swiss cantons forbade the grazing of goats in coppice stands of a minimum age, varying from five to twelve years. By decree of the king, the royal forests of France were closed to sheep and goats in 1515, but the decree was never effectively enforced. The later Ordinance of 1669 was similarly disregarded, as were all other attempts to control grazing. It has been estimated that in some parts of the Alps, the timber line has been lowered as much as 1,000 feet by overgrazing within historic times.

The French Revolution was the immediate cause of the modern movement for forest conservation in France. The law of Sept. 29, 1791, gave the villagers full freedom of the royal forests and game preserves. By 1803, the government was forced to limit the use of the forests because of increasing losses of herds, crops, and houses as a result of floods. The herdsmen complained and resisted with force all efforts at reforestation, because limitation of the grazing area decreased their immediate earnings. But the prevalence of floods in 1840 was sufficient to focus public attention on problems of torrent control. The flood of 1856 caused unprecedented damage in the basins of the Rhine, Loire, Rhône, and Garonne.[2] The French forestry law, passed in 1859, is still in force.

Realizing that floods are natural phenomena that cannot be controlled completely by reforestation, the French Government initiated in 1882 an active campaign for pasture improvement and the construction of engineering works, such as dams and reinforcements along river banks.

The success of torrent control in France led to the adoption of similar methods in Austria.[3] Less effort was made, however, to acquire additional lands for reforestation, and more reliance was placed on the use of control works downstream in areas directly subject to flooding. In prewar Austria-Hungary, erosion was most destructive on the lower slopes where unstable beds of sedimentary sandy, gravelly, and clayey materials fre-

[1] Sparhawk, William N. Forest Rights in Foreign Countries with Special Reference to Grazing Rights, U. S. Dept. Agr. *Circ.* 456, 1937.

[2] "Restauration et Conservation des Terrains en Montagne." France. Direction Générale des Eaux et Fôrets. Paris. 1911.

[3] Strele, G. Fifty Years Experience with Mountain Stream Control in Austria. Mimeographed trans. from *Wasserkraft und Wasserwirtschaft* by W. Pundt, and J. G. Woodburn. 1936.

quently predominate. Although the floods were lower than along the Alpine torrents, erosion attacked some of the most productive farm land in northwestern Bohemia, southern Moravia, and Slovakia. Because the construction of a sufficient number of large dams for flood control was prohibitively expensive, and horizontal ditching proved inadequate, reforestation has been accepted as the most practical, large-scale control measure for these regions.

In Germany, most of the rivers subject to flooding have been canalized, and scientifically managed forests further contribute to regulation of the runoff from the mountains (Fig. 345). Today, the value of these forests in reducing erosion seldom is mentioned; but in the past, control through the maintenance of ground cover and the construction of small dams contributed to the evolution of a highly developed forest policy. Forestry is now focused primarily on the problem of increasing the production of existing stands, since any considerable extension of the forests would mean the curtailing of valuable farm or meadow land.

## THE AGRICULTURAL REVOLUTION

Some erosion occurred even on the comparatively smooth fields of northern Europe, but it probably was not sufficient to attract attention except in areas, like Scotland and the Rhine valley, where level land was at a premium. Because the population of Europe was comparatively sparse until the middle of the thirteenth century, there was little incentive to improve crop yields. The growth of city population that accompanied the development of the weaving industry in Flanders, about that time, increased the demand for food and raw material. The immediate result was an effort to increase crop yields in Flanders and wool production in England. By the seventeenth century, the urban population of England also had increased, and with it the need for agricultural improvement. The English experimental farmers of that period derived their agricultural inspiration directly or indirectly from Flanders.

Because the application of newly developed agricultural improvements was contingent on complete landownership, the traditional open-field system of farming gradually gave way to field enclosure. Before the end of the eighteenth century, England had become the agricultural leader of Europe; and from there, the agrarian revolution spread to the Continent. In France, the enclosure of fields began with the French Revolution, when the peasants became a dominant political force and the abolition of primogeniture prevented the accumulation of land by individuals.

In Germany, the abandonment of the open-field system, which accompanied the freeing of the peasants and the subdivision of large estates, occurred first in Bavaria and Hesse, where French influence was felt

Fig. 345.—Upper, overgrazing following clearing off the beech forest about 1810 seriously damaged this and neighboring areas, near Riedenburg, Germany. Some of this land now being reforested at cost of approximately $50.00 per acre. Center, gully healed in well-managed state forest by construction of loose-rock, drop-notch spillway dams (installed twenty years ago) near Tharandt-by-Dresden, Germany. Lower, one of the type of structures used in torrent control in state forest near Oberammergau, Germany. (*Courtesy, M. H. Cohee.*)

most strongly. The liberation of the peasants between 1801 and 1848, in western Germany, marked the beginning of an agricultural movement focused on the increase in crop yields through the restoration and maintenance of soil productivity, known as the *Theorie der Statik*.

In the east, the large estates of the Junker nobility not only continued but even increased in size during the first half of the nineteenth century. The free landless class, created by the liberation of the peasants, was no longer bound to the land, and the agricultural population declined. Subdivision of the estates proceeded slowly until the 1870's, when landowners were on the verge of bankruptcy as a result of declining yields, labor shortage, and cheap grain from the newly opened prairies of the United States.

Throughout northern Europe, erosion control has attracted little attention. Flanders, where the agricultural revolution had its inception, has maintained a high standard of cultivation; but today dairy and garden products are of greater economic importance than turnips and grain. But the Netherlands, like Denmark, is constantly fighting the landward movement of coastal dunes. In southwestern France, large-scale dune stabilization by vegetative methods was first undertaken in 1787 by Brémontier. The completion of this work made it possible to reclaim Les Landes, the swamplands lying behind the dunes and covering an area of 2½ million acres. The justification of the financial expenditure was demonstrated during the American Civil War, when the export of resin from southeastern United States was cut off. Les Landes then became the principal source of resin for Europe.

In eastern Europe, exploitive farming has created erosion problems that resemble more closely those of America. Moving dunes are a menace to farming in the plains of Poland, East Prussia, and Hungary. In Russia, the freeing of the serfs in 1861 marked the beginning of extensive grain farming, encouraged by the government in order to produce a favorable balance of trade. The situation is roughly comparable to the exploitation of the Great Plains of the United States, but destruction was slower because at that time agriculture was not highly mechanized. The small landowners, unable to compete with the large estates, lacked the capital necessary to develop methods for controlling erosion; and because of debts and high taxes, the peasants were forced to exploit the land. This situation continued until the Russian Revolution, in spite of the high development of soil science in Russia.

After the revolution, the land was divided into long, narrow holdings, usually running up- and downhill, so that each peasant might have his share of both good and bad land. Particularly in the Dnieper basin and the Ukraine, the land has been badly cut by gullies directly attributable to excessive subdivision of land. Under the circumstances it appears that only through collective farming could the holdings be reunited in such a way that erosion could be controlled without the reestablishment of large-scale capitalistic farms. But collectivization has also been accompanied by mechanization which has led to rapid expansion of agriculture into the semiarid plains of the south and east.

Erosion did not completely escape notice during the imperial regime in Russia. In the report of the Tula Hydrographic Expedition of 1908, a distinction was made between normal geologic and culturally induced erosion. The area covered lies in the heart of the forest steppe of the central Russian plateau and is one of the oldest areas of cultivation in the country. Because it was less well adapted to extensive grain farming than the southwest, much of it had been subdivided into small farms. From 63 to 75 per cent of the land under cultivation was situated on slopes of sufficient steepness to be in danger of rapid erosion, depending on the locality, and from 10 to 18 per cent already had been ruined by gullying. The most serious erosion occurred on the lower slopes where runoff accumulated. Cultivation then shifted to the upper slopes, and new land was exposed to sheet washing and gullying.

Reclamation of the lower slopes, commonly described as *spent lands*, has been a major problem at the Novosilsk Experiment Station since its establishment in 1921. Rotations of 6 years or less, depending on the stage of deterioration, are recommended. All rotations include 2 years of grass or leguminous crops to improve the soil by addition of needed nitrogen and organic matter. One year of clover usually supplies sufficient nitrogen for a subsequent grain crop, but at least 2 years are required to develop adequate improvement in the soil structure.

From the standpoint of national economy, the wheat-producing *chernozem* area of southwestern Russia is more valuable than the central plateau. Pankov has estimated that 1,350,000 acres of chernozem soil have been ruined within the last 200 or 300 years. Much of the destruction has occurred since the introduction of extensive grain farming following 1861. As lands became eroded, forests were cut to increase the wheat-producing area.

Fertility studies of chernozem soils have indicated that cultivation tends to break down the aggregated structure and that, although increasing the depth of the humus layer, it causes a reduction of the actual humus content. To restore the favorable structure, it has been recommended that long rotations, including grass or clover, be adopted. Professor W. R. Williams, of the Timiriasev Academy of Moscow, has recommended that the divides between drainage systems be planted in forests as a means of preventing floods and erosion. He is equally positive on the subject of grass, which he regards as the soil's chief protection against erosion.

Sheet washing and gullying, prevalent in the Transcaucasus region, are under investigation at the research institute at Tiflis.[1] Since the Russian Revolution, the increased number of tea, tobacco, and citrus plantations has been accompanied by accelerated erosion. It is estimated that at the current rate, all the topsoil may be removed from the cultivated land within 10 to 20 years.

Although it is recognized that wind erosion is severe not only in Asiatic Russia but also in the southern part of European Russia, satisfactory data concerning its extent are not available. The planting of shelter belts is the chief method adopted for its control, but this is still in an experimental stage. With the extension of mechanized cooperative farms into the semiarid sections, the problem of wind erosion is increasing; but Russia, forewarned of the danger by disasters that have overtaken other countries, is organizing surveys preliminary to the development of adequate control measures.

## The Far East

Before it became necessary to cultivate the sloping land of northern Europe, the discovery of the New World provided additional sources of food supply. China had no such outlet for its increasing population, and it was necessary to cultivate all available land. The skill developed by the "farmers of forty centuries" is justly praised, but China is also the "land of famine." Within a single country are to be found some of the most effective examples of erosion control as well as the most convincing evidence of destructive erosion caused by man. That erosion has been recognized for generations is illustrated by two proverbs quoted by Lowdermilk[2]: "Mountains exhausted of forests are washed bare by torrents" and "Mountains empty—rivers gorged." Chinese legends[3] also record a classification of soils according to color and structure that was developed about 4,000 years ago—nearly 1,000 years before the burning of Troy.

In southern China, where the rainfall averages from 60 to 80 inches annually, rice is the staple food crop. Because the fields must be flooded, rice can be grown only on level land or terraces banked to retain the water;

[1] Gussak, V. B. Experimental Study of Sheet Erosion and Runoff on Red Soils in the Wet Sub-tropics, *Pedology* 1, pp. 35–56, 1935.

[2] Lowdermilk, W. C. Forestry in Denuded China, Am. Acad. Political and Social Sci. *Pub.* 2433, 1930.

[3] Thorp, James. "Geography of the Soils of China." Published by the National Geologic Survey of China in cooperation with the Institute of Geology of the National Academy of Peiping and the China Foundation for the Promotion of Education and Culture. Nanking, China. 1936.

consequently, erosion is practically nonexistent. But where the slopes are steep, and upland crops are grown in rows customarily running up- and downhill, erosion is so severe that fields frequently are abandoned after 2 or 3 years of cultivation. But very little thought has been given to its control. It is probable that most of the widely distributed red and yellow soils of south China originally were covered with a topsoil of yellowish or brownish color. Complete reclamation is not possible in all instances, but some areas could be converted into usable farm land, if terracing, contour plowing, strip cropping, and rotations were employed. One of the major requirements of adequate rotations is the introduction of leguminous cover crops. In other areas, reforestation seems to be the only practical method of erosion control; but where soil wastage is far advanced, even this is difficult.

Central China has a lower rainfall (40 to 60 inches annually); natural revegetation is slower than in the south; and rice cultivation is less important. As in southern China, however, erosion is completely controlled on rice fields but otherwise neglected. On an experimental farm near Nanking, where fruit trees have been planted in rows running up- and downhill, and in an experimental mulberry grove, planted without regard for slope, gullies are forming between the rows, and sheet wash is affecting the entire surface. At the current rate of erosion, the soil remaining will soon be insufficient to support tree growth.

Northern China presents a more complex problem. The rainfall, which is less than that farther south, becomes progressively lower toward the west. The soft loess soil that predominates is easily washed away by rains and blown away by the winds during the frequently recurring droughts. In the northwest, erosion that has been in progress for centuries has completely altered the nature of the country (Fig. 346); and in the northeast, sediment brought down by the Yellow River has created one of the world's most serious flood problems.[1]

That many of the mountains in the northwest were once wooded is evident from accounts of early travelers in the Orient and by the presence today of occasional isolated woodlands that have escaped destruction because of religious veneration. The Buddhist temples, built in the first century A.D., were surrounded by forests. Those forests which are left are still free from erosion. They are composed of various hardwood species which reproduce naturally. At Chung Pu, Lowdermilk[2] noticed an isolated forest-covered mountain in the midst of the barren, eroded landscape. Here lies the tomb of Hwang Ti, a monarch of 30 centuries ago.

[1] Todd, O. J. "Two Decades in China." Association of Chinese and American Engineers. Peiping, 1938.

[2] Lowdermilk, W. C. Erosion in the Orient as Related to Soil Conservation in America, *Jour. Am. Soc. of Agronomy*, Vol. 21, pp. 404–414, 1929.

The unprotected forests were cut to make room for farming or, where population remained sparse for a longer period, to protect the inhabitants against wild animals. As late as the fifteenth century, Marco Polo referred to the ferocious tigers and wolves of the forests of Cathay.

One of the most stupendous engineering feats of the world has been the gradual building of terraces in northwest China to conserve the soil. As the soil began washing downhill, the farmers built embankments or terraces. Their height was increased year by year until the landscape

FIG. 346.—Formerly cultivated land on watershed of Yellow River, China, destroyed by erosion. Note remnants of terraces with which the attempt was made to conserve these lands. (*Courtesy, W. C. Lowdermilk.*)

has been converted into a series of giant steps. But even terraces provide insufficient protection. In dry periods, the loess shrinks, and deep cracks are formed, which permit rainwater to enter the soil and form underground watercourses which carry away the loess from the lower levels. Underground caverns increase in size until there is no longer sufficient support for the overlying soil. In this way, sinkholes have formed in many places, even on level land.

During each famine, farmers by the thousands migrate to the cities farther south. Those who return after an absence of a few years frequently find that their terraces have been cut by gullies or ruined by sinkholes; but the population is so dense that the land cannot be abandoned until advancing erosion forces the people off it. Farther west, along the margin of the Ordos Desert, cultivation has pressed outward toward more arid lands. The soil, loosened by the plow, has become more susceptible to wind erosion, thus artificially increasing the area of drifting sand. But here, again, population is so dense that land abandonment or a shift from cultivation to grazing is not practical.

The Yellow River flowing from the northwestern highlands carries with it each year about 2,500,000,000 tons of silt,[1] a quantity sufficient to increase by 5 feet the elevation of an area 400 square miles, or 256,000 acres, in extent. During the flood of 1933, the north dike of the Yellow River, which had been built to restrain the flood waters, was completely buried by silt. Since a principal cause of the floods of the Yellow River has been the heavy burden of silt carried by flood waters, the prevention of floods by means of engineering works alone is impossible.

Although the two countries have a common agricultural tradition, the problems of erosion control in China and Japan are striking in their contrast. Chinese agriculture developed first on the extensive delta plains where erosion proceeded slowly. Japan, on the other hand, is a mountainous region with comparatively small areas of plains country, where the major agricultural problem is the protection of the limited farming areas from flooding and sedimentation caused by the removal of the mountain forests.

The annual rainfall of Japan is heavy, ranging from 40 to 80 inches annually, and intense storms are frequent. Because of the steepness of the slopes and the nature of the soil, derived chiefly from coarse-grained granite and volcanic ash and lava, erosion is rapid where the slopes are denuded. This fact was the basis of the old Japanese proverb "To control the river is to control the mountain."[2] By the seventeenth century, reforestation was practiced widely for torrent control, but subsequent political and religious unrest prevented general reforestation of critical slopes.

About 400 years ago, the temples of central Honshiu were burned by a shogun who was hostile to Buddhism, but they were later rebuilt at the expense of the mountain forests of the Shiga Prefecture. Erosion proceeded without restraint until 1871, when reclamation was started under the direction of the Dutch engineer Johann Dorehk. "Now one may walk up the valleys, past check dams overgrown with vines and bracken over which torrential waters no longer charge, then into a carpet of . . . thick forest litter, under a full canopy of pine forest. . . . "[3] This probably marks the beginning of the modern period of flood and erosion control.

In 1897, the Forestry Act of Japan was passed; Article 14 includes regulations for erosion control. Provision is made for: the prevention of

[1] Eliassen, Sig. Soil Erosion and River Regulation with Special Reference to the Yellow River, Assoc. Chinese and Am. Eng. *Jour.*, Vol. 17, pp. [22]–38, 1936.

[2] Lowdermilk, W. C. Erosion Control in Japan, *Oriental Eng.*, Vol. 8, No. 3, pp. 3–13, 1927.

[3] *Ibid.*

soil denudation, shifting sand, floods, and avalanches and the maintenance of a constant water supply (Fig. 347). The law was not fully enforced until some years later.

FIG. 347.—Upper, type of soil and water conservation work to protect valuable cultivated lands below. Some of this work exceeds the value of the land by ten times. Lower, torrent control in the Zaoh Valley, Japan. (*Courtesy, S. Sakurai, Tokyo Imperial University.*)

The forests near the headwaters of the Kirisumi River had been subject to unrestricted cutting for centuries. Runoff after the exceptionally heavy rains of 1911 started new gullies and increased the length and depth of those already existing. Houses and bridges were washed away by floods; mud flowed from the slopes and covered the fields. In 1916, the Forest Act of 1897 was brought into operation, and a *protection forest* was established. Within 10 years, the headwater area of the Kirisumi, with the exception of two small tributary valleys, was reforested; and as the result, erosion and floods were effectively controlled. The cost of the reclamation, 120,000 yen, was more than the market value of the land restored and protected. However, the price of the land is an arbitrary figure based on its sale value rather than on the benefits derived from the maintenance of agriculture over a period of centuries.

Today, about 8.5 per cent of the total area of Japan is in protection forests; and of this, nearly 93 per cent serves the dual purpose of erosion and flood control. Within a period of forty years, 33 million yen have been spent for reclamation work by the Departments of Agriculture and the Interior, and an equal or larger sum by prefectures and private landowners. In 1925 alone, the Tokyo Forestry Board spent 170,000 yen for erosion control on an area of about 375 acres—nearly ten times the market value of the land.

The methods of torrent and erosion control applied resemble, in general, those of other countries having analogous problems. Japan, however, has introduced one interesting new feature.[1] Plants used for erosion control have been reduced to a minimum number of species, which not only bind the soil but also have a wide range of adaptability to climate, soil, slope, and erosion conditions. In this way, growing and planting have been standardized, and the cost of revegetation materially reduced. Through systematic effort and legislation, Japan has now reached the stage where accelerated erosion is practically nonexistent on farm lands.

### India and Netherlands India

From the standpoint of erosion, India occupies an intermediate position between the Old and New Worlds. Like China, it has an old culture, dense population, and great variations in climatic and physiographic conditions. The Aryan invaders, who displaced the earlier Negroid and Dravidian populations, probably entered India about 2000 B.C. Some of the forests escaped the depredations of these agricultural and pastoral invaders; and in the northern Punjab, Alexander the Great found *sal* forests of sufficient density to conceal his entire army. The Mohammedan invaders after the eleventh century A.D. had no respect for forests; and under the Mogul Empire,[2] forest destruction proceeded apace.

Although the role of forests in maintaining stream flow, preventing erosion and sedimentation, and preserving the level of the water table was recognized by early English foresters, such as Gibson and Cleghorn, and although conservation was urged repeatedly in the strongest terms, the

[1] Detwiler, S. B. Fifty Years' Experience Gives Japan Simple, Effective Program, *Soil Cons.*, Vol. 3, pp. 9–10, 1937.

[2] Stebbing, E. P. "The Forests of India." London. 1922.

Indian forest policy was not established until the latter part of the nineteenth century.

Exploitive agriculture, by contrast, developed too rapidly to permit gradual adaptation to environment. The Ganges and the Indus were the only river systems that were navigable to a large extent, and roads were little more than cart trails, passable only in dry weather. Because the era

Fig. 348.—Famine laborers reclaiming land wasted by erosion. Punjab Plains, India. (*Courtesy, E. A. Smythies.*)

of road and railroad building under the British Empire coincided with the Civil War in the United States, it was accompanied by rapid exploitation of the wheat- and cotton-producing areas of India.

Erosion, however, had been in progress for centuries but was local rather than general. Along the Jumna River, destruction of the native vegetation by overgrazing was followed by gullying severe enough to necessitate the abandonment of entire villages. Both gullying and sheet wash had ruined sections of the Chota Nagpur hills, and the silt-laden waters from the slopes had clogged the pores in the soil of the plain below and caused them to become waterlogged.

Nearly a century ago, Sleeman[1] observed that, as a result of exploitation, "the most productive parts of the surface of Bundelkhand, like that

[1] Sleeman, William Henry. "Rambles and Recollections of an Indian Official." London. 1844.

of some of the districts of the Narbada territories, which repose upon the back of the sandstone of the Vindhya chain, is fast flowing off to the sea. . . ." He also had seen "valuable estates reduced in value to almost nothing, in a few years, by some new *antennae* . . . thrown out from the tributary streams of great rivers into their richest and deepest soils." This he attributed to deforestation and to plowing and cross plowing which increased the tendency of the soil to wash. As erosion progressed, methods for its control became more difficult to apply, because under British dominion the population increased rapidly, and with it the need for cash crops and food (Fig. 348).

From the Indus in the northwest to the southern extremity of the peninsula, *bunding* has been practiced for hundreds of years for controlling erosion and conserving moisture. In general, bunds are low, earthen ridges built on the contour along the lower edges of fields. Simple bunds are characteristic of the sloping lands of the Madras and Bombay Presidencies. In the upper Narbada valley, where the land is nearly level, fields are completely surrounded by bunds that retain water on them until shortly before plowing, when it is released. Near Jubbulpore, the building of the bunds is attributed to Raja Man, who lived about 500 years ago.[1]

In the hill country of northeastern India, erosion is of recent origin, and bunding has heretofore been unnecessary. Until a few years ago, the eastern Himalayas were covered with dense forests. Small garden patches were cleared, cultivated for 2 or 3 years, and abandoned. Because of recent and rapid growth in population, the rest periods have been shortened; and runoff, erosion, and floods have increased.

The Punjab is bordered on the north by the Siwalik hills. The seasonal torrents from the hills disgorge their water and sand upon the plains, forming sandy *cho* beds. About a hundred years ago, the *chos* ran between well-defined banks, and a few perennial streams flowed from the hills. As early as 1852[2] the *cho* acreage in three areas near Hoshiapur amounted to 48,206 acres; by 1884 it had increased to 80,057 acres; and by 1897 to 94,236 acres.

In 1900, a Chos Act, passed by the Punjab Government, resulted in the establishment of a few small forest preserves and partial closure of some of the most eroded grazing lands in Hoshiapur. But, in general, overgrazing continues. Effort is being made to increase the capacity of the pastures through education of the natives, rotational closure of seriously eroded areas, the introduction of stall feeding and fodder production, and adoption of proper pasture management. Of far greater value would be a

[1] Howard, Albert. Soil Erosion and Surface Drainage, Agr. Research Inst. Pusa *Bull.* 53, 1916.

[2] Hamilton, A. F. T. Siwalik Erosion, *Himalayan Jour.*, Vol. 7, pp. 87–102, 1935.

drastic reduction in the size of the herds, which would be difficult to accomplish because of the religious scruples of the natives.

A technique for volumetric measurement of runoff and erosion from controlled plots was developed in 1936 at the Punjab Irrigation Research Institute at Madhopur.[1] The losses show marked acceleration of both runoff and erosion as the result of close clipping of the grass cover, to simulate, in some degree, the effects of overgrazing.

According to estimates based on experience and intelligent observation, some of those who have lived in the country fear that if erosion is permitted to continue at the present rate, the situation in India may, in a few decades, be as serious as that in northwestern China.

To an even greater extent than in India, soil erosion in Ceylon and Netherlands India has been caused by commercial exploitation. The building of the first road into the interior in 1824 was followed by rapid extension of coffee cultivation, and coffee remained the chief commercial product until the appearance of the leaf disease about 1880, when it was supplanted by tea.

The relation between the maintenance of mountain forests and the prevention of erosion on the tea plantations of Ceylon was pointed out as early as 1873 by Hooker. In 1885, an ordinance was passed for the protection of forests on the higher slopes; but 10 years later, the reason given for railroad extension was that transportation on the Kelani, the longest river in Ceylon, had been impaired by silting which had resulted from the extension of tea planting.

A committee appointed in 1929 to investigate erosion in Ceylon recommended that an extensive campaign of education be initiated and that after about 5 years, the subject be reviewed in order to determine whether or not legislative control would be needed. In 1935, it was found[2] that erosion control had been introduced on 70,000 acres of tea land and that 9,500 additional acres had been placed under control on rubber plantations. Further control has been assured through the Ordinance of 1935. Reserves are to be established wherever necessary to prevent erosion, and no Crown Lands above 5,000 feet are to be alienated except upon the advice of the Executive Committee on Agriculture and Lands and with the approval of the governor. Land orders were laid down for the guidance of revenue officers, and all new landowners are required to install and maintain all recommended erosion-control works at their own expense.

[1] Gorrie, R. M. Soil Erosion in India, *Nature*, Vol. 142, pp. 560–561, London, 1938; Soil Losses from Indian Forest, Grasslands and Farms, *Indian Forester*, Vol. 64, pp. 327–329, 1938.

[2] Park, Malcolm. The Soil Erosion Questionnaire—1935, *Tropical Agriculturist*, Vol. 85, pp. 369–380, 1935.

In his account of the coffee industry in Java, Gillett[1] commends to the settlers in English possessions the soil conservation practices of the Dutch planters:

FIG. 349.—Above, bench terracing for control of erosion and conservation of water—rubber plantation, East Coast of Sumatra. Below, sloping face of bench terraces (same as above) stabilized with viny legumes. (*Courtesy, C. B. Manifold.*)

"No rush methods are here employed; the planting requirements for the following year are prepared in good time. The land is cleared and dug, and then without exception, some suitable method is taken to prevent soil erosion. The method

[1] Gillett, S. Report on a Visit to Southern India and Java; II, The Coffee Industry of Java, *East African Agr. Jour. of Kenya, Tanganyika, Uganda and Zanzibar*, Vol. 2, pp. 149–163, 1936.

adopted depends according to circumstances, but whatever system is carried out it may be relied upon to be 100 per cent efficient."

Under the Dutch system, the holes for planting are dug at least a year in advance, after which the whole area is sown with some green manure crop. To reduce erosion further, contour terraces and silt pits are commonly employed, either separately or jointly (Fig. 349). The silt accumulating in the trenches is cleared out and spread over the field, or new trenches are dug to replace those which have become filled. In all cases, provision is made for drainage of excess water.

FIG. 350.—Intensive utilization of sloping land for production of rice. Java. (*Courtesy, W. A. Foot and Travelers' Official Inf. Bureau of Batavia, Java.*)

In Java, the first of the hill crops requiring scientific assistance was coffee. In order to maintain profits by reducing the cost of production, L. J. Tijl (1890–1896) developed the *weed system* to replace the earlier clean cultivation of the land. This may be regarded as the beginning of modern *hill culture* in Java, which led to the development of analogous research in the United States.[1]

On rubber plantations, a form of applied ecology known as the *forestry method* has been employed of recent years to improve yields, reduce costs, and prevent erosion by the scientific maintenance of the humus content of the soil. Under ideal circumstances, clean cultivation is replaced by selective weeding in which desirable species are retained, the less desirable eliminated, and the rubber trees allowed to reproduce naturally. On badly eroded lands, even universally condemned weeds, such as stag-horn or

[1] Hopp, H., and Cunniff, H. A. "History of Bergcultuur and Bergcultures in Netherlands East Indies." Manuscript. 1937.

bracken, may temporarily serve the useful purpose of producing a cool, damp soil that favors the growth of more desirable species.

Although Java has been outstanding in the development and application of methods for controlling erosion, these have been employed chiefly on European estates. In the rice lands, both European and native-owned, terraces are bunded to retain water, and erosion is almost nonexistent (Fig. 350). However, where upland crops are grown on native farms, erosion is general but has not as yet reached catastrophic proportions. In the future, they will require the same careful attention that has been given to the European-owned estates.

## The New World

### South Africa

In a recent address, General J. C. Smuts expressed the opinion that "erosion is the biggest question before the country today, bigger than any politics." He spoke for the Union of South Africa, but his remark is applicable to many of the countries of the New World. In these regions, ready-made social and economic structures were imposed upon lands that had never before contributed to the wealth of the world. Changes came abruptly, and economic progress was rapid, but the civilization of the Old World was seldom completely adapted to the new lands. Political maladjustments were the first to receive attention, whereas agricultural problems were regarded as matters for individual rather than national concern. The new countries at first failed to realize that rapid development of agriculture would be accompanied by early and accelerated land deterioration.

At the time of initial white settlement, the native populations of America and Australia were sparse, and nations developed unhampered by indigenous population. In Africa, by contrast, the native population was denser and increased rapidly with the establishment of peace and the control of disease. As long as population growth was held in check, soil wastage was not serious. The traditional practices of shifting cultivation and grazing were not permanently destructive to the land if sufficient time could be allowed for recuperation. Even the clearing of land by burning at first improved the quality of the vegetation. As the influx of white settlers reduced the amount of land available to the natives and restricted their mobility, primitive agriculture became more destructive.

The Union of South Africa, like the southern Great Plains of the United States, is a young region; and in both areas, climate varies from subhumid in the east to desert in the west. Periods of drought and heavier than average rainfall occur at irregular intervals and may last for a few months or for a number of years. During the periods of heavier rainfall, the area devoted to agriculture and grazing increases. Retrenchment and economic loss accompany the droughts.

In the early 1890's, rainfall was above average, and South Africa had hopes of becoming a major wheat-producing area, but droughts had been recognized as a limiting factor as early as 1684, when settlers were advised to keep on reserve a 2 years' supply of grain. A century later, Sparrman[1] warned the settlers that, if the misuse of the land continued, "the presage of the country people may chance to prove true, *viz.*, that many spots now inhabited and cultivated must be relinquished and suffered to lie waste." Today, it is recognized that the native pastures, or *veld*, are the chief source of wealth in South Africa.

Preservation of the veld was the chief aim of the drought studies initiated during the first quarter of the twentieth century. As investigations continued, it was found that erosion was integrally associated with problems of drought and pasture management. The coordinated government attack on erosion began as a result of the publication of the final report of the Drought Investigation Commission,[2] in 1923.

Veld burning and the assembling of cattle in *kraals* at night, generally employed throughout the Union, have been direct causes of accelerated erosion and have prevented the natural recovery of pasture vegetation after droughts. The custom of veld burning was traditional among the natives long before the arrival of Europeans and was readily adopted by the white settlers. As early as 1687, pasture deterioration as a result of repeated burning had been observed by Simon van der Stel, who issued the first *placaat* prescribing punishment for offenders. Severe scourging was the penalty for the first offense; and "the cord until death do follow," for the second. Subsequent generations have passed laws to regulate burning, but no government of the past has been strong enough to enforce them. "The most pressing and necessary change in farm methods" in South Africa, according to the Drought Investigation Commission, however, is the abolition of *kraaling*. This practice not only precludes natural recovery of overgrazed pastures but also results in the formation of cattle trails that serve as starting points for *donga*, or gully erosion. It has been calculated that the income of the Union from wool alone could be doubled if fencing and paddocking (in place of kraaling) were generally adopted.

As a part of the program of veld management, experiments are now in progress to determine the pasture values of selected native and foreign grasses, methods of propagation, breeding possibilities, and the relative

[1] Hall, T. D. South African Pastures: Retrospective and Prospective, *South African Jour. Sci.*, Vol. 31, pp. 59–97, 1934.

[2] Union of South Africa Drought Inv. Comm. *Final Rept.*, October, 1923, Cape Town, 1923.

values of different grasses to serve as soil binders and to retard runoff. The Union of South Africa also recognizes the fact that the preservation of soil and natural vegetation is an obligation of the government, but financing such work presents a serious problem.

> "We know that these undeveloped areas of natural veld are deteriorating rapidly under man's management. To what extent is it in the farmer's hands to remedy this? As a business man, and a small one at that, he must make his 5 per cent to 10 per cent each year on capital invested. Thus, before we can advise him to sink more money in fences, water facilities, artificial pastures, etc., we must be sure that it will be worth his while financially, in the immediate present."[1]

To encourage the installation of erosion-control works, South Africa has adopted a program of government subsidy for farmers. Under Scheme *A*, the control works are installed by the farmer at his own expense; and on their completion, the state pays a bonus of $33\frac{1}{3}$ per cent on the final valuation. Scheme *B* provides for loans from the Department of Agriculture and Forestry at $3\frac{1}{2}$ per cent interest, the capital to be repaid over a period of thirty years. Scheme *C* was established to encourage the use of unemployed European workers during the depression. Since the beginning of economic recovery, it has suffered from lack of available labor, and farmers are now being encouraged to shift to Scheme *A*. As supplementary measures, Schemes *D* and *E* have been adopted to provide loans for the purchase of equipment and rebates on the cost price of fencing to be used for erosion control.

By the conclusion of the fiscal year ending Aug. 31, 1937,[2] over 18,000 dams, at an estimated cost of £2,163,260, were approved under Schemes *A*, *B*, and *C*. These serve the immediate need of conserving water and increasing fodder production and, by facilitating proper veld management, help prevent future erosion. Because the reclamation of eroded areas has lagged, the government is considering new methods for encouraging and subsidizing gully control and the construction of contour banks.

In spite of governmental effort,[3] many farmers still fail to realize the gravity of the erosion problem; others have insufficient capital to install the necessary control measures; and still others feel that the government is trying to obtain control of their farms.

[1] Pasture Research in South Africa, Union of South Africa Dept. Agr. and Forestry, *Progress Rept.* 1, Pretoria, 1938.

[2] Viljoen, P. R. Strengthening Agriculture: Improved Agricultural Conditions in the Union, Annual Report of the Secretary of Agriculture and Forestry for the Year Ending 31 August, 1937, *Farming in South Africa*, Vol. 12, pp. 471–517, 1937.

[3] Nöthling, I. J. The Attitude of Farmers towards the Soil-erosion Scheme, *Farming in South Africa*, Vol. 13, pp. 113–114, 1938.

"Whether South Africa will become another uninhabitable desert or a prosperous pastoral people will depend on how quickly the mass of the people react to the Government's extensive efforts in trying to arouse a national realization of the seriousness of the position with regard to erosion control and the necessity for keeping fewer but better animals on more productive pastures."[1]

## East Africa

The countries of East Africa,[2] from Southern Rhodesia to Somaliland, have essentially the same erosion problems (Fig. 351) because of similarity of relief, climate, and native customs. The low coastal plain is backed on the west by mountains and semiarid plateaus whose elevation is sufficient to make the temperature suitable for white settlement.

Agricultural and grazing customs are as widespread as are the racial groups. The Bantus, from Uganda to the Union of South Africa, practice shifting agriculture, employ fire for clearing the land and hunting game, and have an almost religious veneration for livestock. Before the era of white settlement, disease and intertribal war prevented rapid increase in population, and a delicate balance was maintained between the land and the number of people and animals that it supported. Enforced peace destroyed this balance, and the settlers brought with them methods of farming that were new to the natives. When these did not interfere with the traditional way of living, they were readily adopted. But practices that conflicted with the ingrained social and agricultural structure of native society, such as the limitation of the size of herds, were rejected even though necessary to maintain the productivity of the land.

Of the East African countries, Northern Rhodesia approximates primitive conditions most closely. Although burning and shifting agriculture have been practiced for centuries, most of the country is still forested, and erosion is negligible. But near the roads and railroads, plowing has become general; corn is supplanting finger millet; less land is devoted to ground nuts; and sweet potatoes are being planted flat instead of in hills. In spite of accelerated erosion, reversion to primitive methods of farming is not advisable if the land is to support an increasing population. The natives rather must be educated to use the plow in conjunction with mixed farming, green manuring, and the application of fertilizers.

Even more dangerous is the premature growth of commercial farming. In Uganda, where the natives are producing cotton for export on large

[1] Hall, T. D. South African Pastures: Retrospective and Prospective, *South African Jour. Sci.*, Vol. 31, pp. 59–97, 1934.

[2] Stockdale, Sir Frank. Report . . . on his visit to East Africa, January–March, 1937. Great Britain Colonial Advisory Council of Agriculture and Animal Health. C. A. C. 345. [London.] 1937.

communal fields, year after year, gullies 2 or 3 feet deep frequently are formed during a single rain.

As population increases, native methods of farming become more destructive. Hobley[1] estimates that more than 1,000 square miles of "magnificent primeval forest" in Uganda have been destroyed by the Southern Kikuyu tribes during the last 85 years. In sections of the Taita ranges of northern Tanganyika, 90 per cent of the cultivable soil has been removed since the forests were destroyed. Similar complaints come from Nyasaland, where immigration from Portuguese East Africa augments

FIG. 351.—Land waste by erosion. Rhodesia, Africa.

the natural increase in population. According to Dr. Robert Law (quoted by Hobley), the mountains of the north were well wooded a hundred years ago and drained by perennial streams. Within his memory, 20 such streams have ceased to flow, and the agricultural capacity of the land has been reduced 50 per cent, chiefly as a result of sheet washing.

Today, however, the most seriously eroding areas within East Africa are the overgrazed native pastures. Cattle tracks serve as foci for gully formation, and the pastures near the waterholes are overgrazed and trampled to such an extent that rainfall does not readily penetrate the soil. From these centers, overgrazing spreads outward until it may be necessary for the cattle to travel 20 miles to secure water. On exhausted pastures, cattle are replaced by the more destructive sheep and goats.

[1] Hobley, C. W. Soil Erosion: A Problem in Human Geography, *Geog. Jour.*, Vol. 82, pp. 139–150, 1933. Paper read at the afternoon meeting of the Society on May 8, 1933, immediately after Mr. Champion's paper.

When the ground cover is gone, the agile goats stand on their hind legs and browse on tree and shrub vegetation to heights of 5 or 6 feet. Near Lake Baring, in Kenya, Champion[1] observed a line some 5 feet above the ground level below which no foliage had survived; it represented the upper limit of goat grazing. When goats have exhausted all pasture possibilities, nothing remains but an eroding, desertlike waste.

Where education has overcome the native antipathy for reducing herds, the problem of adequate compensation remains. Scrub cattle can be used for beef extract, but only the skins of goats have commercial value—usually not more than 6 or 8 cents apiece because of their poor quality. Since other natives provide the only market for live goats, their sale represents merely the shifting of overgrazing from one area to another.

Within the last decade, overgrazing has become especially acute in Kenya because of a plague of locusts lasting from 1929 through 1931, followed by a prolonged drought. In the northwest, where the rainfall is low and high winds prevail, it has been estimated that the Turkana Desert is advancing at a rate of 6 or 7 miles a year. In 1937, the Native Authority Ordinance made illegal the burning of live grass without the permission of the tribal headman, and the Crop and Livestock Rules authorized the scheduling of areas for controlled stocking, at the discretion of the District Commissioner. Grazing regulations have proved effective within the Machakos Reserve, where the surplus cattle are sold to the local Liebig beef extract factory, with a capacity for handling 30,000 head of cattle annually.

In a zone 2,000 miles wide, straddling the equator (Map 17), the tsetse fly, the carrier of the fatal sleeping sickness, provides an effective check on the animal population. The infested area, comprising most of central and about half of the total area of East Africa, has been saved by the tsetse fly from erosion contingent on animal-drawn plows and overgrazing.

The late C. F. M. Swynnerton, director of the Tanganyika Tsetse Research Department and the leader in this field of investigation, questioned the right of our generation to eradicate the fly, saying[2]:

"The tsetses are the most potent preservers of the natural flora and fauna. Drive out the tsetse and the whole landscape changes. The close settlements of the pastoral agriculturalists press in, very soon every tree and shrub is cut down, the ground is denuded of grass and for part of the year looks like desert, cultivation covers much of it during the rains, and during the dry season its coarse and

[1] Champion, A. M. Soil Erosion in Africa, *Geog. Jour.*, Vol. 82, pp. 130–139, 1933.

[2] Swynnerton, C. F. M. The Tsetse Flies of East Africa: A First Study of Their Ecology, with a View to Their Control, Royal Entomological Soc. London *Trans.*, Vol. 84, pp. 464–465, 1936.

untidy stubble adds to the ugliness of the scene that has replaced the beauties of nature. . . . Scrub cattle replace the fat game animals and, annually destroying their own grazing, die in large numbers from starvation. The ground, laid bare to the winds and rains, undergoes sheet or gully erosion according to its soil and gradient and becomes more devastated yearly."

Were it not for the tsetse fly, Tanganyika could support a far denser population, provided trees were preserved, soil-conserving methods of

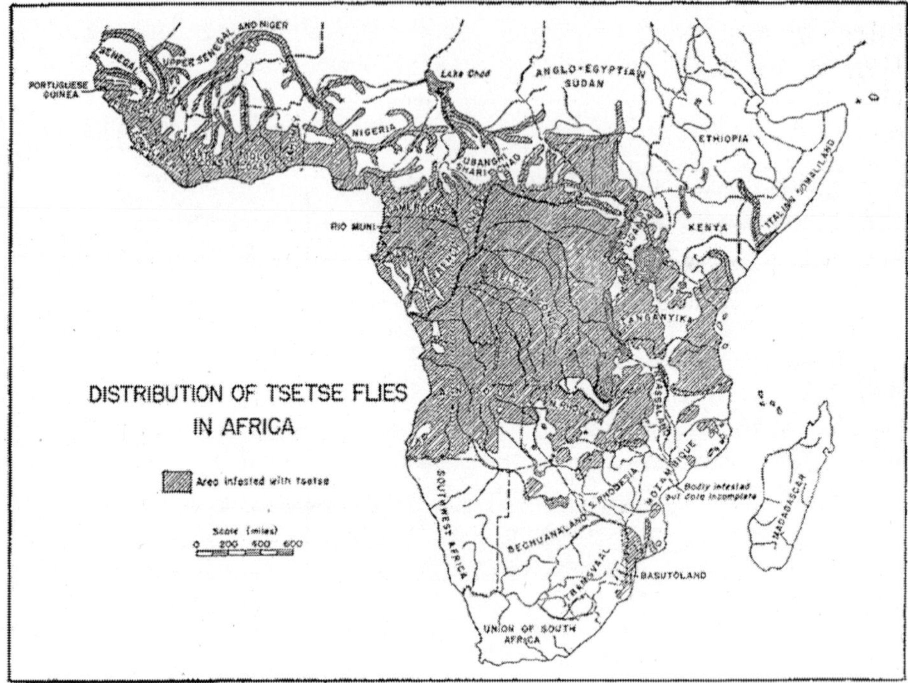


MAP 17.—Showing areas infested with tsetse flies. Africa. (*Adapted after Swynnerton.*)

farming adopted, and the number of livestock closely regulated. But Swynnerton continues:

"I do not believe that we have in the past shown fully our fitness to reclaim; but I do think I see the cloud, as yet no bigger than a man's hand, that indicates that the pleas of the apostles of anti-erosion are at last producing an impression, and that the land we reclaim in the future will be subject to regulated settlement and limitation of the numbers of its cattle.

"If I am right, we are justified in continuing to reclaim for the relief of existent erosion and the fuller development of the country."

On European estates, soil wastage has been less serious in pastures than on cultivated land. At first, no provision was made for erosion control, because plantation owners had had little experience with the problem in Europe, and because effort was devoted primarily to the rapid expansion of commercial farming. Uganda, in 1916, had but 133,000 acres in cotton; by 1926, the cotton area had increased to 533,000 acres; and by 1936, to 1,500,000 acres. A part of the Serere Station was practically denuded of soil within 20 years. Similarly, at the Makwapala Cotton Experiment Station in Nyasaland, erosion became so serious that, three years after its establishment, it was necessary to ridge-terrace the entire station.

Following the example of Ceylon, Tanganyika inaugurated an intensive educational campaign to be followed by a systematic survey to determine whether or not erosion-control legislation was needed. Much the same policy has been adopted in other East African countries. In all of them, advice on erosion control is available, and most of the countries provide financial aid for work. Throughout East Africa, experimentation is in progress to determine, on a quantitative basis, the amount of erosion occurring under different types of land use.

The reduction of crop yields as a result of erosion and the general world depression were felt almost simultaneously by the European landowners. This is one of the major reasons for the success of erosion-control education. In Nyasaland, for example, control measures are in use on practically all of the European-owned tea estates. To the East African native, however, erosion control often means nothing more than an increase in the work needed to produce food. More frequently, it requires a reduction of stock and an actual curtailment of visible wealth.

The most successful approach to the erosion problem has been through education of native leaders who, in turn, can influence their fellow tribesmen more effectively than can foreign agriculturists. Practically all of the hill tribes of Tanganyika and a large number of the local native councils of Kenya now have erosion-control regulations, passed and enforced by the tribal authorities. The use that may be made of visual education is shown by a recent bulletin[1] of the Soil Conservation Service of the Department of Agriculture, Kenya, East Africa, whose attractive pen sketches illustrate the meaning of erosion and methods for its control on native farms.

The increasing interest in erosion control, among both Europeans and natives, provides hope that the little "cloud" visualized by Swynnerton may grow until it effectively covers all of the countries of East Africa.

## West Africa

Erosion has been less serious in West than in East Africa, although shifting cultivation and burning are characteristic of both regions. Equatorial West Africa still remains in almost its natural state; but in the

[1] The Evils of Soil Erosion and Ways of Preserving the Land, Kenya Department of Agriculture Soil Conservation Service. Nairobi, Kenya.

countries to the north of the Gulf of Guinea, European influence is greater. Along the coast are tropical lowlands, bordered on the north by a transitional zone of mixed evergreen and deciduous trees occupying the highly dissected margin of the interior plateau. The interior is semiarid, and its vegetation ranges from savanna on the south to thorn scrub along the desert margin farther north.

The widely criticized indolence and primitive farming methods of the natives, combined with the prevalence of the tsetse fly and a climate ill-suited to white settlement, have preserved much of the tropical zone from erosion. The native garden plots are never clean-weeded; and under shifting agriculture, the land is not abandoned completely, as in Ceylon and the East Indies, but allowed to lie idle with a definite expectation of recultivation when the soil has recovered.

Within the last half century, cacao has become an important export, particularly from the Gold Coast and Nigeria; but unlike commercial crop production in East Africa, it is grown chiefly by the natives on farms seldom larger than 10 acres. Although methods of cultivation are primitive, Van Hall[1] advises against the introduction of improvements, such as wider spacing or clean cultivation, that would lead to accelerated soil wastage, unless sufficient care were given cultivation, ground cover, and shade trees.

Because of steeper slopes and the prevalence of easily eroded soils in the central zone, the destruction of the forests is followed by rapid acceleration of erosion. Ainslee[2] has described the eroded lands of the Udi plateau as "unsurpassed in ugliness and uselessness." Much of the topsoil has been removed by sheet wash, and gullies have cut to depths of 100 feet.

Erosion control was neglected until 1928, when the Public Works Department found that headward-cutting gullies were approaching dangerously near the main Udi-Enugu highway. A reserve was then established; and within it burning, grazing, and farming were forbidden. Rows of contour ditches were constructed in such a way that the intervening spaces of one row were directly opposite the ditches of the next. After partially leveling the walls of the steeper gullies, a few small earthen dams were built near their heads. The ridges bordering the ditches were seeded, and the remaining area forested. By 1936, the natural vegetative succession was completely restored, and further planting seemed unnecessary.[3]

[1] Van Hall, C. J. J. "Cacao." London. 1932.

[2] Ainslee, J. R. Soil Erosion in Nigeria. British Empire Forestry Conference, South Africa, 1935. Kaduna, Nigeria. 1935.

[3] Grasovsky, A. A World Tour for the Study of Soil Erosion Control Methods, Oxford Univ. Imp. Forestry Inst. *Paper* 14, Oxford, 1938. (Mimeographed.)

The natives of central and northern Nigeria have evolved a system of ridge and basin cultivation that effectively checks erosion. Contour trenches, dug with a hoe and interrupted at intervals by cross ridges,[1] collect the runoff and retain the soil washed from above. Although independently developed, the system resembles basin listing, which is now used successfully in California orchards and in other parts of the United States.

In northern Nigeria and the adjacent French colonies, wind erosion already has become critical. Prophecies of increasing desiccation, however, have awakened general interest in the problem of land deterioration. Outstanding authorities, including E. P. Stebbing,[2] have stated that the area of blowing sand is spreading, that the water table has been lowered and rainfall reduced as a result of excessive clearing and cultivation, and that general southward migration of peoples from the edge of the Sahara into northern Nigeria has been caused by progressive desiccation or desert expansion.

But all scientists do not concur in these opinions. Well borings indicate that the lowering of the water table is local[3] and that the failure of wells usually is caused by neglect. Before the opening of the West African coast, trade crossed the Sahara; and by the establishment of towns as far north as possible, the length of the desert trip was shortened. When the trade shifted to the sea routes, southern locations were more desirable, and the population shifted accordingly. It is claimed by some observers that the more recent migration from the French Niger Colony to Nigeria has been caused by the digging of additional wells and tax reduction in the British territory rather than by desiccation.

Tree planting is the most widely practiced method for reducing wind erosion. In the Katsina Emirate of northern Nigeria, a belt of dum palms about $\frac{1}{4}$ mile wide is being planted along the international boundary. Trees set out in 1934 were growing well two years later. Supplementary strips of *Dalbergia sissoo*, oriented at right angles to the prevailing winds, proved less successful; and as a substitute, all uncultivated areas of more than a square mile in extent are being protected from burning, cultivation, and grazing and are allowed to revert to their natural condition. Following the example of Nigeria, the governments of the Gold Coast[4]

[1] Stamp, L. Dudley. Land Utilization and Soil Erosion in Nigeria, *Geog. Rev.*, Vol. 28, pp. 32–45, 1938.

[2] Stebbing, E. P. "The Forests of West Africa, and the Sahara: A Study of Modern Conditions." London and Edinburgh. 1937.

[3] Jones, Brymor. Desiccation and the West African Colonies, *Geog. Jour.*, Vol. 91, pp. 401–423, 1938.

[4] Marshall, R. C. Report on the Forestry Department for the year 1935–1936, Accra, Gold Coast, 1936.

and French West Africa are protecting existing trees and planting shelter belts to prevent further extension of areas of mobile sand.

In northern Nigeria, the introduction of mixed farming, dairying, leguminous cover crops, and fertilization has been comparatively successful. The plow, however, is recognized as a dangerous introduction which may accelerate erosion unless Nigeria can profit by experience in other parts of the world.

### Australia

Land exploitation in Australia and the consequent erosion are comparatively recent. In the days of sailing vessels, distance alone was a deterrent, and the accounts of the early explorers did little to overcome prejudice against settlement in a region so remote. During the first half of the nineteenth century, colonization spread slowly along the eastern coast; and although the discovery of gold in the 1860's stimulated settlement west of the mountains, the population of Australia seventy years ago was still under 1,500,000. Australia, long separated from the rest of the world by its surrounding seas, had retained many archaic forms of plant and animal life. Colonization brought not only new methods of cultivation but also new crops and animals.

In any consideration of erosion in Australia, it must be recognized that the continental character of the country results in such great diversity of climate, topography, and land utilization that generalization is impossible. Australia is as large as the United States. The northern part, lying within the tropics, has summer monsoonal rains, whereas the southern coast has a marked winter maximum. Along the eastern coast, the rainfall is more evenly distributed. Toward the interior, the precipitation becomes progressively lower; a fourth of the continent has rainfall averaging less than 8 inches a year. Correspondingly, the eastern coast and mountains originally were covered with dense eucalyptus forests, which give way toward the west first to *mallee* vegetation, consisting of isolated eucalyptus trees and a fair cover of grass; then to saltbush and bluebush scrub; and finally to *mulga* (scrub acacia) and the spinifex vegetation of the desert.

The first settlers attacked the eastern forests with fire and axe to open farms and pastures and to supply the markets with eucalyptus, a wood hard enough to be used for cogwheels by the British shipyards. In the more arid and remote interior, wood was cut locally for fuel. With the cutting of the forests, the flow of rivers became less regular, sediment blocked the channels, and floods increased. Since 1886, the bridge over the Avon at Stratford, in Victoria, has had to be lengthened 600 feet because of siltation and river shifting; and on the Tambo flats, the productivity of some of the lower fields has been seriously reduced by sand deposits.[1]

The siltation of the Hume Reservoir, however, is of national concern because it is the largest source of irrigation water in the Murray drainage basin. This area, including the tributary Darling Valley, comprises one-seventh of the mainland of Australia and, together with the narrow coastal strip between it and the sea, contains three-fourths of the population. Although the River Murray Commisson reported in 1928 that erosion was not serious nor silting excessive, it recognized the danger of future forest destruction. Further alienation of Crown Lands was prohibited, and lessees were forbidden to interfere with the native timber. In spite of this, extensive burning continues, and both erosion and sedimentation have increased since 1928.

The high mountain pastures are burned annually to remove old growth and improve the grass. Fires spreading to the forests are not checked, because they are believed to "do no harm." In 1930, forest burning, combined with unusually heavy rains, caused a landslide along the Geehi River,[2] a tributary of the Murray, damming it with an earthen wall 20 feet high. The course of the stream was diverted; and for two weeks thereafter, the river, for a distance of 15 miles, flowed mud instead of water.

The enforcement of erosion control in the foothills and the lowlands is more difficult, because the land is privately owned and has been cleared of all tree growth. In the granite areas, gullies 50 feet deep have cut into the soil and bedrock; and at their mouths, fans of sand spread out over fields and pastures. Some of the gullies are discharging their sediment directly into the Hume Reservoir.

In the semiarid lands west of the mountains, farms are large; and because of mechanized agriculture, the cost of wheat production is low. For a few years after the land was broken, crop yields were good. Thereafter, the story is much like that of the Great Plains of America. High prices combined with a succession of years with heavier than average rainfall, from 1915 to 1926, caused an overexpansion of the wheat area. The soil, pulverized by frequent plowing and reduced in content of binding organic matter, became susceptible to blowing during the recurrent droughts and to sheet washing and gullying during the rains. As much as 6 feet of soil have been removed by wind in some of the dry, sandy areas since settlement.

[1] Report of Committee Appointed to Investigate Erosion in Victoria. Victoria Department of Lands and Survey, Soil Erosion Committee. Melbourne, 1938.

[2] Byles, B. Y. Report on a Reconnaissance of the Mountainous Part of the River Murray Catchment in New South Wales, Commonwealth Forestry Bur. *Bull.* 13, 1932.

Farmers and herdsmen, advancing westward, encountered a more arid climate, characterized by frequent droughts. During the thirty-six years from 1900 through 1935,[1] the northwestern section of New South Wales experienced eighteen years of desert climate. To the east, the number of desert years decreased at an irregular rate. There were eighteen desert years at Wanaaring; but Walgren, 150 miles farther east, experienced but six during the thirty-six-year period. In 1929, the eastern interior of Australia was still suffering from a drought that had lasted for several years, and 1929 was the third successive year in which the South Australian Parliament had extended drought relief to farmers. Western Australia, however, had no drought that year.

The difficult problem of adjusting herds to such unpredictable climatic conditions is further complicated by the presence of hordes of rabbits, incidentally introduced by the early settlers. The rabbits devour the succulent young vegetation and force the sheep to overgraze the perennial saltbush and bluebush, even in good years. Although their number decreases rapidly during droughts, rabbits multiply with equal rapidity when the rains begin and consequently prevent natural regeneration of vegetation. Before settlement, only the annual vegetation died during droughts; and although some wind erosion occurred, the saltbush, bluebush, and mulga prevented widespread destruction. When the perennial vegetation is destroyed, no reserve of food remains for the herds, and wind erosion is unrestrained.

The effects of wind erosion are not limited to the livestock industry. Between 1929 and 1931, the Victorian Sand Drift Committee, which controls 4,000 miles of river channels, spent over £300,000 clearing sand drift from watercourses. The Railway Department spends large sums annually to maintain service in areas of drifting sand.[2]

Among the suggestions for the control of erosion on pastures are the resting of paddocks and the fencing of small areas to permit natural regeneration and seed dispersal, but regeneration is slow in areas of scant rainfall. The control of rabbits would undoubtedly help, but rabbit-proof fencing, the digging out of burrows, or even the plowing of strips across pastures may be too expensive for land of low value. A better remedy probably would be the regulation of grazing and the rejuvenation and preservation of indigenous perennial vegetation. For the Western Division of New South Wales, where the land is still owned by the Crown, the suggestion has even been made that, in the long run, it would

[1] Holmes, Macdonald. The Meaning of Soil-erosion, Univ. Sydney *Pub. Geography* I, Sydney, Australia, 1938.

[2] Richardson, A. E. V. The Problem of Soil Erosion in Pastoral Areas, *Pastoral Rev.*, Vol. 45, pp. 1055–1057, 1935.

be cheaper to pension off the 18,500 pastoral inhabitants than to allow the land to degenerate and then spend vast sums on reclamation.[1]

The Australian Council of Agriculture has decided that erosion has become a sufficiently important national problem to justify investigation by the Council for Scientific and Industrial Research.[2] A preliminary survey of the nature, extent, and possible control of erosion already has been initiated under the direction of F. N. Ratcliffe. In this way, it will be possible to evaluate the divergent opinions on erosion, some of which tend to stimulate interest in the national menace, and others to underestimate the extent of the problem, often from the standpoint of national pride.

The new act establishing the Soil Conservation Service in New South Wales provides for procedures that in many ways are similar to the provisions of the soil conservation districts acts recently passed by 36 states of the United States (see *A Standard State Soil Conservation Districts Law*, prepared, at the suggestion of representatives of a number of states, by the Soil Conservation Service, United States Department of Agriculture, 1936).

The following statement with reference to this soil conservation bill is given below[3]:

"Until a few months ago, it was possible for most of us to read with complacent interest of enormous losses of soil in the American 'dust bowl', and of the spectacular effects of water erosion in other parts of the United States. But Australians have now been shocked into the realization that the devastating forces which have ruined, or are even now attacking a sixth of the area of the United States, are at work here. Wind in the West and water in the eastern areas are stealing our soil. Pasture improvement, crop rotation, improved cultivation and manuring have bolstered up production against the natural drain upon the earth's fertility. But erosion, the greatest depleting agency of all, has worked unhampered. It has even been encouraged by some of the recognized methods of farming, until the damage has become so obvious that the State has been forced to act.

"Investigations had been proceeding in New South Wales for about five years, but until the introduction last month of the Soil Conservation Bill public interest had been aroused only spasmodically. It is perhaps unfortunate that we shun the spectacular methods of American publicity for national questions of this

[1] Holmes, Macdonald. The Erosion-pastoral Problem of the Western Division of New South Wales, Univ. Sydney *Pub. Geography* 2, Sydney, Australia, 1938.

[2] Richardson, A. E. V. Shifting Sands—The Growth of the Menace in Australia, *Proc.* Royal Geog. Soc. Australasia, South Australian Branch, Sess. 1934–1935, Vol. 36, pp. 43–51, 1936.

[3] Anonymous. The N.S.W. Soil Conservation Bill, Australian Inst. Agr. Sci. *Jour.*, Vol. 4, pp. [121]–123, 1938.

kind. A wider and more vigorous public appeal might have saved some of our valuable wheat lands years ago. No doubt some areas were saved through the co-operation of farmers with departmental experts, for the control of water erosion in its early stages is a simple matter. But action on a wider scale needs legislative support and the establishment of an authority to deal with this matter and nothing else. That is what the new Bill will do. It has passed both Houses of Parliament, and will shortly become law. It is the first legislation of its kind in the Commonwealth, and it is to be hoped that it will inspire similar legislation in other States.

"The basis of the Bill is co-operation between the landowner and a special Soil Conservation Service. It is not intended to deal with what might be termed normal, geologic erosion, but with the accelerated erosion which is due, largely, to the efforts of man. It is in this sense a measure of insurance against further loss of capital, for the surface soil of a country is its richest asset. It has become the custom in Parliament to attach to each important Bill a short explanatory note. We can do no better than quote the note attached to this Bill for a description of the things it sets out to correct:

"'The continued losses of the surface soils of our rural lands as a consequence of erosional action, the depreciation of vast tracts of grazing areas from the same causes, the siltation of water storages, the increasing liability to floods following abnormal run-off through vegetative denudation, the erosion of river banks and associated damage to adjacent lands, demand that a rational soil conservation policy be initiated and vigorously pursued in the interests of every person in the State.'

"The new Service will thus have a vast field in which to operate. Its first work probably will be a reconnaissance of the damage caused by the erosional forces of wind and water. The areas particularly liable to serious erosion will be indicated, and the landowners concerned will be invited to enter into voluntary agreements with the Service with the object of mitigating erosion. They may receive advances out of moneys provided by Parliament to have the work done. Owners will not be forced to mitigate erosion purely in their own interests. Only when damage originating from an owner's land is causing deterioration of the lands of adjoining owners, or of existing national works, will there be any element of compulsion. At the outset, there was a good deal of apprehension among farmers about this compulsory provision, but fuller discussion, in Parliament and at various farmers' conferences, appears to have satisfied most people that 90 per cent of the Soil Conservation Service's work will be based on voluntary co-operation. The method to be adopted with private lands is interesting. Section 17 of the Bill states:

"'Where the Minister is of opinion that any tract of land is subject to erosion, or is liable or likely to become liable to erosion, and that such tract of land should be notified as an area of erosion hazard, he may, by notification published in the Gazette, and in a newspaper circulating in the locality in which such tract of land is situated, give notice that such tract of land should be notified as an area of erosion hazard.'

"Owners concerned may then lodge objections, and if these are rejected by the Minister they may appeal to the Land and Valuation Court. When the area has been notified finally, the Director of the Service may be authorised to enter upon any specified land therein and initiate any work considered necessary, but not unless (a) the Minister is satisfied that work on the land is necessary to protect other land in the area, (b) the owner has neglected or refused to enter into an agreement, and (c) the owner has been given notice, against which, again, he may lodge an objection. A further provision empowers the Minister to demand contributions from such owner towards the cost of the work done on his land, and again there is a right of appeal. Thus the compulsory provisions of the measure appear to be hedged about with adequate safeguards.

"The Government has been careful to point out that it would be impossible for the State to carry out anti-erosion work on every farm requiring it. It is intended to limit the main activities in agricultural and grazing lands to districts where requests for assistance originate, and where urgent action is most necessary. Whatever work is done on private properties, according to a Ministerial statement, will be done in districts where the voluntary co-operation of the farmers is assured.

"It appears, then, that precept and example will be the two big factors in the work of combating erosion. The need for soil protection is so urgent that the Conservation Service must become almost at once one of the most important of all Government agencies."

## Canada

In Canada, the problem of erosion ceases to be "foreign." The boundary between the United States and Canada is purely political; and erosion, like climate and vegetation, takes no cognizance of this arbitrary line. Similarly, settlement spread from lower, or French, Canada to New England. In both areas, the farms were long and narrow so that each settler might have direct access to river or highway transportation. The shape of the farms was in itself conducive to erosion; and in addition, French agriculture was notably backward. Wheat was grown year after year; and during the winter, the manure from the barnyard was dumped on the ice of the St. Lawrence so that this "useless encumbrance" might be washed away by the spring thaws. By the beginning of the nineteenth century, yields were so reduced that wheat lands were being abandoned.

Although the British settlers in upper Canada were better and more progressive farmers, their lands gradually deteriorated, and floods are said to have become more frequent as the forests were removed. On the impoverished fields, dairying and mixed farming superseded wheat. In Ontario, cheese manufacture, which made possible the agricultural change without financial loss to the farmers, was introduced via the Mohawk Valley by westward-moving pioneers and British loyalists from the newly established United States.

The Prairie Provinces remained under control of the Hudson's Bay Company until 1870, when they were purchased for a nominal sum by the Dominion of Canada. Although the land had been surveyed before it was opened to settlement, population advanced westward in spite of the advice of surveyors like Palliser.[1]

As in the Great Plains of the United States, profits from newly opened wheat fields were at first large; but within a few years, yields began to decline, and wind erosion made its appearance. Near Indian Head, Saskatchewan, drifting was recorded as early as 1887; the first intensive effort to control it was initiated in 1918 in the vicinity of Monarch, in southern Alberta. Because farmers had noticed that the soils within 10 or 20 yards of the western margins of their fields were the last to show signs of blowing, strip cropping was introduced. So successful were their efforts that today the Monarch area is considered the outstanding example of erosion control on land susceptible to blowing.[2]

Of the 50 million acres of grain land in the Prairie Provinces, about a fourth is left fallow and exposed to wind erosion each year. Although drifting attracts greatest attention during the dry periods, the soil recuperates slowly, and reduced crop yields are said to continue long after the conclusion of the droughts. By 1935, erosion was so widespread and the reduction of crop yields so serious that the Prairie Farm Rehabilitation Act was passed by the Dominion Government to assist in the reclamation of eroded farm lands and to develop and promote systems of farming, tree planting, water conservation, and land settlement that would afford greater economic security to the farmers. During 1935–1936 and 1936–1937, appropriations of $780,000 a year were granted for rehabilitation; and for 1937–1938, an appropriation of $2,000,000.

In order to forestall damage to farm land by drift from neighboring fields, the Alberta Control of Sand Drifting Act of 1936 holds each farmer responsible for damage to neighboring land if it is a result of drifting caused by his own negligence. No claim for damage, however, may be made unless the claimant has installed on his own farm the precautionary measures prescribed by the Act.

On the short-grass range lands of southwestern Saskatchewan, southern Alberta, and the foothills of British Columbia, erosion did not become serious until after 1915, when a series of dry years caused overgrazing and resulted in general range deterioration. This led to the establishment of the Dominion Range Experiment Station at Manyberries, Alberta; and following the drought of the early 1930's, a branch

[1] Palliser, John. Journals, detailed reports, and observations relative to the exploration of that portion of British North America which in latitude lies between the British boundary line and the height of land of the Northern or Frozen Ocean, respectively, and in longitude between the Western Shore of Lake Superior and the Pacific Ocean, during the years 1857, 1858, 1859, and 1860. British Parliamentary Papers. 1863.

[2] Hopkins, E. S., Barnes, S., Palmer, A. E., and Chepil, W. S. Soil Drifting Control in the Prairie Provinces, Canada Dept. Agr. *Bull.* (New Series), No. 179, 1937.

station was opened at Kamloops, B. C., in 1935. The pasture and range management investigations at both stations are directed toward the prevention of overgrazing, a major cause of soil erosion.

In both the northern Great Plains of the United States and the Prairie Provinces of Canada, the causes of erosion and the techniques employed for controlling it on farm and range lands are essentially the same. Canada, however, still has one opportunity that no longer exists in the United States. The Peace River Valley is today a land of agricultural pioneers where the application of control measures may forestall the need for later reclamation.

# Chapter XLII. Research, an Arm of Coordinated Land Use

A growing realization of the menace of erosion during the past few decades has given rise to numerous investigations into the scientific aspects of soil and water conservation and land use by agriculturists of the states and of the Federal Department of Agriculture. *Most reassuring, however, is the purpose of the nation behind the Soil Conservation Act of 1935. Congress, in this Act, adopted the policy to provide for a permanent agriculture in the prevention and control of erosion as it menaces the national welfare. This Act of 1935 marks the beginning of a new era in the relationship of the American people to its land, a transition from exploitative to conservative use of land.* Only a beginning has been made, however, in the direction of permanent conservation of the soil resource. A long road of discovery, adjustments, and refinements in the *science and practice of soil conservation* lies ahead. Advance along this road must not be delayed, for the ravages of erosion proceed uncompromisingly with every rain and dust storm.

Even with past attention to this problem, the science and practice of soil conservation have lagged behind other phases of agricultural science. The effect of soil wastage has been obscured or nullified for the time by other gains, such as the development and large-scale use of fertilizers, improved agronomic practices, and improved varieties and hybrids of crops. An urgent need in agricultural science today is to bring up into line the science and practice of soil conservation.

As an aid to planning and work on the land, a continuing program of integrated research is essential to the success of a national program for soil and water conservation and correct land use. It will not be enough merely to find new and more efficient methods for supplying the physical needs of the great diversity of land types in the interest of such pressing problems as conservation of soil, water, grass, forests, and wildlife. These are tremendously important problems and must be solved. But it is also important that better methods be found through which these ends can be accomplished so that the results will accord with the economic and social needs of those who live on the land. Conservation of soil and water, control of floods, and establishment of wildlife refuges and recreational and forest areas through methods that force people off the land, although necessary in some localities, will, if pursued without due regard for the needs of the entire nation, lead eventually to insurmountable difficulties. Accomplishment of proper land use will, of course, necessitate the removal of some people from some areas, chiefly submarginal; but mathematically, the process cannot go beyond the productive capabilities of the land available for population transfer.

## Problems Involved

Defense of lands upstream may be necessary for the protection of a reservoir or of farm land downstream, on which the welfare of those living in the lower part of a valley may depend. Availability of farm land; tax delinquency; the debt situation; prices, transportation, market, and school facilities; and even the character of the people may all bear upon the complicated problem of soil and water conservation and properly balanced types of land use. Soil conservation technicians, geologists, climatologists, economists, sociologists, educators and other specialists are collectively confronted with the necessity of finding practical solution to these complexly related problems.

The problems involved in a national program of proper land use are almost endless; many of them are inseparably interrelated and cannot be solved independently. Some of them lead into fields of conservation not ordinarily thought of as pertinent to the land. To illustrate: Because of the development of marshes and stagnant pools of water in some localities through the filling of stream channels with the products of erosion, malaria has become a menace to the population, where formerly the disease was unknown. Mosquito control by draining the marshes would be only a temporary palliative, if erosion should be permitted to continue over the uplands. Coordinated research and work in drainage, soil conservation, and entomology are needed here to help solve the problem.

Education needed to awaken the rural and urban population of wind-erosion areas and adjacent territory to the need for control of soil blowing undoubtedly could be helped through the results of research into the health hazards involved with dust storms. Accordingly, the assistance of research in matters pertaining to the science of medicine is needed in connection with conservation and land-use programs.

General acceptance of water and soil conserving measures, such as introduce radical changes in the details of surface configuration—terraces, diversion ditches, contouring, basin listing, etc.—involves the development of farm machinery that will operate efficiently under man-imposed alterations of the topography of the land.

Long-established practices of reducing cultivated soil to the finest possible condition of tilth have so increased the hazards of runoff and erosion in many localities that the physicist must be asked to help with the problem of determining what can be done to improve cultural operations of this nature.

Some types of stream-borne sediments are deposited much more rapidly than others of the same textural class because of the chemical nature of the solid-liquid flow. In other instances, sedimentation is more rapid from summer floods than from cold-weather floods, probably because of bacteriological conditions. Accordingly, the help of the chemist, the bacteriologist, and the climatologist is needed in that phase of erosion effect which has to do with the shoaling of streams, reservoirs, harbors, drainage ditches, and irrigation canals.

Plans for a comprehensive and properly coordinated national program of land use calls not only for the assistance of continuing research but for research in many specialized fields—in economics, sociology, ethnology, education, health, sanitation, medicine, engineering, climate and so on—all so coordinated as to meet the numerous, intricately involved problems pertinent to proper land use. Our experience on the land must be capitalized, of course; but this will not be enough. Nor will it be enough merely to find out what one specialty has to offer. The combined forces of these specialties must be brought to bear on our land difficulties, and this can be done only through integration—through the collective efforts of all pertinent specialists looking at the composite aspects of the problem.

## Function of Research

Research in soil conservation involving causes and control of erosion must begin with the soil itself, including the normal processes of its formation and removal and its interrelations with plant and water resources. It must explore the nature, causes, extent, and effects of soil and water wastage under various types of agricultural use. It must determine and test fundamental, practical, and economically feasible means of conserving these resources by preventing undue wastage and by restoring those conditions conforming with needful and sustained land use. It must bring to bear on the problems involved several fields of science and practice and call into effective cooperation the agencies concerned with these special fields; it must develop, in cooperation with state agricultural experiment stations and other appropriate scientific and technical agencies, a forward-looking program of basic and applied research for the various problem areas of the country. It must take part in fitting the findings of research to the land in accord with its needs and adaptabilities. It must assist in determination of the effects of land use on stream flow and sedimentation. It must determine the economics of erosion and of its control from the viewpoint of the farmer as well as from that of the community and the public.

## Research Program of the Soil Conservation Service

The research program of the Soil Conservation Service covers a variety of investigations pertinent to the execution of the responsibility of the Department of Agriculture in developing a sound and practical operating program of soil and water conservation on nearly a billion acres of land where erosion and water wastage are serious problems. The problems attacked are those which arise in the conduct of field work now under way on 150 million acres of land in soil and water conservation demonstration projects and CCC camp areas, soil conservation districts, flood-control districts, and drainage developments in watersheds throughout the 48 states. The results and findings of research under this program, carried on in cooperation with other agencies of the Federal Government and states, are immediately tested in the field in order to facilitate effective application in the action programs. The research program includes: investigations of soil characteristics basic to the development of measures for conserving soil and water; determination of the effect of cultivation and cropping systems on erosion and runoff; development of water conservation practices for agricultural lands in the arid and semiarid regions; control of wind erosion; determination of the effect of land use on flood flows and the characteristics of runoff from farm fields in the principal agricultural areas of the United States as a basis for the design of erosion-control structures, channels, and outlets; development of economical systems for safely removing and disposing of runoff water from farm lands; determination of the factors affecting the processes, rates, and control of reservoir, channel, and bottomland silting; climatic factors as they directly affect erosion; the relation of former land-use practices to the present erosion problem; the economic appraisal of erosion damage and soil and water conserving practices; testing of the efficacy of erosion-control plants; development of erosion-control practices for farm lands unsuited to ordinary methods of cultivation, and development of more efficient and effective practices for irrigation and drainage of agricultural lands; and improvement of equipment used on land treated for soil and water conservation.

# Chapter XLIII. Soil Conservation Surveys

In determining the major physical characteristics of the numerous kinds of land that must be dealt with in soil and water conservation work and land-use readjustments, a new technique has been developed by the Soil Conservation Service under the soil conservation surveys. These surveys give an accurate inventory of the physical characteristics of the land, without which it is not possible to plan effectively for the proper installation of the various measures that must be employed to control erosion and conserve rainfall in fields, on farms, and over entire watersheds. These surveys show the dominant conditions of slope, soil, and erosion as well as the present vegetative cover, or the use now being made of the land. With these facts before us, it is generally possible, within a sound economic pattern, to make effective use of the various measures required to meet the needs and adaptabilities of the different types of land, so applied that what is done in one field contributes protection to an adjacent field, or an adjacent farm, or to some area or property downstream, and so also that one control measure supplements another, where several are used on the same area. Without this information, obviously, no such treatment is possible. Therefore, accurate farm and ranch surveys are essential to the execution of the best possible job of land protection, water conservation, and land-use adjustments.

Needed readjustments in the use of land may sometimes call for the removal of people from one section to another—from land too severely washed for practical rehabilitation to land of productive value. Such movements, however, cannot go beyond the productive capacity of the land available for the transfer. Difficulties involved with transfers of this kind indicate the importance of the relationship of land of one locality or farm or range to the land of some other locality. Without adequate land surveys, efficient planning and assistance for those who are on the point of being forced off the land by advancing erosion cannot be achieved.

The necessary readjustments in the physical patterns of land use referred to under "Interior West-Gulf Coastal Plain," Chap. XXX,

Part 2, were carried out on the basis of very careful surveys of each farm, showing not only the kind and distribution of the soils, slope gradients, erosion, and land use but such additional features as field boundaries, fences, drainageways, and farm buildings (see Fig. 261, Chap. XXX, Part 2). Additional information was acquired also—information with respect to the economic situation of the land operators.

## Soil Conservation Surveys

The basic physical factors involved with sound planning for conserving soil and water, whether by farms or by large watersheds, are soil type, slope, land use or cover, and degree of and susceptibility to erosion.[1]

The land conservation survey, in other words, is a field inventory of these important physical conditions—a preliminary stocktaking to accumulate and classify the information necessary for efficient guidance of programs of erosion control, water conservation, and land-use readjustments. Other factors, of course, must be taken into consideration in the development of these programs. Economic and social conditions, including market and transportation facilities, land tenure, size of farm unit, indebtedness, taxes, prices, farm requirements, school facilities, and so on, influence the manner of using land and the practicability of soil defense measures; and they must be considered along with the basic framework of the physical environment.

### Grouping Physical Factors

Soil character, degree of slope, and condition of erosion may vary widely within a very small area, even within the limits of a single farm field. For practical purposes, then, all physical variations that definitely affect erosion potentials and adaptable types of land use must be indicated on a map. In order to avoid confusion through multiplicity of detail, it is necessary to group the numerous variations entering into the physical complex. For instance, it is necessary to establish definite slope groups for particular types of soils, in order to express the correct meaning of the dominant slope factor in relation to the practical use of the land and to avoid mapping every slight change in gradient. The same holds true with respect to the soil and the degree of erosion. Not only must each land factor or group of factors be classified separately, but the several factors affecting a given area must be classified for each parcel of land mapped, in order to express, in a practical way, distinctive land conservation significance.

[1] Soil Conservation Survey Handbook. U. S. Department of Agriculture *Misc. Pub.* 352, 1939.

### System of Mapping

What is actually done in showing the areal distribution of soil types, slopes, and erosion conditions on a soil conservation survey map is the superimposing of three separate classifications one upon another in such a way as to develop a readable three-factor chart. Each delineated area on the map carries a composite symbol showing the soils, slope gradients, and conditions of erosion (see Fig. 356). The use that is being made of the land—pasture, cultivation, forest, idle—is indicated separately, usually by color or hachures.

## Major Physical Factors Affecting Erosion

Soil. Since soil is the thing actually affected by erosion, and since its character greatly influences rates of infiltration, runoff, and erosion, it is necessary to give full consideration to its principal characteristics for every area shown on the map. It is necessary to present these characteristics not only for the topsoil but also for the successive layers of the profile down to and including the parent or geological materials. Among the more important characteristics affecting infiltration of rainfall, retention of moisture, susceptibility to erosion, and productivity are the texture, structure, content of organic matter (in the upper layers), prominent chemical features, and thickness of the main soil horizons or layers. Soils are classified in accordance with these and other pertinent considerations, such as color, origin, and climatic environment, into types and series or groups.

Slope. Both gradient (steepness) and length of slope profoundly affect rates of runoff and soil removal and, indirectly, the amount of moisture absorbed by the soil. Where the land is steep, and the cover thin, the combined effects of erosion and mass movement often prevent the development of deep soils with well-defined profile features. Also, steepness of slope affects the feasibility of cultivation and adaptability to grazing.

For purposes of practical mapping, slopes are classified under five broad groups of gradient ranges, indicated by the letters *A*, *B*, *C*, *D*, and *E*. *Group A* includes level to gently sloping land where active water erosion is not in evidence; *groups B* and *C* cover moderately sloping areas on which erosion is either active or a hazard but controllable under annual cultivation—with simple practices for *B* slopes and complex practices for *C* slopes; *group D* comprises those steep slopes where erosion is active or so hazardous that the land should be plowed only to seed a protective crop of grass or other dense cover; and *group E* includes those very steep slopes where the land is so erodible that it should never be plowed.

These group definitions apply to all soils, but the range of gradient within any group may vary for different soil types or groups of types according to variations in degree of erodibility. For example, the absorptive Cecil sandy loam, occurring extensively over the southern Piedmont, normally can be cultivated with safety on slopes as steep as 8 to 10 per cent, with proper treatment; whereas the Iredell sandy loam, an associated soil, because of physical characteristics that make it less absorptive of rainfall, cannot be cultivated safely on slopes exceeding 5 or 6 per cent.

EROSION. Most of the major land ills, where the soils are not inherently poor, such as decreasing productivity, increasing runoff, and increased resistance to tillage, are directly traceable to erosion losses. The degree to which the soil has suffered from erosion as the result of past treatment is of great importance in planning the future use and treatment of a given tract of land. For example, a productive soil having a depth of 10 to 12 inches of friable, loamy topsoil over a raw, clayey subsoil and capable of being cultivated readily on slopes up to 10 or 12 per cent presents a very different farming problem from an area, originally the same, that has lost half or three-fourths of the productive surface layer.

The classification of erosion is so designed that this factor can be correlated with other important physical characteristics of the land and so aid in determining adaptable and necessary land-use and conservation practices. The facts determined with respect to the characteristics of soil, slope, and erosion also make it possible to appraise, at least approximately, the susceptibility of a given area to erosion and so to devise necessary protective treatment, even for a newly-cleared, uneroded area.

## Classification of Erosion

Erosion, in general, is classified into two broad categories: normal erosion and accelerated erosion.

### NORMAL, OR GEOLOGIC, EROSION

Normal erosion pertains to soil movement under natural environmental conditions, as under a cover of grass or trees. Rates of soil removal are generally imperceptibly slow where the land is thickly covered with vegetation (Fig. 352). Naturally bare or slightly vegetated areas, such as occur in some parts of low-rainfall areas, erode more rapidly. Erosion is also more rapid in many depressions where water concentrates and on precipitous slopes.

Normal erosion is caused by water or wind or the two combined. The principal classes are: sheet erosion by water, sheet and gully erosion by water, and wind erosion.

## Accelerated Erosion

Accelerated erosion pertains to speeded-up removal of soil by water or wind following removal or impairment of a stabilizing cover of vegetation. It is separable into two general classes: naturally accelerated erosion and man-induced, or artificial, erosion.

Fig. 352.—Above, no erosion, class (0)—never cultivated. Below, stabilized erosion, class (38)—formerly cultivated and abandoned because of erosion. Stabilized by second-growth pine. (*Photographs by Soil Conservation Service.*)

*Naturally accelerated erosion* results from extreme changes in the environment, such as those produced by severe drought, lightning, plant disease, snow slides, and rock avalanches. Erosion thus started is indicated on the land conservation maps by combining the symbol for normal erosion with that for accelerated erosion. For example, naturally accelerated water erosion would be shown as *W2*, *W3*, etc.; normally accelerated sheet and gully erosion would be indicated by the symbols *W27*, *W38*, or *W9*; normally accelerated wind erosion would be indicated by such symbols as *WP* and *WR*.

Much of the *advanced geological erosion*, such as the badlands and "painted deserts" of the West, existed when the white man came to America.

FIG. 353.—Above, sheet erosion (as classed in *conservation surveys*), class (2)—*B* slope. Tillage up and down slope. Rills can be obliterated by cultivation. Below, sheet erosion, class (3)—*B* slope. Incipient gullies between rows. Exposes the subsoil. (*Photographs by Soil Conservation Service.*)

Erosion accelerated as the result of man's activities is classified as follows:

## WATER EROSION

SHEET EROSION. The constant removal of soil by rain and wind in more or less thin sheets from considerable areas is designated *sheet erosion* (see Chap. IV, Part 1). The classifications of sheet erosion are as follows (Fig. 353):

| | Class | Per Cent of Topsoil Removed |
|---|---|---|
| Degree of soil removal under the various classifications of water erosion | *0* No perceptible erosion | |
| | *1* Slight erosion | 0 to 25 |
| | *2* Moderate to moderately severe erosion | 25 to 75 |
| | *3* Severe erosion | 75 to 100 |
| | *4* Very severe erosion | All topsoil and 25 to 75 per cent of subsoil |
| | *5* Exceedingly severe erosion | More than 75 per cent of subsoil removed |
| | *6* Miscellaneous erosion | Slips, catsteps, etc. |

Gully Erosion. Gully erosion (see Chap. IV, Part 1) is caused by the concentration of runoff water in unstable depressions that cannot be obliterated by tillage methods alone. They are classified according to size and frequency, as follows (Fig. 354):

Shallow gullies that can be crossed by ordinary tillage and harvesting machinery.

| | Class | Frequency |
|---|---|---|
| Shallow gullies | *7* | Occasional gullies—over 100 feet apart |
| | *8* | Frequent gullies—less than 100 feet apart |
| | *9* | Very frequent or large gullies—75 per cent or more of area actually in gullies |

Deep gullies that cannot be crossed by ordinary farm machinery:

Resistant substratum:

| | | |
|---|---|---|
| Deep gullies—resistant substratum | ⑦ | Occasional |
| | ⑧ | Frequent |
| | ⑨ | Very frequent |

Unstable substratum:

| | | |
|---|---|---|
| Deep gullies—unstable substratum | *7V* | Occasional |
| | *8V* | Frequent |
| | *9V* | Very frequent |

Accumulation of the Products of Erosion. Accumulations of the products of erosion are indicated by the symbol + (plus) if less than 12 inches in depth. If the deposits are deeper than 12 inches, classes such as *1*+, *2*+, and *3*+ are established to indicate variations in depth.

## Wind Erosion

The effects of wind erosion are indicated by the symbols *O*, *P*, *R*, *S*, *T*, and *U* for soil removal and *F*, *H*, *K*, *L*, *M*, and *N* for soil accumulation (Fig. 355).

The degrees of removal are expressed as follows:

| | Class | | Per Cent of Topsoil Removed |
|---|---|---|---|
| Degree of soil removal under the various classifications of wind erosion | *O* | No discernible erosion | |
| | *P* | Slight erosion | 0 to 25 |
| | *R* | Moderate to moderately severe erosion | 25 to 75 |
| | *S* | Severe erosion | 75 to 100 |
| | *T* | Very severe erosion | All topsoil and 25 to 75 per cent of the subsoil |
| | *U* | Exceedingly severe erosion | More than 75 per cent of the subsoil |

Fig. 354.—Above, very active gully eating back into farm land, class (9*V*) erosion. Below, destructive erosion on a moderately steep slope. All topsoil has been washed away and gullies have riddled the substratum, class (9) erosion. (*Photographs by Soil Conservation Service.*)

Degree of accumulation (materials recently blown onto affected areas) is expressed as follows:

| | Class | | Depth of Accumulation, Inches |
|---|---|---|---|
| Degree of soil accumulation under the various classes of wind erosion | *F* | Shallow accumulation | 0 to 6 |
| | *H* | Moderate accumulation, evenly distributed | 6 to 12 |
| | *K* | Moderate accumulation, hummocky distribution | 6 to 12 |
| | *L* | Deep accumulation | 12 to 36 |
| | *M* | Dune accumulation (small) | 36 to 72 |
| | *N* | Dune accumulation (large) | 72 plus |

FIG. 355.—Above, wind erosion, class (*F*)—shallow accumulations. Soil drifting from untreated neighboring field, filling listed furrows of cultivated land. Below, severe wind-erosion removal, subsoil exposed, class (*T*). Close-up of subsoil showing plow marks of former cultivation. Topsoil all blown off. (*Photographs by Soil Conservation Service.*)

The addition of a numeral suffix 1, 2, or 3 to the class symbol indicates the percentage of the surface area affected by the various classes of accumulation:

(1) *F1, K1, L1* indicates one-third or less of area affected.
(2) *F2, K2, L2* indicates one-third to two-thirds of area affected.
(3) *F3, K3, L3* indicates two-thirds or more of area affected.

### UNDIFFERENTIATED EROSION

Various classes of erosion not shown separately on the land conservation maps are indicated by such symbols as the following:

Soil scouring by floods is indicated by placing − (minus) before the symbol normally used; bank cutting is indicated by hachures; stabilization of erosion is shown by the use of a circumflex over the appropriate symbol for partial control (as $\hat{7}$ or $\hat{R}$) and by overscoring the appropriate symbol for complete control (as $\overline{7}$ or $\overline{R}$).

## Present Type of Land Use

Four major classes and one miscellaneous class are employed in mapping present types of land use, as follows: (1) cultivated land; (2) pasture and grassland; (3) woodland; (4) idle land; and (5) urban areas, home sites, etc. These classes are defined as follows:

Cultivated land:

*a.* Land used for cultivated crops
*b.* Fallow land
*c.* Orchards
*d.* Land seeded to grass, alfalfa, or other forage crops grown in regular rotation for hay; and meadows cut for hay and grazed in late season

Pasture or grassland:

Natural grazing areas and cultivated grass and legume lands devoted primarily to grazing. Wild grassland used solely for hay, land used for pasture in rotation, and pastured woodlands are not included

Woodlands:

Areas with 40 per cent or more of the ground covered by the spread of trees of all ages, and land devoted to forest plantations

Idle lands:

Idle land, with cover too scant for adequate protection or economic use other than possible food source for wildlife

Miscellaneous:

Urban areas, home sites, golf courses, etc.

## Kinds of Land Conservation Surveys

The two major types of surveys are detailed and reconnaissance.

Detailed Surveys. This kind of survey develops a detailed inventory of all the dominant land factors. It is particularly applicable to those lands where intensive and diversified agricultural operations are carried on over a wide variety of physical land conditions, and where intensive planning is required. The scale of mapping is usually 4 or 8 inches to the mile; the base maps are enlargements of aerial photographs. Each area outlined on the field sheet is fully identified by a composite symbol showing the soil, slope, and degree of erosion. Cultural features are shown in colors. Examples of the variations shown by detailed mapping are given in Fig. 356.

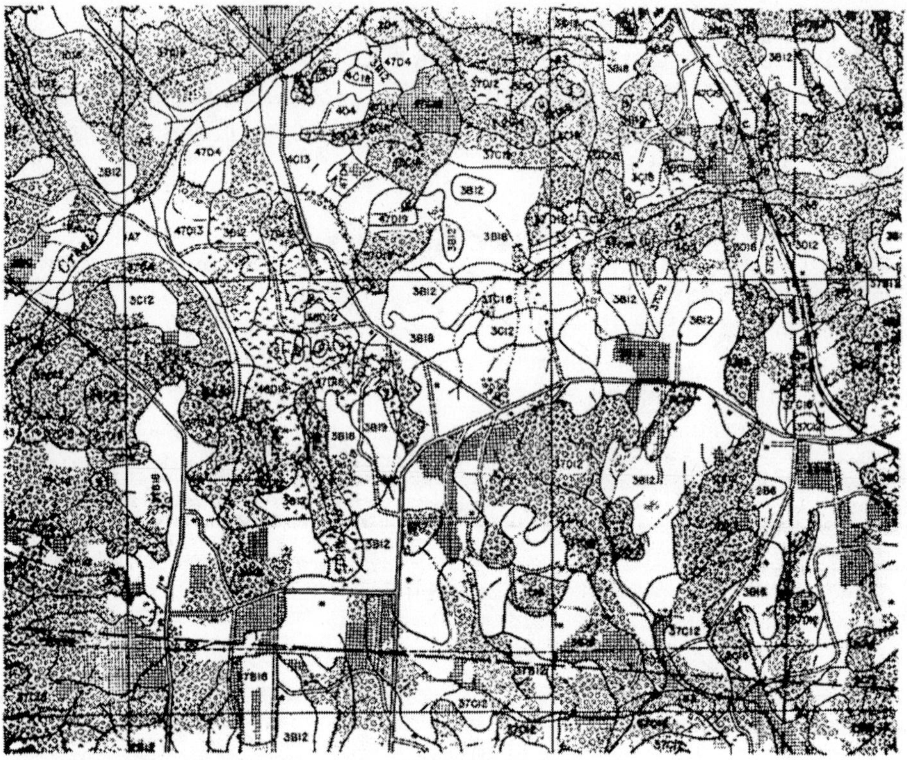

Fig. 356.

### Erosion

*Sheet erosion*

2 Slight. Less than 25 per cent of surface soil removed.
3 Moderate. 25–75 per cent of surface soil removed.
4 Severe. More than 75 per cent of surface soil, or all of surface soil and some subsoil removed.
5 Very severe. Surface soil and subsoil removed, with erosion of parent material.

*Gully erosion*

7 Occasional gullies. More than 100 feet apart.
8 Frequent gullies. Less than 100 feet apart, but less than 75 per cent of area included in gullies.
9 Very frequent or large gullies. Applied to a single deep gully or to an intricately dissected area.

*Miscellaneous*

1 No apparent erosion.
\+ Recent alluvial or colluvial deposits

*Stabilized erosion*

____over symbol indicates stabilization ($\overline{37}$).

### Land Use

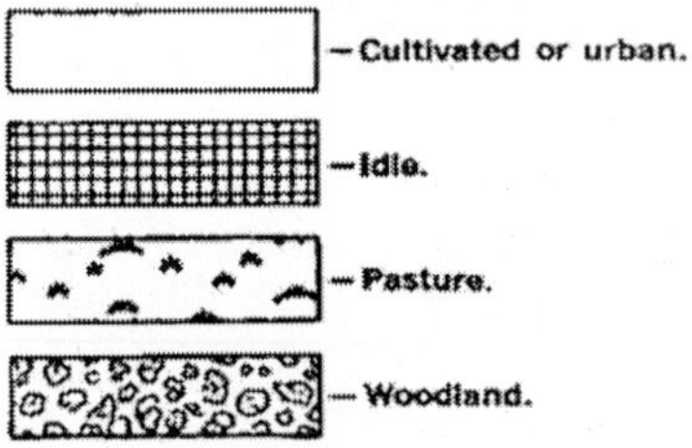

—Cultivated or urban.

—Idle.

—Pasture.

—Woodland.

Explanation of Symbol 37B12

37 = moderate sheet erosion with occasional gullies.
*B* = slope, 3–7 per cent.
12 = soil, Appling sandy loam.

(Since publication of this survey, slight revisions have been made in the symbols designating the erosion and slope classes. The revised symbols are those used in the text.)

### Slope

| | | Dominant per cent |
|---|---|---|
| *A* | Nearly level areas with little or no erosion. | 0–3 |
| *B* | Subject to erosion when cultivated; control practicable. | 3–7 |
| *C* | Too steep for practical erosion control under clean cultivation; suitable for close-growing crops. | 7–12 |
| *D* | Too steep for practical erosion control if cultivated; suitable only for permanent cover. | |

### Soils

3 Alluvial soils, undifferentiated
4 Wilkes sandy loam
6 Helena sandy loam
7 Congaree silt loam
12 Appling sandy loam
13 Appling fine sandy loam
16 Cecil sandy loam
17 Cecil fine sandy loam
18 Cecil sandy loam and sandy clay undifferentiated
19 Cecil clay loam

In mapping, the land is examined, on foot, with sufficient detail to determine the dominant characteristics of every acre. The large-scale base maps used for mapping farms usually enable the technician to keep his location with accuracy. Where this cannot be done, the mapping is executed by plane-table traverse. Slope is determined largely by use of the Abney hand level. The degree of erosion is determined at the time the soil type is identified, chiefly by comparing the profiles of eroded areas with uneroded or only slightly eroded lands, originally the same. Inspections of field work are made from time to time by regional inspectors in order to maintain uniformity and accuracy in the classification and mapping.

Reconnaissance Surveys. In these more generalized surveys, base maps of much smaller scale are used, usually 1 or ½ inch to the mile. The procedure is to make only enough examinations to determine the predominant characteristics of erosion, slope, soil, and land use. In some areas of rough, broken country, base maps of still smaller scale are used, and the classifications shown are more inclusive. Such generalized maps are used for broad planning purposes and to indicate where detailed mapping is necessary.

An example of how surveys of physical land factors help with the development of conservation programs for large areas is found in the *Soil Conservation Problem-area Groups of Land* map, covering 96,900,000 acres in the Southern Plains, prepared by H. H. Finnell.[1] This problem-area groups map is not a soils map nor a rainfall, topographic, or land-use map; it represents, rather, a map that isolates and labels those bodies of land where soil, climate, and appropriate farming practices are similar. This map was developed, however, from information shown on the *Soil Conservation Reconnaissance Survey of the Southern Great Plains Wind-erosion Area*,[2] together with information supplied by climatic records, crop yields, and the results of actual work on the land.

Adjustments in land use based on the *Problem-area Groups* map for the entire area surveyed are indicated in Fig. 357.

## Land-use Capability

The physical facts determined by the land conservation survey can be evaluated in terms of *land-use capability*. Such capability is determined by appraisal of the soil, slope, erosion, drainage, etc., in their collective relation to the capacity of the soil to produce under those practical methods of intensive use conforming with permanent main-

[1] Finnell, H. H. "Problem-area Groups of Land in the Southern Great Plains." U. S. Department of Agriculture. 1939.

[2] Joel, A. H. U. S. Dept. Agr. *Tech. Bull.* 556, 1937.

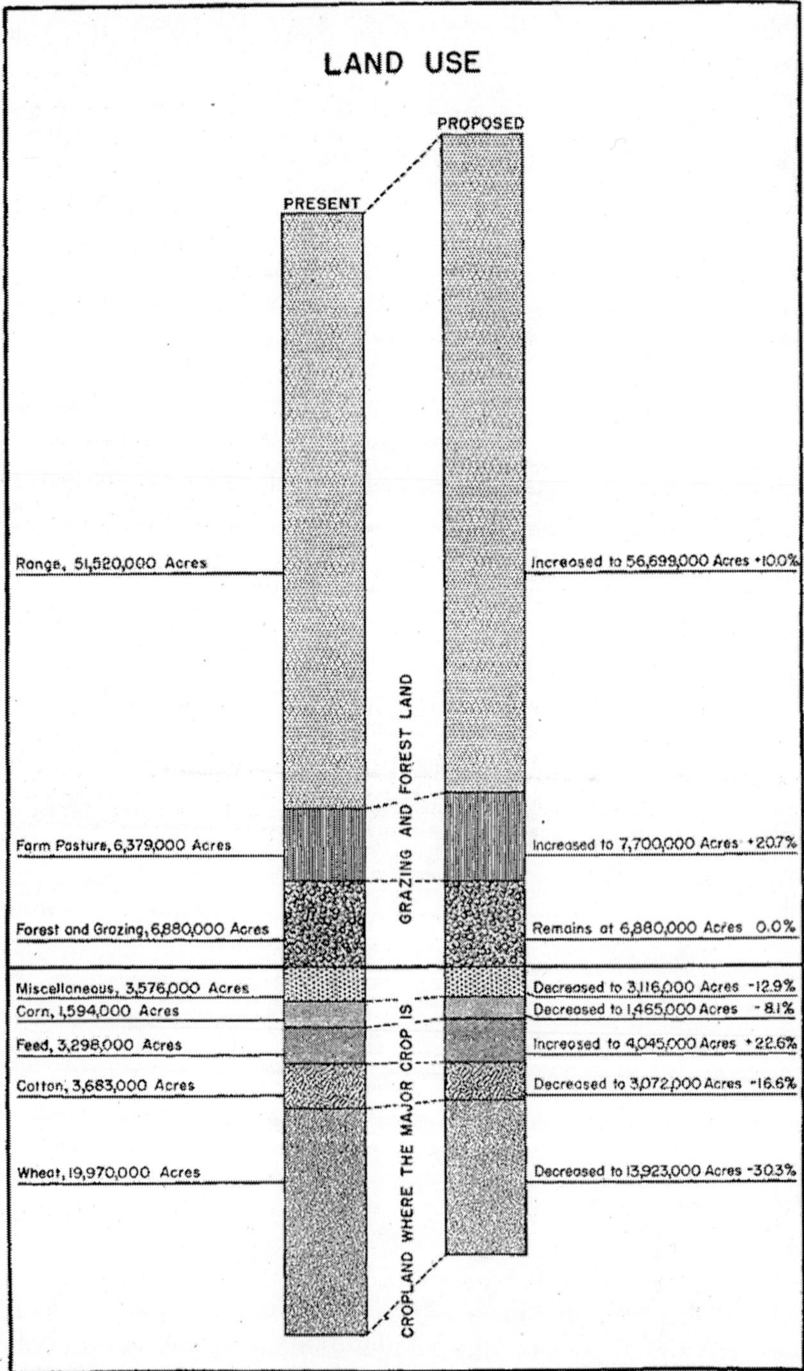


Fig. 357.—Needed adjustments as between cultivated and noncultivated land in the Southern Plains, based on survey of 96,900,000 acres. Column on left shows the way this land is now used; column on right shows the proposed improved use to meet the physical adaptabilities and needs of the land and the economic conditions. (*Soil Conservation Service.*)

tenance of the integrity of the soil. The classification, however, does not attempt to establish methods for applying the practices, measures, and treatments that may be required in any capability class. That determination is a function of the operating technician collaborating with the farmer.

This capability evaluation answers in a practical way the need for simplifying complicated survey data by grouping numerous technical separations into a small number of general classes adequately expressing

Fig. 358.—Land-use capabilities, by classes. (*Photograph by Soil Conservation Service.*)
Class II, *B* slope (5–8 per cent)—class 2 erosion; requires simple conservation practices.
Class III, *C* slope (8–15 per cent)—class 3 erosion; requires intensive conservation practices.
Class IV, *D* slope (15–25 per cent)—class 37 erosion; suitable for legumes and hay only.
Class V, *E* slope (25 per cent plus)—forest land only.

the principal kinds of adaptable land use (Fig. 358). (Also see Fig. 44, Chap. IV, Part 1.)

The following land-use capability classes have been established:

*Class I.* Lands that can be cultivated safely on a sustained-production basis for at least moderate to good yields of adaptable crops, without the need of special practices or treatments. Commonly recognized methods of good farming for the region are assumed.

*Class II.* Lands that cannot be cultivated safely on a sustained-production basis for at least moderate to good yields of adaptable crops without the application of some special practice or treatment of simple character. Such practices might be contour cultivation, strip cropping with wide strips, improvement of drainage, removal of rocks or troublesome debris, soil amendments, simple rotations, or preservation of crop residues.

*Class III.* Lands that cannot be cultivated safely on a sustained-production basis for at least moderate to good yields of adaptable crops without the application of intensive practices or measures, such as contour cultivation, strip cropping in narrow bands, terracing, tile drainage, fertilization, or systematic rotations.

*Class IV.* Lands that cannot be cultivated safely under any plan of continuous use but that can be used safely for pasture grasses and tame hay with such limited cultivation as may be necessary to maintain a good cover.

*Class V.* Lands that cannot be cultivated safely at any time and that are suitable only for permanent cover.

# INDEX

Pages on which illustrations appear are in **boldface** type.

C

D

F

G

H

N

O



P

U



V